Stochastic Planning and Modeling for Energy Systems

Stochastic Planning and Modeling for Energy Systems

Methods, Applications, and Developments

Edited by

Miadreza Shafie-khah

Research and Innovation Division, Nowocert, Dublin, Ireland

ELSEVIER

Elsevier
Radarweg 29, PO Box 211, 1000 AE Amsterdam, Netherlands
50 Hampshire Street, 5th Floor, Cambridge, MA 02139, United States

ISBN: 978-0-443-45298-7

For Information on all Elsevier publications visit our website at https://www.elsevier.com/books-and-journals

Publisher: Priscilla Braglia
Acquisitions Editor: Dan Egan
Editorial Project Manager: Arushi Rawat
Production Project Manager: Fizza Fathima
Cover Designer: Bakyalakshmi S

Working together to grow libraries in developing countries
www.elsevier.com • www.bookaid.org

Typeset by Aptara, New Delhi, India

Contents

Contributors

Ali-Akbar Ahmadi, Industrial Automation Laboratory, Department of Electrical and Computer Engineering, Kharazmi University, Tehran, Iran

Muzammil Ahmed, Indian Institute of Technology Guwahati, Guwahati, Assam, India

Mohammad Hossein Alizadeh, Department of Aerospace Engineering, Amirkabir University of Technology, Tehran, Iran

Hossein Ameliaiiz, EEE department, Control and Power Group, Imperial College, London, United Kingdom

Mohammadtaghi Ameli, Department of Electrical Engineering, Shahid Beheshti University, Tehran, Iran

Fatemeh Aminizade, Department of Mathematics & Computer Science, Amirkabir University of Technology, Tehran, Iran

Hamid Reza Baghaee, Department of Electrical Engineering, Tarbiat Modares University, Tehran, Iran

Seyed Mohammad Taghi Bathaee, Department of Electrical Engineering, K. N. Toosi University of Technology, Tehran, Iran

Frede Blaabjerg, Department of Energy Technology, Aalborg University, Aalborg East, Denmark

Enrico De Santis, Department of Information Engineering, Electronics and Telecommunications, Sapienza University of Rome, Italy

Mostafa Eidiani, Consultant for Equipment Research & QC Office, Khorasan Regional Electric Company, Mashhad, Razavi Khorasan, Iran

Gianluca Ferro, Department of Information Engineering, Electronics and Telecommunications, Sapienza University of Rome, Italy

Rachana Garg, Electrical Engineering, Delhi Technological University, New Delhi, India

Lukas Gebremariam, Department of Electrical and Computer Engineering, Wachemo University, Hosanna, Ethiopia

Gevork B. Gharehpetian, Department of Electrical Engineering, Amirkabir University of Technology, Tehran, Iran

Atiye Ghasemzadeh, Azarakhsh Energy Maham Shargh Engineering Company, Mashhad, Razavi Khorasan, Iran

Seyed Mehdi Hakimi, Department of Electrical Engineering, Dam.C., Islamic Azad University, Damavand, Iran; Renewable Energy Research Center, Dam.C., Islamic Azad University, Damavand, Iran

Hamid HassanzadehFard, Department of Electrical Engineering, Mi.C., Islamic Azad University, Miyaneh, Iran

Md. Rabiul Islam, School of Engineering, University of Wollongong, Wollongong, NSW, Australia

Farha Khan, Department of Electrical Engineering, Aligarh Muslim University, Aligarh, Uttar Pradesh, India

Jamil Ahmad Khan, Electrical Engineering Department, Abbottabad Campus, COMSATS University Islamabad, Abbottabad, Khyber Pakhtunkhwa, Pakistan

Mathewos Lolamo, Department of Electrical and Computer Engineering, Wachemo University, Hosanna, Ethiopia; Research and Innovation Division, Nowocert, Dublin, Ireland

Priya Mahajan, Electrical Engineering, Delhi Technological University, New Delhi, India

Mohsen Masoumi, Azarakhsh Energy Maham Shargh Engineering Company, Mashhad, Razavi Khorasan, Iran

Seyed Pouria Miri, Department of Electrical Engineering, Amirkabir University of Technology, Tehran, Iran

Manvi Mishra, Electrical Engineering, Delhi Technological University, New Delhi, India

Sambeet Mishra, Department of Electrical engineering, Information Technology and Cybernetics, University of South-Eastern Norway

Farzaneh Mohammadi, Department of Information Technology Management, Tarbiat Modares University, Tehran, Iran

Rozhina Mohammadi, Department of Electrical Engineering, Amirkabir University of Technology, Tehran, Iran

Shaghayegh Nobakht, Industrial Automation Laboratory, Department of Electrical and Computer Engineering, Kharazmi University, Tehran, Iran

Hamid Reza Pilehvar Javid, Department of Electrical Engineering, Amirkabir University of Technology, Tehran, Iran

Nazanin Pourmoradi, Department of Electrical Engineering, Shahid Beheshti University, Tehran, Iran

Homa Rashidizadeh-Kermani, Research and Innovation Division, Nowocert, Dublin, Ireland; Department of electrical and computer engineering, University of Birjand, Birjand, Iran

Antonello Rizzi, Department of Information Engineering, Electronics and Telecommunications, Sapienza University of Rome, Italy

Mohammad Hossein Alizadeh Roknabadi, Department of Aerospace Engineering, Amirkabir University of Technology, Tehran, Iran

Kumars Rouzbehi, Department of Systems Engineering and Automatic Control, University of Seville, Sevilla, Spain

Buddhadeva Sahoo, Department of Electrical and Electronics Engineering, SR University, Warangal, Telangana, India

Niusha Mohammadi Sedgh, Department of Electrical Engineering, Amirkabir University of Technology, Tehran, Iran

Mohammad Hossein Shaabani, Department of Electrical Engineering, Amirkabir University of Technology, Tehran, Iran

Miadreza Shafie-khah, Research and Innovation Division, Nowocert, Dublin, Ireland

Syed Ayaz Ali Shah, Electrical Engineering Department, Abbottabad Campus, COMSATS University Islamabad, Abbottabad, Khyber Pakhtunkhwa, Pakistan

Preeti Sharma, Department of EE, SOET, Pandit Deendayal Energy University, Gandhinagar, Gujarat, India

Neelofar Shaukat, Electrical Engineering Department, Abbottabad Campus, COMSATS University Islamabad, Abbottabad, Khyber Pakhtunkhwa, Pakistan

Chandan Kumar Shiva, Department of Electrical and Electronics Engineering, SR University, Warangal, Telangana, India

Uday Chandra Sikkam, Department of Electrical and Electronics Engineering, SR University, Warangal, Telangana, India

Saeed Tavakkoli, Azarakhsh Energy Maham Shargh Engineering Company, Mashhad, Razavi Khorasan, Iran

Aschalew Tirulo, Department of Software Engineering, Wachemo University, Hosaena, Ethiopia; Research and Innovation Division, Nowocert, Dublin, Ireland

Mostafa Vahedipour-Dahraei, Department of Electrical & Computer Engineering, University of Birjand, Birjand, South Khorasan, Iran

Mahmoud Mollayousefi Zadeh, Department of Electrical Engineering, K. N. Toosi University of Technology, Tehran, Iran

Sonia Zakeri, Department of Electrical and Computer Engineering, University of Tabriz, Tabriz, Iran

Danial Zendehdel, Department of Information Engineering, Electronics and Telecommunications, Sapienza University of Rome, Italy

About the Editor

Prof. Miadreza Shafie-khah earned two PhDs, his first in Electrical Engineering from Tarbiat Modares University, Tehran, Iran and his second in Electromechanical Engineering from the University of Beira Interior, Portugal, where he also held his first postdoctoral position. He subsequently completed his second postdoctoral appointment at the University of Salerno, Italy. He was an Assistant Professor at the University of Beira Interior and a Full Professor in Sustainable and Flexible Energy Systems at the University of Vaasa, Finland. Currently, he is a Scientific Director for Energy Business eMBA at the Executive Education and Head of Research and Innovation Division at Nowocert, Dublin, Ireland. He is the Editor-in-Chief of *Smart Energy Systems*, and an Editor of the *IEEE Transactions on Sustainable Energy, IEEE Transactions on Intelligent Transportation Systems, IEEE Systems Journal, IEEE Access, IEEE Open Access Journal of Power and Energy, IEEE Power Engineering Letters,* and *IET Generation, Transmission and Distribution.* He is a Top Scientist in Research.com in Electronics and Electrical Engineering, and was ranked number 1 in Finland among the World's Rising Stars of Science in 2022-2024. His research interests include smart grids, electricity markets, demand response, and electric vehicles.

Acknowledgments

As an editor, I am grateful to the colleagues and contributors whose expertise and collaboration shaped this volume. I would like to express my deepest gratitude to my wife for her unwavering love, patience, and encouragement throughout the preparation of this book.

AI and data-driven methods in scenario generation and reduction

1

Aschalew Tirulo[a], *Mathewos Lolamo*[b], *Lukas Gebremariam*[b] *and Miadreza Shafie-khah*[c]
[a]Department of Software Engineering, Wachemo University, Hosanna, Ethiopia, [b]Department of Electrical and Computer Engineering, Wachemo University, Hosanna, Ethiopia, [c]Research and Innovation Division, Nowocert, Dublin, Ireland

1.1 Introduction

1.1.1 Comprehensive overview

Uncertainty is intrinsic to modern energy systems. Renewable generation exhibits temporal variability, demand forecasts contain errors, and market prices fluctuate unpredictably. A stochastic program is a mathematical program that involves some uncertain data. These parameters may be most accurately described as continuous random variables, but, except in some trivial situations, it is difficult to optimize directly in terms of those distributions. Hence, in most applications, the continuous distributions are approximated by discrete distributions with a finite number of scenarios, or possible realizations, for the random variables. This discretization procedure is often called scenario generation. Stochastic planning methods such as stochastic programming, chance-constrained optimization, and robust optimization rely on a set of scenarios representing possible realizations of the uncertain variables. The quality of these scenarios strongly influences planning outcomes: If the scenario set fails to capture extreme events or joint dependencies, the resulting solution may be sub-optimal or even infeasible. Conversely, an excessively large scenario set is computationally prohibitive. Scenario generation and reduction thus aim to strike a balance between fidelity and tractability.

Classical scenario generation techniques include sampling from parametric distributions, moment-matching methods (Latin hypercube sampling), dynamic factor models, generalized autoregressive score models, vine copulas, principal component analysis, and Gaussian mixture models. While these approaches are grounded in probability theory and offer tractable sampling schemes, they may struggle to capture complex cross-correlations or multimodal behavior inherent in energy data. In the last decade, data-driven approaches have emerged that leverage the representational power of neural networks to learn implicit distributions from historical data. Similar machine-learning approaches have also been used to generate realistic synthetic datasets for validating healthcare applications (Arvanitis et al., 2022). Generative adversarial networks (GANs) and variational autoencoders (VAEs) can replicate the joint distribution of high-dimensional variables without assuming a particular parametric form. Moreover,

Stochastic Planning and Modeling for Energy Systems: Methods, Applications, and Developments.
DOI: https://doi.org/10.1016/B978-0-443-45298-7.00004-5

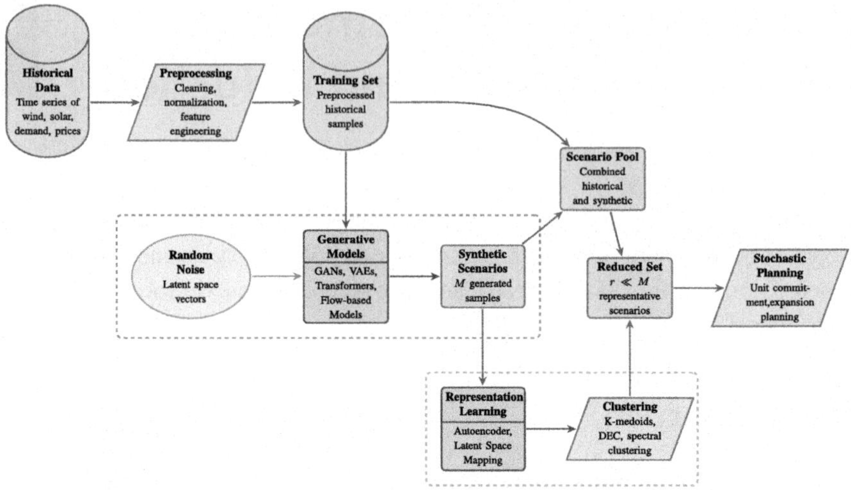

Figure 1.1 Pipeline for AI-driven scenario generation and reduction. Historical data are fed into generative models (GANs or VAEs) to produce synthetic scenarios. These are then reduced via clustering (e.g., K-medoids) to obtain a compact set for stochastic planning. *DEC*, Deep embedded clustering; *GAN*, generative adversarial networks; *VAE*, variational autoencoders.

deep clustering methods reduce large scenario pools by learning latent representations that preserve temporal and spatial structure.

This section of the chapter provides a comprehensive overview of AI-based scenario generation and reduction methods tailored for stochastic planning in energy systems. The remaining sections of the chapter are outlined as follows. Section 1.2 reviews the State of the Art from 2019 to 2025, highlighting seminal contributions in GAN-based scenario generation, deep clustering, and hybrid frameworks. Section 1.3 develops the mathematical foundations of stochastic energy planning, scenario quality metrics, GAN formulation, and clustering. Section 1.4 describes the practical implementation framework, including data preparation, pseudocode, complexity analysis, and benchmarking. Section 1.5 presents a worked case study on wind-power scenarios using Wasserstein-GANs and K-medoids reduction. Section 1.6 deals with additional cases, Section 1.7 discusses future directions and open research challenges, and Section 1.8 offers theoretical questions, coding exercises, and mini-projects for further exploration. Fig. 1.1 provides a visual summary of the methodological pipeline. Fig. 1.2 illustrates the adversarial learning architecture used for GAN-based scenario generation.

1.1.2 Motivation and learning objectives

The increasing penetration of renewable energy sources introduces unprecedented variability and uncertainty into power systems (Khalid, 2024). Traditional deterministic planning, which relies on single forecasts, cannot account for the wide

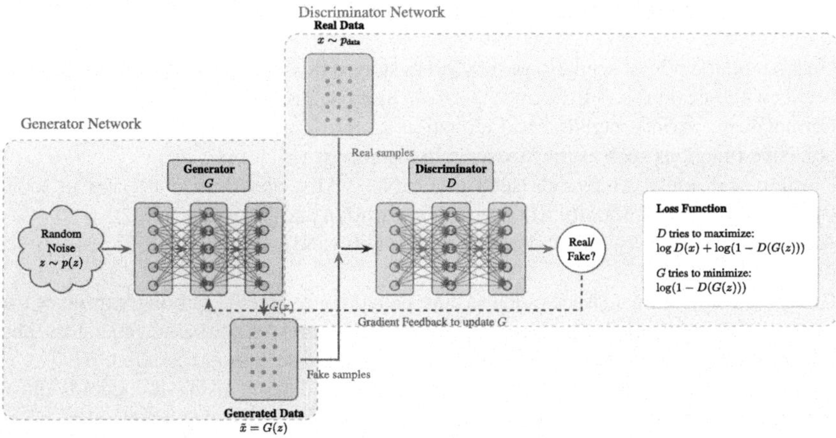

Figure 1.2 Architecture of a generative adversarial network. A generator transforms random noise into synthetic scenarios, while a discriminator attempts to distinguish real from generated samples. The two networks are trained adversarially. Convolutional or recurrent layers can be used depending on data modality. The GAN architecture used for scenario generation is illustrated in Fig. 1.2.

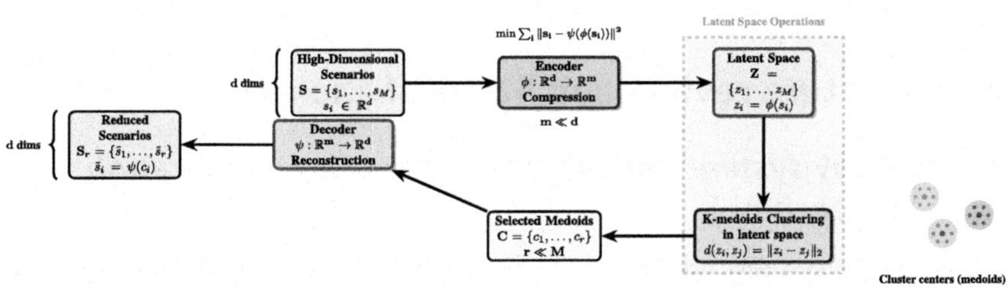

Figure 1.3 Autoencoder and clustering workflow. The encoder compresses high-dimensional scenarios into a latent space; clustering is performed in the latent space to select representative scenarios (medoids). The decoder reconstructs scenarios if needed. The autoencoder-based clustering and scenario reduction workflow is shown in Fig. 1.3.

range of possible outcomes. Scenario-based stochastic programming addresses this challenge by optimizing over a discrete set of possible futures (Kammammettu & Li, 2023). However, generating a scenario set that accurately reflects the joint distribution of uncertain variables, including their temporal dynamics and spatial correlations, while remaining computationally manageable, is nontrivial. Data availability has grown exponentially, and AI techniques offer an opportunity to learn complex distributions directly from data. Fig. 1.3 presents the autoencoder-based clustering workflow used to reduce high-dimensional scenarios into representative medoids.

After studying this chapter, the reader will be able to:

- Understand the role of scenario generation in stochastic energy planning and the differences between classical probabilistic and AI-driven approaches.
- Formulate scenario generation and reduction as optimization problems and derive relevant objective functions such as the Wasserstein distance.
- Develop and apply AI models such as GANs, VAEs, and deep clustering in tools like PyTorch/TensorFlow for efficient scenario generation and reduction.
- Analyze trade-offs between model accuracy (fidelity) and computational efficiency (tractability).
- Understand and mitigate issues such as bias, data privacy, and the carbon footprint of training large models, while exploring emerging methods like transformer-based generators, physics-informed GANs, and federated learning for responsible scenario generation.
- Apply AI-based methods to a realistic wind-power dataset, using K-medoids clustering for scenario reduction; ensure models achieve Wasserstein distance below 0.05 and reduce scenario sets by at least 90%.
- Design AI models and reduction pipelines that reduce scenario computation time by at least 50% while preserving more than 95% of distributional fidelity, as validated by divergence metrics; design and benchmark pipelines accordingly.
- Evaluate performance using metrics such as mean absolute error (MAE), Kullback-Leibler divergence (KL), and Wasserstein distance, and interpret benchmark tables.
- Benchmark AI methods to achieve Wasserstein distance below 0.05 on standard datasets while reducing scenario sets by at least 90%.

1.2 Literature review and State-of-the-Art

1.2.1 Energy system uncertainty and scenario generation approaches

Modern energy systems are subject to a rich tapestry of uncertainties. Renewable generation sources such as wind and solar exhibit strong variability driven by weather conditions and are difficult to forecast accurately beyond a few hours 3. Market prices fluctuate with supply-demand balance, fuel costs, and regulatory interventions; extreme price spikes or negative prices may occur during rare events. Infrastructure unreliability introduces another layer of uncertainty: Component failures, maintenance schedules, and unplanned outages affect the availability of generation and transmission assets. Long-term planning is further complicated by climate change, which alters wind patterns, cloud cover, and temperature distributions over decades. Properly characterizing these diverse sources of uncertainty, both their marginal distributions and their cross-correlations, is essential for robust planning and real-time operations.

Various approaches have been reported in the literature for scenario generation in an uncertain energy system, ranging from traditional to modern AI-driven methods. Classical scenario generation relies on Monte Carlo sampling, moment matching, and statistical distance preservation techniques. Latin hypercube sampling and quasi-Monte Carlo sequences aim to cover the input space uniformly; copula models capture pairwise dependencies but struggle with high-dimensional correlations. Parametric time-series models autoregressive integrated moving average (ARIMA), vector autoregression (VAR), generalized autoregressive conditional heteroskedasticity (GARCH) extrapolate trends and volatilities but require strong stationarity assumptions. Scenario reduction is often performed by greedy forward or backward selection algorithms based on Kantorovich distance or probability re-weighting (Gomes et al., 2020). While these methods have been used extensively in energy planning, they may fail to reproduce nonlinear patterns, heavy tails, or regime shifts present in real data. AI-driven approaches leverage deep learning to learn complex, nonlinear mappings from noise to data distributions. GANs and VAEs are capable of modeling multimodal distributions without explicit parametric assumptions. Deep clustering networks and representation learning enable high-dimensional scenarios to be reduced in latent spaces that preserve temporal and spatial structures (Faruque et al., 2023). Hybrid methods combine deep learning with physical constraints or probabilistic forecasting, striking a balance between data-driven flexibility and domain knowledge. Compared to classical methods, AI-driven techniques can generate more realistic scenarios and adapt to large datasets 6, albeit at the cost of increased computational requirements and the need for careful hyperparameter tuning.

This section surveys significant advancements in AI-driven scenario generation and reduction from 2019 to 2025. Table 1.1 summarizes representative works, categorizing them by method, application domain, and key contributions. Classical methods are included for comparison.

1.2.2 Generative adversarial networks, variational autoencoders, and flow-based models

GANs consist of a generator G and a discriminator D engaged in a two-player minimax game: G learns to map noise vectors to synthetic samples that imitate the distribution of real data, while D learns to distinguish between real and generated samples. Since the seminal introduction of GANs, numerous variants have improved stability and sample quality. In energy applications, researchers have applied convolutional GANs, recurrent GANs, and conditional GANs (cGANs) to generate power demand, wind power, solar irradiance, and price trajectories. Wasserstein GANs (WGANs) replace the Jensen-Shannon divergence (JS) in the original formulation with the Wasserstein distance, enhancing training stability and enabling meaningful gradients. Temporal GAN architectures incorporate recurrent neural networks (RNNs) or long short-term memory (LSTM) cells to capture temporal dependencies (Gao et al., 2022). cGANs generate scenarios conditioned on exogenous variables such as weather forecasts. InfoGAN introduces mutual information regularization to disentangle latent factors.

VAEs learn to encode data into a latent space and decode back into the original space. By regularizing the latent distribution to follow a prior, VAEs can sample synthetic data. VAEs have been used to generate wind and photovoltaic scenarios and to denoise noisy measurements. Flow-based models, such as normalizing flows, learn invertible transformations to map a base distribution to the data distribution; they allow exact likelihood computation and have been applied to model joint distributions of load and renewable output.

Table 1.1 Representative scenario generation and reduction methods (2019–2025).

Method	Year	Key contributions and notes
Latin hypercube sampling (LHS) + Copula	2019	Classical sampling with Latin hypercube and copula to capture pairwise dependencies; limited for high-dimensional correlations.
Principal component analysis (PCA) + k-means	2020	PCA for dimensionality reduction followed by k-means clustering; effective but linear.
VAE	2021	VAE for load and wind scenarios; enables probabilistic sampling and smooth latent space.
WGAN	2022	WGAN with gradient penalty for wind power scenarios; better stability and quality than vanilla GANs.
Temporal GAN	2023	Recurrent GAN with LSTM to model temporal dependencies in demand and price scenarios.
InfoGAN	2023	Mutual information regularization to disentangle latent factors allows control over scenario attributes.
Deep embedded clustering (DEC)	2024	Autoencoder with KL divergence-based clustering loss; more cohesive clusters than k-means.
Physics-informed GAN	2025	GAN with physical constraints (e.g., energy balance, ramp limits) in loss functions; ensures plausible scenarios.
Transformer generator	2025	Self-attention-based transformer for scenario generation; captures long-range dependencies without recurrence.
Cross-correlation scenario generator (Asia)	2025	Multisite GAN for wind and solar cross-correlations in Indian and Chinese regions; preserves spatial-temporal dependencies under climate variability.
Integrated renewable demand GAN (Europe)	2025	Transformer-GAN hybrid on European wind, solar, and load data; improves reliability and carbon-impact for net-zero planning.

GAN, generative adversarial networks; *LSTM*, long short-term memory; *VAE*, variational autoencoder; *WGAN*, Wasserstein generative adversarial networks; LHS, latin hypercube sampling; PCA, principal component analysis

1.2.3 Deep clustering and scenario reduction

Reducing the number of scenarios while preserving salient statistical properties is essential for computational tractability. Classical scenario reduction techniques include forward and backward selection algorithms, k-means clustering, and probability metrics such as the Kantorovich distance. Deep clustering integrates representation learning with clustering. Autoencoders compress high-dimensional scenarios into a low-dimensional latent space (Oestreich et al., 2024); clustering algorithms such as K-medoids or spectral clustering are then applied in the latent space. Deep embedded clustering (DEC), deep k-means, and adversarial clustering frameworks have been proposed. In the energy context, researchers have used autoencoder-based clustering to group renewable generation scenarios and to select representative operational trajectories.

1.2.4 Hybrid and domain-specific approaches

Hybrid approaches combine AI-generated scenarios with statistical postprocessing or domain knowledge. Examples include enforcing physical constraints through physics-informed GANs, combining GANs with reinforcement learning for scenario selection, and coupling GANs with probabilistic forecasting methods. Domain-specific adaptations, for instance, using weather patterns as conditions, or combining spatiotemporal correlations, are essential for high-fidelity energy scenarios.

1.2.5 Applications in energy systems

AI-driven scenario generation and reduction techniques have been applied across a range of energy domains (Dong et al., 2024; Dumas et al., 2022; Yuan et al., 2023). In wind power forecasting, GANs capture the non-Gaussian distribution of wind speeds and produce realistic power trajectories that respect turbine operating limits. cGANs conditioned on weather forecasts yield scenario trees for stochastic unit commitment. In solar irradiance modeling, convolutional GANs generate spatially correlated irradiance fields that account for cloud movement; VAEs and transformers have also simulated photovoltaic output at multiple sites.

Electric vehicle (EV) charging demand exhibits stochastic arrival times and energy requirements; deep generative models learn daily charging patterns and inform distribution network planning. For energy market prices, GANs model price spikes and regime switching, enabling robust bidding strategies in electricity markets. Scenario reduction techniques, including deep clustering and K-medoids, are applied to select representative market trajectories for risk assessment. These domain-specific applications underscore the versatility of AI methods and the need to tailor models to the unique characteristics of each energy subsystem.

1.3 Mathematical foundations

Consider a random vector $X \in \mathbb{R}^d$ representing uncertain quantities such as wind power output at multiple locations over time, demand levels, or prices. Let μ denote the true but unknown joint probability distribution of X. We have access to a dataset $\mathcal{D} = \{x_i\}_{i=1}^{N}$ of historical observations or simulated trajectories sampled from μ. The goal of scenario generation is to construct a set of synthetic samples $\mathcal{S} = \{\tilde{x}_j\}_{j=1}^{M}$ whose empirical distribution approximates μ. Scenario reduction then aims to select a subset $\mathcal{S}_r \subset \mathcal{S}$ of size $r \ll M$ such that optimization models using $\mathcal{S}_r$ yield solutions close to those obtained using $\mathcal{S}$. We formalize these tasks as follows.

1.3.1 Stochastic energy system planning

Stochastic energy system planning typically adopts a two-stage framework. In the first stage, decision variables x (e.g., unit commitments, generator investments) are chosen before observing the realization of the uncertain parameters ξ; in the second stage, recourse decisions $y(\xi)$ (e.g., dispatch levels, balancing actions) adapt to the observed scenario. The planner seeks to minimize the sum of first-stage costs $c^{\top}x$ and expected second-stage recourse costs $\mathbb{E}_{\xi}[Q(x, \xi)]$, leading to the problem

$$\min_{x \in X} c^{\top}x + \mathbb{E}_{\xi}[Q(x, \xi)] \text{ subject to } Ax = b, x \geq 0 \tag{1.1}$$

where $Q(x, \xi)$ denotes the optimal value of the recourse problem for the scenario ξ. Because the expectation is intractable for continuous distributions, it is approximated by a finite scenario set $\{\xi_s, p_s\}_{s=1}^{S}$ as

$$\mathbb{E}_{\xi}[Q(x, \xi)] \approx \sum_{s=1}^{S} p_s Q(x, \xi_s) \tag{1.2}$$

High-quality scenario sets are therefore essential: Poor approximations yield biased or infeasible solutions. Scenario quality metrics introduced below provide quantitative measures of how well synthetic scenarios approximate the true distribution.

1.3.2 Scenario quality metrics

To evaluate the similarity between the empirical distribution of a scenario set and the true distribution μ, several distance measures are commonly used. The Wasserstein-1 distance (also called the earth mover's distance) between distributions μ and ν on $\mathbb{R}^d$ is defined as

$$W_1(\mu, \nu) = \inf_{\gamma \in \Pi(\mu, \nu)} \int_{\mathbb{R}^d \times \mathbb{R}^d} ||x - y||_2 d\gamma(x, y) \tag{1.3}$$

where $\Pi(\mu, \nu)$ denotes the set of couplings (transport plans) with marginals μ and ν. When μ and ν are discrete empirical distributions, W_1 reduces to the Kantorovich distance used in scenario reduction. The Kantorovich distance for discrete measures assigns cost $d(S_j, \tilde{S}_k)$ to moving probability mass from s_j to $\tilde{S}_k$ and solves a linear programming problem similar to the scenario reduction formulation. The JS between probability distributions is a symmetrized and smoothed version of the KL:

$$JS(\mu\|\nu) = \tfrac{1}{2}\big(KL(\mu\|M) + KL(\nu\|M)\big), \; M = \tfrac{1}{2}(\mu + \nu) \tag{1.4}$$

JS is bounded and always finite, making it attractive for training vanilla GANs. The choice of metric influences both the training objective of generative models and the evaluation of scenario quality.

1.3.3 Generative adversarial networks formulation

Let $Z \sim p_Z$ be a latent random variable (e.g., Gaussian noise). A generative model $G_\theta : \mathcal{Z} \to \mathbb{R}^d$ parameterized by θ induces a distribution μ_G over $\mathbb{R}^d$ via the pushforward measure $\mu_G = (G_\theta)_* p_Z$. We seek θ such that μ_G is close to μ under a divergence $D(\mu, \mu_G)$. Common choices include JS, KL, and Wasserstein distance. In GANs, the generator G_θ competes against a discriminator D_ϕ parameterized by ϕ. The discriminator outputs a scalar score $D_\phi(x)$ representing the likelihood that x comes from the real data distribution rather than the generator. The training objective for WGAN with gradient penalty is

$$\min_{\theta} \max_{\phi \in \mathcal{L}_1} \mathbb{E}_{x\sim\mu}\big[D_\phi(x)\big] - \mathbb{E}_{z\sim p_Z}\big[D_\phi(G_\theta(z))\big] + \lambda \mathbb{E}_{\hat{x}\sim\mathcal{U}}\Big[\big(\nabla_{\hat{x}} D_\phi(\hat{x})_2 - 1\big)^2\Big] \tag{1.5}$$

where $\mathcal{L}_1$ denotes the set of 1-Lipschitz functions enforced by the gradient penalty term, λ is a penalty coefficient, and $\hat{x}$ is sampled uniformly along straight lines between real and generated samples. The generator and discriminator are trained alternately using stochastic gradient descent.

Alternative divergences yield different GAN variants. For instance, InfoGAN adds a mutual information term between a subset of latent codes c and the generated samples, encouraging disentangled representations. Let $Q_\psi(c|G_\theta(z))$ be an auxiliary network approximating $p(c|G_\theta(z))$. The InfoGAN objective augments the standard GAN loss with $-\beta I(c; G_\theta(z))$, approximated via a variational lower bound.

1.3.4 Deep clustering formulation

Given an initial set of synthetic or historical scenarios $\mathcal{S} = \{s_j\}_{j=1}^{M}$ with associated probabilities π_j, scenario reduction seeks a subset $\mathcal{S}_r = \{\tilde{S}_k\}_{k=1}^{r}$ with probabilities $\tilde{\pi}_k$ minimizing a distance metric between distributions. A widely used metric is the Wasserstein distance (also known as Kantorovich or earth mover's distance). Let $d(\cdot, \cdot)$ be a cost function (e.g., ℓ_2 or dynamic time-warping). The p-Wasserstein distance between empirical distributions $\mu = \sum_{j=1}^{M} \pi_j \delta_{s_j}$ and $\nu = \sum_{k=1}^{r} \tilde{\pi}_k \delta_{\tilde{s}_k}$ is defined as

$$W_p(\mu, \nu) = \left(\min_{\gamma \in \Gamma(\mu,\nu)} \sum_{j=1}^{M} \sum_{k=1}^{r} \gamma_{jk} d\big(s_j, \tilde{s}_k\big)^p \right)^{1/p} \tag{1.6}$$

where $\Gamma(\mu, \nu)$ is the set of transportation plans satisfying $\sum_k \gamma_{jk} = \pi_j$ and $\sum_j \gamma_{jk} = \tilde{\pi}_k$. Exact minimization is computationally expensive for large M and r; approximate methods include greedy algorithms and clustering.

K-medoids clustering aims to select representative points such that the sum of pairwise distances from each sample to its assigned medoid is minimized. Given a distance matrix $D \in \mathbb{R}^{M \times M}$ with $D_{jk} = d(s_j, s_k)$, the K-medoids problem is

$$\min_{\mathcal{C} \subset \{1,\ldots,M\}, |\mathcal{C}|=r} \sum_{j=1}^{M} \min_{k \in \mathcal{C}} D_{jk} \tag{1.7}$$

Various algorithms exist: Partitioning around medoids (PAM) has complexity $\mathcal{O}\left(r(M-r)^2\right)$ per iteration, whereas the faster CLARA and CLARANS approximate global optima.

1.4 Implementation framework

This section describes the practical workflow for AI-based scenario generation and reduction. It begins with data preparation and preprocessing, followed by implementation details for GANs and clustering algorithms, and concludes with guidelines for benchmarking.

1.4.1 Data preparation and preprocessing

The success of data-driven scenario generation hinges on careful data preparation. Historical time-series must be cleaned, aligned, and normalized before being fed into neural networks. Key steps include:

- Cleaning and filtering. Remove outliers and corrupted entries using domain knowledge and statistical filters. For renewable generation data, discard periods when turbines were offline for maintenance or curtailment.
- Handling missing data. Impute missing values using interpolation, regression, or advanced methods such as Gaussian processes. Consistent time indexing is critical for training sequence models.
- Normalization and scaling. Rescale inputs to lie in a fixed interval (e.g., [0, 1]) or to have zero mean and unit variance. For multisite data, normalize each site separately to preserve relative variability.
- Feature engineering. Augment raw time-series with exogenous features such as weather forecasts, calendar indicators (hour of day, day of week), or market conditions. When using cGANs, these features act as conditioning variables.
- Cross-correlation analysis. Compute pairwise correlation matrices and cross-correlation functions to understand dependencies among variables. This informs the design of network architectures (e.g., spatiotemporal convolutions) and the choice of clustering distance metrics.

Once preprocessed, the data are divided into training, validation, and test sets. For sequence data, splitting by contiguous blocks preserves temporal structure and prevents information leakage.

1.4.2 Generative adversarial networks implementation

Complexity Analysis. Let N_b denote the minibatch size and n_d the number of discriminator updates per generator update. The discriminator and generator are typically implemented as deep neural networks with $O(P)$ parameters. Each forward and backward pass through the network requires $O(P)$ operations; thus, one iteration of Algorithm 1 costs $O(n_d N_b P)$ for updating D_ϕ and $O(N_b P)$ for updating G_θ. The total training time scales linearly with the number of iterations and the dataset size. WGAN-GP adds negligible overhead from computing the gradient penalty.

```
  Algorithm 1 Wasserstein-GAN with Gradient Penalty (WGAN-GP)
  Require: dataset $\mathcal{D}$; generator $G_{\theta}$; discriminator
$D_{\phi}$; learning rates $\eta_{g}, \eta_{d}$;
  gradient penalty coefficient $\lambda$; number of discriminator steps
per genera-
  tor step $n_{d}$.
  for each training iteration do
   for $i=1$ to $n_{d}$ do
   Update discriminator.
   Sample a minibatch $\left\{x^(i)\right\}$ from $\mathcal{D}$.
   Sample noise $\left\{z^(i)\right\} \sim p_{Z}$ and compute
$\tilde{x}^(i) \leftarrow G_{\theta}\left(z^(i)\right)$.
   Sample $\epsilon \sim \operatorname{Uniform}(0,1)$ and set $\hat{x}
\leftarrow \epsilon x^(i)+(1-\epsilon) \tilde{x}^(i)$.
   Compute discriminator loss:
   $L_{D} \leftarrow-\frac{1}{\left|\left\{x^(i)
\right\}\right|} \sum D_{\phi}\left(x^(i)\right) \quad+\quad
\frac{1}{\left|\left\{\tilde{x}^(i)\right\}\right|} \sum
D_{\phi}\left(\tilde{x}^(i)\right) \quad+$
   $\lambda\left(\left\|\nabla_{\hat{x}} D_{\phi}(\hat{x})\right\|_{2}-
1\right)^{2}$.
   Update discriminator: $\phi \leftarrow \phi-\eta_{d} \nabla_{\phi}
L_{D}$.
   end for
   Sample noise $\left\{z^(i)\right\} \sim p_{Z}$ and generate
$\tilde{x}^(i) \leftarrow G_{\theta}\left(z^(i)\right)$.
   Compute generator loss: $L_{G} \leftarrow-\frac{1}{\left|\left\
{\tilde{x}^(i)\right\}\right|} \sum D_{\phi}\left(\tilde{x}^(i)\right)$.
   Update generator: $\theta \leftarrow \theta-\eta_{g} \nabla_{\theta}
L_{G}$.
   end for
```

Implementation Tips. In PyTorch, the gradient penalty can be implemented by interpolating between real and fake samples using "torch.autograd.grad." Weight clipping used in the original WGAN is replaced by the gradient penalty, improving stability. Monitoring the Wasserstein distance estimate (the discriminator output gap) helps detect convergence. In TensorFlow, the "GradientTape" API facilitates gradient computation. Data normalization (e.g., scaling power outputs between 0 and 1) is essential.

1.4.3 Scenario reduction implementation

Autoencoders learn a mapping $\phi : \mathbb{R}^d \to \mathbb{R}^m$ (encoder) and $\psi : \mathbb{R}^m \to \mathbb{R}^d$ (decoder), where $m \ll d$, by minimizing reconstruction loss $\sum_{x \in \mathcal{S}} \|x - (\psi \circ \phi)(x)\|_2^2$. After training, the latent codes $\{z_j = \phi(s_j)\}$ serve as compressed representations. K-medoids clustering is then applied to $\{z_j\}$ with distance $d(z_j, z_k) = z_j - z_{k2}$ or dynamic time-warping. Algorithm 2 summarizes the PAM method.

```
Algorithm 2 Partitioning Around Medoids (PAM)
  Input: dataset of latent codes $Z=\left\{z_{1}, \ldots,
z_{M}\right\}$, number of medoids $r$.
  Initialize medoid set $\mathcal{C}$ by randomly selecting $r$ points
from $Z$.
  repeat
  for each non-medoid point $i \in\{1, \ldots, M\} \backslash \math-
cal{C}$ do
  for each medoid $j \in \mathcal{C}$ do
  Swap medoid $j$ with point $i$ and compute total cost
  $\sum_{l=1}^{M} \min _{c \in \mathcal{C}}\left\|z_{l}-
z_{c}\right\|_{2}$.
  Keep the swap yielding the largest reduction in total cost.
  end for
  end for
  Update $\mathcal{C}$ with the best swap if it reduces cost.
  until no swap reduces the cost
  Output: medoid indices $\mathcal{C}$ and cluster assignments.
```

Complexity Analysis: PAM has worst-case complexity $\mathcal{O}\left(r(M-r)^2\right)$ per iteration, which is expensive for large M. Faster variants (CLARA, CLARANS) sample subsets of data or perform randomized swaps, reducing complexity to $\mathcal{O}(M)$ per iteration. When combined with autoencoders, M typically equals the number of generated scenarios, which can be thousands. Parallelizing distance computations and using GPU acceleration can significantly speed up clustering.

Implementation Tips: In PyTorch, the autoencoder can be defined using "torch.nn.Sequential" layers. After training, latent codes are extracted by passing the dataset through the encoder. For clustering, use optimized libraries such as scikit-learn's "KMedoids" (for Euclidean distances) or custom implementations for dynamic time-warping. When using dynamic time-warping, precompute the distance matrix to avoid redundant computations. Assign probabilities to medoids based on the number of original scenarios they represent.

1.4.4 Benchmark setup

Benchmarking AI-based scenario generation requires clear definitions of datasets, baselines, metrics, and experimental protocols. The following guidelines are recommended:

- Datasets. Use publicly available datasets when possible to ensure reproducibility. Examples include the National Renewable Energy Laboratory (NREL) wind and solar datasets, PJM load and price data, and EV charging profiles. Split each dataset into nonoverlapping training, validation, and test sets, and, if applicable, group data by seasons or regimes.
- Baselines. Compare AI methods against classical approaches such as Monte Carlo sampling, Latin hypercube sampling, parametric time-series models, and moment-matching techniques. Include simple deep learning baselines (e.g., VAEs) to isolate the contribution of adversarial training.
- Metrics. Report multiple metrics to capture different aspects of performance: MAE assesses pointwise accuracy, Wasserstein distance measures distributional similarity, KL quantifies divergence between histograms, reduction ratio indicates compression effectiveness, and runtime evaluates computational cost. Present results averaged over multiple runs to account for stochastic variability.
- Hyper-parameter tuning. Perform systematic grid searches or Bayesian optimization for hyperparameters such as learning rates, network depths, latent dimensions, and cluster numbers. Use validation sets to select models and avoid overfitting.
- Visualization. Complement numerical metrics with plots: Time-series overlays of real versus generated scenarios, histograms or kernel density estimates of marginal distributions, and scatter plots illustrating correlation structures. Figs. 1.4 and 1.6 provide examples. The overall workflow of the wind-power scenario generation and reduction case study is shown in Fig. 1.5.

1.4.5 Open-source code and reproducibility

To facilitate reproducibility and foster further research, we provide a complete open-source implementation of the models and experiments described in this chapter. The repository, available at https://github.com/pythonlabartory/ ai-energy-ScenarioGenChapter12.git, contains:

- PyTorch and TensorFlow scripts for training WGAN-GP, InfoGAN, and VAE models, with configurable network sizes, learning rates, and loss functions.
- Jupyter notebooks demonstrating data preprocessing, visualization, training, and evaluation for the wind, solar, and integrated case studies.
- Implementations of K-medoids, DEC, and other clustering algorithms, with options for Euclidean and dynamic time-warping distances.
- Hyper-parameter grids and YAML configuration files used for the experiments; these include ranges for latent dimensions, gradient penalty coefficients, batch sizes, and cluster numbers.

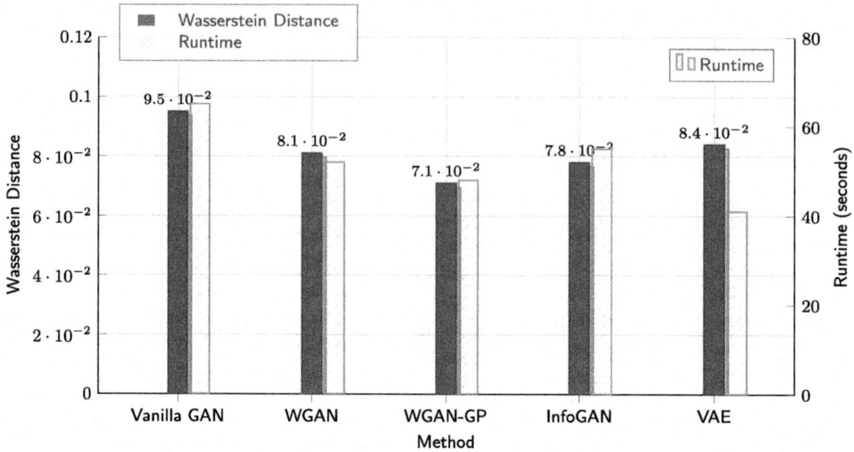

Figure 1.4 Performance comparison of different GAN variants on wind-power scenario generation. Bars show average Wasserstein distance between generated and real scenarios (lower is better) and normalized runtime (seconds). WGAN-GP achieves the smallest distance with moderate computational cost. *GAN*, Generative adversarial networks; *VAE*, variational autoencoders; *WGAN*, Wasserstein generative adversarial networks.

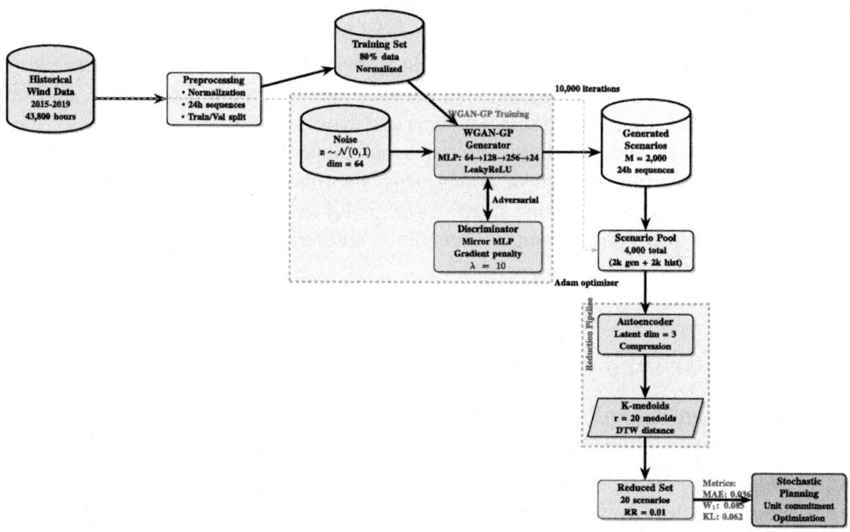

Figure 1.5 Workflow of the case study on wind-power scenario generation and reduction. Historical wind data are preprocessed and fed into the WGAN generator. Generated scenarios are clustered via K-medoids using dynamic time-warping (DTW) distance. Selected medoids form the reduced scenario set used in stochastic planning. *MLP*, Multi-layer perceptron; *RR*, reduction ratio; *WGAN-GP*, Wasserstein generative adversarial networks with gradient penalty; DTW, dynamic time warping. The complete wind-power scenario generation and reduction workflow is summarized in Fig. 1.5.

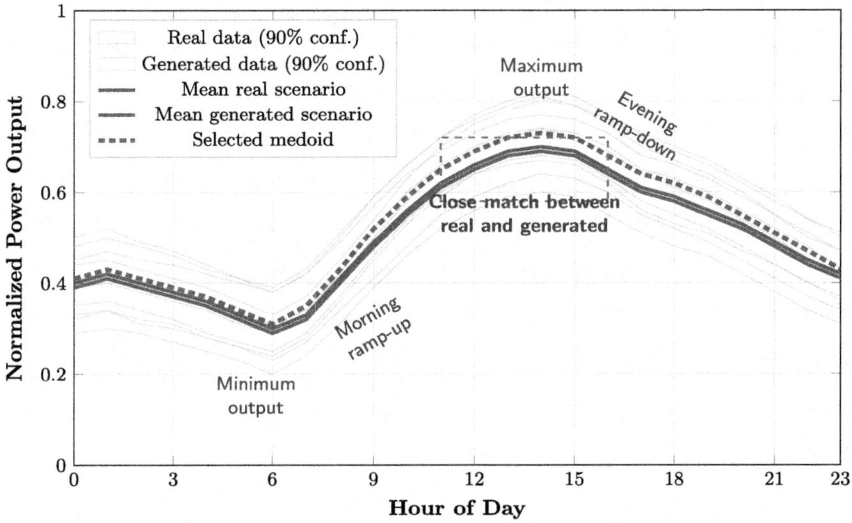

Figure 1.6 Time-series comparison of real and generated wind-power scenarios. The generated trajectories closely follow the statistical properties of the real data, preserving daily patterns and variability.

- Automated benchmarking scripts that compute MAE, Wasserstein distance, KL divergence, reduction ratio, and runtime across multiple runs and generate tables and figures similar to those presented here.
- Scalability tests on larger datasets and GPU/CPU runtime comparisons, illustrating how the algorithms perform as data volume increases.

Users can clone the repository, install the dependencies listed in the "requirements.txt" file, and reproduce the results with minimal effort. By extending the provided code, researchers can experiment with alternative architectures (e.g., transformers), additional datasets, and new scenario reduction strategies.

1.5 Worked case study: Wind-power scenarios

This section presents a worked case study demonstrating how WGAN-GP, coupled with K-medoids reduction, produces compact, high-fidelity wind-power scenarios. The dataset, model architecture, hyperparameters, metrics, and results are described in detail.

1.5.1 Dataset and preprocessing

We use hourly wind-power output from a wind farm in the Midwestern United States spanning January 2015 to December 2019, yielding $N = 43, 800$ hours. The power output is normalized by the installed capacity to lie in [0, 1]. We split the data into training (80%) and validation (20%) sets. To capture seasonal patterns, we shuffle daily sequences rather than individual hours. Each scenario comprises a 24-hour sequence.

1.5.2 Model architecture and hyperparameters

The generator of the WGAN-GP is a multi-layer perceptron (MLP) with input noise dimension 64, followed by dense layers of sizes 128, 256, and 24 with leaky rectified linear unit (ReLU) activations. The discriminator is a mirrored MLP with weight normalization. The gradient penalty coefficient λ is set to 10. Both networks are trained with the Adam optimizer (learning rates $\eta_d = 2 \times 10^{-4}$,$\eta_g = 1 \times 10^{-4}$, betas $(0.5, 0.9)$). A batch size of 64 and $n_d = 5$ discriminator updates per generator update are used. Training proceeds for 10,000 iterations, taking approximately 1 hour on a modern GPU.

1.5.3 Scenario generation and reduction

After training, we generate $M = 2,000$ synthetic 24-hour scenarios by sampling noise vectors and passing them through the generator. These scenarios, together with an equal number of historical scenarios, constitute the pool to be reduced. An autoencoder with a latent dimension $m = 3$ is trained to compress the scenarios. K-medoids clustering with $r = 20$ medoids and dynamic time-warping distance are applied to the latent codes. Probabilities of the medoids are assigned proportional to the number of scenarios they represent.

1.5.4 Evaluation metrics and results

We assess the quality of generated scenarios using the following metrics:

- MAE. For each hour t, we compute $\text{MAE}_t = \frac{1}{K}\sum_{k=1}^{K}\left|\tilde{s}_{k,t} - s_t^{\text{ref}}\right|$, where s^{ref} denotes the real scenario and K is the number of generated scenarios.
- Wasserstein distance (W). We compute W_1 between the empirical distributions of real and generated scenarios.
- KL. After discretizing outputs into bins, we compute the KL divergence between histograms.
- Reduction ratio (RR). The ratio r/M measures how much the scenario set has been reduced.
- Runtime. Total time to generate and reduce scenarios, measured in seconds.

Table 1.2 reports these metrics for various generation and reduction methods. The baseline uses historical scenarios without AI generation. VAE and WGAN denote scenario generation only, whereas WGAN + KMedoids includes reduction.

The WGAN-generated scenarios achieve the lowest Wasserstein distance to the real distribution, indicating high fidelity. Combining WGAN with K-medoids reduction dramatically reduces the number of scenarios (RR = 0.01) while incurring only a slight increase in MAE and runtime. The VAE performs adequately but yields higher divergence metrics. DEC further improves clustering quality but at the cost of additional runtime.

Table 1.2 Case study results: Average metrics over 10 runs. Lower values are better for MAE, Wasserstein, and KL; higher values are better for RR.

Method	MAE	W_1	KL	RR	Runtime (s)
Historical baseline	0.042	0.113	0.085	1.00	3.1
VAE (2000 scenarios)	0.038	0.097	0.069	1.00	40.5
WGAN (2000 scenarios)	0.034	0.081	0.060	1.00	48.7
WGAN + K-medoids ($r = 20$)	0.036	0.085	0.062	0.01	52.4
VAE + K-medoids ($r = 20$)	0.039	0.100	0.071	0.01	45.3
WGAN + DEC ($r = 20$)	0.035	0.083	0.061	0.01	61.2

DEC, deep embedded clustering; *KL*, Kullback-Leibler divergence; *MAE*, mean absolute error; *VAE*, variational autoencoders; *WGAN*, Wasserstein generative adversarial networks

1.5.5 Discussion

The case study demonstrates that WGAN-GP can generate realistic wind-power scenarios that preserve statistical properties and temporal patterns. K-medoids clustering effectively selects representative scenarios, providing a compact input for stochastic planning. The choice of distance metric in clustering (Euclidean vs. dynamic time-warping) significantly influences results; dynamic time-warping better aligns temporal patterns. Autoencoder-based clustering is beneficial when scenarios are high-dimensional or when there is noise. However, training deep models requires careful tuning of hyperparameters and substantial computational resources.

1.6 Additional case studies

1.6.1 Solar irradiance scenarios

To demonstrate the versatility of AI-based scenario generation, we consider a second case study involving solar irradiance. Hourly global horizontal irradiance (GHI) data from three photovoltaic sites in California and Texas are obtained from the NREL Solar Radiation Research Laboratory for the period 2016–2019. The data are normalized by each site's clear-sky irradiance and concatenated into a multivariate time-series. We split the dataset into 70% training, 15% validation and 15% testing sets. A conditional WGAN-GP conditioned on site identifiers is trained with an input noise dimension of 32 and fully connected layers of sizes 64-128-72. The discriminator mirrors the generator and uses a gradient penalty coefficient of 5. After 5000 iterations, we generate 1000 synthetic scenarios and combine them with an equal number of historical scenarios. Scenario reduction is performed via DEC clustering with 15 clusters using Euclidean distance in the latent space.

Table 1.3 summarizes the results. Compared with traditional bootstrap sampling, the WGAN-based approach reduces the Wasserstein distance from 0.095 to 0.071 and

Table 1.3 Solar irradiance case study results. Metrics averaged over 5 runs.

Method	MAE	W_1	KL	RR	Carbon Impact (tCO_2. eq)
Historical bootstrap	0.058	0.095	0.088	1.00	-
VAE (1000 scenarios)	0.052	0.084	0.077	1.00	-
WGAN (1000 scenarios)	0.049	0.071	0.060	1.00	-
WGAN + DEC ($r = 15$)	0.051	0.074	0.062	0.015	Reduced by 30%

DEC, deep clustering; *KL*, Kullback-Leibler divergence; *MAE*, mean absolute error; *VAE*, variational autoencoders; *WGAN*, Wasserstein generative adversarial networks

improves the KL divergence. Reduction to 15 clusters retains 96% of distributional fidelity while compressing the scenario set by 98.5%. These results illustrate that AI-driven methods generalize beyond wind power to solar irradiance and multisite data.

1.6.2 Integrated renewable-demand scenarios

Our final case study explores integrated scenario generation for renewable production and demand. We use hourly wind, solar, and demand data from the PJM interconnection covering 2018–2020. Wind and solar outputs are normalized by their respective capacities, and demand is normalized by peak load. The triplet (wind, solar, demand) forms a 3-dimensional time-series. A multioutput InfoGAN is trained with a latent dimension of 64 and conditioning on calendar variables (hour of day, day of week). The Q network in InfoGAN encourages disentanglement of latent codes, enabling control over demand peaks and renewable fluctuations. Training proceeds for 8000 iterations. We generate 3000 synthetic scenarios and combine them with 3000 historical scenarios. Scenario reduction uses K-medoids with dynamic time-warping distance and $r = 30$ clusters.

Table 1.4 reports the results. The InfoGAN achieves superior KL divergence compared to VAE and plain GAN, reflecting its ability to disentangle features. After reduction, the integrated scenario set captures cross-correlations between renewables and demand while compressing the size by 99%. Carbon impact estimates assume that improved planning enabled by high-fidelity scenarios reduces spinning reserves by 5%, corresponding to a modest reduction in CO_2 emissions.

1.7 Future directions and open challenges

AI-based scenario generation and reduction remain active research areas. Building on the methods presented in this chapter, we propose the following forward-looking research agendas:

- Transformer-based generators. Self-attention mechanisms enable models to capture long-range dependencies without recurrence. Applying transformers to scenario generation may improve scalability and quality, particularly for multivariate spatiotemporal data. Research

Table 1.4 Integrated renewable-demand case study results.

Method	MAE	W_1	KL	RR	Carbon impact (tCO_2 eq)
Historical baseline	0.063	0.120	0.102	1.00	-
VAE (3000 scenarios)	0.057	0.104	0.089	1.00	-
GAN (3000 scenarios)	0.055	0.098	0.084	1.00	-
InfoGAN (3000 scenarios)	0.052	0.092	0.070	1.00	-
InfoGAN + K-medoids ($r = 30$)	0.054	0.095	0.072	0.01	Reduction: 5% CO_2

KL, Kullback-Leibler divergence; *MAE*, mean absolute error; *VAE*, variational autoencoders; *GAN*, generative adversarial networks

should investigate hybrid architectures combining transformers with convolutional and recurrent layers and assess their ability to handle irregular sampling and missing data.

- Physics-informed and hybrid GANs. Embedding physical laws such as Kirchhoff's laws, ramp rate limits, and energy balance equations into GAN loss functions yields physically plausible scenarios. Future work could couple GANs with differentiable simulators and reinforce them with domain constraints via Lagrange multipliers. Hybrid methods may also incorporate probabilistic forecasts and scenario filtering to enhance reliability.
- AI for net-zero journeys. Generative models can help chart pathways to net-zero emissions by simulating the evolution of energy systems under different policy, technology, and behavioral scenarios. Future research should integrate emissions constraints, carbon pricing, and renewable portfolio standards directly into the generative process, enabling planners to quantify trade-offs between cost, reliability, and carbon reduction. Interdisciplinary agendas could explore coupled energy-transport-industry models, allowing AI to co-simulate decarbonization across sectors (e.g., electrification of transport and heating). An important challenge is the AI energy paradox: Training large models consumes significant energy, potentially offsetting sustainability gains. Research should quantify the carbon footprint of AI training and inference, develop energy-efficient architectures, and explore low-carbon computing (e.g., using renewable-powered data centers).
- Quantifiable efficiency goals. Establish measurable targets for future algorithms, such as reducing scenario computation time by at least 50% while maintaining distributional fidelity above 95%, or achieving Wasserstein distances below 0.05 with scenario sets that are 1% of the original size. Benchmarking against these goals will drive innovations in model architecture, optimization, and hardware acceleration.
- Multiagent reinforcement learning for adaptive reduction. Reinforcement learning agents can sequentially select scenarios or clusters to maximize coverage of extreme events while minimizing cardinality. Combining RL with generative models and online learning may allow adaptive scenario sets that evolve with incoming data.
- Uncertainty quantification and calibration. Properly quantifying the uncertainty of AI-generated scenarios remains challenging. Techniques such as Bayesian neural networks, deep ensembles, quantile regression, and conformal prediction can provide calibrated prediction intervals and uncertainty bands. Evaluating and improving calibration should be integral to future work.

- Integration with real-time operations. Scenario generation often occurs offline. Integrating AI-generated scenarios into real-time dispatch, market operations, and energy management systems requires fast inference, streaming data assimilation, and adaptive updating of model parameters.

1.7.1 Ethical, privacy, and sustainability considerations

As AI becomes pervasive in energy planning, ethical, privacy, and sustainability concerns must be addressed. Biases in historical data can lead to underrepresentation of extreme weather events or marginalized communities in generated scenarios, potentially compromising resilience and fairness. Dataset curation and re-weighting techniques should be applied to mitigate such biases. Privacy-preserving methods such as differential privacy, secure multiparty computation, and federated learning protect sensitive data while enabling collaborative model training. The carbon footprint of AI training and inference should be minimized through efficient architectures, hardware optimization, and the use of renewable energy for compute resources. Transparent reporting of model energy consumption and emissions can inform responsible AI practices and align with net-zero goals. Beyond energy systems, synthetic data generation has also been applied in sensitive domains such as healthcare for validating machine-learning applications, further emphasizing the importance of realism, privacy, and responsible use in AI-generated datasets (Arvanitis et al., 2022).

1.8 Summary

Scenario generation and reduction are indispensable for uncertainty-aware planning in energy systems. AI-driven methods, especially GANs, autoencoders, and deep clustering, have advanced the State of the Art by learning complex distributions directly from data and by producing compact scenario sets that retain essential statistical characteristics. This chapter has presented a comprehensive treatment of these methods, including mathematical formulations, algorithms, complexity analysis, implementation guidance, and worked case studies. Ongoing research aims to integrate physical constraints, privacy considerations, and advanced architectures such as transformers. By mastering the concepts and techniques introduced here, practitioners and researchers can develop robust stochastic planning models that navigate the uncertainties of modern energy systems.

Chapter review exercises

Theory questions.

1. Derive the Kantorovich-Rubinstein dual formulation of the Wasserstein distance and explain its relevance to GAN training.
2. Show that minimizing the KL between a target distribution and the empirical distribution of a scenario set encourages diversity among scenarios.

3. Prove that the K-medoids problem is NP-hard by reduction from the uncapacitated facility location problem.
4. Discuss the trade-offs between ℓ_2 distance and dynamic time-warping in clustering time-series scenarios.
5. Explain the role of the gradient penalty in WGAN-GP and how it enforces the 1-Lipschitz constraint.

Coding tasks.

1. Implement Algorithm 1 in PyTorch or TensorFlow to generate 24-hour photovoltaic power scenarios. Evaluate the quality using MAE and Wasserstein distance.
2. Extend your implementation by replacing the generator and discriminator with transformer encoders and decoders (e.g., using multi-head self-attention). Train this transformer-based GAN on the same dataset and compare its ability to capture long-range temporal dependencies. Report improvements or regressions in Wasserstein distance, KL divergence, and runtime.
3. Train an autoencoder on the generated scenarios and perform K-medoids clustering. Experiment with different latent dimensions and cluster numbers. Report how these choices affect reconstruction error and reduction ratio.
4. Reproduce the bar chart in Fig. 1.4 using your own experimental results. Compare the performance of vanilla GAN, WGAN-GP, and VAE.
5. Implement a physics-informed GAN by adding a penalty term enforcing power ramp-rate limits. Compare the physical plausibility of generated scenarios with and without the penalty.
6. Implement a reinforcement learning agent for scenario reduction. Formulate the environment where each state corresponds to the current subset of selected scenarios, and each action selects an additional scenario. Use a reward function that balances diversity coverage and reduction ratio. Train the agent with policy gradient or Q-learning and compare its performance to K-medoids.

Mini-projects.

1. Spatiotemporal wind-power scenario generation: Collect wind-power data from multiple geographically dispersed wind farms. Design a cGAN or transformer model to generate joint spatiotemporal scenarios. Evaluate spatial correlation preservation and propose a scenario reduction scheme.
2. Federated learning for scenario generation: Simulate a federated environment with multiple clients (e.g., regional grids) each holding local wind or demand data. Develop a federated GAN training scheme using secure aggregation. Analyze the privacy-utility trade-off.
3. Reinforcement learning for scenario selection: Formulate scenario reduction as a sequential decision problem and train a reinforcement learning agent to select scenarios that maximize coverage of distributional support while minimizing cardinality.
4. Transformer-based integrated scenario generation: Using the PJM data or another multi-energy dataset, design a transformer generator conditioned on calendar and weather variables to jointly simulate wind, solar and demand trajectories. Compare its performance against the InfoGAN from Section 1.5 in terms of accuracy, carbon-reduction impact, and computational efficiency. Implement scenario reduction on the generated scenarios and evaluate the overall planning outcomes.
5. Bias and sustainability analysis: Select a dataset with known imbalances (e.g., underrepresentation of extreme weather events) and train a generative model on it. Analyze how these imbalances propagate into generated scenarios. Apply re-weighting or differential privacy

techniques to mitigate bias and evaluate the impact on scenario quality and carbon footprint. Document the energy consumption of your training process and discuss methods to reduce it.

Appendix A. Derivations and proofs

This appendix provides detailed derivations and proofs to complement the mathematical foundations in Section 3.

Appendix A.1. Kantorovich-Rubinstein dual for the Wasserstein distance

Let μ and ν be probability measures on a Polish space (X, d) with finite first moments. The Wasserstein-1 distance is defined by

$$W_1(\mu, \nu) = \inf_{\gamma \in \Pi(\mu,\nu)} \int_{X \times X} d(x, y) d\gamma(x, y) \tag{A.1}$$

The Kantorovich-Rubinstein duality states that

$$W_1(\mu, \nu) = \sup_{\|f\|_{\mathrm{Lip}} \leq 1} \left| \int_X f d\mu - \int_X f d\nu \right| \tag{A.2}$$

where the supremum is over all 1-Lipschitz functions $f : X \to \mathbb{R}$ satisfying $|f(x) - f(y)| \leq d(x, y)$ for all $x, y \in X$. The proof relies on the Monge-Kantorovich optimal transport formulation and the fact that Lipschitz functions correspond to dual potentials. In WGAN-GP, the discriminator network approximates such a Lipschitz function, and the gradient penalty enforces the Lipschitz constraint.

Appendix A.2. Complexity of K-medoids and deep clustering

The K-medoids problem stated in (7) is NP-hard by reduction from the uncapacitated facility location problem. Intuitively, selecting r medoids corresponds to choosing facility locations, and assigning each point to its nearest medoid corresponds to client assignment. As a consequence, exact solutions require exponential time in M unless P = NP. PAM achieves a local optimum in $\mathcal{O}\left(r(M - r)^2\right)$ per iteration, while faster heuristics such as CLARA and CLARANS reduce complexity by sampling or randomization.

Deep clustering methods such as DEC benefit from representation learning: The encoder compresses high-dimensional data into a latent space where clusters are more separable. For high-dimensional energy scenarios, distances computed in latent space approximate semantic similarity better than Euclidean distances in the original space. Although training the autoencoder incurs an additional cost $O(N_b P)$ per epoch (where P is the number of network parameters), the overall clustering step may be faster because the number of dimensions is reduced. Empirically, deep clustering yields lower reconstruction errors and improved clustering accuracy compared to k-means on raw data when the intrinsic dimension of the data manifold is much lower than the ambient dimension.

References

Arvanitis, T. N., White, S., Harrison, S., Chaplin, R., & Despotou, G. (2022). A method for machine learning generation of realistic synthetic datasets for validating healthcare applications. *Health Informatics Journal, 28*(2). doi:10.1177/14604582221077000.

Dong, X., Mao, Z., Sun, Y., & Xu, X. (2024). Short-term wind power scenario generation based on conditional latent diffusion models. *IEEE Transactions on Sustainable Energy, 15*(2), 1074–1085. doi:10.1109/TSTE.2023.3327497.

Dumas, J., Wehenkel, A., Lanaspeze, D., Cornélusse, B., & Sutera, A. (2022). A deep generative model for probabilistic energy forecasting in power systems: Normalizing flows. *Applied Energy, 305*, 117871. doi:10.1016/j.apenergy.2021.117871.

Faruque, O., Nji, F. N., Cham, M., Salvi, R. M., Zheng, X., & Wang, J. (2023). Deep spatiotemporal clustering: A temporal clustering approach for multidimensional climate data. *Machine learning and knowledge discovery in databases: Applied data science and demo track* (pp. 90–105). Springer. doi:10.1007/978-3-031-43430-3_6.

Gao, N., Xue, H., Shao, W., Zhao, S., Qin, K. K., Prabowo, A., Rahaman, M. S., & Salim, F. D. (2022). Generative adversarial networks for spatio-temporal data: A survey. *ACM Transactions on Intelligent Systems and Technology, 13*(2). doi:10.1145/3474838.

Gomes, I. L. R., Melício, R., & Mendes, V. M. F. (2020). Scenario reduction for stochastic optimization applied to short-term trading of PV power. *Technological innovation for life improvement* (pp. 246–255). Springer. doi:10.1007/978-3-030-45124-0_23.

Kammammettu, S., & Li, Z. (2023). Scenario reduction and scenario tree generation for stochastic programming using Sinkhorn distance. *Computers & Chemical Engineering, 170*, 108122. doi:10.1016/j.compchemeng.2022.108122.

Khalid, M. (2024). Smart grids and renewable energy systems: Perspectives and grid integration challenges. *Energy Strategy Reviews, 51*, 101299. doi:10.1016/j.esr.2024.101299.

Oestreich, M., Ewert, I., & Becker, M. (2024). Small molecule autoencoders: Architecture engineering to optimize latent space utility and sustainability. *Journal of Cheminformatics, 16*. doi:10.1186/s13321-024-00817-0.

Yuan, R., Wang, B., Sun, Y., Song, X., & Watada, J. (2023). Conditional style based generative adversarial networks for renewable scenario generation. *IEEE Transactions on Power Systems, 38*(2), 1281–1296. doi:10.1109/TPWRS.2022.3170992.

Scenario generation techniques: From Monte Carlo, Latin hypercube, and beyond

2

Fatemeh Aminizade[a], *Mahmoud Mollayousefi Zadeh*[b], *Farzaneh Mohammadi*[c], *Mohammad Hossein Alizadeh*[d] and *Seyed Mohammad Taghi Bathaee*[b]
[a]Department of Mathematics & Computer Science, Amirkabir University of Technology, Tehran, Iran, [b]Department of Electrical Engineering, K. N. Toosi University of Technology, Tehran, Iran, [c]Department of Information Technology Management, Tarbiat Modares University, Tehran, Iran, [d]Department of Aerospace Engineering, Amirkabir University of Technology, Tehran, Iran

2.1 Introduction

In the past few years, scenario generation has acquired much attention in many different domains, as Fig. 2.1 shows. The bibliometric analysis of publications retrieved from the Web of Science database using the keyword scenario generation reveals a multidisciplinary research landscape with strong representation across both engineering and computational sciences. As illustrated in the figure, the most influential research areas include engineering, electrical and electronic, and energy fuels, indicating that scenario generation techniques are predominantly applied in energy systems modeling, optimization, and power system planning. These fields are followed by domains within computer science, such as artificial intelligence, information systems, and theory and methods, which underscores the integration of algorithmic and data-driven approaches, such as Monte Carlo simulations (MCSs), Latin hypercube sampling (LHS), and machine learning-based scenario generation. Additionally, categories like operations research, management science, and transportation science and technology highlight the methodological applications of scenario generation for decision-making, uncertainty quantification, and system optimization. The relatively high publication counts in these interconnected disciplines reflect the growing interdisciplinary relevance of scenario generation in addressing complex problems related to sustainability, energy transitions, and intelligent infrastructure design. Overall, this distribution suggests that the field is evolving toward a synergistic convergence of computational intelligence and engineering-driven system modeling.

The evolution of modern energy systems has initiated a profound transformation from centralized, fully dispatchable infrastructures toward decentralized, data-rich, and uncertainty-driven networks. The growing integration of renewable energy resources, coupled with the liberalization of electricity markets, has fundamentally changed how energy systems are planned and operated (Infield & Freris, 2020). In this new landscape, randomness is not a marginal phenomenon but an inherent property of the

Stochastic Planning and Modeling for Energy Systems: Methods, Applications, and Developments.
DOI: https://doi.org/10.1016/B978-0-443-45298-7.00007-0

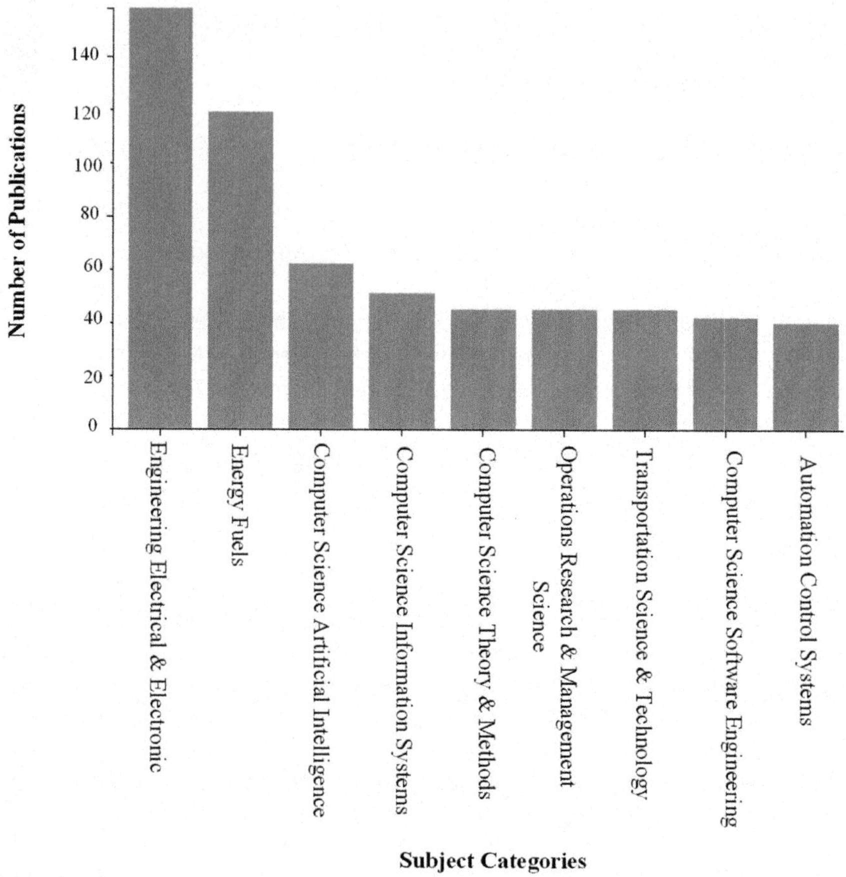

Figure 2.1 Interdisciplinary map showing the central position of scenario generation research among major scientific fields. Source Web of Science database (2025).

system itself. Fluctuating wind speeds, intermittent solar radiation, volatile market prices, and uncertain demand patterns collectively create an environment where deterministic optimization models-once the backbone of energy system analysis-can no longer provide reliable or realistic solutions (Zadeh et al., 2020; Zadeh et al., 2023).

To address these challenges, *stochastic programming* has emerged as a central paradigm for decision-making under uncertainty (Prékopa, 2013). By explicitly embedding random variables within the optimization framework, stochastic programming enables planners and operators to assess and mitigate risks associated with uncertain future conditions. Yet, the quality of any stochastic optimization model is fundamentally tied to the accuracy with which uncertainty is represented (Marti, 2024). This representation is captured through a finite set of scenarios, each describing one plausible evolution of the uncertain parameters over time.

Consider a finite planning horizon

$$\mathcal{T} = \{1, 2, \ldots, T\} \tag{2.1}$$

where T denotes the number of time periods under consideration. The uncertain parameters influencing system operation are collectively modeled as a multivariate stochastic process

$$\xi_t = \left(\xi_t^1, \xi_t^2, \ldots, \xi_t^d\right), \quad t \in \mathcal{T}, d \in \mathbb{N}, \tag{2.2}$$

where each component ξ_t^i represents a random variable corresponding to a particular source of uncertainty, and d denotes the dimensionality of the process-that is, the number of distinct uncertain factors considered. The symbol $\mathbb{N} = \{1, 2, 3, \ldots\}$ denotes the set of natural numbers, ensuring that d is a positive integer. The process ξ_t is defined on a probability space $(\Omega, \mathcal{F}, P)$, where Ω is the sample space, $\mathcal{F}$ is a σ-algebra of measurable events, and P is a probability measure satisfying $P(\Omega) = 1$.

In the context of energy systems, the vector ξ_t typically includes multiple correlated random quantities, such as the electrical load L_t, the renewable generation $G_t^{\text{ren}} = \left(G_t^{\text{wind}}, G_t^{\text{pv}}\right)$, the electricity market price λ_t, and the fuel price ϕ_t:

$$\xi_t = \left(L_t, G_t^{\text{ren}}, \lambda_t, \phi_t\right) \tag{2.3}$$

Each of these quantities evolves over time following its own stochastic behavior, and their joint probability distribution defines the uncertain operating environment of the energy system.

The process of *scenario generation* aims to construct a finite and computationally tractable approximation of the continuous stochastic process ξ_t. Let

$$\mathcal{S} = \{1, 2, \ldots, S\} \tag{2.4}$$

denote the index set of scenarios, where S is the number of discrete realizations used to represent uncertainty. Each scenario $s \in \mathcal{S}$ corresponds to a complete trajectory of the stochastic process over the planning horizon:

$$\xi_s = \left(\xi_{s,1}, \xi_{s,2}, \ldots, \xi_{s,T}\right) \tag{2.5}$$

where $\xi_{s,t} \in \mathbb{R}^d$ denotes the vector of realizations at time t for scenario s. Each scenario is associated with a probability weight π_s satisfying

$$\pi_s > 0, \quad \sum_{s=1}^{S} \pi_s = 1 \tag{2.6}$$

The discrete probabilistic approximation of the true stochastic process is therefore given by the finite set

$$\mathcal{X} = \left\{\left(\xi_s, \pi_s\right) | s \in \mathcal{S}\right\} \tag{2.7}$$

An ideal scenario set $\mathcal{X}$ should reproduce the essential statistical properties of the continuous process, such as the expected values, variances, correlations, and temporal dependencies, while keeping the model computationally tractable. Moreover, decisions derived from this discrete approximation should remain near-optimal when evaluated against the true continuous distribution. Balancing these two objectives, namely *statistical fidelity and computational efficiency*, represents the central challenge of scenario generation (Alonso-Travesset et al., 2022; Wedel, 2024).

Scenario generation thus serves as the bridge between abstract probability theory and actionable decision-making. It transforms complex random phenomena into a structured set of representative outcomes, enabling the application of optimization and simulation tools in uncertain environments (Dellino & Meloni, 2015). The evolution of this field, from classical Monte Carlo sampling (Dunn & Shultis, 2022) to more structured techniques such as LHS (Shu & Jirutitijaroen, 2011), and further toward Copula-based (Yoo et al., 2023, Jing et al., 2025) and data-driven approaches, reflects the ongoing effort to represent uncertainty both faithfully and efficiently. As energy systems become more interconnected, decentralized, and variable, the development of advanced scenario generation methods will remain fundamental to achieving resilient, adaptive, and data-informed energy planning.

2.2 The principles of statistical sampling

Uncertainty modeling in energy systems relies on the ability to represent a continuous probability distribution by a finite, discrete set of representative samples, or *scenarios*. The fundamental challenge is to approximate the true multivariate probability distribution F_ξ governing the uncertain parameters by a finite discrete distribution that captures its essential statistical properties. This is particularly critical in the context of stochastic optimization, where expectations over random variables must be replaced by sample averages to achieve computational tractability.

2.2.1 Representing distributions with discrete scenarios

Consider a generic two-stage stochastic program, a canonical formulation widely used in energy system planning and operation. The decision-maker seeks to minimize the total expected cost:

$$\min_{\mathbf{x}} \quad \mathbf{c}^T\mathbf{x} + \mathbb{E}_\xi[Q(\mathbf{x}, \xi)] \tag{2.8}$$

subject to the first-stage constraints

$$\mathbf{Ax} \geq \mathbf{b}, \quad \mathbf{x} \in X \ , \tag{2.9}$$

where $\mathbf{x}$ represents the first-stage (here-and-now) decisions, such as investment levels in generation capacity, bilateral contracts, or grid reinforcement. These must be chosen before the uncertain parameters $\boldsymbol{\xi}$ are realized.

The function $Q(\mathbf{x}, \xi)$ in Eq. (2.8) denotes the *recourse function*, which captures the second stage (wait-and-see) decisions made after uncertainty unfolds. It is defined by the following optimization problem:

$$Q(\mathbf{x}, \xi) = \min_{\mathbf{y}} \quad \mathbf{q}(\xi)^T y \tag{2.10}$$

$$\text{s.t.} \quad \mathbf{T}(\xi)\mathbf{x} + \mathbf{W}(\xi)\mathbf{y} \geq \mathbf{h}(\xi), \quad \mathbf{y} \in Y, \tag{2.11}$$

where $\mathbf{y}$ represents second-stage operational decisions such as dispatch levels, real-time balancing, or reserve activation.

The expected value term $\mathbb{E}_{\xi}[Q(\mathbf{x}, \xi)]$ in Eq. (2.8) involves integration over the continuous distribution of ξ, which is often analytically and numerically intractable. A practical and powerful approach is to approximate this expectation through the sample average approximation (SAA) method:

$$\mathbb{E}_{\xi}[Q(\mathbf{x}, \xi)] \approx \sum_{s=1}^{S} \pi_s Q(\mathbf{x}, \xi_s) = \frac{1}{S} \sum_{s=1}^{S} Q(\mathbf{x}, \xi_s), \text{ (for equiprobable scenarios)} \tag{2.12}$$

where $\{\xi_s\}_{s=1}^{S}$. are the sampled scenarios drawn from F_{ξ}, and π_s are their associated probabilities satisfying $\sum_{s=1}^{S} \pi_s = 1$.

Through this substitution, the stochastic program Eqs. (2.8)–(2.11) is converted into a large but finite deterministic equivalent problem (DEP), which can be solved using conventional optimization techniques. The quality of the solution obtained through this sample-based approximation critically depends on the statistical properties of the scenario set $\{(\xi_s, \pi_s)\}_{s=1}^{S}$.

2.2.2 Desired properties: Representativeness and convergence

The reliability of stochastic optimization results is only as strong as the statistical quality of the scenarios employed. A well-designed scenario set must exhibit the following key properties.

(1) Representativeness and distributional accuracy. The empirical distribution of the generated scenarios, denoted F_S, should converge to the true underlying distribution F_{ξ} as the number of samples $S \to \infty$. Formally, under independent and identically distributed (i.i.d.) sampling, this convergence is guaranteed by the Glivenko-Cantelli theorem, which ensures:

$$\sup_{\xi} |F_S(\xi) - F_{\xi}(\xi)| \to 0 \quad \text{almost surely as } S \to \infty$$

This property ensures that the empirical distribution faithfully reproduces the probabilistic structure of the underlying process.

(2) Accuracy and statistical fidelity. The discrepancy between F_S and F_ξ can be quantified using probability metrics such as the *Wasserstein distance*, defined as:

$$W_p(F_S, F_\xi) = \left(\inf_J \int \left\|\xi - \xi'\right\|^p dJ(\xi, \xi')\right)^{1/p} \tag{2.13}$$

where J denotes a joint distribution with marginals F_S and F_ξ, and $p \geq 1$ specifies the order of the metric. A small Wasserstein distance implies that the generated scenario set closely matches the statistical features of the true distribution, including its mean, variance, and higher-order dependencies.

(3) Correlation structure preservation. In energy systems, uncertainties are rarely independent. Spatial, temporal, and cross-variable correlations must be preserved to ensure physical realism:

- Spatial correlations: Wind speeds or solar irradiance across neighboring sites are strongly correlated.
- Cross-variable correlations: Demand and temperature or solar power and load often exhibit direct statistical coupling.
- Temporal correlations: The stochastic process ξ_t exhibits serial dependencies, where ξ_t is correlated with ξ_{t-1}.

Ignoring these dependencies leads to unrealistic operational outcomes and underestimation of system risks.

(4) Preservation of extreme events. Energy systems must be robust against high-impact, low-probability events such as prolonged wind droughts, cold snaps, or price spikes. Therefore, the generated scenarios must include tail realizations of F_ξ, ensuring the optimization model internalizes risk associated with rare but critical situations.

(5) Computational tractability. Finally, the number of scenarios S must remain manageable to keep the resulting deterministic problem solvable within practical computational limits. Excessive scenario counts increase problem dimensionality, leading to the well-known "curse of dimensionality." Consequently, scenario reduction or clustering techniques are often applied postgeneration.

Table 2.1 summarizes the main requirements that high-quality scenario sets must fulfill in energy applications.

Statistical sampling lies at the heart of stochastic energy modeling. Scenario generation is not merely a random draw from data; it is a disciplined, mathematically grounded process of constructing a finite approximation that balances fidelity to the true probability distribution with computational feasibility. The next section will explore how these principles are realized in practice through classical and advanced sampling methods, beginning with the Monte Carlo approach and its structured extensions.

2.3 Monte Carlo simulation for scenario generation

MCS is one of the most fundamental and widely used techniques for uncertainty quantification and scenario generation in energy systems. Its intuitive nature and

Table 2.1 Comparison of scenario generation requirements.

Property	Description and implication for energy models
Representativeness	Scenarios must capture the full range of possible outcomes (e.g., from wind droughts to stormy periods, from low demand to peak load). Inadequate coverage leads to operational and planning vulnerabilities.
Accuracy	The statistical properties (mean, variance, tails) of the scenario set must align with historical or simulated data. Inaccurate representations bias investment and dispatch outcomes.
Correlation structure	Scenarios must preserve interdependencies among variables and across time to ensure physical and temporal realism.
Extreme events	Plausible but rare events (e.g., extreme weather or market shocks) must be represented to avoid systemic underpreparedness.
Computational tractability	The number of scenarios must be limited so that the resulting stochastic optimization model can be solved efficiently. This often necessitates a trade-off between accuracy and tractability.

minimal assumptions about the underlying distributions make it a cornerstone of stochastic modeling. The basic idea of MCS is to approximate the expected value of a function of random variables by repeatedly sampling from the associated distributions and averaging the resulting outcomes.

Let $z \in \mathbb{R}^d$ be a random vector representing uncertain parameters, such as hourly wind speeds, solar irradiance, or fuel prices, with a joint cumulative distribution function (CDF) $F_z(\bullet)$. The goal is to compute the expected value of a cost or performance function $C(z)$, defined as:

$$\mu = \mathbb{E}[C(z)] = \int_{\mathbb{R}^d} C(z) f_z(z) dz \tag{2.14}$$

where $f_z(z)$ is the joint probability density function (PDF). Analytical computation of Eq. (2.14) is often intractable for high-dimensional systems or nonlinear cost functions, which motivates the use of Monte Carlo methods.

2.3.1 Standard Monte Carlo algorithm

The standard MCS procedure involves generating N i.i.d. samples $\{z^1, \ldots, z^N\}$ from F_z, evaluating the deterministic function $C(\bullet)$ for each realization, and computing the sample mean as an unbiased estimator of μ (Algorithm 1).

Algorithm 1 Standard Monte Carlo simulation (MCS) for cost estimation.

Input: Number of scenarios N, cost function $C(\cdot)$, cumulative distribution function $F(\cdot)$
Initialize: Total cost $T \leftarrow 0$
for $i = 1$ to N do
  Sample a random vector $u^i \sim U[0,1]^d$ ▷Draw from the d-dimensional uniform Distribution
  Transform to target distribution: $z^i \leftarrow F^{-1}(u^i)$ ▷Inverse Transform Sampling
  Evaluate cost for this scenario: $c^i \leftarrow C(z^i)$
$T \leftarrow T + c^i$
end for
$\hat{\mu}_N \leftarrow 1/NT$ ▷Compute the sample mean estimator
Return: Estimated expected cost $\hat{\mu}_N$

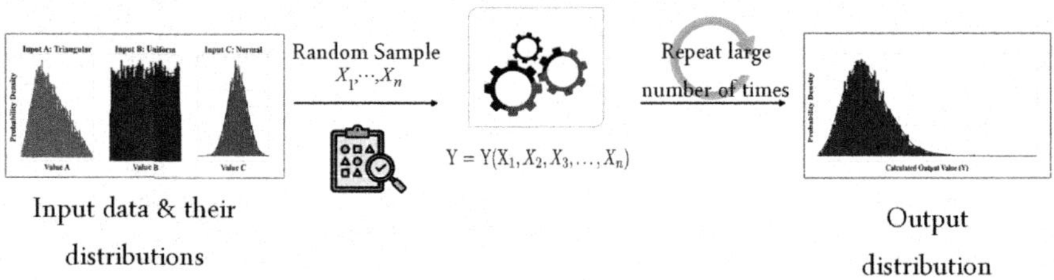

Figure 2.2 Conceptual overview of the Monte Carlo method.

The estimator $\hat{\mu}_N$ is mathematically expressed as:

$$\hat{\mu}_N = \frac{1}{N} \sum_{i=1}^{N} C(z^i), \tag{2.15}$$

and by the Law of Large Numbers, it converges almost surely to the true expected value μ as $N \to \infty$:

$$\lim_{N \to \infty} \hat{\mu}_N = \mu. \tag{2.16}$$

Fig. 2.2 illustrates the process of repeated random sampling from a probability distribution to approximate an unknown expected value in the MC method.

2.3.2 Convergence and limitations

The primary advantage of MCS is its simplicity and universality. The estimator $\hat{\mu}_N$ is unbiased, i.e., $\mathbb{E}[\hat{\mu}_N] = \mu$. Its convergence rate, however, is relatively slow. According to the Central Limit Theorem (CLT), if $\sigma^2 = \mathrm{Var}[C(z)] < \infty$, then:

$$\sqrt{N}(\hat{\mu}_N - \mu) \xrightarrow{d} N(0, \sigma^2), \tag{2.17}$$

implying that the standard error of the estimator decreases proportionally to $O(N^{-1/2})$. To reduce the estimation error by a factor of 10, the number of samples N must be increased by a factor of 100. Although this rate is independent of the dimensionality d, it makes standard MCS computationally expensive for high-dimensional, high-accuracy energy system problems.

2.3.3 Energy system example: Wind farm portfolio

To illustrate MCS in practice, consider a simple wind farm portfolio with two independent farms. Consider two independent wind farms with distinct wind resource characteristics. Farm 1 experiences wind speeds following a Weibull distribution with scale parameter $\lambda_1 = 10\,\text{m/s}$ and shape parameter $k_1 = 2.0$, while Farm 2 exhibits more energetic conditions with $\lambda_2 = 12\,\text{m/s}$ and $k_2 = 2.2$. Both farms utilize identical wind turbines characterized by cut-in, rated, and cut-out speeds of $v_{ci} = 3\,\text{m/s}$, $v_r = 12\,\text{m/s}$, and $v_{co} = 25\,\text{m/s}$, respectively, with a rated power output $P_r = 2000$ kW. The power output of farm i is a function of wind speed v_i:

$$P_i = f_i(v_i) \tag{2.18}$$

where $v_1 \sim \text{Weibull}(\lambda_1 = 10, k_1 = 2)$ and $v_2 \sim \text{Weibull}(\lambda_2 = 12, k_2 = 2.2)$. The Weibull CDF and its inverse are:

$$F(v; \lambda, k) = 1 - \exp\left(-(v/\lambda)^k\right), \tag{2.19}$$

$$F^{-1}(u; \lambda, k) = \lambda(-\ln(1-u))^{1/k}. \tag{2.20}$$

For each scenario s, wind speeds $v_{s,1}$ and $v_{s,2}$ are generated via inverse transform sampling, and the power output is computed using the turbine power curve:

$$P_{s,i} = \begin{cases} 0, & v_{s,i} < v_{ci} \\ a v_{s,i}^3 + b, & v_{ci} \le v_{s,i} < v_r \\ P_r, & v_r \le v_{s,i} < v_{co} \\ 0, & v_{s,i} \ge v_{co} \end{cases} \tag{2.21}$$

where v_{ci}, v_r, and v_{co} are cut-in, rated, and cut-out wind speeds, respectively. The total power output for a single scenario s is the sum $P_{s,\text{total}} = P_{s,1} + P_{s,2}$. This is a single sample from the random variable $\mathbf{P}_{\text{total}} = \mathbf{P}_1 + \mathbf{P}_2$, whose expected value $\mu_{\text{total}} = \mathbb{E}[\mathbf{P}_{\text{total}}]$ is the quantity we wish to estimate. The Monte Carlo estimate of total energy production is then:

$$\hat{E}_{\text{total}} = \frac{1}{S}\sum_{s=1}^{S} (P_{s,1} + P_{s,2}) \ , \tag{2.22}$$

Implementing the MCS framework with $S_{\max} = 10,000$ scenarios reveal several key characteristics of the Monte Carlo method. A reference "true" expected value of

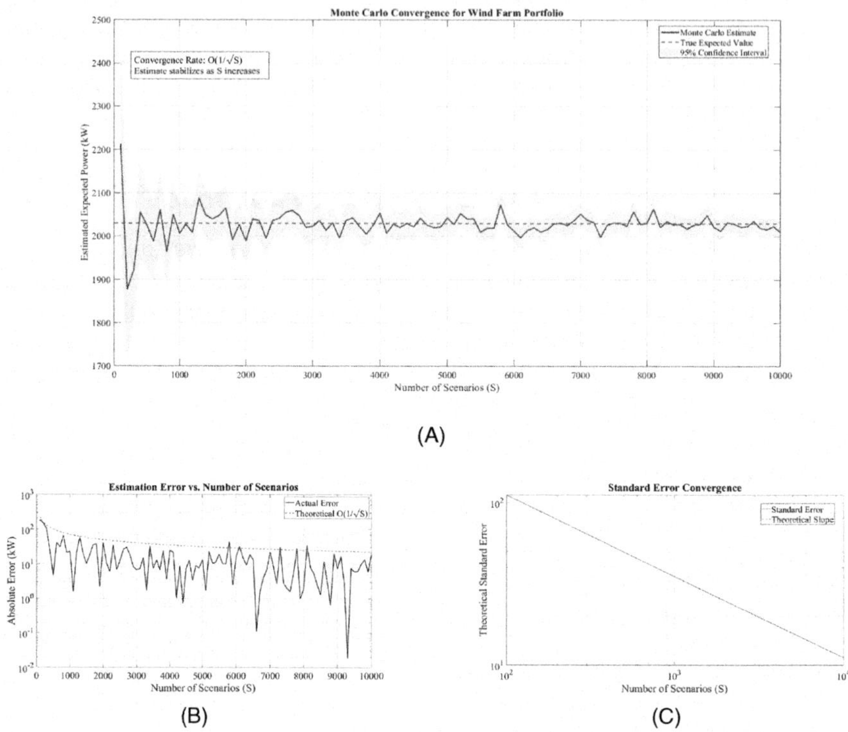

Figure 2.3 Monte Carlo convergence analysis for the wind farm portfolio. (A) Convergence of the estimator toward the true expected value $\mu_{\text{total}} = 1842.36$ kW with 95% confidence intervals, demonstrating stabilization with increasing scenarios. (B) Absolute estimation error showing $O\left(1/\sqrt{S}\right)$ behavior, where error reduction follows the theoretical inverse square root relationship. (C) Standard error convergence with the theoretical slope, validating the variance reduction principle $\text{Var}\left(\hat{E}_{\text{total}}\right) = \text{Var}(\mathbf{P}_{\text{total}})/S$.

$\mu_{\text{total}} = 1842.36$ kW was established using 100,000 samples to provide a benchmark for convergence analysis.

Fig. 2.3A illustrates the fundamental convergence property of the Monte Carlo estimator. The sample mean $\hat{E}_{\text{total}}$ initially exhibits significant variability for small sample sizes ($S < 1000$), with oscillations of approximately $\pm$ 50 kW around the true mean. As the number of scenarios increases beyond 2000, the estimator stabilizes markedly, converging toward the theoretical expected value. The 95% confidence interval, derived from the theoretical standard error $\sigma/\sqrt{S}$, narrows progressively, reflecting the increasing precision of the estimate. This visual convergence pattern empirically validates the almost sure convergence guaranteed by the Strong Law of Large Numbers.

The error convergence behavior, depicted in Fig. 2.3B, provides compelling evidence for the theoretical $O\left(1/\sqrt{S}\right)$ convergence rate. The absolute error $\left|\hat{E}_{\text{total}} - \mu_{\text{total}}\right|$

decreases from approximately 15 kW at $S = 100$ to less than 2 kW at $S = 10,000$. The logarithmic scale clearly reveals the characteristic linear decay with slope $-1/2$ when plotted against $\log(S)$ confirming the inverse square root relationship. This error reduction pattern demonstrates the fundamental trade-off in Monte Carlo methods: Each quadrupling of computational effort yields only a halving of estimation error.

Fig. 2.3C examines the standard error $\sigma/\sqrt{S}$, which quantifies the uncertainty in the Monte Carlo estimator. The estimated population standard deviation ≈ 890 kW, combined with the $1/\sqrt{S}$ scaling, produces the observed convergence pattern. The standard error diminishes from 89 kW at $S = 100$ to approximately 9 kW at $S = 10,000$, representing a tenfold improvement in precision. The parallel alignment between the empirical standard error and the theoretical slope line confirms the proper implementation of the sampling methodology and validates the variance reduction principle underlying Monte Carlo methods.

The numerical results yield a final estimate of $\hat{E}_{\text{total}} = 1841.28$ kW with a relative error of 0.06% after 10,000 scenarios. This high precision demonstrates the efficacy of Monte Carlo methods for renewable energy portfolio assessment. The convergence analysis provides practical guidance for determining appropriate sample sizes: For preliminary assessments requiring $\pm$ 5% accuracy, a few hundred scenarios may suffice, while high-precision engineering applications may require several thousand simulations.

The demonstrated $O(1/\sqrt{S})$ convergence, while fundamental to Monte Carlo methods, also highlights the computational limitations for ultra-high-precision requirements, motivating consideration of variance reduction techniques for applications demanding extreme accuracy with constrained computational resources.

2.4 Stratified sampling and Latin hypercube sampling

The slow convergence of MCS motivates the use of variance reduction techniques. Stratified sampling is a canonical variance reduction method that partitions the input domain Ω into M disjoint strata (sub-regions) $\Omega_1, \ldots, \Omega_M$. A predetermined number of samples n_j is allocated to each stratum (where $\sum n_j = N$), and MCS is performed within that subregion.

The final estimate is a weighted average of the strata-specific estimates. By enforcing a representative number of samples from all parts of the distribution, including low-probability, high-impact tail regions, stratification prevents the sample clustering that can occur in pure MCS and reduces the overall variance of the estimator: $\text{Var}(\hat{\mu}_N^{\text{Strat}}) < \text{Var}(\hat{\mu}_N)$.

2.4.1 Latin hypercube sampling

While effective in low dimensions, partitioning the domain Ω becomes intractable as the dimension K increases (the curse of dimensionality). LHS is a sophisticated technique that achieves the benefits of stratification in high-dimensional spaces (Shields & Zhang, 2016).

Algorithm 2 Standard Latin hypercube sampling (LHS).

Input: Number of scenarios N, Number of dimensions K
Initialize: $N \times K$ sample matrix S
for $j = 1$ to K **do**
    Generate a random permutation P_j of $\{1, \ldots, N\}$
    for $i = 1$ to N **do**
        Sample a random number $U_{ij} \sim U[0,1]$
$S_{ij} \leftarrow \frac{P_j(i) - U_{ij}}{N}$ ▷Create stratified sample in [0,1]
    end for
end for
Initialize: $N \times K$ output matrix $\hat{X}$
for $i = 1$ to N **do**
    for $j = 1$ to K **do**
    $\hat{X}_{ij} \leftarrow F_{xj}^{-1}(S_{ij})$ ▷Apply Inverse Transform Sampling
    end for
end for
Return: Scenario matrix $\hat{X}$

LHS operates by stratifying the marginal distribution of each of the K input variables. For N samples:

1. The range of each of the K variables is divided into N equiprobable intervals.
2. N sample points are placed in the K-dimensional space such that there is exactly one sample in each of the N intervals for all K variables.

This design guarantees full stratification across all 1-D marginals, ensuring a much more even and uniform coverage of the parameter space than standard MCS.

2.4.2 Algorithmic solution: Latin hypercube sampling

The LHS algorithm (Algorithm 2) generates this stratified sample matrix. It creates K permutations of the integers 1, …, N and uses them to define the bins from which to sample, adding a uniform random jitter within each bin (Olsson et al., 2003).

The primary advantage of LHS is its efficiency. By spreading samples more evenly, it reduces variance, and its convergence rate is often faster than $O(N^{-1/2})$, making it a superior choice to MCS for many engineering applications (Olsson et al., 2003). The Fig. 2.4 demonstrates the LHS method on samples. The left panel shows a bivariate distribution of two correlated random variables, X_1 and X_2, where sample points (white circles) are drawn according to the joint probability density (blue contours). To perform LHS, each variable is first transformed into a uniform distribution on the interval [0, 1] using its CDF, as shown in the middle panel. In this uniform space, the range of each variable is divided into equal, nonoverlapping intervals (strata), and one sample is randomly selected from each interval, ensuring that the entire range of each variable is well represented. The red point labeled "1. Plan Point" is an example of such a sample in the uniform space. In the final step, shown in the right panel, these uniform samples are mapped back to the original correlated variable space using the

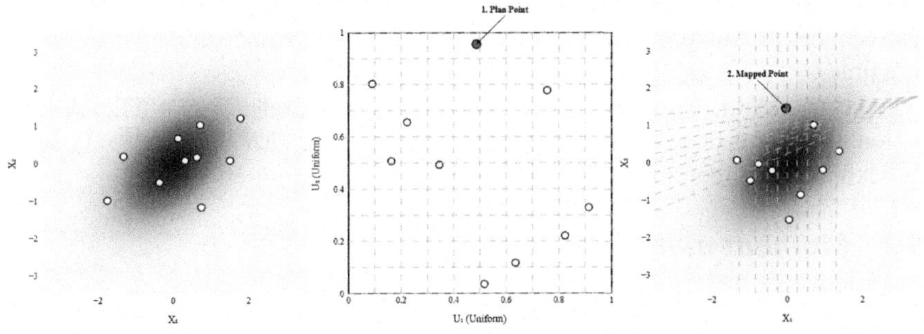

Figure 2.4 Demonstration of Latin hypercube sampling (LHS).

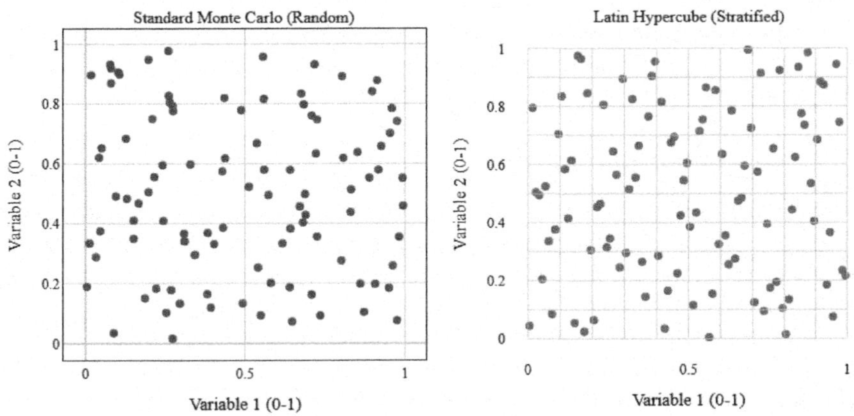

Figure 2.5 Schematic comparison of sampling efficiency in a 2D space. (*left*) Standard Monte Carlo (MCS) exhibits random clustering and gaps, leading to slower convergence. (*right*) Latin Hypercube Sampling (LHS) stratifies the domain, ensuring a more uniform coverage of the sample space and faster convergence rates.

inverse CDF, producing the "2. Mapped Point." This process preserves the correlation structure while ensuring good coverage of the input space, making LHS an efficient and systematic method for generating representative samples for uncertainty analysis, sensitivity studies, and simulation experiments.

The difference in sampling efficiency is visually compared in Fig. 2.5.

2.5 Other relevant models

2.5.1 Quasi-Monte Carlo

Quasi-Monte Carlo (QMC) methods take the idea of uniform coverage a step further. Instead of using pseudo-random numbers, QMC uses deterministic quasi-random sequences, also known as low-discrepancy sequences (e.g., Sobol, Halton, Faure). These sequences are specifically engineered to fill the K-dimensional space as evenly as possible.

The key benefit is a superior theoretical convergence rate. For sufficiently smooth functions, the error of QMC converges at approximately $O(N^{-1}(\log N)^{K-1})$, which is significantly faster than the $O(N^{-1/2})$ rate of MCS (Niederreiter, 1992).

2.5.2 Moment matching

Moment Matching methods generate a set of discrete scenarios and their associated probabilities by solving an optimization problem. The goal is to create a discrete distribution whose first few statistical moments (typically the first four: Mean, variance, skewness, and kurtosis) exactly match the moments of the target continuous distribution.

This approach is computationally efficient but has a major drawback: By focusing only on the first few moments, it may fail to capture the behavior in the tails of the distribution, potentially missing critical extreme events.

2.5.3 Optimal quantization

Optimal quantization (or vector quantization) is a more advanced technique that formalizes the idea of optimal scenario generation. It seeks to find the best discrete approximation Q_n (a set of N scenarios) to a continuous distribution P by minimizing a probability distance, most commonly the Wasserstein distance.

This is essentially a clustering problem. Finding the optimal N scenarios is equivalent to finding the N cluster centroids that best represent the distribution P. This problem is often solved using variants of the k-means algorithm or stochastic gradient descent.

2.5.4 Copula functions

A significant challenge is modeling the dependence between multiple uncertain variables (e.g., wind, solar, and load). Copula functions are a powerful statistical tool for this task.

According to Sklar's Theorem, any multivariate joint distribution can be decomposed into its marginal distributions (describing each variable individually) and a copula, which describes the dependence structure that joins them.

This allows a plug-and-play approach:

1. Model the marginal distribution of wind (e.g., with a Weibull distribution).
2. Model the marginal distribution of solar (e.g., with a Beta distribution).
3. Fit a copula (e.g., a Gaussian or Student's t Copula) to the historical data to capture the correlation between them.

This method is extremely flexible and is highly effective at capturing the complex, nonlinear dependencies found in energy systems (de Carvalho Silva et al., 2025). The Fig. 2.6 demonstrates the other relevant models in the mind map chart. Furthermore,

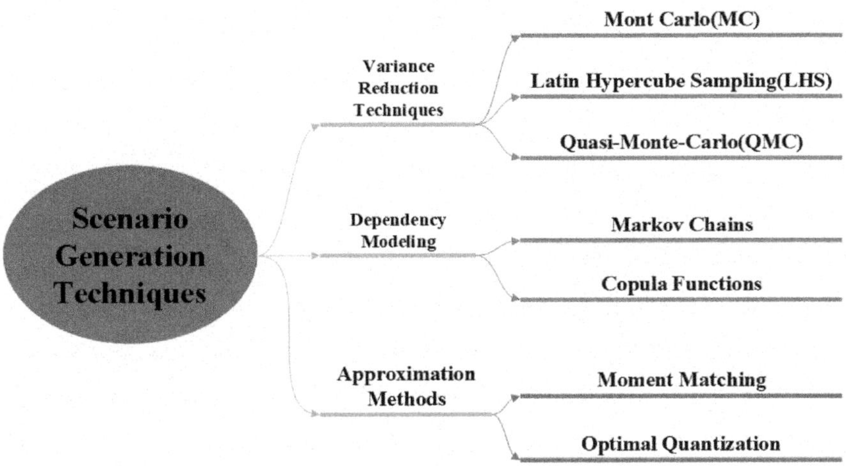

Figure 2.6 A mind map classifying advanced generation models.

the capacity of copulas to capture dependence structures compared to independent sampling is illustrated in Fig. 2.7.

2.5.5 Problem statement and dataset

To demonstrate the practical application of these sampling methods, this case study addresses the problem of probabilistic forecasting for stochastic energy variables. The public dataset suitable for this analysis is the "Pennsylvania-New Jersey-Maryland (PJM) Hourly Energy Consumption Data" (Mulla, 2018). This dataset contains hourly power consumption data (in megawatts, MW) from PJM Interconnection Llimited liability company (LLC), a regional transmission organization (RTO) in the United States. This data allows for the analysis of stochastic load, a different but equally critical uncertainty for system operators.

This case study will proceed using the PJM Hourly Energy Consumption Data (Mulla, 2018). The objective is to move beyond a simple, single-value point forecast and generate a full probabilistic forecast. This is formulated as a deterministic-stochastic decomposition problem.

2.5.6 Methodology

The methodology proceeds in two stages.

1. Deterministic Modeling: A base forecasting model (Facebook's Prophet) is trained on three years of historical price data. This model's purpose is to capture the predictable, cyclical components of the time series. Fig. 2.8 shows the model's successful decomposition, isolating the (1) long-term overall trend, (2) distinct weekly seasonality (e.g., weekend vs weekday), (3) broad yearly seasonality (e.g., summer vs winter), and (4) clear daily seasonality (e.g., overnight lulls).

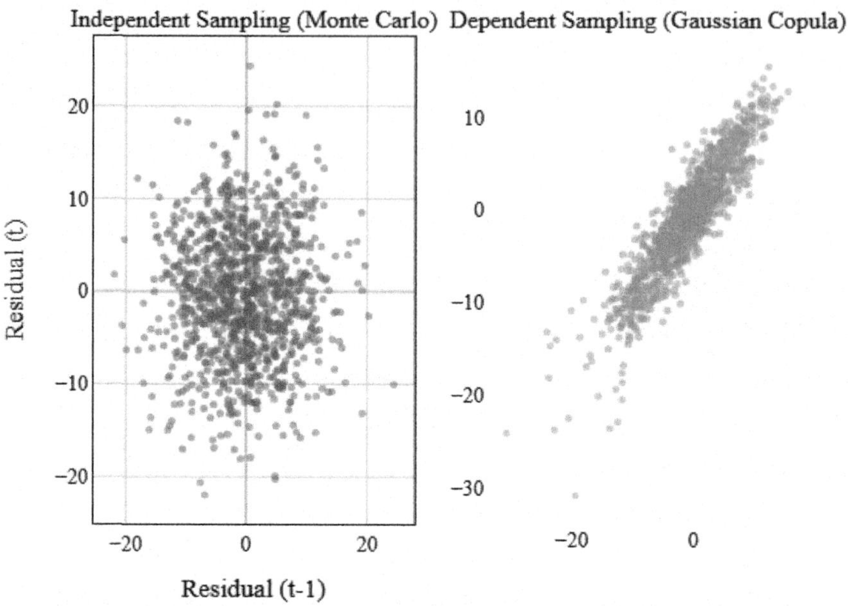

Figure 2.7 Visualization of temporal dependence capture. (*left*) Independent sampling (MCS) produces an uncorrelated "cloud." (*right*) The Gaussian Copula approach successfully reproduces the correlation structure.

2. Stochastic Modeling: The model's in-sample errors, or residuals, are extracted from the training period. These residuals represent the unpredictable, stochastic component of the price. The scenario generation task is thus reformulated as: Price Scenario = Deterministic Forecast + Stochastic Residual Scenario

This case study compares five models from this chapter-Standard MCS, LHS, Bootstrap, a first-order Markov Chain, and an autoregressive Gaussian Copula-to generate $N = 1000$ scenarios for this stochastic residual component over a one-year test horizon. The code implementations for these models are available in (Amini, 2025).

Critical View of the Base Model: The selection of Facebook's Prophet as the deterministic base model is a pragmatic one, valued for its automatic handling of multiple seasonalities (daily, weekly, yearly) and nonlinear trends, as shown in Fig. 2.8. However, this choice is not without criticism. Prophet is largely a black-box decomposer, and its primary limitation in this context is its default assumption that the residuals (the errors) are independent and i.i.d. Gaussian noise. This case study's core hypothesis is that this assumption is flawed. The subsequent analysis of the residuals will demonstrate that they are, in fact, neither independent nor Gaussian, establishing the critical need for the advanced stochastic models this chapter explores.

The Price Scenario Formulation: The decomposition formula is the conceptual heart of this case study. It formalizes a separation of uncertainties:

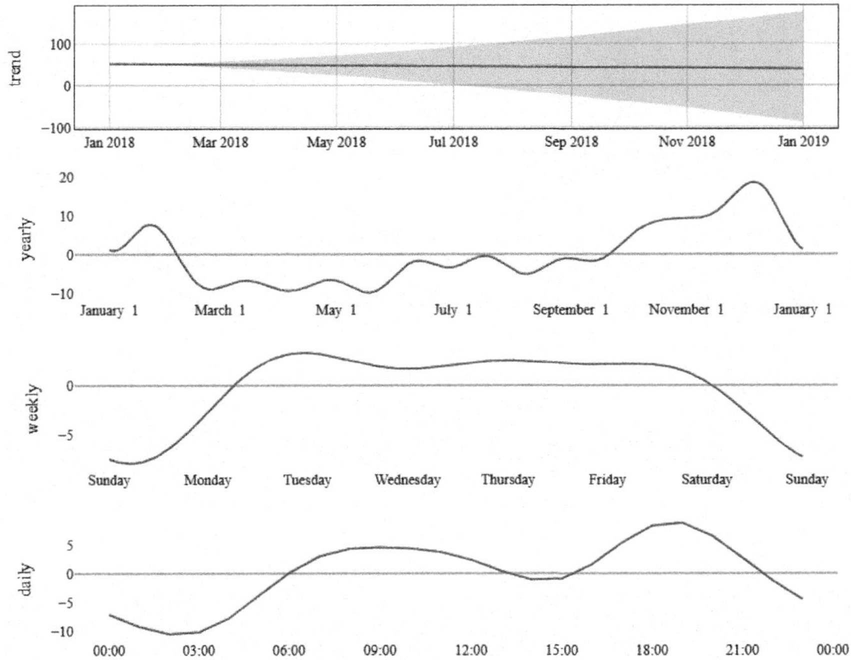

Figure 2.8 Decomposition of deterministic price components by the Prophet model.

- Deterministic Forecast: This term represents the knowable or predictable part of the price signal. It is the center of the probabilistic forecast, capturing the structural, cyclical patterns we expect to repeat (e.g., higher demand on weekdays, lower demand overnight).
- Stochastic Residual Scenario: This term represents the unknowable component-the random shocks, sudden events, and volatile noise that the deterministic model cannot predict.

By separating them, we move from the simple task of predicting the price to the more sophisticated task of modeling the structured uncertainty around the prediction. The quality of the final probabilistic forecast, therefore depends entirely on how realistically we can generate scenarios for these stochastic residuals.

2.5.7 Analysis of stochastic residuals

A critical finding of the analysis is that the extracted residuals are not simple white noise, establishing the core challenge for the scenario generation models. Fig. 2.9 illustrates two key properties of the training residuals: Fig. 2.9 Plot (A) reveals a non-Gaussian, fat-tailed (leptokurtic) distribution, while Fig. 2.9 Plot (B) confirms strong temporal autocorrelation, showing a slow decay and a 24-hour seasonal spike, proving the residuals are not independent.

First, the distribution (Fig. 2.9A) is non-Gaussian. The original statistics in Table 2.2 confirm this: The distribution is slightly left-skewed (Skew = –0.354) and strongly leptokurtic (Kurtosis = 2.232), meaning it has fatter tails and a sharper peak than a

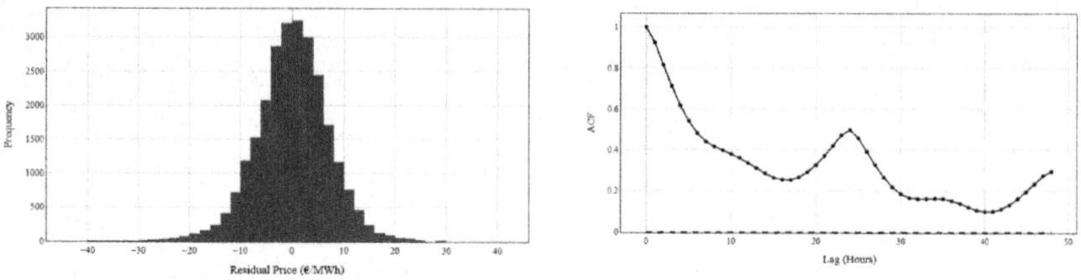

(A) Distribution of stochastic residuals. (B) Autocorrelation Function (ACF) of residuals.

Figure 2.9 Statistical properties of stochastic residuals from training data.

Table 2.2 Statistical fairness results: Comparison of statistical moments.

Model	Residual mean	Residual stddev	Residual skew	Residual kurtosis
Monte Carlo	0.001	7.351	-0.000	-0.000
LHS	-0.000	7.351	-0.000	-0.001
Bootstrap	0.000	7.351	-0.353	2.234
Markov Chain	0.008	10.381	-0.176	2.975
Gaussian Copula	-0.006	7.337	-0.362	2.208
Original	-0.000	7.352	-0.354	2.232

LHS, Latin hypercube sampling

normal distribution. This implies that extreme price events are more common than a simple Gaussian model would predict.

Second, the autocorrelation function (ACF) plot (Fig. 2.9D) reveals significant temporal dependence. The residual at hour t is highly correlated with the residual at hour $t - 1$ (ACF ≈ 0.9), and this memory persists for more than 24 hours, with a secondary peak at the 24 -hour lag.

This finding implies that simple i.i.d. models (MCS, LHS, Bootstrap) are fundamentally misspecified. They will generate scenarios that may match the *distribution* but will fail to model the *temporal dynamics*.

2.5.8 Evaluation of probabilistic forecasts

The five models are evaluated on the unseen, one-year test set. The evaluation is structured to answer two criteria that generate scenarios fair and have statistical properties that match the original data, and scenarios useful, and they accurately model the temporal dynamics and improve probabilistic scores.

2.5.8.1 Statistical fidelity (fairness)

This test assesses how well the generated scenarios, in aggregate, replicate the statistical fingerprint of the original data. The results are shown in Table 2.2 and Fig. 2.10.

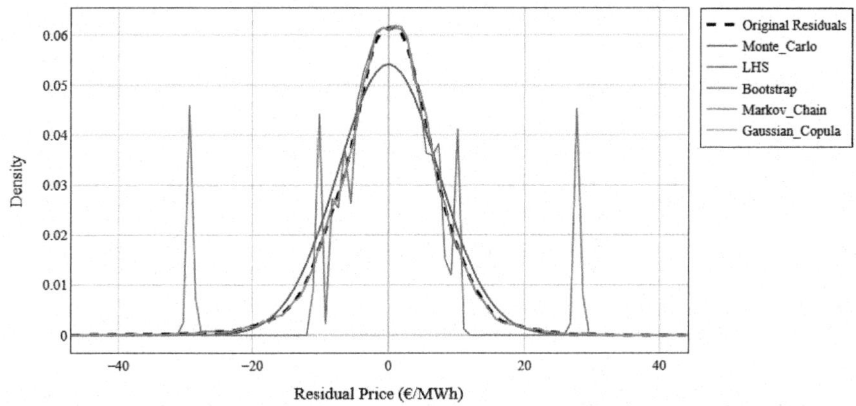

Figure 2.10 Comparison of probability density functions for generated residuals.

Table 2.2 provides a quantitative comparison of the first four statistical moments for the generated residuals versus the original data; the goal is to match the "Original" row, and bolded models indicate a successful match.

Fig. 2.10 provides a visual comparison of the overall probability density, confirming the findings from Table 2.2. As seen in the plot, the Bootstrap (green) and Gaussian Copula (yellow) models perfectly overlay the Original (black dashed) line. In contrast, the Markov Chain (pink) model clearly fails, creating an artificial, spiky distribution (while MCS/LHS, not shown, would form a simple Gaussian bell curve).

The results show a clear divergence:

- MCS and LHS are, by their parametric design, assuming a Gaussian distribution. They are therefore fundamentally incapable of capturing the non-Gaussian Skew (-0.35) and Kurtosis (2.23) and fail this test.
- Markov Chain fails catastrophically on a statistical level. The discretization process (using 20 bins) is the clear culprit, resulting in the spiky, artificial distribution seen in Fig. 2.10. Furthermore, its standard deviation (10.381) is dramatically incorrect compared to the original (7.352), indicating it is generating scenarios with far too much volatility.
- Bootstrap and Gaussian Copula are the only models to pass the fairness test. As nonparametric (or semi-parametric) methods, they flawlessly reproduce all four statistical moments. Their distribution plots (green and yellow) are visually indistinguishable from the original (black), confirming their statistical fidelity.

Based on statistical fidelity alone, Bootstrap and the Gaussian Copula are the only fair and acceptable representations.

2.5.8.2 Temporal and dynamic fidelity (plausibility)

This test evaluates if the models produce plausible dynamics. A fair model (from the previous section) that fails this test would produce scenarios that look right on average but are physically unrealistic hour-to-hour. We test this by comparing the ACF of their generated scenarios.

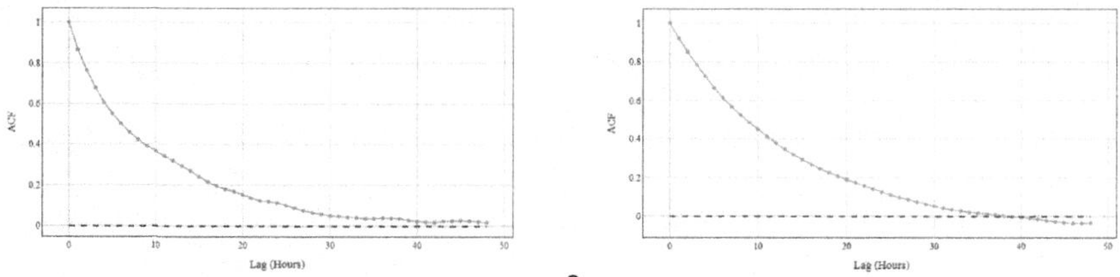

s

(A) ACF of Markov Chain Scenario (B) ACF of Gaussian Copula Scenario

Figure 2.11 ACF comparison for dynamic models (Markov chain and Gaussian Copula). *ACF,* Autocorrelation function.

- MCS, LHS, and Bootstrap are all i.i.d. (independent and i.i.d.) methods. They fail this test by definition. Their scenarios are generated without any memory of the previous time step, resulting in zero autocorrelation (their ACF plots would be flat at zero for all lags >0).
- Markov Chain and Gaussian Copula are, by contrast, autoregressive models designed to capture this very property. As shown in Fig. 2.11, both models successfully replicate the strong, slowly-decaying autocorrelation pattern seen in the original data (Fig. 2.9B), proving their temporal fidelity. This contrasts sharply with i.i.d. models, which would show no correlation at all.

Why the Dynamic Models Succeeded: The success of these two models lies in their internal logic.

- The Markov Chain model is inherently autoregressive. The probability of transitioning to a new state (a residual bin) at time t is explicitly conditional on the state at time $t-1$. This chain of probabilistic dependencies creates a memory that is precisely what autocorrelation measures.
- The Gaussian Copula was implemented as an autoregressive Copula. Its structure explicitly links the value at time t to the value at time $t-1$ (likely via an AR(1) process) within the copula's dependency structure. It was designed to preserve the temporal correlation observed in the original data.

This test demonstrates that only the Markov chain and Gaussian Copula models are dynamically plausible. The i.i.d. models, despite Bootstrap's statistical fairness, are fundamentally flawed for modeling time series.

2.5.8.3 Probabilistic scores and qualitative analysis

This final evaluation is the most important, that illuminate which model is the most useful by combining quantitative probabilistic scores (Table 2.3), which measure overall accuracy, with a qualitative inspection of the scenario fan charts (Fig. 2.12), which reveals their behavior. In Table 2.3, the continuous ranked probability score (CRPS) is the primary metric for overall probabilistic accuracy (where lower is better), and bold text highlights the winning model.

The quantitative results in Table 2.3 reveal a critical and complex trade-off, culminating in a surprising conclusion:

Table 2.3 Probabilistic metric results (lower is better).

Model	CRPS	P80 reliability	Pinball p50
Monte Carlo	14.922	-0.581	9.223
LHS	14.916	-0.581	9.222
Bootstrap	15.039	-0.604	9.141
Markov chain	14.600	-0.596	9.143
Gaussian Copula	15.047	-0.602	9.140

CRPS, continuous ranked probability score

- CRPS winner: The Markov Chain achieves the lowest (best) CRPS (14.600), indicating its overall probabilistic forecast is the most accurate.
- Surprising failure: The Gaussian Copula (15.047) and Bootstrap (15.039) models—the two that were statistically fair-perform very poorly. Their final accuracy scores are among the worst, even underperforming the simple LHS model. This implies that their fairness was misleading.
- Systemic failure (reliability): All models exhibit poor calibration. The P80 Reliability score of approx. -0.59 is dreadful. It indicates that the 80% prediction interval (P10-P90) was far too narrow, capturing only $(80\% - 59\%) \approx 21\%$ of the actual outcomes. This suggests a systemic issue, likely that the deterministic base model is overconfident. This is visually confirmed in Fig. 2.12, which shows the base Prophet forecast (blue) for a sample week capturing the general shape but completely missing the extreme volatility and spikes of the actual price (black), thereby creating large, unpredictable residuals.

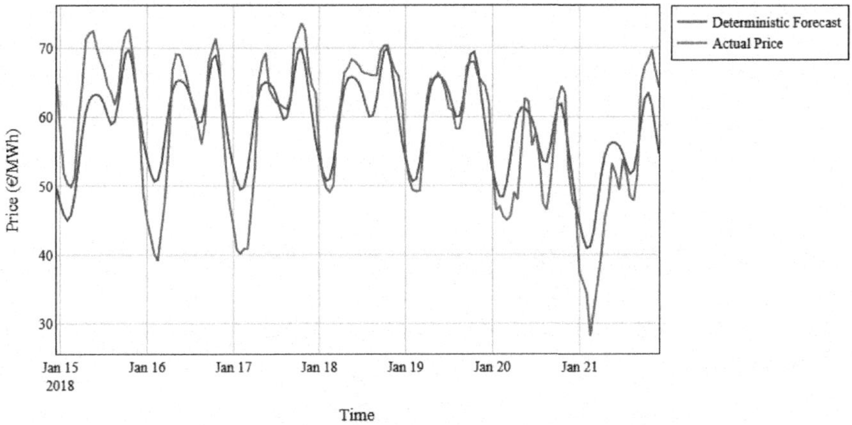

Figure 2.12 Deterministic (Base) Forecast vs. Actuals for a sample week.

This conclusion is reinforced visually in Fig. 2.13, which provides a qualitative comparison of scenario fan charts for the same sample week. This comparison highlights the critical trade-off identified: The models that were statistically fair Fig. 2.13A and B ultimately had poor probabilistic scores, while the model that was statistically flawed Fig. 2.13C achieved the best CRPS score. The Bootstrap scenarios (Fig. 2.13A), while

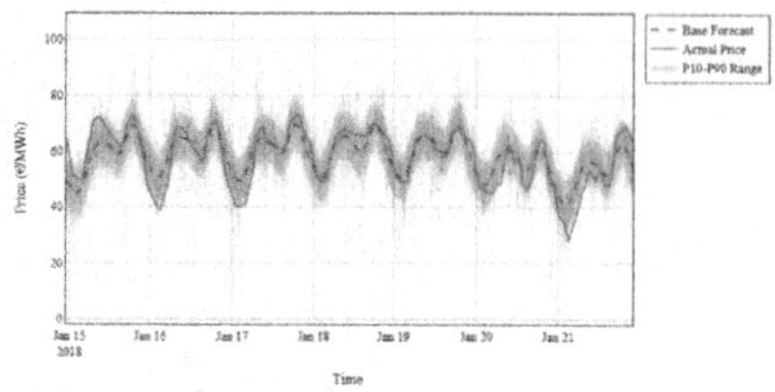

(A) Scenarios from: Bootstrap (Statistically Fair, Bad Score)

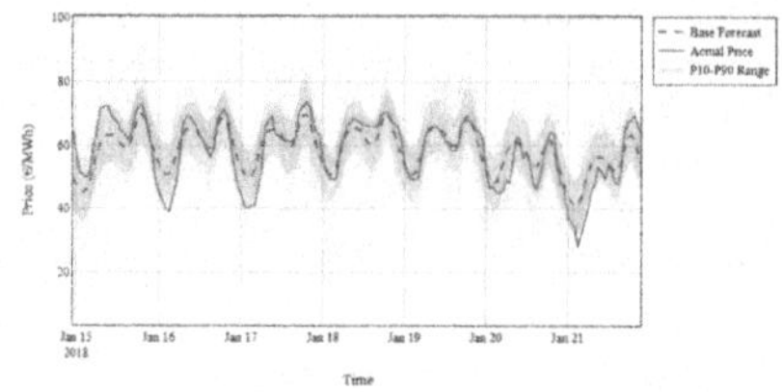

s

(B) Scenarios from: Gaussian Copula (Fair & Dynamic, Bad Score)

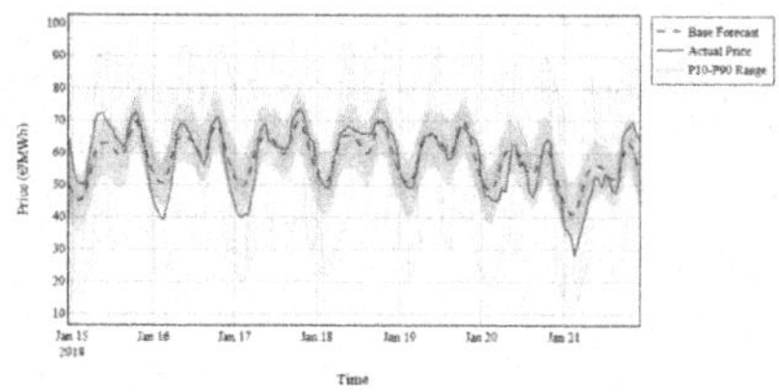

s

(C) Scenarios from: Markov Chain (Statistically Flawed, Best Score)

Figure 2.13 Qualitative comparison of scenario fan charts for the same sample week.

statistically fair in aggregate, are unrealistically spiky and volatile on an hourly basis. The Gaussian Copula scenarios (Fig. 2.13B) are smoother and more plausible, but their poor CRPS score indicates their probabilistic range was ultimately incorrect.

In stark contrast, the Markov Chain scenarios (Fig. 2.12C), despite being built on a flawed statistical distribution, provide the best quantitative score. This suggests that their *dynamic properties* (e.g., realistic ramp rates and temporal dependencies, which are linked to autocorrelation) are more important for overall forecast accuracy than matching the static distribution.

2.5.9 Case study conclusion

This case study successfully demonstrated the formal methodology for generating probabilistic scenarios for a complex, real-world energy time series.

The analysis, which compared five models, revealed a critical and persistent trade-off for practitioners:

1. Statistical versus dynamic fidelity: The "Bootstrap" and "Gaussian Copula" models were the only ones to achieve perfect *statistical fairness* (Fig. 2.10). However, this fairness did not translate to accuracy; both models produced poor CRPS scores (Table 2.3).
2. The superior, flawed model: The "Markov Chain" model failed the statistical fairness test catastrophically (Fig. 2.10, Table 2.2). However, it was dynamically plausible (Fig. 2.11A) and achieved the best probabilistic score (CRPS) by a significant margin.
3. Implications for decision-making: For a system operator scheduling reserves, the i.i.d. models (MCS, LHS, Bootstrap) are dangerously misleading. The "Gaussian Copula," despite its theoretical elegance, also failed to produce an accurate forecast. The "Markov Chain," despite its distributional flaws, was the most accurate and thus the most useful for decision-making.

It is concluded that for this operational energy system analysis, modeling the temporal dynamics was more critical to forecast accuracy (CRPS) than perfectly matching the static statistical moments. The failure of the "Gaussian Copula" suggests that its simple AR-1 structure, while capturing the ACF, was insufficient. The Markov Chain's state-based logic, despite its flaws, provided the most plausible scenarios and achieved the best overall probabilistic score.

2.6 Conclusions

2.6.1 Summary of methodologies

This chapter has detailed the foundational statistical methods for generating scenario sets to represent uncertainty in energy systems. We began with standard MCS, which is simple and robust but suffers from slow $O(N^{-1/2})$ convergence. We then introduced LHS and QMC as powerful variance reduction techniques that provide faster convergence by sampling the parameter space more uniformly. Finally, we discussed advanced methods like Copula Functions, which are essential for modeling the complex, nonlinear dependencies between multiple uncertain variables.

2.6.2 The impracticality of large, raw scenario sets

The case study demonstrated that while these methods can generate fair scenarios, they often need to generate a very large number (=1000 or more) to be accurate. For complex stochastic optimization models used in energy planning (e.g., unit commitment or capacity expansion), solving a problem with 1000 scenarios is computationally infeasible. This creates a curse of dimensionality in stochastic programming.

References

Alonso-Travesset, À., Martín, H., Coronas, S., & de la Hoz, J. (2022). Optimization models under uncertainty in distributed generation systems: A review. *Energies, 15*(5), 1932.

Amini, F., (2025). Probabilistic price forecasting with stochastic scenario generation. Available from: https://github.com/fateme-amini/Probabilistic-Price-Forecasting

de Carvalho Silva, T. E. B., Maçaira, P., Oliveira, F. L. C., & de Almeida Pereira, G. A. (2025). Enhancing energy planning with copula-based dependency models. *Energy Reports, 14*, 43–55.

Dellino, G., & Meloni, C. (2015). *Uncertainty management in simulation-optimization of complex systems*. Springer.

Dunn, W. L., & Shultis, J. K. (2022). *Exploring Monte Carlo methods*. Elsevier.

Infield, D., & Freris, L. (2020). *Renewable energy in power systems*. John Wiley & Sons.

Jing, W., Aksoy, M., Rahman, M. A., Alenezi, A. H., Deriche, M., & Tao, H. (2025). Multi-objective planning and optimal configuration of wind, solar, and energy storage in interconnected microgrid clusters using Vine Copula scenario generation and antlion optimization. *Renewable Energy, 256*, 124313.

Marti, K. (2024). Stochastic optimization methods. *Stochastic optimization methods: Applications in engineering and operations research* (pp. 1–36). Springer.

Mulla, R., (2018). Hourly energy consumption. kaggle Available from: https://www.kaggle.com/datasets/robikscube/hourly-energy-consumption

Niederreiter, H. (1992). *Random number generation and quasi-Monte Carlo methods*. SIAM.

Olsson, A., Sandberg, G., & Dahlblom, O. (2003). On Latin hypercube sampling for structural reliability analysis. *Structural Safety, 25*(1), 47–68.

Prékopa, A. (2013). *Stochastic programming*: 324. Springer Science & Business Media.

Shields, M. D., & Zhang, J. (2016). The generalization of Latin hypercube sampling. *Reliability Engineering & System Safety, 148*, 96–108.

Shu, Z., & Jirutitijaroen, P. (2011). Latin hypercube sampling techniques for power systems reliability analysis with renewable energy sources. *IEEE Transactions on Power Systems, 26*(4), 2066–2073.

Wedel, W. G. (2024). *Energy systems optimization considering the uncertainty of future developments*. BoD-Books on Demand.

Yoo, J., Son, Y., Yoon, M., & Choi, S. (2023). A wind power scenario generation method based on copula functions and forecast errors. *Sustainability, 15*(23), 16536.

Zadeh, M. M., Afshar, Z., Farahani, G., & Bathaee, S. M. T. (2023). An adaptive robust optimization model for microgrids operation using convexified ac power flow equations. *International Transactions on Electrical Energy Systems, 2023*(1), 6483030.

Zadeh, M. M., Afshar, Z., Heydari, R., Bathaee, S. M. T., & Savaghebi, M. (2020). A linear adaptive robust optimization model for day-ahead scheduling of microgrid. In *IECON 2020 The 46th Annual Conference of the IEEE Industrial Electronics Society* (pp. 1501–1506). IEEE. doi:10.1109/IECON43393.2020.9254928.

Scenario reduction methods: Clustering, fast forward selection and distance metrics

3

Mahmoud Mollayousefi Zadeh[a], *Niusha Mohammadi Sedgh*[b], *Rozhina Mohammadi*[b], *Sonia Zakeri*[c], *Mohammad Hossein Alizadeh*[d] *and Seyed Mohammad Taghi Bathaee*[a]
[a]Department of Electrical Engineering, K. N. Toosi University of Technology, Tehran, Iran, [b]Department of Electrical Engineering, Amirkabir University of Technology, Tehran, Iran, [c]Department of Electrical and Computer Engineering, University of Tabriz, Tabriz, Iran, [d]Department of Aerospace Engineering, Amirkabir University of Technology, Tehran, Iran

3.1 Introduction

The comprehensive scenario sets generated in Chapter 2 accurately represent energy system uncertainties but create computationally prohibitive stochastic optimization problems. The complexity of stochastic programming grows exponentially with scenario count:

$$\text{Computational Complexity} = O\left((S)^T\right) \tag{3.1}$$

where S is the number of scenarios and T the number of time stages. For example, a 24-hour stochastic unit commitment with 100 scenarios per hour yields 100^{24} possible paths-far beyond computational feasibility (Mollayousefi Zadeh et al., 2023; Montero et al., 2022). The fundamental trade-off in scenario reduction can be expressed as:

$$\min_{J\subset I} D\left(\mathbb{P}, \tilde{\mathbb{P}}_J\right) \text{ subject to } |J| = N \tag{3.2}$$

where I is the original scenario set, J is the reduced set of size N, $\mathbb{P}$ is the original probability measure, $\tilde{\mathbb{P}}_J$ is the reduced probability measure, and D is a probability distance metric.

The scenario reduction problem is fundamentally a mass transportation problem (Li & Li, 2016). Let $\mathbb{P} = \{(\xi_i, p_i)\}_{i=1}^{S}$ be the original scenario set with probabilities p_i, and $\mathbb{Q} = \left\{\left(\zeta_j, q_j\right)\right\}_{j=1}^{N}$ be the reduced set. As depicted in Fig. 3.1, the scenario reduction process transforms a large scenario set into a computationally tractable reduced set while preserving statistical properties.

Stochastic Planning and Modeling for Energy Systems: Methods, Applications, and Developments.
DOI: https://doi.org/10.1016/B978-0-443-45298-7.00009-4

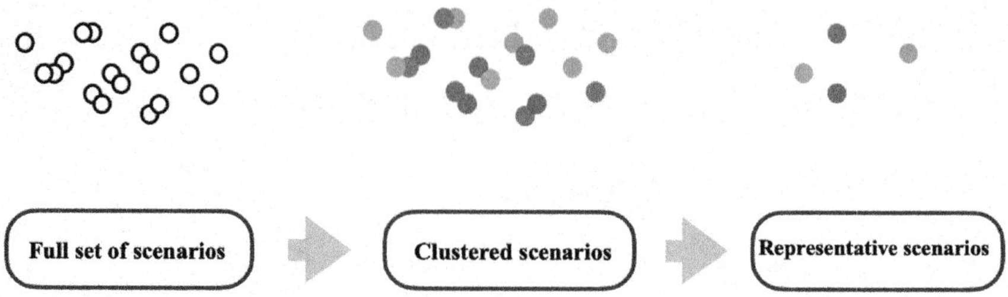

Figure 3.1 The scenario reduction process.

The optimal reduction minimizes the Wasserstein distance:

$$W_p(\mathbb{P}, \mathbb{Q}) = \left(\min_{\pi_{ij}} \sum_{i=1}^{S} \sum_{j=1}^{N} \pi_{ij} d(\xi_i, \zeta_j)^p \right)^{1/p} \tag{3.3}$$

subject to:

$$\sum_{j=1}^{N} \pi_{ij} = p_i \, \forall i = 1, \ldots, S \tag{3.4}$$

$$\sum_{i=1}^{S} \pi_{ij} = q_j \, \forall j = 1, \ldots, N \tag{3.5}$$

$$\pi_{ij} \geq 0 \, \forall i, j \tag{3.6}$$

where π_{ij} represents the probability mass transported from the scenario ξ_i to scenario ζ_j.

Several scenario reduction techniques, such as K-means clustering and others, can be employed to manage computational complexity in stochastic optimization. An effective scenario reduction method must not only accurately represent the original probability density function (PDF) but also significantly reduce the problem's complexity. The key properties that such an ideal method should possess are summarized in Table 3.1.

3.2 Distance metrics

The efficacy of scenario reduction methods is fundamentally governed by the choice of distance metric (Rujeerapaiboon et al., 2022). This metric quantifies the dissimilarity between scenario paths, directly influencing which scenarios are deemed similar and are consequently merged or eliminated. A well-chosen distance metric ensures that the reduced scenario set preserves the essential statistical and geometrical features of the original data.

It is crucial to define the problem's dimensions: Let S be the total number of scenarios in the set, and let T be the number of time periods (or the dimensionality) of each individual scenario. For example, a set of $S = 1000$ daily wind power scenarios,

Table 3.1 Key properties of ideal scenario reduction methods.

Property	Description
Statistical consistency	The reduced set should preserve key statistical properties (mean, variance, covariance) of the original distribution.
Computational efficiency	The reduction algorithm should be computationally feasible for large scenario sets.
Theoretical foundation	The method should have solid mathematical foundations and convergence guarantees.
Adaptability	The approach should handle various types of uncertainties and time dependencies.
Implementation simplicity	The method should be relatively straightforward to implement and use.

each with $T = 24$ hourly values, results in 1000 vectors of length 24. The computational complexity of a distance metric can depend on either S or T, leading to significant differences in performance, as detailed in this section and Table 3.2.

3.2.1 Euclidean distance

The Euclidean distance is the most intuitive metric, measuring the straight-line distance between two scenario vectors $\xi_i = (\xi_i^1, \dots, \xi_i^T)$ and $\xi_j = \left(\xi_j^1, \dots, \xi_j^T\right)$ in a T-dimensional space. It is defined as:

$$d_{\mathrm{E}}(\xi_i, \xi_j) = \sqrt{\sum_{t=1}^{T} \left(\xi_i^t - \xi_j^t\right)^2} \tag{3.7}$$

While computationally efficient, its primary limitation is that it treats all time periods as independent and equally important, and it is sensitive to the scale of the variables. It performs well when uncertainties are uncorrelated and have similar magnitudes (Hu & Li, 2019).

3.2.2 Manhattan distance

Also known as the L^1 norm or taxicab distance, the Manhattan distance sums the absolute differences along each dimension:

$$d_{\mathrm{Mah}}(\xi_i, \xi_j) = \sum_{t=1}^{T} \left|\xi_i^t - \xi_j^t\right| \tag{3.8}$$

This metric is more robust to outliers than the Euclidean distance. If a large deviation in a single time period is considered less critical than small deviations across many periods, Manhattan distance may be preferable (Chiu et al., 2016; Gangil et al., 2025).

Table 3.2 Comparison of distance metrics for scenario reduction in energy systems.

Metric	Computational complexity	Suitability for energy systems
Euclidean	$O(T)$	Good for independent, similarly scaled variables (e.g., load profiles). Simple and fast, but poor for correlated or differently-scaled data.
Manhattan	$O(T)$	Robust to outliers. Suitable for applications where the total absolute deviation is a more relevant measure of error than squared deviation.
Mahalanobis	$O(T^2)$ to $O(T^3)$	Excellent for correlated uncertainties (e.g., wind and solar generation, electricity prices). Requires a well-estimated, invertible covariance matrix.
Wasserstein	$O(S^3)$ for linear programming	Theoretically optimal but computationally expensive. Best for the final reduction of already preprocessed sets.
Energy score	$O(S^2T)$	Proper scoring rule for multivariate distributions. Ideal for evaluating spatial-temporal dependencies in renewable generation.
CRPS	$O(S^2)$ per time period	Standard for univariate probabilistic forecasting. Excellent for wind/solar power and price forecasting applications.
Total variation	$O(KT)$	Measures worst-case probability differences. Useful for risk assessment and applications where extreme events are critical.

CRPS, continuous ranked probability score.

Algorithm 1 Fast forward selection for scenario reduction.

Require: Original scenario set $\{(\xi_i, p_i)\}_{i=1}^{S}$, target size N, distance metric d

Ensure: Reduced scenario set $\{(\zeta_j, q_j)\}_{j=1}^{N}$

Initialize $J=\emptyset$ (selected scenarios), $U=\{1, \ldots, S\}$ (unselected scenarios)

Initialize distances $D_{i}=\infty$ for all $i \in U$

for $k=1$ to N **do**

for each i U **do**

Compute temporary reduction quality if i were added to J

$Qi = \sum_{j\in U\setminus\{i\}} P_j \min m \in J\cup\{i\} d(\xi j, \xi m)$

end for

Select $i^* = \arg\min_{i\in U} Q_i$

$J = J\cup\{i^*\}$, $U = U\setminus\{i^*\}$

Update probabilities: $q_m = p_m + \sum_{j\in U_m} P_j$ for $m \in J$

end for

return $\{(\xi_j, q_j)\}_{j\in J}$

Algorithm 2 Backward reduction for scenario reduction.

Require: Original scenario set $\{(\xi_i, p_i)\}_{i=1}^{S}$, target size N, distance metric d
Ensure: Reduced scenario set $\{(\zeta_j, q_j)\}_{j=1}^{N}$
Initialize $J = \{1, ..., S\}$
for $k = S$ down to $N + 1$ **do**
  for each $i \in J$ **do**
  Compute the degradation if scenario i is removed:
  $\Delta i = \sum_{j \in J \backslash \{i\}} pj\left[\min_{m \in J \backslash \{i\}} d(\xi j, \xi m) - \min m \in Jd(\xi j, \xi m)\right]$
  end for
  Remove $i^* = \arg\min_{i \in J} \Delta_i$
  $J = J \backslash \{i^*\}$
  Redistribute probability pi* to nearest scenario in J
end for
Return $\{(\xi_j, q_j)\}_{j \in J}$

3.2.3 Mahalanobis distance

The Mahalanobis distance accounts for the correlations between variables and their respective variances. For two scenario vectors ξ_i and ξ_j, it is defined as:

$$d_{\text{Mal}}(\xi_i, \xi_j) = \sqrt{(\xi_i - \xi_j)^{\mathsf{T}} \Sigma^{-1} (\xi_i - \xi_j)} \tag{3.9}$$

where Σ is the covariance matrix of the entire scenario set. This metric effectively measures the distance in terms of standard deviations from the mean. A key advantage is that it is scale-invariant and can handle correlated uncertainties, such as the joint behavior of wind speed and solar irradiance at different locations (Kumari & Gupta, 2024; Srinivasaraghavan & Allada, 2006).

3.2.4 Wasserstein distance

The Wasserstein distance, originating from optimal transport theory, provides a powerful framework for comparing probability distributions. For two discrete probability distributions $P = \{(\xi_i, p_i)\}_{i=1}^{S}$ and $Q = \{(\zeta_j, q_j)\}_{j=1}^{S'}$, the pth order Wasserstein distance is:

$$W_p(P, Q) = \left(\inf_{\gamma \in \Gamma(P, Q)} \sum_{i=1}^{S} \sum_{j=1}^{S'} \gamma_{ij} d(\xi_i, \zeta_j)^p \right)^{1/p} \tag{3.10}$$

where $\Gamma(P, Q)$ denotes the set of all joint distributions (couplings) with marginals P and Q, and $d(\bullet, \bullet)$ is a ground metric, often the Euclidean distance. Scenario reduction can be viewed as finding a distribution Q with fewer scenarios that minimize the Wasserstein distance to the original distribution P. While theoretically optimal for preserving the distribution's shape, its computational cost is significant (Dong et al., 2023; Liu et al., 2025).

3.2.5 Kantorovich–Rubinstein metric

In the context of scenario reduction, the Kantorovich–Rubinstein metric, which is dual to the first-order Wasserstein distance, is also highly relevant. It is defined for a class of Lipschitz continuous functions $\mathcal{F}$ as:

$$d_{\mathrm{KR}}(P, Q) = \sup_{f \in \mathcal{F}} \left| \mathbb{E}_P[f(\xi)] - \mathbb{E}_Q[f(\xi)] \right| \tag{3.11}$$

This formulation is particularly useful in stability analysis of stochastic programs, as it bounds the error in the optimal objective value when the original distribution P is replaced by the reduced distribution Q (Giordano, 2023).

3.2.6 Energy score

The energy score is a proper scoring rule that measures the distance between a multivariate predictive distribution and the observed value. It generalizes the concept of the continuous ranked probability score (CRPS) to higher dimensions. For two multivariate distributions P and Q with scenarios $\{\xi_i\}_{i=1}^{S}$ and $\{\zeta_j\}_{j=1}^{S'}$, the energy score is defined as:

$$\mathrm{ES}(P, Q) = \mathbb{E}\left[\|\xi - \zeta\|^{\beta}\right] - \frac{1}{2}\mathbb{E}\left[\|\xi - \xi'\|^{\beta}\right] - \frac{1}{2}\mathbb{E}\left[\|\zeta - \zeta'\|^{\beta}\right] \tag{3.12}$$

where $\xi, \xi' \sim P$ are independent, $\zeta, \zeta' \sim Q$ are independent, $\|\bullet\|$ denotes the Euclidean norm, and $\beta \in (0, 2]$ is a parameter typically set to 1. For scenario sets with probabilities $\{p_i\}$ and $\{q_j\}$, the empirical Energy Score becomes:

$$\mathrm{ES}(P, Q) = \sum_{i=1}^{S}\sum_{j=1}^{S'} p_i q_j \|\xi_i - \zeta_j\| - \frac{1}{2}\sum_{i=1}^{S}\sum_{k=1}^{S} p_i p_k \|\xi_i - \xi_k\| - \frac{1}{2}\sum_{j=1}^{S'}\sum_{l=1}^{S'} q_j q_l \|\zeta_j - \zeta_l\| \tag{3.13}$$

The energy score is particularly valuable for evaluating multivariate probabilistic forecasts in energy systems, where dependencies between different time periods or locations are crucial (Ziel, 2021).

3.2.7 Continuous ranked probability score

The CRPS serves as a comprehensive probabilistic evaluation metric that quantifies the discrepancy between predicted and empirical distributions through their cumulative distribution functions. As a proper scoring rule, CRPS provides a rigorous framework for assessing the quality of probabilistic forecasts by simultaneously valuating calibration, sharpness, and reliability. This metric has gained significant prominence in energy forecasting applications, particularly for renewable generation and electricity prices, where accurate uncertainty quantification is paramount for operational decision-making (Zamo & Naveau, 2018).

The fundamental formulation of CRPS integrates the squared differences between cumulative distribution functions across the entire domain of possible outcomes. For a predictive distribution characterized by its Cumulative Distribution Function (CDF) F_P and an empirical reference distribution F_Q, the CRPS is mathematically defined as:

$$\mathrm{CRPS}\left(F_P, F_Q\right) = \int_{-\infty}^{\infty} \left[F_P(x) - F_Q(x)\right]^2 dx \tag{3.14}$$

This integral representation effectively measures the total area between the two cumulative distribution curves, providing a geometrically intuitive quantification of distributional differences. The metric exhibits several desirable properties: It attains its minimum value of zero only when the distributions are identical, is strictly proper, ensuring honest assessment of probabilistic forecasts, and is sensitive to all moments of the distribution rather than focusing solely on central tendencies.

Within the context of scenario reduction methodologies, where continuous distributions are approximated by discrete scenario sets, the empirical computation of CRPS requires careful consideration. Given an original scenario ensemble $P = \{(\xi_i, p_i)\}_{i=1}^{S}$ and its reduced counterpart $Q = \left\{\left(\zeta_j, q_j\right)\right\}_{j=1}^{S'}$, the empirical CRPS derivation follows from the continuous definition through the construction of empirical cumulative distribution functions. The computational expression emerges as:

$$\mathrm{CRPS}(P, Q) = \sum_{i=1}^{S} p_i \int_{-\infty}^{\infty} \left(\mathbb{1}_{\{\xi_i \le x\}} - F_Q(x)\right)^2 dx - \frac{1}{2} \sum_{i=1}^{S} \sum_{k=1}^{S} p_i p_k |\xi_i - \xi_k| \tag{3.15}$$

In this formulation, the indicator function $0 = \mathbb{1}_{\{\xi_i \le x\}}$ assumes unity when the scenario ξ_i does not exceed threshold x and vanishes otherwise, while $F_Q(x) = \sum_{j=1}^{S'} q_j \mathbb{1}_{\{\zeta_j \le x\}}$ constructs the empirical CDF of the reduced scenario set. The first component measures the fidelity of the reduced set in approximating the original scenarios, whereas the second term represents an intrinsic property of the original ensemble that remains invariant under reduction.

For practical implementation in energy system applications, a more computationally tractable representation proves advantageous:

$$\mathrm{CRPS}(P, Q) = \sum_{i=1}^{S} p_i \cdot \mathbb{E}_Q |\xi_i - \zeta| - \frac{1}{2} \sum_{i=1}^{S} \sum_{k=1}^{S} p_i p_k |\xi_i - \xi_k| \tag{3.16}$$

where the expectation $\mathbb{E}_Q|\xi_i - \zeta| = \sum_{j=1}^{S'} q_j\left|\xi_i - \zeta_j\right|$ computes the probability-weighted absolute distance between the original scenario ξ_i and the reduced distribution Q.

3.2.8 Average total variation distance

The total variation distance (TVD) measures the largest possible difference between probabilities assigned by two distributions. For discrete probability distributions P and Q defined on the same probability space, the TVD is defined as:

$$\text{TVD}(P, Q) = \sup_{A\in\mathcal{F}} |P(A) - Q(A)| \tag{3.17}$$

where $\mathcal{F}$ is the sigma-algebra of measurable events. For finite discrete distributions, this simplifies to:

$$\text{TVD}(P, Q) = \frac{1}{2}\sum_{\omega\in\Omega} |P(\omega) - Q(\omega)| \tag{3.18}$$

where Ω is the sample space. In the context of scenario reduction, where we have scenario sets with associated probabilities, the Average TVD across time periods can be computed as:

$$\text{AvgTVD}(P, Q) = \frac{1}{T}\sum_{t=1}^{T}\frac{1}{2}\sum_{k=1}^{K}\left|F_P^t(I_k) - F_Q^t(I_k)\right| \tag{3.19}$$

where F_P^t and F_Q^t are the empirical CDFs at time t, and $\{I_k\}_{k=1}^{K}$ forms a partition of the range of possible values. TVD provides a strong measure of distributional difference, with values ranging from 0 (identical distributions) to 1 (completely disjoint distributions) (Rahimian et al., 2019).

3.3 Selection-based reduction algorithms

3.3.1 Fast forward selection algorithm

The fast forward selection (FFS) algorithm iteratively selects the scenario that minimizes the increase in the Wasserstein distance at each step (Heitsch & Römisch, 2003) (Algorithm 3.1).

3.3.2 Backward reduction algorithm

The backward approach starts with the full set and iteratively removes the least important scenario (Heitsch & Römisch, 2003) (Algorithm 3.2).

Table 3.3 Distance metrics comparison for wind power scenario reduction.

Metric	FFS	Backward reduction
Euclidean RMS	0.209070	0.212734
Manhattan RMS	0.823928	0.833622
Wasserstein (first-order)	0.000981	0.001000
Kantorovich–Rubinstein	0.000981	0.001000
Variance preserved	0.972876	1.094922
Computation time (s)	0.45	1.23

FFS: fast forward selection; RMS: Root Mean Square.

3.3.3 Energy system example

This section presents a practical application of scenario reduction methods to wind power generation data, comparing the performance of FFS and backward reduction algorithms. The case study involves reducing a set of 200 synthetic wind power scenarios representing 24-hour generation profiles to a more manageable set of 20 scenarios while preserving key statistical properties. A synthetic dataset was generated with a temporal correlation characteristic of real wind patterns, featuring higher generation during daytime hours and lower generation at night. Both reduction algorithms were implemented in MATLAB [available in (Rojinamdii, 2025)] and evaluated using multiple distance metrics to assess their performance in preserving the original distribution's characteristics.

Fig. 3.2 presents a comprehensive comparison of the two reduction methods across multiple evaluation criteria. Both methods successfully captured the variability of the original wind power scenarios, as shown in subplots (A) and (B). The FFS algorithm selected scenarios that provide good coverage of the distribution tails, while the backward reduction method-maintained scenarios that more closely represent the central tendency of the original set. The mean profile comparison in subplot (C) demonstrates that both methods accurately preserved the diurnal pattern of wind generation, with the backward method showing marginally better alignment with the original mean, particularly during the morning ramp-up period between hours 6–10.

The temporal variance analysis in subplot (F) reveals important differences in statistical preservation. The backward reduction method better maintained the variance structure, particularly during hours 12–18 when wind variability is typically highest. This is quantitatively confirmed in Table 3.3, where backward reduction achieved superior variance preservation at 109.5% compared to 97.3% for FFS, indicating slight over-representation of high-variance scenarios. The distribution of selected scenarios in subplot (E) shows that FFS tended to select scenarios from the extremes of the original set, while backward reduction maintained a more uniform distribution across the original scenario indices.

Table 3.3 provides a detailed quantitative comparison of the distance metrics. The FFS algorithm demonstrated slightly better performance in point-wise distance metrics, with Euclidean RMS of 0.209070 and Manhattan RMS of 0.823928 compared to 0.212734 and 0.833622 for backward reduction respectively. This indicates that, on average, the FFS-reduced scenarios are marginally closer to their nearest neighbors

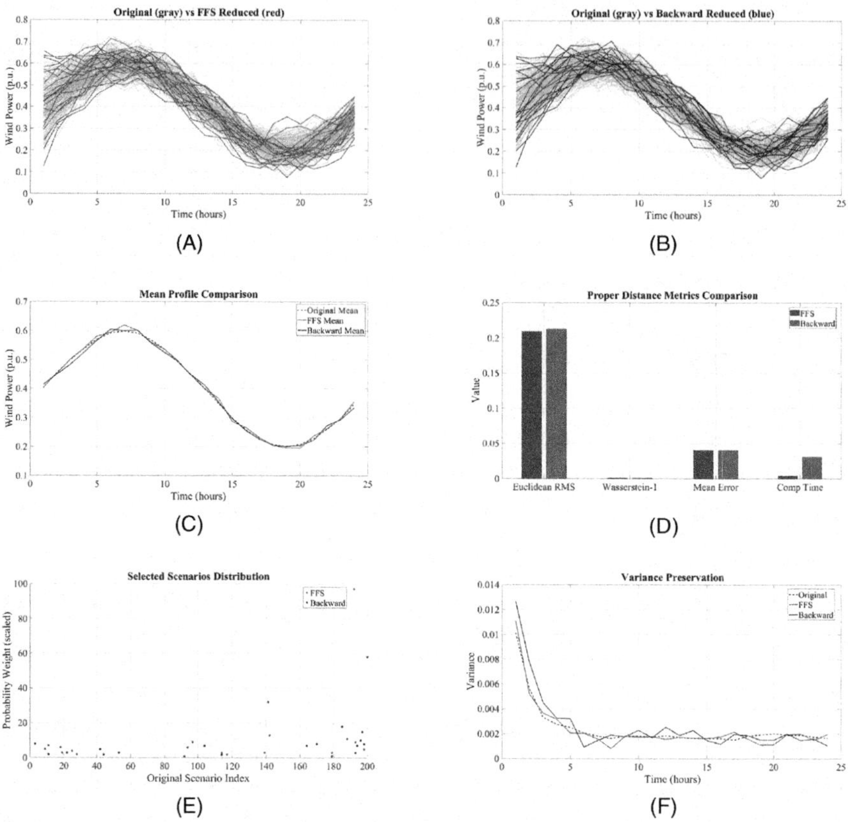

Figure 3.2 Comparison of scenario reduction methods for wind power data: (A) Original scenarios versus FFS-reduced scenarios; (B) original scenarios versus backward-reduced scenarios; (C) mean profile preservation across methods; (D) performance metrics comparison; (E) distribution of selected scenarios in original set; (F) temporal variance preservation over 24-hour period. *FFS*, Fast forward selection.

in the original set. However, both methods achieved nearly identical performance in distribution-based metrics, with Wasserstein and Kantorovich–Rubinstein distances of approximately 0.001, confirming their effectiveness in preserving the overall probability distribution. A significant practical consideration is the computational performance, where the backward reduction method required approximately 2.7 times longer computation time than FFS, making FFS more suitable for applications requiring rapid scenario reduction or dealing with very large datasets.

The results have important implications for stochastic optimization in energy systems. For day-ahead scheduling applications, the backward reduction method's superior variance preservation makes it preferable when capturing extreme wind ramping events is critical for maintaining system reliability. For long-term planning problems, the computational efficiency of FFS makes it more suitable for applications

Table 3.4 Comparison of selection-based reduction methods.

Characteristic	Fast forward selection	Backward reduction
Computational complexity	$O(N \cdot S^2)$	$O(S^3)$
Memory requirements	Moderate	High
Quality of solution	Good heuristic	Theoretically better
Best use case	Large scenario sets	Small to medium sets where optimality is crucial

Algorithm 3 k-means scenario reduction.

Require: Original scenario set $\{\xi_i\}_{i=1}^{S}$, number of clusters k, distance metric d

Ensure: Reduced scenario set $\{\zeta_j, q_j\}_{j=1}^{k}$

Randomly initialize cluster centroids $\mu_1, \mu_2, \ldots, \mu_k$

repeat

  Assignment Step: Assign each scenario to nearest centroid

  $C_j = \{i: d(\xi_i, \mu_j) \leq d(\xi_i, \mu_m) \forall m\}$

  Update Step: Recompute centroids

  $\mu_j = \frac{1}{|C_j|} \sum_{i \in C_j} \xi i$

until centroids stabilize

Compute cluster probabilities: $q_j = \frac{|C_j|}{S}$

return$\{(\mu_j, q_j)\}_{j=1}^{k}$

requiring frequent scenario generation or dealing with multiple uncertainty sources simultaneously. Both methods adequately preserved the mean wind profile, ensuring accurate estimation of expected wind generation and consequent reserve requirements for power system operations. The choice between methods should consider the specific application requirements: Backward reduction for maximum statistical fidelity when computational resources permit, and FFS for large-scale or time-sensitive applications where near-optimal performance is sufficient. While this case study demonstrates the effectiveness of both methods, it should be noted that the evaluation used synthetic wind data, and performance validation on real historical wind generation data would strengthen these findings. Further investigation is also warranted to establish the relationship between distance metric performance and the quality of downstream operational decisions in unit commitment and economic dispatch problems. Table 3.4 compares selection-based reduction methods.

3.4 Clustering-based reduction algorithms

3.4.1 k-means clustering for scenario reduction

The k-means algorithm partitions scenarios into k clusters and uses cluster centroids as representative scenarios (Algorithm 3.3).

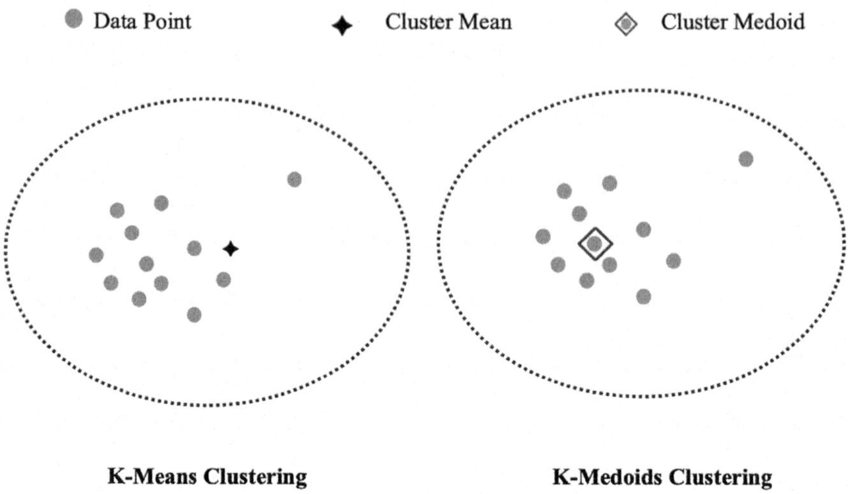

Figure 3.3 Comparison of k-means (centroids) and k-medoids (actual scenarios).

3.4.2 k-medoids clustering

Unlike k-means, k-medoids selects actual scenarios from the set as cluster representatives (see Fig. 3.3).

$$\text{Medoid} = \arg\min_{\xi \in C_j} \sum_{\xi_i \in C_j} d(\xi, \xi_i) \tag{3.20}$$

K-means clustering K-medoids clustering

3.4.3 Energy system example

In this study, a real-world wind turbine Supervisory control and data acquisition (SCADA) dataset (Erisen, 2020) was employed to evaluate the performance of various scenario-generation and scenario-reduction techniques. The original dataset included several operational variables such as wind speed, wind direction, and theoretical power curves. To focus the analysis on the actual power output, all irrelevant columns were removed and the primary variable of interest- Low Voltage (LV) ActivePower (kW), was converted to a numeric format to ensure consistency and enable valid mathematical operations. Missing or zero readings, which could bias descriptive statistics and distort clustering outcomes, were corrected using a local averaging approach in which each zero entry was replaced by the mean of its immediate neighbors within a predefined temporal window. This method preserved short-term temporal structure while mitigating measurement gaps. To further streamline the dataset and reduce redundancy while retaining essential temporal behavior, every sixth record was preserved, generating a manageable down sampled series. Time information extracted from timestamps enabled assignment of each observation to its corresponding season—Spring, Summer,

Table 3.5 First 12 rows of clustered scenarios.

Date/time	LV activepower (kW)	Season	Load level
1/1/2018 0:00	380.0478	Winter	Low
1/1/2018 1:00	447.6057	Winter	Low
1/1/2018 2:00	710.5873	Winter	Low
1/1/2018 3:00	736.6478	Winter	Low
1/1/2018 4:00	1493.808	Winter	Medium
1/1/2018 5:00	1164.893	Winter	Medium
1/1/2018 6:00	1114.027	Winter	Medium
1/1/2018 7:00	909.8878	Winter	Low
1/1/2018 8:00	585.2594	Winter	Low
1/1/2018 9:00	725.8681	Winter	Low
1/1/2018 10:00	881.0621	Winter	Low
1/1/2018 11:00	233.9906	Winter	Low

Autumn, or Winter-based on the month of the year. This preprocessing yielded a clean, structured dataset suitable for subsequent clustering, scenario reduction, and statistical evaluation.

Following preprocessing, the intra-seasonal variability of active power was analyzed to identify natural load-level groupings. To accomplish this, K-means clustering was independently applied to each seasonal load-level subset, partitioning the data into three clusters corresponding to low, medium, and high-power levels. The cluster centers were then sorted in ascending order to provide consistent and interpretable labels across seasons. The resulting labeled dataset allowed each observation to be characterized simultaneously by its season and its load level. Table 3.5 illustrates the first twelve rows of this clustered dataset, showing the combined temporal, seasonal, and load-level classifications.

To obtain a reduced yet representative set of scenarios suitable for use in stochastic optimization, the FFS algorithm was applied within each seasonal subset. FFS iteratively selects scenarios that minimize the weighted distance to all other points, thereby ensuring that the chosen representatives capture both the central tendency and structural variability of each cluster. From each of the three clusters per season, one representative scenario was selected, resulting in a reduced scenario set containing twelve scenarios overall. Table 3.6 lists the selected representatives, categorized by season and load level.

The fidelity of the reduced scenario set was evaluated by comparing descriptive statistics from the original seasonal data to those computed for the representative scenarios. Table 3.7 reports the seasonal means and variances for both datasets. The close correspondence between the two sets indicates that the representative scenarios preserve the central tendencies and dispersion characteristics of the original data.

To provide a visual comparison of the full and reduced distributions, Fig. 3.4 presents seasonal histograms with overlaid kernel density estimates (KDEs). The representative scenarios align well with the dominant modes and general shapes of the seasonal distributions. Minor secondary peaks present in multimodal seasons are

Table 3.6 Reduced scenarios.

Date/time	LV activepower (kW)	Season	Load level
2/10/2018 18:50	97.2951	Winter	Low
1/15/2018 19:40	1617.1610	Winter	Medium
12/22/2018 8:20	3502.7361	Winter	High
5/24/2018 5:30	69.5924	Spring	Low
3/15/2018 21:00	1666.9301	Spring	Medium
3/17/2018 10:00	3473.0090	Spring	High
6/5/2018 14:30	136.8365	Summer	Low
7/14/2018 20:30	1427.0540	Summer	Medium
8/4/2018 1:00	3132.3701	Summer	High
10/25/2018 20:20	302.5698	Autumn	Low
11/30/2018 13:30	1785.6930	Autumn	Medium
10/9/2018 22:20	3421.1479	Autumn	High

Table 3.7 Statistical comparison between original and reduced scenarios.

Season	Mean (Orig.)	Var (Orig.)	Mean (Red.)	Var (Red.)
Winter	1377.7181	1922650.855	1739.0641	1940268.179
Spring	1225.8738	1746740.192	1736.5105	1932961.515
Summer	1161.3792	1373696.262	1565.4202	1505109.548
Autumn	1574.8626	1632712.951	1836.4702	1622210.783

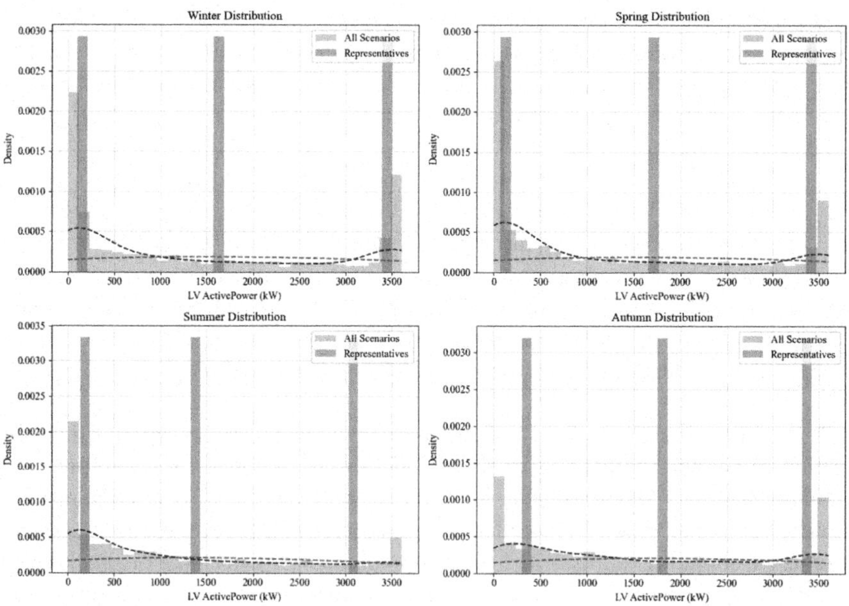

Figure 3.4 Seasonal distributions (Histogram + KDE).

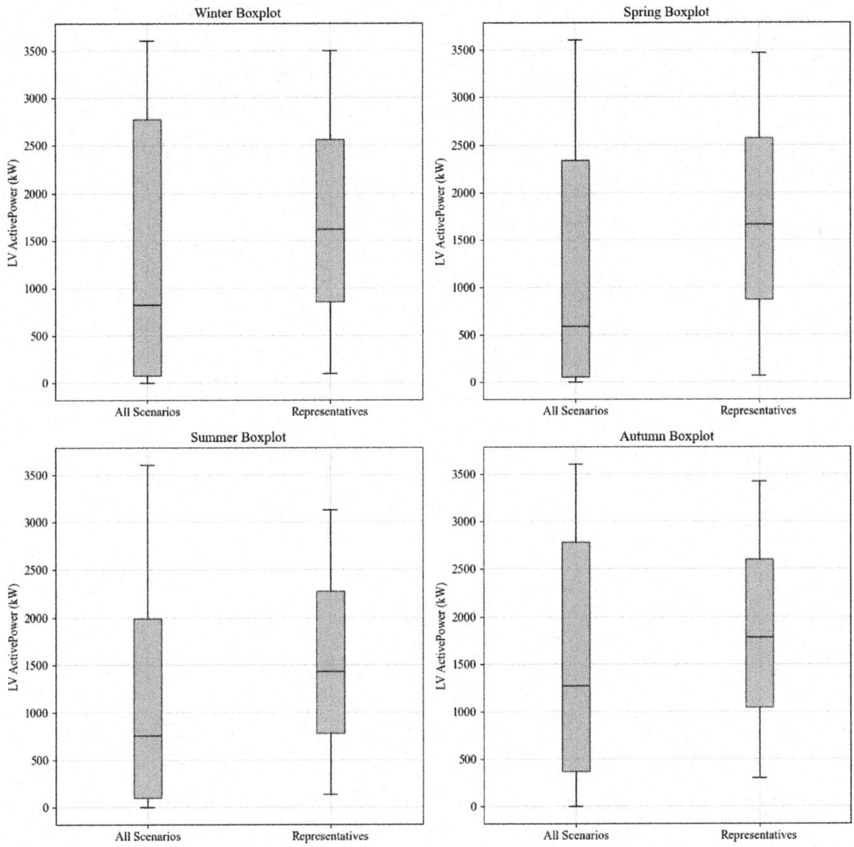

Figure 3.5 Seasonal boxplots.

naturally omitted due to the small number of representative points, but the principal seasonal characteristics are preserved.

A complementary visualization is provided in Fig. 3.5, which compares boxplots of the full and reduced datasets. Across all seasons, the median and interquartile ranges of the representative scenarios remain closely aligned with those of the original data. As expected, the reduced dataset exhibits narrower whiskers because of its reduced size, yet it still captures the key seasonal variability patterns, including the relative magnitudes of extreme values.

Beyond descriptive statistics, several distributional metrics were employed to quantify the similarity between the original and reduced datasets, including the Wasserstein distance, energy distance, CRPS, and TVD. These metrics evaluate different aspects of distributional similarity, ranging from spatial alignment (Wasserstein) to histogram overlap (TVD). The results, presented in Table 3.8, confirm that the representative scenarios remain close to the full datasets across all seasons.

Table 3.8 Advanced metric analysis between original and reduced scenarios.

Season	Wasserstein	Energy	CRPS	TVD
Winter	407.7035	139.9298	831.1352	0.012129
Spring	538.8989	220.1518	824.1992	0.009301
Summer	482.5069	181.0894	731.8309	0.012895
Autumn	377.3083	112.3582	782.0978	0.013186

CRPS, continuous ranked probability score; *TVD,* total variation distance.

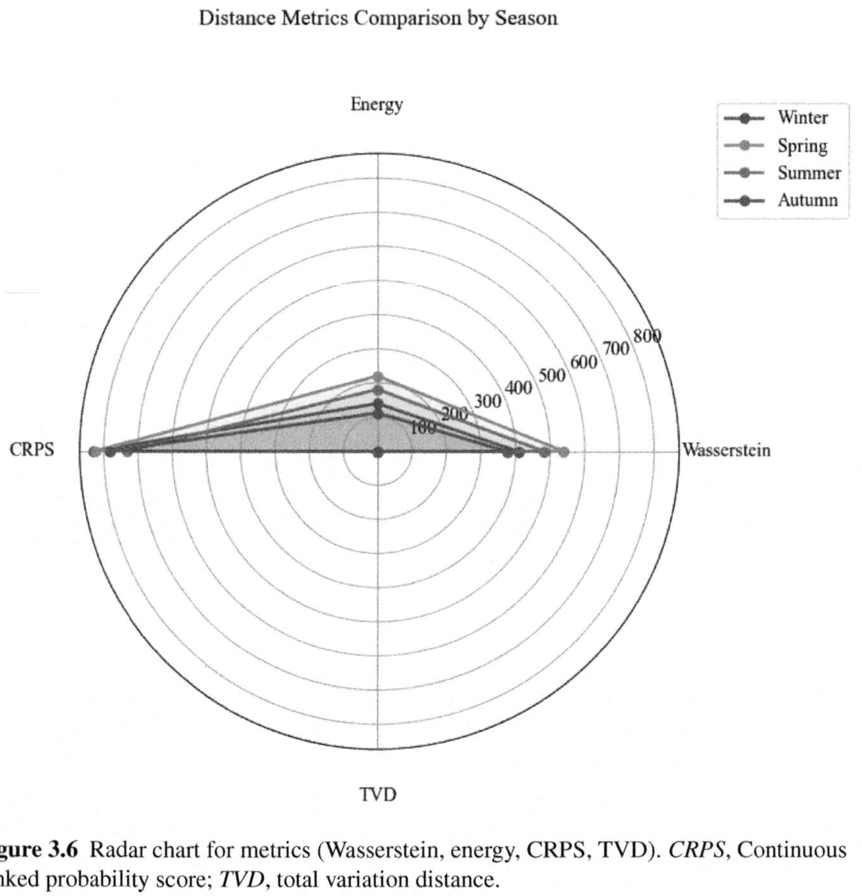

Figure 3.6 Radar chart for metrics (Wasserstein, energy, CRPS, TVD). *CRPS*, Continuous ranked probability score; *TVD*, total variation distance.

Fig. 3.6 summarizes these four metrics using a radar chart. The compact areas covered by each season indicate that the reduced scenarios faithfully approximate the original distributions, with no season showing substantial divergence along any metric axis.

Finally, to justify the selection of three clusters per season, several cluster-determination methods were evaluated. Table 3.9 summarizes standard approaches such as the Elbow method, Silhouette analysis, and Gap statistic, along with

Table 3.9 Determining optimal number of clusters.

Method	Description and application
Elbow method	Plots the within-cluster sum of squares (WCSS) against the number of clusters; the "elbow" point suggests the optimal *k*.
Silhouette analysis	Assesses cluster quality based on similarity within clusters versus between clusters; higher silhouette width indicates better-defined clusters.
Gap statistic	Compares WCSS of the observed data to that of a reference null dataset; the optimal *k* maximizes the gap.
Energy-specific criteria	Selects *k* based on the computational requirements of downstream optimization models in energy systems.

energy-system-specific considerations where computational tractability motivates a smaller number of representative clusters.

3.5 Optimal *k* selection methods

Selecting the optimal number of clusters k is a fundamental step in scenario reduction using K-means clustering. Different statistical and geometric criteria may identify different optimal values of k, and each choice affects the trade-off between model accuracy and computational tractability in downstream stochastic optimization. This section presents four widely used methodologies-elbow method, silhouette analysis, gap statistic, and energy-based criteria-together with their clustering results and comparative performance metrics.

3.5.1 Elbow method

The elbow method is one of the most widely adopted heuristic techniques for identifying the appropriate number of clusters. The method evaluates the within-cluster sum of squares (WCSS), which quantifies the internal compactness of clusters (Cui, 2020). For k clusters, the WCSS is defined as:

$$WCSS(k) = \sum_{i=1}^{k} \sum_{x \in C_i} x - \mu_i{}^2, \tag{3.21}$$

where C_i denotes the set of data points in the cluster i, and μ_i is the corresponding centroid.

As k increases, WCSS decreases due to reduced intra-cluster variance. However, beyond a certain value of k, the incremental improvements become marginal. The point where the reduction in WCSS begins to plateau is referred to as the elbow. Fig. 3.7 illustrates this behavior for the resampled wind turbine SCADA dataset. A noticeable bend occurs at approximately $k = 6$, indicating a balance between cluster compactness and model simplicity.

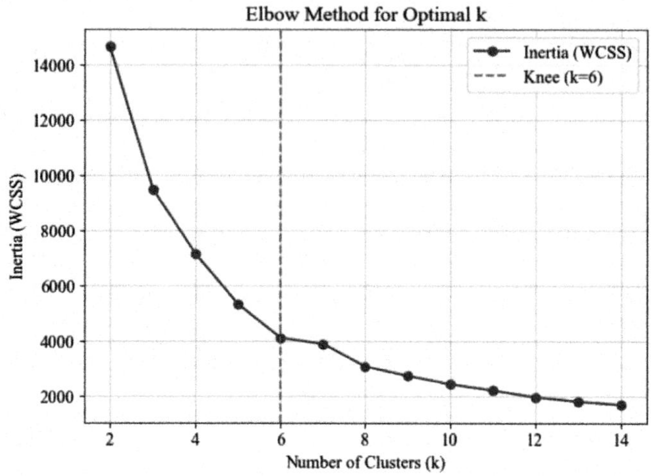

Figure 3.7 WCSS versus k plot. The elbow point indicates the optimal number of clusters. *WCSS,* Within-cluster sum of squares.

To remove the subjectivity inherent in visual identification, the Kneedle algorithm was employed. This algorithm detects points of maximum curvature on a monotonically decreasing convex curve. By normalizing the WCSS sequence and computing deviations from a linear reference, Kneedle identifies the most significant change in slope. Using this automated method, the elbow was detected at $k = k^*$, yielding the same conclusion as the visual inspection.

Fig. 3.8 presents the final clustering output for the elbow-selected value $k = 6$. The resulting clusters capture the main operational regimes of the wind turbine, characterized by hourly wind-speed and power pairs.

When Mahalanobis distance is used instead of Euclidean distance, the WCSS becomes:

$$WCSS(k) = \sum_{i=1}^{k} \sum_{x \in C_i} (x - \mu_i)^T \Sigma^{-1} (x - \mu_i), \tag{3.22}$$

where Σ is the covariance matrix of the dataset, allowing the distance metric to account for correlated features.

3.5.2 *Silhouette analysis*

Silhouette analysis provides a geometric measure of cluster validity by evaluating how similar each point is to its assigned cluster relative to other clusters (Ilyas & Priscila, 2024). For each sample i, the silhouette value is defined as:

$$s(i) = \frac{b(i) - a(i)}{\max(a(i), b(i))}, \tag{3.23}$$

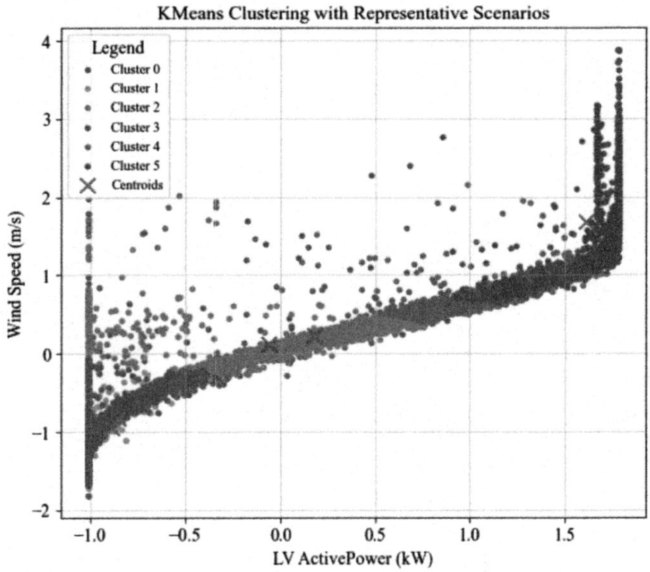

Figure 3.8 Clustering output for elbow method (Optimal = 6).

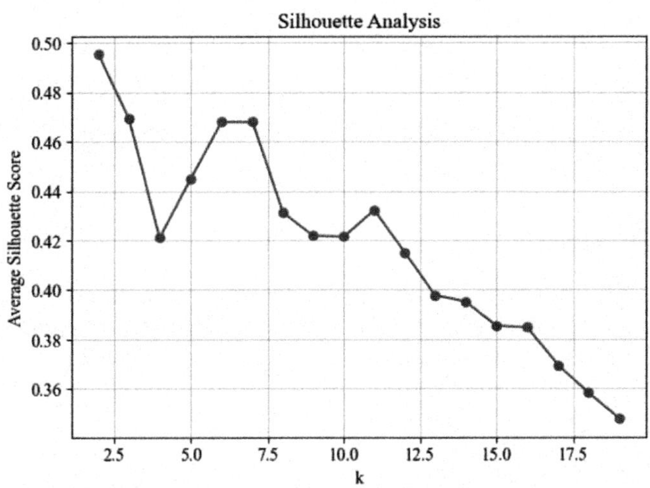

Figure 3.9 Silhouette scores for different *k*. The optimal *k* corresponds to the maximum average silhouette.

where $a(i)$ is the mean distance between i and other points in its own cluster and $b(i)$ is the minimum mean distance from i to any other cluster.

Silhouette values range from -1 (incorrect clustering) to 1 (high-quality clustering). The average silhouette score for different values of k is shown in Fig. 3.9. The

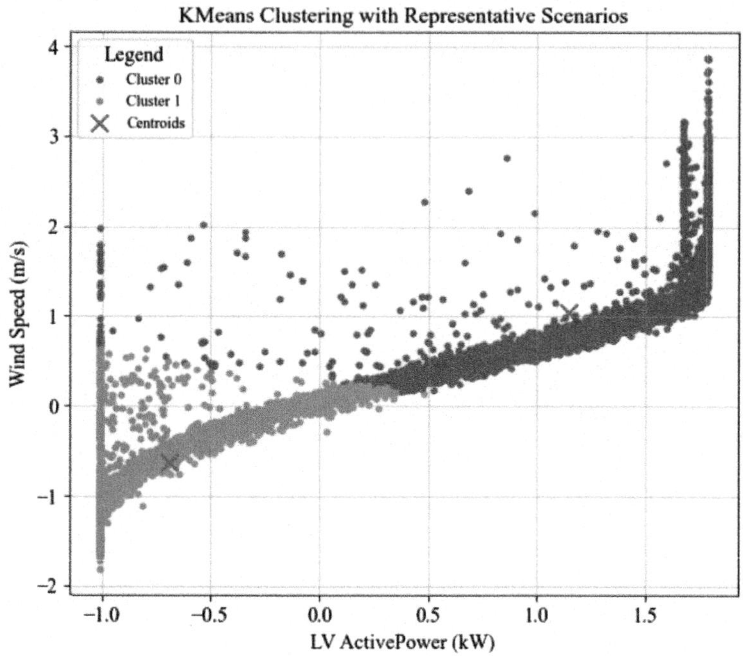

Figure 3.10 Clustering output for silhouette analysis (Optimal = 2).

maximum silhouette value occurs at $k = 2$, suggesting that two clusters achieve the best balance between cohesion and separation. The corresponding clustering result is displayed in Fig. 3.10.

3.5.3 Gap statistic

The gap statistic evaluates clustering quality by comparing the within-cluster dispersion of the empirical data with that of uniformly distributed reference datasets (El-Mandouh et al., 2019). For each value of k, the gap statistic is defined as:

$$\text{Gap}(k) = \frac{1}{B}\sum_{b=1}^{B}\log\left(W_k^{(b)}\right) - \log\left(W_k\right), \tag{3.24}$$

where W_k is the within-cluster dispersion of the observed data, $W_k^{(b)}$ corresponds to the bth reference dataset, and B is the number of such datasets.

The optimal number of clusters is the smallest k satisfying:

$$\text{Gap}(k) \geq \text{Gap}(k+1) - s_{k+1}, \tag{3.25}$$

where s_{k+1} is the standard deviation of the reference dispersions. Fig. 3.11 illustrates

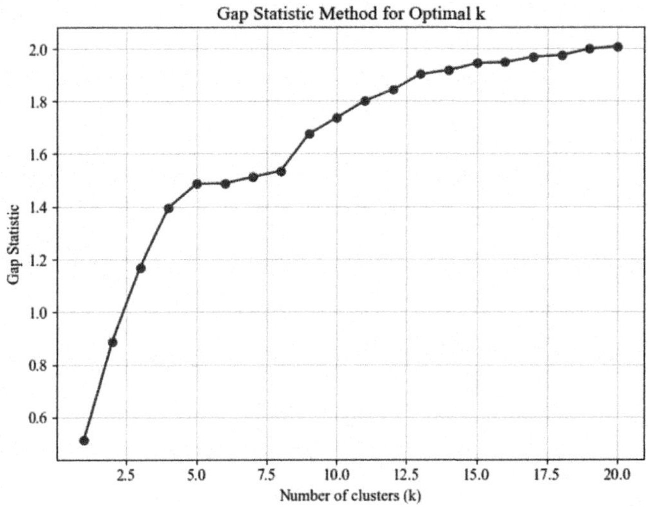

Figure 3.11 Gap statistic for different k. The optimal k is at the point with the maximum gap or the first k satisfying the inequality.

the gap statistic for $k = 1$ to 20. The criterion identifies $k = 5$ as optimal, and the corresponding cluster visualization is shown in Fig. 3.12.

3.5.4 Energy-based criterion

Energy-based criteria quantify the discrepancy between the empirical data distribution and the distribution represented by cluster centroids. A simple form of the energy measure is (Paltsev, 2017):

$$E(k) = \frac{1}{n}\sum_{i=1}^{n} \|x_i - \mu_{c(i)}\|^2, \tag{3.26}$$

where n is the number of samples, and $c(i)$ denotes the cluster assignment of the sample i.

A more rigorous formulation uses the energy distance between the original dataset and the clustered representation:

$$\text{Energy Score}(k) = 2\mathbb{E}\|X - Y\| - \mathbb{E}\|X - X'\| - \mathbb{E}\|Y - Y'\|, \tag{3.27}$$

where X, X' are independent samples from the data distribution and Y, Y' are independent samples from the cluster-centroid distribution.

Fig. 3.13 shows the energy distance for multiple values of k, where the minimum occurs at $k = 12$. Fig. 3.14 provides the corresponding clustering visualization.

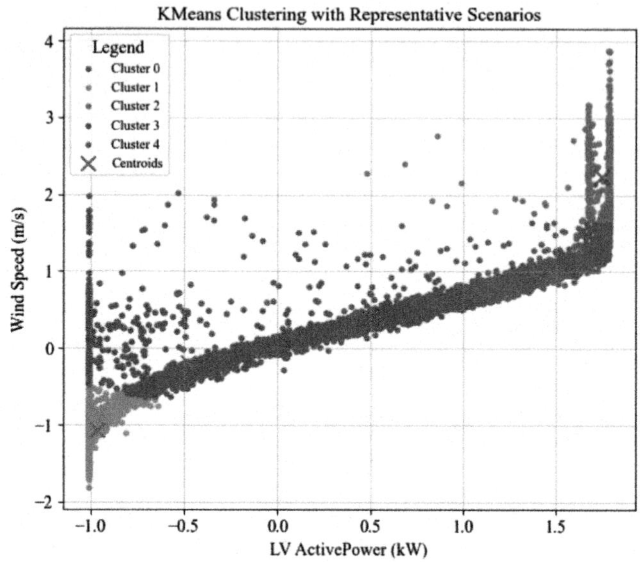

Figure 3.12 Clustering output for gap statistic. (Optimal $k = 5$).

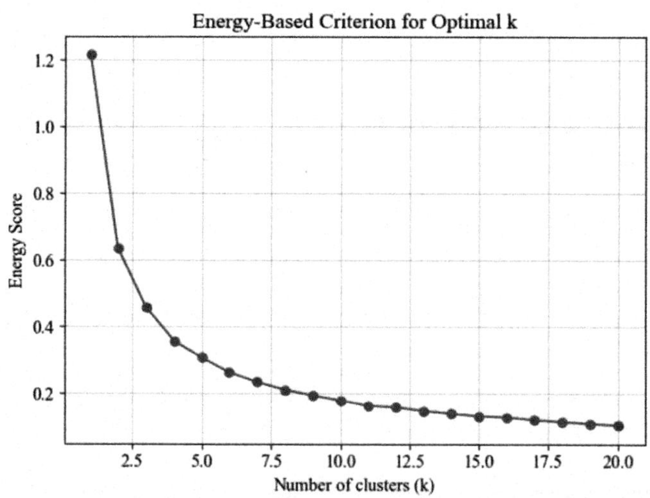

Figure 3.13 Energy distance for different k. The optimal k corresponds to the minimum energy distance.

3.5.5 Comparative evaluation

Table 3.10 summarizes key performance metrics for the reduced scenario sets obtained using the four methods. Among these, the energy score exhibits the lowest value for $k = 12$, indicating the closest multivariate match to the original dataset. Although CRPS

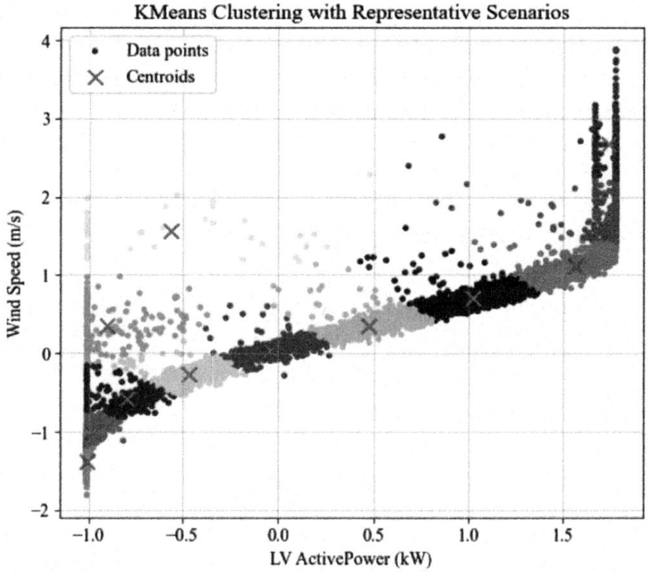

Figure 3.14 Clustering output for energy distance (Optimal = 12).

Table 3.10 Performance metrics for different optimal *k* methods.

Method	Optimal *k*	Wasserstein	Energy score	CRPS	Avg TVD
Elbow	6	858.547	2279.449	0.216	0.75
Silhouette	2	945.665	2083.739	0.23	0.75
Gap statistic	5	958.507	1900.824	0.26	0.5
Energy state	12	758.291	1359.865	0.21	0.458

CRPS, continuous ranked probability score; *TVD,* total variation distance.

values remain similar across methods, they are computed separately for each feature and therefore do not fully capture cross-variable dependencies.

Increasing the number of clusters improves representational accuracy but simultaneously raises computational complexity, particularly in stochastic programming models, where each scenario induces additional branching in the scenario tree. Solver time may grow superlinearly with the number of nodes, making it excessively large *k* values undesirable.

To balance accuracy and efficiency, practitioners may:

- select the smallest *k* for which relative improvements in the chosen metric fall below a threshold (e.g., 1%),
- use scalable clustering methods such as MiniBatch K-Means or BanditPAM for large datasets,
- apply two-stage reduction: Coarse clustering followed by refined representative selection.

Overall, the results show that while methods such as silhouette and gap statistic suggest relatively small values of k, the energy-based criterion identifies $k = 12$ as offering the best fidelity-complexity trade-off. Given the requirements of preserving distributional characteristics for scenario-based optimization, $k = 12$ is adopted as the final choice.

3.6 Advanced and hybrid techniques

As scenario reduction methodologies mature, advanced techniques have emerged to address the complex challenges inherent in energy system applications. These sophisticated approaches recognize that practical problems often demand more nuanced solutions than standard reduction algorithms can provide.

3.6.1 Simultaneous branching reduction

Multistage stochastic programming problems introduce additional complexity through their tree-like structure, where scenario reduction must preserve both statistical properties and the nonanticipativity constraints essential for temporal consistency. Simultaneous branching reduction approaches address this challenge by optimizing across the entire tree structure rather than treating individual scenarios in isolation. The mathematical formulation extends the classical reduction framework:

$$\min_{\tilde{\mathbb{P}}} W_p\left(\mathbb{P}, \tilde{\mathbb{P}}\right) + \lambda \cdot \text{Tree Complexity}\left(\tilde{\mathbb{P}}\right) \tag{3.28}$$

This optimization balances distributional fidelity against computational tractability, where the regularization term penalizes excessively complex tree structures that would undermine the computational advantages of scenario reduction. The approach proves particularly valuable for capacity expansion planning and multistage investment decisions where temporal consistency across decision stages is paramount.

3.6.2 Moment-matching heuristics

In many energy system applications, preserving specific statistical moments may be more critical than minimizing overall distributional distance. Moment-matching heuristics explicitly enforce consistency between original and reduced distributions for key statistical properties:

$$\mathbb{E}\left[\tilde{\xi}\right] = \mathbb{E}[\xi] \tag{3.29}$$

$$\mathrm{Var}\left[\tilde{\xi}\right] = \mathrm{Var}[\xi] \tag{3.30}$$

$$\mathbb{S}kew\left[\tilde{\xi}\right] = \mathbb{S}kew[\xi] \tag{3.31}$$

Table 3.11 Comprehensive comparison of scenario reduction methods.

Method	Speed	Quality	Ease of use	Recommended use
FFS	Fast	Very good	Easy	Large sets, quick prototyping
Backward	Slow	Excellent	Moderate	Small sets, final optimization
k-means	Fast	Good	Easy	Path-dependent problems
k-medoids	Moderate	Very good	Moderate	When actual scenarios are needed
Hybrid	Moderate	Excellent	Difficult	Complex, high-stakes applications

FFS, fast forward selection.

This approach ensures that the reduced scenario set maintains the mean, variability, and asymmetry characteristics of the original distribution, which often correspond directly to physically meaningful quantities in energy systems. For renewable integration studies, preserving variance captures fluctuation magnitudes, while skewness preservation maintains the balance between over-generation and under-generation events.

3.6.3 Hybrid methods

Recognizing that no single reduction method dominates across all performance dimensions, hybrid approaches strategically combine elements from multiple algorithms. A representative hybrid methodology proceeds through three stages:

1. Initial clustering: Partition the original scenario set into moderate-sized groups using k-means or similar clustering techniques, exploiting their computational efficiency for initial coarse reduction.
2. Intra-cluster selection: Apply FFS within each cluster to identify representative scenarios that capture local distributional characteristics.
3. Global integration: Combine selected scenarios from all clusters, with potential final refinement using backward reduction to ensure global optimality.

This layered approach leverages the complementary strengths of different reduction paradigms, typically achieving superior performance compared to any individual method applied in isolation.

3.7 A practical guide for energy system modelers

3.7.1 Comparative analysis

The selection of an appropriate scenario reduction method represents a critical decision point in stochastic optimization workflows. Table 3.11 provides a structured comparison to guide this selection process, evaluating methods across multiple practical dimensions relevant to energy system applications.

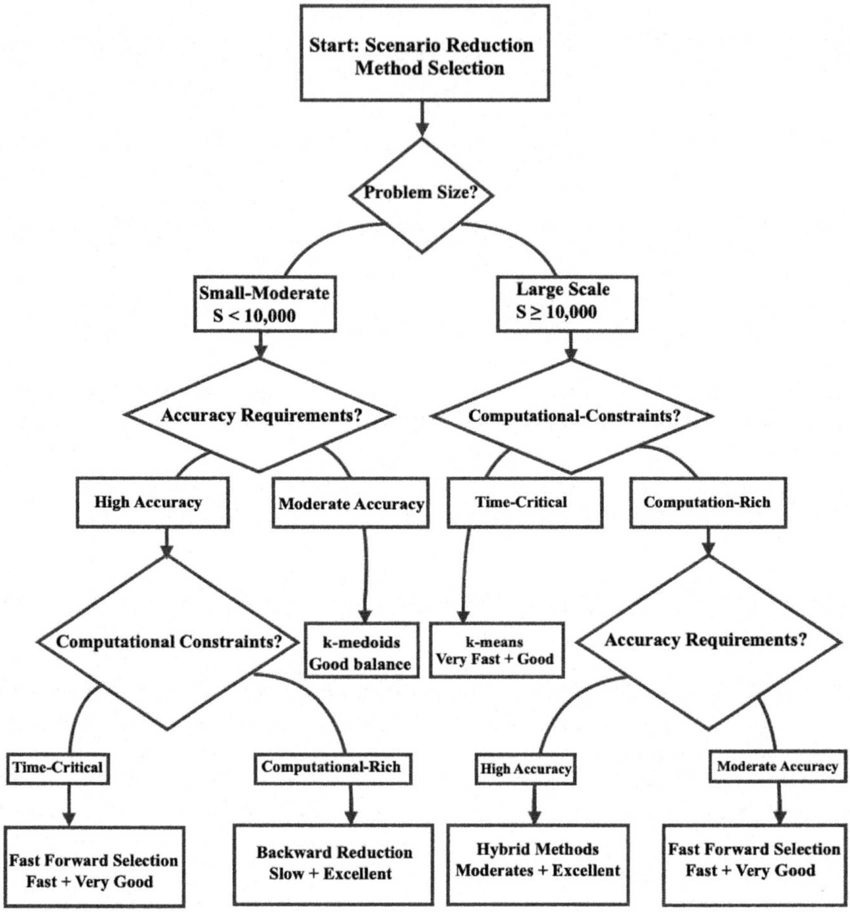

Figure 3.15 Decision tree for selecting the appropriate scenario reduction method based on problem characteristics.

3.7.2 Decision framework

Navigating the methodological landscape requires consideration of problem-specific characteristics, including computational constraints, accuracy requirements, and implementation complexity. Fig. 3.15 provides a structured decision framework that maps problem attributes to appropriate reduction methodologies, serving as a practical guide for method selection.

3.7.3 Quality assessment

The ultimate validation of any reduction methodology lies in its impact on decision quality. Beyond distributional metrics, practical assessment should evaluate how the reduction affects key performance indicators:

$$\text{Relative Error} = \frac{|f(\mathbb{P}) - f(\tilde{\mathbb{P}})|}{|f(\mathbb{P})|} \tag{3.32}$$

Here, f represents domain-specific objective functions such as expected system cost, risk measures like CVaR, or operational reliability metrics. This application-focused evaluation ensures that reduction quality aligns with ultimate optimization objectives rather than abstract statistical measures.

3.8 Conclusion

This chapter has established scenario reduction as an indispensable component of the stochastic optimization workflow, transforming computationally intractable problems into manageable formulations while preserving essential uncertainty characteristics. The methodological progression from basic selection and clustering approaches to advanced hybrid techniques reflects the evolving sophistication of energy system modeling requirements.

The comparative analysis reveals that selection-based methods, particularly FFS and backward reduction, provide theoretically grounded approaches with compelling performance guarantees. Their mathematical foundations in optimal transport theory offer rigorous quality bounds that support their application in high-stakes decision contexts. Concurrently, clustering methodologies deliver computationally efficient alternatives well-suited to problems where path-dependent characteristics dominate system behavior.

A recurring theme throughout this exploration has been the critical importance of distance metric selection, which fundamentally shapes reduction outcomes. The choice between Euclidean, Mahalanobis, Wasserstein, and other metrics should align with both statistical considerations and the physical realities of energy system operations. Furthermore, the emergence of specialized metrics like the energy score and CRPS enables more nuanced evaluation of reduction quality specifically tailored to energy forecasting applications.

Looking forward, the integration of scenario reduction within broader stochastic optimization frameworks suggests several promising directions. Adaptive reduction methods that respond to problem structure and solution progress offer potential efficiency improvements. Tighter coupling between reduction and optimization processes may yield specialized techniques that preserve precisely those distributional aspects most relevant to decision quality. As energy systems incorporate greater uncertainty from renewable integration, electrification, and climate impacts, the role of scenario reduction will only grow in importance.

The complete workflow encompassing scenario generation, reduction, and optimization now provides a mature foundation for addressing complex stochastic problems in energy systems. This methodological triad enables decision-makers to navigate uncertainty with both computational practicality and mathematical rigor, supporting the robust planning and operation of increasingly complex and renewable-dependent energy infrastructures.

References

Chiu, W.-Y., Yen, G. G., & Juan, T.-K. (2016). Minimum Manhattan distance approach to multiple criteria decision making in multiobjective optimization problems. *IEEE Transactions on Evolutionary Computation, 20*(6), 972–985.

Cui, M. (2020). Introduction to the k-means clustering algorithm based on the elbow method. *Accounting, Auditing and Finance, 1*(1), 5–8.

Dong, X., Sun, Y., Malik, S. M., Pu, T., Li, Y., & Wang, X. (2023). Scenario reduction network based on Wasserstein distance with regularization. *IEEE Transactions on Power Systems, 39*(1), 4–13.

El-Mandouh, A. M., Mahmoud, H. A., Abd-Elmegid, L. A., & Haggag, M. H. (2019). Optimized k-means clustering model based on gap statistic. *International Journal of Advanced Computer Science and Applications, 10*(1), 183–188.

Erisen B., Wind turbine SCADA dataset, (Accessed 16 September 2025) (2020). https://www.kaggle.com/datasets/berkerisen/wind-turbine-scada-dataset

Gangil, G., Saraswat, A., & Goyal, S. K. (2025). A new hybrid distance and similarity based scenario reduction approach for stochastic economic operation of microgrid, e-Prime-advances in electrical engineering. *Electronics and Energy, 11*, 100905.

Giordano G. (2023). On the use of the Kantorovich-Rubinstein distance for dimensionality reduction. arXiv preprint arXiv:2309.09442 .

Heitsch, H., & Römisch, W. (2003). Scenario reduction algorithms in stochastic programming. *Computational optimization and applications, 24*(2), 187–206.

Hu, J., & Li, H. (2019). A new clustering approach for scenario reduction in multi-stochastic variable programming. *IEEE Transactions on Power Systems, 34*(5), 3813–3825.

Ilyas, F. M., & Priscila, S. S. (2024). An optimized clustering quality analysis in k-means cluster using silhouette scores. *Explainable AI applications for human behavior analysis* (pp. 49–63). IGI Global.

Kumari, P., & Gupta, S. (2024). Comparative analysis between Euclidean distance metric and Mahalanobis distance metric. *International Journal of Innovative Research in Technology and Science, 12*(2), 271–277.

Li, Z., & Li, Z. (2016). Linear programming-based scenario reduction using transportation distance. *Computers & Chemical Engineering, 88*, 50–58.

Liu, Y., Luo, L., Li, X., Liu, H., Yu, Z., & Wang, Y. (2025). A Wasserstein distance-based distributionally robust optimization strategy for a renewable energy power grid considering meteorological uncertainty. *Symmetry, 17*(10), 1602.

Mollayousefi Zadeh, M., Afshar, Z., Farahani, G., & Bathaee, S. M. T. (2023). An adaptive robust optimization model for microgrids operation using convexified AC power flow equations. *International Transactions on Electrical Energy Systems, 2023*(1), 6483030.

Montero, L., Bello, A., & Reneses, J. (2022). A review on the unit commitment problem: Approaches, techniques, and resolution methods. *Energies, 15*(4), 1296.

Paltsev, S. (2017). Energy scenarios: The value and limits of scenario analysis. *Wiley Interdisciplinary Reviews: Energy and Environment, 6*(4), e242.

Rahimian, H., Bayraksan, G., & Homem-de Mello, T. (2019). Identifying effective scenarios in distributionally robust stochastic programs with total variation distance. *Mathematical Programming, 173*(1), 393–430.

Rojinamdii, Source code for scenario reduction. (2025). Available from: https://github.com/rojinamdii/ScenarioReduction

Rujeerapaiboon, N., Schindler, K., Kuhn, D., & Wiesemann, W. (2022). Scenario reduction revisited: Fundamental limits and guarantees. *Mathematical Programming, 191*(1), 207–242.

Srinivasaraghavan, J., & Allada, V. (2006). Application of mahalanobis distance as a lean assessment metric. *The International Journal of Advanced Manufacturing Technology, 29*(11), 1159–1168.

Zamo, M., & Naveau, P. (2018). Estimation of the continuous ranked probability score with limited information and applications to ensemble weather forecasts. *Mathematical Geosciences, 50*(2), 209–234.

Ziel, F. (2021). The energy distance for ensemble and scenario reduction. *Philosophical Transactions of the Royal Society A, 379*(2202), 20190431.

A synergistic framework for efficient and uncertainty-calibrated solar irradiance forecasting using data compression and optimized neural networks

4

Manvi Mishra, Priya Mahajan and Rachana Garg
Electrical Engineering, Delhi Technological University, New Delhi, India

4.1 Introduction

In the era of the depletion of fossil fuel-based energy resources, policy makers, electrical utilities, and researchers have been working strategically to meet electricity demand along with the target of minimal CO_2 emissions. While promoting the utilization of renewable resources for power generation, India has become the world's third largest producer of renewable energy, with 40% of the installed capacity from nonfossil fuel sources, out of which solar photovoltaics have the highest installed capacity of 48.55 GW (PIB, 2023). Solar irradiance is an important weather parameter that affects solar power generation. However, the solar irradiance pattern shows a transitional behavior with respect to time and geographical location. This inconsistent nature of the parameters may lead to an imbalance between power generation and the demand chain. Accurate forecasting of weather parameters can mitigate high fluctuations between the demand and supply systems.

Solar irradiance forecasting has traditionally been performed using conventional statistical techniques, such as auto regressive moving average (ARMA) (Singh & Pozo, 2019), auto regressive integrated moving average (ARIMA), seasonal auto regressive integrated moving average (SARIMA) (Alsharif et al., 2019), etc. However, these techniques might lack: (1) Seasonality in the time series dataset, for which the differencing of the dataset has to be applied, and (2) a loss of information in the form of residuals. These drawbacks of stationary forecasting techniques can be addressed by machine-learning-based forecasting techniques, such as artificial neural network (ANN), recurrent neural network (RNN), and long short-term memory (LSTM). The authors (Essam et al., 2022) analyzed different machine learning-based forecasting models to predict the solar power output. The study includes several forecasting models, ANN as the ANN16 model, LSTM, Random Forest (RF), Decision Tree (DT), and Extreme Gradient Tree Boosting (XGBoost). They concluded that among all the considered models, the ANN model outperformed the others in predicting the solar

Stochastic Planning and Modeling for Energy Systems: Methods, Applications, and Developments.
DOI: https://doi.org/10.1016/B978-0-443-45298-7.00021-5

power output. The authors (Pereira et al., 2024) suggested the development of ANN models based on prediction data from operational numerical weather prediction (NWP) and aerosol data. The operational solar direct normal irradiance (DNI) prediction datasets received from the European Center for Medium-range Weather Forecasts (ECMWF) and the Copernicus Atmospheric Monitoring Service (CAMS) were used as input to the first ANN model. Another ANN model was developed using the period of the improved DNI forecast. The thorough analysis in this study suggests that the combination of ANN models can improve the quality of the forecasting results. Although ANN models are easy to implement, other versions of neural networks, such as RNN (Pang et al., 2020), and LSTM (Yildirim et al., 2023) are gaining attention owing to their ability to attain higher accuracy.

One of the issues with solar irradiance forecasting is that it requires a large volume of datasets of other meteorological factors, such as temperature, pressure, wind speed, etc. These datasets as inputs to the forecasting algorithms cause a computational burden; hence, the algorithms' processing time may increase. Therefore, preprocessing the raw dataset has become an important step in forecasting the weather parameters. The authors (Ayesha et al., 2020) reviewed various dimensionality reduction techniques, such as Principal Component Analysis (PCA) and Singular Value Decomposition (SVD), which ensure the data compression to reduce the burden on computation and storage memory. Another study (Nespoli et al., 2022) suggested a novel normalization approach to reduce the computational burden of any forecasting technique. They introduced an ensemble forecasting technique for predicting solar PV power along with a new normalization approach. A novel approach (Lyu et al., 2023) has been introduced that consists of data compression and feature selection to obtain the optimal dataset as an input to the forecasting technique, that is, XGBoost. Kernel-based Principal Component Analysis (kPCA) was used for data compression, and a novel Deep Reinforcement Learning (DRL) was used for optimal feature selection. Although various preprocessing techniques are available in the literature, the key to achieving accurate forecasting results with minimum computational burden lies in adopting and integrating appropriate pre-processing techniques with the selected forecasting model.

Moreover, with the exponential advancement in the optimization algorithms domain, especially in the field of metaheuristic optimization algorithms, the integration of these optimization algorithms with forecasting models has become a viable option for enhancing result accuracy. The fundamentally known Particle Swarm Optimization (PSO) (Shafqat et al., 2021) has been widely integrated with forecasting models to obtain accurate results. The authors integrated the ANN model with the PSO algorithm to forecast daily global solar radiation, and performed an analysis against a traditional backpropagation ANN model (BP-ANN) (Mohandes, 2012). An in-depth analysis suggests that the ANN-PSO model performs better than the BP-ANN model. Similarly, another extensive study (Mughal et al., 2025) was carried out in which different ANN models trained with various algorithms were compared with the ANN-PSO model to highlight the effectiveness of the integration of PSO with the ANN model. The evolution of mature metaheuristic optimization algorithms with neural networks can further enhance the quality of the results. In a recent study, the authors (Xu et al., 2024) suggested the optimization of a feed-forward ANN model using an Equilibrium

Optimization (EO) algorithm. This developed EO-FFANN model yields the RMSE error of 142.38 W/m^2, which is less than the other ANN models optimized with Wind-Driven Optimization (WDO), Optics-Inspired Optimization (OIO), and Social Spider Algorithm (SOSA). Similarly, another study (Bansal et al., 2023) introduced biogeography-based weights optimization of an ANN model for accurate forecasting of solar irradiance. The superiority of the proposed model was proven by comparing it with the BP-ANN, PSO-based ANN, and GA-based ANN models. Although significant studies have been conducted to optimize the two well-known topologies of neural networks, FF-ANN and BP-ANN, the internal parameter optimization of advanced neural networks, such as RNN and LSTM, requires further in-depth investigation.

Studies have also been conducted to analyze the limitations of optimization algorithms. The authors (Xin et al., 2024) suggested that the biogeography-based optimization algorithm might struggle to deal with areas where the function values shift dramatically. Similarly, the swarm-based optimization algorithm suffers from the drawback of convergence at the local optimal solution. Therefore, there is a need for a robust and efficient optimization algorithm has been observed that can perform effectively under complex search spaces and has a tendency to converge to the best global solution without adhering to the local best solution. In this direction, the authors (MiarNaeimi et al., 2021) proposed a novel Horse Herd Optimization (HHO) algorithm that is capable of attaining the global best solution for higher-dimensional complex problems owing to a better trade-off between the exploration and exploitation phases. In evaluating the performance of the HHO algorithm, it has been observed that it outperforms other optimization algorithms, such as Grey Wolf Optimization (GWO), Dragonfly Algorithm (DA), etc., based on the tendency to find the optimal solutions of different types of objective functions. To the best of our knowledge, the application of the HHO algorithm for optimizing the weights of neural networks has not been thoroughly explored in the field of solar irradiance forecasting. The authors of the present study investigated the probability of performance enhancement of the neural network, particularly the RNN model, by optimizing its parameters using the HHO algorithm.

The present study suggests a novel methodology that proposes a notion to develop a reliable solar irradiance forecasting model, which can potentially deal with any size of meteorological dataset (smaller dataset, i.e., 4 months, to large dataset, i.e., 3 years). The proposed novel model mitigates computational complexities through the application of the kPCA algorithm and enhances the performance of neural networks via parametric optimization using the HHO algorithm. The main contributions of this study are as follows.

1. The novel framework has the underlying capability of compressing all the input data information using a dimensionality reduction technique (kPCA) and integrating the metaheuristic algorithm (HHO) for internal parametric optimization of neural network-based models (ANN and RNN).
2. Performance evaluation of the different test models was performed on different ranges of datasets, including four months, one year, and three years, along with the seasonal variation dataset. The error matrices of the proposed model were compared with those of PSO and GWO-tuned neural network test models.

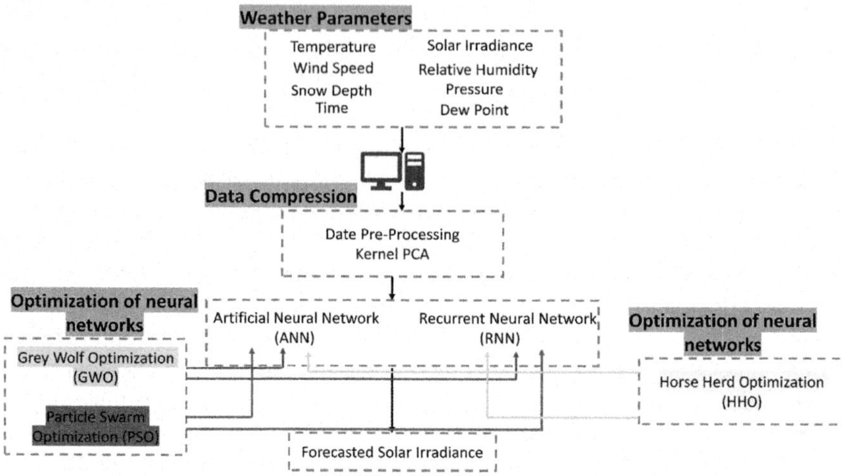

Figure 4.1 Block diagram of the methodology used in the present study.

3. The performance of the novel kPCA-HHO-RNN model outperformed that of the conventional RNN model with significantly less computational time. Additionally, the proposed model also withstands the uncertainty analysis. Moreover, the model has also been compared with other State-of-the-Art models present in the literature.

4.2 Methodology

The present work proposes a novel solar irradiance forecasting model that involves the compression of a large volume of datasets without the loss of any information. Subsequently, the neural networks, ANN and RNN, were tuned using the HHO algorithm to obtain the precise forecasting outcomes. HHO-tuned neural networks were compared with GWO and PSO-tuned neural networks. Fig. 4.1 shows the strategic flow of the methodology applied in this study.

Data pre-processing of the raw dataset was performed using Google Colab, which supports a Python coding environment, and for developing novel neural networks, that is, HHO-based ANN and HHO-based RNN, a MATLAB coding environment was used. The proposed method is explained in detail in the next section.

4.3 Data dimensionality reduction technique—kernel principal component analysis

The dataset for this study was obtained from the open-source website of the National Renewable Energy Laboratory (NREL) (NSRDB, 2024). This website provides a gateway for acquiring the meteorological dataset for any geographical location. The

4 months, 1 year, and 3 years datasets with hourly variations in meteorological parameters, such as dew point, temperature, pressure, relative humidity, wind speed, and solar irradiance, were extracted from this website. The snow depth meteorological parameter has been omitted as the considered location does not receive any snow. The 4-month dataset has 2880 × 9 data samples, 1 year dataset has 8760 × 9 data samples, and the 3-year dataset has 26,280 × 9 data samples.

To develop a robust solar irradiance forecasting model, it is necessary to consider the relevant weather parameters, owing to their temporal and geographical variations. However, consideration of critical information for different ranges and dimensions may cause a computational burden. Therefore, a broad range of datasets was normalized to fit within a specific range. The mathematical expression is as follows:

$$D_{norm} = \frac{D - D_{min}}{D_{max} - D_{min}} \tag{4.1}$$

Although the normalization of the dataset helps in setting the dataset to a particular range of magnitude, it does not ensure a reduction in the computational time of the forecasting technique. Therefore, data dimensionality reduction can be helpful for compressing the volume of data and ensuring less memory allocation for processing the forecasting algorithm. The authors (Davò et al., 2016), suggested the incorporation of PCA with an ANN model for forecasting long term solar irradiance. This dimensionality reduction technique suffers from the setback of losing a certain amount of dataset information during the compression of data. To overcome this issue, kPCA was used to ensure the extraction of 100% of the information from the dataset.

Using a specific kernel function, KPCA ensures the mapping of data to a dot product feature space Ɛ, where the new dataset ø becomes linearly dependent. Thus, the evaluation of eigenvalues and eigenvectors was performed on this linearly dependent dataset ø. For the proposed study, the Radial Basis Function (RBF) was selected as given in Equation (Singh & Pozo, 2019). Compared to other kernel functions, RBF can effectively linearize the weather data used for solar forecasting (Jaffel et al., 2017). The RBF kernel function is given as:

$$K(x_i, x_j) = \exp\left(-\frac{x_i - x_j^{\,2}}{2\sigma^2}\right) \tag{4.2}$$

The original weather dataset $\boldsymbol{X} = \{x_1, x_2, x_3, \ldots, x_N\}, \boldsymbol{X} \in R^{N\times m}$, where N is the number of data in $\boldsymbol{X}$ and m is the dimension of $\boldsymbol{X}$, which is mapped to a higher-dimensional space as:

$$K(x_i, x_j) = \emptyset(x_i)\emptyset(x_j)^T \tag{4.3}$$

where $i, j = 1, 2 \ldots N$. The $M \times M$ kernel matrix has been constructed:

$$K(x_i, x_j) = \emptyset(X)\emptyset(X)^T \tag{4.4}$$

The given equation has $K_{ij} = \kappa(x_i,x_j)$ whereas $\varnothing(X) = \{\varnothing(x_i),\ldots, \varnothing(x_N)\}$ and this is the higher-dimensional dataset, which is defined in the dot product feature space $\mathcal{E}$. The $\varnothing(X)$ should be centered, i.e., $\sum_{i=1}^{t} \not\equiv(x_i) = 0$, to compute the covariance matrix.

$$\emptyset(X) = \emptyset(X) - \frac{1}{N}\sum_{i=1}^{t}\emptyset(x_i) \tag{4.5}$$

The covariance matrix helps to analyze the correlation between variables. The covariance matrix can be calculated as:

$$\sum\overline{\emptyset(X)} = \frac{1}{N}\sum_{i=1}^{N}\bar{\emptyset}(x_i)\bar{\emptyset}(x_j)^T \tag{4.6}$$

The eigen vector $\boldsymbol{A}$ of the given covariance matrix is described in Eq. 4.7, which provides the direction of data variance.

$$\lambda A = \sum\overline{\emptyset(X)}A \tag{4.7}$$

$\boldsymbol{A}$ ϵspan $\{\bar{\emptyset}(x_1), \bar{\emptyset}(x_2), \ldots, \bar{\emptyset}(x_N)\}$. Hence, $\boldsymbol{A}$ can be represented as a linear combination of $\bar{\emptyset}(x_i)$'s with coefficient δ_i's:

$$A = \sum_i^N \delta_i\bar{\emptyset}(x_i) \tag{4.8}$$

Substituting Eqs. (4.6) and (4.7) in Eq. (4.8) yields:

$$\lambda\sum_i^N \delta_i\bar{\emptyset}(X_i) = \frac{1}{N}\sum_i^N \bar{\emptyset}(x_i)\bar{\emptyset}(x_i)^T\sum_i^N \delta_i\bar{\emptyset}(X_i) \tag{4.9}$$

$$\lambda\bar{\emptyset}(X)\delta = \frac{1}{N}\bar{\emptyset}(X)\bar{\emptyset}(X)^T\bar{\emptyset}(X)\delta \tag{4.10}$$

$$N\lambda\bar{\emptyset}(X)\bar{\emptyset}(X)^T\delta = \bar{\emptyset}(X)\bar{\emptyset}(X)^T\bar{\emptyset}(X)\bar{\emptyset}(X)^T\delta \tag{4.11}$$

Applying Eq. (4.4) into Eq. (4.11) yields Eq. (4.12):

$$N\lambda K\delta = K^2\delta \tag{4.12}$$

The finalized form of Eq. (4.12) is given by Eq. (4.13):

$$N\lambda\delta = K\delta \tag{4.13}$$

This is a higher-dimensional eigenvalue problem. Finally, the transformed data were calculated using $X_{trans} = K\delta$. A transformed dataset of three dimensions can be obtained by selecting the first three columns of δ for further study.

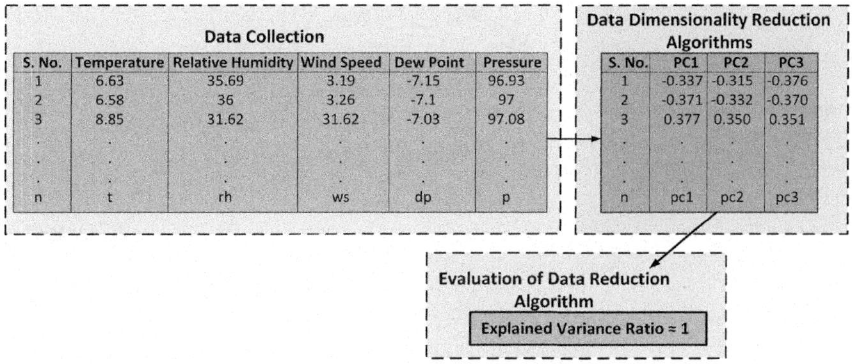

Figure 4.2 Presentation of dimensionality reduction algorithm.

Fig. 4.2 shows that the original dataset was compressed into three columns denoted as components 1, 2, and 3. The accountability of the compression of the original dataset, without any information loss, was checked using a term known as the Explained Variance Ratio. This ratio shows the percentage of the variance attributed to each kPCA component. The explained variance ratio of each of the three kPCA components was 0.414, 0.296, and 0.288, respectively. The summation of these three kPCA component ratios accounts for almost one, which suggests that the given dimensionality reduction technique holds 100% of the useful information from the raw input dataset.

4.4 Neural network-based forecasting models

4.4.1 Artificial neural network

An ANN is a computational strategy that falls under the field of artificial intelligence, which is inspired by the human brain's neural network. The internal configuration of the ANN involves an input layer, a hidden layer, an output layer, weights, and bias (El-Amarty et al., 2023), as shown in Fig. 4.3.

With the advantage of acquiring less memory space, feedforward topology is utilized for many computational applications. It is also a reliable tool for yielding accurate forecasting results by training nonlinear datasets.

The weights and biases are used to make the interconnection of neurons of different hidden layers. Eq. (4.14) defines the output expression of the ANN model by using weights and bias.

$$y = \sum_{i=1}^{n} w_{ij} x_j + b_j \tag{4.14}$$

where w_{ij} are the connected weights, b_j is the bias, the total number of inputs is denoted by n, and x_j are the inputs at the initial stage. The computation of the feedforward

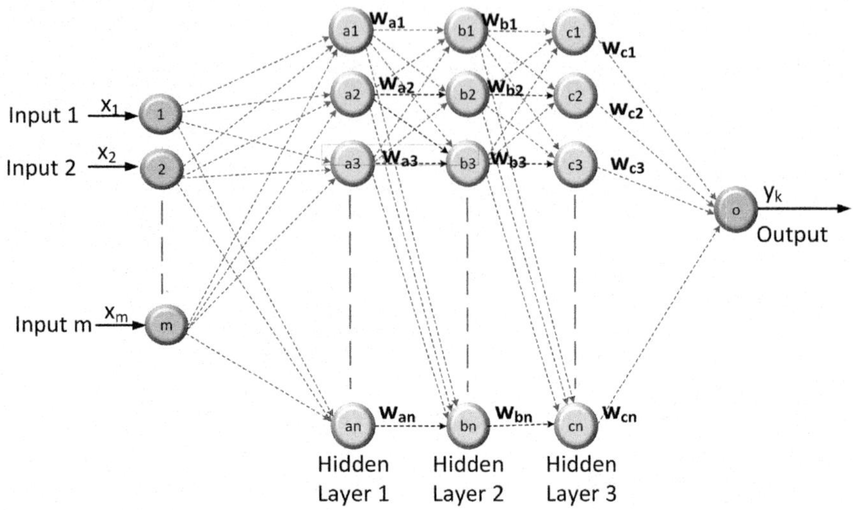

Figure 4.3 Internal structure of ANN feedforward model. *ANN,* Artificial neural network.

ANN model requires a complex gradient algorithm that iterates the associated weights and is biased to achieve the desired results, which are evaluated by calculating the error between the original and forecasted values. Therefore, the objective function for computing the ANN model is the Mean Square Error (MSE), which is defined as:

$$MSE = \frac{1}{n}\sum_{i=1}^{n}\sum_{j=1}^{o}\left[Y_j(i) - T_j(i)\right]^2 \tag{4.15}$$

where n and o denote the overall input and output data, respectively, $Y_j(i)$ is the actual data value, $T_j(i)$ is the output value.

4.4.2 Recurrent neural network

The RNN is an extended network of ANN that constitutes an input layer, a hidden layer, and an output layer with some additional weights are shown in Fig. 4.4. Suppose the input time series is $x = [x_1, x_2,\ldots, x_t]$ at time t, the output h_t of the hidden layer and y_t of the RNN output layer are calculated as follows:

$$h_t = A(W_{hh}h_{t-1} + W_{xh}x_t + b_h) \tag{4.16}$$

$$y_t = \sigma\left(W_{hy}h_t + b_y\right) \tag{4.17}$$

where the output layer is shown by y_t at time t, $h_t$is hidden layer, A () and σ are the activation function of the hidden layer and output layer respectively, input is x_t at time t, h_{t-1} is the output of the previous hidden layer, and $W_{hh}$and W_{xh} are the weight matrix corresponding to the output of the previous hidden layer and corresponding to the input

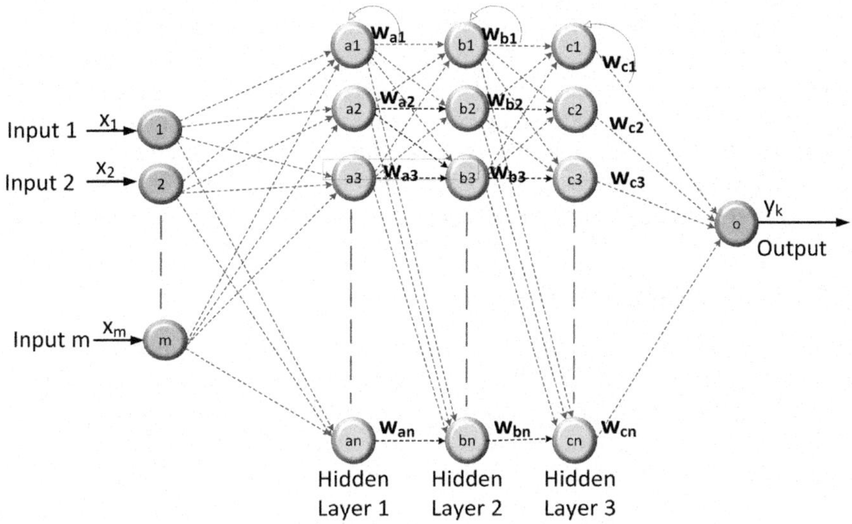

Figure 4.4 Diagrammatic representation of RNN Model. *RNN,* Recurrent neural network.

at time t, respectively. W_{hy} is the weight matrix corresponding to the output layer, and b_h and b_y are the bias corresponding to the hidden layer and output layer, respectively (Liu et al., 2021).

The main difference between conventional ANN and RNN models is that the RNN has a feedback signal. Because of this feedback signal, information from earlier states is utilized in later states. RNNs also operate on the premise of adding a time-delay unit and feedback connection (Granata & Di Nunno, 2021).

4.5 Optimization algorithms used for tuning parameters of neural networks

4.5.1 Particle swarm optimization

The moving tendency of birds inspired the foundation of the PSO algorithm. The entire PSO algorithm process was divided into four sets (Shafqat et al., 2021). First, the PSO optimizer starts with a random particle value. This value is based on the degree of possibility of the optimal solution space. In the trailing step, the fitness values of the previous (P_{bi}) and next (P_{li}) steps were compared to achieve the optimal solution within the current space. Subsequently, the best and global best position (G_{bi}) are compared to obtain the global fitness value. The velocity expression of the algorithm is:

$$V_i^{(t+1)} = wv_i^{(t)} + c_1r_1\left(P_{bi} - x_i^{(t)}\right) + c_2r_2\left(G_{bi} - x_i^{(t)}\right) \tag{4.18}$$

The particle position is iterated as:

$$x_i^{(t+1)} = x_i^{(t)} + V_i^{(t+1)} \tag{4.19}$$

where, $x_t^{(t)}$ is the particle's current position, $v_t^{(t)}$ is the speed, i is the optimization vector, and t is the number of iterations. w represents the inertia weight factor of the speed, c_1 and c_1 are the cognitive and social coefficients, r_1 and r_2 are the random velocity values of the search space in the range of 0 to the maximum velocity. In the final step, the best particle has been suggested by evaluating the fitness function, and based on this function value, the positions of other particles are updated. In the present study, the optimally best particle position helped in updating the weights of the neural network.

4.5.2 Grey wolf optimization

The GWO Algorithm was inspired by the prey hunting and encircling characteristics of grey wolves (Faris et al., 2018). The pack of wolves has been divided according to the hierarchy, that is, the leader is the Alpha wolf, and fellow wolves have been divided into Beta, Delta, and Gamma wolves. The mathematical interpretation of the wolf's characteristic of encircling the prey is represented by Eqs. (4.20) and (4.21):

$$\vec{D} = \left|\overrightarrow{C.X_p}(t) - \vec{X}(t)\right| \tag{4.20}$$

$$\vec{X}(t+1) = \overrightarrow{X_p}(t) - \vec{A}.\vec{D} \tag{4.21}$$

where the current iteration is shown by t, vector coefficients are denoted by A, D, and C, position of the prey shown by $X_{p,}$ and position of the grey wolves shown by X. The calculations for the A and C coefficients were evaluated using:

$$\vec{A} = 2\vec{a}.\overrightarrow{r_1} - \vec{a} \tag{4.22}$$

$$C = 2.\overrightarrow{r_2} \tag{4.23}$$

where a decreases linearly from 2 to 0 during the process, and r_1, r_2 are random vectors in [0,1]. The positions of the wolves were updated in each iteration using Eq. (4.21). The best position was extracted by evaluating the average positions of all age groups of the wolves present in the wolf pack. The optimization of the weights of the neural network was evaluated based on the best position attained.

4.5.3 Horse herd optimization

The HHO algorithm is a metaheuristic optimization algorithm that attempts to follow the behavioral patterns of horses. Horse herd behavior has been divided into six characteristics: Grazing (G), hierarchy (H), sociability (S), imitation (I), defence mechanism (D), and roam (R) (MiarNaeimi et al., 2021). The mathematical equation shown in Eq. (4.24) indicates the movement of horses of different ages:

$$M_n^{iter,age} = V_n^{iter,age} + M_n^{(iter-1),age}, age = \alpha, \beta, \gamma, \delta \tag{4.24}$$

where $M_n^{iter,age}$ is the position of the nth horse, *iter* is the current iteration, *age* is the range of horses' ages, $V_n^{iter,age}$ is the velocity vector of the horse. The total lifespan of horses is almost 25–30 years, which is divided as follows: Ages of 0–5 years are represented as δ, ages 5–10 years are represented as γ, 10–15 years are represented by β, and ages of more than 15 years are denoted by α. This whole herd is divided as 10% of the total horses are α horses, next 20% of the total horses are from the β age group, next 30% horses are from the γ age group, and 40% horses are from the age group of δ. The velocity expressions, including the behavioral pattern according to age group, are represented in Eqs. (4.25)–(4.28):

$$V_n^{iter,\alpha} = G_n^{iter,\alpha} + D_n^{iter,\alpha} \tag{4.25}$$

$$V_n^{iter,\beta} = G_n^{iter,\beta} + H_n^{iter,\beta} + S_n^{iter,\beta} + D_n^{iter,\beta} \tag{4.26}$$

$$V_n^{iter,\gamma} = G_n^{iter,\gamma} + H_n^{iter,\gamma} + S_n^{iter,\gamma} + I_n^{iter,\gamma} + D_n^{iter,\gamma} + R_n^{iter,\gamma} \tag{4.27}$$

$$V_n^{iter,\delta} = G_n^{iter,\delta} + I_n^{iter,\delta} + R_n^{iter,\delta} \tag{4.28}$$

where *G, D, H, S, I,* and *R* are the different characteristics of horse behavior. The comprehensive discussion of behavioral traits and their mathematical expressions is discussed below:

1. Grazing

   With short resting periods, horses have the ability to graze for 16–20 hours a day. The HHO algorithm mathematically models the grazing behavior by implementing the following expression:

$$\vec{G}_n^{iter,age} = g_{iter}(u + r.l)\left[M_n^{(iter-1)}\right], age = \alpha, \beta, \gamma, \delta \tag{4.29}$$

$$g_n^{iter,age} = g_n^{(iter-1),age} \times \omega_g \tag{4.30}$$

where $\vec{G}_n^{iter,age}$ is the motion parameter of the *ith* horse that presents the moving and grazing tendency, which decreases linearly with ω_g. l and u are the lower and upper bounds of the grazing space, respectively, and r is a random number between 0 and 1. The values of l, u, and g used in this work are 0.95, 1.05, and 1.5, respectively.

2. Hierarchy

   Horses tend to follow the hierarchy as one of the most important behavior patterns. The adult horse is responsible for handling the leadership of the herd. The mathematical expression used for hierarchy is:

$$\vec{H}_n^{iter,age} = h_n^{iter,age}\left[M_*^{(iter-1)} - M_n^{(iter-1)}\right], age = \alpha, \beta, \gamma (4.31)$$

$$h_n^{iter,age} = h_n^{(iter-1),age} \times \omega_h \tag{4.32}$$

where $\vec{H}_n^{iter,age}$ indicates the effects of the best horse location on the velocity parameter, $M_*^{(iter-1)}$ shows the location of the best horse.

3. Sociability

The social behavior of horses increases their chances of survival and escape from being hunted by predators. The mathematical expression for sociability is:

$$\vec{S}_n^{iter,age} = s_n^{iter,age}\left[\left(\frac{1}{N}\sum_{j=1}^{N} M_j^{(iter-1)}\right) - M_n^{(iter-1)}\right], age = \beta, \gamma \tag{4.33}$$

$$s_n^{iter,age} = s_m^{(iter-1),age} \times \omega_s \tag{4.34}$$

where $\vec{S}_n^{iter,age}$ is the social motion vector of the *ith* horse, and $s_n^{iter,age}$ indicates the horse's inclination toward the herd, which decreases in each cycle with ω_s. N is the total number of horses in the herd.

4. Imitation

Similar to other mammals, horses are also likely to imitate each other, and in this process of imitation, horses acquire good and bad habits from others. This imitation behavior is not inherent throughout the life of horses. The mathematical equation for the imitation behavior is presented in Eq. (4.35):

$$\vec{I}_n^{iter,age} = i_n^{iter,age}\left[\left(\frac{1}{pN}\sum_{j=1}^{pN} \hat{M}_j^{(iter-1)}\right) - M^{(iter-1)}\right], age = \gamma \tag{4.35}$$

$$i_m^{iter,age} = i_m^{(iter-1),age} \times \omega_i \tag{4.36}$$

In the above equations, $\vec{I}_n^{iter,age}$ is the motion vector of the *ith* horse toward the average of the best horse locations, which is denoted by $\hat{M}$, and pN is the number of horses with the best locations. ω_i is the imitation coefficient that decreases with each iteration.

1. Defence Mechanisms

With respect to Darwin's theory, which suggests the survival of the fittest, the defence mechanism is one of the factors that support the better survival of horses in case of a prey attack. The mathematical expression is as follows:

$$\vec{D}_n^{iter,age} = -d_n^{iter,age}\left[\left(\frac{1}{qN}\sum_{j=1}^{qN} \tilde{M}_j^{(iter-1)}\right) - M^{(iter-1)}\right], age = \alpha, \beta \tag{4.37}$$

$$d_n^{iter,age} = d_n^{(iter-1),age} \times \omega_d \tag{4.38}$$

where $\vec{D}_n^{iter,age}$ shows the vector of the *ith* horse tending to escape from the horses with the worst location, as shown by $\tilde{M}$ vector. qN is the number of horses in the worst location, where the value of q to be suggested is 20% of the total horses. ω_d is the reduction factor. The negative sign in the expression indicates that the horses are prone to move away from the danger of being hunted.

2. Roam

As a social animal, roaming is one of the characteristic behaviors of horses. They are curious to roam from one location to another to search for fresh food. The mathematical expression of roaming behavior has been expressed as:

$$\vec{R}_n^{iter,age} = r_n^{iter,age} \times r \times M^{(iter-1)}, age = \gamma, \delta \quad (4.39)$$

$$r_n^{iter,age} = r_n^{(iter-1),age} \times \omega_r \quad (4.40)$$

where $\vec{R}_n^{iter,age}$ is the random velocity vector and ω_r is the reduction factor. To express the expanded velocity expression of horses in each age group, the mathematical expression of related behavioral traits was incorporated, as shown in Table 4.1

In the above expressions, w_g, w_d, w_h, w_s, w_i, and w_r are the linearly reducing coefficients that are being updated in each iteration, L and U are the lower and upper bounds of the grazing space, r is a random number between 0 and 1; the best and worst locations of horses are represented by pN and qN, respectively.

The inclusion of behavioral pattern in the HHO algorithm is advantageous over the PSO and GWO algorithms as it provides a better window for exploration and exploitation phases, which eventually provides the best solution in lesser computational time (Mehrabi et al., 2024). The pseudocode presented in Table 4.2 indicates the important steps involved in the development of a metaheuristically tuned neural network model for solar irradiance forecasting. The objective function used in the code is the MSE (step 3 in the pseudocode), and the code termination condition is 1×10^6 MSE error or MSE at the end of the 1000th epoch.

4.6 Results and discussion

In this study, neural network-based solar irradiance forecasting models were considered. The meteorological parameters of Jaipur, Rajasthan (26° 55′ 19.4520′′ N and 75° 46′ 43.9860′′ E) were considered for the analysis. The internal configurations of the ANN and RNN-based neural networks are listed in Table 4.3.

The error between the actual and forecasted values is calculated to prove the potency of the proposed solar irradiance forecasting model.

4.6.1 Result and analysis of artificial neural network-based forecasting models

The different types of ANN solar irradiance forecasting models used for analysis are:

1. Conventional ANN Model.
2. kPCA integrated ANN model.
3. kPCA integrated PSO tuned ANN model.
4. kPCA integrated GWO tuned ANN model.
5. kPCA integrated HHO tuned ANN model.

Table 4.1 Velocity expression of each age group.

Age group	Expanded velocity expression
Alpha	$V_n^{iter+1,\alpha} = \Big[w_g \times g_n^{(iter-1),\alpha}(L + r.U)\Big[M_n^{(iter-1)}\Big]\Big] - d_n^{(iter-1),\alpha} \times w_d \times \Bigg[\Bigg(\frac{1}{qN}\sum_{j=1}^{qN} M_j^{(iter-1)}\Bigg) - M_n^{(iter-1)}\Bigg]$ (6.41)
Beta	$V_n^{iter,\beta} = \Big[w_g \times g_n^{(iter-1),\beta}(L + r.U)\Big[M_n^{(iter-1)}\Big]\Big] - h_n^{(iter-1),\beta} \times w_h \times \Big[M_{lbh}^{(iter-1)} - M_n^{(iter-1)}\Big] +$ $s_n^{(iter-1),\beta} \times w_s \times \Big[\Big(\frac{1}{N}\sum_{j=1}^{N} M_j^{(iter-1)}\Big) - M_n^{(iter-1)}\Big] - d_n^{(iter-1),\beta} \times w_d \times \Big[\frac{1}{qN}\sum_{j=1}^{qN} M_j^{(iter-1)} - M_n^{(iter-1)}\Big]$ (6.42)
Gamma	$V_n^{iter,\gamma} = \Big[w_g \times g_n^{(iter-1),\gamma}(L + r.U)\Big[M_n^{(iter-1)}\Big]\Big] + h_n^{(iter-1),\gamma} \times w_h \times \Big[M_{lbh}^{(iter-1)} - M_n^{(iter-1)}\Big] +$ $s_n^{(iter-1),\gamma} \times w_s \times \Bigg[\Bigg(\frac{1}{N}\sum_{j=1}^{N} M_j^{(iter-1)}\Bigg) - M_n^{(iter-1)}\Bigg] + i_n^{(iter-1),\gamma} \times w_i \times \Bigg[\Bigg(\frac{1}{pN}\sum_{j=1}^{pN} M_j^{(iter-1)}\Bigg) - M^{(iter-1)}\Bigg] -$ (6.43) $d_n^{(iter-1),\gamma} \times w_d \times \Bigg[\Bigg(\frac{1}{qN}\sum_{j=1}^{qN} M_j^{(iter-1)}\Bigg) - M^{(iter-1)}\Bigg] + r_n^{(iter-1),\gamma} \times w_r \times r \times M^{(iter-1)}$
Delta	$V_n^{iter,\delta} = \Big[w_g \times g_n^{(iter-1),\delta}(L + r.U)\Big[M_n^{(iter-1)}\Big]\Big] + i_n^{(iter-1),\delta} \times w_i \times \Bigg[\Bigg(\frac{1}{pN}\sum_{j=1}^{pN} M_j^{(iter-1)}\Bigg) - M^{(iter-1)}\Bigg] + r_n^{(iter-1),\delta} \times w_r \times r \times M^{(iter-1)}$ (6.44)

Table 4.2 Pseudocode of HHO optimized NN forecasting model.

Algorithm 1	**HHO optimized NN forecasting model**
Steps	**1.** Initialize neural network internal configurations including number of hidden layers, number of neurons, and activation function **2.** Assign random weights to the NN **3.** Provide mean square error function as to evaluate the performance of NN model **4.** Initialize the constant of optimization algorithm and set *iter* = *1*. **5.** In a random manner initialize the position of the horses in the suitable search space. **6.** Evaluate the fitness values of all the horses as per the objective function and classify the horses in α, β, γ, and δ age groups. **7.** For $iter \leq max^{m}\ iter$ for n = 1: no. of horses **1.** Calculate the velocity vector of horses of each group as per Eqs. (4.29)–(4.32) **2.** Evaluate the new position with Eq. (4.24) **3.** For the given updated position of the horses, evaluate the fitness value of all the horses. N = n + 1 Iter = iter + 1 end end while Return the best position **1.** Run the optimized weights enabled NN model. **2.** Plot a graph between actual value and target values. **3.** Terminate

HHO, horse herd optimization; *NN*, neural network.

Table 4.3 Internal configuration of neural networks.

S. no.	**Hyper-parameter**	**Magnitude**
1	Epochs	1000
2	Goal	1×10^{-6}
3	Learning rate	1×10^{-7}
4	Training function	trainlm
5	Activation function	tansig
6	No. of hidden layers	3
7	No. of neurons (3-cases)	5; 15; 25

Three configurations of ANN models with 5, 15, and 25 neurons in each of the three hidden layers were analyzed. Table 4.4 shows the error parameters of the different ANN models for 4 months dataset. Table 4.5 and Table 4.6 present the evaluated parameters for the 1-year and 3 years dataset, respectively.

Table 4.4 Evaluated error parameters of ANN models on 4-month dataset.

Errors	ANN	kPCA-ANN	kPCA-PSO-ANN	kPCA-GWO-ANN	kPCA-HHO-ANN
5,5,5- Hidden layer neurons					
MSE	0.0307	0.0287	0.0262	0.0251	0.0243
RMSE	0.1757	0.1695	0.162	0.1584	0.1559
MAE	0.0981	0.0887	0.0852	0.0801	0.0794
R^2	0.6721	0.6884	0.7153	0.73	0.7366
15,15,15- Hidden layer neurons					
MSE	0.0199	0.0165	0.0154	0.0062	0.0049
RMSE	0.1414	0.1284	0.1241	0.0841	0.0705
MAE	0.0759	0.0658	0.0628	0.0426	0.0364
R^2	0.7844	0.8212	0.833	0.913	0.9461
25,25,25- Hidden layer neurons					
MSE	5.85×10^{-3}	7.27×10^{-4}	6.79×10^{-4}	5.698×10^{-4}	3.53×10^{-4}
RMSE	0.0765	0.0269	0.0261	0.0239	0.0188
MAE	0.0385	0.0132	0.0133	9.3×10^{-3}	$8.7 \times ^{-3}$
R^2	0.9365	0.9921	0.993	0.994	0.9962

ANN, artificial neural network; *GWO*, grey wolf optimization; *HHO*, horse herd optimization; *kPCA*, kernel principal component analysis; *MSE*, mean square error; *PSO*, particle swarm optimization; MAE, mean absolute error.

Table 4.5 Evaluated error parameters of ANN models on 1-year dataset.

Errors	ANN	kPCA-ANN	kPCA-PSO-ANN	kPCA-GWO-ANN	kPCA-HHO-ANN
5,5,5- Hidden layer neurons					
MSE	0.1648	0.1509	0.1428	0.1377	0.1295
RMSE	0.4059	0.3884	0.3778	0.3711	0.3599
MAE	0.2956	0.2444	0.2316	0.228	0.2234
R^2	0.895	0.9148	0.9194	0.9222	0.9269
15,15,15- Hidden layer neurons					
MSE	0.110	0.096	0.0852	0.0796	0.0803
RMSE	0.3317	0.3098	0.2918	0.2821	0.2834
MAE	0.2546	0.1892	0.1828	0.1810	0.1775
R^2	0.9379	0.9458	0.9519	0.9550	0.9546
25,25,25- Hidden layer neurons					
MSE	0.0992	0.0878	0.0632	0.0522	0.0480
RMSE	0.3150	0.2963	0.2515	0.2285	0.2191
MAE	0.2354	0.1797	0.1558	0.1466	0.1416
R^2	0.944	0.9504	0.9643	0.9705	0.9729

ANN, artificial neural network; *GWO*, grey wolf optimization; *HHO*, horse herd optimization; *kPCA*, kernel principal component analysis; *MSE*, mean square error; *PSO*, particle swarm optimization.

Fig. 4.5 shows a typical forecasted solar irradiance curve for a random day. The curve shows the peaks around 10:20 hour and 14:45 hour with 564.06 W/m^2 and 338.02 W/m^2 of solar irradiance, respectively. For the two peaks considered in the curve, the kPCA-based HHO-integrated ANN model forecasted the magnitude of solar irradiance

Table 4.6 Evaluated error parameters of ANN models on 3-year dataset.

Errors	ANN	kPCA-ANN	kPCA-PSO-ANN	kPCA-GWO-ANN	kPCA-HHO-ANN
5,5,5- Hidden layer neurons					
MSE	0.3106	0.2293	0.2085	0.2061	0.2060
RMSE	0.5572	0.4788	0.4566	0.4540	0.4538
MAE	0.4028	0.3127	0.2912	0.2891	0.2876
R^2	0.8250	0.8708	0.8825	0.8838	0.884
15,15,15- Hidden layer neurons					
MSE	0.2625	0.1785	0.1663	0.1660	0.1656
RMSE	0.5123	0.4225	0.4078	0.4074	0.4070
MAE	0.3881	0.2618	0.2524	0.2540	0.2520
R^2	0.8521	0.8994	0.9061	0.9064	0.9066
25,25,25- Hidden layer neurons					
MSE	0.1668	0.1556	0.1395	0.1386	0.1374
RMSE	0.4085	0.3945	0.3735	0.3722	0.3706
MAE	0.3246	0.2413	0.2334	0.2319	0.23
R^2	0.906	0..9123	0.9215	0.9219	0.9226

ANN, artificial neural network*; GWO,* grey wolf optimization; *HHO,* horse herd optimization*; kPCA,* kernel principal component analysis; *MSE,* mean square error; *PSO,* particle swarm optimization.

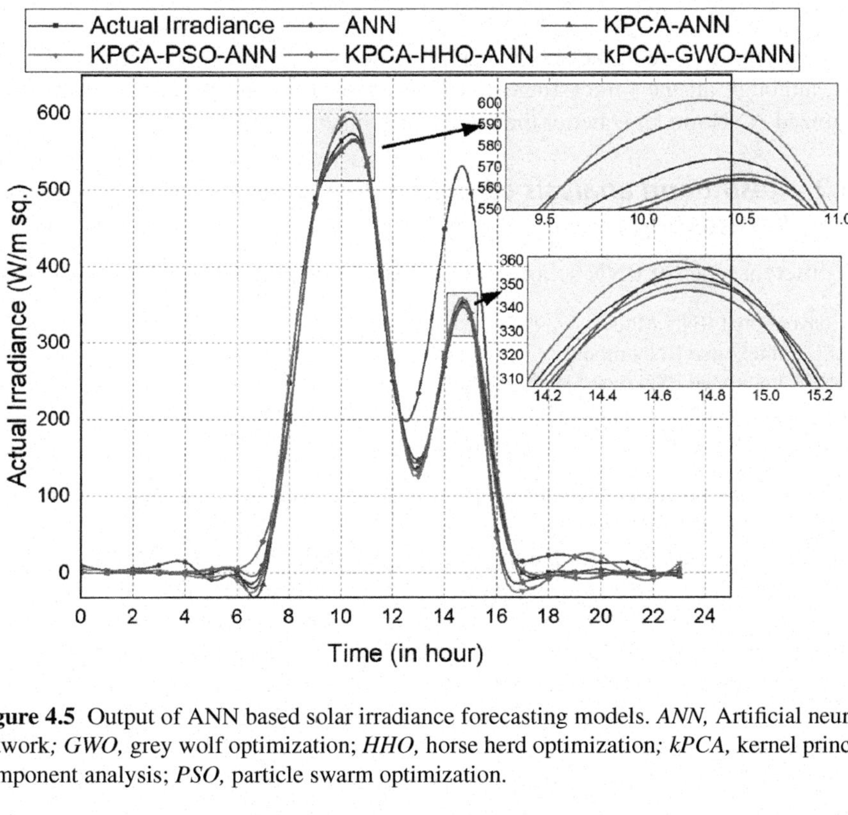

Figure 4.5 Output of ANN based solar irradiance forecasting models. *ANN,* Artificial neural network*; GWO,* grey wolf optimization; *HHO,* horse herd optimization*; kPCA,* kernel principal component analysis; *PSO,* particle swarm optimization.

Table 4.7 Evaluated error parameters of the RNN model on the 4-month dataset.

Errors	RNN	kPCA-RNN	kPCA-PSO-RNN	kPCA-GWO-RNN	kPCA-HHO-RNN
5,5,5-Hidden layer neurons					
MSE	0.0302	0.0279	0.0251	0.0246	0.0235
RMSE	0.1737	0.1669	0.1583	0.1570	0.1534
MAE	0.1039	0.0885	0.08132	0.0806	0.0792
R^2	0.6755	0.6977	0.7325	0.738	0.7449
15,15,15- Hidden layer neurons					
MSE	0.0179	0.0157	0.0136	7.1×10^{-3}	2.55×10^{-3}
RMSE	0.1337	0.1251	0.1166	0.0786	0.0505
MAE	0.0682	0.0628	0.0579	0.0474	0.0234
R^2	0.8064	0.8303	0.8527	0.96	0.9723
25,25,25- Hidden layer neurons					
MSE	3.78×10^{-3}	6.45×10^{-4}	5.402×10^{-4}	4.49×10^{-4}	2.21×10^{-4}
RMSE	0.0615	0.0254	0.0232	0.0212	0.0149
MAE	0.0333	0.0116	0.0092	0.0109	0.0059
R^2	0.9594	0.993	0.9941	0.9963	0.9976

GWO, grey wolf optimization; *HHO*, horse herd optimization; *kPCA*, kernel principal component analysis; *MSE*, mean square error; *PSO*, particle swarm optimization; *RNN*, recurrent neural network.

which was quite close to the actual irradiance magnitude. However, for the rest of the time, although all the forecasting models work well, the performance of the HHO-optimized ANN model is better than that of the other considered models.

4.6.2 Result and analysis of the recurrent neural network-based forecasting model

The different types of RNN solar irradiance forecasting models used for analysis are:

1. Conventional RNN Model.
2. kPCA integrated RNN model.
3. kPCA integrated PSO tuned RNN model.
4. kPCA integrated GWO tuned RNN model.
5. kPCA integrated HHO tuned RNN model.

Three configurations of RNN models with 5, 15, and 25 neurons in each of the three hidden layers were analyzed.

In Tables 4.7–4.9, the error parameters, that is, MSE, MAE, and RMSE, are minimal for the kPCA-based HHO-integrated RNN forecasting model. In addition, there was a significant reduction in errors as the number of hidden layer neurons increased. Although the RNN is a mature form of neural network, the optimized form of the RNN model performs a commendable job in precisely forecasting solar irradiance.

Fig. 4.6 shows the forecasting capabilities of the different RNN-based solar irradiance forecasting models. The peaks have been shown at 10:20 hours and 14:45 hour having 564.06 W/m^2 and 338.02 W/m^2 of solar irradiance, respectively. The enlarged form of the two peaks within the 24-hour variation of the solar irradiance shows

Table 4.8 Evaluated error parameters of the RNN model on the 1-year dataset.

Errors	RNN	kPCA-RNN	kPCA-PSO-RNN	kPCA-GWO-RNN	kPCA-HHO-RNN
5,5,5- Hidden layer neurons					
MSE	0.0556	0.0255	0.0136	0.0111	0.0101
RMSE	0.2358	0.1597	0.1165	0.1054	0.1004
MAE	0.150	0.1025	0.0751	0.066	0.063
R^2	0.9686	0.9856	0.9923	0.9937	0.9943
15,15,15- Hidden layer neurons					
MSE	0.0389	0.0206	0.0115	0.0075	0.0072
RMSE	0.1973	0.1437	0.1075	0.0864	0.0851
MAE	0.1488	0.0914	0.0701	0.0641	0.0599
R^2	0.9780	0.9883	0.9934	0.994	0.9959
25,25,25- Hidden layer neurons					
MSE	0.0297	0.0122	0.0070	0.0058	0.0052
RMSE	0.1732	0.1104	0.0837	0.0764	0.0721
MAE	0.186	0.0687	0.0575	0.0546	0.0545
R^2	0.965	0.9931	0.996	0.9967	0.9968

GWO, grey wolf optimization; *HHO*, horse herd optimization; *kPCA*, kernel principal component analysis; *MSE*, mean square error; *PSO*, particle swarm optimization; *RNN*, recurrent neural network.

Table 4.9 Evaluated error parameters of the RNN model on the 3-year dataset.

Errors	RNN	kPCA-RNN	kPCA-PSO-RNN	kPCA-GWO-RNN	kPCA-HHO-RNN
5,5,5- Hidden layer neurons					
MSE	0.31	0.0278	0.0188	0.0179	0.0171
RMSE	0.5568	0.166	0.1372	0.1338	0.1306
MAE	0.3751	0.1169	0.0875	0.0871	0.089
R^2	0.8253	0.984	0.9894	0.9899	0.9904
15,15,15- Hidden layer neurons					
MSE	0.0672	0.026	0.0111	0.0104	0.010
RMSE	0.2593	0.1634	0.1054	0.1022	0.099
MAE	0.1882	0.1102	0.0688	0.0675	0.0664
R^2	0.9621	0.9850	0.9937	0.9941	0.9944
25,25,25- Hidden layer neurons					
MSE	0.052	0.0211	0.0101	0.0083	0.0078
RMSE	0.2279	0.1452	0.1005	0.0912	0.0885
MAE	0.1721	0.0951	0.0651	0.0623	0.0544
R^2	0.9707	0.9881	0.9939	0.9953	0.9956

GWO, grey wolf optimization; *HHO*, horse herd optimization; *kPCA*, kernel principal component analysis; *MSE*, mean square error; *PSO*, particle swarm optimization; *RNN*, recurrent neural network.

that the kPCA-based HHO-integrated RNN model has the best forecasting capability among other forecasting models, even under fluctuating solar irradiance conditions. Simultaneously, for the rest of the time, the kPCA-based HHO-optimized RNN model outperformed the other RNN-based forecasting models.

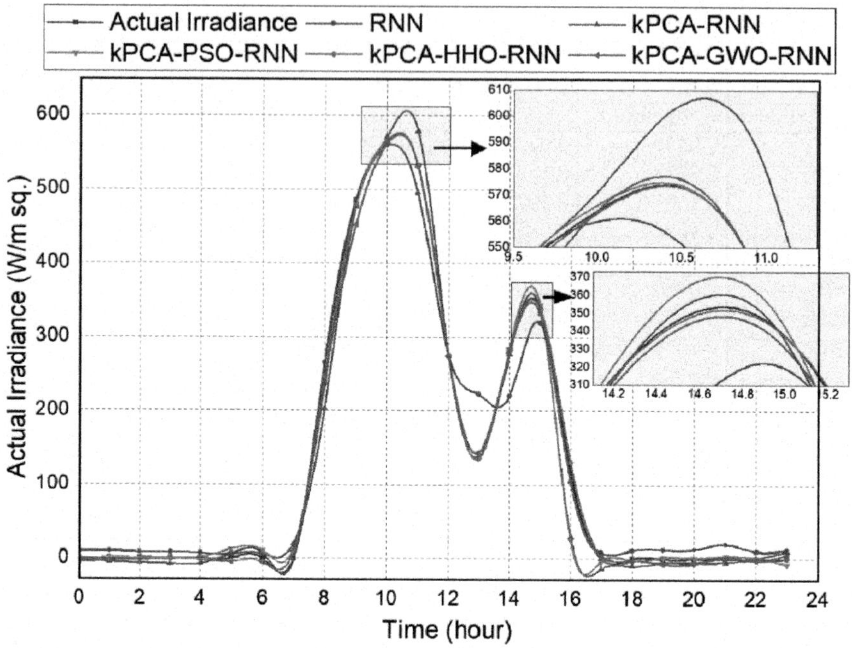

Figure 4.6 Output of RNN based solar irradiance forecasting models. *GWO,* Grey wolf optimization; *HHO,* horse herd optimization; *kPCA,* kernel principal component analysis; *PSO,* particle swarm optimization; *RNN,* recurrent neural network.

4.6.3 Comparison of the solar irradiance forecasting test models

The evaluated error parameters for the four-month, one-year, and three-year datasets have been indicated in the above sub-sections. To visualize the performance of each of the test models, an RMSE comparison was performed as shown in Fig. 4.7.

The bar presentation of the RMSE errors clearly shows that the neural networks with a higher number of hidden layer neurons, that is, 25, perform better than the lower number of neurons. Moreover, the extensive analysis of the RMSE result shows that among all the considered models, the performance of the RNN-based test models is superior. This is due to the inherent memory unit in the RNN model, and the increment in the training dataset significantly improved its training compared to the ANN models.

To highlight the best RNN-based test model, an accuracy evaluation was performed. The authors of the present research work have developed an expression for accuracy measurement using the mean of the target dataset and the mean of the predicted outcomes (Zulfiqar et al., 2023).

$$Accuracy = \frac{mean(predicitedSI)}{mean(actualtargetSI)} \tag{4.45}$$

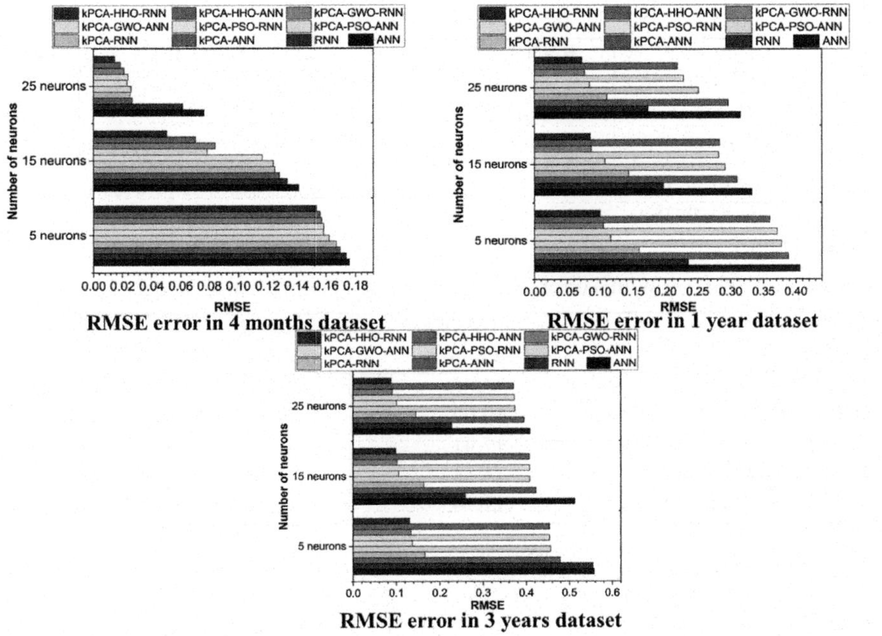

Figure 4.7 RMSE comparison of different test models. *ANN,* Artificial neural network*; GWO,* grey wolf optimization; *HHO,* horse herd optimization*; kPCA,* kernel principal component analysis; *PSO,* particle swarm optimization; *RNN,* recurrent neural network; RMSE, root mean square error.

Table 4.10 Evaluated accuracy of different datasets.

Datasets	RNN	kPCA-RNN	kPCA-PSO-RNN	kPCA-GWO-RNN	kPCA-HHO-RNN
4-month	99.8	99.83	99.90	99.93	99.98
1-year	88.96	92.17	93.45	93.7	94.03
3-year	80.44	89.19	92.59	92.91	93.79

GWO, grey wolf optimization; *HHO,* horse herd optimization*; kPCA,* kernel principal component analysis; *PSO,* particle swarm optimization; *RNN,* recurrent neural network.

Based on the above expression, the accuracy of each type of dataset is measured to compare and validate the efficacy of the proposed model. The RMSE comparison suggests that the RNN neural network consisting of 25 hidden neurons attains lower RMSE errors, therefore to obtain the best solar irradiance forecasting model, further comparison of accuracy was performed for the 25 hidden-layer neuron-based RNN models.

From the tabular information presented in Table 4.10, it is observed that the kPCA-HHO-RNN-based solar irradiance forecasting model attained higher accuracy in all three types of datasets.

Table 4.11 Computational Time of the models.

Dataset	RNN	kPCA-HHO-RNN
4-month	6 min 38 sec	2 min 48 sec
1 year	30 min 35 sec	10 min 26 sec
3 year	79 min 46 sec	28 min 57 sec

HHO, horse herd optimization; *kPCA,* kernel principal component analysis; *RNN,* recurrent neural network.

Table 4.12 RMSE (in W/m^2) error of different considered forecasting models.

Models Dataset	RMSE 4 month	RMSE 1 year	RMSE 3 year
RNN	46.06	129.72	170.69
kPCA+RNN	19.02	82.68	108.75
kPCA+ PSO tuned RNN	17.41	62.69	78.72
kPCA+GWO tuned RNN	15.88	57.22	68.3
kPCA+ HHO tuned RNN	11.14	53.92	66.28

GWO, grey wolf optimization; *HHO,* horse herd optimization; *kPCA,* kernel principal component analysis; *PSO,* particle swarm optimization; *RNN,* recurrent neural network.

To further highlight the advantage of using the kPCA algorithm for data preprocessing, a computational time analysis of the original high-dimensional dataset and kPCA based dimensionally reduced dataset was also performed. The dimensionality reduction algorithm kPCA provides the advantage of minimizing the noise and noncollinearity of the dataset and can also shorten the memory requirement (Das et al., 2016). The configuration of the system used for time analysis has a 64-bit operating system with an Intel Core i5 1035G1 CPU processor at 1.00 GHz. Table 4.11 presents the computational time attained by the model consisting of the original high-dimensional dataset (RNN) and kPCA-based dimensionally reduced dataset with the HHO-RNN model (kPCA-HHO-RNN).

The computational time analysis of the different models suggests that the computational time of the model is directly proportional to the volume of the dataset. For 4 months dataset, the overall computational time was less. However, the optimized model with a dimensionally reduced dataset (kPCA-HHO-RNN) attained the best result in a lesser computational time. For the 1-year dataset, the time taken by the original high-dimensional data with the nonoptimized model (RNN) was greater than that of the kPCA-HHO-RNN model. Simultaneously, the computational time for the RNN model with 3 years original high-dimensional dataset was 79 minute 46 second whereas the proposed kPCA-HHO-RNN model successfully attained better forecasting results in 28 minutes and 57 seconds. The computational time analysis of the test models shows that the combination of an appropriate data preprocessing algorithm (kPCA) and metaheuristically optimized (HHO) parameters of the neural model leads to improvement in both the empirical error result (lesser error matrices and higher accuracy) and computational efficiency (lower computational time).

Furthermore, the RMSE error (W/m^2) of each of the considered neural networks with 25 hidden layer neurons was compared, as presented in Table 4.12

The outcome of the above error evaluation can be concluded as a model trained on larger dataset spanning over 3 years exhibit higher error magnitudes compared to the model trained on a shorter time span (i.e., 4 months), which might be due to the seasonal effect and cloud situations that have inherent stochastic behavior which makes it difficult for the model to forecast with higher accuracy over extended time periods (Kumari & Toshniwal, 2021).

4.7 Performance validation and comparative analysis of the proposed model

To highlight the robustness of the model, the proposed model was analyzed under monthly climatic variation. Moreover, to validate the performance of the proposed model further, a dataset from another geographical location was considered. Finally, to benchmark the performance ability, the RMSE of the proposed model was compared with other recently developed solar irradiance forecasting models.

4.7.1 Monthly climatic variation

In this section, the evaluation of the proposed kPCA-HHO-RNN is performed by analyzing its performance capability under different seasonal variations. The 1-year dataset was divided into four parts according to seasonal climatic variation (Winter, Spring, Summer, and Autumn) to determine the robustness of the proposed algorithm under various dynamic characteristics of the dataset. The months of the year were clustered into four seasons as: Winter (November–February), spring (March–April), summer (May–August), and autumn (September–October). As suggested in the above section, the proposed kPCA-HHO-RNN model outperforms the present case study, which explored the robustness of the proposed model with the kPCA-HHO-ANN test model under various seasonal variations. Table 4.13 shows the evaluated errors from kPCA-HHO-ANN and kPCA-HHO-RNN under different climatic variations.

As shown in Table 4.13, the kPCA-HHO-RNN model outperformed the kPCA-HHO-ANN model with an accuracy of more than 99 % for all the variations in the climatic seasons. A visualization of the forecasted solar irradiance curve is shown in Fig. 4.8, representing the 24-hour solar irradiance curve for each season.

4.7.2 Predictive uncertainty quantification using the Monte Carlo dropout approach

The Monte Carlo Dropout (MC Dropout) approach is a computationally effective method that has been used to approximate Bayesian inference in deep NNs. This approach is quite useful in measuring the predictive uncertainty. By keeping the dropout layers active during the testing phase, the model samples from an approximate posterior distribution. Each step with dropout enabled represents a different configuration of the network, hence yields a unique outcome. This approach is a practical method

Table 4.13 Evaluated errors of the models.

Summer Season		
Matrices	**kPCA-HHO-ANN**	**kPCA-HHO-RNN**
MSE	5.896×10^{-4}	9.76×10^{-5}
MAE	0.0166	0..0064
RMSE	0.0243	0.0099
R^2	0.9995	0.9996
Accuracy	98.26	99.33
Spring Season		
Matrices	kPCA-HHO-ANN	kPCA-HHO-RNN
MSE	8.9514×10^{-5}	1.8675×10^{-5}
MAE	0.0066	0.0028
RMSE	0.0095	0.0043
R^2	0.99996	0.99999
Accuracy	99.32	99.71
Autumn Season		
Matrices	kPCA-HHO-ANN	kPCA-HHO-RNN
MSE	7.8529×10^{-5}	1.3804×10^{-5}
MAE	0.0053	0.0026
RMSE	0.0089	0.0037
R^2	0.99995	0.99999
Accuracy	99.382	99.71
Winter Season		
Matrices	kPCA-HHO-ANN	kPCA-HHO-RNN
MSE	6.6388×10^{-4}	5.9401×10^{-5}
MAE	0.0171	0.0054
RMSE	0.0258	0.0077
R^2	0.9995	0.99997
Accuracy	97.72	99.28

ANN, artificial neural network*; HHO,* horse herd optimization*; kPCA,* kernel principal component analysis; *MSE,* mean square error; *RNN,* recurrent neural network.

of determining the sensitivity of the network toward small changes in its internal configuration.

For determining the uncertainty, two metrics have been used, namely, Prediction Interval Coverage Probability (PICP) and Prediction Normalized Average Width (PINAW). Both of these metrics evaluate the quality of prediction intervals (for this case 90%) in uncertainty quantification (UQ).

$$PICP = \frac{1}{N}\sum_{i=1}^{N} \mu_i \tag{4.46}$$

$$PINAW = \frac{1}{N.R}\sum_{i=1}^{N} (U_i - L_i) \tag{4.47}$$

where N is the number of samples, μ_i is the indicator function, L_i, U_i are the lower

Table 4.14 Comparison of recently developed forecasting techniques.

References	Algorithm	Dataset Time Span	RMSE (W/m^2)
Pereira et al. (2024)	ANN + operational NWP + aerosol data	4.5 years	204.3
Lyu et al. (2023)	kPCA + SC + DQN + XGBoost	3 years	111.04
Xu et al. (2024)	EO-FFANN	1 year	142.38
Vaisakh & Jayabarathi (2022)	GU-DHOA-MLP	4 months	213.54
	GU-DHOA-CNN	4 months	211.39
	GU-DHOA-RNN	4 months	315.16
Ali-Ou-Salah et al. (2022)	Seasonal clustering + ANN	4 years	106.4
Wang et al. (2024)	Multi-physical process and LSTM-BP	1 year	65.49
Proposed models	kPCA+ HHO tuned RNN	4 months	11.13
	kPCA+ HHO tuned RNN	1 year	53.92
	kPCA+ HHO tuned RNN	3 years	66.28

ANN, artificial neural network; *EO*, equilibrium optimization; *HHO*, horse herd optimization; *kPCA*, kernel principal component analysis; *NWP*, numerical weather prediction; *RNN*, recurrent neural network; *XGBoost*, extreme gradient tree boosting; SC, spectral clustering; DQN, deep q-network; FFANN, feed-forward artificial neural network; GU, grey updated; DHOA, deer hunting optimization algorithm; MLP, multilayer perceptron; CNN, convolutional neural network; LSTM-BP, long short-term memory- back propagation.

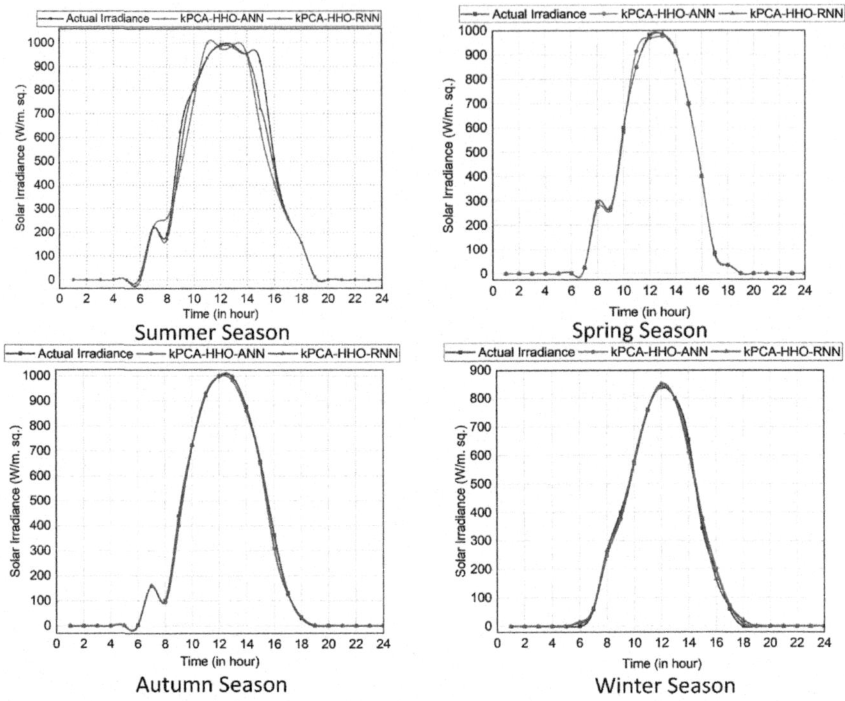

Figure 4.8 Solar irradiance curve representing different seasons. *ANN,* Artificial neural network*; GWO,* grey wolf optimization; *HHO,* horse herd optimization*; kPCA,* kernel principal component analysis; *PSO,* particle swarm optimization; *RNN,* recurrent neural network.

and upper bounds of the prediction interval, and R is the range of actual target data. The UQ was performed on the kPCA-HHO-RNN model, which is identified as the best-performing model across all the considered scenarios. The UQ analysis utilized the MC Dropout approach for determining the model's predictivity when trained with the 1-year dataset and 3 years dataset. For the 1-year dataset, PICP is 91.01%, and PINAW is 0.15, while for 3 years dataset, PICP is 90.62 and PINAW is 0.14. For both datasets, the PICP is within the required 90% confidence level, and PINAW is also on the lower side. Thus, the proposed kPCA-HHO-RNN model provides reliable and well-calibrated probabilistic forecasts, confirming its effective performance.

4.7.3 Comparison with other recently developed solar irradiance forecasting models

Furthermore, the proposed work has been compared with other research work presented in the literature, as shown in Table 4.14. It can be seen from the above discussion that for all three types of datasets, the proposed solar irradiance forecasting model works better than other recently developed forecasting models.

From the above results, it was observed that the present study is the effective integration of a data compression algorithm, that is, kPCA, with a metaheuristically tuned RNN model, which can result in the development of an efficient solar irradiance forecasting model. The kPCA algorithm can effectively compress a dataset of any volume, preserving all the information of the original dataset and providing the benefit of a lower computational burden. However, the optimization of the parameters of the RNN model helps to achieve better solar irradiance forecasting results. Hence, the proposed kPCA-HHO-RNN model can be effectively used in real-world solar irradiance forecasting applications using various dataset volumes.

4.8 Conclusion

This study suggests an advancement in the field of solar irradiance forecasting by providing a comprehensive framework that includes the integration of the dimensionality reduction technique and an optimization algorithm for tuning the parameters of the neural network. The original meteorological datasets (4 months, 1 year, and 3 years) were compressed using the dimensionality reduction technique, viz., kPCA, which significantly compressed all the information into smaller dimensions and reduced the computation time required for the execution of the model. The dimensionally reduced dataset was used to train the neural networks (ANN and RNN) whose parameters were optimized using HHO. The efficacy of the HHO optimized models was compared with that of the PSO and GWO-optimized neural network models. The evaluation of various error matrices and accuracy measurement suggests that the kPCA-HHO-RNN model outperforms the other test models. Additionally, the superiority of the proposed model was highlighted by evaluating the error matrices with monthly climatic datasets. The model effectively withstands the UQ analysis with PICP of more that 90%. Further RMSE comparison with other recently developed models presented in the literature benchmarked the superior performance of the proposed kPCA-HHO-RNN model.

References

Ali-Ou-Salah, H., Oukarfi, B., & Mouhaydine, T. (2022). Short-term solar radiation forecasting using a new seasonal clustering technique and artificial neural network. *International Journal of Green Energy, 19*(4), 424–434. doi:10.1080/15435075.2021.1946819.

Alsharif, M. H., Younes, M. K., & Kim, J. (2019). Time series ARIMA model for prediction of daily and monthly average global solar radiation: The case study of Seoul, South Korea. *Symmetry, 11*(2). doi:10.3390/sym11020240.

Ayesha, S., Hanif, M. K., & Talib, R. (2020). Overview and comparative study of dimensionality reduction techniques for high dimensional data. *Information Fusion, 59*, 44–58. doi:10.1016/j.inffus.2020.01.005.

Bansal, A. K, Swaroop Sangtani, V., Dadheech, P., Aneja, N., & Yahya, U. (2023). Biogeography-based optimization of artificial neural network (BBO-ANN) for Solar radiation forecasting. *Applied Artificial Intelligence, 37*(1). doi:10.1080/08839514.2023.2166705.

Das, S. P., Achary, N. S., & Padhy, S. (2016). Novel hybrid SVM-TLBO forecasting model incorporating dimensionality reduction techniques. *Applied Intelligence, 45*(4), 1148–1165. doi:10.1007/s10489-016-0801-3.

Davò, F., Alessandrini, S., Sperati, S., Delle Monache, L., Airoldi, D., & Vespucci, M. T. (2016). Post-processing techniques and principal component analysis for regional wind power and solar irradiance forecasting. *Solar Energy, 134*, 327–338. doi:10.1016/j.solener.2016.04.049.

El-Amarty, N., Marzouq, M., El Fadili, H., Bennani, S. D., & Ruano, A. (2023). A comprehensive review of solar irradiation estimation and forecasting using artificial neural networks: Data, models and trends. *Environmental Science and Pollution Research, 30*(3), 5407–5439. doi:10.1007/s11356-022-24240-w.

Essam, Y., Ahmed, A. N., Ramli, R., Chau, K. W., Idris Ibrahim, M. S., Sherif, M., Sefelnasr, A., & El-Shafie, A. (2022). Investigating photovoltaic solar power output forecasting using machine learning algorithms. *Engineering Applications of Computational Fluid Mechanics, 16*(1), 2002–2034. doi:10.1080/19942060.2022.2126528.

Faris, H., Aljarah, I., Al-Betar, M. A., & Mirjalili, S. (2018). Grey wolf optimizer: a review of recent variants and applications. *Neural Computing and Applications, 30*(2), 413–435. doi:10.1007/s00521-017-3272-5.

Granata, F., & Di Nunno, F. (2021). Forecasting evapotranspiration in different climates using ensembles of recurrent neural networks. *Agricultural Water Management, 255*. doi:10.1016/j.agwat.2021.107040.

Jaffel, I., Taouali, O., Harkat, M. F., & Messaoud, H. (2017). Kernel principal component analysis with reduced complexity for nonlinear dynamic process monitoring. *International Journal of Advanced Manufacturing Technology, 88*(9-12), 3265–3279. doi:10.1007/s00170-016-8987-4.

Kumari, P., & Toshniwal, D. (2021). Deep learning models for solar irradiance forecasting: A comprehensive review. *Journal of Cleaner Production, 318*, 128566. doi:10.1016/j.jclepro.2021.128566.

Liu, M.-D., Ding, L., & Bai, Y.-L. (2021). Application of hybrid model based on empirical mode decomposition, novel recurrent neural networks and the ARIMA to wind speed prediction. *Energy Conversion and Management, 233*, 113917. doi:10.1016/j.enconman.2021.113917.

Lyu, C., Eftekharnejad, S., Basumallik, S., & Xu, C. (2023). Dynamic feature selection for solar irradiance forecasting based on deep reinforcement learning. *IEEE Transactions on Industry Applications, 59*(1), 533–543. doi:10.1109/TIA.2022.3206731.

Mehrabi, N., Boroujeni, S. P. H., & Pashaei, E. (2024). An efficient high-dimensional gene selection approach based on the binary horse herd optimization algorithm for biological data classification. *Iran Journal of Computer Science, 7*(2), 279–309. doi:10.1007/s42044-024-00174-z.

MiarNaeimi, F., Azizyan, G., & Rashki, M. (2021). Horse herd optimization algorithm: A nature-inspired algorithm for high-dimensional optimization problems. *Knowledge-Based Systems, 213*, 106711. doi:10.1016/j.knosys.2020.106711.

Mohandes, M. A. (2012). Modeling global solar radiation using Particle Swarm Optimization (PSO). *Solar Energy, 86*(11), 3137–3145. doi:10.1016/j.solener.2012.08.005.

Mughal, S. N., Sood, Y. R., & Jarial, R. K. (2025). Techno-economic assessment of photovoltaics by predicting daily global solar radiations using hybrid ANN-PSO model. *Energy Systems, 16*(3), 1233–1261. doi:10.1007/s12667-023-00646-4.

Nespoli, A., Leva, S., Mussetta, M., & Ogliari, E. G. C. (2022). A selective ensemble approach for accuracy improvement and computational load reduction in ANN-based PV power forecasting. *IEEE Access, 10*, 32900–32911. doi:10.1109/ACCESS.2022.3158364.

NSRDB, Weather Parameters 2024 https://nsrdb.nrel.gov/data-viewer

Pang, Z., Niu, F., & O'Neill, Z. (2020). Solar radiation prediction using recurrent neural network and artificial neural network: A case study with comparisons. *Renewable Energy, 156*, 279–289. doi:10.1016/j.renene.2020.04.042.

Pereira, S., Canhoto, P., & Salgado, R. (2024). Development and assessment of artificial neural network models for direct normal solar irradiance forecasting using operational numerical weather prediction data. *Energy and AI, 15*. doi:10.1016/j.egyai.2023.100314.

PIB, Renewable energy in India 2023 Available from: https://pib.gov.in/FeaturesDeatils.aspx?NoteId=151141&ModuleId=2

Shafqat, W., Malik, S., Lee, K.-T., & Kim, D.-H. (2021). PSO based optimized ensemble learning and feature selection approach for efficient energy forecast. *Electronics, 10*(18), 2188. doi:10.3390/electronics10182188.

Singh, B., & Pozo, D. (2019). A guide to solar power forecasting using ARMA models. In *2019 IEEE PES Innovative Smart Grid Technologies Europe (ISGT-Europe)* (pp. 1–4). doi:10.1109/ISGTEurope.2019.8905430.

Vaisakh, T., & Jayabarathi, R. (2022). Analysis on intelligent machine learning enabled with meta-heuristic algorithms for solar irradiance prediction. *Evolutionary Intelligence, 15*(1), 235–254. doi:10.1007/s12065-020-00505-6.

Wang, Z., Zhang, Y., Li, G., Zhang, J., Zhou, H., & Wu, J. i (2024). A novel solar irradiance forecasting method based on multi-physical process of atmosphere optics and LSTM-BP model. *Renewable Energy, 226*, 120367. doi:10.1016/j.renene.2024.120367.

Xin, J., Khishe, M., Zeebaree, D. Q, Abualigah, L., & Ghazal, T. M. (2024). Adaptive habitat biogeography-based optimizer for optimizing deep CNN hyperparameters in image classification. *Heliyon, 10*(7). doi:10.1016/j.heliyon.2024.e28147.

Xu, T., Sabzalian, M. H., Hammoud, A., Tahami, H., Gholami, A., & Lee, S. (2024). An innovative machine learning based on feed-forward artificial neural network and equilibrium optimization for predicting solar irradiance. *Scientific Reports, 14*(1). doi:10.1038/s41598-024-52462-0.

Yildirim, A., Bilgili, M., & Ozbek, A. (2023). One-hour-ahead solar radiation forecasting by MLP, LSTM, and ANFIS approaches. *Meteorology and Atmospheric Physics, 135*(1). doi:10.1007/s00703-022-00946-x.

Zulfiqar, M., Kamran, M., Rasheed, M. B., Alquthami, T., & Milyani, A. H. (2023). A hybrid framework for short term load forecasting with a navel feature engineering and adaptive grasshopper optimization in smart grid. *Applied Energy, 338*, 120829. doi:10.1016/j.apenergy.2023.120829.

Resilient microgrid operation under uncertainty

5

Mostafa Vahedipour-Dahraei[a], *Homa Rashidizadeh-Kermani*[b] *and Miadreza Shafie-khah*[b]
[a]Department of Electrical & Computer Engineering, University of Birjand, Birjand, South Khorasan, Iran, [b]Research and Innovation Division, Nowocert, Dublin, Ireland

5.1 Introduction

The rapid evolution of modern power systems, primarily driven by the increasing penetration of renewable energy sources (RESs), has introduced new layers of complexity and uncertainty into distribution networks (Macmillan et al., 2024; Tabares & Cortés, 2024). Recent studies have highlighted how RES-based systems significantly influence both operational and planning challenges, emphasizing the need for advanced stochastic modeling approaches (Tabares & Cortés, 2024). Renewable-based microgrids (MGs), as decentralized and flexible energy management entities, have become central to enhancing resilience, integrating distributed generation (DG), and improving the overall stability of power systems (Du et al., 2024; Macmillan et al., 2024). Long-term trends—such as climate-induced variations in load and renewable resource availability—have further reinforced the importance of MGs in both short-term operational strategies and long-term planning frameworks (Macmillan et al., 2024).

However, the intermittent nature of renewable generation—especially wind and photovoltaic output—combined with volatile load profiles, real-time market price fluctuations, and unforeseen disturbances, poses substantial challenges to the reliability and economic efficiency of MG operations (Wang et al., 2025). These inherent uncertainties render traditional deterministic scheduling approaches insufficient for meeting real-world operational requirements (Wang et al., 2025). Consequently, there is a heightened need for advanced stochastic decision-making frameworks capable of capturing inherent uncertainties while ensuring system resilience under disruptive events (Du et al., 2024; Lin et al., 2024; Wang et al., 2025). Recent research has proposed multiobjective robust dispatch strategies, stochastic optimization schemes, and uncertainty-aware reinforcement learning algorithms to enhance both the economic performance and operational robustness of MGs under uncertain environments (Lin et al., 2024; Vahedipour-Dahraie et al., 2017).

Resilience has become a critical operational objective in MG scheduling, particularly given the growing frequency of extreme weather events, cyber intrusions, and unexpected contingencies that may isolate MGs from the main grid (Liu et al., 2023; Vahedipour-Dahraie et al., 2022). Resilient MGs must not only maintain the continuity of critical loads during such disturbances but also adapt their operational decisions

Stochastic Planning and Modeling for Energy Systems: Methods, Applications, and Developments.
DOI: https://doi.org/10.1016/B978-0-443-45298-7.00014-8

to minimize economic losses. This requires the integration of advanced scheduling strategies that consider both probabilistic uncertainties and operational risks over multitime horizons (Nakiganda & Aristidou, 2023). The capability of MGs to operate in islanded mode and prioritize essential loads during contingencies underscores the importance of resilience-oriented operational planning (Lari, 2025).

In parallel, demand response (DR) programs have gained substantial attention as an effective mechanism for enhancing system flexibility, reducing peak demands, and improving demand-side participation in energy markets (Babu et al., 2025; Du et al., 2024). DR enables consumers to adjust their energy consumption in response to price signals or operational instructions, thus contributing to real-time balancing and stability enhancement. When integrated into MG scheduling, DR becomes a powerful tool to mitigate the adverse effects of renewable intermittency, reduce reliance on expensive reserves, and enhance operational resilience (Tahere Daemi et al., 2023; Vahedipour-Dahraie et al., 2021). Flexible and consumer-centric DR mechanisms, including interruptible/curtailable (I/C) programs and capacity-market schemes, allow customers to actively participate in both normal and emergency conditions, incentivizing load adjustments to maintain system stability (Guo et al., 2025).

The interaction between uncertainty, risk, and DR participation within resilient MGs is highly complex and requires sophisticated modeling (Kumar et al., 2023; Zhang et al., 2024). MG operators must continuously balance expected operational costs against risk exposure, particularly under high renewable penetration and uncertain contingencies. Traditional deterministic or single-scenario approaches are insufficient to capture these complexities (Chen et al., 2025). Instead, stochastic optimization, especially multistage and scenario-based methods, has become a key paradigm for modeling uncertain renewable generation, load deviations, and market fluctuations (Chen et al., 2025). Nevertheless, stochastic optimization alone does not guarantee adequate risk control. Therefore, MG operators increasingly adopt risk measures such as conditional value-at-risk (CVaR) to limit exposure to low-probability, high-impact events and ensure acceptable operational safety margins (Chen et al., 2025; Kumar et al., 2023; Zhao et al., 2023).

This chapter presents a comprehensive risk-based stochastic scheduling framework for resilient MGs. The primary objective of the chapter is to introduce a methodology that enables MG operators to make optimal and disruption-resilient operational decisions, while explicitly considering uncertainties arising from renewable generation, load fluctuations, and market price variations. In addition, the proposed framework simultaneously addresses system resilience and DR flexibility, allowing the MG to maintain stable operation even under emergency and islanded conditions. The proposed methodology consists of several key components: Modeling uncertainties using multistage scenario-based approaches, employing a CVaR-based risk measure to limit exposure to low-probability, high-impact events, realistically integrating DR programs to enable participation of responsive loads, ensuring the supply of critical loads in islanded mode, and analyzing the dynamic interactions among responsive loads, ESS, and renewable resources. Furthermore, the chapter includes numerical studies and simulations based on real-world data, demonstrating how the proposed framework can reduce operational costs, limit risk exposure, and enhance overall system resilience.

The results clearly highlight the advantages of the proposed methodology over deterministic or nonrisk-aware approaches.

5.2 Resilience concepts and operational requirements in microgrids

In the context of modern power systems, resilience refers to the ability of a system to anticipate, absorb, adapt to, and rapidly recover from disruptive events, ensuring the continuous delivery of critical services (He & Lin, 2025; Zidane et al., 2025). Within MGs, resilience has emerged as a core operational objective, particularly given the increasing frequency of extreme weather events, cyber threats, equipment failures, and unforeseen contingencies that may isolate MGs from the main grid (He & Lin, 2025). Importantly, the resilience of an MG is not solely determined by the reliability of its components; it also depends on the flexibility, adaptability, and reconfigurability of its resources under adverse conditions (Zahraoui et al., 2024).

A resilient MG must be capable of operating both in grid-connected and islanded modes. Under normal conditions, an MG can rely on the upstream network to balance supply and demand. However, when disturbances occur—such as transmission line faults, natural disasters, or significant mismatches between generation and load—the MG must automatically switch to islanded operation and continue supplying critical loads without interruption (Shahravi et al., 2025). Achieving this requires real-time monitoring, rapid reconfiguration, and intelligent control of distributed energy resources (DERs), including dispatchable generators, ESS, and demand-side resources (Khosravi et al., 2025).

Key features that enhance MG resilience include:

- Load prioritization and critical load assurance: MG operators must guarantee the supply of critical loads, such as hospitals, emergency facilities, and essential infrastructure, even during extended islanding periods. Noncritical loads can be selectively curtailed to maintain overall system stability (Liu et al., 2023; Yang et al., 2024).
- Islanding capability and seamless transition: The ability to automatically detect upstream disturbances and switch to islanded operation is vital for resilience. Islanding introduces operational challenges such as power imbalances, frequency deviations, and voltage instability, which must be addressed through robust control strategies and optimized operational planning (Moreno et al. 2020).
- Integration of flexible resources: DERs, responsive loads, and energy storage systems (ESS) play a pivotal role in supporting resilience. Coordinated operation of these resources ensures adequate spinning reserves, load-following capability, and rapid response during disturbances. In particular, DR programs allow flexible loads to reduce or shift consumption, effectively acting as virtual reserves to absorb shocks and maintain system balance (Li et al., 2023).
- Adaptability under uncertainty: RESs such as wind and solar are inherently variable and stochastic. Resilient MGs must employ uncertainty modeling and stochastic optimization techniques to anticipate fluctuations in generation and demand. This ensures that operational decisions are not only economically efficient but also robust against rare and severe scenarios (Elazab et al., 2024).

- Multitime-horizon operational planning: Achieving resilience requires planning across multiple time scales—from minute-level control during islanded events to daily energy and reserve scheduling. This allows MGs to preposition resources, strategically manage energy storage, and coordinate DR actions in anticipation of potential disruptions (Moreno et al. 2020).

Integrating these resilience-enhancing features enables MGs to absorb shocks, limit service interruptions, and rapidly recover after disturbances. Moreover, resilient MGs provide a foundation for risk-aware operational strategies, in which uncertainties and contingencies are explicitly incorporated into the scheduling framework (Wu et al., 2024). This is especially critical when MGs interact with real-time markets or participate in energy trading, as it ensures both economic efficiency and operational reliability under dynamic and uncertain conditions (Yang et al., 2024).

Recent studies emphasize that resilience is a dynamic attribute, dependent on both system design and operational strategies. For example, a three-stage dispatch and reconfiguration strategy has been proposed to enhance resilience under islanding events by optimal scheduling and reconfiguration of resources (Wu et al., 2024). Models that quantify islanding capability probabilistically and embed it into scheduling frameworks further strengthen the MG's ability to operate under uncertainty and supply critical loads. The integration of uncertain renewables, storage, and DR in probabilistic scheduling has also shown significant improvements in system stability and robustness under extreme conditions (Lari et al., 2025; Shahravi et al., 2025).

In operational terms, resilience-oriented MG scheduling aims to maximize the expected profit of the MG operator while ensuring reliable supply through coordinated management of both demand- and supply-side resources. A typical MG may include dispatchable DG units and RESs supplying a local customer base, while responsive loads under DR programs provide additional flexibility. Under normal operation, the MG is grid-connected; upon detection of upstream disturbances, it transitions to islanded mode, ensuring uninterrupted supply of critical loads—with reconnection once disturbances clear (Liu et al., 2023).

Given the stochastic nature of disturbances and renewable generation, scheduling must anticipate uncertainty across a multistage horizon. By combining flexible resources (storage, DR, DERs) with probabilistic scheduling and resilience-oriented constraints, the MG can maintain performance and service continuity even in adverse conditions (Lari et al., 2025; Shahravi et al., 2025).

Responsive loads can also provide spinning reserve, enhancing system reliability while reducing energy costs, as freed-up generation can be allocated to other loads. Effective participation requires responsive loads with storage and control capabilities, along with real-time monitoring and communication to ensure rapid and accurate response (Li et al., 2023).

Furthermore, the MG operator clears day-ahead, real-time, and spinning reserve markets using an optimal power flow (OPF) model that considers network constraints, determines energy credits, and evaluates voluntary/involuntary load reductions while calculating reliability and economic indices. This integrated framework ensures that operational decisions maximize both MG operator profit and customer benefits while maintaining resilience under uncertainty (Wu et al., 2024).

5.3 Demand response programs and their operational role in microgrids

DR has emerged as a pivotal strategy in modern MG operation, enabling system operators to leverage the flexibility of end-users to improve reliability, economic efficiency, and resilience. In essence, DR refers to the ability of consumers to adjust their electricity consumption in response to price signals, incentive programs, or operational commands, thereby assisting the MG in balancing supply and demand under both normal and contingency conditions (Du et al., 2024).

5.3.1 *Classification of demand response programs*

DR programs can be broadly classified into three main categories:

- Price-based DR: These programs incentivize consumers to modify their electricity usage according to dynamic price signals. Examples include time-of-use (TOU) tariffs, critical peak pricing (CPP), and real-time pricing (RTP) schemes. By responding to price fluctuations, consumers can shift loads from high-price periods to lower-cost intervals, thereby flattening demand peaks and reducing overall operational costs (Zeng et al., 2023).
- Incentive-based DR (IBDR): In this category, consumers receive direct incentives for reducing or curtailing their load during specific events. Programs such as I/C loads and capacity market (CAP) participation allow operators to temporarily shed load to relieve system stress, especially during contingencies or periods of high energy prices. Noncompliance often results in penalties, ensuring reliability of DR participation (Shahravi et al., 2025; Liu et al., 2024; Zeng et al., 2023).

DR programs can be utilized across all time horizons associated with power system management. Each program, depending on customer participation mechanisms, pricing policies, and response timing, can be implemented in various electricity markets. For example, DR programs that require day-ahead notification must be considered in the day-ahead market of restructured power systems and within the generation scheduling process of traditional systems. Similar to generating units that deliver their scheduled output in real time, these responsive loads react to network signals during real-time operation. Therefore, different DR strategies—whether incentive-based or price-based—can be executed over different time horizons depending on specific network conditions. The time required to activate each DR program is illustrated in Fig. 5.1 (Wu et al., 2024).

In price-based DR programs, if the system operator has an accurate understanding of load behavioral patterns in response to price variations, these programs can be applied at multiple levels of power system planning. For instance, a direct load control program can be activated in less than one minute after a control signal is issued. Many household devices possess this capability. A typical example is electric water heaters with storage tanks, which account for approximately 11% of total residential electricity consumption on average, and their share can exceed 31% during peak demand periods (Elazab et al., 2024). Therefore, these devices and similar equipment offer significant potential for participation in DR programs requiring response times of less than 15 minutes.

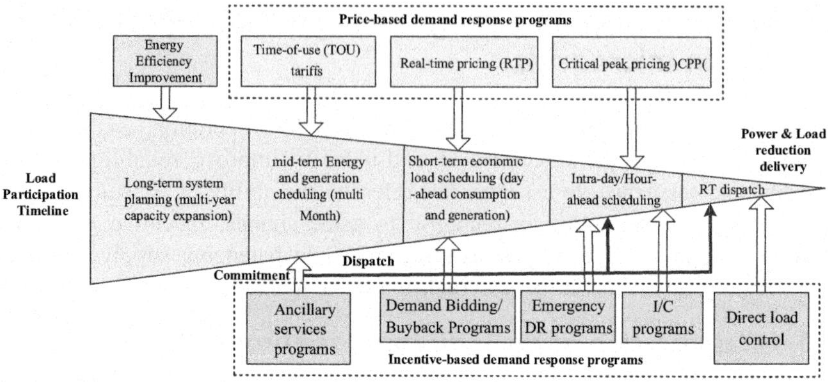

Figure 5.1 Time horizons of various demand response program types.

5.3.2 Role of demand response in risk mitigation and resilience enhancement

DR plays a central role in reducing operational risks and enhancing MG resilience. By enabling flexible load adjustment, DR provides a virtual reserve that can respond rapidly to unforeseen disturbances, such as sudden renewable generation shortfalls or islanding events. In particular:

- DR reduces exposure to high-cost or high-risk operating scenarios by curtailing consumption during periods of scarcity or market volatility. The integration of DR with risk-aware scheduling models, such as CVaR-based frameworks, allows MG operators to explicitly control the probability of extreme economic losses (Li et al., 2023).
- During islanding or other contingencies, DR ensures critical load supply by strategically shedding nonessential loads. When combined with DERs and ESS, DR enables MGs to maintain operational stability even under severe disturbances, contributing to the overall adaptive and flexible operation of the system (Liu et al., 2023; Zeng et al., 2023).

5.3.3 Interaction of demand response with energy storage and renewable generation

The synergistic operation of DR with ESS and RESs significantly amplifies MG performance:

- DR allows consumption patterns to be aligned with the intermittent nature of RESs. Flexible loads can be increased during periods of high solar or wind generation and reduced when renewable output declines, smoothing net demand profiles (Liu et al., 2024).
- Responsive loads participating in DR can act as virtual spinning reserves, complementing ESS and dispatchable DERs. This reduces reliance on conventional generators while ensuring rapid response to contingencies (Elazab et al., 2024; Shahravi et al., 2025).
- By integrating DR, ESS, and RESs in a unified scheduling framework, MG operators can minimize operational costs, reduce peak demand charges, and improve market participation,

while simultaneously maintaining high reliability and resilience (Chen et al., 2024; Lari et al., 2025).

In practice, MG operators implement multistage stochastic optimization models where DR, ESS, and RES outputs are jointly optimized. This ensures that the combined flexibility of supply and demand is fully leveraged to withstand uncertainties in load, generation, and market conditions (Danner et al., 2024; Yang et al., 2024). Notably, higher levels of DR participation have been shown to significantly reduce system risk exposure, improve economic outcomes, and strengthen the resilience of MG operations (Sun et al., 2025; Zhang et al., 2024).

5.4 Uncertainty modeling and stochastic scheduling in microgrid operation

Uncertainty modeling constitutes a foundational component for modern MG operation, enabling optimal, resilient, and risk-aware decision making. MGs naturally face multiple sources of variability—including intermittent renewable energy generation, stochastic load demand, and volatile electricity market prices. Ignoring these uncertainties can lead to suboptimal operational schedules, increased risk exposure, and compromised system resilience (Macmillan et al., 2024).

5.4.1 Sources of uncertainty

The primary categories of uncertainty in MG operations are:

1. Renewable generation uncertainty; wind turbines (WTs) and photovoltaic units exhibit inherent variability due to stochastic wind speed and solar irradiance, leading to unpredictable changes in available power that must be accurately captured to ensure reliability (Macmillan et al., 2024).
2. Load uncertainty: Demand varies with consumer behavior, environmental conditions, and special events. Participation of flexible or responsive loads (e.g., under DR programs) adds further randomness because load reduction or shifting often depends on price signals or control directives (Hu et al., 2024).
3. Market price uncertainty; electricity prices in day-ahead, real-time, and reserve markets are subject to rapid fluctuations, affecting energy trading, scheduling, and reserve allocation decisions (Liu et al., 2025).
4. Islanding and contingency uncertainty: The occurrence, timing, and duration of disturbances that force an MG into islanded mode is stochastic. These uncertainties critically impact resilience planning and operational strategy design (Vahedipour-Dahraie et al., 2020; Zeng et al., 2023).

5.4.2 Operational modes and scheduling horizons

MG operations are typically managed across two modes:

- Normal (grid-connected) mode: The MG is connected to the upstream network. The energy management system (EMS) optimizes scheduling of local DERs and energy exchange with the main grid, considering potential future islanding events.

- Resilient (islanded) mode: Under severe upstream disturbances, the MG operates independently. In this mode, EMS schedules local resources to supply critical and noncritical loads, minimizing load shedding if needed.

Accordingly, uncertainties are categorized as:

- Normal-operation uncertainties: Variability in renewable generation, load demand, and electricity prices.
- Contingency-based uncertainties: Stochastic characteristics related to islanding events (occurrence probability, duration).

5.4.3 Scenario-based and probabilistic modeling

To properly deal with uncertain parameters, probabilistic, scenario-based modeling approaches are widely used:

- Definition of probability distributions (PDFs): Renewable output, load demand, and market prices are modeled using appropriate distribution functions. When historical data is sufficient, probabilistic characterization of islanding durations or contingency events may also be considered (Macmillan et al., 2024).
- Monte Carlo simulation (MCS): Random samples are drawn from PDFs to generate a large number of operational scenarios. MCS helps in assessing reliability, risk, and expected performance under uncertainty (Hu et al., 2024).
- Scenario reduction: To make optimization tractable, algorithms such as clustering or reduction based on statistical similarity are used to select a representative subset of scenarios that still captures the variability space adequately (Vahedipour-Dahraie et al., 2020; Zeng et al., 2023).
- Joint probability calculation: When multiple uncertain parameters (e.g., renewable output, load, price, islanding events) are considered, the joint probability of combined scenarios is computed—allowing stochastic or robust optimization models to evaluate worst-case or expected outcomes.

Such probabilistic and multistage modeling strategies, together with advanced optimization (e.g., stochastic programming, robust optimization, model predictive control), allow MGs to operate efficiently, maintain resilience, and manage risk proactively (Vahedipour-Dahraie et al., 2020).

5.4.4 Stochastic optimization for resilient microgrid scheduling

A multistage stochastic optimization framework is implemented to determine the **optimal scheduling of MG resources under uncertainty**. Key aspects include:

- **Normal operation:** Unit commitment, energy trading with the main grid, and reserve allocation are optimized considering responsive loads and their contribution to spinning reserve.
- **Resilient operation:** During islanded mode, prescheduled reserves and ESS capacity are utilized to minimize load curtailment while guaranteeing supply to critical loads.
- **Economic modeling of DR participation:** Customers participate in DR programs through load curtailment and load shifting, incentivized via dynamic pricing or bill credits. Mandatory load shedding is applied to noncritical loads when necessary to ensure feasibility.

This integrated approach ensures that MG operations are not only economically optimized but also resilient to both predictable and unforeseen disruptions. By modeling uncertainties probabilistically and combining stochastic scheduling with DR and ESS coordination, the MG can maintain critical load supply, reduce operational risk, and enhance overall system reliability in both grid-connected and islanded modes.

5.5 Risk-based stochastic scheduling

Effective operation of modern MGs requires not only consideration of uncertainty but also explicit management of **risk**. In this context, **risk** refers to the potential for low-probability, high-impact outcomes that may compromise the reliability, resilience, or economic performance of the MG (Liu et al., 2025). While stochastic models can capture variability in generation, load, and market prices, they do not inherently provide mechanisms for controlling exposure to extreme events. This is where **risk-based optimization** becomes essential.

Risk in MG scheduling is typically quantified using advanced metrics that go beyond simple expected cost or profit. One of the most widely adopted approaches is CVaR, which evaluates the expected losses in the tail of the distribution beyond a specified probability level, enabling operators to explicitly manage exposure to extreme adverse events (Hu et al., 2024; Liu et al., 2025). Formally, for a confidence level α, CVaR represents the expected cost in the worst $(1 - \alpha)$ fraction of scenarios. Incorporating CVaR into the MG scheduling objective allows operators to balance expected economic performance with protection against high-risk outcomes, providing a robust and risk-aware operational strategy (Zeng et al., 2024).

CVaR-based frameworks are particularly effective when combined with stochastic, scenario-based optimization, as they account for uncertainties in renewable generation, load variations, market prices, and contingency events. This approach ensures that MG operations remain resilient, economically efficient, and capable of handling extreme disruptions without excessive reliance on conservative reserve allocations. Other risk metrics include **VaR, probabilistic reliability indices**, and scenario-based penalty functions for critical load violations. Integrating these risk measures into MG scheduling ensures that operational strategies are robust against both frequent fluctuations and rare but severe contingencies.

5.5.1 Multistage stochastic optimization framework

MG scheduling under risk and uncertainty is naturally modeled as a **multistage stochastic optimization problem**, which considers decisions over multiple time horizons and scenarios:

Stage 1: Day-ahea**d planning:** In this stage, the MG operator schedules energy generation, storage dispatch, and DR commitments based on forecasted renewable generation, load profiles, and market prices. Probabilistic scenarios generated from MCSs or historical data represent potential realizations of uncertainties.

Stage 2: Real-tim**e adjustment:** As real-time measurements and market signals are observed, adjustments to generation, DR participation, and storage dispatch are made to account for deviations from forecasts. This stage ensures that the MG maintains balance and resilience under evolving conditions.

Stage 3: Con**tingency management:** In case of unexpected islanding events or other disruptions, the operator implements resilient strategies to minimize load curtailment, guarantee supply to critical loads, and optimally utilize storage and responsive loads. The probabilistic nature of islanding duration and occurrence is explicitly considered in scenario modeling.

This multistage approach allows MG operators to anticipate uncertainties, prepare flexible responses, and maintain service continuity while controlling operational risks. Ensuring the resilience of MGs requires a holistic approach that integrates risk management, DR, and coordinated resource operation to maintain critical load supply under both normal and islanded operational modes. In this context, resilience refers to the MG's capacity to anticipate, absorb, adapt to, and rapidly recover from disruptive events—thereby minimizing service interruptions and economic losses.

5.6 Formulation of the optimization problem

This section presents the mathematical formulation of the proposed stochastic optimization model for MG scheduling under uncertainty. The formulation incorporates normal grid-connected operation, resilient islanded operation, DR participation, and probabilistic representation of uncertainties. All indices, sets, parameters, and decision variables used throughout the model are defined below.

Nomenclature

	Indices and sets
$(.)._{, t, s}$	Variable at time t in scenario s.
$(.)^{\min}$, $(.)^{\max}$	Minimum and maximum limits of a variable.
t, N_T	Time index and number of scheduling periods.
h,N_H	Index and number of islanding-duration scenarios.
τ	Timeslot index in island mode scheduling problems.
s, N_S	Index and number of normal operation scenarios.
i, N_G	Index and number of dispatchable DG units.
w, N_W	Index and number of wind turbine units.
b, n, r	Indices of system buses.
Parameters and constants	
$P_{j,t}$	Demand of customer group j at time t (kW).
$\Pr_{j,t}$	Selling price of electricity to customer group j (\$/kWh).
$\Pr_{m,t}^{buy,(sell)}$	Market price for buying/selling electricity from/to main grid (\$/kWh).
Β	Risk-aversion parameter.
α	Per unit confidence level.
$\lambda_{i,t}^{R,up}$ $(\lambda_{i,t}^{R,dn})$	Bid price of up/down spinning reserve from DG unit i (\$/kWh).

(continued on next page)

$\lambda_{j,t}^{R,up}$ ($\lambda_{j,t}^{R,dn}$)	Bid price of up/down spinning reserve from loads *j* (\$/kWh).
$\lambda_{m,t}^{R,up}$ ($\lambda_{m,t}^{R,dn}$)	Up/down regulation market price at time *t* (\$/kWh).
$\lambda_{i,t}^{R,non}$	Nonspinning reserve bid from DG unit *i* (\$/kWh).
π_s (ϕ_h)	Occurrence probability of normal-operation scenarios (or islanding-duration scenario h).
CU_i (CD_i)	Start-up/shut-down cost constants for DG unit *i* (\$).
RU_i, (RD_i)	Ramp-up and ramp-down limits of DG unit *i*.
UTi, (DTi)	Minimum up-time / down-time for DG unit *i*.
G^l, (B^l)	Conductance and susceptance of line l.
VOLL	Value of lost load.
EENS	Expected energy not served.
Variables	
P (Q)	Active (reactive) power (kW).
$P_{m,t}$	Power exchange between the MG and the main grid (kW).
$P_{m,t}^{buy,(sell)}$	Active power bought (sold) from (to) the main grid at time *t* (kW).
$fl_{(n,r)}^{P(Q)}$	Active (reactive) power flowing between bus *n* and *r*.
SUC_i, (SDC_i)	Start-up (Shut-down) cost variables of DG unit *i* (\$).
$R_{i,t}^{up}$ ($R_{j,t}^{up}$)	Up-spinning reserve deployed by DG unit *i* (customers in group *j*).
$R_{i,t}^{dn}$ ($R_{j,t}^{dn}$)	Down-spinning reserve deployed by DG unit *i* (customers in group *j*).
$R_{m,t}^{up}$ ($R_{m,t}^{dn}$)	Up (down)-spinning reserve deployed by main grid (kW).
$R_{i,t}^{non}$	Nonspinning reserve deployed by DG unit *i*.
$E_{t,h}^{j}$	Cross elasticity of period *t* to period *h*.
$P_{j,t}^{shed}$, ($Q_{j,t}^{shed}$)	Active and reactive power of emergency load shedding (kW).
$u_{i,t,s}$	Commitment status of DG unit *i*, {0, 1}.
$y_{i,t,s}$ ($z_{i,t,s}$)	Start-up (shut-down) indicator of DG *i*, {0, 1}.
$\upsilon_{t,s}$	Auxiliary variables, {0, 1}.

5.6.1 Objective function

The primary goal of the proposed stochastic scheduling framework is to maximize the MG's expected profit across the scheduling horizon under both normal operating conditions and islanded operation. As previously discussed, the scheduling procedure is updated multiple times throughout the study period to accommodate evolving system conditions and uncertainties.

Let the probability of islanding during hour *t* be represented byϕ_h. Accordingly, the probability that the MG remains in normal operating mode is ($1 - \phi_h$). Based on these probabilities, the expected profit is formulated by combining the objective functions associated with the normal mode (OF_{NORMAL}) and the islanded mode (OF_{ISLAND}) across all possible scenarios representing the stochastic behavior of loads, market prices, renewable generation, and other sources of uncertainty. The resulting objective function for the overall scheduling problem can therefore be expressed as:

$$Max\{OF_{NORMAL} + OF_{ISLAND}\} \tag{5.1}$$

$$OF_{NORMAL} = \sum_{t=1}^{N_T}\sum_{h=1}^{N_H}(1-\phi_h)\sum_{s=1}^{N_S}\pi_s(F_1 - F_2 - F_3 - F_4) + \beta\left[\xi + \frac{1}{(1-\alpha)}\sum_{s=1}^{N_S}\pi_s\eta_s\right] \tag{5.2}$$

$$\begin{aligned} OF_{ISLAND} = &\sum_{t=1}^{N_T}\sum_{h=1}^{N_H}\phi_h\sum_{\tau=t}^{t+\tau-1}\sum_{s=1}^{N_S}\pi_s\left[\sum_{j=1}^{N_J}\left(\text{Pr}_{j,t,s}.P_{j,t,s}\right) - F_2 - F_4\right] \\ &+\beta\left[\xi + \tfrac{1}{(1-\alpha)}\sum_{h=1}^{N_H}\sum_{s=1}^{N_S}\pi_s\phi_h\eta_{s,h}\right] \end{aligned} \tag{5.3}$$

where, functions F_1 to F_4 are defined as follows:

$$F_1 = \underset{m,t,s}{\overset{sell}{\text{Pr}}}.P^{sell}_{m,t,s} - \underset{m,t,s}{\overset{buy}{\text{Pr}}}.P^{buy}_{m,t,s} + \sum_{j=1}^{N_J}\text{Pr}_{j,t,s}.P_{j,t,s} \tag{5.4}$$

$$\begin{aligned} F_2 = &\sum_{i=1}^{N_G}\left[C_i(P_{i,t,s}) + SUC_{i,t,s} + SDC_{i,t,s}\right] \\ &+\left(\lambda^{R,up}_{i,t}R^{up}_{i,t} + \lambda^{R,non}_{i,t}R^{non}_{i,t} + \lambda^{R,dn}_{i,t}R^{dn}_{i,t}\right) \end{aligned} \tag{5.5}$$

$$F_3 = \lambda^{R,up}_{m,t}R^{up}_{m,t} + \lambda^{R,dn}_{m,t}R^{dn}_{m,t} \tag{5.6}$$

$$F_4 = \sum_{j=1}^{N_J}\left[\left(\lambda^{R,dn}_{j,t}R^{dn}_{j,t} + \lambda^{R,up}_{j,t}R^{up}_{j,t}\right) + P^{shed}_{j,t,s}VOLL_{j,t}\right] \tag{5.7}$$

Function F_1 represents the total revenue obtained from energy transactions with the main grid as well as the income generated from selling electricity to local customers. Function F_2 denotes the operating costs of dispatchable units, including both their energy and reserve provision costs, along with startup and shutdown expenses. Function F_3 captures the cost associated with the reserve capacity supplied to the main grid, whereas F_4 accounts for the cost of reserve provided by responsive loads and the penalty costs related to emergency load shedding.

The second terms in Eqs. (5.2) and (5.3) correspond to the CVaR component of a candidate solution. The risk-aversion parameterβ regulates the tradeoff between maximizing expected profit and reducing the variability of profit under uncertain operating conditions. Additionally, the auxiliary variable ξ is introduced to compute the VaR, while η_s ($\eta_{s,h}$) represents the deviation between the MG's operating cost in scenario s and ξ the VaR level.

It is also important to note that the operating cost of RESs is assumed to be negligible and is therefore excluded from the cost formulation in this study.

5.6.2 Problem constraints

The optimization problem described in Section 5.7.1 is subject to a set of operational, physical, and technical constraints that ensure feasible and reliable MG operation under both normal and islanded modes. These constraints include DR participation, power balance, unit commitment, reserve allocation, DR participation, and network limits.

5.6.2.1 Demand response modeling and constraints

The participation of responsive loads in DR programs is modeled based on the incentives and penalties applied to customers. When customer j participates in a DR program, its hourly demand is adjusted from the initial value $D_{j,t}^{\text{int}}$ to a modified level $D_{j,t}^{DR}$ as follows:

$$D_{j,t}^{DR} = D_{j,t}^{\text{int}} + \Delta D_{j,t} \tag{5.8}$$

In this case, the revenue of customer j participating in the DR program is calculated based on the hourly incentive rate as follows:

$$INC(\Delta D_{j,t}) = \underset{j,t}{\overset{inc}{\Pr}} \times \left(D_{j,t}^{\text{int}} - D_{j,t}^{DR}\right) \tag{5.9}$$

Similarly, the penalty payments for customer j, who fails to respond or comply with the predefined DR contract, are calculated as follows:

$$PEN(\Delta D_{j,t}) = \underset{j,t}{\overset{pen}{\Pr}} \times \left[L_{j,t}^{c} - \left(D_{j,t}^{\text{int}} - D_{j,t}^{DR}\right)\right] \tag{5.10}$$

Here, $L_{j,t}^{c}$ denotes the predetermined contractual consumption level of customer j at time period t. As previously discussed, customer j adjusts its demand profile to maximize its net profit, defined as the difference between the monetary benefits derived from electricity consumption and the associated costs. The corresponding profit function is formulated as follows:

$$PRF_{j,t}^{c} = BEN(\Delta D_{j,t}) - \underset{j,t}{\Pr}\Delta D_{j,t} + INC(\Delta D_{j,t}) - PEN(\Delta D_{j,t}) \tag{5.11}$$

To maximize the profit of customer j, the derivative of Eq. (5.11) with respect to the decision variable must be set to zero.

$$\frac{\partial BEN(\Delta D_{j,t})}{\partial D_{j,t}} - \underset{j,t}{\Pr} + \frac{INC(\Delta D_{j,t})}{\partial D_{j,t}} - \frac{PEN(\Delta D_{j,t})}{\partial D_{j,t}} = 0 \tag{5.12}$$

$$\frac{\partial B(\Delta D_{j,t})}{\partial D_{j,t}} = \underset{j,t}{\Pr} + \underset{j,t}{\overset{inc}{\Pr}} + \underset{j,t}{\overset{pen}{\Pr}} \tag{5.13}$$

Therefore, the total electricity rate in any period t, denoted by (ρ_t), is given by $\mathrm{Pr}_{j,t}+\mathrm{Pr}_{j,t}^{inc}+\mathrm{Pr}_{j,t}^{pen}$. By applying the concepts of self- and cross-elasticity, the following relationship can be expressed:

$$\frac{\partial D_{j,t}}{D_{j,t}} = E_{j,t,h}\frac{\partial \mathrm{Pr}_{j,h} + \partial\, \mathrm{Pr}_{j,h}^{inc} + \partial\, \mathrm{Pr}_{j,h}^{pen}}{\mathrm{Pr}_{j,h} + \mathrm{Pr}_{j,h}^{inc} + \mathrm{Pr}_{j,h}^{pen}} \tag{5.14}$$

By integrating Eq. (5.14) over the entire scheduling horizon, the following expressions are obtained:

$$\int_{D_{j,t}^{\mathrm{int}}}^{D_{j,t}} \frac{\partial D_{j,t}}{D_{j,t}} = \sum_{h=1}^{N_T} E_{j,t,h}\left[\int_{\mathrm{Pr}_{j,t}^{\mathrm{int}}}^{\mathrm{Pr}_{j,t}} \frac{\partial \mathrm{Pr}_{j,h}}{\mathrm{Pr}_{j,h} + \mathrm{Pr}_{j,h}^{inc} + \mathrm{Pr}_{j,h}^{pen}} + \int_{0}^{\mathrm{Pr}_{j,t}^{inc}} \frac{\partial\, \mathrm{Pr}_{j,h}^{inc}}{\mathrm{Pr}_{j,h} + \mathrm{Pr}_{j,h}^{inc} + \mathrm{Pr}_{j,h}^{pen}} + \int_{0}^{\mathrm{Pr}_{j,t}^{pen}} \frac{\partial\, \mathrm{Pr}_{j,h}^{pen}}{\mathrm{Pr}_{j,h} + \mathrm{Pr}_{j,h}^{inc} + \mathrm{Pr}_{j,h}^{pen}}\right] \tag{5.15}$$

$$\ln\left(\frac{D_{j,t}}{D_{j,t}^{\mathrm{int}}}\right) = \sum_{h=1}^{N_T} E_{j,t,h}\left[\ln\left(\frac{\mathrm{Pr}_{j,h} + \mathrm{Pr}_{j,h}^{inc} + \mathrm{Pr}_{j,h}^{pen}}{\mathrm{Pr}_{j,h}^{\mathrm{int}} + \mathrm{Pr}_{j,h}^{inc} + \mathrm{Pr}_{j,h}^{pen}}\right) + \ln\left(\frac{\mathrm{Pr}_{j,h} + \mathrm{Pr}_{j,h}^{inc} + \mathrm{Pr}_{j,h}^{pen}}{\mathrm{Pr}_{j,h} + \mathrm{Pr}_{j,h}^{inc}}\right) + \ln\left(\frac{\mathrm{Pr}_{j,h} + \mathrm{Pr}_{j,h}^{inc} + \mathrm{Pr}_{j,h}^{pen}}{\mathrm{Pr}_{j,h} + \mathrm{Pr}_{j,h}^{pen}}\right)\right] \tag{5.16}$$

By simplifying Eq. (5.16), the postparticipation demand of responsive loads in the IBDR programs is calculated as follows:

$$D_{j,t} = D_{j,t}^{\mathrm{int}} \times \left\{\prod_{h=1}^{N_T}\left[\frac{\left(\mathrm{Pr}_{j,h} + \mathrm{Pr}_{j,h}^{inc} + \mathrm{Pr}_{j,h}^{pen}\right)^3}{\left(\mathrm{Pr}_{j,h}^{\mathrm{int}} + \mathrm{Pr}_{j,h}^{inc} + \mathrm{Pr}_{j,h}^{pen} c\right)\left(\mathrm{Pr}_{j,h} + \mathrm{Pr}_{j,h}^{inc}\right)\left(\mathrm{Pr}_{j,h} + \mathrm{Pr}_{j,h}^{pen}\right)}\right]^{E_{j,t,h}}\right\} \tag{5.17}$$

$$0 \le P_{j,t,s}^{shed} \le P_{j,t}^{\max} \tag{5.18}$$

Additionally, by assuming a given power factor for load j ($\cos\theta_j$), the amount of curtailed reactive power can be calculated as follows:

$$Q_{j,t,s}^{shed} = P_{j,t,s}^{shed} \tan\left(\cos^{-1}\theta_j\right) \tag{5.19}$$

5.6.2.2 Network constraints

These equations describe the real-time operational behavior of the MG by representing the AC power flow for each scenario and each time interval. Specifically, they ensure that the electrical balance between supply and demand is maintained at every network node. Eqs. (5.20) and (5.21) formulate the active and reactive power balance at node n, respectively, capturing the relationship between generation, consumption, and network exchanges. These linearized expressions provide a tractable approximation of AC power flow, enabling the stochastic scheduling problem to be solved efficiently while preserving acceptable operational accuracy.

$$P^n_{i,t,s} + P^n_{w,t,s} - P^n_{j,t,s} + P^{n,shed}_{j,t,s} = \sum_{r=1}^{N_B} fl^P_{(n,r),t,s} \tag{5.20}$$

$$Q^n_{i,t,s} + Q^n_{w,t,s} - Q^n_{j,t,s} + Q^{n,shed}_{j,t,s} = \sum_{r=1}^{N_B} fl^Q_{(n,r),t,s} \tag{5.21}$$

By designating bus 1 as the slack bus, which is directly connected to the main grid, the $P_{m,t,s}$ and $Q_{m,t,s}$ should be included on the left-hand side of Eqs. (5.20) and (5.21), respectively. Furthermore, $fl^P_{(n,r),t,s}$ and $fl^Q_{(n,r),t,s}$ represent the active and reactive power flows between buses n and r at time t under scenario s. In this study, the linearized form of (5.22) and (5.23) is used with the following assumptions (Vahedipour-Dahraie et al., 2017; Vahedipour-Dahraie et al., 2022): (1) over a typical range of voltage amplitude ($0.95pu \leq |V| \leq 1.05pu$), it can be assumed $(|V_{n,t,s}| - |V_{r,t,s}|)^2 \approx 0$, and (2) over a typical range of difference in voltage phase angle across branch n and r, i.e.,$|\delta_{n,t,s} - \delta_{r,t,s}| \leq 10°$, it is assumed that $\sin(\delta_{n,t,s} - \delta_{r,t,s}) \approx \delta_{n,t,s} - \delta_{r,t,s}$ and $\cos(\delta_{n,t,s} - \delta_{r,t,s}) \approx 1$.

$$lf^P_{(n,r),t,s} = G_{n,r}(V_{n,t,s} + V_{n,t,s} - 1) + B_{n,r}(\delta_{n,t,s} - \delta_{r,t,s}) \tag{5.22}$$

$$lf^Q_{(n,r),t,s} = -B_{n,r}(V_{n,t,s} + V_{n,t,s} - 1) + G_{n,r}(\delta_{n,t,s} - \delta_{r,t,s}) \tag{5.23}$$

To ensure network congestion constraints are respected, the active and reactive power flows on each line must be limited as follows:

$$-lf^{P,\max}_{(n,r),t,s} \leq lf^P_{(n,r),t,s} \leq lf^{P,\max}_{(n,r),t,s} \tag{5.24}$$

$$-lf^{Q,\max}_{(n,r),t,s} \leq lf^Q_{(n,r),t,s} \leq lf^{Q,\max}_{(n,r),t,s} \tag{5.25}$$

Moreover, to guarantee safe operation with respect to permissible voltage magnitudes and phase angles, the following constraints must be satisfied:

$$\left|V^{\min}_{n,t}\right| \leq |V_{n,t,s}| \leq \left|V^{\max}_{n,t}\right|, \text{ and } -\pi \leq \delta_{n,t,s} \leq \pi \tag{5.26}$$

Additionally, the following constraints on exchange power capacity must be considered to govern the purchase and sale of electricity from and to the main grid at each time interval.

$$P_{m,t,s} = P^{buy}_{m,t,s} - P^{sell}_{m,t,s} \tag{5.27}$$

$$0 \leq P^{buy}_{m,t,s} \leq P^{\max,buy}_{m,t}.\upsilon_{t,s} \tag{5.28}$$

$$0 \leq P^{sell}_{m,t,s} \leq P^{\max,sell}_{m,t}(1 - \upsilon_{t,s}) \tag{5.29}$$

5.6.2.3 Dispatchable distributed generator constraints

These constraints encompass the power capacity limits of DGs (5.18), ramp-up and ramp-down limits as given in (5.31)–(5.32), startup and shutdown cost limitations in (5.33), as well as minimum up and down time requirements specified in (5.34)–(5.35).

$$P^{\min}_{i}.u_{i,t,s} \leq P_{i,t,s} \leq P^{\max}_{i}.u_{i,t,s} \tag{5.30}$$

$$P_{i,t,s} - P_{i,t-1,s} \leq RU_{i}.(1 - y_{i,t,s}) + P^{\min}_{i}.y_{i,t,s} \tag{5.31}$$

$$P_{i,t-1,s} - P_{i,t,s} \leq RD_{i}.(1 - z_{i,t,s}) + P^{\min}_{i}.z_{i,t,s} \tag{5.32}$$

$$SUC_{i,t,s} \leq CU_{i}.y_{i,t,s}; \quad SDC_{i,t,s} \leq CD_{i}.z_{i,t,s} \tag{5.33}$$

$$\sum_{h=t}^{t+UT_i-1} u_{i,t,s} \geq UT_{i}.y_{i,t,s} \tag{5.34}$$

$$\sum_{h=t}^{t+DT_i-1} (1 - u_{i,t,s}) \geq DT_{i}.z_{i,t,s} \tag{5.35}$$

5.6.2.4 Wind Power Constraints

The utilized wind power in each scenario and at every considered time interval is restricted by the maximum available wind power.

$$0 \leq P_{w,t,s} \leq P^{\max}_{w} \tag{5.36}$$

5.6.2.5 Reserve constraints

The reserve capacity provided by dispatchable units and responsive loads is bounded by the limits specified in constraints (5.37)–(5.41).

$$0 \leq R^{up}_{i,t} \leq P^{\max}_{i} u_{i,t} - P_{i,t} \tag{5.37}$$

Table 5.1 Technical data of dispatchable DG units.

DGs Type	P^{min} (kW)	P^{max} (kW)	Operation cost ($/kWh)	Start-up cost ($)	Shut-down cost ($)
MT_1	25	150	0.9	0.09	0.08
MT_2	25	150	1	0.09	0.08
FC_1	20	100	2.4	0.16	0.09
FC_2	20	100	2.6	0.16	0.09
DG	35	150	3.1	0.12	0.08

DG, distributed generation, *FC*, fuel cells; *MT*, microturbines.

$$0 \le R_{i,t}^{dn} \le P_{i,t} - P_i^{\min} u_{i,t} \tag{5.38}$$

$$0 \le R_{i,t}^{non} \le P_i^{\max}(1 - u_{i,t}) \tag{5.39}$$

$$0 \le R_{j,t}^{up} \le P_{j,t} - P_{j,t}^{\min} \tag{5.40}$$

$$0 \le R_{j,t}^{dn} \le P_{j,t}^{\max} - P_{j,t} \tag{5.41}$$

Moreover, it is assumed that the MG can provide reserve services to the main grid. The available reserve capacity is limited by:

$$0 \le R_{m,t}^{up} \le P_m^{\max} \upsilon_t - P_{m,t} \tag{5.42}$$

$$0 \le R_{m,t}^{dn} \le P_{m,t} - P_m^{\min} \upsilon_t \tag{5.43}$$

5.7 Case studies and numerical analysis

5.7.1 Test system and main assumptions

To validate the effectiveness of the proposed **risk-based stochastic scheduling framework for resilient MGs**, comprehensive case studies and numerical analyses are conducted. These studies aim to demonstrate the operational, economic, and resilience benefits of integrating **DR, ESS**, and **RESs** under uncertainty and risk considerations. The case studies are designed to highlight how the framework can optimize MG operations, enhance resilience during islanding events, and mitigate the impact of uncertainties in generation, load, and market prices.

The proposed methodology is implemented on a typical MG system, as illustrated in Fig. 5.2. The MG configuration comprises five dispatchable DG units, three WTs, and eight aggregated groups of responsive loads. The technical characteristics of the dispatchable DG units—specifically microturbines (MT), fuel cells (FC), and diesel engines (DE)—are detailed in Table 5.1 (Vahedipour-Dahraie et al., 2017). To account for system variability, the hourly forecasted profiles for the MG's base load, wind power generation, and wholesale electricity market prices are adopted from and, as

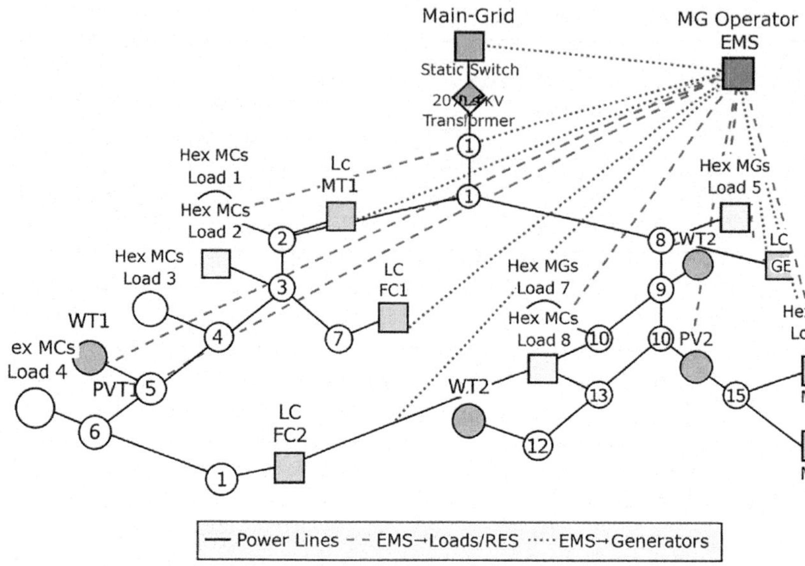

Figure 5.2 Structure of the studied MG. *MG*, Microgrid.

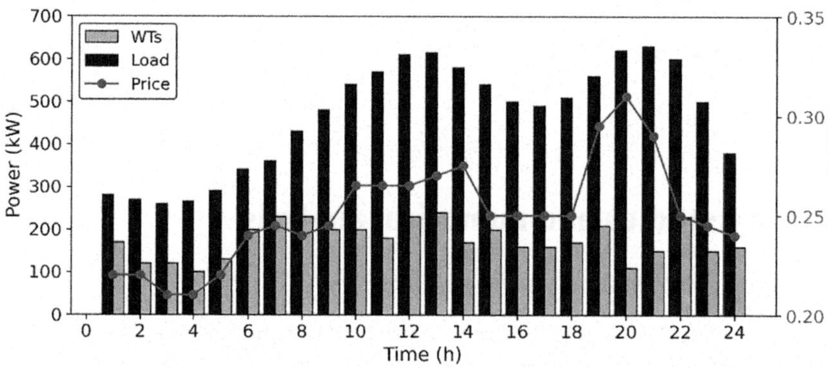

Figure 5.3 The hourly forecasted values of MG load, WTs output power, and electricity price. *MG*, Microgrid; *WTs*, wind turbine.

shown in Fig. 5.3 (Vahedipour-Dahraie et al., 2017; Vahedipour-Dahraie et al., 2022). Realistic uncertainty modeling is incorporated by assuming that the forecast errors for load demand, wind power output, and electricity prices follow normal probability distributions. The standard deviations (SDs) for these uncertainties are set to 8%, 5%, and 10% of their respective forecasted values (Vahedipour-Dahraie et al., 2022; Vahedipour-Dahraie et al., 2021). Furthermore, to capture the economic behavior of end-users participating in DR programs, the price elasticity of responsive loads is modeled based on data provided in (Vahedipour-Dahraie et al., 2021).

Table 5.2 Different islanding duration scenarios

Hours	9	10	11	12	13	14	15
ϕ_h	0.006	0.061	0.242	0.382	0.242	0.061	0.006

Islanding events are also incorporated into the case studies to evaluate the MG's resilience. The **islanding duration** is modeled as a stochastic variable following a normal distribution with a mean of **12 hours** and varying SDs to represent different levels of uncertainty. The **probabilities of different islanding durations** are summarized in Table 5.2 (Vahedipour-Dahraie et al., 2021). During islanded operation, the MG EMS schedules available local resources to supply critical and noncritical loads while minimizing load curtailment. By integrating DR and ESS, the MG can adjust consumption and energy storage to balance supply and demand, thus maintaining resilience under varying islanding durations.

The **stochastic scheduling framework** uses multistage scenario generation to capture the combined effect of uncertainties in load, renewable generation, market prices, and islanding events. MCS is applied to generate a large set of scenarios, which are then reduced using the K-means clustering algorithm to a manageable number of representative scenarios (Zeng et al., 2024). Each scenario represents a combination of normal operation uncertainties and contingency-based uncertainties, allowing the EMS to optimize resource allocation in both **grid-connected** and **islanded** modes. The probability of occurrence for each combined scenario is determined as the product of the probabilities of the normal operation scenario and the associated islanding duration.

Accordingly, the reduced scenario set is applied to the proposed optimization problem to maximize the expected profit of the MG operator. The required coding and optimization algorithm is carried out on a PC with 4 GB of RAM and an Intel Core i7 at 2.60 GHz processor with GAMS software and CPLEX solver, considering an optimality gap of 0.0. The computation time in different cases is less than two minutes, which further illustrates the practical merits of the proposed strategy.

5.7.2 Numerical results and discussion

To evaluate the performance and practical applicability of the proposed risk-based stochastic scheduling framework for resilient MGs, four distinct case studies are defined. Each case examines the impacts of DR integration, uncertainty management, and resilient operation on MG performance. The scheduling horizon is set to one day, divided into 24 hourly intervals. In all scenarios, the value of lost load (VOLL) is assumed to be 1 \$/kWh, and the confidence level for risk assessment is set at 0.95.

Case 1 represents the optimal scheduling of the MG under normal conditions without DR participation. In this scenario, the MG operator seeks to maximize expected profit while considering only normal operation uncertainties, including variability in load, renewable generation, and market prices.

Case 2 extends Case 1 by incorporating DR programs into the scheduling process. Here, responsive loads are allowed to adjust their consumption based on electricity prices or incentive signals, enabling the MG to better manage peak demands, reduce operational costs, and partially hedge against uncertainties.

Case 3 examines the resilient scheduling of the MG by considering credible islanding duration scenarios derived from historical data (as summarized in Table 5.2). In this scenario, DR actions are not included. The focus is on ensuring the continuity of critical loads during potential islanding events, while also evaluating the operational cost increase associated with providing resilience.

Case 4 combines the resilient scheduling of Case 3 with the active participation of customers in DR programs. In this scenario, DR-enabled loads contribute to mitigating the additional costs of islanded operation, enhancing both economic and operational performance.

Fig. 5.4 illustrates the efficient frontiers for these four cases, generated by varying the risk-aversion parameter β from 0 (risk-neutral) to 50 (highly risk-averse) (Vahedipour-Dahraie et al., 2021). The figure demonstrates two key observations:

1. As the operator becomes more risk-averse, the expected profit decreases, reflecting a deliberate trade-off to reduce exposure to low-probability, high-impact scenarios.
2. Simultaneously, the CVaR index increases, indicating that the MG's exposure to worst-case scenarios is mitigated and profit volatility is reduced.

Comparing across the four cases reveals several important insights:

- Impact of DR: When customers participate in DR programs (Cases 2 and 4), both the expected profit and CVaR improve relative to scenarios without DR. DR participation helps reduce operational costs, particularly by decreasing the need for imported energy from the main grid and optimizing dispatchable DG unit operation.
- Impact of resilient scheduling: Introducing credible islanding scenarios (Cases 3 and 4) generally reduces expected profit compared to normal operations (Cases 1 and 2) due to increased operational costs and potential mandatory load shedding. Nevertheless, Case 4 shows a smaller reduction in profit than Case 3, demonstrating that DR can effectively alleviate some of the additional costs associated with islanded operation.
- Trade-offs between risk and reward: The results highlight the value of risk-aware decision-making, where adopting a slightly risk-averse strategy reduces extreme losses without excessively sacrificing expected profits. The integration of DR in resilient operations further shifts the trade-off curve upward, offering both higher expected returns and better risk mitigation.

Overall, the case studies validate that the combination of stochastic optimization, risk management, and DR integration provides a robust framework for MG scheduling. This approach enables MGs to maintain operational resilience, optimize economic performance, and efficiently handle uncertainties in load demand, renewable generation, market prices, and islanding events. The findings also emphasize that DR programs are essential for enhancing both economic and operational resilience in future smart grid systems.

Since the SD of energy price forecasts is typically higher than that of responsive loads, trading energy with the upstream network may introduce additional exposure to unfavorable scenarios. Consequently, when customers actively participate in DR programs, the frequency of scenarios with negative profits decreases. This effect leads

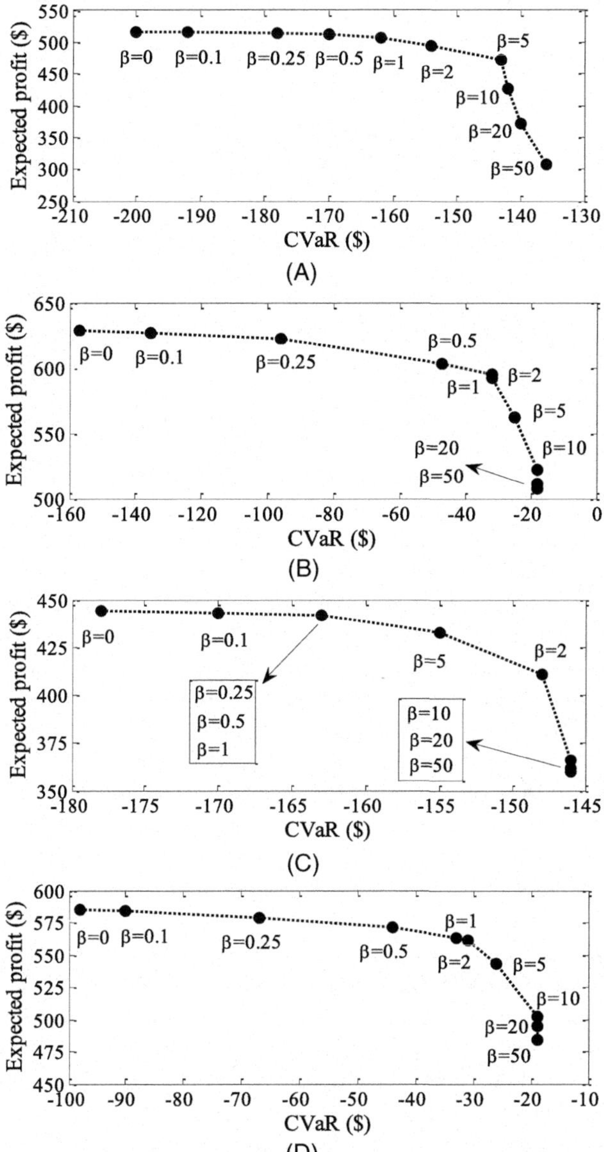

Figure 5.4 Operator's expected profit versus CVaR for different values of β, (A) Case 1, (B) Case 2, (C) Case 3, and (D) Case 4. *CVaR*, Conditional value-at-risk.
From Vahedipour-Dahraie, M., Rashidizadeh-Kermani, H., & Anvari-Moghaddam, A. (2021). Risk-based stochastic scheduling of resilient microgrids considering demand response programs. *IEEE Systems Journal, 15* (1), 971–980.

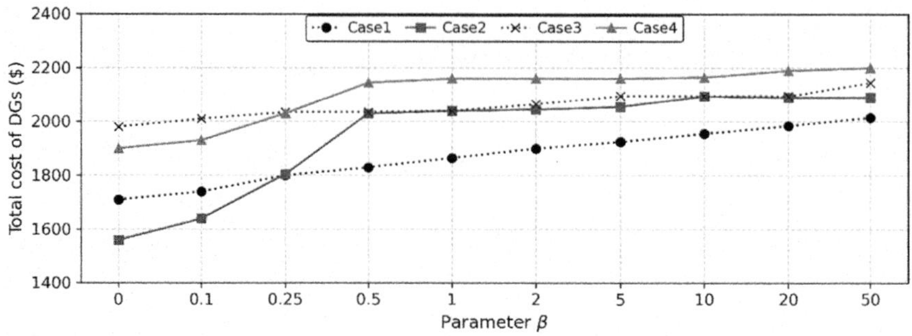

Figure 5.5 Total operational cost of dispatchable DG units across different cases versus risk-aversion. *DG*, Distributed generation.

to higher CVaR values in Cases 2 and 4 compared to Cases 1 and 3, respectively, demonstrating the risk-mitigating impact of DR integration.

By varying the risk-aversion parameter β from 0 to 50, the expected profits of Cases 1, 2, 3, and 4 decrease by 39%, 18%, 19%, and 16%, respectively. In contrast, their corresponding CVaR values increase by 32%, 88%, 18%, and 71%, highlighting the trade-off between risk exposure and expected profitability. These results indicate that implementing DR reduces the dependency between the operator's profit and their risk-aversion level, both in normal and resilient MG operations. Moreover, the influence of risk aversion on CVaR is substantially higher in cases with DR, emphasizing the protective role of responsive loads against worst-case outcomes.

In scenarios where resilient operation is considered (Cases 3 and 4), CVaR values increase compared to normal operation Cases (1 and 2). This behavior arises because ensuring feasible islanding following a disturbance requires rescheduling both energy and reserve resources to accommodate the worst possible contingencies, thereby increasing the perceived risk in the operator's planning horizon.

Fig. 5.5 illustrates the total operational cost of dispatchable DG units across different cases with respect to varying levels of risk aversion. At low risk-aversion values ($\beta < 0.25$), operational costs in cases with DR actions are lower than in those without DR. This outcome occurs because, at lower β, the operator prioritizes maximizing expected profit by relying more on the main grid to supply loads. Additionally, in DR scenarios, customers shift consumption to off-peak hours, reducing peak-period generation from DG units, which lowers operational costs. Conversely, at higher risk-aversion levels, operational costs for DG units in DR cases may exceed those in non-DR cases, reflecting the operator's strategy to sacrifice potential high profits to avoid losses in worst-case scenarios.

Fig. 5.6 presents the total expected energy not served (EENS) and scheduled reserve costs across varying risk-aversion levels. As shown in Fig. 5.6(A), DR participation reduces the total cost of scheduled reserves, as DR, DG units, and the main grid collectively provide reserves in a competitive manner, decreasing overall reserve expenditure. Notably, reserve costs in resilient operation Cases (3 and 4) are higher than in normal operation Cases (1 and 2) due to increased uncertainties associated with

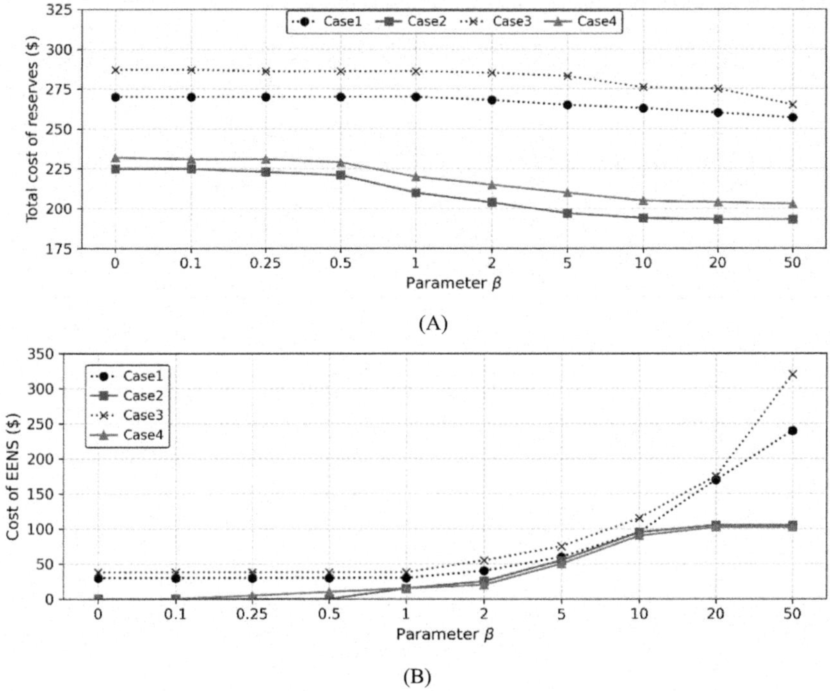

Figure 5.6 Costs of different cases versus risk aversion, (A) cost of EENS, and (B) cost of scheduled reserves. *EENS*, Expected energy not served.

islanding events. Furthermore, the operator's risk attitude influences reserve allocation: Higher risk aversion reduces the probability of supply-demand mismatches, thereby decreasing the required reserve.

Fig. 5.6(B) shows that EENS costs generally increase with higher risk-aversion, although not monotonically. Cases 2 and 4 exhibit lower EENS costs because responsive loads contribute additional reserve capacity, reducing the need for mandatory load shedding. During unscheduled islanding events, EENS costs further decrease due to preallocated reserves being higher in these scenarios compared to normal operations.

Table 5.3 details the impact of risk-aversion on scheduled reserve costs provided by dispatchable DGs, DR, and the main grid. It can be observed that at higher risk-aversion levels, resources are scheduled to mitigate potential mismatches between supply and demand, effectively lowering the required reserve. In essence, a risk-averse operator prioritizes resilience over maximum profit, reducing exposure to extreme losses. By decreasing the number of worst-case scenarios, the cost of scheduled reserves across all resource types diminishes, highlighting the interplay between DR, reserve management, and risk-aware MG operation.

In practice, when the islanding duration extends, the MG operator tends to rely more on local resources rather than the main grid. Consequently, the amount of energy

Table 5.3 Effects of risk aversion on the cost of scheduled reserve in different cases.

β	Case1			Case2			Case3			Case4		
	C_R^{DGs}	C_R^m	C_R^{DR}	C_R^{DGs}	C_R^m	C_R^{DR}	C_R^{DGs}	C_R^m	C_R^{DR}	C_R^{DGs}	C_R^m	C_R^{DR}
0	176	49	0	127	38	24	191	40	0	118	29	20
0.5	176	34	0	107	3	24	190	32	0	113	4	20
1	176	30	0	95	0	23	189	30	0	103	2	19
5	171	23	0	82	0	22	188	23	0	92	2	19
10	168	20	0	76	0	21	184	12	0	88	0	18
20	164	16	0	75	0	21	182	11	0	86	0	18
50	161	12	0	75	0	20	172	8	0	85	0	17

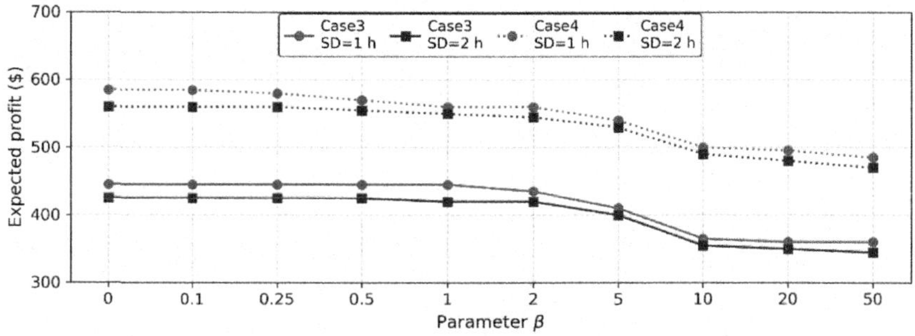

(A)

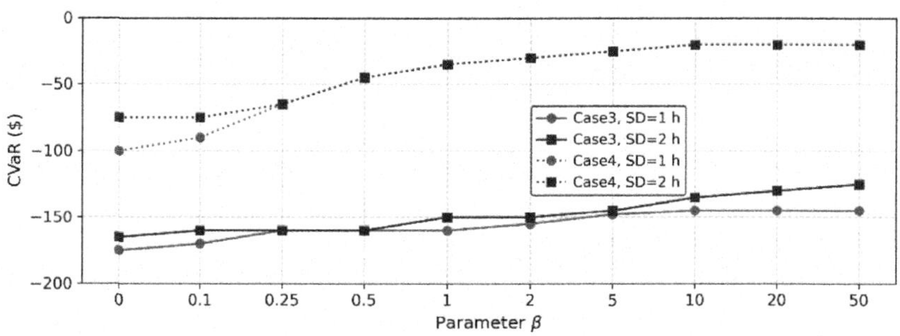

(B)

Figure 5.7 The effect of SD on the expected profit and CVaR in different risk aversion, (A) expected profit, (B) CVaR. *CVaR*, Conditional value-at-risk; *SD*, standard deviation.

purchased from the upstream network decreases. To analyze the influence of the SD of the PDF associated with islanding duration on the MG's expected profit and CVaR, the scheduling problem is solved for two different SD values, namely SD = 1 hour and SD = 2 hours, and the results are presented in Fig. 5.7.

The total energy exchanged between the MG and the main grid over the 24-hour scheduling horizon is summarized in Table 5.4. Here, E1 and E2 denote the energy purchased from and sold to the main grid, respectively, while Enet represents the net

Table 5.4 Exchanged energy (kWh) between the MG and the main grid versus risk aversion in different cases

Operation state	β	Case1			Case2		
		E_1	E_2	E_{net}	E_1	E_2	E_{net}
Normal operation	0	1281	120	1161	1485	87	1308
	1	781	110	661	100	302	-202
	10	487	107	380	0	365	-365
	50	296	98	198	0	366	-366
Operation with resiliency (SD = 1 hour)		Case3			Case4		
	0	991	82	909	791	152	639
	1	791	82	709	100	311	-211
	10	391	82	309	0	355	-355
	50	192	79	113	0	357	-357
Operation with resiliency (SD = 2 hours)	0	891	82	809	491	152	339
	1	691	82	609	100	310	-210
	10	390	82	308	0	353	-353
	50	190	78	112	0	355	-355

MG, microgrid; *SD*, standard deviation.
From Vahedipour-Dahraie, M., Rashidizadeh-Kermani, H., & Anvari-Moghaddam A. (2021). Risk-based stochastic scheduling of resilient microgrids considering demand response programs. *IEEE Systems Journal, 15* (1), 971–980.

traded energy (i.e., E1 – E2). To assess the impact of islanding duration variability on resilient MG operation, the two SD values are considered. The results indicate that as the risk-aversion parameter (β) increases, E1 consistently decreases across all cases. At the same time, unlike in Cases 1 and 3, E2 increases, reflecting more frequent energy exports to the main grid.

These observations can be explained as follows. During normal operation, DR programs help reduce hourly peak loads and shift consumption to off-peak periods, particularly when energy from the main grid is cheaper. Consequently, in a risk-neutral scenario (β = 0), the operator tends to purchase more energy blocks from the main grid in Case 2 than in Case 1, where DR is not utilized. Conversely, as the risk-aversion level rises, the operator prefers to satisfy MG loads using more reliable local DG units rather than relying on the upstream grid. This approach reduces energy imports and often leads to exporting surplus energy to maximize overall profit.

Furthermore, it is observed that increasing the SD of islanding durations from 1 to 2 hours reduces the amount of energy procured from the main grid, particularly at lower risk-aversion levels. This behavior highlights that higher uncertainty in islanding duration prompts the operator to rely more on local resources, thereby enhancing the MG's resilience and reducing dependency on external supply during potential disruptions.

The results in Fig. 5.7(A) show that with higher SD, due to the increased operation cost in longer islanding duration, the expected profit decreases. Moreover, as can be observed from Fig. 5.7(B), varying the SD parameter does not have a substantial effect on the CVaR in most values of β. However, in low risk aversion, when the SD of islanding duration is considered higher, the local resources are scheduled in the optimization process in such a way as to decrease demand-supply mismatch. In this condition, a part of the profit associated with the undesired scenarios is reduced, and therefore the CVaR index increases.

5.8 Discussion and insights

The proposed risk-based stochastic scheduling framework provides a unified perspective on the operational and planning challenges of resilient MGs. By integrating DR, ESS, and distributed renewable generation under stochastic uncertainties, the model demonstrates how operational efficiency, reliability, and resilience can be jointly enhanced.

A key insight is the pivotal role of DR programs in improving MG flexibility and reducing operational risks. Through incentive-driven participation, responsive loads help mitigate peak demand, support reserves, and dynamically react during islanding events, thereby lowering reliance on costly emergency resources. Another important lesson is the necessity of scenario-based stochastic modeling to capture uncertainties arising from renewable variability, load fluctuations, price volatility, and islanding events. Multistage stochastic optimization enables the EMS to anticipate diverse operating conditions and balance cost and risk effectively.

The framework also highlights the importance of risk-aware decision-making tools such as CVaR, which limit exposure to rare but disruptive events and ensure the continuous supply of critical loads. Among its advantages are enhanced resilience, operational cost savings via optimal DR and ESS usage, comprehensive risk mitigation, and adaptability to different MG configurations. Nevertheless, limitations include high computational burden, strong dependency on accurate data, and the need for advanced EMS and real-time monitoring infrastructures.

Looking ahead, DR and resilience will be fundamental pillars of future MGs as renewable penetration grows and decentralized power systems become more dynamic. DR serves as a fast and cost-effective "virtual reserve," complementing ESS and DG. Integrating DR with stochastic scheduling and risk-aware optimization promotes adaptive, intelligent MG operation—supporting peer-to-peer trading, participation in ancillary service markets, and multi-MG coordination.

Overall, the proposed framework shows that effective DR integration, robust uncertainty modeling, and risk-aware optimization are essential for achieving resilient, flexible, and cost-efficient MG operation. Future research should aim to improve computational efficiency, expand the approach to interconnected MGs, and develop practical strategies that leverage real-time data for dynamic decision-making.

5.9 Conclusion

This chapter presents a comprehensive framework for the risk-based stochastic scheduling of resilient MGs, integrating DR and DERs under both normal and contingency scenarios. The proposed methodology demonstrates that combining stochastic modeling with risk-aware optimization can significantly enhance the resilience, reliability, and economic efficiency of MG operations. By leveraging DR programs, MGs can adapt load profiles in response to price signals or operational events, providing a flexible mechanism to mitigate peak demand, support spinning reserves, and respond effectively during islanding events. The key findings of the study include:

- MGs can maintain supply to critical loads under both grid-connected and islanded conditions, minimizing involuntary load shedding during contingencies.
- Incorporating risk measures such as CVaR allows operators to proactively manage low-probability, high-impact events, ensuring robust performance under uncertainty.
- The integrated scheduling of DR and DERs optimizes resource allocation, reduces operational costs, and mitigates the effects of renewable intermittency and market price volatility.
- A multistage, scenario-driven optimization approach captures the probabilistic nature of uncertainties in generation, load, and market prices, providing a realistic basis for operational planning.

References

Babu, K. V. S., Chakraborty, P., & Pal, M. (2025). Demand response optimization MILP framework for microgrids with DERs. *arXiv:2502.08764*.

Chen, J., Zhang, X., & Zhao, H. (2024). Integrated energy management of microgrids with DR and storage under uncertainty. *Energy, 277*, 126901.

Chen, L., Sun, Y., & Zhang, X. (2025). Multi-stage stochastic optimization for microgrids integrating renewable energy and demand response program. *Renewable Energy, 222*, 1213–1231.

Danner, P., Volkova, A., & de Meer, H. (2024). Two-step blackout mitigation by flexibility-enabled microgrid islanding. arXiv:2406.09805.

Du, J., Hu, W., Zhang, S., Liu, W., Zhang, Z., Wang, D., & Chen, Z. (2024). A multi-objective robust dispatch strategy for renewable energy microgrids considering multiple uncertainties. *Sustainable Cities and Society, 116*, 105918.

Du, X., Yang, Y., & Guo, H. (2024). Optimal scheduling of a microgrid based on renewable resources and demand response program using stochastic and IGDT-based approach. *Journal of Energy Storage, 86*, 111306.

Elazab, A., Abdelnaby, A. T., & Ali, A. A. (2024). Impacts of multiple demand-side management strategies on microgrids planning: A literature survey. *Clean Energy, 8*(1), 36–54.

Guo, S., He, J., Ma, K., Yang, J., Wang, Y., & Li, P. (2025). Demand response strategy for microgrid energy management integrating electric vehicles, battery energy storage system, and distributed generators considering uncertainties. *Sustainable Energy, Grids Networks, 41*, 101594.

He, J.-H., & Lin, J.-H. (2025). Review of microgrids to enhance power system resilience. *Engineering Proceeding., 92*(1), 82.

Hu, J., Shan, Y., Yang, Y., Parisio, A., Li, Y., Amjady, N., Islam, S., Cheng, K. W., Guerrero, J. M., & Rodríguez, J. (2024). Economic model predictive control for microgrid optimization: A review. *IEEE Transactions on Smart Grid, 15*(1), 472–484.

Hu, J., Shan, Y., Yang, Y., Parisio, A., Li, Y., Amjady, N., Islam, S., Cheng, K. W., Guerrero, J. M., & Rodríguez, J. (2024). Risk-aware microgrid scheduling using conditional value-at-risk: A review and future directions. *IEEE Transactions on Smart Grid, 15*(2), 890–905.

Khosravi, N., Çelik, D., Bevrani, H., & Echalih, S. (2025). Microgrid stability: A comprehensive review of challenges, trends, and emerging solutions. *International Journal of Electrical Power & Energy Systems, 170*, 110829.

Kumar, P., Srivastava, S. K., & Gupta, R. (2023). Risk-aware scheduling of microgrids considering DR and CVaR constraints. *IEEE Transactions on Smart Grid, 14*(3), 1987–1998.

Lari, D. S., Nafar, M., Abbasi, A. R., & Bahmani-Firouzi, B. (2025). Probabilistic scheduling of microgrid resilience: Integrating renewables, storages and demand response in unit commitment and reconfiguration. *International Journal of Electrical Power & Energy Systems, 168*, 110710.

Li, J., Yang, M., Zhang, Y., Li, J., & Lu, J. (2023). Micro-grid day-ahead stochastic optimal dispatch considering multiple demand response and electric vehicles. *Energies, 16*(8), 3356.

Lin, F.-J., Lu, S.-Y., Hu, M.-C., & Chen, Y.-H. (2024). Stochastic optimal strategies and management of electric vehicles and microgrids. *Energies, 17*(15), 3726.

Liu, G., Ferrari, M., Ollis, B., & Sundararajan, A. (2023). Resilient microgrid scheduling considering stochastic chance-constrained islanding capability. In *2023 IEEE Power and Energy Society General Meeting (PESGM)*.

Liu, H., Zhang, Y., & Li, P. (2024). Dynamic pricing and peak shaving in smart microgrids with renewable integration. *Applied Energy, 375*, 121974.

Liu, X., An, L., Liang, X., & Tan, Z. (2025). CVaR-based stochastic scheduling of resilient microgrids with demand response and storage integration. *Applied Energy, 380*, 126478.

M Vahedipour-Dahraie, H. R. - K., Anvari-Moghaddam, A., Siano, P., & Catalão, J. P. S. (2022). Short-term reliability and economic evaluation of resilient microgrids under incentive-based demand response programs. *International Journal of Electrical Power & Energy Systems, 138*, 1–10.

Liu, X., Liang, X., An, L., & Tan, Z. (2025). Robust optimization model for micro-energy grid accounting for demand response and carbon-green certificate market transactions. *Frontiers in Energy Research, 13*, 1636892. https://doi.org/10.3389/fenrg.2025.1636892.

Macmillan, M., Zolan, A., Bazilian, M., & Villa, D. L. (2024). Microgrid design and multi-year dispatch optimization under climate-informed load and renewable resource uncertainty. *Applied Energy, 368*, 123355.

Moreno, R., Panteli, M., Mancarella, P., Rudnick, H., Lagos, T., Navarro, A., Araneda, J.C., (2020). From Reliability to Resilience: Planning the Grid Against the Extremes. IEEE Power & Energy Magazine, 18, 41–53.

Nakiganda, A. M., & Aristidou, P. (2023). Resilient microgrid scheduling with secure frequency and voltage transient response. *IEEE Transactions on Power Systems, 38*(4), 3580–3592.

Shahravi, H., Olamaei, J., Kheradmandi, M., & Sadr, A. (2025). Enhancing the small-signal stability of the island microgrids under the virtual impedances-gray wolf optimization model: Two distinct case studies. *International Journal of Low-Carbon Technologies, 20*, 1024–1035.

Sun, H., Wang, X., & Li, F. (2025). Synergistic operation of demand response, storage, and renewable energy in resilient microgrids. *Renewable Energy, 225*, 1245–1262.

Tabares, A., & Cortés, P. (2024). Using stochastic dual dynamic programming to solve the multistage energy management problem in microgrids. *Energies, 17*(11), 2628.

Tahere Daemi, A. N., Pourfarzin, S., & Akbari, H. (2023). Microgrid system energy management with demand response program for clean and economical operation. *Applied Energy, 334*, 120717.

Vahedipour-Dahraie, M., Najafi, H. R., Anvari-Moghaddam, A., & Guerrero, J. M. (2017). Study of the effect of time-based rate demand response programs on stochastic DA energy and reserve scheduling in islanded residential microgrids. *Applied Sciences, 7*(4), 1–19.

Vahedipour-Dahraie, M., Rashidizadeh-Kermani, H., & Anvari-Moghaddam, A. (2021). Risk-based stochastic scheduling of resilient microgrids considering demand response programs. *IEEE Systems Journal, 15*(1), 971–980.

Vahedipour-Dahraie, M., Rashidizadeh-Kermani, H., Anvari-Moghaddam, A., & Guerrero, J. M. (2020). Stochastic risk-constrained scheduling of renewable-powered autonomous microgrids with demand response actions: Reliability and economic implications. *IEEE Transactions on Industry Applications, 56*(2), 1882–1895.

Vahedipour-Dahraie, M., Rashidizadeh-Kermani, H., Anvari-Moghaddam, A., & Siano, P. (2020). Risk-averse probabilistic framework for scheduling of virtual power plants considering demand response and uncertainties. *International Journal of Electrical Power and Energy Systems, 121*, 106126.

Wang, T., Liu, H., & Su, M. (2025). Energy optimization for microgrids based on uncertainty-aware deep deterministic policy gradient. *Processes, 13*(4), 1047.

Wu, Y.-S., Liao, J.-T., & Yang, H.-T. (2024). Three-stage resilience enhancement via optimal dispatch and reconfiguration for a microgrid. *Frontiers in Energy Research, 12*. doi:10.3389/fenrg.2024.1461383.

Yang, L., Yang, H., Cao, X., & Guan, X. (2024). Distributionally robust frequency-constrained microgrid scheduling towards seamless islanding. arXiv:2401.03381.

Zahraoui, Y., Korõtko, T., Rosin, A., Mekhilef, S., Seyedmahmoudian, M., Stojcevski, A., & Alhamrouni, I. (2024). AI applications to enhance resilience in power systems and microgrids — A review. *Sustainability, 16*(12), 4959.

Zeng, F., Wang, S., & Wang, X. (2023). Price-based demand response in microgrids: Optimization and scheduling. *IEEE Transactions on Smart Grid, 14*(2), 1250–1262.

Zeng, F., Wang, S., & Wang, X. (2024). Stochastic CVaR optimization for resilient energy management in microgrids. *IEEE Transactions on Industrial Informatics, 20*(3), 2150–2162.

Zhang, M., Li, J., & Wang, H. (2024). Stochastic optimization of resilient microgrids with demand response under renewable uncertainty. *Applied Energy, 374*, 120895.

Zhang, Y., Li, J., Yang, M., & Liu, Q. (2024). DR-based multi-stage stochastic optimization for microgrid resilience. *IEEE Transactions on Industry Applications, 60*(6), 5250–5263.

Zhao, F., Li, H., & Wang, M. (2023). Conditional value-at-risk based operational planning of resilient microgrids with DR participation. *Energy, 283*, 125614.

Zidane, T. E. K., Muis, Z. A., Ho, W. S., Zahraoui, Y., Aziz, A. S., Su, C.-L., Mekhilef, S., & Campana, P. E. (2025). Power systems and microgrids resilience enhancement strategies: A review. *Renewable and Sustainable Energy Reviews, 207*, 114953.

Case studies in renewable-dominant and islanded microgrids

6

Farha Khan[a] *and Preeti Sharma*[b]
[a]Department of Electrical Engineering, Aligarh Muslim University, Aligarh, Uttar Pradesh, India, [b]Department of EE, SOET, Pandit Deendayal Energy University, Gandhinagar, Gujarat, India

6.1 Introduction

The energy transition in the world is being influenced by an increasing focus on sustainable and low-carbon alternatives. One of the emerging technologies is the microgrid, which is central to the provision of reliable, local, and affordable power (Guerrero et al., 2014). They are especially needed in remote and islanded areas where extension of the grid is technically or economically infeasible. Dominated by renewable energy sources like solar photovoltaics (PV) and wind power, such systems are referred to as renewable-dominant microgrids, and they can lead to less fossil fuel dependence, resiliency, and long-term sustainability (Parhizi et al., 2015). Nevertheless, renewable-dominant microgrids have operational issues with the variable and intermittent nature of the renewable resources. All these problems are forcing the inclusion of energy storage technology, backup generation systems, and intelligent control measures.

6.1.1 Evolution of the microgrid concept

Historically, microgrids have been proposed as the local solution to an uninterrupted power supply, and most of them have used only diesel generators (DGs). Although these diesel-powered systems were reliable, they were expensive to run, harmful in terms of their emissions, and logistically demanding when it came to ferrying fuel (Niknam et al., 2012). In the long term, purely fossil-based hybrid microgrids were replaced by hybrid microgrids that combined renewables and small-scale storage with diesel backup.

There are three phases of evolution:

1. Diesel-heavy microgrids: Display dispatchable power, and are very costly and polluting.
2. Hybrid microgrids: A Hybrid of PV/wind with backup diesel and batteries.
3. Renewable-dominant microgrids: Supplied with renewables as the main source of energy, but with advanced energy storage and innovative control measures ensuring regulated operations.

Fig. 6.1 shows the evolution from diesel dominant microgrids to hybrid and renewable dominant microgrids.

Stochastic Planning and Modeling for Energy Systems: Methods, Applications, and Developments.
DOI: https://doi.org/10.1016/B978-0-443-45298-7.00012-4

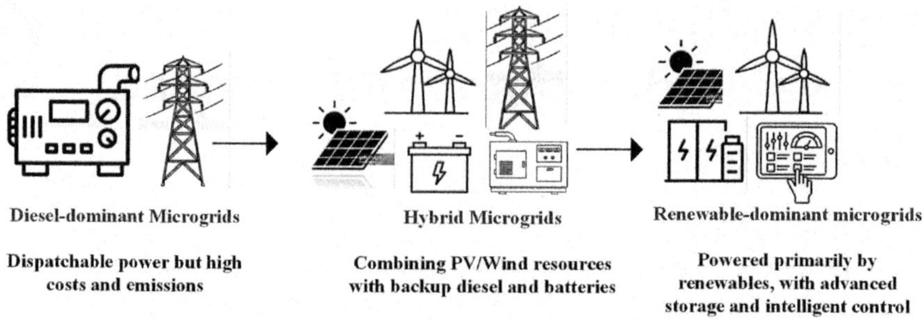

Figure 6.1 Evolution from diesel dominant microgrids to hybrid and renewable dominant microgrids.

This evolution can also be summarized as follows in Table 6.1, with the main features, benefits, disadvantages, and applications of each type of microgrid. It highlights the transition between the high-emission diesel-based system to a renewable-based dominant system, indicating the advantages and the limitations at each level.

6.1.2 Renewable penetration and its implications

The rate at which renewable energy is integrated in a microgrid is the most important factor that dictates design and operation strategies within the microgrid (Bhattacharyya et al., 2021). At lower penetrations, renewables are substitutes, but when it is at medium, it decreases the fuel consumption significantly, yet still relies on backup diesel generation. The high penetration (>70%) level leads to the possibility of relying almost entirely on renewables, with robust storage integration and advanced control (Aldaouab & Daniels, 2018).

The advantages of high levels of renewables are as follows:

- Environmental sustainability: Cleaner environment and air.
- Economic feasibility: Reduced cost of operations in the long-term due to fuel substitution.
- Resilience: Increased system capacity to resist disturbances and disasters.

Islanded systems, which do not have the support of an external grid, cannot compensate for these disadvantages with the same advantages, and therefore, frequency instability, voltage fluctuations, and power quality degradation are more pronounced in such systems (Saha et al, 2023). Therefore, high-renewable penetration not only necessitates storage and intelligent control but also requires probabilistic forecasting, multistage stochastic scheduling, and risk-aware planning tools that can anticipate variability and uncertainty in both supply and demand.

6.1.3 Role of energy storage and hybridization

The key to renewable-dominant microgrids is energy storage. It smoothed intermittent renewable generation, provides frequency regulation, and maintains load following

Table 6.1 Evolution of microgrid architectures.

Microgrid type	Primary energy sources	Advantages	Limitations	Typical applications
Diesel-dominant	Diesel generators	High reliability, dispatchable	High fuel cost, high emissions, and logistical challenges	Remote islands, military bases
Hybrid	Diesel + PV/wind + small storage	Lower fuel use, better economics, improved sustainability	Still reliant on fossil backup, storage cost	Rural electrification, small industries
Renewable-dominant	PV, wind, energy storage (diesel minimal/none)	Very low emissions, low long-term costs, scalable	Variability, high storage, and control requirements	Remote villages, 100% RE islands

PV, photovoltaics.

Table 6.2 Key performance metrics for renewable-dominant microgrids.

Metric category	Indicators/measures	Purpose
Technical performance	Reliability index (% uptime), storage utilization efficiency, converter efficiency	Evaluate the ability to maintain a continuous supply with stable power electronics and optimized storage integration.
Economic viability	LCOE, net present value (NPV), cost of storage per kWh	Assess affordability, investment feasibility, and competitiveness against centralized grid supply.
Environmental sustainability	CO_2 reduction (tons/year), renewable penetration (%), lifecycle emission intensity	Quantify environmental benefits and alignment with decarbonization targets.
Social impact	Energy access index (EAI), Job Creation index (JCI), Participation satisfaction factor (PSF)	Measure inclusivity, community benefits, and long-term social acceptance.
System resilience and security	Black-start capability, recovery time after disturbance, cybersecurity robustness	Ensure adaptability to grid disturbances, natural disasters, and cyber threats.

LCOE, levelized cost of electricity.

capability (Alipour et al., 2020). Such technologies include lithium-ion and flow batteries, which are extensively applied, and flywheel and hydrogen technologies, which are increasingly used in more sophisticated designs (Cohen et al., 2023).

Hybridization is also used to increase system robustness. With biomass-based units, small-scale hydro, or a limited capacity of diesel, it is more reliable as the system provides backup in the case of a long-term resource deficit without threatening its long-term sustainability (Savitz, 2025).

6.1.4 Performance metrics for microgrid evaluation

The dynamics of performance assessment of renewable-dominant microgrids must be multidimensional with technical, economic, environment-related, and resiliency metrics (Etemadi et al., 2018; Gomez Anccas et al., 2025; Tang et al., 2016). Table 6.2 displays the key performance metrics for renewable-dominant microgrids.

6.1.5 Importance of case studies

Even though modeling and simulations offer theoretical guidance, it is important to validate these through case studies (Atinkut, 2025). Case studies also provide practical experiences in operating renewable-dominant microgrids in terms of resource variability, the effectiveness of controls, the socioeconomic impact, and the acceptance in the community (Tahir, 2025; Simon, 2019).

The chapter uses a wide variety of case studies that include PV-based rural microgrids in India, hybrid wind PV systems in the Mediterranean, and close to 100%

renewable systems in Alaska. These cases demonstrate the adaptability of renewable-dominant microgrids subject to changing circumstances and their ability to become repeatable designs across the globe.

6.1.6 Chapter objectives and structure

The aim of this chapter is to:

1. Introduce detailed case studies in renewable-dominant and islanded microgrids.
2. Assess the network operation with technical, economic, environmental, and resilience indicators.
3. Identify challenges, lessons, and directions for future research.
4. Place stochastic modeling, probabilistic forecasting, and uncertainty-aware optimization at the center of renewable-dominant microgrid analysis, showing how risk management frameworks strengthen technical, economic, and policy outcomes.

The rest of this chapter is organized to present a justification of renewable-dominant microgrids. Section 6.2 introduces the conceptual framework of the renewable-dominant microgrids, including their main principles and features of operation. Section 6.3 presents mathematical representations and governing equations of the major system components that provide the analytical framework of performance assessment. Section 6.4 presents geographical regional case studies displaying practical solutions and field experience.Section 6.5 includes a comparative analysis of the performance of each energy-scale technology on technical, economic, environmental, and resilience attributes. Section 6.6 includes the main difficulties found, which can be learned by the existing projects, and Section 6.7 discusses the future research prospects. Lastly, Section 6.8 sums up the chapter and establishes the importance of a renewable-dominant microgrid to sustainable energy systems.

6.2 Conceptual framework of renewable-dominant microgrids

Microgrids based on renewable energy sources are a radically new way of producing and delivering power, and allowing communities and industrial plants to become almost independent of fossil fuels. In contrast to traditional microgrids with diesel- or gas-based generators dominating, renewable-dominant systems focus on a strong engagement in solar PV energy, wind power, and energy storage facilities. This design approach is especially important in microgrids on islands and remote off-grid sites when the supply of energy must be local and when design autonomy is a concern.

The conceptual framework details a holistic view of the system architecture, dynamics of energy flows, classification to be used, and evaluation of performance indicators needed to understand the design and functioning of these microgrids.

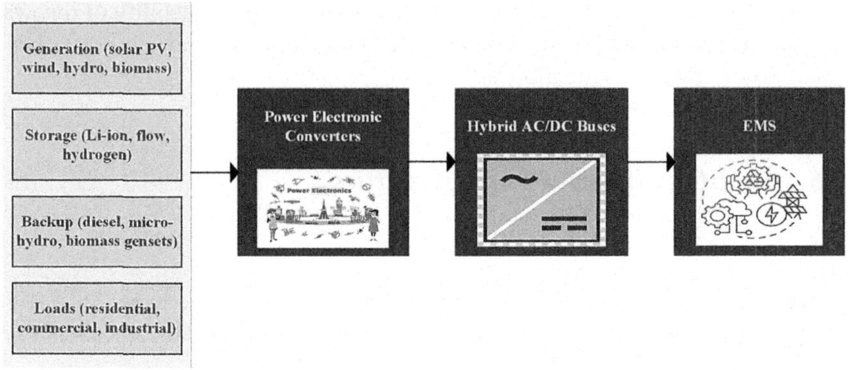

Figure 6.2 General architecture of a renewable-dominant microgrid.

6.2.1 Definition and scope of renewable-dominant microgrids

Renewable-dominant microgrid can be described as a localized power supply in which the source of renewable energy (solar PV, wind, biomass, or small hydro) has a majority stake in the power supply, usually over 70% the overall energy mix (Wikipedia, n.d.) shown in Fig. 6.2. These systems have been designed to work autonomously in islanded mode or through coordination with the main grid. The major difference with the traditional microgrids is the low reliance on fossil-based generation, hence guaranteeing the long-term viability, low carbon emissions, and lower cost of operations. The range of renewable-based microgrids spans a wide range of applications (Niknam et al., 2012). They conduct electrification in remote villages where a centralized transmission is not possible. They substitute diesel generation in island regions and decrease logistical burdens, which come with the transportation of fuels. They add dependability and provide stability during grid disruption in industrial or commercial clusters. Moreover, microgrids that are dominated by renewables are now commonly deployed to military bases and disaster-prone areas where energy security is very important.

6.2.2 Structural components of renewable-dominant microgrids

The renewable-dominant microgrids architecture may be divided into four fundamental subsystems:

- Generation Subsystem: The generation block is mainly made up of renewable sources like PV arrays, wind turbines, biomass gasifiers, or small-scale hydro units. The power electronic converters that provide grid-forming or grid-following capabilities integrate these units. The noncontinuous character of renewables makes it necessary to combine several sources to manage the differences in the day and the season.
- Energy Storage Subsystem: The key to renewable-dominant microgrids is energy storage. The most common ones deployed are lithium-ion batteries, which are highly efficient, and their costs are decreasing. Flow batteries and hybrid hydrogen-based storage are being considered more often as long-duration applications. The storage subsystem allows load

shifting, frequency regulation, peak shaving, and black-start capability, while providing a smooth integration of renewable sources (Alipour et al., 2020).

- Backup and Auxiliary Sources: The long-term objective should be the operation of the system without fossil fuels, but there can be a certain amount of backup, which can be supported by diesel or natural gas gensets, biomass-based units, or micro-hydro turbines. Such dispatchable units are started in periods of sustained renewable shortages or during periods of critical load support. They are intended to play an auxiliary role and will guarantee the attainment of reliability goals without compromising the microgrid nature of the renewable-dominant system (MDPI Editorial Board, 2023).
- Load Subsystem and Interfaces: Loads of renewable-dominant microgrids are generally a combination of residential, commercial, and industrial loads. The application of smart meters, demand response programmes, and load prioritization methods is becoming increasingly common in order to streamline consumption patterns. All of these are synchronized by power electronic interfaces that include inverters, DC–DC converters, and hybrid AC/DC buses that regulate energy flow and ensure the quality of power.

6.2.3 Operating modes and functional characteristics

Microgrids that are renewable-dominant have two main modes of operation: Grid-connected and islanded. They dynamically interact with the utility network in grid-connected operation, providing excess renewable power and receiving assistance during deficit periods. Contrastingly, with islanded operation, the microgrid operates independently, with generation and demand completely within its area of operation (Sarker et al., 2026). The key attribute of such systems is the possibility of a smooth transition between grid-connected and islanded operation. This demands high-level synchronization systems, protection coordination, and black-starts. In practice, operating modes determine the control philosophy, storage size, and extent of reserve capacity that have to be held.

6.2.4 Hierarchical control framework

The stability and coordination of renewable-dominant microgrids rely on a hierarchical control framework shown in Fig. 6.3 (Shrivastava & Subudhi, 2020; Joshal & Gupta, 2023; Shahid et al., 2018).

- Primary Control: This provides instant stability of voltage and frequency based on droop properties and decentralized control legislation. It can support real-time load sharing between distributed generators without the need for rapid communication.
- Secondary Control: The secondary controllers correct deviations caused by the primary droop control so that the voltage and frequency are returned to nominal values. They may be centralized, based upon a supervisory controller, or they may be distributed, based upon consensus algorithms.
- Tertiary Control: In the upper level, tertiary control is in charge of the economic dispatch, load scheduling, and communication with the main grid when possible. It allows rationalization of the flow of energy, minimization of costs, and incorporation of market-related transactions in energy.

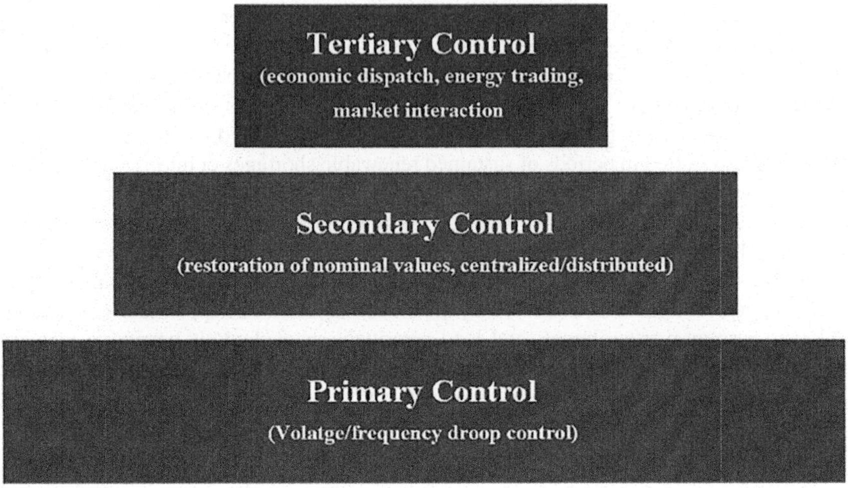

Figure 6.3 Hierarchical control framework of renewable-dominant microgrids.

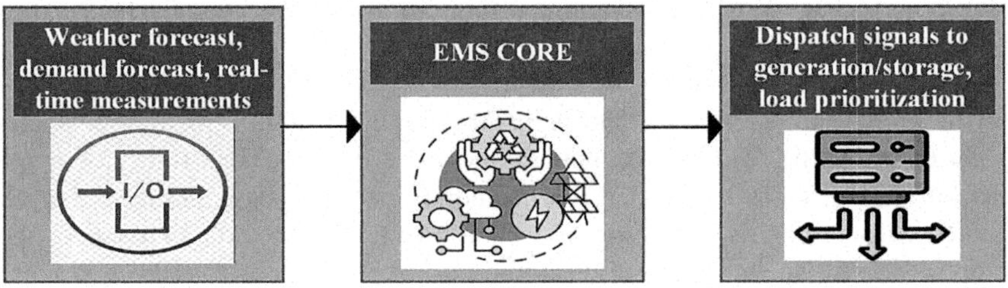

Figure 6.4 Workflow of the Energy Management System in renewable-dominant microgrids.

Emerging control solutions combine model predictive control (MPC), reinforcement learning, and multiagent systems to control the variability of renewables and load uncertainties in an adaptive manner.

6.2.5 Energy management strategies

The intelligence of renewable-dominant microgrids is energy management. It ensures that generation, storage, and demand are synchronized in the most efficient way to achieve technical and economic goals. The workflow of the Energy Management System (EMS) in renewable-dominant microgrids shown in Fig. 6.4. Demand-side management (such as load forecasting and demand response) are significant factor in balancing variability (Atef et al., 2026). Predictive scheduling, resting on weather prediction of solar irradiance and the speed of wind, is essential in the optimization of the utilization of renewable energy. Energy management systems also optimize the storage to reduce cycling losses, maximize battery life, and have sufficient reserve

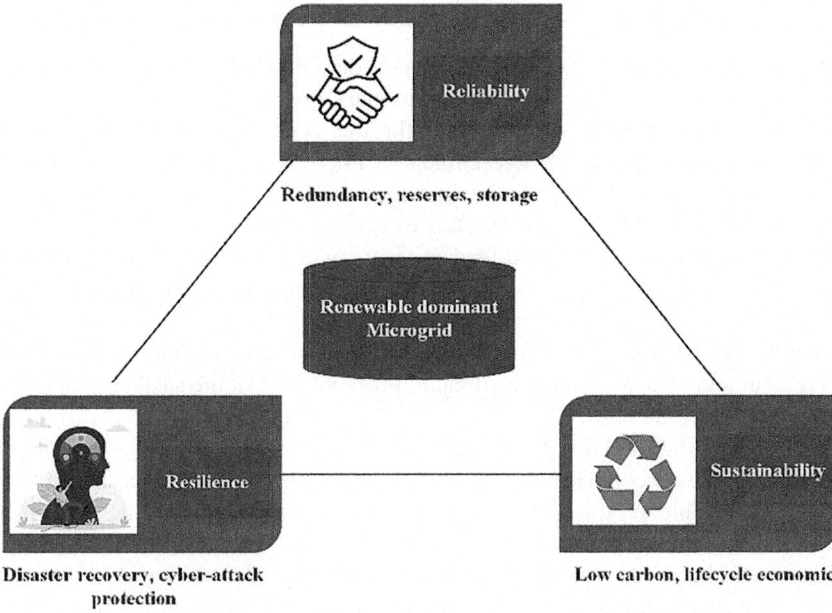

Figure 6.5 The reliability–resilience–sustainability triad in renewable-dominant microgrid design.

margins (Sarwar et al., 2024). The use of real-time dispatch strategies means that the high-priority loads are never missed, and the noncritical loads can be curtailed during supply shortages.

6.2.6 Principles of reliability, resilience, and sustainability

The principles of renewable-dominant microgrid design go beyond technical efficiencies to include more general aims of reliability, resiliency, and sustainability (Ghosh & Das, 2019; Akorede et al., 2010).

- Redundancy, spinning reserves, and strong storage capacities ensure reliability by keeping the supply continuous even in times of fluctuation of generation.
- Resilience focuses on the resistance to, and recovery of, extreme incidents, like a storm or cyber-attack, with fault-tolerant design and fast restoration policies.
- Sustainability is realized through the emphasis on the integration of renewable energy, minimization of lifecycle carbon emissions, and economic affordability over the long-term operation.

All these are the guiding principles of renewable-dominant systems. Fig. 6.5 represents the reliability–resilience–sustainability triad in renewable-dominant microgrid design.

6.2.7 Socioeconomic and policy dimensions

Socioeconomic and policy-based factors also determine the success of renewable-dominated microgrids. They have a revolutionary impact on the rural electrification of villages, empowering the communities through accessibility to clean and reliable energy (Marnay et al., 2025). Economically, microgrids have the effect that local industries are stimulated, dependence on imported fossil fuels is minimized, and jobs are created. The deployment is boosted by policy frameworks, such as subsidies, renewable portfolio standards, and feed-in tariffs, which make investments financially viable (Peterson et al., 2021; Asmus, 2010). Nonetheless, the speed of adoption is frequently dictated by administrative issues, including interconnection standards, tariff frameworks, and grid codes. Community acceptance is also essential, and stakeholder involvement and education must be done to promote trust in the sustainability of such systems over the long term.

To conclude, the modular architecture, smart control, and operation-oriented sustainability are the conceptual pillars of renewable-dominant microgrids. These systems combine renewable production with high-energy storage, ensure the flexibility of operations of hybrids, and utilize hierarchical control to stabilize and optimize. In addition to technical factors, they represent such principles as resilience, economic affordability, and socioeconomic development. This conceptualization forms the basis of the following parts of the chapter, which explore mathematical models, case studies, and performance analysis of renewable-dominant microgrids.

6.3 Mathematical models and equations for system components under uncertainty

Mathematical modeling of the components of a renewable energy system is the core of microgrid analysis and control. A microgrid with renewable dominance or an isolated microgrid has a variety of resources, including PV arrays, wind energy conversion systems (WECS), Energy Storage System (ESS), backup generators (usually diesel), and power electronic interfaces. All these units have nonlinear features, environmental implications, and dynamics that need to be represented in the proper design, simulation, and control. The requirements of proper modeling of the components are motivated by three core requirements. First, renewable systems with high penetration will need to be capable of balancing the variability of the generator with a storage and backup system that keeps operations stable. Second, the controllers (local at the converter level and more global at the hierarchical microgrid control level) require the use of predictive models to satisfy the voltage, frequency, and power-sharing levels. Finally, performance analysis, optimization, and reliability assessment require mathematical models of efficiency and state of charge (SOC) dynamics, and power quality performance.

Besides deterministic modeling, uncertainty-sensitive formulations need to be included to reflect stochastic characteristics of renewable resources and demand. Solar irradiance, wind speed, temperature fluctuations, fuel price fluctuations, and load fluctuations are inherently random, and therefore their neglect can lead to an overly optimistic system performance. Thus, probabilistic forecasting, scenario generation,

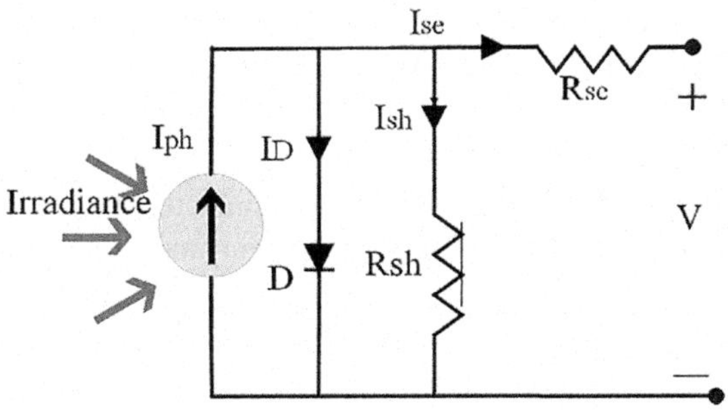

Figure 6.6 Equivalent single-diode model of the PV cell.

and stochastic optimization techniques are incorporated into the subsequent component model, which makes sure that the variability and uncertainty of the microgrid operation are tackled directly.

6.3.1 Solar photovoltaics system modeling

Solar PV systems convert incident sunlight into electricity in the form of electrical energy on a semiconductor surface. The PV generation source needs to be modeled accurately to analyze the performance of the system, design power conditioning stages, and the development of efficient control strategies, as maximum power point tracking (MPPT) (Seane et al., 2022).

6.3.1.1 Basic photovoltaics cell equation

The electrical properties of a PV cell are highly nonlinear and dependent on the environment, especially the irradiance and temperature. To model these dynamics, the single-diode equivalent circuit has been popular because it is both simple and accurate enough (Rawa et al., 2022). This model forms the foundation of obtaining the current voltage (*I–V*) characteristics of a PV cell, which can then be used to represent PV cell modules and arrays used in the practical world.

The governing current–voltage (*I–V*) relationship of the single-diode model is shown in Fig. 6.6 and given as:

$$I_{se}=\mathrm{I_{ph}}-\mathrm{I_S}\left\{e^{\frac{qV_d}{nk_T}}-1\right\}-\frac{\mathrm{V}+\mathrm{IR}_{se}}{\mathrm{R_{sh}}} \tag{6.1}$$

where:

I_{se}= output current
$\mathrm{I_S}$= diode saturation current
$\mathrm{I_{ph}}$= light-generated current
V= output voltage
I_{sh}= shunt current
V_d= potential difference of diode

q= electron charge (1.602176×10^{-19} C)
k= Boltzmann constant (1.380650×10^{-23} J/K)
T= temperature
n= diode ideality factor

This nonlinear equation defines the output characteristics of the PV cell and is extended to a module or array by scaling current and voltage with series/parallel connections.

Under uncertain irradiance and temperature, the PV output can be modeled as a random process described in Eq. (6.2):

$$P_{PV}(t, \omega) = V_{mpp}(t, \omega) \cdot I_{mpp}(t, \omega) \tag{6.2}$$

where $\omega \in \Omega$ represents a scenario sampled from probabilistic solar forecasting models (e.g., Gaussian process regression or Copula-based joint distributions). To ensure reliability, chance constraints are applied by Eq. (6.3):

$$\mathbb{P}(P_{PV}(t, \omega) \leq P_{PV}^{max} \geq 1 - \in \tag{6.3}$$

with $\in$as the acceptable risk level.

6.3.1.2 Photocurrent expression

The photocurrent depends on irradiance (G) and cell temperature (T) and is given by Eq. (6.4):

$$I_{ph} = \left[I_{sc,ref} + \alpha\left(T - T_{ref}\right)\right]\frac{G}{G_{ref}} \tag{6.4}$$

where:

$I_{sc,ref}$ = short-circuit current in reference condition
α = temperature coefficient of short-circuit current (A/°C)
G = incident irradiance (W/m^2)
G_{ref} = reference irradiance (1000 W/m^2)
T_{ref} = reference temperature (25°C)

6.3.1.3 Maximum power point tracking

An important facilitator in a microgrid based on renewable energy and islanded is the implementation of advanced MPPT techniques. The consistency of maximum PV generator power extraction, as well as the stability of microgrid operation, can be guaranteed because of the fact that MPPT operates PV generators at their optimal operating point, which is directly affected by the variability of solar irradiance and wind speed, which affects the available energy (Sundararajan & Ferrari Maglia, 2024). Fig. 6.7 represents the flowchart that illustrates the Perturb and Observe (P&O) MPPT principle, where the duty cycle of the DC–DC converter is periodically perturbed,

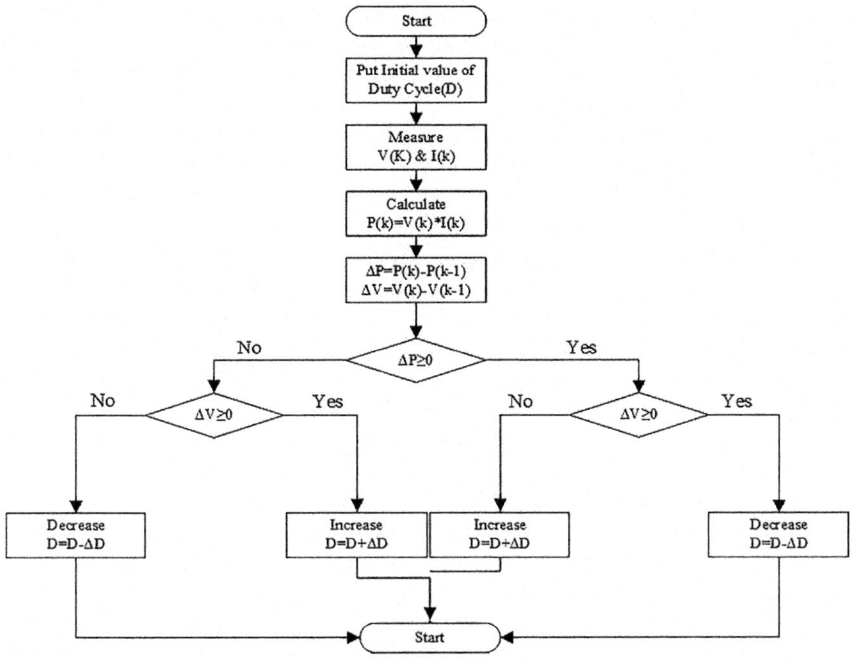

Figure 6.7 Flowchart of the P&O MPPT algorithm.

and the resulting change in PV power is observed. If the power increases ($\Delta P \geq 0$), the perturbation direction is maintained; otherwise, the direction is reversed. The algorithm uses voltage and current measurements to calculate instantaneous PV power and compares it with the previous cycle to determine adjustments in the duty cycle. This iterative process ensures that the operating point converges toward the Maximum Power Point (MPP) of the PV array. In renewable-dominant microgrids, P&O is favored for its simplicity, ease of implementation, and fast dynamic response, though it may exhibit small oscillations around the MPP under steady-state conditions. In hybrid microgrid power systems where PV is combined with storage and backup power units, intelligent MPPT can further improve efficiency, diminish dependence on storage units that are oversized to meet intermittent requirements, decrease the levelized cost of electricity (LCOE), and achieve better voltage and frequency controls. High-efficiency MPPT methods that may overcome the intermittency challenge (e.g., MPPT through a neural network-based, reinforcement-learning-based, and bio-inspired MPPT) are becoming adopted at a higher rate, and are thus critical to furthering the economic and environmental values of decentralized energy infrastructure.

The maximum power output of PV is given by Eq. (6.5) under certain conditions:

$$P_{max} = V_{mpp} \cdot I_{mpp} \tag{6.5}$$

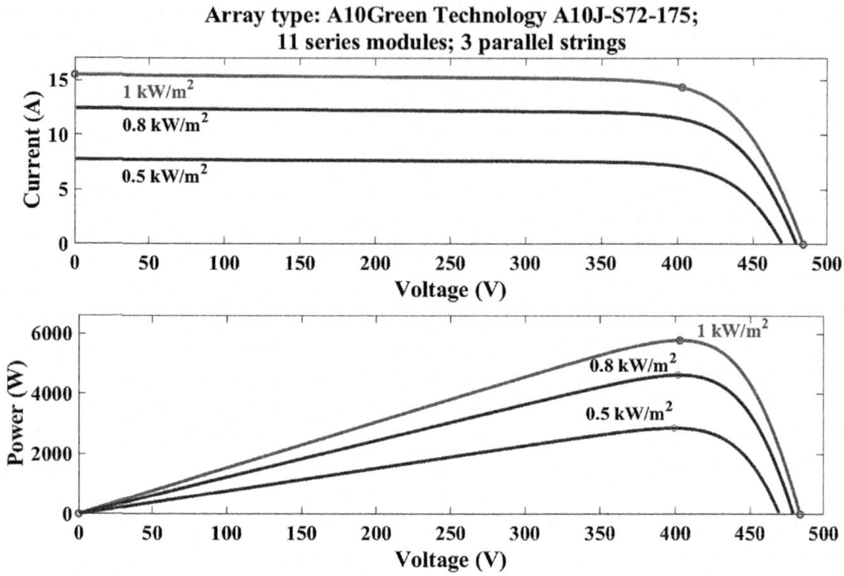

Figure 6.8 *I–V* and *P–V* characteristics of PV panel at varying irradiations.

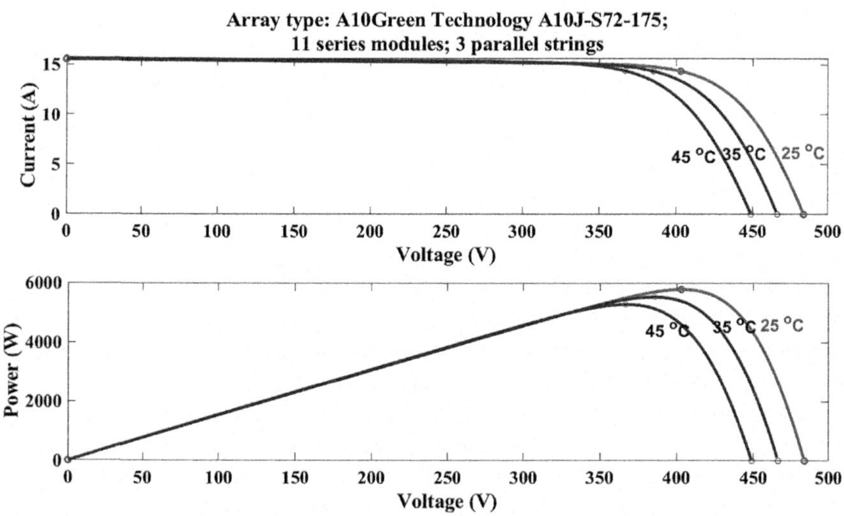

Figure 6.9 *I–V* and *P–V* characteristics of PV panel at varying temperatures.

where V_{mpp} and I_{mpp} are the voltage and current at MPPT. MPPT algorithms are applied to ensure that the PV operates at P_{max}.Figs. 6.8 and 6.9 show the *I–V* and *P–V* curves at varying solar irradiance and temperature shifts, respectively, of the PV panel A10Green Technology A10J-S72-175 with 11 in series and 3 parallel strings with a power rating of 5 kW, highlighting the movement of the MPP.

6.3.2 Wind energy conversion system

One of the most promising renewable resources to invest in microgrid implementation is wind energy because it is abundant, scalable, and has minimal environmental impact. A WECS uses the moving air to generate mechanical torque, then in turn electrical power through a generator by converting the kinetic energy of moving air (He et al., 2021). Precise mathematical modeling of WECS components, such as aerodynamic conversion, turbine drive-train dynamics, and generator electrical models, is a critical part of studying the performance of the WECS, designing control methods, and the secure integrations of the WECS coupled with microgrids. The basic equations of the behavior of WECS are provided later in subsections.

6.3.2.1 Wind power equation

The power available in wind is described by Eq. (6.6):

$$P_{wind} = \frac{1}{2}\rho A V_{\omega}^{3} \tag{6.6}$$

where:

ρ = air density (kg/m^3)
A = swept area of turbine blades (m^2)
V_w = wind speed (m/s)

This equation captures the theoretical power content of the wind stream as a function of air density (ρ), rotor swept area (A), and wind velocity (V_w). The power varies with the cube of the speed of wind, and the consequence is that any slight change in wind speed produces a large change in the available energy. Nevertheless, this is the total kinetic energy within the airflow; the practically usable amount is limited by the air turbine aerodynamics and efficiency.

Wind speed is uncertain and generally modeled using a Weibull probability distribution, illustrated by Eq. (6.7):

$$f(V_w) = \frac{k}{c}\left(\frac{V_w}{c}\right)^{k-1} e^{-\left({V_w}/{c}\right)^{k}} \tag{6.7}$$

Monte Carlo sampling from this distribution generates wind scenarios used in stochastic optimization. The extracted turbine power given by Eq. (6.8):

$$P_{turbine}(t,\omega) = C_p(\lambda(t,\omega), \beta(t,\omega)) \cdot \frac{1}{2}\rho A V_w(t,\omega)^3 \tag{6.8}$$

6.3.2.1.1 Extracted turbine power

The mechanical power extracted is given by Eq. (6.9):

$$P_{turbine} = C_p(\lambda, \beta).P_{wind} \tag{6.9}$$

where:

C_p= power coefficient (function of tip-speed ratio λ and pitch angle β)
$\lambda = \frac{R.\omega_t}{V_\infty}$ the tip-speed ratio
R = turbine radius (m)
ω_t = turbine angular speed (rad/s)

The maximum C_p is limited by Betz's law ($C_{p,max}$593).

The coefficient of performance C_p defines the efficiency with which wind power is converted to shaft power and depends on the tip-speed ratio λ and blade pitch angle β. According to Betz's law, the maximum theoretical value of $C_p$is 0.593, meaning that no turbine can extract more than 59.3% of the wind's kinetic energy. Losses occur in practical turbines, which have a range of 0.35–0.45 due to friction losses and aerodynamic losses. It emphasizes the necessity of blade design optimization and pitch control to capture maximum energy at a variety of wind speeds.

6.3.2.2 Drive train dynamics

The turbine is connected to the generator through a gearbox. Eq. (6.10) is the equation of motion:

$$J\frac{d\omega}{dt} = T_m - T_e - B\omega \tag{6.10}$$

where:

J= inertia constant
T_m = mechanical torque
T_e= electrical torque
B= friction coefficient

This second-order differential equation models the rotational dynamics of the turbine–generator system. The inertia constant J accounts for stored kinetic energy in the rotating mass, which helps smooth fluctuations in wind input (Sebastián, 2021). The friction term $B\omega$ represents mechanical losses, while T_m and T_e describe the mechanical input torque from the turbine and the opposing electromagnetic torque from the generator, respectively. The stability of the WECS microgrid depends on the balance of these rotor torques, and the drive train modeling is therefore very important in the design of control in a microgrid, as well as in transient stability studies in a microgrid featuring WECS-based interconnection.

6.3.2.2.1 Generator models

For a permanent magnet synchronous generator (PMSG), the *d–q* axis model is used and represented by Eqs (6.11) and (6.12):

$$V_d = R_s i_d + L_d\frac{di_d}{dt} - \omega_e L_q i_q \tag{6.11}$$

Figure 6.10 Block diagram of the WECS.

$$V_q = R_s i_q + L_q \frac{di_q}{dt} + \omega_e (L_d i_d + \psi_f) \tag{6.12}$$

where:

i_d, i_q = d–q axis currents
V_d, V_q = d–q axis voltages
L_d, L_q = inductances
ψ_f = flux linkage
ω_e = electrical angular velocity

The d–q axis representation simplifies the analysis of three-phase machines under rotating reference frames. Here, i_d, i_q are the d–q axis currents, V_d, V_q the corresponding voltages, and ψ_f the permanent magnet flux linkage. The model has the advantage that active and reactive power can be decoupled and thus controlled through field-oriented control strategies. This is because in the modern WECS, PMSGs are highly desirable because of their high efficiency, reliability, and compatibility with variable-speed flow. In addition, no external excitation is required. Power generator modeling is therefore essential to designing the power electronic interface and implementing MPPT in wind systems. Fig. 6.10 shows the block diagram of the WECS.

6.3.3 Battery energy storage system (BESS)

BESS coupled with a DG imparts stability, reliability, and continuity of supply in the hybrid microgrid system (Nazir et al., 2022). The BESS can be a crucial contributor to interfacing fluctuations in renewable generation by storing surplus energy during peak periods and releasing energy stored during a deficit. BESS should be accurately modeled, taking into account its charge/discharge profile, voltage profile, and internal losses that can directly impact system efficiency and battery life (Miao et al., 2014). Meanwhile, on the other hand, the DG serves as a dispatchable backup supply that gives firm power when a prolonged shortage of renewables occurs. Its work is affected by the fuel conversion and governor dynamics, which maintain a balance of the frequency and power. The subsections later discuss the SOC mathematical model, battery voltage, and DG modeling, the basis of the hybrid system programming and control.

6.3.3.1 State of charge

The SOC specifies the available energy level in the battery as a fraction of its rated capacity and is one of the most important indices of the work of BESS (Hu et al., 2012). For the dynamic SOC model in Eq. (6.13), both the charging and discharging are captured along with their efficiencies. High charging efficiency (η_{ch}) implies that a greater proportion of the charging current will go to stored energy; comparatively, the discharging efficiency (η_{dis}) will account for the conversion losses in the delivery of energy (Long et al., 2020). This model allows tracking the health of batteries, which helps to avoid overcharging or deep discharging, which negatively impact the battery life cycle and safety. The SOC is modeled by Eq. (6.13):

$$SOC(t) = SOC(t_0) + \frac{1}{C_{bat}} \int_{t_0}^{t} \eta_{ch} I_{ch} dt - \frac{1}{C_{bat}} \int_{t_0}^{t} \frac{I_{dis}}{\eta_{dis}} dt \tag{6.13}$$

where:

C_{bat} = battery capacity (Ah)
I_{ch}, I_{dis} = charging and discharging currents (A)
η_{ch}, η_{dis} = charging/discharging efficiencies

Charging and discharging efficiencies are uncertain due to aging and environmental effects. Hence, the SOC under uncertainty is expressed as Eq. (6.14):

$$SOC(t,\omega) = SOC(t_0) + \frac{1}{C_{bat}} \sum_{\tau=t_0}^{t} (\eta_{ch}(\omega) I_{ch}(\tau,\omega) - \frac{I_{dis}(\tau,\omega)}{\eta_{dis}(\omega)} \Delta\tau \tag{6.14}$$

with efficiency values modeled as random variables. A chance constraint ensures battery safety:

$$\mathbb{P}(SOC^{min} \leq SOC(t,\omega) \leq SOC^{max}) \geq 1 - \delta \tag{6.15}$$

6.3.3.2 Battery voltage model

The conduction of the battery is modeled by an Equivalent Circuit Model (ECM) as in Eq. (6.16). A nonlinear potential versus SOC [open circuit voltage (E_{oc})] determines the operation baseline potential of the cell. The internal resistance (R_{int}) is associated with an ohmic loss that increases in conditions of high loads and as the battery ages. The polarization voltage (V_{pol}) represents both transient diffusion and the electrochemical reactions and has implications for dynamic response in the case of rapid changes in current. Collectively, these parameters present a realistic description of the end-of-battery voltage, which allows the battery to be successfully simulated in grid-connected and islanded applications (Shah et al., 2022).

$$V_{bat} = E_{oc} - IR_{int} - V_{pol} \tag{6.16}$$

where:

E_{oc}= open circuit voltage (function of SOC)
R_{int} = internal resistance
V_{pol}= polarization voltage due to diffusion effects

6.3.3.3 Diesel generator model

The DG can be used as a dispatchable, controllable source of energy to augment variable renewable energy resources. Its steady-state power output is regulated by the mass flow rate of fuel and its efficiency, which can be represented with Eq. (6.17). The efficiency factor *(η_{DG})* is used to characterize the efficiency of the conversion of mechanical and electrical energy, and the lower heating value (LHV) of diesel is used to specify the energy content of diesel (Droop speed control, 2025). The equation permits determining the fuel consumption at specific power requirements, which is a critical feature in an analysis of costs and emissions. The DG serves as a backup in renewable-dominant islanded systems.

$$P_{DG} = \eta_{DG} \cdot \dot{m}_f LHV \tag{6.17}$$

where:

η_{DG} = efficiency
$\dot{m}_f$ = fuel mass flow rate (kg/s)
LH = LHV of diesel (MJ/kg)

In practice, the DG's availability and fuel cost are uncertain. A stochastic cost model is defined as Eq. (6.18):

$$C_{DG}(\omega) = P_{DG}(t, \omega) \cdot C_{fuel}(\omega) \tag{6.18}$$

where $C_{fuel}(\omega)$ reflects fuel price volatility. Generator outages can be represented by a Bernoulli random variable with failure probability *pfail*.

To have stable frequency and load distribution in microgrids, the DG is fitted with a governor mechanism as shown by Eq. (6.19). The droop characteristic (*R*) is the measure of how responsive power output is to frequency changes, which allows various generators to work in parallel without having to communicate directly (Tayab et al., 2017). A smaller droop constant means greater frequency regulation at the expense of load-sharing flexibility. The tuning of droop control must thus be done properly to ensure both system stability and power quality in hybrid renewable-DG systems.

Governor dynamics are modeled as Eq. (6.19):

$$\frac{\Delta P_m}{\Delta f} = -\frac{1}{R} \tag{6.19}$$

6.3.3.4 Load modeling

This model will be realistic because there are devices that are modeled as resistive *(Z)*, current driven *(I)*, and fixed power *(P)*. The coefficients *(α, b, c)* can be optimized to represent the type of load and its sensitivity to changes in voltages; thus, the ZIP model can be applied to stability, control, and planning studies on renewable-dominated microgrids (Mehranvand & Sarrami, 2023).

Microgrid loads are often modeled as a combination of constant power, constant impedance, and constant current components (ZIP model):

$$P = P_0\left(a_p\left(\frac{V}{V_0}\right)^2 + b_p\left(\frac{V}{V_0}\right) + c_p\right) \tag{6.20}$$

$$Q = Q_0\left(a_q\left(\frac{V}{V_0}\right)^2 + b_q\left(\frac{V}{V_0}\right) + c_q\right) \tag{6.21}$$

To represent demand variability, load is expressed by Eq. (6.22) as a stochastic process:

$$L(t,\omega) = \mu_L(t) + \sigma_l(t) \cdot \xi(w) \tag{6.22}$$

where $\mu_L(t)$is the mean load forecast, $\sigma_l(t)$ is the standard deviation, and $\xi(w)$is a random variable. This enables probabilistic reliability indices such as Loss of Loads Probability (LOLP) and Loss of Load Expectancy (LOLE).

6.3.4 Power electronic converter models

Power electronic converters are the backbone of renewable-dominant microgrids, which have the further beneficial capability of facilitating the energy conversion, regulation, and bidirectional flow of power. They serve as an interface between renewable sources, storage, and load, and ensure that voltage and frequency are stable. The conversion modeling must be realistic to have switching dynamic profiles, losses, and interactions with control. System-level studies are frequently done with simplified averaged models, whereas switching models are built for controller design and transient analysis (Espín-Sarzosa et al., 2023). The models play essential roles in the evaluation of efficiency, stability, and dynamic response of the microgrids under changing renewable and loading conditions.

6.3.4.1 Inverter model

The model in the *d–q* frame Eqs. (6.23) and (6.24) represent the interaction between active and reactive current dynamics in the presence of DC-link voltage and modulation indices. This representation finds widespread application because it simplifies the AC signals representation into steady-state variables and makes control design simpler. The inverter dynamics in the *d–q* frame are expressed as:

$$V_d = L\frac{di_d}{dt} - \omega L i_q + R i_d + V_{dc} m_d \tag{6.23}$$

$$V_q = L\frac{di_q}{dt} - \omega L i_d + R i_q + V_{dc} m_q \tag{6.24}$$

where:

V_{dc} = DC-link voltage
m_d, m_q = modulation indices

Converter efficiency is subject to device aging, switching losses, and thermal variations. Hence, efficiency can be modeled as Eq. (6.25):

$$\eta_{conv}(\omega) = \eta_{nominal} - \Delta_\eta(w) \tag{6.25}$$

where $\Delta_\eta(w)$follows a probability distribution.

6.3.4.2 Droop control

In islanded microgrids, the inverters employ droop control Eqs. (6.26) and (6.27) with frequency adjustment against active power and voltage adjustment against reactive power. This allows the sharing of decentralized power among multiple inverters without the exchange of information, and this makes the systems stable. Although primary droop can provide proportional loading, secondary and tertiary controls are applied over the former to correct frequency/voltage deviations and to generate optimal economic dispatch (Morstyn et al., 2017). Inverters in islanded operation follow droop characteristics:

$$f = f_0 - K_p(P - P_{ref}) \tag{6.26}$$

$$V = V_0 - K_q(Q - Q_{ref}) \tag{6.27}$$

where K_p and K_q are active and reactive power droop coefficients.

The mathematical models of the main renewable-dominant microgrid components presented in this section include PV arrays, wind turbines, batteries, diesel generators, loads, and power converters. The models include models of steady-state and dynamic behavior and provide the basis of performance analysis, optimization, and design of controls in the later sections of this chapter. The integration of stochastic extensions into the deterministic models ensures that component behavior is represented not only under nominal conditions but also under realistic uncertain operating environments. These models provide the foundation for the stochastic optimization and risk-aware evaluation presented in the subsequent case studies.

6.4 Case studies from different geographical regions

Renewable-dominant and islanded microgrids have significantly different performance and feasibility in diverse regions and depend on the availability of resources, policy regime, and socioeconomic dynamics, as well as how they adopt the technology (Jena et al., 2021). Case studies offer realistic evidence on how high penetration PV, wind, and hybrid systems work across a variety of settings. The analysis of considerations of microgrids in Asia, Africa, Europe, North America, Australia, China, and South America may be used to derive lessons that focus on scalability, reliability, affordability, and resilience (Gómez-Expósito et al., 2021). The case studies also show how the storage systems, operation strategies, and community acceptance are integrated together, thus joining the theoretical models with practices.

6.4.1 Case study 1—India: Remote villages and renewable-dominant microgrids

India hosts one of the most diverse renewable microgrid landscapes, as there is a dual need to electrify the rural population, coupled with decarbonization needs. In remote villages in states like Rajasthan, Uttar Pradesh, and Ladakh, the villages have been disconnected from the central grid, and decentralized renewable-based microgrids are very viable and sustainable options (Bhattacharyya et al., 2021). The Indian geographic map describing the renewable dominant regions is shown in Fig. 6.11.

Although these case studies have provided evidence of deterministic values like LCOE, reliability, and renewable fraction (RF), the reality on the ground is heavily subject to stochastic changes in the solar irradiance, wind speed, fuel prices, and load demand. Therefore, the regional case studies are further elaborated with uncertainty-sensitive models, such as LOLP, LOLE, Conditional Value-at-Risk (CVaR) of costs, and probabilistic resilience indices. This enables evaluation of performance not only based on average values, but also based on realistic uncertainty distributions.

6.4.1.1 System architecture

Typical Indian renewable-dominant microgrids employ:

- PV arrays (20–200 kWp) as the main source, since India has a high solar insolation (4.5–6 kWh/m^2/day).
- Small-scale (10–50 kW) wind turbines (hybrid arrangements, mainly in the coastal regions (Tamil Nadu, Gujarat).
- BESS, initially lead-acid (early installations) but increasingly with lithium-ion chemistry.
- Diesel seasonal back-up generators.

The system is modeled using the fundamental power balance expressed by Eq. (6.28):

$$P_{PV} + P_{biomass} + P_{BESS} = P_{Load} + P_{Loss} \tag{6.28}$$

6.4.1.2 Mathematical representation

The penetration of renewables can be quantified as Eq. (6.29):

$$RPR = \frac{E_{RES}}{E_{Total}} * 100\% \tag{6.29}$$

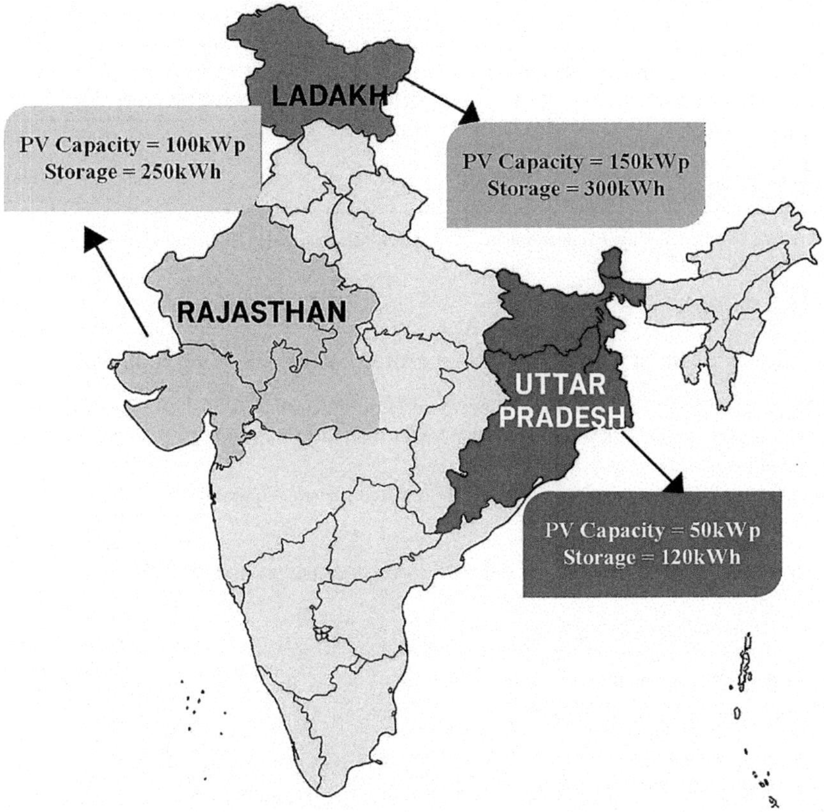

Figure 6.11 Map of the renewable dominant region of the India.
Map lines delineate study areas and do not necessarily depict accepted national boundaries.

where:

E_{RES} = annual energy supplied by renewable sources (kWh)
E_{Total} = total annual energy demand (kWh)

For high-performing microgrids in Rajasthan, values of Renewable Penetration Ratio (RPR) exceed 85%, reducing diesel dependence significantly.

To capture variability, a stochastic renewable penetration ratio (SRPR) can be expressed by Eq. (6.30):

$$SRPR = \mathbb{E}\left(\frac{E_{RES}(\omega)}{E_{Total}(\omega)}\right) \times 100\% \tag{6.30}$$

where ω denotes a scenario drawn from probabilistic solar/wind forecasts. For Rajasthan microgrids, deterministic RPR ≈ 85%–90%, but stochastic analysis shows a 10%–15% probability of falling below 70% during extended cloudy periods.

Table 6.3 Performance indicators of selected indian microgrids.

Location (state)	PV capacity (kWp)	Storage (kWh)	RPR (%)	LCOE (USD/kWh)	CO_2 reduction (%)
Rajasthan	100	250	88	0.20	78
Uttar Pradesh	50	120	82	0.19	72
Ladakh	150	300	90	0.22	80

LCOE, levelized cost of electricity; *PV*, photovoltaics; *RPR*, Renewable Penetration Ratio.

6.4.1.3 Performance metrics

The efficiency of India's microgrids is measured by the following indicators:

- LPSP: 0.05–0.1, which is within the acceptable range for the rural and residential loads.
- LCOE: USD 0.18–0.22/kWh, nearly half lower than the generation costs of diesel-only units (>USD 0.35/kWh).
- Reduction of CO_2 emission: 70%–80% relative to diesel standards, which strengthens environmental gains.

The performance indicators of the Indian region are shown in Table 6.3

6.4.1.4 Socioeconomic impact

In addition to the high technical performance, renewable-dominant microgrids bring significant socioeconomic benefits:

- Enabled 24/7 electricity access in previously unelectrified villages.
- Facilitated productive loads such as irrigation pumps, small enterprises, and cold storage.
- Enhanced educational and healthcare services through reliable lighting and refrigeration.

6.4.2 Case study 2—Africa: Islanded and off-grid microgrids for electrification

The electrification rates across Sub-Saharan Africa remain among the worst in the world, with more than half a billion still not electrified (Howells et al., 2017). In this regard, microgrids based on renewable dominance have imposed themselves as breakthrough solutions in rural electrification. Islanded microgrids have been established in countries like Kenya, Tanzania, and Nigeria that use PV, wind, and hybrid systems made up of diesel and renewable energy (Eder et al., 2022; Riva et al., 2022). Fig. 6.12 shows the African country highlighting the renewable-dominant microgrid region.

6.4.2.1 System architecture

- PV-dominated systems (20–500 kWp) for community clusters.
- Wind-PV hybrids in regions with strong coastal winds (e.g., Kenya's Lake Turkana region).
- Biomass integration where agricultural residues are abundant (e.g., Tanzania).
- Battery systems are generally lithium-ion, although cost is a major barrier.

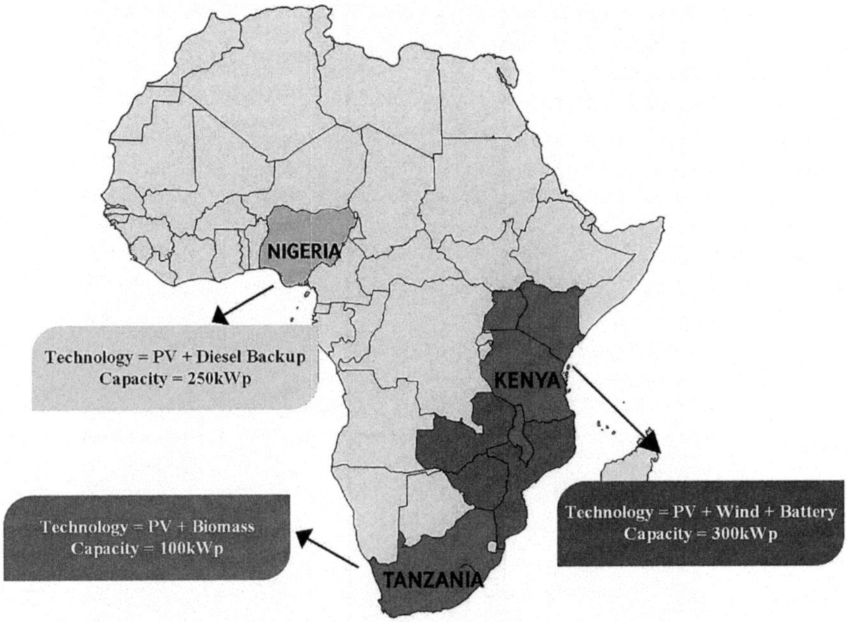

Figure 6.12 African region of the renewable-dominant microgrid.

6.4.2.2 Mathematical model

Reliability is often assessed using LOLE:

$$LOLE = \sum_{t=1}^{T} P(LOLP_t) \cdot \Delta t \tag{6.31}$$

where:

$P(LOLP_t)$= probability of loss of load at time$t$
Δt= time interval

In high-performing microgrids in Kenya, LOLE values are <20 hours/year, ensuring >99.7% reliability. In addition to mean LOLE, scenario-based LOLE distributions can be generated using a Monte Carlo simulation of solar irradiance and load. This enables quantifying the probability that annual LOLE exceeds a policy threshold (e.g., 24 hours/year), which is crucial for designing subsidy frameworks.

6.4.2.3 Performance metrics

- LCOE: USD 0.22–0.28/kWh (still cheaper than kerosene or diesel-based lighting).
- Reliability: >99% availability with hybrid storage.
- Social benefits: Supports education, healthcare, and women-led enterprises.

Table 6.4 demonstrates the performance of renewable microgrids in the sub-Saharan Africa region.

Table 6.4 Performance of renewable microgrids in sub-Saharan Africa.

Country	Technology mix	Capacity (kWp)	LCOE (USD/kWh)	Reliability (%)	Beneficiaries
Kenya	PV + wind + battery	300	0.25	99.5	1200 people
Tanzania	PV + biomass	100	0.22	98.7	700 people
Nigeria	PV + diesel backup	250	0.28	99.1	950 people

LCOE, levelized cost of electricity; *PV*, photovoltaics.

6.4.2.4 Challenges

- Affordability remains an issue as initial connection costs are high.
- Policy uncertainty slows scaling of private sector-led microgrid programs.
- Maintenance capacity in rural areas is limited, leading to downtime.

6.4.3 Case study 3—Europe: Renewable microgrids for energy transition and resilience

The European community has been one of the global leaders in terms of renewable integration, with many countries installing microgrids not only in remote sites but also in cities and industrial areas to enhance resilience and decarbonization. Microgrids have been implemented in countries such as Germany, Denmark, and Spain, which have integrated renewable-dominant and community-based microgrids. Examples include island systems in Greece and the Canary Islands, which have significantly reduced their reliance on imported diesel by combining wind and PV with storage (Bello & Bayod-Rújula, 2020; Rodríguez et al., 2007). These projects also illustrate how advanced energy management systems optimize dispatch to ensure reliability. The European renewable-dominant regions are shown in Fig. 6.13.

6.4.3.1 System architecture

- Solar PV (50–500 kWp) integrated with rooftops and community plants.
- Onshore/offshore wind farms (100 kW–multi-MW) connected to microgrids, especially in Denmark and Germany.
- Battery storage systems with capacities ranging from 200 kWh to 2 MWh.
- Hydrogen-based storage and fuel cells for seasonal balancing in pilot projects.

6.4.3.2 Mathematical representation

To evaluate energy self-sufficiency, the RF is used and expressed by Eq. (6.32):

$$RF = \frac{E_{RES}}{E_{Load}} * 100\% \tag{6.32}$$

where:

E_{RES} = renewable energy supplied (kWh)
E_{Load}= total load demand (kWh)

Community projects in Denmark report RF > 95%, enabling near-complete independence from fossil fuels. Since both renewable generation and demand are stochastic, RF can also be expressed as a distribution. Risk-averse planning uses CVaR-based RF, ensuring that renewable contribution remains above a minimum level in 95% of scenarios.

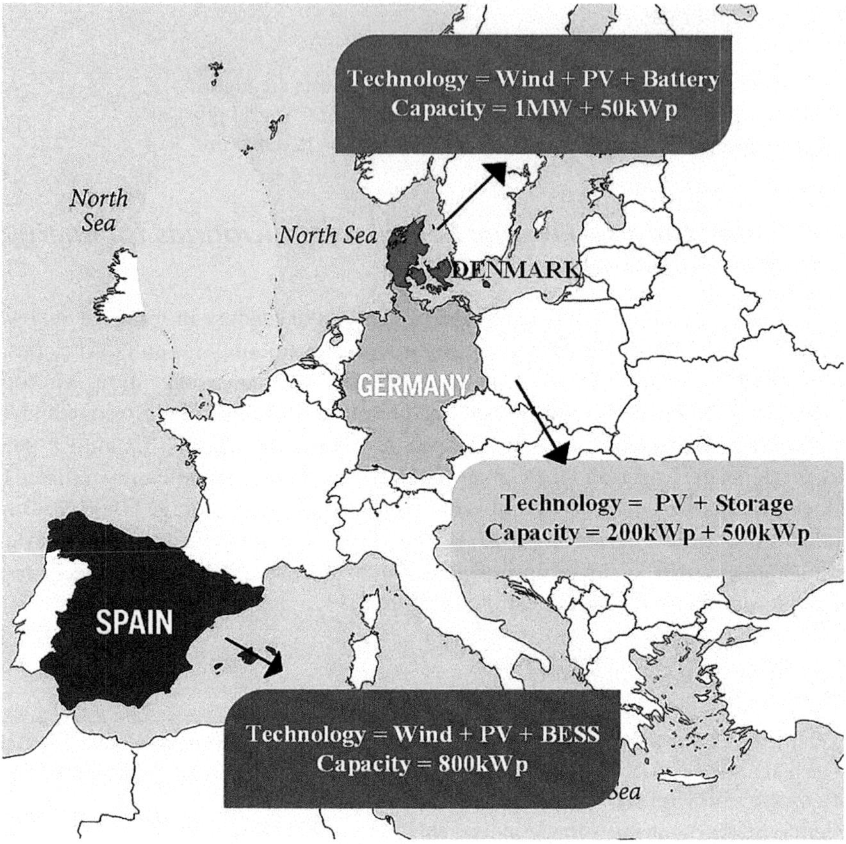

Figure 6.13 European renewable-dominant regions. *PV*, Photovoltaics.

6.4.3.3 Performance metrics

- LCOE: USD 0.15–0.20/kWh, competitive with grid prices in Europe.
- Grid resilience: Able to provide islanded operation during outages.
- Decarbonization: Up to 90% reduction in CO_2 emissions compared to natural gas-based systems.

Table 6.5 depicts the European renewable microgrid projects.

6.4.3.4 Applications

- Smart communities (Germany) with demand-response and smart appliances.
- Industrial microgrids in Spain using wind + PV + battery for peak shaving.
- Island systems in Greece and the Canary Islands, reducing reliance on imported diesel.

Table 6.5 Selected European renewable microgrid projects.

Country	Technology mix	Capacity (kWp/MW)	RF (%)	LCOE (USD/kWh)	CO_2 reduction (%)
Denmark	Wind + PV + battery	1 MW + 500 kWp	97	0.16	90
Germany	PV + storage (urban)	200 kWp + 500 kWh	92	0.18	85
Spain	Wind + PV + BESS	800 kWp	95	0.17	88

LCOE, levelized cost of electricity; *RF*, renewable fraction.

6.4.4 Case study 4—North America: Resilience-oriented microgrids

Resilient microgrids in California have been deployed on university campuses and community centers, reporting Expected Unserved Energy (EUE) close to zero even during wildfires and grid blackouts (California Energy Commission 2021). Similarly, remote communities in Alaska are adopting hybrid PV–wind–battery systems to reduce dependence on costly diesel generation (McLellan et al., 2020). In Puerto Rico, post-hurricane microgrids based on PV and storage provide critical services to hospitals and emergency facilities, showcasing resilience against extreme climate events (Figueroa & O'Neill-Carrillo, 2019). The North American renewable-dominant regions are shown in Fig. 6.14.

6.4.4.1 System architecture

- PV systems (500 kWp–5 MW) integrated in campuses and communities.
- Wind turbines (100 kW–multi-MW) in hybrid settings.
- Battery storage systems (1–10 MWh) for resilience.
- Natural gas CHP (combined heat and power) in hybrid low-carbon systems.

6.4.4.2 Mathematical representation

Resilience is quantified using EUE, and it is quantified by Eq. (6.33):

$$EUE = \sum_{t=1}^{T} (L_t - G_t).P(LOLP_t) \tag{6.33}$$

where:

L_t = load demand at time t
G_t = generation at time t
$LOLP_t$ = probability of loss of load

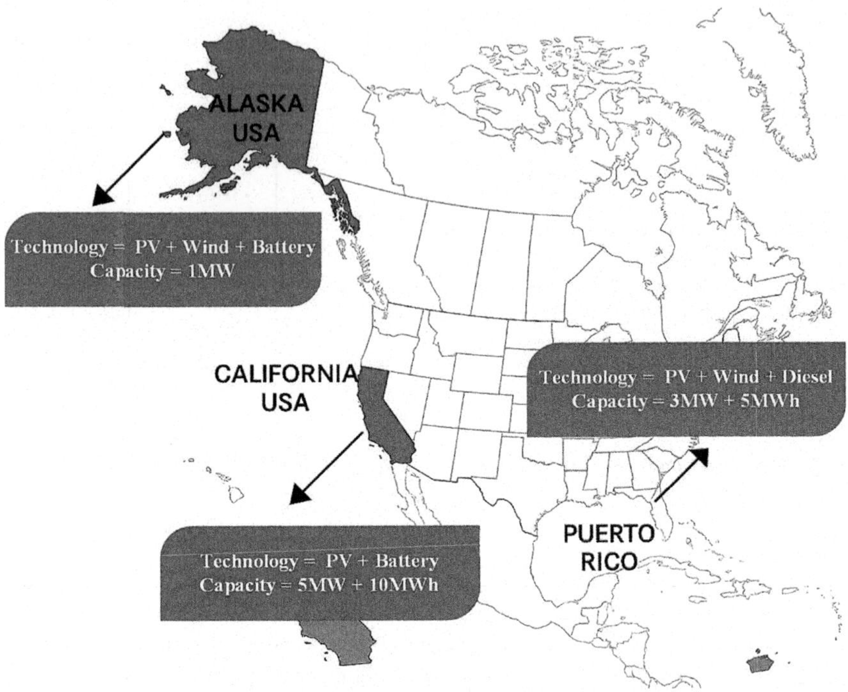

Figure 6.14 North American renewable dominant regions. *PV*, Photovoltaics.

Resilient microgrids in California campuses report EUE ≈ 0, maintaining critical loads during grid outages. Instead of only reporting mean EUE, uncertainty-aware resilience is evaluated using scenario-based distributions of EUE. For example, wildfire-prone California sites report deterministic EUE ≈ 0, but stochastic modeling reveals a 5% probability of unmet demand during multi-day cloud cover and simultaneous grid outages.

6.4.4.3 Performance metrics

- LCOE: USD 0.18–0.24/kWh (higher due to storage cost).
- Reliability: >99.9% for critical infrastructure (hospitals, data centers).
- CO_2 reduction: 70%–85% compared to the conventional grid.

Table 6.6 presents the performance indicators of North American microgrids.

6.4.4.4 Applications

- Hurricane-resilient microgrids in Puerto Rico combining PV + storage.
- Campus microgrids at universities in California and New York.
- Remote communities in Alaska are integrating PV + wind + storage to replace costly diesel.

Table 6.6 Performance indicators of North American microgrids.

Location	Technology mix	Capacity (MW)	LCOE (USD/kWh)	Reliability (%)	Application
California, USA	PV + battery	5 MW + 10 MWh	0.20	99.95	University campus
Puerto Rico	PV + battery + diesel	3 MW + 5 MWh	0.22	99.9	Community resilience
Alaska, USA	PV + wind + battery	1 MW	0.24	99.8	Remote community

LCOE, levelized cost of electricity; *PV*, photovoltaics.

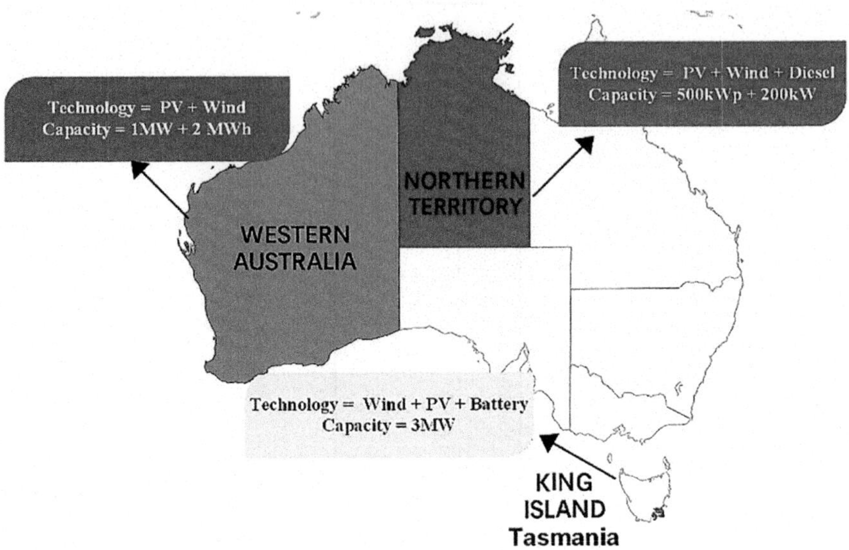

Figure 6.15 Australian region of the renewable-dominant regions. *PV*, Photovoltaics.

6.4.5 Case study 5—Australia: High renewable penetration in remote communities

Australia has pioneered renewable-dominant microgrids in remote communities. King Island in Tasmania is a notable example, where wind, solar PV, and battery storage are integrated with advanced control systems to achieve high renewable penetration and low reliance on fossil fuels (Gamble, Piekutowski, & Willems, 2011). These experiences underline the role of microgrids in transitioning isolated communities toward sustainable energy independence. Hence, Fig. 6.15 highlights the Australian renewable-dominant regions.

6.4.5.1 System architecture

- PV-dominant systems (200 kWp–multi-MW) deployed in mining towns and indigenous communities.
- Wind-PV hybrids in coastal regions and islands.
- Large-scale lithium-ion batteries (1–5 MWh) for stability.
- Diesel backup minimized to <10% contribution.

6.4.5.2 Mathematical representation

System autonomy is often defined using the Self-Sufficiency Ratio (SSR):

$$SSR = \frac{E_{RES} + E_{Storage}}{E_{Load}} * 100\% \tag{6.34}$$

In Australian microgrids, SSR exceeds 90%, indicating strong renewable independence. An SSR can be computed by integrating the distributions of solar/wind and load. For King Island, while deterministic SSR is ~94%, stochastic SSR simulations reveal a 90% confidence interval of 86%–95%, useful for resilience assessment.

6.4.5.3 Performance metrics

- LCOE: USD 0.17–0.21/kWh (competitive with diesel >0.35).
- Reliability: >99% with advanced battery management.
- Emission reductions: >80% CO_2 savings compared to diesel baselines.

The performance metrics of Australian microgrids are shown in Table 6.7

6.4.5.4 Applications

- Mining sector in Western Australia using PV + storage for 24/7 operations.
- Remote indigenous communities are benefiting from clean and affordable energy.
- Island microgrids are reducing diesel imports significantly.

6.4.6 China: Renewable-powered eco-villages and hybrid systems

China is now the global leader in adopting renewable energy and has invested in solar, wind, and battery storage. In recent years, China has implemented several eco-villages and microgrid demonstration projects aimed at showcasing the incorporation of renewable-dominant hybrid energy systems into the rural and semi-urban populations. These eco-villages are also expected not only to decrease reliance on coal-generated power but also to support local economic development and achieve carbon footprint reduction and sustainable rural electrification.

By contrast, Chinese eco-villages are meant to be self-sustainable, intelligent, and scalable energy systems, with the capability of operating in islanded mode, yet having access to the national grid. This section summarizes technical composition, performance, and socioeconomic consequences of these projects.

6.4.6.1 Technical configuration of eco-village microgrids

Typical Chinese eco-village microgrids consist of:

- Solar PV: Rooftop solar panels installed on households and community buildings, typically in the range of 500 kW–2 MW for a village.
- Wind turbines: Small to medium-sized wind turbines (100 kW–1 MW) to complement PV and provide night-time power generation.
- BESS: Lithium-ion or flow batteries sized between 1 and 4 hours of storage capacity to maintain stability.
- Diesel Generators (as backup): Retained for resilience but operated minimally.

Table 6.7 Performance of Australian microgrids.

Location	**Technology mix**	**Capacity (kWp/MW)**	**SSR (%)**	**LCOE (USD/kWh)**	**Diesel reduction (%)**
Western Australia	PV + battery	1 MW + 2 MWh	92	0.18	85
Northern Territory	PV + wind + diesel	500 kWp + 200 kW	90	0.20	80
King Island (Tas.)	Wind + PV + battery	3 MW	94	0.17	88

LCOE, levelized cost of electricity; *PV*, photovoltaics; *SSR*, self-sufficiency ratio.

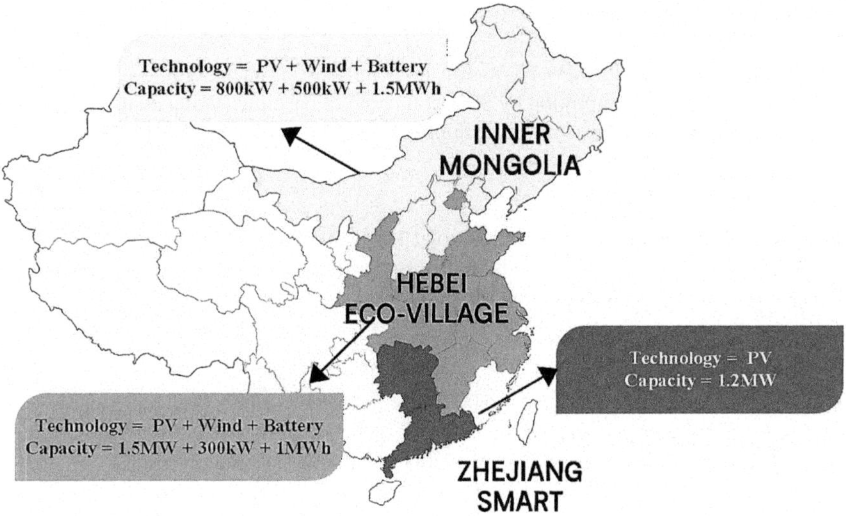

Figure 6.16 China renewable dominant eco-village regions. *PV*, Photovoltaics.

- Advanced Controllers: Supervisory controllers using MPC and AI-based demand-side management to optimize energy flows.

6.4.6.2 Mathematical modeling of hybrid systems

For such hybrid systems, optimal sizing and performance evaluation are critical. The total power balance at any instant can be represented as Eq. (6.35):

$$P_{PV}(t) + P_{WT}(t) + P_{BESS}(t) + P_{DG}(t) = P_{Load}(t) + P_{Loss}(t) \tag{6.35}$$

where:

$P_{PV}(t)$ = power from solar PV
$P_{WT}(t)$ = power from a wind turbine
$P_{BESS}(t)$ = charging/discharging power from batteries
$P_{DG}(t)$ = diesel generator backup power
$P_{Load}(t)$ = demand from households and community centers
$P_{Loss}(t)$ = line and conversion losses

In eco-village projects, scenario generation for solar/wind profiles is conducted using time-series stochastic models. DRO is then applied to ensure that the hybrid system can meet load demand under worst-case probability distributions.

6.4.6.3 Case examples from China

The Chinese eco-villages powered by renewable energy are highlighted in Fig. 6.16 and described these projects below:

Hebei eco-village project:

- Installed capacity: 1.5 MW PV + 300 kW wind + 1 MWh battery.
- Reduced diesel consumption by 85% compared to baseline (Wang et al., 2020).
- Reliability indices: Loss of Power Supply Probability (LPSP) < 0.01.

Zhejiang smart village:

- Rooftop PV across 200 households (total 1.2 MW).
- BESS of 2 MWh with AI-based demand forecasting.
- Achieved annual CO_2 reduction of 2000 tons.

Inner Mongolia hybrid system:

- A combination of 800 kW solar, 500 kW wind, and 1.5 MWh flow batteries.
- Designed for extreme weather resilience.
- Reduced blackouts from 14 hours/month to less than 1 hour/month.

6.4.7 Lessons learned from global implementations

Global renewable-dominant microgrid analysis has identified a number of lessons to be learnt in future deployments. One, the storage of energy is a key component in all applications, be it in batteries, flywheels, or pumped hydro, and as such, there is a need to have adaptable storage systemsto balance variability (Hossain et al., 2020). Second, these systems often depend on bespoke solutions to make visible the availability of local resources, demand, and socioeconomic factors, such as community-owned solar in India or a hybrid wind-hydro system in Spain. Third, favorable policy frameworks, subsidies, and financing mechanisms are still needed to counter high initial capital expenditures, especially in developing areas. Lastly, social acceptance and community participation prove to be fundamental forces behind long-term sustainability, proper maintenance, tariff adherence, and fair gains. A key insight from global experiences is that uncertainty-aware design—using stochastic optimization, scenario reduction, and risk metrics such as CVaR—results in more resilient and cost-effective microgrids than purely deterministic planning, especially in climate-vulnerable and fuel-volatile regions. Altogether, these lessons illustrate how renewable-dominant and islanded microgrids are not only technically feasible but also scalable avenues toward resilient and decentralized energy systems globally (REN21 2023).

6.5 Comparative performance analysis of renewable-dominant microgrids

The comparative evaluation of renewable dominant microgrids offers an integrated picture of the effectiveness of their operations, economic sustainability, environmental and socio-political acceptability. The performance of these systems can be examined in connection not only with technical parameters, but also with financial, environmental, and social aspects, as they are highly context-dependent. This part will provide an evidence synthesis of the current projects and research studies to achieve a systematic evaluation framework, strengths and weaknesses, and critical success factors that can be used to guide future implementation and policy intervention.

6.5.1 Technical performance attributes

Microgrid evaluation is still based on technical performance because the resilience of the system and operational reliability determine the practicality of microgrids in real-life scenarios. System reliability, power quality, efficiency, and disturbance resilience are some of the critical indicators. Reliability metrics such as System Average Interruption Duration Index (SAIDI) and System Average Interruption Frequency Index (SAIFI) are used to quantify the reliability performance of power supply on microgrids in comparison with centralized grids. When combined with storage and intelligent control, renewable-dominant microgrids exhibit significant gains in the outage frequency and duration in remote and islanded areas (Hirsch & Guerrero, 2018; Lasseter et al., 2004).

Another important parameter is power quality, which can be measured by voltage regulation, frequency stability, and harmonics. The most notable contribution to power quality in a hybrid microgrid dominated by solar and wind is advanced converter topologies, digital signal processing, and real-time control schemes. Energy losses can be reduced by implementing efficiency measures, either as round-trip efficiency in storage-coupled systems or as energy conversion efficiency in general (Kanase-Patil et al., 2010). Although generation based on renewable sources will necessarily lead to a lower thermal efficiency concern, storage integration and bidirectional converters dictate the net efficiency (Wang & Wang, 2015). Finally, resilience (i.e., capacity to withstand and recover from natural disasters or cyber-physical attacks) of the system is increasingly quantified with resilience indices that measure restoration speed, load prioritization, and adaptive control strategies (Moreira et al., 2015).

$$SAIDI = \frac{\sum_{i=1}^{N} U_i N_i}{\sum_{i=1}^{N} N_i} \tag{6.36}$$

$$SAIFI = \frac{\sum_{i=1}^{N} \lambda_i N_i}{\sum_{i=1}^{N} N_i} \tag{6.37}$$

where U_i is the outage duration, λ_i is the outage frequency, and N_i is the number of affected customers. SAIDI/SAIFI can be modeled stochastically with probability distributions of outage duration and frequency.

- Example: European microgrids achieve SAIDI < 0.5 hours/year, whereas rural African systems record SAIDI > 30 hours/year due to weak infrastructure.

1. Power Quality is assessed by Total Harmonic Distortion (THD):

$$THD = \frac{\sqrt{\sum_{n=2}^{\infty} V_n^2}}{V_1} \times 100\% \tag{6.38}$$

where V_1 is the fundamental RMS voltage and V_n the *nth* harmonic. THD can be represented as a probabilistic index due to inverter switching randomness.

- IEEE 519 standard recommends THD < 5%.
- Smart inverters in Germany and China achieve THD ≈ 2%–3%, but low-cost setups in remote Asia report THD > 8%.

2. Voltage Deviation Index (VDI):

$$VDI = \frac{|V_{measured} - V_{nominal}|}{V_{nominal}} \tag{6.39}$$

- Voltage deviations > ±5% can damage appliances.
- Microgrids with advanced droop control (e.g., Japan, United States) maintain VDI < 2%, while developing regions often experience VDI up to 10%.

3. Efficiency of renewable-storage integration is expressed as Eq. (6.40):

$$\eta_{rt} = \frac{E_{discharge}}{E_{charge}} \times 100\% \tag{6.40}$$

where E_{charge} and $E_{discharge}$ denote battery input and output energies.

- Lithium-ion batteries typically achieve 90%–95%, while lead-acid systems remain at 70%–80%.
- Flow batteries in pilot projects (United States, China) report 65%–75% efficiency.

4. Overall System Efficiency (η_{SYS}):

$$\eta_{SYS} = \frac{E_{delivered}}{E_{input}} * 100\% \tag{6.41}$$

Hybrid solar-wind-storage microgrids in India and Chile achieve 65%–75%, while diesel-dominated systems fall below 45%.

5. Resilience Index (RI) can be expressed by Eq. (6.42):

$$RI = \frac{\int_0^T L_{served}(t)dt}{\int_0^T L_{demand}(t)dt} \tag{6.42}$$

where $L_{served}(t)$is the load successfully served after a disturbance and $L_{demand}(t)$is the total demand.

6. Loss of Power Supply Probability (LPSP): A measure of how often the system fails to meet the demand. Its lower values indicate higher reliability.

$$LPSP = \frac{\sum E_{unserved}}{\sum E_{demand}} \tag{6.43}$$

Developed regions (e.g., Germany, United States) achieve LPSP < 1%, while isolated island systems (Pacific Islands, Africa) often report LPSP between 5%–15%. CVaR-based LPSP as a risk measure for planning. Table 6.8 represents the technical performance comparison across the region.

6.5.2 Economic performance attributes

The economic viability of renewable-dominant microgrids is mainly measured in terms of capital expenditure (CAPEX), operational expenditure (OPEX), LCOE, internal rate of return (IRR), and payback periods. LCOE is still used to compare microgrids to traditional generation because it considers installation, maintenance, fuel costs (where applicable), and system life span (Ahmed et al., 2019). Empirical data indicate that renewable-intensive microgrids, despite the increased upfront CAPEX associated with storage systems and sophisticated controllers, tend to have competitive or even lower LCOE than systems based on diesel in remote environments. This is associated with insignificant fuel expenses and the declining prices of solar PV and battery technologies (Irena, 2020).

Microgrids are also shaped by finance mechanisms that determine their economic viability, including public-privatization partnerships, community cooperatives, and pay-as-you-go (PAYG) option schemes. An example of this is that community-owned systems typically enjoy better payment compliance and quicker capital recovery. Subsidies, carbon credits, and special loans based on policy further lower the effective LCOE, and renewable-based microgrids become even more appealing in regions of development (Vásquez et al., 2016). The sensitivity analysis of projects shows that the most important parameters in determining economic success are the cost of the battery and the interest rate of financing (Louie, 2016; Mandelli et al., 2016).

Across global microgrid projects:

- India and Africa: CAPEX was relatively low (~USD 1200–1800/kW) due to smaller-scale PV and hybrid installations. However, OPEX remained high (USD 200–300/kW/year), primarily because of battery replacements and diesel backup.
- Europe and North America: CAPEX was higher (~USD 2000–2500/kW), but OPEX was comparatively lower (USD 80–150/kW/year) due to advanced control systems and reliable storage technologies.
- Australia and Chile: Costs were moderate to high (~USD 1800–2300/kW), with OPEX highly dependent on remoteness and logistics of component delivery.

1. The LCOE is the most widely adopted measure and expressed by Eq. (6.44):

$$LCOE = \frac{\sum_{t=0}^{N} \frac{I_t+M_t+F_t+R_t}{(1+r)^t}}{\sum_{t=1}^{N} \frac{E_t}{(1+r)^t}} \tag{6.44}$$

I_t : investment (CAPEX, replacements)
M_t : fixed + variable OandM
F_t : fuel (diesel/biogas)
R_t : residual costs (e.g., compliance) minus salvage value if included as negative in year N

Table 6.8 Technical performance comparison across regions.

Region / use case	Reliability (SAIDI, SAIFI, hrs/year)	Power quality (THD, VDI)	Round-trip efficiency (η_rt %)	Overall system efficiency (η_SYS %)	Resilience index (RI)	LPSP (%)	Key notes
Europe (Germany, Denmark – community PV-wind-BESS)	SAIDI < 0.5; SAIFI < 1	THD ≈ 2%–3%; VDI < 2%	Li-ion BESS: 92–95	70–75	RI > 0.9	< 1	Strong regulatory framework, smart inverters widely deployed.
USA (Campus microgrids, PV-BESS-CHP)	SAIDI ≈ 1–2; SAIFI ≈ 1–2	THD 2%–4%; VDI < 2%	90–93	68–72	RI ≈ 0.85–0.9	< 1	CHP improves resilience; emphasis on critical load support.
Japan (Urban microgrids with droop control, PV-wind-BESS)	SAIDI < 1; SAIFI < 1	THD < 3%; VDI < 2%	90–94	70–78	RI ≈ 0.9	< 1	Advanced droop control ensures stability during disasters.
India (Rural PV-BESS-diesel hybrid)	SAIDI 15–25; SAIFI 20–40	THD 5%–7%; VDI ≈ 5%–8%	Li-ion: 85–90; Lead-acid: 70–80	55–65	RI ≈ 0.6–0.75	5–10	Diesel share raises OPEX; microgrid programs expanding.
Africa (Village PV-diesel-BESS)	SAIDI > 30; SAIFI > 40	THD 6%–10%; VDI 8%–10%	70–80	45–55	RI ≈ 0.5–0.65	10–15	Weak grid infrastructure; resilience is low due to fuel reliance.

(*continued on next page*)

China (Eco-villages PV-wind-biogas-BESS)	SAIDI ≈ 2–5; SAIFI ≈ 2–5	THD 3%–4%; VDI ≈ 3–4%	88–92	65–70	RI ≈ 0.8–0.85	2–5	Biogas integration improves sustainability; government-backed pilots.
Australia (Remote/island PV-wind-BESS)	SAIDI 5–10; SAIFI ≈ 5–10	THD 3%–5%; VDI ≈ 3–5%	85–90	60–70	RI ≈ 0.75–0.85	3–7	High penetration of renewables; storage key for reliability.

CHP, combined heat and power; *LPSP*, loss of power supply probability; *OPEX*, operational expenditure; *PV*, photovoltaics; *SAIDI*, system average interruption duration index; *SAIFI*, system average interruption frequency index; *THD*, total harmonic distortion; *VDI*, voltage deviation index.

E_t: energy delivered
r: discount rate (WACC)
N: analysis life (typically 15–25 years)

In stochastic analysis, LCOE is no longer a single value but a probability distribution, since fuel prices, demand growth, and component lifetimes are uncertain. A CVaR-based LCOE ensures that the expected cost in the worst 5%–10% of scenarios remains within acceptable bounds.

6.5.2.1 Levelized cost of storage

$$LCOS = \frac{\sum_{t=0}^{N} I_t^{BESS} + M_t^{BESS} + R_t^{BESS}}{\sum_{t=0}^{N} \frac{E_{dis,t}^{BESS}}{(1+r)^t}} \tag{6.45}$$

It focuses solely on storage economics and is useful for comparing Li-ion versus flow batteries or hybrid storage stacks.

1. Capital Recovery Factor (CRF) and Annualized CAPEX

$$CRF = \frac{r(1+r)^N}{(1+r)^N - 1}, \; I_{annual} = I_0.CRF \tag{6.46}$$

It allows annualizing up-front CAPEX when cash flow detail is limited.

2. Net Present Value (NPV) for investment decision-making:

$$NPV = \sum_{t=0}^{N} \frac{B_t - C_t}{(1+r)^t} \tag{6.47}$$

where B_t are benefits (avoided diesel/grid purchases, tariff revenue, carbon credits), C_t are costs.

Key sensitivities that drive LCOE/NPV:

- Fuel price pf (diesel, LNG): Most material for hybrids; +20% diesel price can improve PV-BESS NPV by 15%–25%.
- Battery replacement year and cost curve: Shifting replacement from year 8→10 can lower LCOS 5%–10%.
- Discount rate r: Higher risk (islands, new developers) → higher r → higher LCOE.
- Utilization (capacity factor/curtailment): Under-utilized assets inflate LCOE.
- Soiling/derate (Atacama dust): Unmanaged soiling can add 3%–6% LCOE.

1. The payback period (PP) indicates how quickly the investment cost can be recovered:

$$PP = min\left\{ t \middle| \sum_{k=1}^{t} (R_k - C_k) \geq I_o \right\} \tag{6.48}$$

with I_o being an initial investment.

- India and Africa: PP typically ranged from 6 to 10 years, heavily influenced by subsidies and diesel price volatility.
- Europe and North America: Shorter payback of 4–6 years, supported by feed-in tariffs, renewable credits, and tax incentives.
- Australia and Chile: 5–7 years, with strong dependence on solar availability and government incentives.

Return on investment (ROI) was higher in Chile and Europe (15%–20%) compared to India and Africa (8%–12%). The economic performance comparison across the region is shown in Table 6.9.

6.5.3 Environmental and sustainability performance attributes

The microgrids are assessed based on their performance in the reduction of greenhouse gas, the fraction of renewable energy, the size of the land used, and the sustainability of materials. However, in most cases, for a given setup size, replacing diesel with renewable-based plant results in very high CO_2 emission factors of over 60%–80% of RF and storage capacity. RF is the percentage of total energy requirement supplied by renewable sources. It is a simple but effective measure of sustainability. Hybrid systems that combine solar, wind, and storage can typically reach RF over 70%, although some projects on islands reach nearly 100% (Tascikaraoglu & Erdinc, 2012).

Lifecycle assessment (LCA) places trade-offs between lower operational emissions and environmental costs of battery manufacturing and disposal. This highlights the significance of recycling processes, second-life applications of batteries, and use of environmentally friendly chemistries such as lithium iron phosphate (LFP) (Alam et al., 2017). There is also land use implication, notably where large solar PV arrays exist, where dual-use solar PV concepts like agrivoltaics, are agricultural production and generation of renewable energy (Palit & Malhotra, 2013). The environmental profile of renewable-dominant microgrids is therefore an indicator of the benefits of operation as well as material challenges that will need continuous innovation.

1. The renewable energy fraction (REF) can be expressed as Eq. (6.49):

$$REF = \frac{E_{RES}}{E_{TOTAL}} * 100\% \tag{6.49}$$

where:

E_{RES} = annual energy from renewable sources (MWh), E_{TOTAL} = total annual microgrid energy supply (MWh)

a. India and Africa: RF ~60%–75% (solar-diesel-battery hybrids).
b. Europe and North America: RF ~85%–100% (high wind, hydro, and PV penetration).
c. Australia and Chile: RF ~70%–90%, often backed by lithium-ion batteries for stability.

Uncertainty in renewable generation also translates into uncertainty in RF and CO_2 reduction. Probabilistic LCA incorporates distributions of battery lifetime and recycling efficiency, yielding confidence intervals for GHG reductions rather than single-point estimates.

Table 6.9 Economic attributes comparison across the regions.

Region / case study	CAPEX (USD/kW)	OPEX (USD/kW/year)	LCOE (USD/kWh)	Payback period (years)	ROI (%)	Financing model
India (Village MG)	1200–1500	220–280	0.12–0.18	7–9	8–12	Community + subsidy
Africa (Hybrid MG)	1400–1800	250–300	0.14–0.18	8–10	9–12	Community / PPP
Europe (Island MG)	2000–2500	100–150	0.08–0.12	4–6	15–18	PPP/utility partnerships
North America (Resilient MG)	2200–2400	80–120	0.08–0.10	5–6	15–20	PPP/private utility
Australia (Remote MG)	1800–2300	150–200	0.10–0.15	5–7	12–16	Utility/private sector
Chile (Atacama MG)	1900–2200	90–140	0.07–0.09	5–6	16–20	Green bonds/climate funds

CAPEX, capital expenditure; *LCOE*, levelized cost of electricity; *OPEX*, operational expenditure; *PPP*, public and/or private partnerships; *ROI*, return on investment.

2. Carbon Emission Reduction (CER)

One of the most critical contributions of microgrids is the mitigation of greenhouse gas (GHG) emissions by displacing fossil fuels.

a. Diesel-based systems emit ~0.7–0.9 kg CO_2/kWh.
b. Renewable-based microgrids (solar/wind + storage) emit as low as ~0.05–0.15 kg CO_2/kWh over their lifecycle.

Case observations:

a. India and Africa: Hybrid solar-diesel systems reduced CO_2 emissions by 40%–55% compared to diesel-only.
b. Europe and North America: 100% renewable/islanded microgrids achieved 80%–95% emission reductions.
c. Australia and Chile: High solar penetration cut emissions by 70%–85%, particularly in mining and desert regions.

$$CER = \left(\frac{E_{fossil}.EF - E_{RES}.EF_{RES}}{E_{total}} \right) \times 100\% \tag{6.50}$$

where EF and $EF_{RES}$are emission factors for fossil-based and renewable energy.

3. Life Cycle Assessment (LCA) GHG balance:

$$GHG_{net} = GHG_{op} + GHG_{manuf} - GHG_{avoided} \tag{6.51}$$

where GHG_{op} = operational emissions, GHG_{manuf} = embodied emissions, and $GHG_{avoided}$ = displaced emissions by renewables.

Table 6.10 demonstrates the environmental attributes across the region.

6.5.4 Social and policy performance indicators

In addition to the technical and economic aspects, the sustainability of microgrids over the long term will also rely on their social acceptance and conformity to policy frameworks. Social performance indicators are improvements in energy access, community involvement in planning and operation, employment, and inclusivity by gender. Effective projects have always proved that community participation in design and decision-making processes increases the adoption of systems, decreases vandalism, and increases payment dependability. Microgrids can also be an economic activity locally, where they can be used, for example, for small businesses, agricultural processing, and digital connections (Howells et al., 2017).

Regulatory certainty on ownership models, tariff regimes, and interconnection standards is critical on the policy front. In regions where good policies like net metering, renewable portfolio standards, or even renewable feed-in tariffs are in place, microgrids have shown excellent financial and operational outcomes. In addition, policies that accelerate the process of adopting resilience and renewable-heavy designs encourage incentives to integrate storage and smart control systems (Rehmani et al., 2018). On the other hand, the absence of regulatory certainty, standard models, and

Table 6.10 Environmental attributes across the regions.

Region / case study	Renewable fraction (REF, %)	CO_2 emission reduction (CER, %)	Typical emission intensity (kg CO_2/kWh)	Land use / footprint	Recycling and end-of-life strategy	Key sustainability notes
India (solar-diesel-battery villages)	60–70	40–55	0.25–0.35 (vs. 0.8 diesel)	High (farmland used for PV)	Limited battery recycling, emerging second-life reuse	Strong Sustainable Development Goals (SDG) contribution, rural electrification
Africa (hybrid solar-diesel MGs)	65–75	45–55	0.30–0.40	Moderate (savannah land)	Weak recycling frameworks	Key driver for rural access, social benefits
Europe (islanded renewable MGs)	90–100	80–95	0.05–0.15	Low (agrivoltaics mitigate land stress)	Strong PV and Li-ion recycling	Climate leadership, policy-driven
North America (resilient MGs)	85–100	80–90	0.05–0.20	Low (diverse land availability)	Advanced recycling, circular economy models	Disaster resilience, grid reliability
Australia (remote desert MGs)	75–85	70–85	0.10–0.20	Low (desert installations)	Moderate recycling practices	Mining sector sustainability
Chile (Atacama Desert MGs)	80–90	75–85	0.08–0.18	Very Low (abundant desert land)	Moderate PV recycling systems	Extreme solar resource, exportable lessons

PV, photovoltaics.

political resistance to decentralized systems are factors that affect performance results negatively.

a) Energy Access Index (EAI):

$$EAI = \frac{H_{served}}{H_{total}} \tag{6.52}$$

where H_{served} are households connected and H_{total} total in the region.

1. Job Creation Index (JCI)

$$JCI = \frac{J_{microgrid}}{P_{capacity}} \tag{6.53}$$

where $J_{microgrid}$ = direct and indirect jobs created, and $P_{capacity}$= installed capacity (MW).

2. Policy Support Factor (PSF): (qualitative-to-quantitative scale)

$$PSF = \alpha \cdot S + \beta \cdot T + \gamma \cdot I \tag{6.54}$$

where

S = subsidy intensity, T = tariff regulation score, I = incentive/financing support, α, β, γ = weighting coefficients.

Table 6.11 depicts the social and policy performance indicators across the regions.

6.5.5 *Integrated assessment and comparative insights*

A combined approach based on technical, economic, environmental, and social aspects creates a complete picture of renewable-dominant microgrids. Comparative studies show that despite the dominance of technical and economic indicators in design considerations, the long-term sustainability and acceptance by the community is usually determined by environmental and social indicators. The projects that have a balance in all four dimensions will perform better compared to projects that are concentrated on a narrow scope of financial or technical feasibility. To illustrate this, increased levels of renewable energy with optimized storage by microgrids, beyond mitigating levels of emissions, may also reveal greater reliability and low LCOE. Equally, community-based initiatives tend to do well in attaining superior collection and system durability with moderate technical complexity.

Finally, there are the renewable microgrids, a revolutionary approach to energy infrastructure integrating highly modern technology with a socioeconomic novelty. Their relative performance by region and configuration proves that, despite the challenges, especially in storage costs and regulatory hurdles, the model has achieved better results in resilience, sustainability, and inclusiveness. a) A Multi-Criteria Decision-Making (MCDM) approach, such as Analytic Hierarchy Process (AHP) or Technique for Order Preference by Similarity to Ideal Solution (TOPSIS), is often applied to integrate these dimensions.

Table 6.11 Social and policy performance indicators across the regions.

Region	EAI	JCI	PSF	Community participation	Gender inclusivity	Productive use enablement
India	0.72	38	0.68	Medium–high (community-based cooperatives are active)	22% women in Self-Help Groups (SHGs) are linked to microgrids	Agro-processing, rural Small and Medium-sized Enterprises (SMEs), digital kiosks
Sub-Saharan Africa	0.54	45	0.52	High (NGOs and local councils heavily engaged)	28% women in operation/beneficiary roles	Irrigation, cold storage, mobile charging hubs
Southeast Asia	0.61	34	0.63	Medium (community boards partially engaged)	20%	Fisheries, rural electrification for Information and Communication Technology (ICT) services
Europe (remote islands)	0.89	25	0.81	Medium (policy-driven rather than community-driven)	32%	Tourism services, water desalination
Latin America	0.67	41	0.70	High (cooperatives and local participation models are strong)	27%	Agro-industry, education, e-mobility charging
North America (indigenous / remote)	0.84	29	0.78	Medium–high (tribal councils involved in planning)	24%	Digital inclusion, local manufacturing support

EAI, energy access index; *JCI*, job creation index; *PSF*, policy support factor.

For TOPSIS, the performance score of a microgrid project is:

$$C_i^* = \frac{D_i^-}{D_i^+ + D_i^-} \tag{6.55}$$

where:

D_i^+ = Euclidean distance from positive-ideal solution,
D_i^- = Euclidean distance from negative-ideal solution.

This enables ranking of alternative microgrid designs based on their relative closeness to the optimal system. MCDM can also be extended with stochastic inputs, where each criterion (technical, economic, environmental, social) is treated as a random variable. Monte Carlo-based TOPSIS allows ranking of alternatives under uncertainty, providing policymakers with probabilistic confidence levels in ranking results.

The comparative metrics across the region are shown in Table 6.12.

6.5.6 Worked example: Application of performance metrics—PV+BESS versus diesel-only

To illustrate how the economic, environmental, and resilience indicators developed in the previous sections are applied in practice, we compare two alternative microgrid configurations supplying the same annual demand of 9000 MWh:

- Case A—A PV + BESS island microgrid with modest diesel backup (~15% of demand).
- Case B—A diesel-only microgrid.

Common assumptions

- Lifetime $T = 20$ years, discount rate $r = 9\%$.
- Annuity factor:

$$AF = \frac{1-(1+r)^{-T}}{r} \approx 9.133$$

- Present-value energy (denominator for LCOE):

$$E_{PV} = E_{annual} \times AF = 9000 \times 9.133 = 82,197 MWh$$

1. **Case A—PV + BESS**

Inputs:

- Initial CAPEX I0 = \$6.0M
- O&M annual = \$220,000
- Fuel (diesel backup ≈ 8%) annual = \$80,000
- Battery replacement in year 10 = \$1,000,000

Table 6.12 Comparative metrices across the region.

Region	**Technical score (reliability and Renewable Energy Sources (RES) penetration)**	**Economic score (LCOE, payback)**	**Environmental score (CO_2 reduction, RE share)**	**Social/policy score (EAI, JCI, PSF)**	**TOPSIS performance score (C*_i)**	**Ranking**
India	0.72	0.65	0.68	0.70	0.69	3
Sub-Saharan Africa	0.60	0.58	0.74	0.66	0.64	4
Southeast Asia	0.67	0.63	0.70	0.62	0.66	5
Europe (remote islands)	0.84	0.78	0.81	0.76	0.80	1
Latin America	0.71	0.67	0.75	0.72	0.71	2
North America (indigenous/remote)	0.78	0.74	0.76	0.74	0.76	2

EAI, energy access index; *JCI*, job creation index; *LCOE*, levelized cost of electricity; *PSF*, policy support factor.

Present-value costs:

$$PV(O\&M) = 220,000 \times AF \approx \$;2.01\text{M}$$

$$PV(Fuel) = 80,000 \times AF \approx \$;0.73\text{M}$$

$$PV(Replacement) = \frac{1,000,000}{(1.09)^{10}} \approx \$;0.42\text{M}$$

$$C_A = 6.0 + 2.01 + 0.73 + 0.42 \approx \$;9.16\text{M}$$

LCOE:

$$LCOE_A = \frac{C_A}{E_{PV}} = \frac{9.16M}{82,197} \approx 111.5\$;/\text{MWh} \approx 0.112\$;/\text{kWh}$$

Emission (with RF = 85%)

$$RESemissions = 0.85 \times 9,000,000 \times 0.1 = 765,000kg$$

$$Dieselemissions = 0.15 \times 9,000,000 \times 0.8 = 1,080,000kg$$

$$Total = 1,845,000\ kg = 1845tCO_2$$

Baseline diesel-only emissions = 7200 tCO_2 → reduction = **74.4%**.

Resilience: During a 24 hour outage (demand 24 MWh), storage + PV supplies 20 MWh →

$$RI = \frac{20}{24} = 83.3\%$$

1. **Case B—Diesel Only**

Inputs:

- Fuel + variable O&M = \$0.28/kWh

Annual cost:

$$9,000,000 \times 0.28 = \text{w}\$;2.52\,M/yr$$

LCOE:

$$LCOEB \approx 0.28\$;/kWh$$

Emissions:

$$9,000,000 \times 0.8 = 7,200,000\,kg = 7200\,tCO_2$$

Table 6.13 Comparative insights between case A and case B.

Metric	PV + BESS (case A)	Diesel only (case B)
LCOE ($/kWh)	0.112	0.280
CO_2 emissions (t/year)	1845	7200
Emission reduction	74%	–
Renewable fraction (RF)	85%	0%
Resilience index (RI)	83%	0%

LCOE, levelized cost of electricity; *PV*, photovoltaics.

The comparison between case A and case B is demonstrated by Table 6.13

- Economic: PV+BESS reduces LCOE by ~60% compared to diesel.
- Environmental: ~74% CO_2 cut at 85% renewable penetration.
- Resilience: Significant improvement in outage coverage.
- Payback: Annualized cash flows suggest a simple payback ≈ of 3.8 years versus the diesel benchmark.

Sensitivity illustration

Diesel price +20%: Diesel LCOE rises from 0.28 → 0.336 $/kWh, PV + BESS barely changes (0.112 → 0.113 $/kWh).

Battery deferred to year 12 with 20% cheaper cost: PV + BESS LCOE drops further to 0.110 $/kWh.

6.6 Challenges and lessons from existing projects

Implementation of renewable-dominant and islanded microgrids all over the world has proved to have enormous potential and enduring challenges. Pilot projects in islands, rural villages and industrial campuses have proved that high levels of renewable penetration are possible, but these projects also reveal serious technical, financial, policy and social limitations. The insights gained due to these ongoing projects can be invaluable to scale microgrids sustainably and efficiently (Saleh, 2024). Another important issue regarding islanded renewable microgrids is the uncertainty management. Solar irradiance, wind speed, load profiles cannot be perfectly predicted and add stochastic risks to energy balance. There is also the issue of the uncertainty of the parameters in battery degradation models, which causes less accuracy in lifecycle assessment. Further, flexible diesel prices and policy incentives are also contributing to the risks in planning. Future microgrid models must then have robust and distributionally robust optimization (DRO) techniques that directly consider such uncertainty as opposed to deterministic assumptions.

6.6.1 Technical and operational challenges

Renewable intermittency is a key technical challenge. One such case is an Alaska microgrid where Kodiak Island, which is a hydro-wind-battery combination, has been facing significant operational challenges in stabilizing the grid in case of a sudden wind change. Complex control systems such as stabilizers that controlled frequency changes were needed (Clamp, 2020; Mountain & Room, n.d.). In a similar manner, the Bornholm Smart project in Denmark also proposed that inverter-based systems, even though allowing the high level of penetration of renewables, can make the synchronization and fault management issues more difficult and require more vigorous hierarchical control procedures (Haug et al., 2025). Those examples demonstrate that relying operation requires strong forecasting and strong control architecture.

6.6.2 Economic and financial barriers

It is economically viable particularly in low-income areas. This can be seen in the case of Uttar Pradesh Solar Microgrid Program in India. Even though the solar PV systems assisted to a very large extent in breaking the dependence on diesel, the initial capital expenditure and irregular tariff compensations undermined the sustainability of the economic system. In comparison, the Brooklyn Microgrid in New York has proven that new sources of revenue, other than the standard tariff, could be achieved through innovative financing infrastructure, including peer-to-peer (P2P) energy trading, facilitated by blockchain. These examples highlight the fact that, although renewable microgrids can be cost-competitive, success requires innovative financing systems, long-term incentives, and flexible tariff systems.

6.6.3 Policy, regulatory, and institutional gaps

There is uncertainty in the form of unclear policies regarding the ownership and operation of microgrids in most countries. As an example, ambiguous regulatory frameworks concerning the tariffs and the integration of utilities severely affected the functioning of several microgrids in Sub-Saharan Africa, which resulted in conflicts between the developers and the public utilities. On the other hand, El Hierro Island in Spain project on a hybrid system of wind and pumped hydro storage was successful due to favourable regulation and government finances. This example demonstrates that properly developed policies and strong institutional functions may contribute to the continuity of projects and their scalability (Frydrychowicz-Jastrzębska, 2018).

6.6.4 Social and community engagement challenges

The success of the microgrids in the long term is often defined by the participation of the community. Poor reliability and frequent downtime were observed in Nepal with rural microgrids that hydropower was operated by local people and lacked training. Tariff collection and maintenance were always a problem in communities and this weakened faith in the system. Conversely, the JUMEME Solar Hybrid Microgrid project in

Tanzania demonstrated that long-term sustainability increases when communities are consulted throughout the planning process and they are aware of the benefits of the system (Fatma, 2019). It was run under the scheme of SEFA and initially provided services to 16 villages and more than 82,000 people (Pueyo et al., 2022). The transparent model of governance and ownership by local people facilitated good operation through active participation and trust in the community.

6.6.5 Key learnings from global implementations

The evaluation of the case studies in various geographical locations demonstrates some of the lessons that can help design, operate, and scale up renewable-dominant and islanded microgrids. These lessons are not only technological, but also economic, policy, and socio-cultural societal processes.

6.6.5.1 Technical lessons

- Hybridization is essential: No renewable source alone can effectively operate to handle the load with these dynamic requirements. Solar-wind-battery hybrids supplemented by diesel or bioenergy back-up provide the best answers in remote and islanded systems.
- Storage is the most important element: Energy storage systems-lithium-ion batteries and their emerging version, flow batteries-are critical to managing intermittency. Case studies on India and Australia also demonstrated that systems that fail to achieve adequate storage have LOLP over 20%.
- Smart control algorithms: Multilevel and adaptive controllers, on top of enhancing stability and power quality, these models have been tested in European and North American installations to diminish voltage drops. Combination with droop control, predictive controls, and AI-based forecasting is becoming more and more common (Holdmann et al., 2019).
- Robustness over efficiency: To withstand disasters in the high-disaster-risk areas (North America, Caribbean), gravitation in the microgrid design leans towards fault tolerance, black-start, and modularity, rather than on its maximum efficiency (Bass, 2026).

6.6.5.2 Economic insights

- The LCOE declines precipitously with increased renewable distribution, yet becomes stable after 80% because of storage. An example is provided in microgrids in Chile, in which LCOE = 0.08 USD/kWh in the Atacama area was reported, as opposed to LCOE = 0.24 USD/kWh under diesel-only systems.
- In Africa and India, community ownership models have achieved greater sustainability compared to those that are purely privately or government-owned. The power of collective investment is maintenance and long-term participation.
- The significance of policy support is that feed-in tariffs, net-metering, and capital subsidies have promoted faster deployment in Europe and China, but in parts of Sub-Saharan Africa, the lack of policy support has impeded adoption.

6.6.5.3 Social and environmental lessons

- Community acceptance is critical: Projects with active stakeholder involvement (training, employment, participatory planning) reported higher success rates.

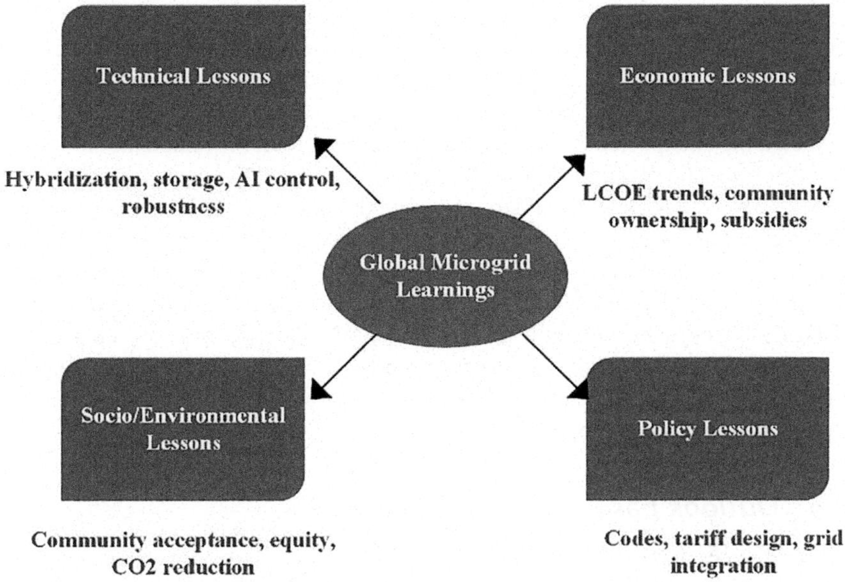

Figure 6.17 Lessons learned map for the renewable-dominant microgrids. *LCOE,* Levelized cost of electricity.

- Environmental benefits can be measured: In Australia and Chile, substitution of diesel-based systems with renewable microgrids cut CO_2 emissions by 50%–80%.
- Fair distribution of energy: In South Asia and rural Africa, the power distribution through microgrids minimized energy poverty as not only households were energized but also schools, health clinics, and small businesses.

6.6.5.4 Policy and regulatory lessons

- Standardization of codes: Lack of unified microgrid standards delays interconnection approvals. Europe's standardized codes accelerate deployment compared to fragmented frameworks in developing countries.
- Tariff design: Flat tariffs discourage efficiency, while time-of-use pricing and demand-response encourage better load management.
- Integration with national grid: Transition pathways must be clear—whether microgrids will remain islanded permanently or integrate into the central grid. Uncertainty often hampers investment.

Fig. 6.17 Lessons learned map for renewable-dominant microgrids

6.6.5.5 Uncertainty lessons

Among the most valuable lessons developed by the current microgrid projects is that uncertainty should be explicitly modeled and addressed instead of being considered as a secondary variable. Most unsuccessful or poor projects were because variability in the generation of renewable, stochastic load patterns or fluctuating fuel prices had not been forecasted during the planning stage. To illustrate, various African and South African microgrids based on solar power significantly underestimated the average solar availability, resulting in a high frequency of reliance on diesel and a higher cost when irradiance was not as predicted. Likewise, in Alaska and islanded systems in the Caribbean, the unexpected loads burst and weather extremes revealed the shortcomings of the deterministic planning tools.

The experience indicates that stochastic modeling, scenario-based simulation, and risk-conscious approaches to optimization are better designs. The methods of chance-constrained optimization, CVaR analysis, and DRO allow the planner to consider extreme events and tail risks. The approaches guarantee that reliability goals like LOLP or LOLE can be achieved with a high likelihood, in objectionable circumstances. Microgrids whose frameworks consider uncertainty are more resilient, economically viable, and socially accepted than those based entirely on deterministic baselines, as demonstrated by global experience.

6.6.6 Outlook based on lessons learned

The experiences from Alaska, Spain, Denmark, India, and Africa demonstrate that renewable-dominant microgrids are not only feasible but also increasingly cost-competitive, as shown in Table 6.14 demonstrates comparative metrices across the region. However, their success depends on combining technological advancements, supportive policy frameworks, and strong community engagement. Future projects should leverage hybrid architectures, digital platforms for energy trading, and AI-driven optimization while ensuring that policies and financing mechanisms align with long-term sustainability.

As renewable microgrids expand, these lessons provide a blueprint for overcoming barriers and accelerating the transition to decentralized, resilient, and community-centered energy systems (Alshahrani et al., 2024).

6.7 Future directions in global microgrid

The shift to renewable-dominated and islanded microgrids is rising faster, but there are research gaps that still require to be filled in, which will help make them more viable technically, economically, and socially. Although there is evidence of the practicality of hybrid microgrid deployments in existing case studies, the lack of solutions to control, storage, interoperability, and regulatory frameworks continues to impose limitations on scalability. The section describes the future research paths that can facilitate the development of renewable-intensive microgrids, with an emphasis on technical innovations, cost-efficiency, digitalization, resilience measures, and socioeconomic frameworks.

6.7.1 Uncertainty-aware planning and optimization

An important research direction within renewable-dominant microgrids is the clear inclusion of uncertainty in the planning and operational processes. The changes in the

Table 6.14 Consolidated lessons from global microgrid case studies.

Challenge category	Case study example	Observed difficulty	Key lesson learned
Technical and operational	Kodiak Island, Alaska	Wind intermittency caused grid instability; frequency fluctuations required stabilization.	Hybrid systems with flywheels and advanced control strategies enhance reliability.
	Bornholm smart grid, Denmark	Inverter-dominated systems complicate synchronization and protection.	Hierarchical and AI-driven control architectures ensure stability under high-RES penetration.
Economic and financial	Solar microgrid, Uttar Pradesh (India)	High upfront cost and irregular tariff payments reduced financial sustainability.	Flexible tariffs, subsidies, and innovative financing mechanisms (e.g., community funds) are vital.
	Brooklyn microgrid, USA	High cost of deployment in urban areas.	Peer-to-peer blockchain trading creates new revenue streams beyond tariffs.
Policy and Regulatory	Sub-Saharan African microgrids	Regulatory ambiguity in tariffs and utility integration caused project delays.	Clear policies and utility coordination frameworks are essential for scalability.
	El Hierro Island, Spain	Complex hybrid wind-hydro integration required strong institutional support.	Government backing and defined policies ensure long-term stability.
Social and Community	Nepal rural hydropower microgrids	Lack of training and weak tariff collection led to poor reliability.	Community training and local ownership are essential for sustainability.
	JUMEME Tanzania solar hybrid	Initially low trust in microgrids among villagers.	Transparent governance and early community involvement improve acceptance and trust.

solar irradiance, wind speed, load demand, and market signals like diesel prices and tariff incentives cannot be predicted exactly. Deterministic methods are easy to use, but do not represent these changes accurately and thus over-size a system, causing unnecessary cost increases, or under-size it, causing reliability and resilience failures. To address this, uncertainty-sensitive strategies like stochastic programming, chance-constrained optimization, and DRO are being actively studied in the literature.

Stochastic models allow the representation of various possible scenarios of the renewable output and demand on the load, with decisions being optimized with respect to minimizing the expected costs or maximizing reliability when there is uncertainty. A two-stage stochastic model can be expressed as Eq. (6.56):

$$\min_{x} C^T x + \mathbb{E}_{\xi}[Q(x, \xi)] \tag{6.56}$$

subject to system constraints, where $x$represents first-stage decisions (e.g., sizing of PV, BESS, and diesel), ξ denotes uncertain variables (irradiance, wind speed, load), and $Q(x, \xi)$is the resource cost capturing operational feasibility under each realization of uncertainty.

Chance-constrained formulations ensure that operational limits (e.g., unmet load, frequency deviations) are satisfied with high probability:

$$P\{g(x, \xi) \leq 0\} \geq 1 - \epsilon \tag{6.57}$$

where ϵ is the acceptable risk tolerance (e.g., 5%). Such formulations are critical in islanded systems where unmet demand translates directly into outages.

DRO extends this by considering not only sampled scenarios but also ambiguity in the probability distributions themselves. The generic form is defined by Eq. (6.58):

$$\min_{x} \underbrace{sup}_{\mathbb{P}\in\mathcal{U}} \mathbb{E}_{\xi\sim\mathbb{P}}[C(x, \xi)] \tag{6.58}$$

where $\mathcal{U}$ is an uncertainty set of probability distributions around empirical data. DRO provides stronger guarantees against rare but high-impact deviations such as prolonged cloudy days or extreme wind variability.

6.7.2 Worked example: Deterministic versus stochastic microgrid design

To illustrate, consider a hybrid PV–BESS–diesel microgrid sized under deterministic assumptions versus a stochastic chance-constrained framework. Table 6.15 compares the deterministic versus stochastic Strategies.

The deterministic approach results in a significantly higher probability of unmet load (~50%) due to unaccounted variability, while the stochastic model, though requiring larger diesel capacity, achieves much higher reliability (>95%) with far

Table 6.15 Comparison of the deterministic versus stochastic strategies.

Strategy	Diesel capacity (kWh)	Diesel energy used (kWh)	Unmet load (kWh)	Total cost ($)	CVaR95_{95}95 cost ($)	Probability of unmet load
Deterministic (means)	100	63.5	56.9	581.4	2939.3	50.2%
Stochastic (chance-constrained)	335	117.6	2.8	52.0	636.4	4.9%

lower expected cost and risk exposure. The CVaR metric further demonstrates that the stochastic model shields the system against extreme cost scenarios.

6.7.3 Research outlook

Future work on uncertainty-aware optimization should extend beyond generation and demand variability to include:

- Battery degradation uncertainty: Electrochemical models with uncertain parameters that affect lifecycle cost and replacement scheduling.
- Policy and market uncertainty: Volatility in diesel prices, carbon credit valuations, and subsidy timelines.
- Resilience under extreme events: Probabilistic frameworks to evaluate recovery after climate-induced blackouts.
- Integration with forecasting: Coupling machine learning-based probabilistic forecasts with stochastic optimization to achieve real-time adaptive scheduling.

Incorporating such uncertainty-aware frameworks will ensure that renewable-dominant microgrids are not only technically feasible but also resilient, cost-optimal, and sustainable under diverse real-world conditions.

6.7.4 Technological advancements

- The technological transformation of microgrids will focus on optimization, intelligence, and advanced storage.
- AI and ML: AI-based predictive controls will enable better real-time, demand-supply balancing, real-time demand forecasting, and proactive fault detection, hike system reliability, and lead to less system downtime.
- Blockchain-facilitated Energy Trading: Distributed ledger technologies will enable secure P2P energy transactions, which will support local energy markets and reduce reliance on centralized utilities.
- New-Generation Batteries: Outmanned are lithium-ion, possibly other technologies, with greater energy density, safety, and life-cycle cost benefits: Solid-state batteries, sodium-ion cells, and redox-flow batteries.
- Hybrid Energy Storage Systems (HESS): Integrating hydrogen storage, supercapacitors with electrochemical storage (batteries) allows multi-timescale energy balancing, increasing microgrid flexibility and resilience.

6.7.4.1 Integration with smart grids

- The potential of next-generation microgrids is to act as modular building blocks of smart grids, where a fully independent microgrid may be powered and operated, or interfaced transparently with the central grid.
- Bidirectional Power Flows: Grid-connected and isolated mode operations will be exchangeable while improving the resilience of the system during grid events.
- Interoperability and Standards: Deployment will necessitate a harmonized communication regime and cyber-secure systems amongst Distributed Energy Resources (DERs).

- Examples of global references, as evident in the post-Fukushima smart community initiatives, can rely on integrated microgrids to ensure urban resilience and diminish dependence on centralized thermal systems.

6.7.4.2 Policy and market innovations

- Microgrid projects as a whole will require emerging regulatory and market processes to be sustainable.
- Business models beyond donor support: Moving to sustainably viable business models, including PAYG programs, cooperatives, and public and/or private partnerships (PPP).
- Community ownership/energy democracy: The use of decentralized ownership systems activates communities, committing them to the lasting adoption of the system and social acceptance.
- Carbon market integration: By measuring the emission reductions, microgrids can enter world global carbon market, emerging new revenue streams, and as an added incentive, renewable penetration.

6.7.4.3 Social dimensions

- Microgrids are socio-technical infrastructures as well as technical systems. The future implementation should lay the values of fairness and inclusion.
- Energy justice: Making electricity available, reliable, and equitable to disadvantaged populations.
- Gender inclusivity in workforce: The active encouragement of women in the domains of operation and management of the microgrid.
- Capacity building: Having local training on the operation and maintenance of microgrid so that they can remain sustainable even after the withdrawal of a donor or the developer.

6.7.5 Case study insights for the future

The local experiences generate some information about localized paths:

- India and Africa: Hybrid solar, battery, and diesel systems are growing at a rapid pace in bringing electrification to the rural population. The future trends of development involve community-based cooperatives with digital intervention.
- Europe: Energy communities under the EU Green Deal will push the adoption of microgrids, and it will be facilitated by robust policies and prosumer involvement.
- North America: Incorporation of resilience-oriented microgrids will be given priority in disaster-prone areas (wildfires, hurricanes).
- Australia: Microgrids based on renewables and storage in mining will increase as renewable costs reduce.
- Chile: Solar-hydrogen-powered and desert-based microgrids will provide testbeds for the transition to Latin America.

Fig. 6.18 represents the future research directions for renewable-dominant and islanded microgrids, and Table 6.16 provides a structured overview of the key challenges

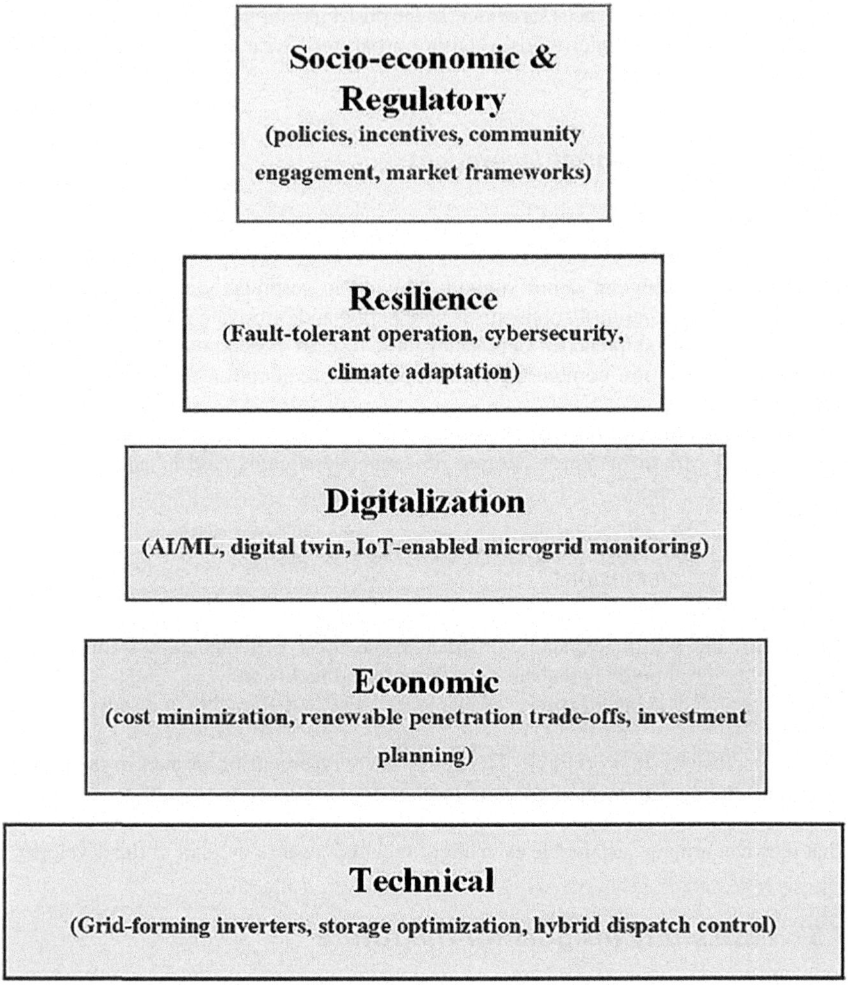

Figure 6.18 Future research directions for renewable-dominant and islanded microgrids.

and corresponding future directions for global microgrids across five major dimensions: technical, economic, policy, social, and environmental. The table highlights how each category presents unique barriers to microgrid development while also identifying possible pathways to overcome them.

6.8 Conclusion

The microgrids gained the centre stage among the global transition to sustainable, reliable, and decentralized energy systems due to the accelerated rate of renewable energy

Table 6.16 Challenges and future directions in global microgrids.

Challenge dimension	Current barriers	Future directions
Technical	Interoperability, intermittency, limited storage, and cybersecurity risks	AI, blockchain, hybrid storage, standardization
Economic	High capex, limited financing, and low affordability	Hybrid financing, carbon markets, and new business models
Policy	Regulatory uncertainty, utility resistance	Clear frameworks, energy community laws, and feed-in tariffs
Social	Affordability, lack of trust, and gender imbalance	Community ownership, energy justice, gender mainstreaming
Environmental	Battery recycling, PV waste	Circular economy, green supply chains

PV, photovoltaics.

technologies development. The conceptual frameworks, mathematical foundations, and various case studies of renewable-dominated microgrid and islanded microgrid have been introduced in this chapter, and the comparative performance and lessons learned are discussed. Following these discussions, it is then evident that microgrids are not a niche application but rather a scalable and flexible solution to energy problems that are urgent in developed and developing environments.

As the case studies demonstrated, high-penetration PV and wind-based systems, combined with advanced energy storage and smart control, can offer the level of stability required and even the resilience to the power supply in geographically constrained or resource-limited locations. The technical performance as a measure of the voltage stability, frequency control, power quality, and reliability indicators improved significantly after hybrid resources were incorporated into microgrids, in which the control strategies were optimized. Similarly, the economic literature demonstrated that, in any case, in the long-term, the level of fossil fuel addiction, LCOE, and positive local jobs make the investment worth it.

Renewable-dominant microgrids have a beneficial impact on the environment, both in terms of preventing the effects of climate change directly by reducing the degree of greenhouse gas emissions and in terms of treating the energy infrastructure as a circular economy. At the social front, microgrids have also been found to have fulfilled the energy justice role of giving the poor communities a reliable, low-cost energy source, improving medical and educational access, and livelihood opportunities. The accessibility of models of community ownership and participatory planning has been vital since it has helped in the provision of local trust and long-term sustainability. Along with these accomplishments, several challenges still exist. Other technical challenges, such as intermittency, small storage, and interoperability barriers, remain as impediments to mass deployment. Without innovative policy structures and integrated financing mechanisms, financial risks involving financing, tariff regulation, and investment dangers cannot be identical. In addition, the absence of governance, underregulation, and the unavailability of technical protocols are critical points of contention. These issues will require a multi-stakeholder initiative with innovative technology, proper policy support, community engagement, and environmental protection. The renewable-dominant microgrids have a bright future. The emerging trends that will accelerate their usage include AI-based predictive control, P2P trading enabled through blockchain, new innovative hybrid storage, and modular plug-and-play structures. In addition, the global energy system is to be decarbonized, and sustainable development policy will support the role of microgrids as one of the building blocks of the next generation of power.

Renewable-dominant and islanded microgrids in this manner represent the convergence of technological progress, environmental sustainability, and social equality. These systems can become the basis of sustainable energy in future, of low-carbon and equitable energy relying on the experience of prior deployments, and on the combination of new solutions to problems. Another aspect that this chapter has highlighted is that uncertainty management plays a central role in assuring the technical and economic feasibility of microgrids. Deterministic models of planning can be useful as tools to aid in the analysis of the baseline but fail when faced with the stochastic variability of solar irradiance, wind velocity, load demand, and market conditions like fluctuation of fuel prices. The use of probabilistic forecasting, scenario generation, chance constraints, CVaR, and DRO frameworks allow microgrid designers to trade-off risk and performance in different operating conditions. Experiences of global projects indicate that microgrids that are developed based on uncertainty-conscious models are more resilient, more economical, and socially sustainable, particularly in the context of islanded situations where the margin of failure is low. With future deployments increasing in velocity, the integration of uncertainty in the very fabric of microgrid design and operation is not only going to enhance their reliability but also render them a real pillar of the energy transition that takes place on a sustainable basis.

References

Ahmed, M. H., Mekhilef, S., & Seyedmahmoudian, R. (2019). Grid-connected renewable microgrids: Economic and technical performance analysis. *Renewable Energy, 131*, 475–485.

Akorede, M. F., Hizam, H., & Pouresmaeil, E. (2010). Distributed energy resources and benefits to the environment. *Renewable and sustainable energy reviews, 14*(2), 724–734.

Alam, M., Mohamed, F. A., & Benbouzid, M. (2017). Microgrids energy management systems: State-of-the-Art and future challenges. *Energies, 10*(9), 1352–1369.

Aldaouab, I., & Daniels, M. (2018). Renewable energy microgrid design for shared loads. *Smart Microgrids*. London, UK: IntechOpen https://www.intechopen.com/chapters/60785.

Alipour, M., Zare, A., & Mohammadi-Ivatloo, B. (2020). Energy storage systems in renewable-based microgrids: A review on technical challenges and optimization approaches. *Journal of Energy Storage, 30*, 101518.

Alshahrani, S., Khan, K., Abido, M., & Khalid, M. (2024). Grid-forming converter and stability aspects of renewable-based low-inertia power networks: Modern trends and challenges. *Arabian journal for science and engineering, 49*(5), 6187–6216.

Asmus, P. (2010). Microgrids, virtual power plants and our distributed energy future. *The Electricity Journal, 23*(10), 72–82.

Atef, M., Alahakoon, S., Mumtahina, U., Wolfs, P., Khatib, T., & Uddin, M. (2026). A review of microgrid energy management systems: methods, challenges, and future directions. *International Journal of Ambient Energy, 47*(1), 2595117.

Atinkut, H.B. (2025). Breakthroughs in Hydrogen and Storage Technologies for a Resilient Grid.

Bass, F. (2026). Renewable Energy in Alaska: A Past and Future History *(Master's thesis.* University of Alaska Fairbanks.

Bello, M. D. C., & Bayod-Rújula, A. A. (2020). Renewable energy microgrids in European islands: Techno-economic analysis and policy implications. *Energies, 13*(14), 3648.

Bhattacharyya, Mandal, M., & Pal, S. (2021). A review of hybrid renewable energy systems for off-grid power supply in developing countries. *Renewable and Sustainable Energy Reviews, 148*, 111276.

Bhattacharyya, P., Singh, A. K., & Kaushik, S. C. (2021). Solar microgrids for rural electrification: Case studies from India. *Energy Reports, 7*, 3272–3283.

California Energy Commission. (2021). Microgrids for resilience: Lessons learned from California deployments (Report No. EPIC PON-14-301). California Energy Commission.

Clamp, A. (2020). Microgrids with energy storage: Benefits, challenges of two microgrid case studies. *Business & Technology Surveillance.*

Cohen, J., Kane, M. B., Marriott, A., Ollivierre III, F., & Govertsen, K. (2023). Economic controls co-design of hybrid microgrids with tidal/PV generation and lithium ion/flow battery storage. *Energies, 16*(6), 2761.

Droop speed control, Wikipedia. https://en.wikipedia.org/wiki/Droop_speed_control.

Eder, Mugge, F., & Kluge, J. (2022). Hybrid microgrids for rural electrification in Sub-Saharan Africa: Renewable integration and business models. *Energy Policy, 164*, 112912.

Espín-Sarzosa, D., Palma-Behnke, R., Cañizares, C. A., Annakkage, U., Elizondo, M., Espina, E., … Reilly, J. T. (2023). Microgrid modeling for stability analysis. *IEEE Transactions on Smart Grid, 15*(3), 2459–2479.

Etemadi, H., Farzan, F., & Aliprantis, K. (2018). Control strategies for reliable microgrid operation under high renewable penetration. *IEEE Transaction on Smart Grid, 9*(6), 6849–6858.

Fatma, B.E.N. (2019). African Development Bank Group.

Figueroa, R. L., & O'Neill-Carrillo, E. (2019). Community-based microgrids in Puerto Rico after Hurricane Maria. *IEEE Electrification Magazine, 7*(2), 18–27.

Frydrychowicz-Jastrzębska, G. (2018). El Hierro renewable energy hybrid system: A tough compromise. *Energies, 11*(10), 2812.

Gamble, S., Piekutowski, M., & Willems, R. (2011). Hydro Tasmania—King Island Case Study. *Energy and Power Generation Handbook: Established and Emerging Technologies*: 10. New York, NY: ASME.

Ghosh, E., & Das, B. (2019). A study on the issue of blockchain's energy consumption. In *In* International Ethical Hacking Conference (pp. 63–75). Singapore: Springer Singapore.

Gomez Anccas, E. D., Hans, C. A., Schulz, D., Microgrid operation control with state-of-charge-dependent storage power constraints. 2025 IEEE Kiel PowerTech, 2025 doi: 10.1109/PowerTech59965.2025.11180499.

Gómez-Expósito, Conejo, A., & Cañizares, C. (2021). *Electric energy systems: Analysis and operation.* CRC Press.

Guerrero, J. M., Vasquez, J. C., Miret, J., Gao, L., & Lee, T. (2014). Microgrids: Challenges and opportunities. *Elsevier /Frontiers in Energy, 8*(2), 82–95.

Haug, F., Eklund, G., Hellstrand, O., Häggblom, F., Nanda Jayawardana, K. D. C., & Karlsson, M. (2025). In *Innovative energy laboratory design for hybrid power systems research and education: 16* (pp. 97–105). Stevenage, UK: The Institution of Engineering and Technology.

He, X., Geng, H., & Mu, G. (2021). Modeling of wind turbine generators for power system stability studies: A review. *Renewable and Sustainable Energy Reviews, 143*, 110865.

Hirsch, Y. P., & Guerrero, J. (2018). Microgrids: A review of technologies, key drivers, and outstanding issues. *Renewable and Sustainable Energy Reviews, 90*, 402–411.

Holdmann, G. P., Wies, R. W., & Vandermeer, J. B. (2019). Renewable energy integration in Alaska's remote islanded microgrids: Economic drivers, technical strategies, technological niche development, and policy implications. *Proceedings of the IEEE, 107*(9), 1820–1837.

Hossain, M. A., Pota, H. R., & Mahmud, R. (2020). Renewable energy microgrids with energy storage: Optimization, control, and future trends. *IEEE Access, 8*, 2169–3536.

Howells, M., Holger Rogner, H., Mentis, D., & Broad, O. (2017). Energy access and electricity planning.

Hu, X., Li, S., & Peng, H. (2012). A comparative study of equivalent circuit models for Li-ion batteries. *Journal of Power Sources, 198*, 359–367.

Irena, R.E.S. (2020). International renewable energy agency. Abu Dhabi, 2020.

Jena, P., Pal, S., & Samantaray, S. (2021). Renewable energy based microgrid in India: Policy, status and challenges. *Renewable and Sustainable Energy Reviews, 152*, 111689.

Joshal, K. S., & Gupta, N. (2023). Microgrids with model predictive control: A critical review. *Energies, 16*(13), 4851.

Kanase-Patil, B., Saini, R., & Sharma, M. S. (2010). Integrated renewable energy systems for off-grid rural electrification of remote area. *Renewable Energy, 35*(6), 1342–1349.

Lasseter, R., Akhil, A., Marnay, C., Stephens, J., Dagle, J., Guttromson, R., Meliopoulous, A. S., Yinger, R., & Eto, J. (2004). Integration of distributed energy resources: The CERTS microgrid concept. *IEEE Transactions on Power Systems, 19*(2), 908–914.

Long, Q., Yu, H., Xie, F., Lu, N., Lubkeman, D., Diesel generator model parameterization for microgrid simulation using hybrid box-constrained levenberg-marquardt algorithm. IEEE Transactions on Smart Grid, 12(2), 943–952, 2020.

Louie, H. (2016). Experiences in off-grid rural electrification projects: A review and implications for microgrid design. *Renewable and Sustainable Energy Reviews, 57*, 1187–1205.

Mandelli, S., Barbieri, J., Mereu, R., & Colombo, E. (2016). Off-grid systems for rural electrification in developing countries: Definitions, classification and a comprehensive literature review. *Renewable and Sustainable Energy Reviews, 58*, 1621–1646.

Marnay, C., Xu, T., Hirase, Y., Mendoza-Araya, P., Hatziargyriou, N., & Lu, S. (2025). Microgrids 2025 editorial. *Applied Energy, 381*, 125348.

McLellan, S. O., Tezuka, T., & Ishihara, A. (2020). Applications of renewable energy in remote areas: Case studies from Alaska and Northern Canada. *Renewable Energy, 145*, 1481–1493.

MDPI Editorial Board. (2023). A Review of Microgrid energy management strategies from the energy trilemma perspective. *Energies, 16*(1), 289.

Mehranvand, S., & Sarrami, M. (2023). Droop control optimization for improved power sharing in AC microgrids. *Energies, 16*(24), 7953.

Miao, Z., Xu, L., Disfani, V. R., & Fan, L. (2014). An SOC-based battery management system for microgrids. *IEEE Transactions On Smart Grid, 5*(2), 966–973.

Moreira, L., Pecas Lopes, J. A., & Madureira, A. (2015). Resilient microgrids: Challenges and prospects. *IEEE Power and Energy Magazine, 13*(3), 16–23.

Morstyn, T., Hredzak, B., Aguilera, R. P., Agelidis, V. G., Model predictive control for distributed microgrid battery energy storage systems. IEEE Transactions on Control Systems Technology, 26(3), 1107–1114, 2017

Mountain, R.O. & Room, A. (n.d.). Renewable microgrids: Profiles from islands and remote communities across the globe.

Nazir, M. I., Hussain, I., Ahmad, A., Khan, I., & Mallik, A. (2022). System modeling and reliability assessment of microgrids: A Review. *Sustainability, 14*(1), 126.

Niknam, T., Azizipanah-Abarghooee, R., & Narimani, M. R. (2012). An efficient scenario-based stochastic programming framework for multi-objective optimal microgrid operation. *Applied Energy, 99*, 455–470.

Niknam, T., Azizipanah-Abarghooee, R., & Narimani, M. R. (2012). Multi-objective optimal microgrid operation: scenario-based stochastic programming. *Applied Energy, 99*, 455–470.

Palit, & Malhotra, A. (2013). Renewable energy based rural electrification: The mini-grid experience from India. *Energy for Sustainable Development, 17*(3), 277–283.

Parhizi, S., Lotfi, H., Khodaei, A., & Bahramirad, S. (2015). State of the Art in research on microgrids: A Review. *IEEE Access, 3*, 890–925.

Peterson, C. J., Van Bossuyt, D. L., Giachetti, R. E., & Oriti, G. (2021). Analyzing mission impact of military installations microgrid for resilience. *Systems, 9*(3), 69.

Pueyo, A., Ngoo, G., Daulinge, E., & Fajardo Mazorra, A. (2022). The quest for scalable business models for mini-grids in Africa: Implementing the keymaker model in Tanzania.

Rawa, M., Calasan, M., Abusorrah, A., Alhussainy, A. A., Al-Turki, Y., Ali, Z. M., … Bassi, H. (2022). Single diode solar cells—Improved model and exact current–voltage analytical solution based on Lambert's W function. *Sensors, 22*(11), 4173.

Rehmani, M. H., Reisslein, M., Rachedi, A., Erol-Kantarci, M., & Radenkovic, M. (2018). Integrating renewable energy resources into smart grid: Recent developments in information and communication technologies. *IEEE Transactions on Industrial Informatics, 14*(7), 2814–2825.

REN21. (2023). *Renewables 2023 global status report*. REN21 Secretariat.

Riva, Morrissey, D., & Byrne, S. (2022). Solar hybrid microgrids for energy access in Africa: A review of deployment and impacts. *Renewable Energy, 192*, 719–731.

Rodríguez, P., Vasquez, J. C., Teodorescu, R., Liserre, M., & Blaabjerg, F. (2007). Flexible active power control of distributed power generation systems during grid faults. *IEEE Transactions on Industrial Electronics, 54*(5), 2583–2592.

Saha, S., Saleem, M. I., & Roy, T. K. (2023). Impacts of high penetration of renewable energy sources on grid frequency behaviour. *International Journal of Electrical Power & Energy Systems, 145*, 108701. doi:10.1016/j.ijepes.2022.108701.

Saleh, M. T. (2024). Mini-grids: Empowering Africa's sustainable energy transition. *Advances in Science and Technology, 142*, 67–77.

Sarker, M. T., Hossen, M. S., Ramasamy, G., Al Qwaid, M., & Karim, H. A. (2026). Techno economic and environmental evaluation of second life battery PV hybrid charging stations for sustainable e-mobility in tropical regions. *Scientific Reports*.

Sarwar, F. A., Hernando-Gil, I., & Vechiu, I. (2024). Review of energy management systems and optimization methods for hydrogen-based hybrid building microgrids. *Energy Conversion and Economics, 5*(4), 259–279.

Savitz, G. (2025). Illuminating Inequality: Going Residential Solar Neutral to Narrow the Energy Equity Gap. *Energy LJ, 46*, 255.

Seane, T. B., Samikannu, R., & Bader, T. (2022). A review of modeling and simulation tools for microgrids based on solar photovoltaics. *Frontiers in Energy Research, 10*. https://doi.org/10.3389/fenrg.2022.772561.

Sebastián, R. (2021). Review on dynamic simulation of wind diesel isolated microgrids (WDIMs). *Energies, 14*(7), 1812.

Shah, C., Wies R. W., Hansen, T. M., Tonkoski, T., Shirazi, M., Cicilio, P., High-fidelity model of stand-alone diesel electric generator with hybrid turbine-governor configuration for microgrid studies. IEEE Access, 10, 110537–10547 2022.

Shahid, M. U., Khan, M. M., Xu, J., Hashmi, K., Habib, S., Mumtaz, M. A., & Tang, H. (2018). A hierarchical control methodology for renewable DC microgrids supporting a variable communication network health. *Electronics, 7*(12), 418.

Shrivastava, S., & Subudhi, B. (2020). Comprehensive review on hierarchical control of cyber-physical microgrid system. *IET Generation, Transmission & Distribution, 14*(26), 6397–6416.

Simon, M. (2019). This Casino's microgrid might be the future of energy. *Wired* Discussing Blue Lake Rancheria solar microgrid with Tesla battery.

Sundararajan, A., & Ferrari Maglia, M. (2024). Data-Driven Model for Photovoltaic Generation: Comparison with Physical Models Using a Microgrid in Puerto Rico. Oak Ridge, TN (United States): Oak Ridge National Laboratory (ORNL).

Tahir, F. (2025). *Exploring the social dimensions of rural and remote microgrids in Australia's shift towards sustainable energy* (Doctoral dissertation).

Tang, X., Hu, X., Li, N., Deng, W., & Zhang, G. (2016). A novel frequency and voltage control method for islanded microgrid based on multienergy storages. *IEEE Transaction on Smart Grid, 7*(1), 410–419.

Tascikaraoglu, S., & Erdinc, O. (2012). Economic and operational benefits of renewable-based microgrids: A case study from Turkey. *Energy, 44*(1), 336–346.

Tayab, U. B., Roslan, M. A. B., Hwai, L. J., & Kashif, M. (2017). A review of droop control techniques for microgrid. *Renewable and Sustainable Energy Reviews, 76*, 717–727.

Vásquez, J. C., Guerrero, J. M., & Luna, A. (2016). Hierarchical control of intelligent microgrids. *Proceedings of the IEEE, 104*(4), 724–745.

Wang, Y., Zhao, J., & Chen, C. (2020). Hybrid renewable energy microgrids in China: Case studies and future perspectives. *Applied Energy, 280*, 115981.

Wang, Z., & Wang, J. (2015). Self-healing resilient distribution systems based on sectionalization into microgrids. *IEEE Transactions on Power Systems, 30*(6), 3139–3149.

Wikipedia, Microgrid, n.d. Available at: https://en.wikipedia.org/wiki/Microgrid

Modeling electric vehicle uncertainty: Charging behavior & grid impact

7

Scope: Aggregate EV load scenarios and distribution impacts

Shaghayegh Nobakht and Ali-Akbar Ahmadi
Industrial Automation Laboratory, Department of Electrical and Computer Engineering, Kharazmi University, Tehran, Iran

7.1 Introduction

Transportation is a major contributor to global greenhouse gas emissions, accounting for an estimated 22% of energy-related CO_2 emissions worldwide (International Energy Agency, 2023). In 2022, emissions from this sector rose by 2.1%, equivalent to 137 million tons, driven largely by the rebound of passenger and freight activities following COVID-19 (Our World in Data, 2020). This reliance on fossil fuels not only accelerates climate change but also poses serious threats to air quality and energy security. Electric vehicles (EVs) have emerged as a key solution for sustainable transportation, significantly reducing carbon footprints by eliminating tailpipe emissions and enabling the use of renewable energy sources such as wind and solar (Van Fan et al., 2022). Moreover, technologies like vehicle-to-grid (V2G) allow EVs to serve as distributed energy resources, enhancing grid resilience and flexibility (Hannan et al., 2017).

EV acceptance in the whole world is at an unprecedented level. The International Energy Agency (IEA) states that EV sales were expected to exceed 20 million units in 2025, not the 25%, though, as projected by the model. EV sales will have reached over 17 million units in 2024 (International Energy Agency, 2025). This boom is being felt in major economies such as China, Russia, and India, with almost 60% of the market likely to be dominated by EVs by 2025, and even in smaller economies such as Austria and Iceland due to the heavy policy incentives given and a slump in costs (Du & Ouyang, 2017; Gass et al., 2014; Shafiei et al., 2018). Nonetheless, the fast penetration of EVs into power systems presents challenges of complexities because the behavior of charging is stochastic; this is dependent on several factors, including location, timing, and period, among others (Kunj & Pal, 2020). Such uncertainties may cause grid, voltage instability, and rise on the cost of operations (Karmaker et al., 2019; Singh et al., 2023). This chapter has used stochastic models to deal with these issues with the aim of offering the methodology for dealing with the uncertainties in EV charging, with the primary objective of ensuring that the grids remain reliable and sustainable to the environment.

Stochastic Planning and Modeling for Energy Systems: Methods, Applications, and Developments.
DOI: https://doi.org/10.1016/B978-0-443-45298-7.00011-2

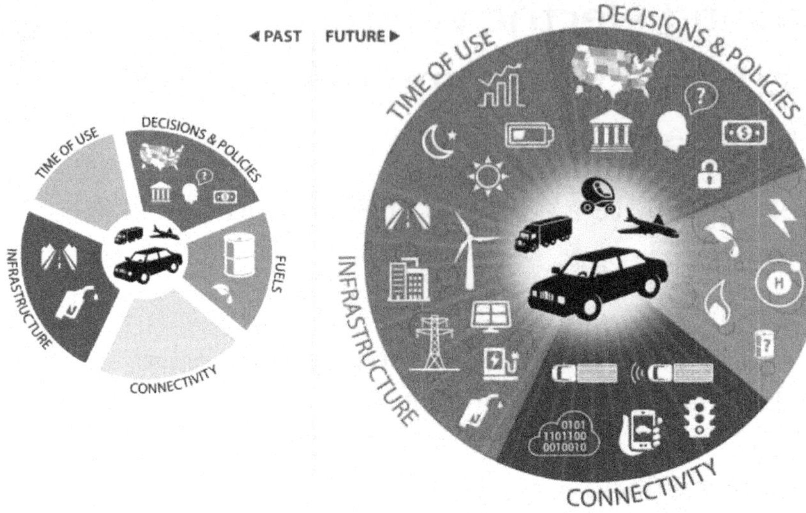

Figure 7.1 EV-grid integration (Muratori et al., 2020). *EV*, Electrical vehicle.

7.1.1 Emergence of electric vehicles in future grid planning

The global EV fleet is projected to grow dramatically over the next 20 years. Current estimates of nearly 3 million units are expected to rise to 66 million by 2040, representing a twenty-two-fold increase (De Lima et al., 2022). Recent projections suggest that annual passenger EV sales could reach 73 million by 2040, accounting for 73% of global car sales, highlighting the transformative shift toward sustainable transportation (BloombergNEF, 2023). Such rapid growth underscores the need for strategic grid planning to manage the increased load and maintain system stability.

EVs present both technical and operational challenges to the grid. Uncontrolled charging can increase peak loads, potentially causing grid congestion (Singh et al., 2023). In contrast, smart charging and V2G technologies allow EVs to serve as flexible grid resources, providing stored energy during periods of high demand. For example, V2G systems enable EVs to function as mobile storage, increasing grid flexibility and reducing reliance on fossil fuel-based generation (Hannan et al., 2017). Furthermore, combining EVs with renewable energy sources creates synergies that enhance the sustainability of the energy system (De Lima et al., 2022). Stochastic modeling is essential for effective grid planning, ensuring that the variable nature of EV charging is accurately represented, resources are properly allocated, and network reliability is maintained (Choudhary et al., 2025; Muratori et al., 2020). Fig. 7.1 illustrates the interaction between EVs and the power grid.

7.1.2 Key challenges posed by electric vehicle uncertainty

EV charging behavior is inconsistent and happens very rarely, which presents a serious challenge to the state of power systems. The location of charging (e.g., residential, commercial, or public station), the timing (peak or off-peak), and the length of the session depend on the behavior of the driver, the battery capacity, and infrastructure availability (Kunj & Pal, 2020). This combination of these variables will cause unpredictable load profiles that are capable of putting a strain on the distribution networks, causing problems such as grid congestion, voltage distortions, and high operational expenses (Karmaker et al., 2019; Singh et al., 2023).

Domestic sites, such as homes, often experience peak electricity demand during the evening, which can overload local transformers (Fischer et al., 2019). Conversely, public and commercial charging stations tend to have higher demand during the daytime, leading to imbalances in grid loads (Kunj & Pal, 2020). These challenges are further amplified by high-capacity batteries and fast-charging systems, which increase both peak demand and the frequency of charging (Muratori et al., 2020). Uncoordinated charging can also degrade power quality by introducing harmonic distortions and reactive loads to grid-connected equipment (Karmaker et al., 2019). To address these issues, advanced stochastic modeling methods—including Monte Carlo simulations (MCSs) and machine learning—are essential for incorporating the probabilistic nature of EV charging and predicting load conditions (Choudhary et al., 2025; Muratori et al., 2020). Additionally, real-time monitoring and control systems can help manage charging loads dynamically, further mitigating grid impacts (Liu et al., 2022b).

7.1.3 Chapter objectives and structure overview

The scope of this chapter is to conduct an overall review of the uncertainties caused by EV charging behavior and its impact on power distributed systems. Through stochastic modeling, we aim to come up with powerful strategies of controlling such uncertainties as they fall so very well as integrating the EV, maintaining grid reliability, and environmental objectives. The following is how the chapter is structured:

- Section 7.2: EV charging behavior and sources of uncertainty: Examines the variables that affect EV charging variability, such as infrastructure limitations and driver behavior.
- Section 7.3: Modeling approaches for EV load uncertainty: Talks about stochastic modeling methods for EV charging variability, including machine learning and MCSs.
- Section 7.4: EV load aggregation and scenario design: This section looks at ways to combine different EV loads and create scenarios to evaluate the effects on the entire system.
- Section 7.5: Impacts of EV charging on the distribution grid examines how EV charging affects capacity needs, power losses, and voltage regulation.
- Section 7.6: Conclusion: Provides a summary of the results and suggestions for additional study.

7.2 Electric vehicles charging behavior and source of uncertainty

The fast uptake of EVs is one of the pillars of international strategies to mitigate the emission of greenhouse gases and fight climate change. Nevertheless, there are serious difficulties with the introduction of large-scale EVs into the current energy systems. It has been projected that the number of passenger battery EVs will demand greater electricity in 2030 and 2050, between 250–480 terawatt-hours (TWh) and 1140–2840

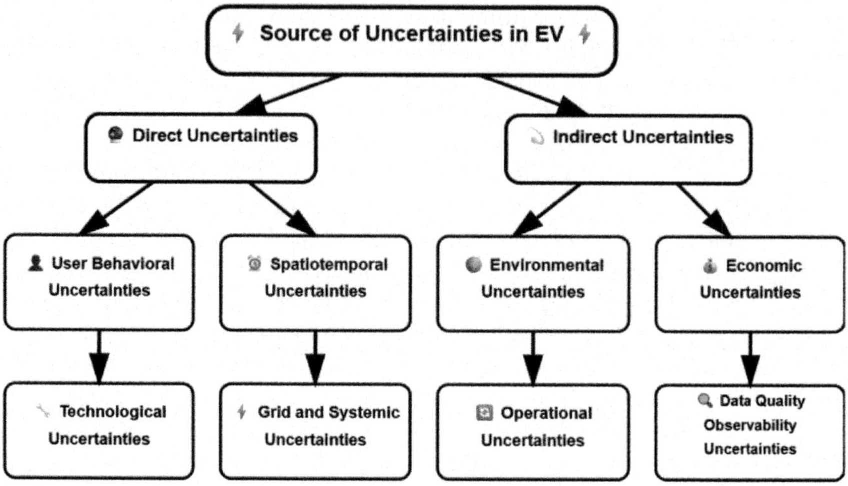

Figure 7.2 Categories of uncertainties in EVs. *EV*, Electrical vehicles.

TWh respectively (Hou et al., 2021). Such a significant influx, when not adequately handled, may cause grid jamming, higher expenses of operations, and unreliability (Li et al., 2022; Saxena et al., 2023).

One of the primary activities in this process is the definition of EV charging behavior, which involves the pattern and the choice taken by EV users in the charging of their vehicles. In the case of EV charging, it is stated that charging behavior is the spatial and temporal variation of charging events that are related to each EV (Daina, 2014). This principle is also extended in the research, which also focuses on the decision-making of EV users who have some freedom with the time, place, and duration of charging (Helmus & Van Den Hoed, 2015). These decisions may be impacted by many factors, such as personal preference, price, and other contacts with other EV users (Shariatzadeh et al., 2024). Besides, EV user behavior is classified into two categories: Charging behavior that includes duration of charging, battery capacity, and charging mode (fast and slow), and mobility behavior that encompasses arrival and departure time, trip time, and modal choice (Baumgartner et al., 2023). An example is that some users can choose to charge at a slow pace, overnight at home, and others can use the rapid charge in the station to charge their vehicles according to the time they have.

The uncertainties involved in EV charging behavior are so complex that they require a systematic classification of the sources of uncertainty. This classification separates uncertainties into two major categories: Direct and indirect uncertainties. A hierarchical system of the uncertainties in EVs is presented in Fig. 7.2.

7.2.1 Direct uncertainties

Direct uncertainties are those that have a direct effect on the charging decisions and patterns by causing immediate effects on the charging pattern, such as the user

Table 7.1 Categories of direct uncertainties in EV charging behavior.

Category	Description	Examples
User behavioral	Variability in how users charge, driven by individual preferences and needs.	Individual variability, range anxiety, cost sensitivity, adaptive decision-making.
Spatiotemporal	Uncertainties in when and where charging occur, tied to user mobility.	Time of charging, charging location, special conditions, external conditions.
Technological	Variations in EV and charging infrastructure capabilities.	Diversity of EV types, charging infrastructure development.
Grid and systemic	Uncertainties related to grid capacity and EV-grid interactions.	Grid capacity, renewable integration, power quality, V2G, aggregate load variability.

EV, electrical vehicle; *V2G*, vehicle-to-grid.

behavioral patterns, technological characteristics, spatial and temporal variations, and direct grid effects. Table 7.1 summarizes these.

7.2.1.1 *User behavioral uncertainties*

EV user behavior is a significant source of uncertainty in the process of modeling aggregate load and system effects. Socio-demographic profiles, pro-environmental behavior, individual preferences, psychological responses to range and charge anxiety, cost sensitivity, attitudes, experience, and knowledge can be identified as the causes of heterogeneity in user behavior. These factors work dynamically in changing circumstances and affect the magnitude of the charging demand as well as the temporal and spatial distribution of the demand. Fig. 7.3 is a detailed diagram that shows the EV user behavioral uncertainties.

7.2.1.1.1 Socio-demographic factors

The socio-demographic factors influencing the probability of EV adoption and the ensuing charging patterns have been demonstrated to be age, gender, income, education, and geographical location (Bae & Yun, 2024; Priyam et al., 2024). As an example, it has been revealed by researchers that younger, better-income, and more-educated users are early EV adopters and have more frequent charging habits than traditional fueling habits (Daina, 2014), whereas it has been shown that charging versus traditional refueling habits vary according to user background (Philipsen et al., 2018). Recent large-scale research confirms that the infrastructure and patterns of charging usage do not vary equally between income groups and place types, thus influencing aggregate demand and the charger deployment strategy (Zheng et al., 2024).

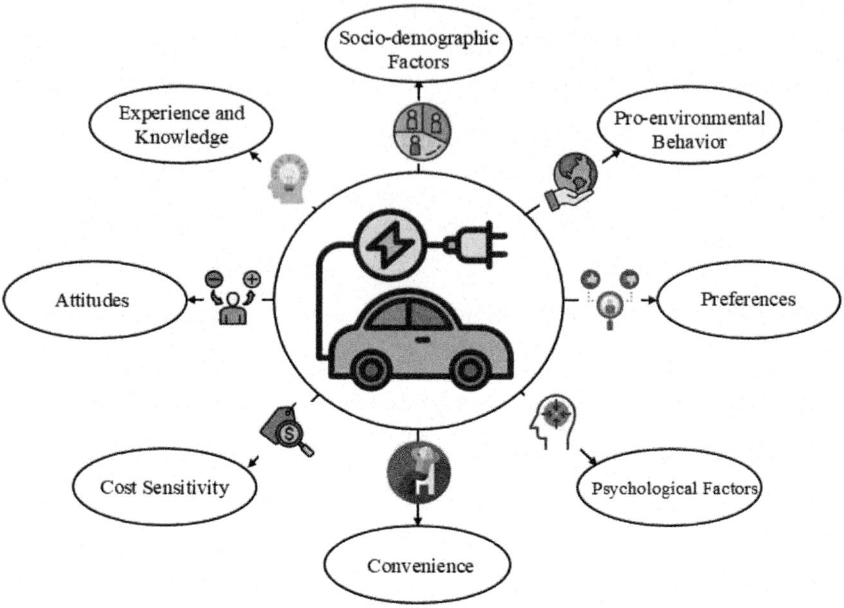

Figure 7.3 Diagram of the EV user behavioral uncertainties. *EV*, Electrical vehicle.

7.2.1.1.2 Pro-environmental behavior

Attitudes that are pro-environmental determine charging decisions, and consumers do participate in green charging when they have a choice (Kramer et al., 2023). As authors emphasized, behavioral interventions (e.g., environmental impact feedback) can be effective to increase the chances of off-peak and renewable-based charging. More eco-friendly users are more willing to use flexible and smart charging solutions.

7.2.1.1.3 Preferences

The practice of charging and choice of place is based on user preference, based on satisfaction, convenience, sensitivity to cost, and knowledge of the protocol. As an example, Wang et al. (2021) discovered that satisfaction and queue times were some of the most important factors that determined preferred charging strategies. Research findings also present that interfaces and protocol transparency standardization go a long way in enhancing user satisfaction and flexibility (Andrenacci & Valentini, 2023).

7.2.1.1.4 Psychological factors

The range anxiety is one of the most significant barriers to EV adoption and a frequent element in the uncertainty about the timing and location of the charging (Noel et al., 2019; Pei, et al., 2025). It is proven by psychological modeling, which involves cumulative prospect theory, that risk-averse personalities exhibit more charge at state of charge (SOC) levels, which leads to higher grid demand variability (Hu et al., 2019). This aspect stimulates the users to charge the batteries to a SOC that is more than

20 V, and in many cases, to 100 V, which may shorten the battery life. This trend is heightened by convenience factors such as fear of a drained battery or the unavailability of chargers, particularly by new or less experienced users (Franke & Krems, 2013).

7.2.1.1.5 Convenience

Convenience—defined by proximity to chargers, access, charging speed, and scheduling flexibility—directly affects when and where users choose to charge their EVs. Model-based research shows that users make trade-offs between cost and convenience, with some willing to pay more for faster or more accessible charging (Chung et al., 2018a). Inconvenience factors, such as crowding or difficulty finding a charger, can disrupt habitual behavior, increasing overall unpredictability in charging patterns.

7.2.1.1.6 Cost sensitivity

The role of cost sensitivity is an important factor in the determination of the decision to charge in both routine and responsive charging behavior. More cost-sensitive users keep track of dynamic tariffs, respond to time-of-use (ToU) price indicators, and often seek cheaper options, whereas the rest are more responsive to convenience or less to price differentials (Li & Jenn, 2022). Put together, such heterogeneity is reflected in the fact that individuals respond unequally to grid management plans and incentive plans.

7.2.1.1.7 Attitudes

Attitudinal development towards technology, flexibility, and smart charging contributes to the uptake and flexibility service uptake. Research indicates that perceived value, risk tolerance, and automation trust are vital toward adaptive responses (Libertson, 2022). Experience may influence knowledge and awareness, which both affect the desire to engage in advanced services (e.g., V2G, demand response [DR]) (Pan et al., 2019).

7.2.1.1.8 Experience and knowledge

Established users would have higher chances of maximizing the charge, implementing new standards, and reacting effectively to live signals. Absence of experience or inadequate knowledge has remained an important impediment to efficient or grid-beneficial behavior (Chamberlain & Al-Majeed, 2021).

7.2.1.2 Spatiotemporal uncertainties

Spatial and temporal factors are very powerful in determining charging behavior. The uncertainty of the time of charging event initiation and the location of said events makes the process of aggregating system loads and grid management rather uncertain. Fig. 7.4 shows the spatiotemporal uncertainties of charging electric cars.

7.2.1.2.1 Time of charging

Empirical research findings have shown that charging events, particularly residential ones, are very concentrated during specific times, mostly in evening peak hours or during the night (Henriksen et al., 2021). The pattern of charging loads shows doubled-peak in weekdays (morning and evening), a single-peak in evenings on weekends, with

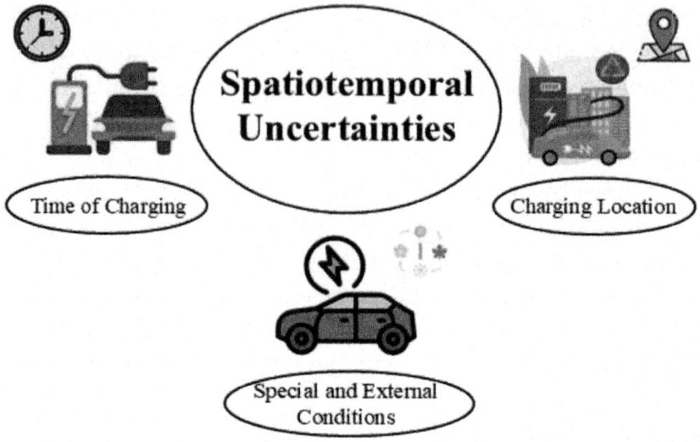

Figure 7.4 Diagram of the EV spatiotemporal uncertainties. *EV*, Electrical vehicle.

residential areas as high as 15:00–24:00 and commercial areas as high as 09:00–24:00 (Yang et al., 2025a). Differences in working schedule, individual schedule, and grid signals (e.g., price, renewable share) bring dissimilarity across individual users as well as across regions. Notably, despite the presence of TOU tariffs, there is a risk that even with this method, clustering can only cause a change in time slot in the maximum; the effectiveness of responses depends on penetration and heterogeneity among users (Bons et al., 2021).

7.2.1.2.2 Charging location

The availability, trip purpose, mobility routines, and infrastructure characteristics (speed, reliability) determine the charging location. Home charging is used mostly on the applications of the private EVs, whereas commercial applications of the taxis make use of the public stations. In urban locations, slow chargers (22 kW) are used, whereas fast (>150 kW) chargers are found in commercial and highway locations (ten Have et al., 2020). The aggregate demand profiles of the various nodes within the grid are influenced by proximity to home/work, existence of enroute fast chargers, and density of charging points (Pan et al., 2020). According to new research, it is essential to consider the user activity model and user satisfaction as part of the location planning because they directly affect deployment efficiency and the risk of congestion (Cui et al., 2022).

7.2.1.2.3 Special and external conditions

Unpredictable incidents like pandemics, climatic extremes, or special events may redefine mobility and charging habits within a few pounds (Shahriar & Al-Ali, 2022). Such events lead to a break if it is usual, in both demand forecasting and the real-time stability of grids. Furthermore, there is the seasonal variation and unexpected system disruptions that cause additional uncertainty (Dominguez-Jimenez et al., 2020).

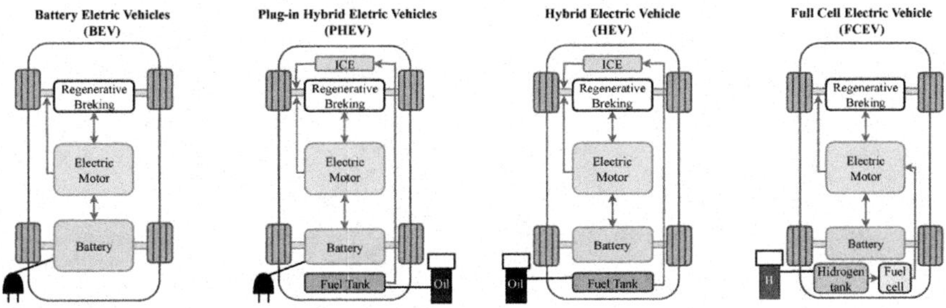

Figure 7.5 Classification of EV types (Huaman-Rivera et al., 2024). *EV*, Electrical vehicle.

7.2.1.3 Technological uncertainties

The technological environment, including the number of EVs on the road, the implementation of chargers, and the development of standards, also brings more ambiguity to the projection of impacts.

7.2.1.3.1 Diversity of electric vehicle types

To date, there are four broad categories of EVs, each of which has a different implication on the charging demand and grid interaction:

- Hybrid electric vehicles (HEVs): It is also equipped with a standard combustion and an electric motor, with regenerative braking to charge the onboard battery. Since HEVs do not need grid charging, they do not directly affect electric grids.
- Plug-in hybrid electric vehicles (PHEVs): Have the ICEs and batteries, which can be charged either using external sources or through the process of regenerative braking. Mostly with smaller batteries than BEVs, PHEVs have a moderate impact on the grid and could have a more substantial impact at high market penetration.
- Battery electric vehicles (BEVs): This is a complete EV, which can be charged only by battery; the charging characteristics are the primary focus of determining aggregate grid effects due to increased battery capacity and wider range. BEVs imply both increased risk and the negative impact of the system in the future, and the ability to offer grid services like V2G.
- Fuel-cell electric vehicles (FCEVs): Using hydrogen fuel instead of electricity as a power source, therefore do not pose direct pressure to the electrical system.

Charging gear is also very varied, and has different standards and speed profiles (AC Levels I-III, DCFC). The vehicle-charger combinations generate unique load signatures, which affect the overall grid modeling and the flexibility of demand (Huaman-Rivera et al., 2024). Fig. 7.5 represents the classification of different types of EV.

7.2.1.3.2 Charging infrastructure development

The different charging needs of these vehicles bring confusion to the demand forecast. The infrastructural development of charging related to the presence and format of chargers (e.g., Level I, II, or III according to SAE J1772 standard) determines the

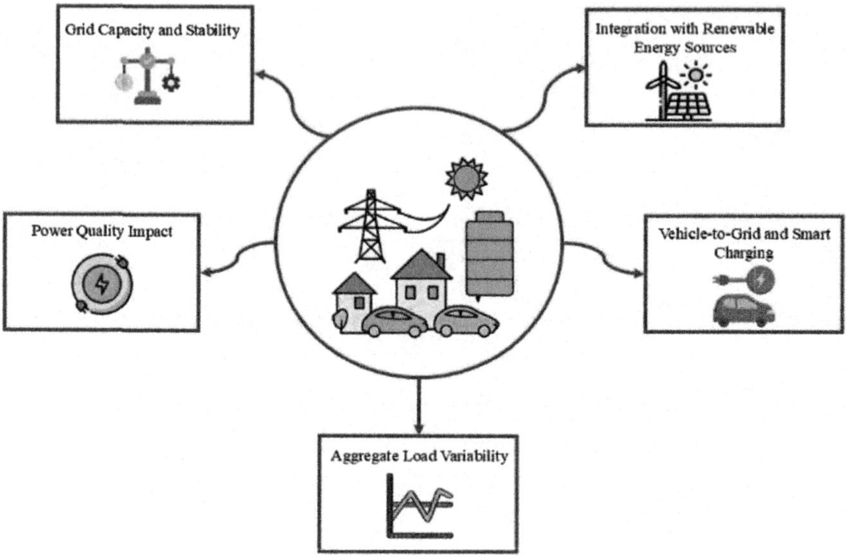

Figure 7.6 Diagram of the EV grid and systemic uncertainties. *EV*, Electrical vehicle.

impact of charging on users and the grid. As an example, fast-charging stations can stimulate frequent charging but higher peak loads, whereas slow chargers can be in line with the off-peak demand (Yang et al., 2018; Zhao et al., 2020b). The recent research highlights the importance of strategic infrastructure planning to handle various user requirements and counter grid stress (Lagomarsino et al., 2022).

7.2.1.4 Grid and systemic uncertainties

The overall effect of EV charging is in addition to the already existing power system and poses uncertainties to stability and power quality, as well as the effectiveness of integration strategies. Fig. 7.6 illustrates the grid and systemic uncertainties of EV integration.

7.2.1.4.1 Grid capacity and stability

Uncoordinated high-power charging, particularly due to clustered DC fast chargers or high EV density in a local area, may be above the thermal limits of distribution transformers and cables, causing overloads, faster wear and tear, and possible failures (Shariatzadeh et al., 2024). Of serious concern is the voltage deviations, especially at the end of the feeders, where the load is concentrated in an increase (Sun & Qiu, 2021).

7.2.1.4.2 Integration with renewable energy sources

EVs have the promise of improved use of RES, but they are hazardous. The unregulated charging can have amplified the evening peak demand when the solar generation declines, and this complicates the problem of the duck curve (Mangipinto et al., 2022).

The success of such strategies as the use of solar charging in the working places is linked to the availability of the local RES, the presence of EVs during the period of their generation, and the desire of the users to take part. There is downside uncertainty in harnessing intermittent generation of RES with stochastic EV demand.

7.2.1.4.3 Power quality impact

In addition to current deviations under voltage, EV chargers, especially fast DC chargers, may cause damage to the grid by injecting harmonic currents into the grid courtesy of the power electronics. Sensitive equipment can be disrupted by high harmonic distortion, lose more, and capacitor bank failures occur (Sevdari et al., 2022). The cumulative harmonic effect of many chargers is unknown and it varies with charger technology, control measures, and grid impedance.

7.2.1.4.4 Vehicle-to-grid and smart charging

Although V2G has major grid advantages (frequency regulation, peak shaving, congestion management (Nagel et al., 2024)) large-scale performance of V2G is uncertain: Batteries degrade (Everoze, 2023), communication and control systems are not always reliable, there is standardization, the market design is complex, and most importantly, people will accept it and participate in its deployment (Ensslen et al., 2018). User flexibility and responsiveness of control signals are also factors that determine the effectiveness of V1G smart charging (shifting time/power).

7.2.1.4.5 Aggregate load variability

All of the multiple summed uncertainties contribute to high stochasticity in the aggregate EV load profile at the various levels of the grid (transformer, feeder, substation, system). The scale of the aggregate EV demand, when it occurs, and how long it lasts, as well as its location, is difficult to predict, making it hard to plan generation, reserve, and network distribution (Kamana-Williams et al., 2024).

7.2.2 *Indirect uncertainties*

The indirect uncertainties are caused by exogenous factors not directly managed by EV users or grid operators but have a greatly important impact on charging patterns and grid stability. These include environmental, economic, operational, and data-related aspects, which form intricate interdependencies that make it difficult to forecast and plan infrastructure. These are outlined in Table 7.2.

7.2.2.1 Environmental uncertainties

The environmental uncertainty has a major effect on EV charging behavior and grid impacts due to variable weather and climatic conditions, changes in policies and regulations, and fluctuations in fuel prices. The factors bring uncertainty to the adoption rates and charging behavior of EVs, which makes stochastic modeling activities difficult. Fig. 7.7 shows the environmental uncertainties with respect to EV integration.

Table 7.2 Categories of indirect uncertainties in EV charging behavior.

Category	Description	Examples
Environmental	External conditions affect driving and charging patterns.	Weather, climate conditions, policy changes, fuel price volatility.
Economic	Financial factors influencing charging decisions.	Charging costs, electricity price fluctuations, incentives, parking fees.
Operational	Real-time challenges in managing charging demand.	Demand forecasting, station availability, cybersecurity, data privacy risks.
Data quality/observability	Issues with data used to model charging behavior.	Sampling bias, privacy-driven data limitations, algorithmic opacity.

EV, electrical vehicle.

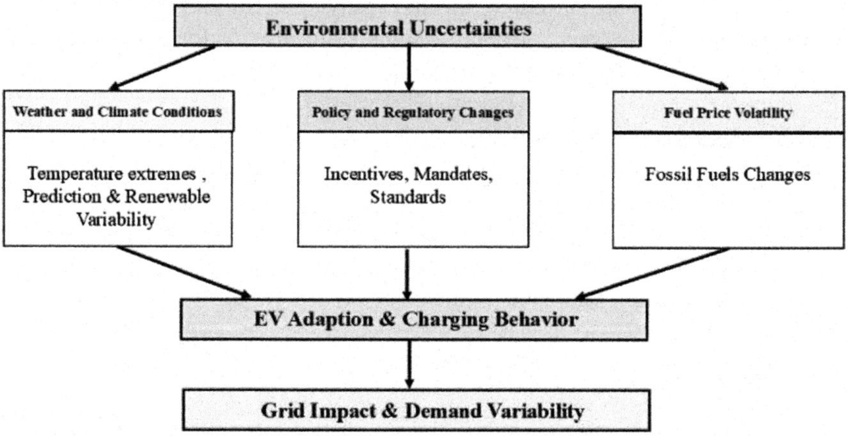

Figure 7.7 Diagram of the EV environmental uncertainties. *EV*, Electrical vehicle.

7.2.2.1.1 Weather and climate conditions

The decisive contribution to the demand for EV charging and grid impact is ambient weather and climate. High and low temperatures affect battery performance and mobility patterns of users and shift aggregate load directly (Hiesl et al., 2021). Experiments that use actual traffic and weather data prove that cold weather may increase the energy consumption up to 30 times, which provokes the peaks of charging demand during winter. Moreover, the conditions related to precipitation and the snowfall have also been found to adjust the traveling and the frequency of the plug-in process, which enhances the stochasticity on a daily basis. Moreover, the changes in the variability

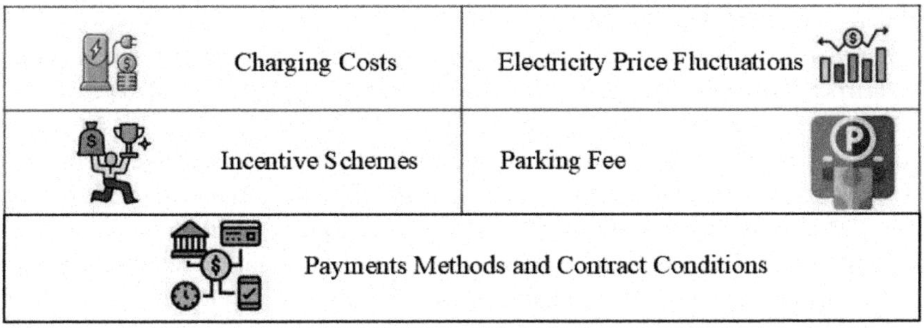

Figure 7.8 Diagram of the EV economic uncertainties. *EV*, Electrical vehicle.

of renewables caused by climate provide another plane: Congruence or incongruence between EV charging schedules and renewable energy availability (Supponen et al., 2016). Weather features, like solar irradiance or temperature differentials, are now included in machine learning and ensemble forecast models, with a significant increase in accuracy of charging load forecasting (Huang et al., 2020).

7.2.2.1.2 . Policy and regulatory changes

The EV regulatory environment is quite dynamic, with the policy rapidly changing and influencing the behavior of charging and the planning of the system. Subsidies (subsidies), mandates (ZEV quotas, ICE bans), and standards (interoperability, flexibility requirements) bring about uncertainty about the rate and heterogeneity of EV adoption and charging responsiveness (LaMonaca & Ryan, 2022). Aggregator participation and flexibility market regulatory frameworks are capable of significantly changing both user and system behavior on short time scales. It has been recently demonstrated that predictable policymaking is less prone to uncertainty and fosters long-term flex investments, and sudden changes can ruin demand projections, infrastructure planning, and user behavior (Hildermeier et al., 2022).

7.2.2.1.3 Fuel price volatility

The volatility in fuel prices around the globe indirectly impacts the rate of EV uptake and driving patterns, thereby impacting the overall charging demand. The price hikes in fossil fuels have already been known to boost EV adoption and shift travel behavior towards frequency charging and home charging (Bushnell et al., 2022). Nevertheless, this volatility also brings modeling difficulties in the projection of a scenario pathway of EV penetration, load development, and possible demand peaks (Sun et al., 2019).

7.2.2.2 Economic uncertainties

Economic uncertainties within EV charging are associated with the costs of charging, electricity prices, incentive programs, payment methods, contract terms, and parking charges, which affect the user behavior and distribution of grid loads. Fig. 7.8 shows the diagram of the EV economic uncertainties.

7.2.2.2.1 Charging costs

EV charging prices are opaque and dynamic. Recent reviews demonstrate that although ToU tariffs and dynamic pricing encourage off-peak charging, they also provide a heterogeneity of behavior, and even counter-intuitive effects, including the shift of peaks to new times. The device of demand charges, which should be used to control the local peak loads, is in most cases not very effective and may lead to unexpected adaptations of users, including those who are inflexible or of low income (El Gohary et al., 2023). Also, cost structure is interrelated with activity time and mobility requirements, which makes it even harder to make predictions about aggregate load (Wang et al., 2022).

7.2.2.2.2 Electricity price fluctuations

The volatility of wholesale and retail prices also affects user choice and grid planning. Unpredictable electricity pricing, whether due to renewables, grid congestion, or market design, can lead to users changing their charging time or investing in stationary storage, although the user price elasticity varies and is not predictable. Uncertainty in the forecasting of electricity prices is transferred to aggregate EV load and infrastructure use uncertainty (Powell et al., 2020).

7.2.2.2.3 Incentive schemes

Incentives provided by the government and by the private sector, such as the provision of rebates, free public charging, and managed charging participation rewards, are key factors that encourage the adoption of smart charging and flexibility services by the users. Yet, their presence (or absence) has a direct and not always linear impact on charging load, since users can flock around supported windows or technologies (Chakraborty et al., 2019; Daziano, 2022; Gong et al., 2021). Recent studies emphasize the design of such incentives, and awareness of users has a dramatic influence on aggregate and local grid effects.

7.2.2.2.4 Payment methods and contract conditions

The uncertainty of contracts and payments comes in due to the fragmented market offerings, the absence of protocol standardization, or complicated fee schemes. According to surveys, users tend to feel disoriented when it comes to the various types of contracts (flat-rate, per-kWh, per-session), the generation of variability in responses to price signals, and restricting the use of smart charging programs (Shell Recharge Solutions, 2023; Huang et al., 2021). Besides, it is evident that the readiness to participate in new models, such as V2G contracts, differs dramatically among user groups, based on their perceptions of fairness and transparency.

7.2.2.2.5 Parking fee

The dwell times, as well as the plug-in decisions, depend on parking policies and particularly the organization of the parking charges at the charging points. In case of high and unpredictable fees, users can decrease charge duration or avoid particular places, which directly influence local load prediction and system use (Pagani et al., 2019). Poorly designed parking pricing may also shift the demand to charging instead of decreasing demand, which becomes an additional spatial complication.

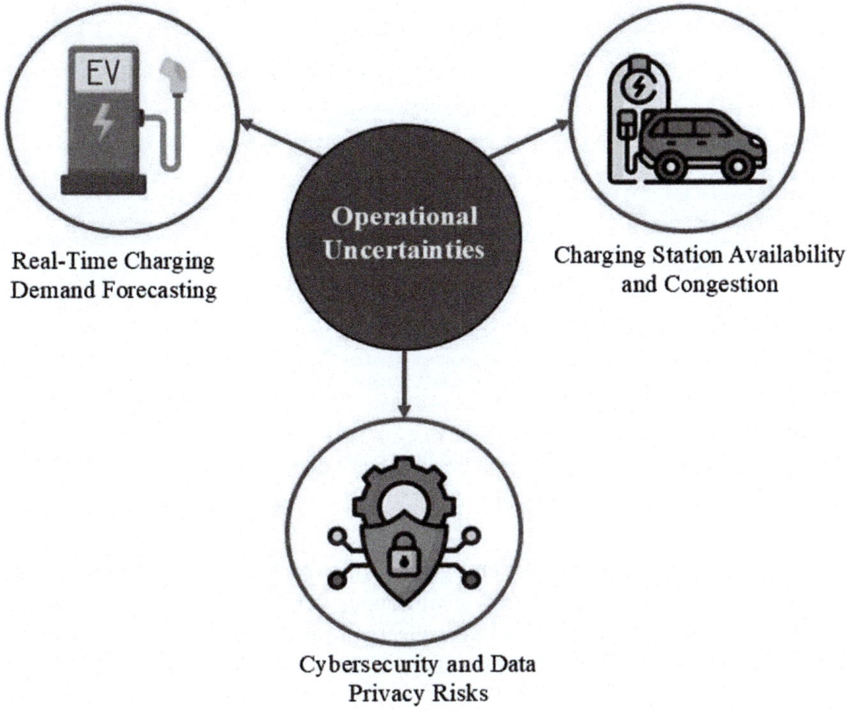

Figure 7.9 Diagram of the EV operational uncertainties. *EV*, Electrical vehicle.

7.2.2.3 Operational uncertainties

The operational risks associated with EV charging include the real-time forecasting of charging demand, availability, and congestion of the charging station, and cybersecurity and data privacy, which affect the effectiveness and reliability of the charging infrastructure. The scheme of the EV operational uncertainties is presented in Fig. 7.9.

7.2.2.3.1 Real-time charging demand forecasting

In real-time forecasting, aggregate or node-level charging demand is highly uncertain because the mobility patterns, weather, and system events of individuals are stochastic. Machine learning and ensemble methods, such as seasonal auto-regressive integrated moving average (SARIMA) and multivariate long short-term memory (LSTM) networks, gain more accuracy when adding additional factors, including weather and user behavior (Akshay et al., 2024; Sanami et al., 2025). This operational uncertainty also confronts utility management and the distribution system reliability, requiring probability-based scenario management.

7.2.2.3.2 Charging station availability and congestion

The uncertainty in the availability of charging stations can be caused by local congestion, equipment failure, or unexpected demand spikes, that is, the nontrivial

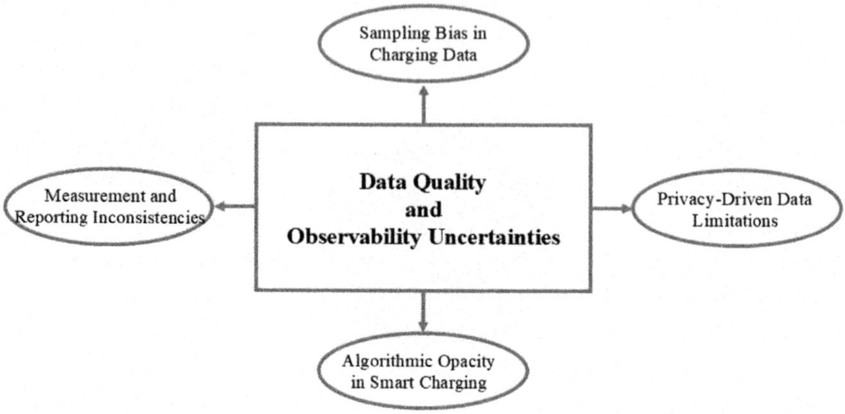

Figure 7.10 Diagram of the EV data quality and observability uncertainties. *EV*, Electrical vehicle.

redistribution in time and space of charging events. As demonstrated in recent agent-based and empirical research, micro-level congestion may drive users to switch stations or time slots, which may increase the load unpredictability on a system-level (Gschwendtner et al., 2023b; Potoglou et al., 2023).

7.2.2.3.3 Cybersecurity and data privacy risks

This is why cybersecurity and privacy issues are becoming more urgent due to the spread of data-driven charging management and V2G solutions. The threats are access to unauthorized data, breach of customer privacy, and the manipulation of charging patterns using digital attack vectors. Recent events and policy reaction points to the fact that the operational consistency and customer confidence may be compromised by improperly implemented cybersecurity or privacy measures, which introduce an additional uncertainty in predicting and controlling smart grids (Pokhrel & Hossain, 2022).

7.2.2.4 Data quality and observability uncertainties

Robust scenario modeling and operation of the system require excellent data that is free of bias and is complete. Nevertheless, there are still a number of lingering uncertainties. Fig. 7.10 shows the diagram of the EV data quality and observability uncertainties.

7.2.2.4.1 Sampling bias in charging data

Empirical charging data sets often have sampling bias: Over- or under-representation of particular groups of users, types of chargers or regions. This affects the generalization of the results because the models used to be trained based on biased data can result in the misrepresentation of the actual load diversity or even flexibility potential in the real world (Austmann, 2021).

7.2.2.4.2 Privacy-driven data limitations

More stringent privacy policies are limiting access to fine-grained user information. This increases the differences between observability of individual mobility and charging decisions, which are important in demand-side management, and potentially leads to using synthetic or aggregate data, which are not behaviorally realistic (Aghapour et al., 2022).

7.2.2.4.3 Algorithmic opacity in smart charging

There is a new source of uncertainty in the smart charging management or V2G dispatch that uses opaque, black box algorithms. The absence of interpretability can destroy user confidence, hinder their regulation, and hind future model improvement, especially in cases where feedback loops are poorly understood or conveyed (Sovacool et al., 2018).

7.2.2.4.4 Measurement and reporting inconsistencies

The noise and systematic error in measurements used in control of operations and long-term planning are caused by inconsistencies in charger metering, error-prone reporting systems, and heterogeneous data standards. The need to have strong data auditing and aligned reporting protocols has been emphasized by the modern reviews (Guo et al., 2025). Table 7.3 gives an overview of various types of direct and indirect uncertainties and their effect.

7.3 Electric vehicles load uncertainty modeling

In this section, the approaches to modeling uncertainties in EVs are discussed. Such uncertainties, as elaborated in Section 7.2, can be attributed to numerous factors, e.g., the arrival and departure time of the EV, charge time, battery level charge, driving habits, and energy consumption habits, among others. With the help of these modeling methods, the effects of uncertainties on EV demand and grid integration can be measured, predicted, and mitigated.

Uncertainty model building usually takes place in a systematic way that involves the mathematical modeling of uncertainties, as shown in Fig. 7.11. Moreover, this part includes information about the advantages, limitations, and needs of every type of modeling. According to the literature review, the following four categories of uncertainty modeling approaches to EVs can be distinguished: Deterministic load modeling, probabilistic and statistical approaches, machine learning-based approaches, and data-driven modeling approaches.

7.3.1 Deterministic load modeling

Deterministic EV load models are one of the basic methods in which there is a set of certainty in the assumption of specific EV parameters. These techniques make use of mathematical models that directly establish a relationship between power consumption and electrical quantities like voltage and current, at the load connection point

Table 7.3 Summary of direct and indirect uncertainties impacting the electricity grid.

Uncertainty	Description	Impact	References
User behavioral			
Socio-demographic	- Age, income, education, and location influence EV adoption and charging patterns.	Varies adoption and charging regularity; impacts charger deployment.	Daina (2014), Bae & Yun (2024), Priyam et al. (2024), Philipsen et al. (2018), Zheng et al. (2024)
Pro-environmental	- Green attitudes encourage off-peak or renewable-based charging with feedback mechanisms.	Promotes flexible, smart charging adoption.	Kramer et al. (2023)
Preferences	- Satisfaction, access, price, and protocol familiarity shape charging choices.	Influence's location and strategy; standardization aids adaptability.	Wang et al. (2021), Andrenacci & Valentini (2023)
Psychological	- Range anxiety leads to frequent charging at high state-of-charge (SOC) levels.	Increases grid demand variability; may reduce battery lifespan.	Noel et al. (2019), Pei, et al. (2025), Hu et al. (2019), Franke & Krems (2013)
Convenience	- Proximity, charger access, and speed drive charging decisions.	Affects timing and location; disruptions increase unpredictability.	Chung et al. (2018a)
Cost Sensitivity	- Cost-conscious users respond to dynamic tariffs and Time-of-Use (ToU) pricing.	Causes varied responses to pricing and grid schemes.	Li & Jenn (2022)

(continued on next page)

Attitudes	- Views on technology and smart charging affect participation in services like Vehicle-to-Grid (V2G).	Impacts uptake of flexibility services; trust is key.	Libertson (2022), Pan et al. (2019)
Experience/ knowledge	- Experienced users optimize charging; lack of knowledge hinders efficiency.	Improves signal response; inexperience limits grid-beneficial behavior.	Chamberlain & Al-Majeed (2021)
Spatiotemporal			
Time of charging	- Peaks in evenings/overnight at homes; weekday double peaks, weekend single peaks. - Varies by schedules and grid signals.	Concentrated loads; ToU tariffs may shift peaks based on user response.	Henriksen et al. (2021), Yang et al. (2025a), Bons et al. (2021)
Charging location	- Home charging for private EVs; public stations for commercial. - Slow chargers in urban areas, fast chargers in commercial/highway settings.	Shapes demand by charger availability and proximity.	ten Have et al. (2020), Pan et al. (2020), Cui et al. (2022)
Special/external conditions	- Pandemics, weather, or events disrupt routines. - Seasonal changes add variability.	Alters patterns, impacting forecasting and grid stability.	Shahriar & Al-Ali (2022), Dominguez-Jimenez et al. (2020)
Technological			
EV types	- HEVs/FCEVs (no grid impact), PHEVs (moderate), BEVs (high, V2G).	Varying needs cause demand uncertainty; BEVs raise peaks.	Huaman-Rivera et al. (2024)
Charging infrastructure	- Charger types (Level I-III, DCFC) affect behavior; fast versus slow.	Impacts forecasting; needs planning for grid stability.	Zhao et al. (2020b), Yang et al. (2018), Lagomarsino et al. (2022)

(*continued on next page*)

Table 7.3 Summary of direct and indirect uncertainties impacting the electricity grid—cont'd

Uncertainty	Description	Impact	References
Grid and systemic			
Grid capacity/stability	- Uncoordinated fast charging causes overloads and voltage deviations.	Risks transformer aging, failures, and instability.	Shariatzadeh et al. (2024), Sun & Qiu (2021)
RES integration	- EV demand may misalign with intermittent Renewable Energy Sources (RES), worsening peak issues.	Complicates solar charging and duck curve management.	Mangipinto et al. (2022)
Power quality	- Fast chargers inject harmonics, disrupting equipment.	Increases losses and risks capacitor failures.	Sevdari et al. (2022)
V2G/smart charging	- V2G offers benefits but faces battery, user, and system challenges.	Uncertainty in feasibility and user participation.	Nagel et al. (2024), Everoze (2023), Ensslen et al. (2018)
Aggregate load variability	- Stochastic EV demand complicates load prediction and planning.	Challenges scheduling, reserves, and network planning.	Kamana-Williams et al. (2024)
Environmental			
Weather and climate	- Temperature and precipitation affect range and charging frequency. - Climate impacts renewable energy availability.	Load shifts, higher winter demand, forecasting challenges.	Hiesl et al. (2021), Supponen et al. (2016), Huang et al. (2020)
Policy and regulatory changes	- Incentives and mandates drive adoption; stability aids planning, abrupt changes disrupt.	Uncertain demand forecasts, need for flexible grid planning.	LaMonaca & Ryan (2022), Hildermeier et al. (2022)

(*continued on next page*)

Fuel price volatility	- Fuel price swings boost EV uptake and home charging; complicates modeling.	Hard to predict load growth, peak demand.	Bushnell et al. (2022), Sun et al. (2019)
Economic			
Charging costs	- Dynamic pricing (ToU tariffs, demand charges) affects behavior.	Behavior heterogeneity, peak shifting, potential penalties for inflexible users.	El Gohary et al. (2023), Wang et al. (2022)
Electricity price fluctuations	- Volatile electricity prices influence charging choices and planning.	Altered charging windows, storage investments, forecasting challenges.	Powell et al. (2020)
Incentive schemes	- Incentives motivate smart charging and flexibility.	Immediate load effects, shaped by design and user awareness.	Chakraborty et al. (2019), Daziano (2022), Gong et al. (2021)
Payment methods and contract conditions	- Fragmented offerings and complex fees cause confusion.	Variable price responses, limited smart charging participation, varied V2G engagement.	Shell Recharge Solutions (2023), Huang et al. (2021)
Parking fees	- Parking fees at charging points influence dwell times and choices.	Reduced charging time, location avoidance, demand displacement.	Pagani et al. (2019)
Operational			
Real-time charging demand forecasting	- Stochastic mobility, weather, and events make forecasting uncertain. - Machine learning (SARIMA, LSTM) improves accuracy.	Challenges utility scheduling and network reliability; requires probabilistic approaches.	Akshay et al. (2024), Sanami et al. (2025)

(continued on next page)

Table 7.3 Summary of direct and indirect uncertainties impacting the electricity grid—cont'd

Uncertainty	Description	Impact	References
Charging station availability and congestion	- Congestion, failures, and surges cause availability uncertainties. - Leads to charging redistribution.	Users switch stations/times, increasing system-level load unpredictability.	Potoglou et al. (2023), Gschwendtner et al. (2023b)
Cybersecurity and data privacy risks	- Threats from unauthorized access and privacy breaches; critical for data-driven systems.	Undermines operational reliability and user trust; adds uncertainty to forecasting and control.	Pokhrel & Hossain (2022)
Data quality and observability			
Sampling bias in charging data	- Over/under-representation of user groups, charger types, or locations.	Reduces model generalizability, misrepresents load diversity.	Austmann (2021)
Privacy-driven data limitations	- Strict privacy laws limit access to detailed user data.	Reduces observability, reliance on less accurate synthetic data.	Aghapour et al., 2022
Algorithmic opacity in smart charging	- Nontransparent algorithms in smart charging and V2G systems.	Undermines trust, hinders oversight, limits model refinement.	Sovacool et al. (2018)
Measurement and reporting inconsistencies	- Inconsistent metering and reporting standards.	Introduces errors, affects operations and planning; needs auditing and standardization.	Guo et al. (2025)

EV, electrical vehicle; *LSTM*, long short-term memory; *SARIMA*, seasonal auto-regressive integrated moving average.

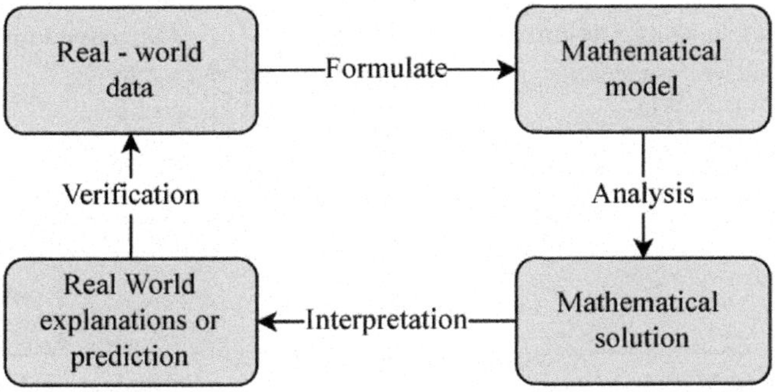

Figure 7.11 Mathematical modeling of EV uncertainties (Huaman-Rivera et al., 2024).

(Arif et al., 2017). A deterministic approach is the basis of more complicated modeling methods and gives fundamental information about the steady-state and dynamic behavior of EV loads operating under different conditions (Tian et al., 2021).

The fundamental concept of deterministic modeling is to assume that EV charging patterns can be forecasted using a set of definite mathematical functions, without taking the stochastics of charging patterns. This method is especially useful when the initial planning study of a system is performed and when no or little historical information is available. Deterministic models can be described as being computationally efficient and easy to implement, and hence appealing in large-scale grid integration studies where computational resources might be a limiting factor (Gil-Aguirre et al., 2019). The deterministic models are usually divided into static and dynamic models.

7.3.1.1 Static modeling

The load characteristics are expressed in the form of algebraic functions called the static modeling, which illustrate the dependency between power consumption and voltage at any moment in time. Such models find this useful in determining steady-state characteristics of EV loads, and have wide application in power flow analysis, infrastructure planning, and voltage stability analysis. Fig. 7.12 presents the methodology of the deterministic simulation of the EV charging demands.

The mathematical basis of the modeling of the static EV loads is based on the formation of the functional relationships between the active power (P), reactive power (Q), and bus voltage (V). The exponential load models (EXP) and polynomial load models (ZIP) are the most used static models in EVs.

The exponential model offers a simplified form of voltage-dependent loads as in Eq. (7.1):

$$P = P_0\left(\frac{V}{V_0}\right)^{\alpha}, Q = Q_0\left(\frac{V}{V_0}\right)^{\beta} \tag{7.1}$$

where P_0, Q_0, and V_0 are rated values, α and β are the voltage sensitivity exponents of active and reactive power, respectively, which are unknown model parameters. The

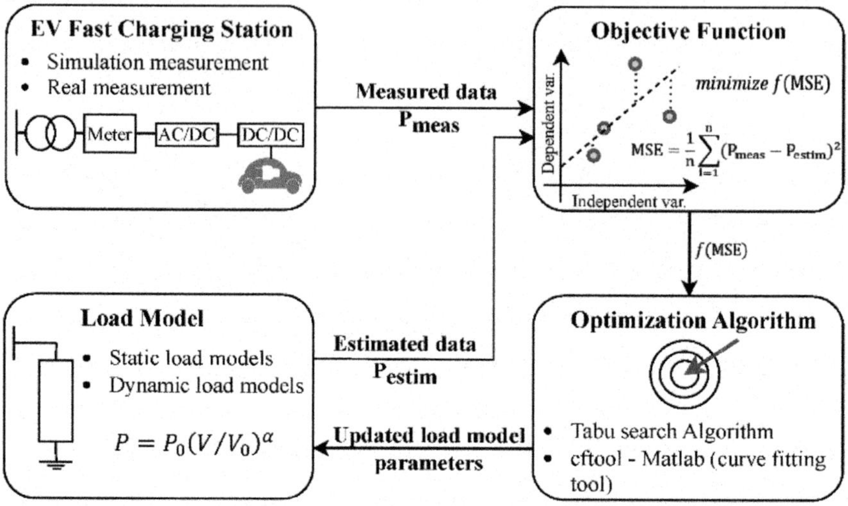

Figure 7.12 Deterministic EV load modeling (Huaman-Rivera et al., 2024).

model is specifically well adapted to EV loads with the power electronic converters having exponential voltage characteristics. As it has been proven, the EXP model can be successfully used to model some kinds of EV chargers, especially those that have a simple control system (Kongjeen et al., 2019). The simplicity of this model, however, might restrict its ability to reflect the charging behavior of a complicated system, like in a fast-charging station with a complicated power management system.

More recent findings (Kongjeen & Bhumkittipich, 2016) have proposed more powerful models in the form of static models, combining the advantages of the methods of polynomials and exponential models. Constant power plus exponential (PEXP) model overcomes the limitations of pure exponential models by adding a constant power term:

$$P = P_0\left[p_1 + p_2\left(\frac{V}{V_0}\right)^{\alpha}\right] \tag{7.2}$$

Equally, the linear plus exponential (LEXP) model is more accurate, it integrates the linear and exponential terms:

$$P = P_0\left[p_1\left(\frac{V}{V_0}\right) + p_2\left(\frac{V}{V_0}\right)^{\alpha}\right] \tag{7.3}$$

It has been demonstrated that in the case of EV charging stations with diverse load compositions, LEXP models can be more accurate than the conventional exponential models (Hannan et al., 2017).

The ZIP model can be illustrated in the form of:

$$\begin{aligned} P &= P_0\left[p_1\left(\frac{V}{V_0}\right)^2 + p_2\left(\frac{V}{V_0}\right) + p_3\right] \\ Q &= Q_0\left[q_1\left(\frac{V}{V_0}\right)^2 + q_2\left(\frac{V}{V_0}\right) + q_3\right] \end{aligned} \tag{7.4}$$

where p_1, p_2, p_3 and q_1, q_2, q_3 are the model parameters are the proportion of each component. The ZIP model has been used extensively in EV load modeling as it is versatile in modeling various forms of charging behavior and can be used in conjunction with standard power flow analysis packages.

7.3.1.2 Dynamic modeling

Dynamic load modeling is used to model phenomena whose effects change over time and system response to disturbances that are considered dynamic, extending the scope of the static formulations, and is therefore a vital component of transient stability studies and transient grid management software. Even with the fact that the description of the load behavior at static models provides a single snapshot at a given time, dynamic models are used to include the use of differential equations, which demonstrate how the load behavior changes with time.

The exponential recovery dynamic load model (ERL), a dynamic model of the active and reactive power response to the step change of the bus voltage, is the most prominent dynamic model used to model EV loads. The model can be used effectively to model loads that have slow recovery properties, with time constants between a few seconds and tens of minutes. The mathematical formulation of the ERL model is expressed as:

$$\begin{aligned} T_p\frac{dP_r}{dt} + P_r &= P_0\left(\frac{V}{V_0}\right)^{a_s} - P_0\left(\frac{V}{V_0}\right)^{a_t} \\ P &= P_r + P_0\left(\frac{V}{V_0}\right)^{a_t} \end{aligned} \tag{7.5}$$

in which P_r is a state variable of power dynamics, T_p is the exponential recovered response time constant, a_s and a_t are the exponent of transient load responses, respectively.

The ERL model incorporates the critical dynamic behavior of EV chargers, such as how they react to voltage disturbance and the recovery process. This ability will be vital in determining the effect of EV charging on power system stability, especially in networks with large coverage of EV penetration. As reference (Tian et al., 2021) showed, the control parameters of EV chargers have a strong impact on their dynamic response, making it hard to capture the entire behavior of EVs when the system is disturbed.

The choice of the type of model used depends on the application needs and tolerance of the computational complexity between the use of static and dynamic models.

Table 7.4 Comparison of key characteristics for dynamic modeling categories.

Characteristic	Static modeling	Dynamic modeling
Mathematical complexity	Low (algebraic equations)	High (differential equations)
Computational requirement	Minimal	Moderate to high
Temporal resolution	Instantaneous	Time-dependent
Parameter sensitivity	Low	High
Grid integration studies	Power flow, steady-state	Transient stability, control
Model validation	Straightforward	Complex
Implementation effort	Simple	Moderate

Table 7.4 gives a detailed comparison of the major attributes of both categories of models.

7.3.2 Probabilistic and statistical models

The statistical characterization models have the objective to produce a distributional analysis of EV charging outputs, either based on data obtained using empirical records of charging stations or based wholly on external variables, including the travel survey records and statistical hypotheses. EV load modeling methods are used to understand and predict the dynamics of the EV load demand depending on any uncertain factors, including the charging time, distance, SOC, and accessibility of charging stations (Huaman-Rivera et al., 2024). Such techniques apply statistical models such as probabilistic distributions and stochastic processes to examine historical data and draw demand models that are useful in planning EV charging infrastructure and grid operations.

7.3.2.1 Probability density functions

The variability and randomness of EV-related parameters are modeled by a probability density function (PDF) and thus allow modeling aggregate demand in the case of uncertainty. The different types of statistical distributions have been used to describe the different types of behaviors that are evident in empirical data. Probably most frequently used distributions are Normal (Nadarajah, 2005), Lognormal (Crow & Shimizu, 1987), Weibull (Khoo et al., 2014), Exponential (Balakrishnan, 2019), Gamma (Stacy, 1962), Beta (McDonald & Xu, 1995), Poisson (Consul & Jain, 1973), Birnbaum–Saunders (fatigue life (Balakrishnan & Kundu, 2019), Inverse Gaussian (Folks & Chhikara, 1978), Logistic (Gupta & Gnanadesikan, 1966), generalized logistic (Gupta et al., 2010), Burr Type XII (Muhammad, 2017), generalized exponential (Gupta & Kundu, 2001), exponentiated Weibull (Pal et al., 2006), and Nakagami (Yacoub et al., 1999). The different types of distributions provide varying flexibility to model skewness, heavy tails, and multimodality, and hence are applicable to patterns of EV behavior. Table 7.5 provides a brief description of the most used PDFs in the modeling of EV loads.

There is empirical research supporting the use of these distributions with an accent on the circumstance-related strengths and limitations. As an illustration, Cheng &

Table 7.5 Most frequently applied PDFs in modeling EV loads.

Normal	Log-normal	Beta
$f(x;\mu,\sigma^2)=\frac{1}{\sigma\sqrt{2\pi}}e^{-\frac{(x-\mu)^2}{2\sigma^2}}$ **Normal** $f(x;\mu,\sigma^2)=\frac{1}{\sigma\sqrt{2\pi}}e^{-\frac{(x-\mu)^2}{2\sigma^2}}$ 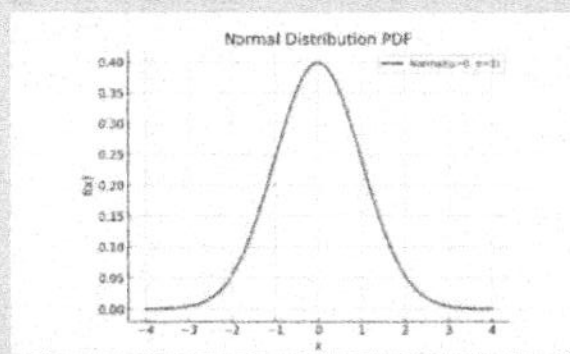	$f(x;\mu,\sigma^2)=\frac{1}{x\sigma\sqrt{2\pi}}e^{-\frac{(\ln(x-\mu))^2}{2\sigma^2}}$ **Log-normal** $f(x;\mu,\sigma^2)=\frac{1}{x\sigma\sqrt{2\pi}}e^{-\frac{(\ln(x-\mu))^2}{2\sigma^2}}$ 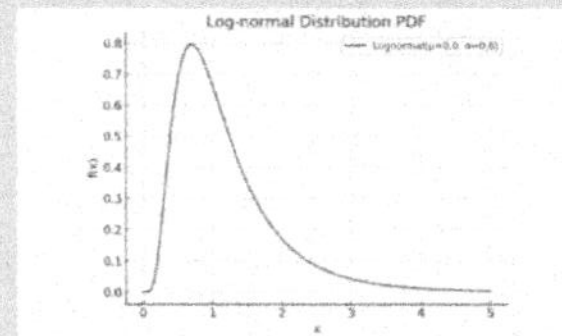	$f(x;\alpha,\beta)=\frac{x^{\alpha-1}(1-x)^{\beta-1}}{B(\alpha,\beta)}$ **Beta** $f(x;\alpha,\beta)=\frac{x^{\alpha-1}(1-x)^{\beta-1}}{B(\alpha,\beta)}$ 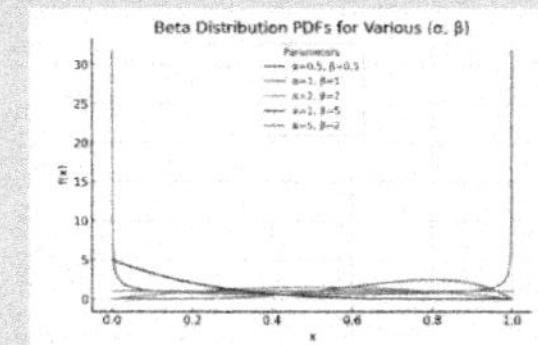
Weibull	**Binomial**	**Poisson**
$f(x;\ \lambda,\ k)=\frac{k}{\lambda}\left(\frac{x}{\lambda}\right)^{k-1}e^{-(x/\lambda)^k}$ **Weibull** $f(x;\lambda,k)=\frac{k}{\lambda}\left(\frac{x}{\lambda}\right)^{k-1}e^{-(x/\lambda)^k}$ 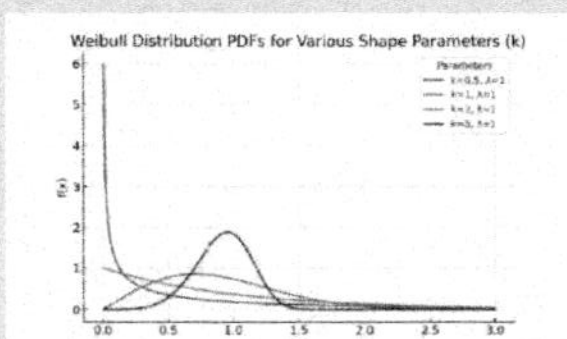	$(k;\ x)=\binom{n}{k}p^k(1-p)^{n-k}$ **Binomial** $(k;\ x)=\binom{n}{k}p^k(1-p)^{n-k}$ 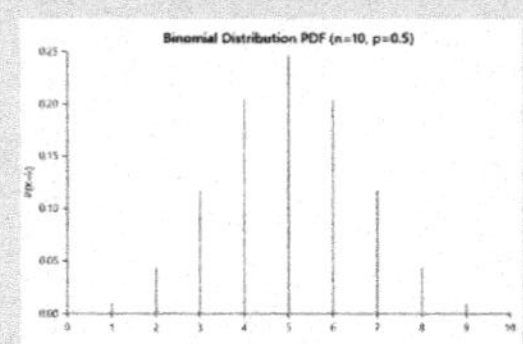	$f(k;\ \lambda t)=\frac{(\lambda t)^k e^{-\lambda t}}{k!}$ **Poisson** $f(k;\ \lambda t)=\frac{(\lambda t)^k e^{-\lambda t}}{k!}$ 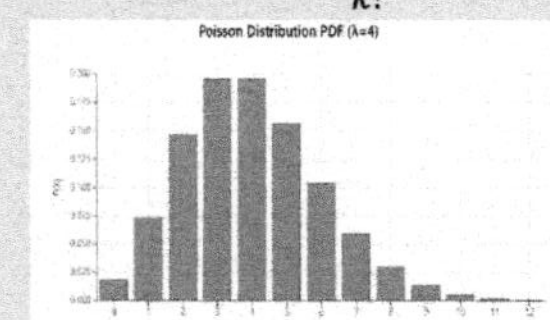

EV, electrical vehicle; *PDFs*, probability density functions.

Liu (2022) fitted normal distributions to the daily travel distance and charging start time, and found that these models generally gave good predictivity of power demand, though they were known to give unrealistic negative values. On the same note, Sun et al. (2015) used the NHTS (National Household Travel Survey, 2021) 2009 ICEV data to model EV fleets and discovered that the NHTS can be used to simulate actual demand patterns using the normal distribution, but only strictly positive distributions can provide more physical consistency, such as charging time. Conversely, (Singh et al., 2023) found that the Weibull distribution was most effective to use in charging duration in a 33-EV experiment in Victoria, Australia, and was better than the exponential and lognormal fits. They also parameterized inter-charging intervals as a combination of two lognormal distributions, but even these parametric forms continue to have problems with highly irregular driver behavior.

A multimode analysis (Flammini et al., 2019) used beta mixture models, which are optimized by reducing root mean square errors (RMSE) to the data of Elaad NL (Elaad n.d.) and performing Kolmogorov-Smirnov tests, which are useful in identifying multiple peaks of charging in the morning and afternoon. Generalizing this, Helmus et al. (2020) performed Gaussian mixture clustering of Dutch G4 city data to determine 13 charging profiles, and Lee et al. (2019);Lahariya et al. (2020) used multivariate Gaussian mixtures to simultaneously predict arrival time, charging duration, and energy consumption, but with less integration of exogenous effects.

Nonparametric methods, including Gaussian kernel density estimation (GKDE) (Chen et al., 2020a; Liang et al., 2015), do not have strict distributional requirements and are more flexible in capturing complicated behavioral patterns. It has been demonstrated by comparative studies (Khaki et al., 2018, 2019) that diffusion KDE (DKDE) has a higher outlier-resistant load estimation accuracy due to optimized bandwidth selection, and Chung et al. (2018b) have proposed a hybrid KDE (HKDE) that combines GKDE and DKDE to reduce outlier influence without impairing fidelity.

The shaping of the simple parametric PDFs into more modern and sophisticated nonparametric and hybrid techniques is a testament to the increased awareness of the behavioral complexity of EV charging. Although simple distributions (e.g., normal, exponential) are computationally friendly and applicable to large-scale MCSs, they are not able to adequately reflect the heterogeneous user populations or multimodal load distributions. Mixture models, KDE variants, and joint distributions based on copulas are more faithful at the expense of higher computational costs. Practically, PDF-based models can be seen as a first step toward uncertainty models, to which the parameters of input are provided to stochastic simulation, scenario generation and probabilistic analysis of grid impacts, consequently, creating a link between behavioral description and power system planning that is critical.

7.3.2.2 Monte Carlo simulation

MCS is a stochastic sampling tool that involves random sampling of inputs in order to create synthetic situations and estimate their statistical features (Rubinstein & Kroese, 2016). Repeated sampling of probability distributions on important variables (e.g., arrival times, daily travel distance, initial SOC, charger power)

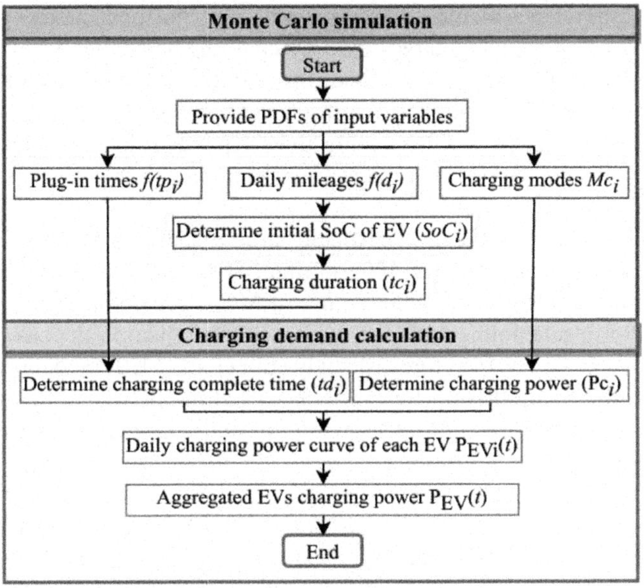

Figure 7.13 Schematic of MCS for EV charging load estimation (Huaman-Rivera et al., 2024). *EV*, Electrical vehicle; *MCS*, Monte Carlo simulation.

in EV charger studies, to generate a large sample of possible charging events, is also known as MCS (Cui et al., 2023; Kilic et al., 2021;Zhang et al., 2024a). Such random draws are then translated into single EV charging loads, which are summed to give aggregate demand profiles and uncertainty limits. The method is common in cases when it is not possible to find analytical solutions, or when the data is sparse, given that it can easily be adapted to work around a number of uncertainties (Li et al., 2023).

The MCS approach to the modeling of EV charging demand consists of the definition of PDFs of the most prominent input variables, such as plug-in times, daily plug-in mileages, and charging modes. These distributions are then used to create random samples of the initial SoC of each EV and, later, the necessary charging time. Based on this information, the simulation continues to determine the charging demand with the charging completion time and charging power of each vehicle. These steps make it possible to derive the individual daily charging power curves and then combine them with all EVs to get the overall demand profile of charging. Through this procedure repeated many times, MCS gives a probabilistic model of EV charging load profiles, which describes variability and uncertainty in user behavior and operational conditions. Fig. 7.13 shows the schematic of the MCS used in estimating the EV charging load.

The recent literature has shown that Monte Carlo methods can be used in the modeling of the uncertainty of EV across various scales and contexts. MCS is mentioned in a review of plug-in EV load modeling methods (Ahmadian et al., 2020) as estimating demand based on the probabilistic sampling of home arrival time, departure time, daily travel distance, vehicle type, and battery capacity to produce stochastic load profiles with high accuracy, but is identified as computationally the most intense. One of the few examples of MCS in modeling large-scale EV charging demand is a New Zealand case study of large-scale EV charging demand (Su et al., 2019), which uses MCS to model the daily driving and charging times of each vehicle and charge fleet makeup (e.g., personal or commercial EVs), producing time-dependent load profiles that indicate peak overlaps and can be used to inform smart charging strategies to prevent network congestion.

Coordinated operation studies of vertically structured power systems (Adianto et al., 2021) model daily travel distance, plug-in time, SOC, and charging time of different types of EV (cars, buses, taxis) with the use of MCS, generating load curves that can be used to inform scheduling and lead to up to 13% reduction in peak loads in high-penetration cases. An EV load forecast study (Liu et al., 2020) models the MCS to the traffic flow and starting driving time, which will allow the spatial-temporal demand to be predicted, with mitigation measures to system overloading measures taken. A model of office building scheduling (Ding et al., 2023), which incorporates MCS of arrival/departure times, SOC, and battery capacity, has a stronger load prediction, and it relocates startup loads by 7.57%–61.67% to off-peak periods. Lastly, an analysis of EV effects on residential distribution (Alsharif et al., 2023) takes up stochastic MCS of arrival/departure times and SOC, which models power flows in hybrid systems and gives zero loss of power supply probability at optimized costs.

7.3.2.3 Markov chains theory

Markov chains: A Markov chain is a stochastic process in which the future state of a system only depends on the present state of the system and follows the memoryless property, and has transition probabilities determined by historical observations. It is a framework that allows probabilistic dependencies to be encapsulated and sequential events predicted in dynamical systems without exhaustive historical information. Markov chains prove useful in the context of EV uncertainty modeling to model the probabilistic characteristics of charging patterns, including arrival time, dwell time, and energy requirements, which are based on irregular mobility patterns and infrastructure accessibility; this will allow resilient forecasting of grid effects, such as load shifts and peaks, and help to support resilient power systems planning (Al-Muhaini, 2020).

In Markov chain modeling, the basic unit is the transition probability matrix (TPM) in which every value is the likelihood of going out of state (i) to state (j) in a single time step, formulated as:

$$p_{ij} = P(X_{n+1} = j | X_n = i) \tag{7.6}$$

where X_n is the state of the system at step n. The sum of every row of this matrix is equal to 1, which becomes clear by the fact that the probability of going through one state to all possible states is equal to 1.

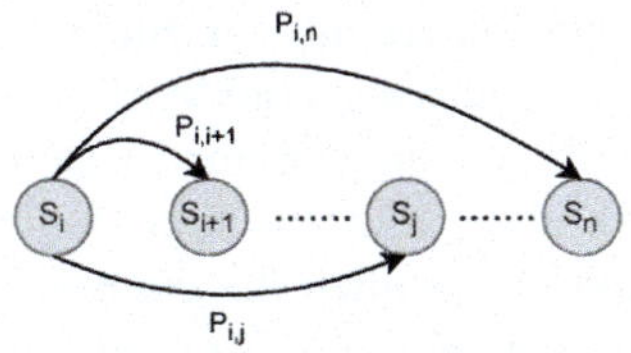

Figure 7.14 State transfer probability diagram (Huaman-Rivera et al., 2024).

Furthermore, the probability of reaching the state j from the initial state i in n steps are given by the (i, j) entry of the n-th power of the transition matrix:

$$p_{ij}^{(n)} = P(X_n = j | X_0 = i) \tag{7.7}$$

Markov chain theory has a three-step approach, which is applied practically in a structured manner. To begin with, a set of states is described to illustrate all possible behaviors of an EV, including its location, activity, SOC. Second, Eq. (7.8) is used to construct TPM with the aim of containing the probability of transitioning between states.

$$P = \left(P_{ij}\right)_n = \begin{bmatrix} P_{11} & P_{12} & \dots & P_{1n} \\ P_{21} & P_{22} & \dots & P_{2n} \\ \vdots & \vdots & \vdots & \vdots \\ P_{n1} & P_{n2} & \dots & P_{nn} \end{bmatrix} \tag{7.8}$$

Such probabilities are generally based on actual or simulated EV data. In changing situations, a nonhomogeneous Markov chain can be adopted, in which the transition matrix varies with time to capture changing trends. Lastly, simulation and analysis of the established TPM is frequently done by producing synthetic EV profiles by random walk over the state space (Liu et al., 2023). The probability diagram of state transfer is presented in Fig. 7.14.

The modern literature illustrates advanced use of Markov chain theory in various areas of charging electric vehicle infrastructure. The (Ahmadian et al., 2020) review concentrated on plug-in EV load methods, and Markov chains were emphasized on probabilistic simulation, including the user preferences and infrastructure to enhance predictivity. The adequacy model of microgrids as a Markov-based algorithm was proposed in (Al-Muhaini, 2020), which discretizes the states of EVs to analyze the reliability in charging uncertainties, including the contribution of the V2G. In (Liu et al., 2023), researchers have developed higher-order Markov chains in load forecasting to capture long-term historical dependencies to provide better temporal accuracy. In (Xiang et al., 2019), the authors reviewed charging load modeling, with more focus on the semi-Markov model of mobility and time uncertainty, embedding dwell times and nonexponential distribution to better understand the impact of the grid. The contribution of Yan et al. (2022) combined Markov chains with the theory of information gap decision in a many-objective model, maximizing charging under the conditions of DR and renewable uncertainty. In (Shepero & Munkhammar, 2018), the authors have come up with a GIS-based spatial Markov model that simulates location-dependent loads by transitions between charging locales.

In (Varshney et al., 2025), under a multiserver station, the authors used queuing-Markov analysis as a method to optimize the controls in the presence of arrival as well as service uncertainty. Puech et al. (2024) used semi-Markov processes coupled with stochastic programming to model the behavior of the users within large stations, dealing with early disconnection. As an extension of queuing-Markov models with feedback to the EV networks, Varshney et al. (2024) investigated the extension and facilitated scalable planning under load variability conditions. Stochastic models of scenarios based on these Markov transitions of vehicle availability and demand were modeled and optimized schedules (Wang et al., 2020). A driving uncertainty based on inhomogeneous Markov chains is modeled with an MDP-based optimal charging framework where cost-minimizing policies with extensions to V2G are derived (Vandael et al., 2015). The (Chen et al., 2020b) approach combines dynamic information to predict the spatial-temporal loads of traffic-informed Markov chains to be integrated with the energy system. The Markov chains of station occupancy and consumption of discrete states were suggested in (Sokorai et al., 2018), which assisted in the simulation of operator revenue.

7.3.2.4 Fuzzy logic

In contrast to the conventional probabilistic approaches, which assume that statistical distributions are accurate, fuzzy logic employs linguistic variables and membership in order to model uncertainties by using terms like low, medium, and high, which are of great importance in reflecting the subjectivity and imprecision inherent in EV user charging decisions. The triangular and trapezoidal fuzzy numbers developed in the methodology are used to model the important parameters, which are SoC, parking time, and travel distance, so that the network planners can undertake the infrastructure evaluation without any assumption of exact data, and the computational complexity of the method remains small in large-scale applications (Jingwei et al., 2014).

A triangular fuzzy number is represented as $\tilde{a}_t = (a_{t1}, a_{t2}, a_{t3})$, with a_{t1},a_{t2} and a_{t3} indicating the lowest, mean, and highest values derived from the projected scenarios. The highest degree of membership is attributed to a_{t2}, as it signifies the most likely condition. Likewise, a_{t1} and a_{t3} define the span of potential power levels. Additionally, the average value may be calculated by considering all available scenarios (Ahmadian et al., 2020). The plug-in EV power can then be modeled with the help of a fuzzy number, as shown in Fig. 7.15.

Recent research has studied the combination of fuzzy logic and different complementary methodologies in detail to increase the accuracy of EV load modeling. A fuzzy inference system (FIS) in (Shahidinejad et al., 2011) represents the recharging choice of drivers in terms of triangular fuzzy numbers of SoC, parking time, and charging likelihood, and estimates hourly loads in different types of vehicles when at home and when at work.

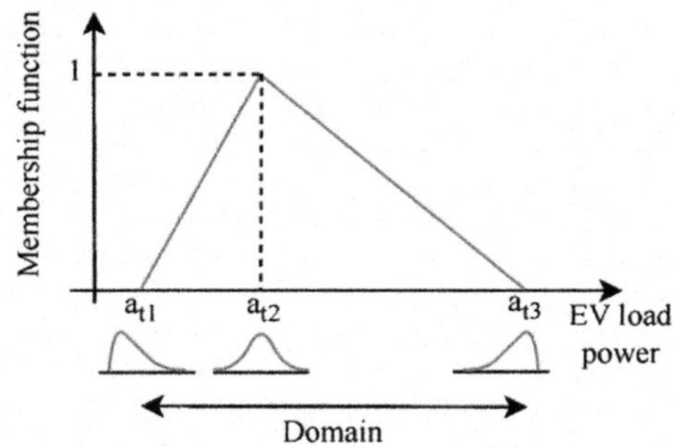

Figure 7.15 EV load demand as a fuzzy number (Huaman-Rivera et al., 2024). *EV*, Electrical vehicle.

A core component of this FIS is the firing strength of each fuzzy rule, which aggregates input uncertainties (e.g., parking time t (t), SOC (s), and threshold factor (th)) by taking the minimum membership degree across inputs:

$$\mu_r = min\left(\mu_{parking}(t),\ \mu_{SOC}(s),\ \mu_{there\,shold}(th)\ \right) \tag{7.9}$$

where μ_r is the firing strength of the rule r, and the membership functions are typically triangular for fuzzy sets.

In (Ahmadian et al., 2020), the authors go further and use fuzzy triangular numbers to represent variability in EV power profiles so that planning can be accomplished based on scenarios without strict data constraints and prove to be more robust in simulations. Equally, Bian et al. (2025); Wan et al. (2019) generalize fuzzy inference to predicting, in which the membership functions of travel-related variables can be used to generate dynamic load curves, relying on the gaps in traditional probabilistic models by considering the human-related vagueness.

More recent advances, such as Gholami et al. (2022); James & Jasmin (2022), incorporate optimization methods, such as genetic algorithms, into refining fuzzy systems to improve real-time responsiveness to price and delay uncertainties; and hybrids such as Lin et al., 2023; Bouallaga & Doumbia, 2019) combine fuzzy qualitative treatment and Monte Carlo quantitatively sampling to model spatial-temporal dynamics in greater detail. The study in Han et al. (2020); Tang & Wang (2015) depicts transitional methods, which combine fuzzy with stochastic or Markov techniques to concern behavioral ambiguity and state transitions, but give more emphasis to probabilistic cores. Table 7.6 at the comparing view presents the common fuzzy applications used in EV modeling.

7.3.3 Machine learning-based techniques

Machine learning has become a revolutionary way of simulating uncertainty in EV work that provides significant benefits over traditional statistical tools. Machine learning, as a subfield of artificial intelligence, is concerned with the development of

Table 7.6 Representative applications of fuzzy logic in EV uncertainty modeling.

References	Primary focus	Key fuzzy inputs	Methodology	Output
Ahmadian et al. (2020)	DG and storage planning	Daily distance, arrival time	Type-2 FIS	Probabilistic load profiles
Shahidinejad et al. (2011)	Charging decision	SoC, parking duration	FIS + driving cycles	Hourly charging load
Wan et al. (2019)	Spatiotemporal load	Travel chain variables	FIS + MCS	Load distribution curves
Gholami et al. (2022)	Smart charging control	SoC, electricity price	Optimized FIS	Charging power set-point
Lin et al. (2023)	Charging station planning	Demand, solar irradiance	Multiobjective FIS	Optimal siting and sizing
Tang & Wang (2015)	Nodal charging demand	SoC, trip purpose	MCS + FIS	Spatial-temporal demand map

DG, distributed generation; EV, electrical vehicle; *FIS,* fuzzy inference system; *MCS,* Monte Carlo simulation; *SoC,* state-of-charge.

algorithms that can allow computers to learn and make decisions based on data they are given, without explicit programming, and discover patterns in data collected through trial-and-error methods. The greatness of the machine learning methods is rooted in their capability to address nonlinear relationships, adapt to novel data trends, and capture irregularities that are common among EV drivers, but they do not assume rigid distributional assumptions that traditional statistical techniques must assume (Zhao et al., 2020a).

These uncertainty modeling machine learning approaches to EVs can be divided into two main groups, namely, the traditional machine learning approaches and contemporary machine learning approaches. The different machine learning-based modeling techniques are depicted in Fig. 7.16.

7.3.3.1 Traditional machine learning-based methods

Traditional machine learning techniques, which are often praised for their interpretability and computational efficiency, can provide strong support for the initial modeling. These techniques fall into three primary groups:

- Linear models (LMs)
- Support vector machines (SVMs)
- Random forest regression (RFR)

7.3.3.1.1 Linear models

LM approaches are a basic base in the machine learning toolkit to solve the uncertainty in EVs. Their simplicity, efficiency in computation, and inherent interpretability can be especially useful in preliminary explorations and baseline evaluations in EV charging behavior modeling.

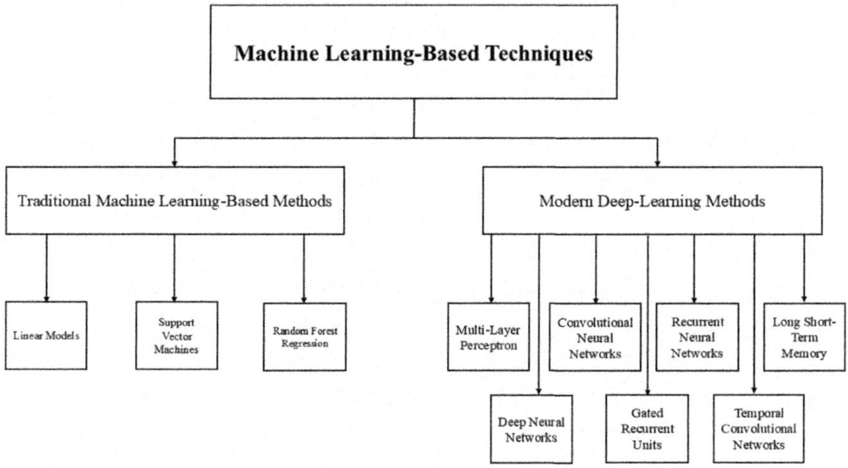

Figure 7.16 Classification of machine learning-based modeling techniques.

Training a linear regression model is going to be the process of identifying the best set of coefficients that will explain the linear relationship between a number of independent variables and a dependent variable in the best way possible. The closed-form solution of the parameter vector θ, which is a cornerstone of the linear regression theory, can be used to accomplish this. This parameter can be determined by the following equation, as explained in Xiong et al. (2017):

$$\theta = \left(X^T X\right)^{-1} X^T y \tag{7.10}$$

where $X = \left[t_{stay}^1, t_{stay}^2, \ldots, t_{stay}^M\right]$ is the pool of sample charging duration and $y = [E^1, E^2, \ldots, E^M]$ contains corresponding energy consumption.

In Xiong et al. (2017), a linear regression was used to predict the user behavior with the prediction of the charging start time, charging end time, and the energy consumed using mean estimation methods. They used a standard linear regression framework using 30 years of history of charging data of 30 EV users. Frendo et al. used linear regression models to estimate the EV departure times to prioritize charging to smart charging, and in the workplace charging case, the mean absolute error was 82 minutes. Their findings revealed that temporal variations in employee charging patterns could be well represented with linear models, and this has a direct impact on enhancing smart charging behavior and optimization of energy delivery (Frendo et al., 2020a). Equally, LM examined variables affecting the charge profile of individuals in Dutch public stations, regressing variables such as temperature and peak hours against average speed until full state-of-charge, and found that the relationship between variables such as voltage deviations caused speeds to reduce by up to 10% (Mies et al., 2018).

LM variations such as multiple linear regression were used in load forecasting systems that adjusted hourly weather and economic variables to normalize probabilistic

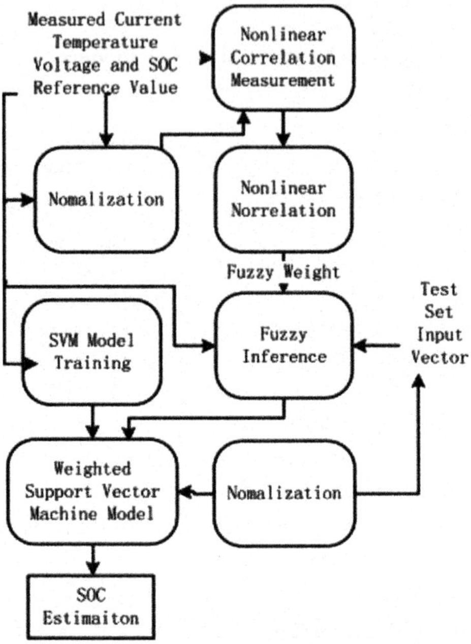

Figure 7.17 Operation flow chart of the SVM model (Sheng & Xiao, 2015). *SVM*, Support vector machines.

EV-influenced loads, and showed significantly lower mean absolute percentage errors than their classical monthly counterparts in ex-post forecasts (Hong et al., 2013). Their inability to capture complex interactions among weather conditions, time-of-day effects, and individual user preferences that have a major impact on influencing the choice of charging is, however limited by the assumption of a linear relationship.

7.3.3.1.2 Support vector machines

SVMs are a powerful methodology of computation that was initially designed by Vapnik (2013) in pattern recognition and regression modeling. SVMs are effective in the context of EVs because of their capability to work with high-dimensional feature spaces and nonlinear relations between charging behaviors, as well as provide stable solutions with regularization characteristics in nature that reduce overfitting of features in uncertain environments.

Xydas et al. (2013) became the first to use SVMs to forecast the demand of charging EVs, and it outperforms MCS approaches on hypothetical charging session models. The paper determined the capacity of SVM to acquire temporal patterns of complex charging behaviors, but it proved ineffective in extending to larger charging station networks because of greater computational complexity. The SVM algorithm modeling process is shown in a flowchart in Fig. 7.17.

This work was later expanded by authors in Majidpour et al. (2016) who applied support vector regression (SVR) to compare customer-centric and station-centric forecasting models using real-world data of the UCLA campus parking lot. Their research found that customer profile-based SVR models had higher prediction speeds, station measurement-based approaches had higher privacy protection with the same accuracy estimates, and projected the trade-off between prediction speed and data sensitivity of SVM applications.

The recent developments have involved improving the performance of SVM by using hybrid approaches as well as preprocessing. The reference Jiang et al. (2020) suggested a hybrid electricity demand forecasting model that combines data-processing schemes with the optimal choice of SVM parameters, and its accuracy was optimal in comparison with isolated machine learning-based models. A novel VMD-SSA-SVR methodology was proposed in Wu et al. (2023), which involves optimization of variational mode decomposition and sparrow search algorithm using SVR to predict the EV charging load. This approach improved the accuracy of forecasting by creating a good signal decomposition and feature extraction. In estimating the state-of-charge of EVs, Sheng & Xiao (2015) applied the fuzzy support vector machine method that involved the use of nonlinear correlation values and fuzzy inference systems to help minimize the sensitivity of the noise to the results and also gave confidence ranges of estimated regression outcomes.

Real-time use in dynamical charging networks may be constrained by the significant disadvantage of computational scalability in large-scale charging networks and longer training times when working with high-dimensional datasets.

7.3.3.1.3 Random forest regression

The first algorithm is the random forest (RF) described by Breiman (2001), which is an ensemble learning algorithm that builds many decision trees using randomly selected subsets of the data and averages their results to enhance predictive accuracy and stability. It is particularly suitable in modeling the stochastic quality of EV charging behavior due to its innate capacity to model nonlinear relationships and measure the amount of prediction variance. Recent research has utilized RF to model these uncertainties well. To illustrate, in one study, harmony search was used to optimize RF parameters to predict spatio-temporal EV load profiles that included charging behaviors of different EV types and were found to be much more accurate in the regional demand distribution than support vectors and neural networks (Ge et al., 2020). The RFR flowchart is presented in Fig. 7.18.

In another study, RF was compared to RF with time-series models, such as seasonal autoregressive integrated moving average in predicting aggregate EV loads in 1700 Dutch sites and found RF to outperform persistence models but outperform autoregressive models because of constraints in the data used (Buzna et al., 2019). To predict the loads of groups of charging stations, a hybrid model that used generative adversarial networks and RF as well as augmented data, was formed to achieve better forecasting within the variability of behaviors, and testing on 15-day half-hourly data within a Chinese region revealed that the hybrid model can be well-predicted (Gang et al., 2021).

Moreover, RF regression was augmented with the sparrow search algorithm to simulate and predict EV charging loads through the MCS, with the emphasis laid

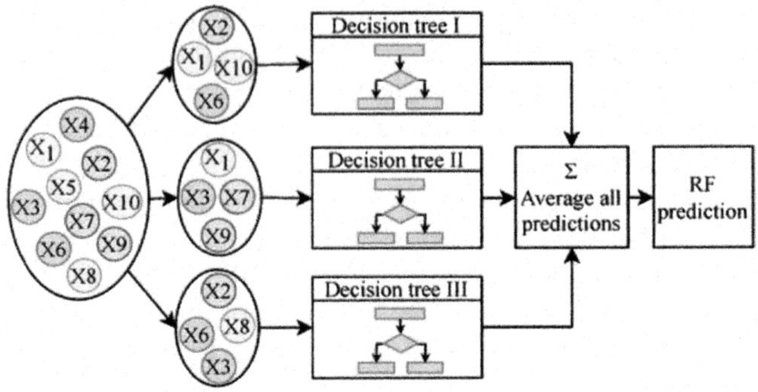

Figure 7.18 Flowchart of RFR algorithm (Huaman-Rivera et al., 2024). *RFR*, Random forest regression.

Table 7.7 Comparison of performance characteristics in traditional machine learning methods.

Aspect	LMs	SVMs	RFR
Interpretability	High	Medium	Low
Computational cost	Low	High	Medium
Nonlinear modeling	Poor	Excellent	Excellent
Uncertainty qualification	Basic	Medium	Advanced

LMs, linear models; *RFR*, random forest regression; *SVMs*, support vector machines.

on tuning parameters to address the load randomness in large fleets and perform better than standard RF in terms of accuracy (Wang et al., 2023). RF was further compared with load forecasting between customer profiles and station measurements, which indicates its effectiveness with station measurements in modeling profile-based uncertainties, and competitive results with k-nearest neighbors and SVR (Pokhrel & Hossain, 2022). RF was used to predict daily charging blocks in a bottom-up model with historical and meteorological inputs in probabilistically forecasting individual EV consumption, where scenario-based aggregation of the fleet outperforms gradient boosting in measures of uncertainty such as continuous ranked probability score (Gerossier et al., 2019). A better RF model, feature selection, and data partitioning to predict short-term loads in single and grouped stations in Shenzhen adequately addressed the effect of randomness and produced lower mean absolute percentage errors compared to SVR and single decision trees (Lu et al., 2018).

The capability of the algorithm to generate uncertainty estimation using the prediction intervals renders the algorithm especially useful in grid planning applications. Nevertheless, drawbacks are low interpretability in comparison with individual decision trees, possible memory demands when the ensemble size is huge, and sensitivity to data quality and size as shown in comparative investigations with conventional time series techniques. Table 7.7 gives the major performance points of conventional machine learning techniques.

7.3.3.2 Modern deep learning methods

Although conventional machine learning methods are simple in modeling EV charging behaviors, they are usually weak in dealing with highly complicated and nonlinear uncertainty. These are the inability to handle temporal dependencies and high-dimensional data with proper accuracy, resulting in low forecasting accuracy on probabilities. Conversely, the contemporary machine learning techniques, which constitute a subdivision of the artificial neural networks (ANNs), offer potent representations of uncertainty modeling through the acquisition of complex patterns on extensive datasets.

ANNs can be traced back to the first calculating models that were based on biological neural systems. The foundational perception, published in 1958 (Rosenblatt, 1958) as a probabilistic brain-model of information processing, was a one-layer network, which could respond to binary classification by changing weights. This idea became multilayer perceptron (MLP) in the 1960s and was forgotten until the resurgence of the 1980s due to the introduction of backpropagation algorithms to efficiently reduce errors over many layers and the introduction of specialized models such as convolutional networks to perform pattern recognition tasks (Rumelhart et al., 1988).

Modern ANN applications in EV uncertainty modeling include a number of specialized architectures, each of which is focused on factors of the prediction and uncertainty quantification problem and may be generally divided into the following types:

- Multilayer perceptron (MLP)
- Convolutional neural networks (CNNs)
- Recurrent neural networks (RNNs)
- Long short-term memory (LSTM)
- Deep neural networks (DNNs)
- Gated recurrent units (GRUs)
- Temporal convolutional networks (TCNs)

7.3.3.2.1 Multilayer perceptron

The use of MLP applications has shown great success in probabilistic EV load forecasting. In the study by Huber et al. (2020), probabilistic forecasting of EV load was done by using an MLP with a tilted loss function; the authors compared their results to those of kernel density estimators and quantile regression. They reported better performance with the same input–output parameters, and this is in line with the overall direction of ANNs, showing optimal performance in a wide variety of scientific tasks. Nevertheless, the main shortcoming of these models is their interpretability, and this poses a challenge toward their implementation on critical infrastructure applications (Chakraborty et al., 2017;Zhang & Zhu, 2018). In Fig. 7.19, the MLP architecture of load forecasting is represented.

7.3.3.2.2 Convolutional neural networks

CNNs have been widely studied on the aspects of spatial and temporal pattern recognition in EV load modeling. In Li et al. (2018), the authors designed a short-term load forecasting that is applied to EV charging stations based on CNNs with niche immunity

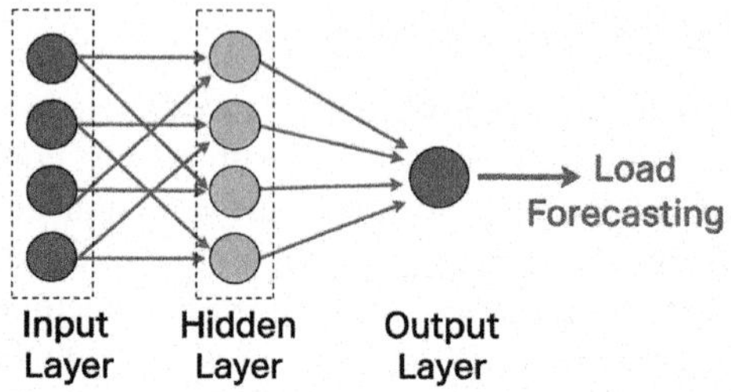

Figure 7.19 MLP architecture for load forecasting. *MLP*, Multilayer perceptron.

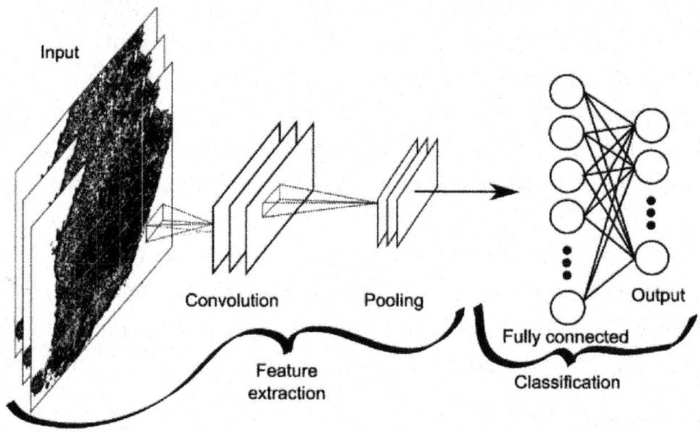

Figure 7.20 Simplified illustration of a CNN (Tikka et al., 2022). *CNN*, Convolutional neural network.

lion algorithm and showed that this method yielded better prediction results due to the ability to extract features. Recently, Zhang et al. (2022) suggested a hybrid CNN-TCN architecture for short-term load forecasting, and Tikka et al. (2022) applied CNNs to estimate the spatial distribution of EVs to aid electricity grid planning, which allows emphasizing the utility of CNNs in various EV-related prediction-related tasks. The architecture of CNN is simplified into a schematic as shown in Fig. 7.20.

7.3.3.2.3 Recurrent neural networks

RNNs are effective in the time-dependent relationship that is important when determining EV charging patterns. Zhu et al. (2019) used RNN-based solutions in short-term load prediction in EV charging stations, proving that the architecture is able to use past charging trends to make better predictions. Moreover, the effectiveness of RNN in forecasting EV charging demand was confirmed by researchers in Shanmuganathan et al. (2022), and they noted the significance of the recognition of time patterns in the uncertainty model application. The RNN model looks like Fig. 7.21.

7.3.3.2.4 Long short-term memory

The operation of LSTM networks has proven to be outstanding in the activities of EV charging prediction that demands the modeling of long-term dependencies. In Ma & Faye (2022), a multistep EV charging station occupancy prediction system based on

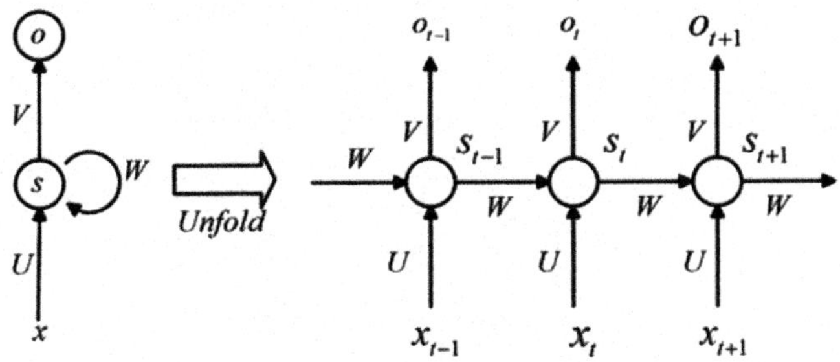

Figure 7.21 Basic recurrent neural networks structure (Zhu et al., 2019).

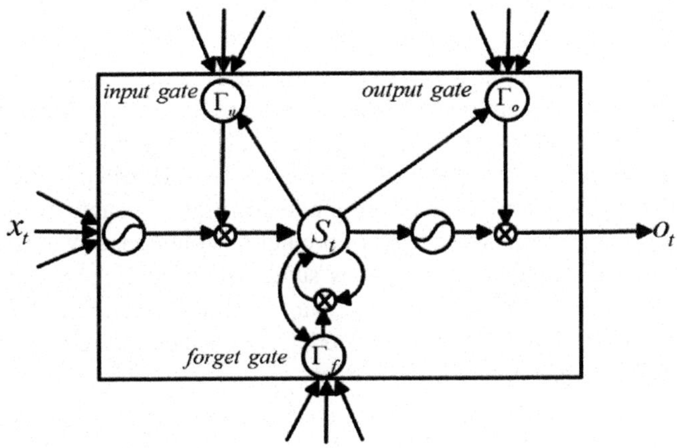

Figure 7.22 LSTM model structure (Zhu et al., 2019). *LSTM*, Long short-term memory.

hybrid LSTM networks was suggested to use a combination of both historical data and time characteristics to make efficient predictions of short periods of time, with diminishing accuracy with increasing prediction horizons, but it is still accurate enough within a prediction window of less than 60 minutes. Zhou et al. (2022) proposed a Bayesian deep learning model that uses the LSTM networks to deal with the uncertainty in their predictions, which outperforms the traditional methods when they are put to test on actual charging station data on the Caltech campus. In Qin et al. (2022), the authors employed a complex model that integrates k-means clustering with LSTM to predict EV charging and used a multilevel discrete Wavelet transform as a feature extractor, and verified their model in more than 1300 charging stations in Zhejiang province. Fig. 7.22 represents the architecture of the LSTM model.

7.3.3.2.5 Deep neural networks

In EVs, DNNs have been used in intricate probabilistic predictions. Zhang et al. (2021) proposed probabilistic forecasting systems using deep learning with the

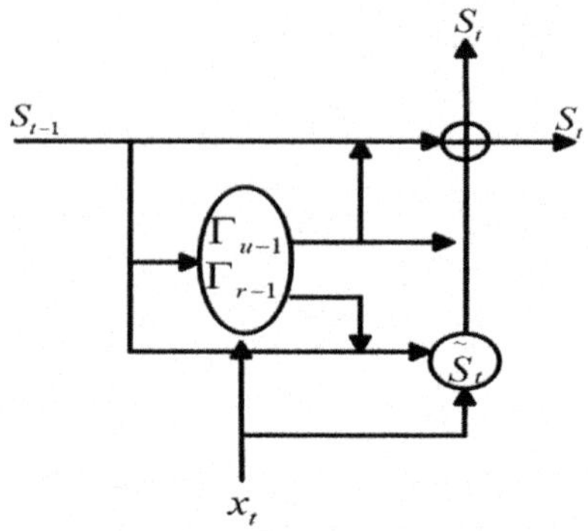

Figure 7.23 Basic gated recurrent units' structure (Zhu et al., 2019).

incorporation of new queuing models to handle the stochastic nature of EV charging behavior using elaborate architectural designs that can capture time as well as the probabilistic nature of charging demand. Recent research points to the fact that point predictions are not sufficient to be used to assess the risks in grid integration cases, and the relevance of probabilistic forecasting models is high to enable grid operators and energy suppliers to understand the level of prediction confidence (Van Kriekinge et al., 2021). Prediction of EV charging loads is suggested in reference Xin et al. (2024), and it relies on MCS to supplement past data, spectral clustering to combine similar load patterns, and custom CNN-LSTM models, per cluster, to model spatiotemporal information and enhance prediction accuracy.

7.3.3.2.6 Gated recurrent units

GRUs have the benefits of computational efficiency and have the ability to model over time. Zhu et al. (2019) compared the performance of GRUs with other deep learning methodologies in short-term EV charging station load forecasting, whereas in Zamee et al. (2023), the authors used self-supervised online learning algorithms based on GRUs to predict the demand and events of an EV charging station, which showed that the architecture can be adjusted to real-time learning cases. The GRU model has an architecture as shown in Fig. 7.23.

7.3.3.2.7 Temporal convolutional networks

TCNs are current developments in EV temporal pattern recognition. In Tian et al. (2025), a TCN-LSTM hybrid network was proposed to forecast short-term EV charging load with consideration of the completely similar day identification methods, which would help to achieve a better prediction accuracy by enriching the ability to extract time-related features and identify patterns. Fig. 7.24 is the TCN-LSTM model.

Table 7.8 provides an overview of the main neural network structures used in the EV uncertainty modeling, their main peculiarities, and applications.

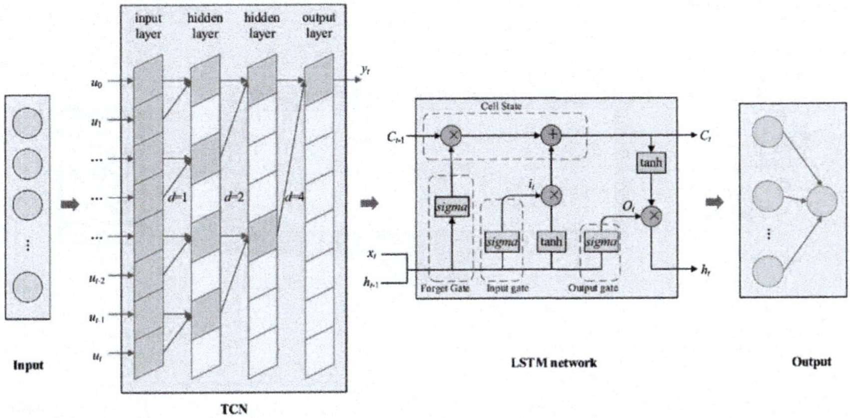

Figure 7.24 TCN-LSTM forecasting model framework (Tian et al., 2025). *TCN-LSTM*, Temporal convolutional network-long short-term memory.

Table 7.8 Neural network architectures for EV uncertainty modeling.

Architecture	Key application	Primary advantage	Limitation
MLP	Probabilistic load forecasting	Simplicity and robust performance	Limited interpretability
CNNs	Grid planning and demand distribution	Spatial-temporal pattern recognition	Require large datasets
RNNs	Short-term load forecasting	Sequential dependency modeling	Gradient vanishing issues
LSTM	Occupancy and demand prediction	Long-term dependency capture	Computational complexity
DNNs	Probabilistic forecasting	Complex nonlinear relationships	Black-box nature
GRUs	Real-time prediction systems	Computational efficiency	Limited long-term memory
TCNs	High-frequency forecasting	Parallel processing capabilities	Architecture complexity

CNNs, convolutional neural networks; *DNNs*, deep neural networks; *EV*, electrical vehicle; *GRUs*, gated recurrent units; *LSTM*, long short-term memory; *MLP*, multilayer perceptron; *RNNs*, recurrent neural networks; *TCNs*, temporal convolutional networks.

7.3.4 Data-driven modeling method

The modeling of EV charging-loads has moved towards data-driven models more. Large charging and mobility datasets are becoming a staple for researchers who are not using synthetic profiles as more EVs are adopted. These approaches use statistical, machine learning, or agent-based models to forecast patterns of loads, optimize charging, and assist in managing demand in real-time (Xydas et al., 2016). The data-driven models toward EV charging optimization are demonstrated in Fig. 7.25.

Figure 7.25 Data-driven modeling methods for EV charging optimization. *EV*, Electrical vehicle.

The success of the uncertainty modeling based on data is essentially determined by the variety and quality of the datasets available, which can be organized into three key sources of data that include the data gathered through the operation of the charging infrastructure, self-reporting information gathered by users, and dynamic movement sensors of vehicles.

The charging infrastructure operational records record the real-time measures of public and commercial charging points, including the length of session, the energy transferred, the timing, and power levels. This kind of data emphasizes the space and time variations of the charging actions between the networks. In one case, the authors of a study by Hecht et al. (2020) used over 20,000 stations in Germany to compare such measures as the initiation of a session, its completion, its duration, and its usefulness as a whole. Although this solution offers a detailed view of the real-time charging behavior, it is not so effective to trace the individual user experience, which is more difficult to identify the long-term personal habits or overall energy requirements related to everyday movement.

Self-reported user observations are received through surveys, discussions, etc., and incorporate information about charging behavior, habits, choice of location, and general experience. These are sources that would be useful in exploring subjective processes such as attitudes, motivation, and satisfaction toward charging options (Yan et al., 2020; Zhang et al., 2020). In one example, a multinomial logit was applied to explain the decision-making process in various conditions depending on the answers to questionnaires (Wang et al., 2021). Nevertheless, it has problems regarding increased collection costs, reduced time intervals, limited geographical coverage, and possible selection biases of participants that would not be as representative as diverse EV owners and reflect long-term behavioral patterns.

Vehicular dynamic movement tracking is based on the GPS or other such technologies, and positional data can be recorded over time to deduce trips, stopovers, and possible areas of charge requirement through patterns of location (Baghali et al., 2022). As an example, Yang analyzed an example of a private EV dataset in Beijing, examining such aspects as the length of trip, time, frequency, speed, and destinations, revealing how urban patterns influence regular patterns (Yang et al., 2022). Similarly, the study by Deng on the preferences of routes within the same city also focused on routine routes based on individual preferences, timing, and facilities (Deng et al., 2024). The main limitation is that most such datasets do not provide information on live battery status because of constraints in the sensors, which means that models have to approximate such features as charge levels or range. Therefore, it is more appropriate when it comes to mobility analysis, but the data tends to place the emphasis on travel rather than on the direct connections with charging choices.

Various types of modeling innovations have been possible through the application of these datasets. The use of data-driven methods has been widely used in the prediction of EV charging load at different levels. Real charging session data in the UK were clustered to characterize the demand profiles based on the type of users and location to explain the variability in arrival time and energy requirements (Xydas et al., 2016). Kang et al. applied the navigation app data to map the dynamics of the demand to charge in Beijing and combined it with the spatial analysis to forecast the variations depending on the urban structure and daily habits (Kang et al., 2022). In Liu & Liu (2023), the stochastic user equilibrium model on the trajectory data was developed, which predicts the spatial-temporal charging distributions based on the uncertainties in the traffic and user preferences. The authors of the study by Powell et al. used large-scale adoption data to optimize the infrastructure placement, and they simulated the uncertainties in access by modeling a variety of varying charging operations (Powell et al., 2022b).

In Li et al. 2025), the authors proposed a multicriteria equity measure based on tract-level data, using artificial intelligence in planning stations and dealing with the uncertainty in equity of demand among demographics. Tao et al. (2022) suggested an agent-based approach with historical fast-charging data, and user agents are simulated to deal with the uncertainty in the use of the stations and time. Travel and demographic data were used to produce probabilistic profiles, which produced synthetic charging conditions to measure the scale of uncertainties (Callanan et al., 2025). Liu and colleagues combined forecasting algorithms and sparse charger information and made use of deep learning to cope with uncertainties in demand in real-time operations (Liu et al., 2022a). Kara et al. examined nonresidential charging sessions through empirical data, calculating the benefits of smart charging by estimating variability in the patterns of sessions (Kara et al., 2015). Frendo et al. applied heterogeneous EVs data through fleet specifics, creating reinforcement learning-based charging schedules to address uncertainties of various types of vehicles (Frendo et al., 2020b).

7.3.5 Comparative review of approaches

A comparative study concluded that deterministic methods are suitable for initial steady-state assessments in data-limited environments; however, they demonstrated low fidelity in uncertain cases. Similarly, methodologies that are probabilistic provide a balance between flexibility and computational performance for larger scales of analysis—but necessitate a certain level of statistical knowledge and professional experience since probabilistic methods can output unrealistic values without careful selection of the statistical distribution.

Table 7.9 Comparative overview of electric vehicle load uncertainty modeling approaches.

Method	Strengths	Weaknesses	Data requirements	Computational cost	Accuracy level
Deterministic	Simple, efficient	Ignores randomness	Minimal	Low	Low
Probabilistic/statistical	Captures variability, flexible	Data-dependent, complex	Moderate-high	Moderate	High
Machine learning	Adapts to trends, nonlinear	High resources, interpretability issues	High	High	Very high
Data-driven	Realistic, scalable	Biases, omissions	High	Moderate	High (data-dependent)

Conversely, the field of machine learning and data-driven approaches provides increased flexibility and accuracy for complex real-time and large-scale applications, for example, load forecasting and optimization of GNE. Machine learning methods require a lot of resources to implement and are generally classified as black-box methods. Recent literature focused on hybridization to improve robustness in the analysis process, as well as the study of the impact of the transition of high penetration EV within the context of uncertainty with respect to EV integration into the grid. This is particularly relevant to the ongoing stress, given the growing load of data from various sources.

Probabilistic methods have the largest proportion of applications across verifier levels, accounting for 19% and 15%, respectively. However, trends suggest a growing use of machine learning methods for probabilistic-based times into the future, with data availability considered to be the key driver of this acceleration. All of this then highlights a requirement for method selection to be specific to context as a means of motivating the EV integration and physiologically prepare to emergent impacts that permeate the sustainability narrative at a grid aspect. Table 7.9 highlights the main categories of EV load modeling methodologies and their features.

7.4 Aggregation of electric vehicles load and scenario design

EVs in distribution networks pose new problems in the prediction and management of load uncertainty, especially given that the EV charging demand is highly fluctuating. To solve this complexity, EV load aggregation, which can be defined as the systematized approach to aggregating individual or clustered EV charging needs into a unified, representative model that assesses overall impacts on power grid characteristics, including voltage shifts, current flows, and system stability as a whole (Nazari-Heris et al., 2022).

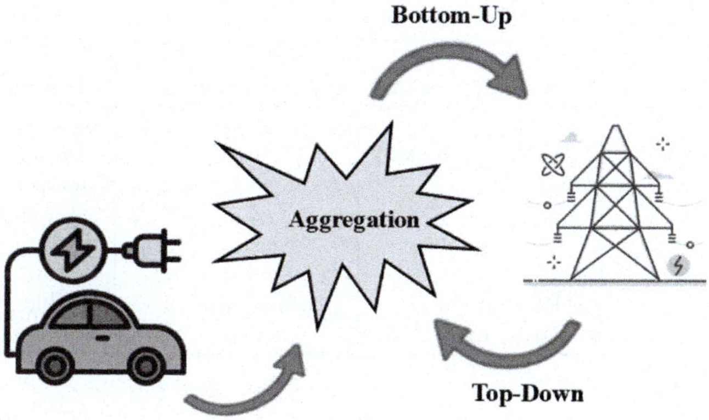

Figure 7.26 Aggregation framework for individual EV integration with power grid. *EV*, Electrical vehicle.

Such aggregation would be needed in stochastic programming, where it would convert the differences in uncertainty to manageable, smaller-dimensional representations, improving predictive accuracy and operational efficiency. Aggregation reduces the complexity of computations by combining all these aspects, yet maintains important probabilistic properties, which allow planners to model realistic scenarios in which EV integration can cause transformer overload or voltage drop (Nelson et al., 2024).

We shall begin in this section, then, upon the grass-root, with bottom-up approaches that start with individual behavior; and on the panoramic plane with top-down ball-estimates that start with general trends; and combine the two in hybrids that may be called the most balanced. We will further explore charging control and management scenario designs. Fig. 7.26 depicts the framework for integrating EVs with the power grid through aggregation.

7.4.1 Bottom-up load construction

In stochastic planning of energy systems, bottom-up load construction is a conceptual framework for aggregating EV charging loads and explicitly removing uncertainties. It is a synthesis of full load profiles where the starting point is the micro-level -individual EV charging events- and the macro-level aggregates, including substation or regional scales. It is especially skillful to integrate direct uncertainties, such as patterns of user behavior, spatiotemporal relations, and indirect ones, such as environmental conditions that influence battery performance (Bartolozzi et al., 2017).

The methodology generally starts with the process of obtaining granular data over empirical sources, including the National Household Travel Survey (NHTS) or real-time telematics data, which will have variables, such as arrival time (a_i), charging duration (d_i), and energy demand (e_i) for each session i.

Probabilistic distributions are estimated to quantify uncertainties, such as in the case of arrival times, a Weibull distribution (or other distributions) is used, and a joint

Table 7.10 Key parameters in bottom-up models.

Uncertainty type	Modeling technique	Example equation/method	Advantages/limitations
Behavioral	PDFs (Weibull/Gaussian)	$p(a_i) = Weibull\ (k, \lambda)$	High accuracy; data-dependent
Spatiotemporal	Monte Carlo simulation	$L(t) = \sum_i l_i(t)$	Captures variability; computationally heavy
Technological	GMM for joint distributions	$\sum_{k=1}^{K} \pi_k \mathrm{N}(\mu_k, \Sigma_k)$	Flexible; requires calibration

PDFs, probability density functions.

variable is modeled with Gaussian Mixture Models (GMM):

$$p(a_i, d_i, e_i) = \sum_{k=1}^{K} \pi_k \mathrm{N}(\mu_k, \Sigma_k) \tag{7.11}$$

where μ_k, and Σ_k denote the mixture weights, means, and covariances, respectively (Amara-Ouali et al., 2025). MCSs then generate multiple load realizations: An elementary load curve for the session i is $l_i(t) = \frac{e_i}{d_i}$ for $t \in [a_i, a_i + d_i]$, and the aggregate $L(t) = \sum_i l_i(t)$ (El-Hendawi et al., 2022).

The procedural steps that are relevant are: (1) Individual session modeling, technology uncertainties (for example, battery efficiency as defined by lithium-ion degradation models); (2) temporal and spatial aggregation, using SARIMA to model counts of sessions, and scenario trees to model variability; (3) validation against grid measures, such as peak amplification, or voltage variation under uncoordinated charging.

The strengths of such an approach are high fidelity resolution and being able to plan the distribution networks, as well as identify bottlenecks, for example, overloaded transformers in residential areas. However, there are also limitations, which include computational intensity and a requirement for high data quality in high resolution, which can limit its usability in low data contexts. More recently, hybrid integrations are suggested with machine learning, such as using PSO-optimized LSTM to make adaptive uncertainty decisions, to minimize forecast error in dynamic situations (Yang et al., 2025b). Table 7.10 shows important parameters in bottom-up models to demonstrate how to handle uncertainties in a comparative manner.

7.4.2 Top-down scenario estimation

In contrast to bottom-up constructions, which pool individual behaviors together, top-down estimation begins with macro-aggregates, such as regional EV fleet forecasts, aggregate energy demand, or adoption, and de-aggregates those to develop graphical load profiles and scenarios. It is an artful paradigm in high-level planning for grids, where systemic uncertainties and economic parameters are absorbed by historical or predicted datasets. Statistically, models are used (e.g., regression and time-series modeling, etc.) to compute aggregate curves, supplemented by stochastic models (for example, scenario trees and direct ensemble models) to reflect variability. Markov chains, advanced transitions including charging-to-idle states, which give considerable probabilistic predictive capacity.

Important components include: (1) Defining baseline conditions (e.g., business-as-usual) and perturbations (e.g., extreme weather driving up demand); (2) model loads using data-driven paradigms, such as grey models in disaggregation; and (3) the analysis of distribution implications, such as voltage instability, transformer overloads, and peak shifts. The virtues of the methodology lie in its sparsity and reduced data demand, which make this approach suitable for long-term stochastic projects in which granular data is limited. However, it can also smooth off micro-absolutes, giving rise to 10%–15% error in local stress calculations unless hybridized (Fischer et al., 2024). Scalability to large-scale energy ecosystems is supported by empirical assertions (Muessel et al., 2023) that alignments in 12% of observed loads are validated by empirical tests, including the calibration of real data.

Artistically, contemporary innovations bring a fresh perspective when harnessing the strengths of a virtual battery, using EV aggregates as dispatchable storage to act as a surrogate flexibility to uncertainties observed (Ebrahimi et al., 2024). It enables the sharing of chargers and demand-response differentiation to better adapt the grid to proliferation; Fused diffusion models assist in scaling to the substitution dynamics of vehicle-vehicle archetypes (Domarchi & Cherchi, 2023). For example, with regional simulations, starting with national projections, demographics enable empirical disaggregation of lines to resolve evening peaks unmanaged charging, and under these policies in high penetrations, demand can escalate by 20%–30% without active DR (Powell et al., 2022a). Deep learning hybridization is a hierarchical model fusion to enhance adaptive nondeterministic uncertainty, eliminating balancing as needed, promoting proactive stewardship.

These archetypes will prioritize extensibility, and new integrations, such as AI-kernel-based flexibility, should add to managed under-frequency shed fears related to uncertainty (Renhai et al., 2025). Combining top-down development scenario architectures illustrate how managed flexibility limits enhancement on system upgrades, but these will all coexist with extraterrestrial adoption and, in the future, bold and innovative grid coordination.

In macro-level modeling, the focus is on using scalable load estimation through probabilistic or statistical models. As an example, the aggregate EV load, L_{total}, is calculated based on higher-level indicators such as vehicle miles travel or adoption forecasts, and allocates these totals with a function, $f(d, t)$, that combines both spatial (d) and temporal (t) variables by Eq. (7.12):

$$L_{top-down} = L_{total} \times f(d, t) \tag{7.12}$$

This distribution commonly uses PDFs provided by national data sets (like Gaussian mixtures of arrival times or Weibull curves of charging seconds) in order to construct simulated regional distributions. The modern applications go through virtual battery conceptualization of batteries whereby EVs are bundled together as single stores (Ebrahimi et al., 2024).

Top-down techniques are good where scalability and efficiency are required, where planning of larger scales; however, where micro-level variations in EV actions are needed, they can also easily fail when increasing the inaccuracy between local grid stress such as voltage imbalance or transformer overload are required, unless refined by integration. Because of these drawbacks, hybrid techniques are becoming more popular and actively used, especially given that they offer a smooth merging of the ground-up rigor of bottom-up techniques and the precision of the top-down methods at a level of analysis to get a middle state, how both the fine uncertainties and large-scale challenges will thus be handled. This renders them especially critical to a broad, realistic base of grids and renewables, variable necessities, supplying such important benefits as high forecast value, expense decreases of operations, and solid assistance in the integration of distributed vitality resources.

The combination of top-down and bottom-up concepts can be best described as a paradigmatic shift in EV load modeling. The methods are good at the interpretational levels between user-specific behavioral responses within an individual vehicle and grid-integration necessities within a system since they combine micro-level choices that occur at the user level with macro-level constraints that occur at the infrastructure level (Nelson et al., 2024). The basic idea of hybrid aggregation is the strategic interrelation between heterogeneous modeling methods to get rid of the limitations of singular methods.

Nowadays, hybrid systems usually combine machine-learning algorithms with optimization models, including the gradient boosting decision tree-temporal convolutional network (GBDT-TCN) sophistication, which works with higher forecasting accuracy, with the root mean square error (RMSE) values of less than 6.23 while predicting discharging (Zhang et al., 2023). The approach is a combination of probabilistic distribution with neural network structures to model stochastic user behavior, as well as temporal dynamics in charging process trends. Likewise, a combination of probability distributions and neural networks has demonstrated potential to embrace the stochastic characteristics of EV charging and at the same time maintain the computational session (Chen et al., 2023). The hybrid schemes are usually followed by a multiphase design; the initial data processing and feature mining are done by machine learning methods, as the next step, then characterized by optimization-based aggregation and scenario generation. Fig. 7.27 shows the hybrid scheme for EV aggregation.

Hybrid methods demonstrate superior flexibility and accuracy compared to traditional techniques, particularly in the integration of renewable energy sources and staged planning. Their main advantages include strong support for renewable and distributed sources, improved forecasting, and a significant reduction in operational costs. As the system of AI and IoT advancements opportunities, these approaches become more flexible and scalable, making it possible to have more solid and safe aggregate load conditions. Disadvantages, however, include the complexity of implementation due to orchestrating algorithmic building blocks and data streams, dependence on local grid

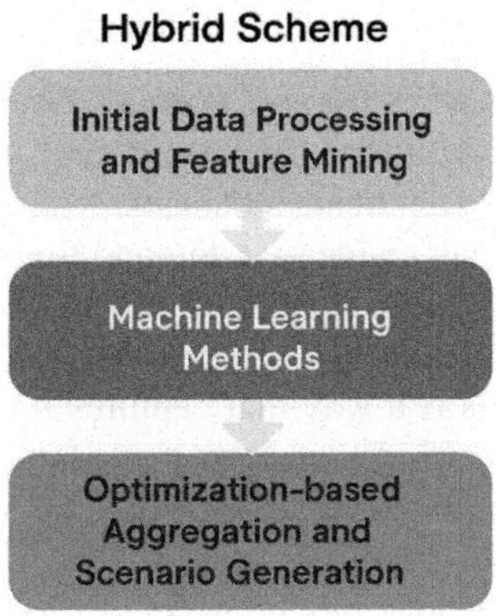

Figure 7.27 Hybrid scheme for EV aggregation. *EV*, Electrical vehicle.

Table 7.11 Comparison of aggregation methods.

Method	Strengths	Limitations	Applications
Bottom-up	High resolution, detailed uncertainties	High computation, data-intensive	Local planning
Top-down	Scalable, low data needs	Overlooks micro-variations	Regional forecasting
Hybrid	Balances accuracy and efficiency	Complex implementation	Renewables integration

connectivity, possible absence of transparency in decision-making, and or use of high-quality data that might impact reproducibility and comparability. All summarized as discussed in Table 7.11, have their own drawbacks and advantages, and best areas of implementation.

7.4.3 Charging control and management scenarios

Effective modeling of EV charging control and management scenarios is the key to preventing the negative effects of grids and increasing the stability of stochastic energy systems. To model such intricate interactions between EV adoption and grid stability, the charging control and management strategies are typically divided into three different cases:

- Uncontrolled charging scenarios and peak demand impact
- Smart and time-shifted charging strategies
- V2G bidirectional power flow scenarios

Each scenario denotes a new technological and coordination stage, such as fully user-controlled grid charging that harms grid-interactive models.

7.4.3.1 Uncontrolled charging scenarios and peak demand impact

EVs start its charging process when they are connected to the power grid without regard to the conditions of the grid. Although this case is easy to model and does not provide extra infrastructure, it creates significant stochastic strain on distribution systems since charged prices are through peak residential demand times.

Uncontrolled charging is the purest example of EV charging, which operates according to user behavior and is not controlled by the grid, which allows charging the cars once connected. This is a way that captures direct uncertainties, such as the spatiotemporal variability, and indirect ones, such as economic demand peaks. It is based on historical data, like the prices of charging and telematics charging durations, and stochastic background (e.g., Monte Carlo) how the load profile is generated to evaluate the risk of peak demand (El-Hendawi et al., 2022). Research further suggests that peak loads may rise up to 20%–40% by uncontrolled charging, which places undue stress on distribution transformers and leads to voltage variations, particularly in residential lines and mixed lines (Electric Vehicle Grid Impact Analysis & Smart-Charge Management, 2025). Using the example of a simulation of 10 real-world feeders, an outlook of 2030 showed that the residential location registered the highest impact of 11.7% during home-dominated conditions and a rise of 16.9 during work-dominated conditions during the midday shift of commercial loads (Jones et al., 2021).

It is done through aggregation of the individual load curves (7.13):

$$L = \sum_{i} c_{ev,i}(t) \tag{7.13}$$

where $c_{ev,i}(t)$ is the charging rate and of investigating the analysis effects, such as sags in voltages or overloads. Its benefits are simplicity and low infrastructure requirements, but countermeasures are also factors that make it unpredictable and tend to cause grid instability, particularly at high penetration levels (Rizvi et al., 2018). Recent studies indicate that there can be inclusion of fuzzy logic to deal with inaccurate human data and provide a new way to forecast using peak occurrences with 15% higher accuracy (Schmalzl et al., 2025). The literature reviews point out that the unregulated charging increases the grid congestion with empirical evidence of Denmark indicating a deviation in bus voltages and congestion in the line (Tayri & Ma, 2025). Conversely, Malaysian network studies have indicated thermal loading, and this has registered that controlled techniques must be discontinued and replaced with controlled techniques (Nutkani et al., 2024). A different study of the agent-based model revealed that without control, EV integration damages equipment, and the central point is the transition to controlled charging (Gschwendtner et al., 2023a).

7.4.3.2 Smart and time-shifted charging strategies

Smart and time-shifted charging, or management charging, is also called smart charging is the introduction of coordination between the system constraints and the EV demand. This approach distributes charging operations both during off-peak times and during times of elevated renewable generation, resulting in enhanced system reliability and lowered cost. Smart charging strategy in EVs is illustrated in Fig. 7.28.

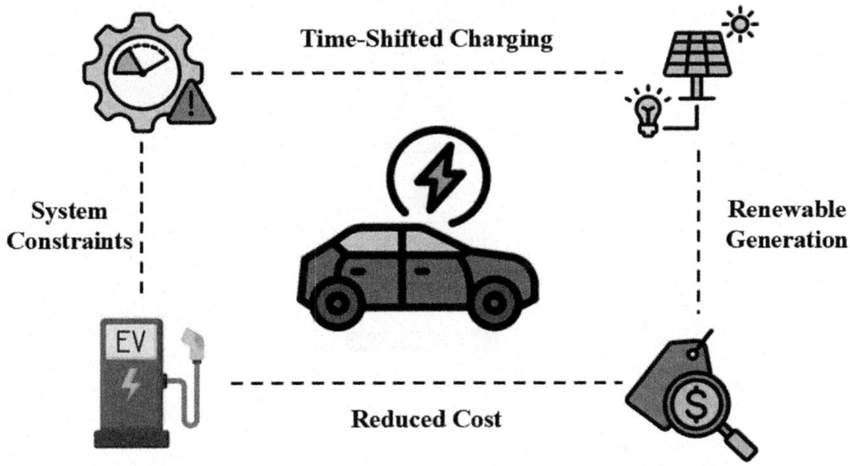

Figure 7.28 Smart charging strategy in EVs. *EV*, Electrical vehicles.

This method employs DR signals to shift EV loads to off-peak periods, optimizing grid utilization. This scenario addresses operational uncertainties by aligning charging with renewable generation or low-demand hours. Techniques include linear programming for load scheduling and machine learning for predictive control, reducing peak demand by up to 30% in simulated networks (Alfaverh, 2023).

Advantages are a boost in grid flexibility and a cut in utility costs, and practical tests indicate that energy costs decreased by 10%–15% (Yang et al., 2025b). Challenges include the issue of user acceptance and smart meters, which might compromise scalability. Novel, hybrid-based real-time data adapting Markov chains over schedules have been found to substantially optimize early scheduling by a quarter in urban environments (Brinkel et al., 2024). A thorough examination has divided smart strategies into architecture, data, intelligence, and methods, and linear programming and heuristics are common. ToU and real-time pricing, as witnessed in the case of Denmark, became effective, which reduced costs and handled grid requirements (Christensen et al., 2025).

Empirical research shows that smart charging overcomes a problem of paused charging whereby EVs respond to the signals after pause, leading to lesser grid pressure (Motlagh et al., 2025). One more analysis has indicated that it postpones the infrastructure enhancements through load control in the distribution systems (Jamil et al., 2025).

7.4.3.3 Vehicle-to-grid bidirectional power flows scenarios

Bidirectional charging, or V2G, extends the benefits of managed charging by enabling EVs to discharge back into the grid, effectively functioning as distributed storage systems. This paradigm allows EV fleets to provide ancillary services such as frequency regulation, renewable balancing, and peak shaving. This scenario tackles systemic

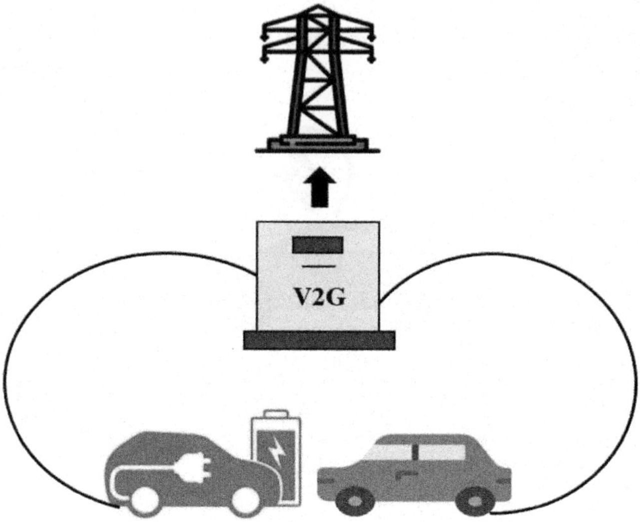

Figure 7.29 V2G System for EV energy exchange. *EV*, Electrical vehicle.

uncertainties by leveraging EV batteries as distributed resources. The V2G system for EV energy exchange is shown in Fig. 7.29.

Stochastic optimization models, such as scenario-based programming, simulate V2G impacts, with aggregate power defined as:

$$P_{V2G}(t) = P_{dis}(t) - P_{ch}(t) \tag{7.14}$$

and $P_{dis}(t)$ and $P_{ch}(t)$ are discharge and charge rates, respectively (Kempton & Tomić, 2005).

Key benefits include enhanced renewable integration, grid support services, and potential revenue streams for EV owners. However, central challenges include battery degradation effects, bidirectional converter costs, regulatory uncertainty, and cybersecurity concerns (Mojumder et al., 2022). Despite these restrictions, V2G has demonstrated the most substantial progression in establishing EVs as active participants in forthcoming power systems to ensure economic efficiency and resilience under a grid with abundant renewables. A deep dive review reframes V2G as a means for grid flexibility that emphasizes bidirectional converters and artificial intelligence, so that real-time control strategies are applicable (Alamgir et al., 2025).

7.5 Distribution grid impacts of electrical vehicle charging

The extensive use of EVs presents radical changes to the operation of distribution grids, both operational difficulties and strategic prospects of power system planning. This part

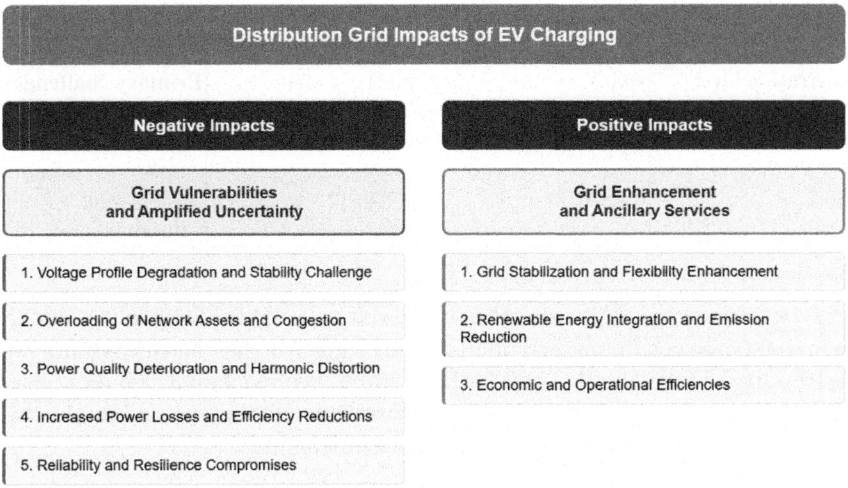

Figure 7.30 Dual impact analysis of EV charging on distribution grids. *EV*, Electrical vehicle.

explores the two-sidedness of EV integration effects, which involves considering how the issue of charging uncertainties presents as grid vulnerability and, at the same time, creates new grid improvement services.

Fig. 7.30 illustrates the bidirectional impacts of EV charging on electrical distribution grid infrastructure, categorizing both detrimental effects and beneficial contributions to grid performance.

7.5.1 Negative impacts: Grid vulnerabilities and amplified uncertainty

Stochasticity of EV charging behavior is a fundamental challenge to the paradigm of traditional design of distribution grids, adding operational uncertainty that occurs via various degradation pathways. The erratic charging patterns, especially those in peak demand periods, generate synchronized loading states that are beyond the infrastructure design requirements and jeopardize the reliability of the system.

7.5.1.1 *Voltage profile degradation and stability challenge*

EV charging uncertainties have been found to cause great interference to voltage profile degradation in the distribution networks, especially in radial feeders where the impedance properties enhance voltage distortion (Khalid et al., 2019).

7.5.1.1.1 Voltage deviation patterns and penetration effects

It has been shown that small penetration rates of EV below 1%–2% cause voltage deviations, which increase nonlinearly to 12.7%–43.3% between nominal values as penetration rates go to 20%–80% (Schlee et al., 2009). There are notably more severe

Table 7.12 EV penetration impacts on distribution network voltage.

Penetration level	LV voltage change (p.u.)	MV voltage change (p.u.)	Primary challenge
≤10%	0.01–0.02	Minimal	Transient flicker onset
25%–50%	0.10–0.20	0.02–0.05	Phase unbalance issues
≥80%	0.25–0.30	0.08–0.12	System-wide instability

EV, electrical vehicle; *LV*, low voltage; *MV*, medium voltage.

effects on low-voltage (LV) networks, where voltage variation is up to 0.10–0.30 per-unit in worst-case conditions, and often should be much higher than the statutory limits of ±10% specified by the IEC 60038 (IEC, 2002). Such violations are more likely at feeder endpoints, where the effects of impedance are most severe, and rural networks are more severely affected due to extended feeder lengths.

7.5.1.1.2 Transient phenomena and power quality issues

Fast charging systems that pull up to 275 A cause significant transient disturbances on the connection and disconnection event, producing flicker in the voltages that can be larger than IEEE standards (Akhavan-Rezai et al., 2012). Residential chargers operating on single-phase add further complexity to the asymmetrical loading conditions, introducing factors of voltage unbalance (VUF) that exceed acceptable limits at penetration of even 25% when the phases are distributed unevenly (Ul-Haq et al., 2015). The effect is especially severe in secondary systems of distribution, where the effect of unbalance is enhanced by lower voltage, and a higher ratio of impedances.

7.5.1.1.3 Load modeling and stability margin impacts

EV loads' representation has a great impact on the results of the voltage stability assessment. The EVs simulated as constant power loads exhibit a behavior of decreasing small-scale stability margins with the constant representations (Das & Aliprantis, 2008). This difference in modeling is important in the definition of the stability limits of the systems, where BEVs would present a more significant voltage load than plug-in hybrid cars owing to higher battery capacities and higher charging power demands (Green et al., 2010).

7.5.1.1.4 Bidirectional power flow complexity

V2G operations also present greater stability issues in the form of the bidirectional power flows that, although they might also be useful to control the voltage during low-demand, can pose a destabilizing effect during mode switching, unless they are synchronized (Da Silva et al., 2021). These operations have a temporal uncertainty that adds voltage regulation challenges, especially in cases where several vehicles are changing between charging and discharging operations at the same time.

Table 7.12 illustrates the nonlinear increase in voltage effects with penetration levels of EV, where low-voltage networks are affected more than medium-voltage systems in all penetration conditions.

7.5.1.2 Overloading of network assets and congestion

Uncontrolled charging of EVs is a significant problem facing the distribution network assets management since it stochastically amplifies the peak loads (Papadopoulos et al., 2013).

7.5.1.2.1 Transformer thermal overloading

Stochastic EV loading patterns cause acceleration in the aging of distribution transformers as a result of thermal stress. The probability of having simultaneous vehicle connections during peak periods causes the loading of the transformer to be far larger than the design specifications. Studies suggest that half EV penetration overloads half of the 50 kVA transformers and a third of the 50 kVA transformers as Level 2 charging, which develops 10% more overloads than Level 1 charging does (Fernandez et al., 2010). Prolonged thermal overloads generate unnecessary heat in the transformer windings, which increases the speed of insulation degradation and oil degradation, essentially decreasing the asset life. Research proves that approximately one out of every five distribution transformers needs replacement/upgrade with 10% Level 2 penetration (Visakh & Selvan, 2023).

7.5.1.2.2 Network congestion dynamics

Urban distribution feeders are disproportionately stressed with centralized EV adoption and low redundancy. The stochastic charging generates load clustering effects, which result in high penetration of EVs that causes a 6% points increment on the peak demand in medium density networks (Nutkani et al., 2024). Fast-charging stations impose a sudden high demand for power, which causes excessive thermal loading of local feeders. Probabilistic models show that only 15% of the networks can support the fast-charging rate or 25% of the slow-charging rate before congestion is likely to occur (Nour et al., 2018).

7.5.1.2.3 Probabilistic risk assessment framework

To quantitatively measure the risks of overloading, probabilistic modeling with time-related uncertainty, area-wise correlation, and variability of charging behavior is needed. Level 2 charging dominates, and the distribution networks can normally support 10% EV penetration before becoming overloaded (Tong et al., 2019).

Table 7.13 shows how critical penetration thresholds where overloading probability becomes significant in the various network configurations, and this shows that the city dense network is most likely to be subjected to a high risk of overloading at comparatively high penetration rates, and the fast-charging case of suburbs gives the most restrictive conditions.

7.5.1.3 Power quality deterioration and harmonic distortion

EV charging infrastructure also presents challenges of power quality in the form of harmonic distortions caused by converters, as well as power factor deterioration. EV chargers use nonlinear power electronics, which distort grid waveforms, resulting

Table 7.13 Probabilistic overloading risk thresholds for distribution networks.

Network type	Asset category	Critical penetration	Overload probability (%)	Mitigation strategy
Urban dense	25 kVA Transformers	30% (Level 2)	50	Immediate upgrade
Urban dense	50 kVA Transformers	30% (Level 2)	35	Load management
Suburban	Distribution lines	15% (Fast charging)	40	Network reinforcement
Rural	Secondary transformers	25% (Slow charging)	20	Enhanced monitoring

in stochastic interactions between charging dynamics and power quality indicators (Richardson et al., 2011).

7.5.1.3.1 Harmonic generation mechanisms and statistical characterization

EV chargers utilize power electronic converters, which in turn produce harmonic currents as a result of the switching process, and the harmonic spectrum is directly proportional to charger architecture and control strategies (Möller et al., 2015). The total harmonic distortion (THD) is reportedly much higher in the various charging technologies, with commercial chargers ranging between 7% and 99% in terms of harmonic injection patterns, with architectural diversity being the key factor that defines harmonic injection patterns (Richardson et al., 2011). The mathematical expression of the harmonic distortion is given as:

$$THD_i = \frac{\sqrt{\sum_{h=2}^{H} I_h^2}}{I_1} \times 100\% \tag{7.15}$$

where h represents the harmonic order, H denotes the maximum harmonic number, I_1 is the fundamental frequency, RMS current, and I_h is the RMS current at the h-th harmonic (Orr et al., 2008).

The stochastic nature of EV charging behavior presents harmonic injection variation in a probabilistic manner. Empirical data show that the existing harmonic distortion is between 2.36 and 5.26 at the commencement of charging, up to 28% at the end of charging. This change of time is associated with the SOC levels, in that the harmonic distortion increases with increasing battery SOC. The mode of charging has a strong impact on the phenomena of distortion, and the constant voltage (CV) mode has a greater harmonic content than the constant current (CC) mode (Fuchs & Masoum, 2015).

7.5.1.3.2 Probabilistic harmonic impact assessment

The cumulative harmonic effect is non-Gaussian given the variability in charging rates, vehicle type, and charging infrastructure. Lower-order odd harmonics (3rd, 5th, 7th, etc.) tend to dominate their harmonic spectrum, with the 3rd harmonic, in particular,

Table 7.14 Statistical distribution of harmonic emissions for EV charger types.

Charger type	Power level	THD_i Range (%)	Dominant harmonics	Key uncertainty factors
Level 1	1.4 kW	15–50	3rd, 5th, 7th	Battery SOC, grid voltage
Level 2	7–19 kW	8–35	3rd, 5th, 7th, 11th	Topology, control strategy
DC fast	50+ kW	5–15	5th, 7th, 11th, 13th	Power level, fleet diversity

EV, electrical vehicle *SoC*, state-of-charge.

a major issue in three-phase systems because it algebraically adds in the neutral conductors (Kütt et al., 2013). The MCS shows that the beneficial cancellation by harmonic diversity can occur when multiple EV chargers are present at the same time with varying phase shifts and magnitudes, thus the distribution systems can be affected up to 19% EV penetration in the condition of summer without breaching 5% voltage THD thresholds (Möller et al., 2015).

Fast charging conditions are of special concern, where the voltage THD values were 11.4% in rapid fast charging conditions, significantly higher than the EN 50160 standard of 8% (Nguyen et al., 2013). Most residential chargers are single-phase, which also leads to phase imbalance in the system, which can cause the degradation of power quality in the form of negative-sequence currents and uneven load distributions (Kütt et al., 2013).

7.5.1.3.3 Power quality degradation and equipment impact

When converting to power, harmonic injection is a factor that lowers the power quality of the line current, and the aging of distribution transformers. Correlation between present THD and transformer life is quadratic, which requires the THD to be restricted to 25%–30% to allow satisfactory equipment life. Analysis of distribution network capacity shows that harmonic effects are taken into account to reduce the acceptable EV penetration to 27.25% and 30.74% when the cable is overloaded earlier (Nguyen et al., 2013).

Table 7.14 is a summary of the common harmonic emission pattern of the various EV charger types, including their power levels, distortion ranges, dominant harmonic/harmonic ranges, and the uncertainty causes.

7.5.1.4 Increased power losses and efficiency reductions

The intermittent charging patterns that lead to variability in currents and increased technical losses due to stochastic integration of EVs increase power losses in the distribution networks (Clement-Nyns et al., 2009). This is in accordance with the law of Joule (($P_{loss} = IR$)) in which even small rises in current (I) lead to quadratic increases in power losses P_{loss} through line resistances (R) and transformer windings (Yunus et al., 2011).

7.5.1.4.1 Stochastic loss mechanisms and probabilistic assessment

The extent of power loss increase is very sensitive to the spatial and temporal concentration of EV loads under uncertain conditions. Simulations based on Monte Carlo show that the worst loss conditions occur in those cases when vehicles start charging the moment they arrive (Schlee et al., 2009). Experiments of realistic distribution networks indicate that uncontrolled charging of moderate penetration (40%–60%) can increase the loss of energy by 40% in the off-peak and 66% in the peak periods. Such rises are especially noticeable in distribution transformers, whereby under high EV penetration conditions, copper losses may shoot by more than 300% as a result of high winding currents (Masoum et al., 2011).

7.5.1.4.2 Reactive power and bidirectional flow considerations

The new EV charging systems bring in complex reactive power demands, which further aggravate the loss conditions past the case of simple active power consumption. Fast charging stations exhibit high reactive power levels, which lead to higher loss in the transmission lines and distribution transformers due to high apparent currents (Kisacikoglu et al., 2010). Moreover, the capabilities to allow the bidirectional power flow provide benefits in grid support at the expense of adding other loss mechanisms via the inefficiency of power electronic converters. V2G operation involves several power conversion steps, each of which adds conversion losses of 2%–5%, which accumulate with stochastic operation conditions (Sortomme et al., 2010).

7.5.1.4.3 Economic implications

The higher the power losses, the higher the utility costs, and in daily energy losses could increase by 66% at the 40% EV penetration, which would necessitate extra generation to replace this loss (Chang et al., 2019). Losses increase the strain on infrastructure, which may require expensive upgrades and have a ripple effect on the economy on a system-wide scale.

The effects of different EV charging strategies on power losses in the system at different penetration levels are summarized in Table 7.15 and provided with the percentage increase and the main properties.

7.5.1.5 Reliability and resilience compromises

The incorporation of EVs in the distribution networks disrupts the established concept of reliability by creating cyber-physical susceptibility, in which digital attacks can lead to immediate physical grid effects, contrary to the traditional equipment-based anxieties (Chandwani et al., 2020). Fig. 7.31 illustrate the cyber-attack on EV charging infrastructure.

Table 7.15 EV charging impact on system power losses.

Charging strategy	Penetration level	Loss increase (%)	Key characteristics
Uncoordinated (peak)	50%	40–66	Maximum loss escalation
Uncoordinated (off-peak)	60%	40	Transformer stress
Coordinated/delayed	50%	10–20	Valley filling optimization
V2G operations	Variable	36 (one-way)	Conversion inefficiencies

EV, electrical vehicle; *V2G*, vehicle-to-grid.

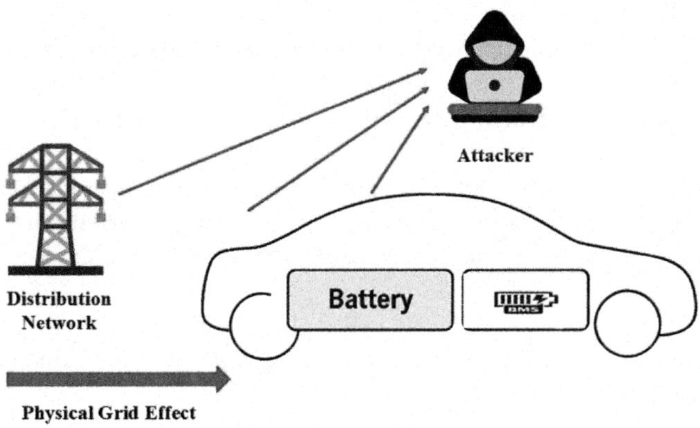

Figure 7.31 Cyber-attack on EV charging infrastructure. *EV*, Electrical vehicle.

7.5.1.5.1 Cyber-physical reliability threats

Smart charging infrastructure exponentially increases the attack surfaces, with data from 2024 indicating that utility cyberattacks have increased 70% to reach 1162 incidents (Hamdare et al., 2023). The authentication of charging stations based on open charge point protocol (OCPP) has critical vulnerabilities, which allow organizing thousands of charging events at a time and coordinating them. Dynamic load altering attacks (D-LAAs) are advanced threats as attackers use EV loads to disrupt grid frequency without having to substantially access the system (Najafi et al., 2025). In 2024, 116,000 records of global charging networks were breached, which confirmed centralized vulnerabilities of the systems (Upstream, 2025).

7.5.1.5.2 Grid resilience under high electrical vehicle penetration

Conventional N-1 contingency planning is not sufficient in the presence of synchronized EV charging that surpasses the capacities of transformers and generates protective operations (Qian et al., 2022). Peak load temporal clustering of EV loads undermines security margins and diminishes the ability of the system to recover due to reduced flexibility in operations. V2G systems create bi-directional interdependencies

Table 7.16 EV-grid cyber-physical threats and mitigation.

Threat category	Risk factors	Impact	Mitigation	Complexity
Dynamic load attacks	Coordinated EV manipulation	Critical	Observer detection	High
Communication failures	Network dependencies	High	Redundant paths	Medium
Data integrity attacks	Protocol vulnerabilities	High	Blockchain auth	High
Synchronized charging	Peak demand clustering	High	Smart scheduling	Medium

EV, electrical vehicle.

with the capability to spread failures across network segments that were originally independent of one another (Sayarshad, 2025).

Communication infrastructure dependencies are single points of failure, preventing smart charging coordination in cases where grid support is most needed during extreme weather events. Areas with a high density of EVs have very much longer recovery times because of complicated coordination needs among charging resumption and recovery procedures, which disproportionately impact underserved communities (Zhang et al., 2024b).

Table 7.16 summarizes significant cyber-physical threats due to EV integration, their risk factors, likely grid effects, and their complexities of implementation to guide uncertainty modeling of load conditions.

7.5.2 Positive impacts: Grid enhancement and ancillary services

Nonetheless, operational difficulties can be turned into useful grid resources with the help of strategic EV integration, sophisticated control measures, and aggregation policies. The current V2G technologies and advanced systems of DR allow EVs to deliver the required grid services and meet the demands of user mobility.

7.5.2.1 Grid stabilization and flexibility enhancement

V2G technologies and sophisticated control systems can turn EVs into distributed energy assets to enhance the stability and flexibility of power systems and to convert the unpredictability of user behaviors and spatiotemporal distributions in charging into grid resources due to DR and aggregation (Rana et al., 2025).

7.5.2.1.1 Frequency regulation services

The most important grid stabilization service in which EVs perform better than traditional generation resources is frequency regulation. Sub-second response times in power electronic converters of EVs allow a much higher response time relative to the traditional synchronous generators in maintaining nominal grid frequency (50/60 Hz) with load-generation imbalance (Nour et al., 2020). Aggregator systems that

Table 7.17 EV-based grid stabilization services.

Service category	Response time	Management approach	Key benefits
Frequency regulation	<1 s	Centralized/decentralized aggregation	50% lower frequency deviation
Peak load management	15 min–4 h	DR with V2G/valley filling	20%–30% peak load mitigation
Voltage regulation	Continuous	Coordinated charging/discharging	Enhanced distribution stability

EV, electrical vehicle; *DR*, demand response; *V2G*, vehicle-to-grid.

are centralized to coordinate the fleet-wide discharge patterns facilitate frequency deviations reduction to achieve stabilization in a microgrid simulation of coordinated EV fleets within a range of ±0.5 Hz (Pinthurat & Hredzak, 2020). Responsive to local frequency signals, decentralized control methods offer extra advantages to individual car owners and can reduce the communications infrastructure needs, having been shown to divide frequency nadir excursions by half with no big coordination protocols (Ramadan et al., 2018).

7.5.2.1.2 Peak load management

Planned EV charging and discharging help bring significant advantages in peak load control, responding to evening demand peaks which ordinarily stress peaking generation units and transmission systems. Valley-filling methods and smart V2G discharge can result in changes in charging demand during off-peak hours, whereas peak shaving in valley filling can be done without overloading at 15%–27%, as seen in field experiments, reducing evening loads, 3000 –2200 kW (Fu et al., 2021). High-end optimization methods, such as particle swarm optimization, decision-tree techniques, confirm 20%–30% flattening of peaks in commercial grid scenarios and obey user-specified restrictions, such as minimum departure SOC limits (Khan et al., 2024; Zainab et al., 2023).

7.5.2.1.3 Voltage regulation and reactive power support

Bidirectional EV chargers at a distribution level offer voltage regulation services and reactive power compensation services, reducing voltage profile distortions due to high loading conditions or large penetration of renewable generation. The power electronic interfaces adjust active power flows together with the DC-link capacitors, inject reactive power support without battery cycling, holding voltage deviations within a range of ±5% of nominal values (Kisacikoglu et al., 2010). When used with traditional voltage control equipment, like on-load tap changers, the systems will reach a power loss reduction of up to 15% in photovoltaic-integrated distribution feeders.

Table 7.17 outlines the most prominent attributes of the EV-based grid stabilization services, their response times, and management strategies, and the main advantages of improving the reliability of a system.

Aggregating EV batteries together as virtual power plants, these services can reach past renewable intermittency: Any surplus solar or wind power can charge idling fleets during downslope, and V2G discharges peak evening ramps, leading to less curtailment and emissions. This symbiotic integration is able not only to increase grid flexibility but also to provide economic incentives to owners due to service markets, to encourage fair energy transitions (Arévalo et al., 2024).

7.5.2.2 Renewable energy integration and emission reduction

EVs facilitate the integration of renewable energy in the sense that the charging uncertainties are transformed into flexible grid assets so that demand can be matched with the fluctuating generation (Štogl et al., 2024).

7.5.2.2.1 Grid integration and curtailment reduction

Probabilistic optimization models that account for EV charging uncertainties show strong potential for increasing renewable energy utilization by 15%–25% in simulated grid scenarios (Demirci et al., 2024). By coordinating charging loads with periods of excess renewable generation, these models reduce curtailment and enhance grid stability. MCSs, combined with renewable forecasting algorithms, help optimize charging schedules to take full advantage of surplus photovoltaic and wind production intervals (Saadatmandi et al., 2024).

Bidirectional capabilities, V2G capabilities, are an expansion of this integration opportunity that allows EVs to serve as distributed energy stores. EVs can absorb excess energy at times of high renewable generation and can emit back to the grid at times of supply shortages to provide intermittency buffering and sectoral CO_2 reductions of 50%–85% in areas dominated by wind (Kofoed-Wiuff et al., 2015).

7.5.2.2.2 Emission reduction pathways

Extensive lifecycle evaluations show that EVs that are charged through renewable sources have 35%–70% reduced footprint of greenhouse gases than internal combustion engines due to well-to-wheel emissions and grid decarbonization paths (Lee et al., 2025). This reduction of the carbon intensity is especially effective when coordinated over time, with off-peak or midday charging occurring during off-peak CO_2 generation periods, with up to 17.68% reduction of CO_2 emissions in carbon-capture-integrated systems (Xiao et al., 2025).

There is even more optimization opportunities involved in geospatial charging. The difference between workplace and residential charging locations allows the implementation of hybrid optimization models that reduce curtailment, and at the same time, user behavioral patterns are considered. Stochastic scheduling algorithm lowers the emissions per km of a photovoltaic-rich urban setting by 40%–55% by operating excess solar generation (which is usually curtailed) directly into EV batteries via peak-generation-free techniques (Hawkins et al., 2013).

Table 7.18 provides an overview of the main approaches to the integration of EVs with renewable, which involves the CO_2 reduction potentials that these approaches may enable, the main uncertainties that they capitalize on, and the resulting grid advantages.

The EV-renewable integration is associated with the optimization process that demands advanced algorithms and converts charging uncertainties into the strategic

Table 7.18 EV-renewable integration strategies and emission reduction potential.

Integration strategy	CO_2 reduction (%)	Primary uncertainty leveraged	Grid benefit
PV-aligned charging	40–55	Spatio-temporal charging patterns	Curtailment absorption
V2G wind integration	50–85	Arrival/departure variability	Intermittency buffering
Stochastic load optimization	15–70	Grid mix forecasting uncertainty	Lifecycle emission reduction
Carbon-aware scheduling	35–45	Real-time carbon intensity variation	Peak emission avoidance

EV, electrical vehicle; *PV*, photovoltaic; *V2G*, vehicle-to-grid.

advantage of the operational challenges. Uncertainty-conscious scheduling systems give importance to clean generation blocks but are not interested in unintentional production of more emissions during periods of high carbon requirements and these are especially considered in systems with high amounts of fossil fuel. Stabilizing renewable integration, but humanizing the energy transition, these probabilistic approaches can be used to model the actual user behaviors in optimization frameworks in our real world.

7.5.2.3 Economic and operational efficiencies

The uncertainty in the charging of EVs can be modeled stochastically to shift the economic paradigm toward the value creation instead of costs burdening, enabling operators to optimize investments and make money on this paradigm by depending on scenario-based strategies (EY, 2025; Virta Global 2024).

7.5.2.3.1 Infrastructure cost optimization and capital deferral

The charging strategies with a degree of uncertainty create exceptional infrastructure cost deferrals through leveraging current capacities of the grid. Empirical studies have proved that uncontrolled EV charging counts system costs to \$263 per vehicle annually, whereas stochastic optimization based smart charging counts \$36 per vehicle, and results in annual savings of 227 dollars, which is 86% of the cost savings per vehicle (Knezović et al., 2016; Richardson, 2013). This conversion is carried out by means of probabilistic peak demand management, which postpones transformer upgrades and minimizes distribution congestion. California grid analyses suggest that EV-based optimization can potentially save the operating costs of \$120–690 million each year via the scenario-based optimization (Coignard et al., 2019).

7.5.2.3.2 Revenue generation through bidirectional value streams

The V2G technologies provide opportunities to generate new revenue never experienced before due to uncertainty-aware service provisions. According to market analyses, the V2G industry is projected to grow by over \$5 billion by 2024, and individual vehicle owners will obtain net present values of about \$7000 by participating

Table 7.19 Economic parameters of EV integration under uncertainty.

Parameter	Unmanaged charging	Smart charging	V2G implementation
Annual system cost ($/vehicle)	263	36	−150 to −250
Peak demand reduction (%)	0	25–50	40–60
Infrastructure deferral (years)	0	3–7	5–12
Revenue potential ($/vehicle/year)	0	50–150	200–700
Grid operating cost reduction (%)	0	5–8	8–15

EV, electrical vehicle.

in the optimized strategies (Geng et al., 2024). Compensation rates in the European implementations are shown to be at an average of €0.35/kWh of ancillary services and rise to a high of 1.11/kWh in high uncertainty times, making it possible to have profitable models of arbitrage (Abdelmotteleb et al., 2024). These two-way value propositions will turn EVs into dynamic economic assets in the distribution networks as opposed to passive loads.

7.5.2.3.3 Operational efficiency through scenario-based management

Stochastic EV integration will improve the efficiency of the distribution system due to the controlled charging policies that can cut the peak demand by a quarter to half without affecting the quality of the services (Szinai et al., 2020). Fleet-scale implementations demonstrate save money in the tune of $200–300 per vehicle/year by managing uncertainty in charging patterns and DR optimization (Colmenar-Santos et al., 2019). This two-fold advantage will optimize infrastructure usage and generate value to the economy through flexible loads.

7.5.2.3.4 Economic-environmental synergy quantification

Economic maximization and the environment meet each other to produce multiplicative benefits in the form of strategic management of uncertainty. It has been shown that every 1 MW of smartly controlled EV capacity can add 2–3 MW of other renewable integration without stability loss, and will invest more in clean energy assets (Bilal et al., 2024). According to European estimates, the annual savings of 4 billion tend to happen in the organization of EV-renewable optimization, and by 2030, EVs will provide 4% of the total power supply in the region (International Energy Agency, 2024).

Table 7.19 compares the main economic indicators of unmanaged charging, smart charging, and V2G implementation and notes that there are cost reductions, benefits in demand management, and revenue prospects under stochastic conditions.

Stochastic modeling would reduce the possible burden of the EVs on the distribution system to optimized cost-efficient assets. Uncertainty-sensitive EV management has developed sustainable electrification routes in large scale by deferring infrastructure costs, generating revenue, and improving operational efficiency, even as it ensures economic sustainability and the reliability of the service.

7.6 Conclusion

This chapter discusses the main difficulties of modeling EV challenges and impact on the grid, the prospective future trends of research and innovation, and offers a conclusion about the most critical findings and implications of the chapter.

7.6.1 Key challenges

There are several long-standing issues with EV charging uncertainty modeling that span the data, technical, and system domains. First, there is the pervasive issue of data heterogeneity and inaccessibility, whereby datasets are either nonstandardized or extremely granular, making it difficult to develop models that can be applied to different user populations and geographical locations (Yu et al., 2025). For example, shifting charging patterns result in nonstationarities that are difficult to describe using parametric distributions, as well as localized maxima that are undervalued and exacerbate the issue of voltage imbalances or asset overloading (Tayri & Ma, 2025). Empirical findings demonstrate the devastating potential of transformer overloads at high penetration rates, such as 60%–100%, which can reach 33%–85% and cause voltage derangements of up to 5.5 (Zhao et al., 2025).

They are further complicated by cybersecurity vulnerabilities against the property of interconnected charging infrastructure, where sophisticated malware could cause cumulative grid instabilities and damage data integrity (U.S. Department of Energy, 2024). Stochastic programming necessitates significant computational efforts to perform scenario analyses, which is not feasible in dynamic scenarios with fluctuating renewable input changes (He et al., 2023; Tayri & Ma, 2025). This is further compounded by real-time operational issues. Ultra-fast supercharging is one example of new technology that causes harmonic and power quality degradation by increasing short-run bursts in load (Yu et al., 2025). The risks of investing in the strategy and conventions favoring certain communities to roll out fairly are increased by regulatory and economic factors, such as incompatible schedules in infrastructure, which can lead to underserved populations (U.S. Department of Energy, 2024). Additionally, the operation dilemma between long-term forecasts and the likelihood that a user will be willing to use their vision and participate in V2G is created by the degradation of battery functioning in the case of the two-way flow (International Energy Agency, 2024). These complications highlight the necessity of developing models that take into account cross-sectoral dynamics, such as how transportation policies affect the need for electricity, in addition to helping to quantify uncertainties.

7.6.2 Future research directions

To address these constraints, future research should focus on developing hybrid paradigms that leverage recent computational advances. One promising direction is to strengthen privacy-preserving methods, such as federated learning, which enable decentralized data aggregation while handling concept drift and supporting incremental, personalized models (Hallak & Kem, 2025). Such approaches can facilitate scalable spatiotemporal predictions that integrate multimodal behavioral predictors—including weather, traffic, and socioeconomic factors—to better capture the complexity of behavioral patterns (Tayri & Ma, 2025; Zhao et al., 2025).

Another promising direction is the integration of distributionally robust optimization (DRO) with explainable AI (XAI), which can strengthen reliability while addressing supply–demand imbalances caused by renewable intermittency (He et al., 2023). For example, V2G dispatch could be optimized using a graph neural network combined with queueing theory, thereby postponing costly infrastructure investments and potentially avoiding millions of transformer replacements (International Energy Agency, 2024). Blockchain-based architectures also offer improved security for peer-to-peer energy transactions, supporting financially sustainable business models for operators and aggregators. Looking ahead, research should also emphasize multiobjective models that capture techno-economic-environmental synergies, such as agent-based models of equitable charging access and responsive assessments of carbon pricing within renewable energy systems (Hallak & Kem, 2025).

Cross-functional collaboration is essential, as power engineers, data scientists, economists, and policymakers should work out standardized protocols and design markets so that the flexibility offered by distributed EVs can be converted into trustworthy grid resources. Computational burdens could be reduced by considering real-time management through model predictive control or deep reinforcement learning, and enhancing workforce skills is relevant to address skills changes in supporting changing infrastructures (U.S. Department of Energy, 2024). Finally, the objectives of these directions are to exploit EV uncertainties as strategic resources, encouraging adaptive and accurate forecasting and grid-compliant performance.

7.6.3 Conclusions

Compared with deterministic approaches, modern tools based on probabilistic modeling and machine learning are particularly valuable for managing the uncertainties of EV integration in complex, multivariate systems. Probabilistic methods provide robust handling of variability, while machine learning excels at detecting patterns, albeit often at the cost of interpretability and high data demands. Statistical methods quantify distributions reliably but depend on accurate data, whereas deterministic approaches offer efficient evaluations for steady-state scenarios but struggle to capture variability. Choosing the appropriate methodology requires careful consideration of context, including data availability and the scope of the analysis.

Stochastic EV charging patterns need to be leveraged as grid stabilization services through careful management and regulation, taking into account challenges such as data quality, cybersecurity, and real-time responsiveness. These strategies will play a key role in building resilient and sustainable energy systems as EV adoption grows worldwide. Effective outcomes depend on coordinated efforts and a perspective that treats uncertainties as opportunities, enabling shared benefits across sectors while supporting decarbonization goals.

Acknowledgment

The authors would like to thank the developers of AI tools, including GPT, Claude, and Grok, for their assistance in generating initial drafts, refining language, and supporting the organization of ideas in this book chapter. Their contributions helped streamline the writing process and enhance clarity, though all interpretations, analyses, and conclusions remain the sole responsibility of the authors.

References

Abdelmotteleb, I., De Felice, M., & Ruiz, P. *V2G: Transforming electric vehicles into rolling energy storage. RaboResearch, Rabobank.* https://www.rabobank.com/knowledge/d011424799-v2g-transforming-electric-vehicles-into-rolling-energy-storage.

Adianto, Y., Baguley, C., Madawala, U., Hariyanto, N., Suwarno, S., & Kurniawan, T. (2021). The coordinated operation of vertically structured power systems for electric vehicle charge scheduling. *Energies, 15*(1), 27.

Aghapour, R., Zeraati, M., Jabari, F., Sheibani, M., & Arasteh, H. (2022). Cybersecurity and data privacy issues of electric vehicles smart charging in smart microgrids. In V. Vahidinasab, & B. Mohammadi-Ivatloo (Eds.), *Electric vehicle integration via smart charging: Technology, standards, implementation, and applications* (pp. 85–110). Springer International Publishing.

Ahmadian, A., Mohammadi-Ivatloo, B., & Elkamel, A. (2020). A review on plug-in electric vehicles: Introduction, current status, and load modeling techniques. *Journal of Modern Power Systems and Clean Energy, 8*(3), 412–425.

Akhavan-Rezai, E., Shaaban, M. F., El-Saadany, E. F., & Zidan, A. (2012). Uncoordinated charging impacts of electric vehicles on electric distribution grids: Normal and fast charging comparison. In *2012 IEEE Power and Energy Society General Meeting* (pp. 1–7).

Akshay, K. C., Grace, G. H., Gunasekaran, K., & Samikannu, R. (2024). Power consumption prediction for electric vehicle charging stations and forecasting income. *Scientific Reports, 14*(1), 6497.

Alamgir, S., Hassan, S. J. U., Mehdi, A., Abdelmaksoud, A., Haider, Z., Shin, G. S., & Kim, C. H. (2025). A comprehensive review of vehicle-to-grid (V2G) technology as an ancillary services provider. *Results in Engineering, 27*, 106813.

IEC. (2002). IEC 60038: IEC standard voltages (6.2 ed.). International Electrotechnical Commission.

EY. (2025). Smart-charging and V2G critical for cost savings, grid stability and renewables integration [Press release]. https://www.ey.com/en_gl/newsroom/2025/03/smart-charging-and-v2g-critical-for-cost-savings-grid-stability-and-renewables-integration

Alfaverh, F. F. (2023). *Demand response management and control strategies for integrated smart electricity networks* (Doctoral dissertation, University of Hertfordshire.

Al-Muhaini, M. (2020). Electric vehicle Markov-based adequacy modeling for electric microgrids. *IEEE Access, 8*, 132721–132735.

Alsharif, A., Tan, C. W., Ayop, R., Al Smin, A., Ali Ahmed, A., Kuwil, F. H., & Khaleel, M. M. (2023). Impact of electric vehicle on residential power distribution considering energy management strategy and stochastic Monte Carlo algorithm. *Energies, 16*(3), 1358.

Amara-Ouali, Y., Hamrouche, B., Principato, G., & Goude, Y. (2025). Quantifying the uncertainty of electric vehicle charging with probabilistic load forecasting. *World Electric Vehicle Journal, 16*(2), 88.

Andrenacci, N., & Valentini, M. P. (2023). A literature review on the charging behaviour of private electric vehicles. *Applied Sciences, 13*(23), 12877.

Arévalo, P., Ochoa-Correa, D., & Villa-Ávila, E. (2024). Systematic review of the effective integration of storage systems and electric vehicles in microgrid networks: Innovative approaches for energy management. *Vehicles, 6*(4), 2075–2105.

Arif, A., Wang, Z., Wang, J., Mather, B., Bashualdo, H., & Zhao, D. (2017). Load modeling—A review. *IEEE Transactions on Smart Grid, 9*(6), 5986–5999.

Austmann, L. M. (2021). Drivers of the electric vehicle market: A systematic literature review of empirical studies. *Finance Research Letters, 41*, 101846.

Bae, J., & Yun, M. H. (2024). Driving change to electric vehicles: A comprehensive study of electric vehicle user segmentation and charging behaviors. *Proceedings of the Human Factors and Ergonomics Society Annual Meeting, 68*(1)), 928–934.

Baghali, S., Guo, Z., & Hasan, S. (2022). Investigating the spatiotemporal charging demand and travel behavior of electric vehicles using GPS data: A machine learning approach. In *2022 IEEE Power & Energy Society General Meeting (PESGM)* (pp. 1–5).

Balakrishnan, K. (2019). *Exponential distribution: Theory, methods and applications*. Routledge.

Balakrishnan, N., & Kundu, D. (2019). Birnbaum-Saunders distribution: A review of models, analysis, and applications. *Applied Stochastic Models in Business and Industry, 35*(1), 4–49.

Bartolozzi, A., Favuzza, S., Ippolito, M. G., La Cascia, D., Riva Sanseverino, E., & Zizzo, G. (2017). A new platform for automatic bottom-up electric load aggregation. *Energies, 10*(11), 1682.

Baumgartner, N., Weyer, K., Eckmann, L., & Fichtner, W. (2023). How to integrate users into smart charging—A critical and systematic review. *Energy Research Social Science, 100*, 103113.

Bian, H., Tang, X., Ji, K., Zhang, Y., & Xie, Y. (2025). Prediction of electric vehicle charging load considering user travel characteristics and charging behavior. *World Electric Vehicle Journal, 16*, 502.

Bilal, M., Oladigbolu, J. O., Mujeeb, A., & Al-Turki, Y. A. (2024). Cost-effective optimization of on-grid electric vehicle charging systems with integrated renewable energy and energy storage: An economic and reliability analysis. *Journal of Energy Storage, 100*, 113170.

BloombergNEF. (2023). Electric vehicle outlook 2023. Bloomberg New Energy Finance, Retrieved July 27, 2005, from https://about.bnef.com/electric-vehicle-outlook/.

Bons, P. C., Buatois, A., Schuring, F., Geerts, F., & van den Hoed, R. (2021). Flexible charging of electric vehicles: Results of a large-scale smart charging demonstration. *World Electric Vehicle Journal, 12*(2), 82.

Breiman, L. (2001). Random Forest. *Machine Learning, 45*, 5–32.

Brinkel, N., van Wijk, T., Buijze, A., Panda, N. K., Meersmans, J., Markotić, P., van der Ree, B., Fidder, H., de Brey, B., Tindemans, S., AlSkaif, T., & van Sark, W. (2024). Enhancing smart charging in electric vehicles by addressing paused and delayed charging problems. *Nature Communications, 15*(1), 5089.

Bushnell, J. B., Muehlegger, E., & Rapson, D. S. (2022). *Energy prices and electric vehicle adoption*. National Bureau of Economic Research.

Buzna, L., De Falco, P., Khormali, S., Proto, D., & Straka, M. (2019). Electric vehicle load forecasting: A comparison between time series and machine learning approaches. In *2019*

1st International Conference on Energy Transition in the Mediterranean Area (SyNERGY MED) (pp. 1–5).

Callanan, A., Samuelsson, O., & Márquez-Fernández, F. J. (2025). A data-driven probabilistic framework for estimating grid impacts of EV charging at scale. *International Journal of Electrical Power & Energy Systems, 172*, 111204.

Chakraborty, D., Bunch, D. S., Lee, J. H., & Tal, G. (2019). Demand drivers for charging infrastructure-charging behavior of plug-in electric vehicle commuters. *Transportation Research Part D: Transport and Environment, 76*, 255–272.

Chakraborty, S., Tomsett, R., Raghavendra, R., Harborne, D., Alzantot, M., Cerutti, F., Srivastava, M., Preece, A., Julier, S., Rao, R. M., Kelley, T. D., Braines, D., Sensoy, M., Willis, C. J., & Gurram, P. (2017). Interpretability of deep learning models: A survey of results. In *2017 IEEE Smartworld, Ubiquitous Intelligence & Computing, Advanced & Trusted Computed, Scalable Computing & Communications, Cloud & Big Data Computing, Internet of People And Smart City Innovation* (pp. 1–6).

Chamberlain, K., & Al-Majeed, S. (2021). Standardisation of UK electric vehicle charging protocol, payment and charge point connection. *World Electric Vehicle Journal, 12*(2), 63.

Chandwani, A., Dey, S., & Mallik, A. (2020). Cybersecurity of onboard charging systems for electric vehicles—Review, challenges and countermeasures. *IEEE access, 8*, 226982–226998.

Chang, M., Bae, S., Yoon, G. G., Park, S. H., & Choy, Y. (2019). Impact of electric vehicle charging demand on a Jeju island radial distribution network. In *2019 IEEE Power & Energy Society Innovative Smart Grid Technologies Conference (ISGT)* (pp. 1–5).

Chen, L., Huang, X., & Zhang, H. (2020a). Modeling the charging behaviors for electric vehicles based on ternary symmetric kernel density estimation. *Energies, 13*(7), 1551.

Chen, L., Yang, F., Xing, Q., Wu, S., Wang, R., & Chen, J. (2020b). Spatial-temporal distribution prediction of charging load for electric vehicles based on dynamic traffic information. In *2020 IEEE 4th Conference on Energy Internet and Energy System Integration (EI2)* (pp. 1269–1274).

Chen, Y., Pang, B., Xiang, X., Lu, T., Xia, T., & Geng, G. (2023). Probabilistic forecasting of electric vehicle charging load using composite quantile regression LSTM. In *2023 IEEE/IAS Industrial and Commercial Power System Asia (I&CPS Asia)* (pp. 984–989). IEEE.

Cheng, J., & Liu, N. (2022). Electric vehicles charging load prediction based on Monte Carlo method. In *2022 2nd International Conference on Electrical Engineering and Control Science (IC2ECS)* (pp. 846–850).

Choudhary, D., Mahanty, R. N., & Kumar, N. (2025). Demand management of plug-in electric vehicle charging station considering bidirectional power flow using deep reinforcement learning. *Engineering Applications of Artificial Intelligence, 139*, 109585.

Christensen, K., Ma, Z. G., & Jørgensen, B. N. (2025). A scoping review on electric vehicle charging strategies with a technical, social, and regulatory feasibility evaluation. *Renewable and Sustainable Energy Reviews, 211*, 115300.

Chung, H. M., Li, W. T., Yuen, C., Wen, C. K., & Crespi, N. (2018a). Electric vehicle charge scheduling mechanism to maximize cost efficiency and user convenience. *IEEE Transactions on Smart Grid, 10*(3), 3020–3030.

Chung, Y. W., Khaki, B., Chu, C., & Gadh, R. (2018b). Electric vehicle user behavior prediction using hybrid kernel density estimator. In *2018 IEEE International Conference on Probabilistic Methods Applied to Power Systems (PMAPS)* (pp. 1–6).

Clement-Nyns, K., Haesen, E., & Driesen, J. (2009). The impact of charging plug-in hybrid electric vehicles on a residential distribution grid. *IEEE Transactions on power systems, 25*(1), 371–380.

Coignard, J., MacDougall, P., Stadtmueller, F., & Vrettos, E. (2019). Will electric vehicles drive distribution grid upgrades?: The case of California. *IEEE Electrification Magazine, 7*(2), 46–56.

Colmenar-Santos, A., Muñoz-Gómez, A. M., Rosales-Asensio, E., & López-Rey, Á. (2019). Electric vehicle charging strategy to support renewable energy sources in Europe 2050 low-carbon scenario. *Energy, 183*, 61–74.

Consul, P. C., & Jain, G. C. (1973). A generalization of the Poisson distribution. *Technometrics, 15*(4), 791–799.

Crow, E. L., & Shimizu, K. (1987). *Lognormal distributions*. Marcel Dekker.

Cui, D., Wang, Z., Liu, P., Wang, S., Zhang, Z., Dorrell, D. G., & Li, X. (2022). Battery electric vehicle usage pattern analysis driven by massive real-world data. *Energy, 250*, 123837.

Cui, D., Wang, Z., Liu, P., Wang, S., Zhao, Y., & Zhan, W. (2023). Stacking regression technology with event profile for electric vehicle fast charging behavior prediction. *Applied Energy, 336*, 120798.

Bouallaga, A., & Doumbia, B. (2019. Stochastic electric vehicle load modeling for HV/MV substation constraint assessment. In *25th International Conference on Electricity Distribution (CIRED 2019), Madrid* (pp. 3-6).

Daina, N. (2014). *Modelling electric vehicle use and charging behaviour* (Doctoral dissertation, Imperial College London).

Das, T., & Aliprantis, D. C. (2008). Small-signal stability analysis of power system integrated with PHEVs. In *2008 IEEE Energy 2030 Conference* (pp. 1–4).

Da Silva, E. C., Melgar-Dominguez, O. D., & Romero, R. (2021). Assessment of distributed generation hosting capacity in electric distribution systems by increasing the electric vehicle penetration. In *2021 IEEE PES Innovative Smart Grid Technologies Europe (ISGT Europe)* (pp. 01–05).

Daziano, R. A. (2022). Willingness to delay charging of electric vehicles. *Research in Transportation Economics, 94*, 101177.

De Lima, T. D., Franco, J. F., Lezama, F., & Soares, J. (2022). A specialized long-term distribution system expansion planning method with the integration of distributed energy resources. *IEEE Access, 10*, 19133–19148.

Demirci, A., Tercan, S. M., Ahmed, E. E., Cali, U., & Nakir, I. (2024). A novel electric vehicle charging management with dynamic active power curtailment framework for PV-rich prosumers. *IEEE Access*.

Deng, J., Gao, L., Chen, X., & Yuan, Q. (2024). Taking the same route every day? An empirical investigation of commuting route stability using personal electric vehicle trajectory data. *Transportation, 51*(4), 1547–1573.

Ding, Y., Zhu, Y., Wang, Q., Tian, Z., Yan, R., Yan, Z., & Xia, X. (2023). A comprehensive scheduling model for electric vehicles in office buildings considering the uncertainty of charging load. *International Journal of Electrical Power & Energy Systems, 151*, 109154.

Domarchi, C., & Cherchi, E. (2023). Electric vehicle forecasts: A review of models and methods including diffusion and substitution effects. *Transport Reviews, 43*(6), 1118–1143.

Dominguez-Jimenez, J. A., Campillo, J. E., Montoya, O. D., Delahoz, E., & Hernández, J. C. (2020). Seasonality effect analysis and recognition of charging behaviors of electric vehicles: A data science approach. *Sustainability, 12*(18), 7769.

Du, J., & Ouyang, D. (2017). Progress of Chinese electric vehicles industrialization in 2015: A review. *Applied Energy, 188*, 529–546.

Ebrahimi, M., Shafie-Khah, M., & Laaksonen, H. (2024). Uncertainty-observed virtual battery model for an electric vehicle parking lot enabling charger-sharing modelling. *Journal of Energy Storage, 89*, 111578.

Elaad N.L. (n.d.). Data analytics. Library catalog. Retrieved March 17, 2021, from www.elaad.nl/research/data-analytics/.

Electric Vehicle Grid Impact Analysis and Smart-Charge Management. (2025). National renewable energy laboratory. Retrieved from https://www.nrel.gov/transportation/managed-electric-vehicle-charging

El Gohary, F., Stikvoort, B., & Bartusch, C. (2023). Evaluating demand charges as instruments for managing peak-demand. *Renewable and Sustainable Energy Reviews, 188*, 113876.

El-Hendawi, M., Wang, Z., Paranjape, R., Pederson, S., Kozoriz, D., & Fick, J. (2022). Electric vehicle charging model in the urban residential sector. *energies, 15*(13), 4901.

Ensslen, A., Ringler, P., Dörr, L., Jochem, P., Zimmermann, F., & Fichtner, W. (2018). Incentivizing smart charging: Modeling charging tariffs for electric vehicles in German and French electricity markets. *Energy research & social science, 42*, 112–126.

Everoze. (2023). *V2G global roadtrip: Around the world in 50 projects*. Retrieved from https://d1oyzg0jo3ox9g.cloudfront.net/app/uploads/2023/10/V2G-Global-Roadtrip-Around-the-World-in-50-Projects.pdf

Fernandez, L. P., San Román, T. G., Cossent, R., Domingo, C. M., & Frias, P. (2010). Assessment of the impact of plug-in electric vehicles on distribution networks. *IEEE Transactions on Power Systems, 26*(1), 206–213.

Fischer, D., Harbrecht, A., Surmann, A., & McKenna, R. (2019). Electric vehicles' impacts on residential electric local profiles–A stochastic modelling approach considering socio-economic, behavioural and spatial factors. *Applied Energy, 233*, 644–658.

Fischer, N. L., Beck, J., Bayer, S., Rudion, K., Maksymchuk, N., & Enzenhöfer, R. (2024). Top-down approach for modeling of aggregated electric vehicle charging load based on real data. In *2024 IEEE PES Innovative Smart Grid Technologies Europe (ISGT EUROPE)* (pp. 1–5).

Flammini, M. G., Prettico, G., Julea, A., Fulli, G., Mazza, A., & Chicco, G. (2019). Statistical characterisation of the real transaction data gathered from electric vehicle charging stations. *Electric Power Systems Research, 166*, 136–150.

Folks, J. L., & Chhikara, R. S. (1978). The inverse Gaussian distribution and its statistical application—A review. *Journal of the Royal Statistical Society Series B: Statistical Methodology, 40*(3), 263–275.

Franke, T., & Krems, J. F. (2013). Understanding charging behaviour of electric vehicle users. *Transportation Research Part F: Traffic Psychology and Behaviour, 21*, 75–89.

Frendo, O., Gaertner, N., & Stuckenschmidt, H. (2020a). Improving smart charging prioritization by predicting electric vehicle departure time. *IEEE Transactions on Intelligent Transportation Systems, 22*(10), 6646–6653.

Frendo, O., Graf, J., Gaertner, N., & Stuckenschmidt, H. (2020b). Data-driven smart charging for heterogeneous electric vehicle fleets. *Energy and AI, 1*, 100007.

Fu, K., Lin, H., Wang, Y., Sun, Q., & Wennersten, R. (2021). A rolling global optimization method of ev fleet for peak-valley shaving considering the uncertainty of EV behavior. In *2021 4th International Conference on Energy, Electrical and Power Engineering (CEEPE)* (pp. 946–950).

Fuchs, E. F., & Masoum, M. A. (2015). *Power Quality in Electrical Machines and Power Systems*. Academic Press.

Gang, W., Wu, L., & Xuan, G. (2021). A load forecasting method of electric vehicles charging station group based on GAN-RF model. In *2021 IEEE 5th Conference on Energy Internet and Energy System Integration (EI2)* (pp. 3076–3079).

Gass, V., Schmidt, J., & Schmid, E. (2014). Analysis of alternative policy instruments to promote electric vehicles in Austria. *Renewable Energy, 61*, 96–101.

Ge, X., Shi, L., Fu, Y., Muyeen, S. M., Zhang, Z., & He, H. (2020). Data-driven spatial-temporal prediction of electric vehicle load profile considering charging behavior. *Electric Power Systems Research, 187*, 106469.

Geng, J., Bai, B., Hao, H., Sun, X., Liu, M., Liu, Z., & Zhao, F. (2024). Assessment of vehicle-side costs and profits of providing vehicle-to-grid services. *ETransportation, 19*, 100303.

Gerossier, A., Girard, R., & Kariniotakis, G. (2019). Modeling and forecasting electric vehicle consumption profiles. *Energies, 12*(7), 1341.

Gholami, M., Mehrasa, M., Razi, R., Hajar, K., Hably, A., Bacha, S., & Labonne, A. (2022). Smart charging strategy for electric vehicles using an optimized fuzzy logic system. In *2022 24th European Conference on Power Electronics and Applications (EPE'22 ECCE Europe)* (pp. 01–09).

Gil-Aguirre, J., Perez-Londoño, S., & Mora-Flórez, J. (2019). A measurement-based load modelling methodology for electric vehicle fast-charging stations. *Electric Power Systems Research, 176*, 105934.

Gong, S., Cheng, V. H. S., Ardeshiri, A., & Rashidi, T. H. (2021). Incentives and concerns on vehicle-to-grid technology expressed by Australian employees and employers. *Transportation Research Part D: Transport and Environment, 98*, 102986.

Green, R. C., II, Wang, L., & Alam, M. (2010). The impact of plug-in hybrid electric vehicles on distribution networks: A review and outlook. In *IEEE PES General Meeting* (pp. 1–8).

Gschwendtner, C., Knoeri, C., & Stephan, A. (2023a). Mind the goal: Trade-offs between flexibility goals for controlled electric vehicle charging strategies. *Iscience, 26*(2).

Gschwendtner, C., Knoeri, C., & Stephan, A. (2023b). The impact of plug-in behavior on the spatial–temporal flexibility of electric vehicle charging load. *Sustainable Cities and Society, 88*, 104263.

Guo, Z., You, L., Zhu, R., Zhang, Y., & Yuen, C. (2025). A city-scale and harmonized dataset for global electric vehicle charging demand analysis. *Scientific Data, 12*(1), 1254.

Gupta, R. D., & Kundu, D. (2001). Generalized exponential distribution: Different method of estimations. *Journal of statistical computation and simulation, 69*(4), 315–337.

Gupta, R. R., Zywiel, M. G., Leadbetter, W. B., Bonutti, P., & Mont, M. A. (2010). Scientific evidence for the use of modern patellofemoral arthroplasty. *Expert Review of Medical Devices, 7*(1), 51–66.

Gupta, S. S., & Gnanadesikan, M. (1966). Estimation of the parameters of the logistic distribution. *Biometrika, 53*(3-4), 565–570.

Hamdare, S., Kaiwartya, O., Aljaidi, M., Jugran, M., Cao, Y., Kumar, S., Mahmud, M., Brown, D., & Lloret, J. (2023). Cybersecurity risk analysis of electric vehicles charging stations. *Sensors, 23*(15), 6716.

Han, X., Wei, Z., Hong, Z., & Zhao, S. (2020). Ordered charge control considering the uncertainty of charging load of electric vehicles based on Markov chain. *Renewable Energy, 161*, 419–434.

Hannan, M. A., Hoque, M. M., Mohamed, A., & Ayob, A. (2017). Review of energy storage systems for electric vehicle applications: Issues and challenges. *Renewable and Sustainable Energy Reviews, 69*, 771–789.

Hawkins, T. R., Singh, B., Majeau-Bettez, G., & Strømman, A. H. (2013). Comparative environmental life cycle assessment of conventional and electric vehicles. *Journal of Industrial Ecology, 17*(1), 53–64.

He, S., Zhang, Z., Han, S., Pepin, L., Wang, G., Zhang, D., Stankovic, J. A., & Miao, F. (2023). Data-driven distributionally robust electric vehicle balancing for autonomous mobility-on-demand systems under demand and supply uncertainties. *IEEE Transactions on Intelligent Transportation Systems, 24*(5), 5199–5215.

Hecht, C., Das, S., Bussar, C., & Sauer, D. U. (2020). Representative, empirical, real-world charging station usage characteristics and data in Germany. *ETransportation, 6*, 100079.

Helmus, J., & Van Den Hoed, R (2015). Unraveling user type characteristics: Towards a taxonomy for charging infrastructure. *The 28th International Electric Vehicle Symposium and Exhibition*.

Helmus, J. R., Lees, M. H., & van den Hoed, R. (2020). A data driven typology of electric vehicle user types and charging sessions. *Transportation Research Part C: Emerging Technologies, 115*, 102637.

Henriksen, I. M., Throndsen, W., Ryghaug, M., & Skjølsvold, T. M. (2021). Electric vehicle charging and end-user motivation for flexibility: A case study from Norway. *Energy, Sustainability and Society, 11*(1), 44.

Hiesl, A., Ramsebner, J., & Haas, R. (2021). Modelling stochastic electricity demand of electric vehicles based on traffic surveys—The case of Austria. *Energies, 14*(6), 1577.

Hildermeier, J., Burger, J., Jahn, A., & Rosenow, J. (2022). A review of tariffs and services for smart charging of electric vehicles in Europe. *Energies, 16*(1), 88.

Hong, T., Wilson, J., & Xie, J. (2013). Long term probabilistic load forecasting and normalization with hourly information. *IEEE Transactions on smart grid, 5*(1), 456–462.

Hou, F., Chen, X., Chen, X., Yang, F., Ma, Z., Zhang, S., Liu, S., Zhao, C., & Guo, F. (2021). Comprehensive analysis method of determining global long-term GHG mitigation potential of passenger battery electric vehicles. *Journal of Cleaner Production, 289*, 125137.

Hu, L., Dong, J., & Lin, Z. (2019). Modeling charging behavior of battery electric vehicle drivers: A cumulative prospect theory based approach. *Transportation Research Part C: Emerging Technologies, 102*, 474–489.

Huaman-Rivera, A., Calloquispe-Huallpa, R., Luna Hernandez, A. C., & Irizarry-Rivera, A. (2024). An overview of electric vehicle load modeling strategies for grid integration studies. *Electronics, 13*(12), 2259.

Huang, B., Meijssen, A. G., Annema, J. A., & Lukszo, Z. (2021). Are electric vehicle drivers willing to participate in vehicle-to-grid contracts? A context-dependent stated choice experiment. *Energy Policy, 156*, 112410.

Huang, X., Wu, D., & Boulet, B. (2020). Ensemble learning for charging load forecasting of electric vehicle charging stations. In *2020 IEEE Electric Power and Energy Conference (EPEC)* (pp. 1–5). IEEE.

Huber, J., Dann, D., & Weinhardt, C. (2020). Probabilistic forecasts of time and energy flexibility in battery electric vehicle charging. *Applied Energy, 262*, 114525.

Hallak, K., & Kem, O. (2025). Adaptive federated learning framework for predicting EV charging stations occupancy. International Journal of Transportation Science and Technology. In press.

International Energy Agency. (2023). CO2 Emissions in 2022. https://www.iea.org/reports/co2-emissions-in-2022

International Energy Agency. (2024). *Global EV Outlook 2024*. IEA Publications.

International Energy Agency. (2025). Global EV outlook 2025. https://www.iea.org/reports/global-ev-outlook-2025.

James, J., & Jasmin, E. A. (2022). Fuzzy logic controller for dynamic price based charging of electric vehicle. In *2022 IEEE International Power and Renewable Energy Conference (IPRECON)* (pp. 1–7).

Jamil, U., Alva, R. J., Ahmed, S., & Jin, Y. F. (2025). Artificial intelligence-driven optimal charging strategy for electric vehicles and impacts on electric power grid. *Electronics, 14*(7), 1471.

Jiang, P., Li, R., Liu, N., & Gao, Y. (2020). A novel composite electricity demand forecasting framework by data processing and optimized support vector machine. *Applied Energy, 260*, 114243.

Jingwei, Y., Diansheng, L., Shuang, Y., & Shiyu, H. (2014). Charging load forecasting for electric vehicles based on fuzzy inference. In *Chinese Conference on Pattern Recognition* (pp. 585–594).

Jones, C. B., Lave, M., Vining, W., & Garcia, B. M. (2021). Uncontrolled electric vehicle charging impacts on distribution electric power systems with primarily residential, commercial or industrial loads. *Energies, 14*(6), 1688.

Kamana-Williams, B., Bishop, D., Hooper, G., & Chase, J. G. (2024). Driving change: Electric vehicle charging behavior and peak loading. *Renewable and Sustainable Energy Reviews, 189*, 113953.

Kang, J., Kong, H., Lin, Z., & Dang, A. (2022). Mapping the dynamics of electric vehicle charging demand within Beijing's spatial structure. *Sustainable Cities and Society, 76*, 103507.

Kara, E. C., Macdonald, J. S., Black, D., Bérges, M., Hug, G., & Kiliccote, S. (2015). Estimating the benefits of electric vehicle smart charging at non-residential locations: A data-driven approach. *Applied Energy, 155*, 515–525.

Karmaker, A. K., Roy, S., & Ahmed, M. R. (2019). Analysis of the impact of electric vehicle charging station on power quality issues. In *2019 International Conference on Electrical, Computer and Communication Engineering (ECCE)* (pp. 1–6).

Kempton, W., & Tomić, J. (2005). Vehicle-to-grid power fundamentals: Calculating capacity and net revenue. *Journal of Power Sources, 144*(1), 268–279.

Khaki, B., Chung, Y. W., Chu, C., & Gadh, R. (2018). Nonparametric user behavior prediction for distributed EV charging scheduling. In *2018 IEEE Power & Energy Society General Meeting (PESGM)* (pp. 1–5).

Khaki, B., Chung, Y. W., Chu, C., & Gadh, R. (2019). Probabilistic electric vehicle load management in distribution grids. In *2019 IEEE Transportation Electrification Conference and Expo (ITEC)* (pp. 1–6).

Khalid, M. R., Alam, M. S., Sarwar, A., & Asghar, M. J. (2019). A Comprehensive review on electric vehicles charging infrastructures and their impacts on power-quality of the utility grid. *ETransportation, 1*, 100006.

Khan, M. R., Haider, Z. M., Malik, F. H., Almasoudi, F. M., Alatawi, K. S. S., & Bhutta, M. S. (2024). A comprehensive review of microgrid energy management strategies considering electric vehicles, energy storage systems, and AI techniques. *Processes, 12*(2), 270.

Khoo, Y. B., Wang, C. H., Paevere, P., & Higgins, A. (2014). Statistical modeling of electric vehicle electricity consumption in the Victorian EV trial, Australia. *Transportation Research Part D: Transport and Environment, 32*, 263–277.

Kilic, E., Akil, M., Bayindir, R., Sebati, A., & Malek, R. (2021). Electric vehicles charging management with Monte Carlo simulation. In *2021 10th International Conference on Renewable Energy Research and Application (ICRERA)* (pp. 423–427).

Kisacikoglu, M. C., Ozpineci, B., & Tolbert, L. M. (2010). Examination of a PHEV bidirectional charger system for V2G reactive power compensation. In *2010 Twenty-Fifth Annual IEEE Applied Power Electronics Conference and Exposition (APEC)* (pp. 458–465).

Knezović, K., Martinenas, S., Andersen, P. B., Zecchino, A., & Marinelli, M. (2016). Enhancing the role of electric vehicles in the power grid: Field validation of multiple ancillary services. *IEEE Transactions on Transportation Electrification, 3*(1), 201–209.

Kofoed-Wiuff, A., Hethey, J., Togeby, M., Sawatzki, S., & Persson, C. (2015). *The Danish Experience with Integrating Variable Renewable Energy*. Agora Energiewende.

Kongjeen, Y., & Bhumkittipich, K. (2016). Modeling of electric vehicle loads for power flow analysis based on PSAT. In *2016 13th International Conference on Electrical Engineering/Electronics, Computer, Telecommunications And Information Technology (ECTI-CON)* (pp. 1–6).

Kongjeen, Y., Bhumkittipich, K., Mithulananthan, N., Amiri, I. S., & Yupapin, P. (2019). A modified backward and forward sweep method for microgrid load flow analysis under different electric vehicle load mathematical models. *Electric Power Systems Research, 168*, 46–54.

Kramer, J., Riza, L., & Petzoldt, T. (2023). Carbon savings, fun, and money: The effectiveness of multiple motives for eco-driving and green charging with electric vehicles in Germany. *Energy Research & Social Science, 99*, 103054.

Kunj, T., & Pal, K. (2020). Optimal location planning of EV charging station in existing distribution network with stability condition. In *2020 7th International Conference On Signal Processing and Integrated Networks (SPIN)* (pp. 1060–1065).

Kütt, L., Saarijärvi, E., Lehtonen, M., Mõlder, H., & Niitsoo, J. (2013). A review of the harmonic and unbalance effects in electrical distribution networks due to EV charging. In *2013 12th International Conference on Environment and Electrical Engineering* (pp. 556–561).

Lagomarsino, M., van der Kam, M., Parra, D., & Hahnel, U. J. (2022). Do I need to charge right now? Tailored choice architecture design can increase preferences for electric vehicle smart charging. *Energy Policy, 162*, 112818.

Lahariya, M., Benoit, D., & Develder, C. (2020). Defining a synthetic data generator for realistic electric vehicle charging sessions. In *Proceedings of the Eleventh ACM International Conference on Future Energy Systems* (pp. 406–407).

LaMonaca, S., & Ryan, L. (2022). The state of play in electric vehicle charging services–A review of infrastructure provision, players, and policies. *Renewable and Sustainable Energy Reviews, 154*, 111733.

Lee, Z. J., Li, T., & Low, S. H. (2019). ACN-data: Analysis and applications of an open EV charging dataset. In *Proceedings of the Tenth ACM International Conference on Future Energy System* (pp. 139–149).

Li, S., Hu, W., Cao, D., Zhang, Z., Huang, Q., Chen, Z., & Blaabjerg, F. (2022). A multiagent deep reinforcement learning based approach for the optimization of transformer life using coordinated electric vehicles. *IEEE Transactions on Industrial Informatics, 18*(11), 7639–7652.

Li, X., & Jenn, A. (2022). Energy, emissions, and cost impacts of charging price strategies for electric vehicles. *Environmental Science & Technology, 56*(9), 5724–5733.

Li, Y., Huang, Y., & Zhang, M. (2018). Short-term load forecasting for electric vehicle charging station based on niche immunity lion algorithm and convolutional neural network. *Energies, 11*(5), 1253.

Li, Z., Li, C., Zhang, B., Duan, Q., Liu, L., Zu, G., & Li, Q. (2023). Research on new energy vehicle charging prediction based on Monte Carlo algorithm and its impact on distribution network. *Frontiers in Energy Research, 11*, 1269041.

Li, Z., Smith-Colin, J., & Wang, J. (2025). A unified multi-criteria equity metric for data-driven decision-making: Application to tract-level electric vehicle charging station planning. *Engineering Applications of Artificial Intelligence, 144*, 110070.

Liang, M., Li, W., Yu, J., & Shi, L. (2015). Kernel-based electric vehicle charging load modeling with improved Latin hypercube sampling. In *2015 IEEE Power & Energy Society General Meeting* (pp. 1–5).

Libertson, F. (2022). Requesting control and flexibility: Exploring Swedish user perspectives of electric vehicle smart charging. *Energy Research & Social Science, 92*, 102774.

Lin, J., Qiu, J., Tao, Y., & Sun, X. (2023). Planning of electric vehicle charging stations with PV and energy storage using a fuzzy inference system. *IEEE Transactions on Transportation Electrification, 10*(3), 5894–5909.

Liu, D., Li, Z., Jiang, J., Cheng, X., & Wu, G. (2020). Electric vehicle load forecast based on Monte Carlo algorithm. In *2020 IEEE 9th Joint International Information Technology and Artificial Intelligence Conference (ITAIC): 9* (pp. 1760–1763).

Liu, H., Shen, H., Hu, W., Ji, L., Li, J., & Yu, Y. (2023). Electric vehicle load forecast based on higher order Markov chain. In *2023 5th Asia Energy and Electrical Engineering Symposium (AEEES)* (pp. 1203–1207).

Liu, J., Lin, G., Rehtanz, C., Huang, S., Zhou, Y., & Li, Y. (2022a). Data-driven intelligent EV charging operating with limited chargers considering the charging demand forecasting. *International Journal of Electrical Power & Energy Systems, 141*, 108218.

Liu, K., & Liu, Y. (2023). Stochastic user equilibrium based spatial-temporal distribution prediction of electric vehicle charging load. *Applied Energy, 339*, 120943.

Liu, Y., Liu, W., Gao, S., Wang, Y., & Shi, Q. (2022b). Fast charging demand forecasting based on the intelligent sensing system of dynamic vehicle under EVs-traffic-distribution coupling. *Energy Reports, 8*, 1218–1226.

Lu, Y., Li, Y., Xie, D., Wei, E., Bao, X., Chen, H., & Zhong, X. (2018). The application of improved random forest algorithm on the prediction of electric vehicle charging load. *Energies, 11*(11), 3207.

Ma, T. Y., & Faye, S. (2022). Multistep electric vehicle charging station occupancy prediction using hybrid LSTM neural networks. *Energy, 244*, 123217.

Majidpour, M., Qiu, C., Chu, P., Pota, H. R., & Gadh, R. (2016). Forecasting the EV charging load based on customer profile or station measurement? *Applied Energy, 163*, 134–141.

Mangipinto, A., Lombardi, F., Sanvito, F. D., Pavičević, M., Quoilin, S., & Colombo, E. (2022). Impact of mass-scale deployment of electric vehicles and benefits of smart charging across all European countries. *Applied Energy, 312*, 118676.

Masoum, M. A., Moses, P. S., & Smedley, K. M. (2011). Distribution transformer losses and performance in smart grids with residential plug-in electric vehicles. In *ISGT* (pp. 1–7). IEEE.

McDonald, J. B., & Xu, Y. J. (1995). A generalization of the beta distribution with applications. *Journal of Econometrics, 66*(1-2), 133–152.

Mies, J. J., Helmus, J. R., & Van den Hoed, R. (2018). Estimating the charging profile of individual charge sessions of electric vehicles in The Netherlands. *World Electric Vehicle Journal, 9*(2), 17.

Mojumder, M. R. H., Ahmed Antara, F., Hasanuzzaman, M., Alamri, B., & Alsharef, M. (2022). Electric vehicle-to-grid (V2G) technologies: Impact on the power grid and battery. *Sustainability, 14*(21), 13856.

Möller, F., Müller, S., Meyer, J., Schegner, P., Wald, C., & Isensee, S. (2015). Impact of electric vehicle charging on unbalance and harmonic distortion-field study in an urban residential area. In *23rd International Conference on Electricity Distribution (CIRED)* (pp. 1–5).

Motlagh, S. G., Oladigbolu, J., & Li, L. (2025). A review on electric vehicle charging station operation considering market dynamics and grid interaction. *Applied Energy, 392*, 126058.

Muessel, J., Ruhnau, O., & Madlener, R. (2023). Accurate and scalable representation of electric vehicles in energy system models: A virtual storage-based aggregation approach. *Iscience*, (10), 26.

Muhammad, M. (2017). The complementary exponentiated burrxii poisson distribution: Model, properties and application. *Journal of Statistics Applications & Probability, 6*(1), 33–48.

Muratori, M., Jadun, P., Bush, B., Bielen, D., Vimmerstedt, L., Gonder, J., Gearhart, C., & Arent, D. (2020). Future integrated mobility-energy systems: A modeling perspective. *Renewable and Sustainable Energy Reviews, 119*, 109541.

Nadarajah, S. (2005). A generalized normal distribution. *Journal of Applied statistics, 32*(7), 685–694.

Nagel, N. O., Jåstad, E. O., & Martinsen, T. (2024). The grid benefits of vehicle-to-grid in Norway and Denmark: An analysis of home-and public parking potentials. *Energy, 293*, 130729.

Najafi, F., Nobakht, S., Samimiat, M., Ahmadi, A. A., & Nateghi, A. (2025). Detection and localization of dynamic load altering attacks in power systems. *Computers and Electrical Engineering, 123*, 110207.

Lee, Y., Bencekri, M., Kwak, J., Jeong, I., & Lee, S. (2025). Electrification pathways toward transport decarbonization with a systematic review. *Clean Technologies and Environmental Policy, 27*(12), 7855–7877.

National Household Travel Survey. Retrieved March 17, 2021, from nhts.ornl.gov.

Nazari-Heris, M., Abapour, M., & Mohammadi-Ivatloo, B. (2022). An updated review and outlook on electric vehicle aggregators in electric energy networks. *Sustainability, 14*(23), 15747.

Nelson, K., Mohammadi, J., Chen, Y., Blasch, E., Aved, A., Ferris, D., Cruz, E. A., & Morrone, P. (2024). Electric vehicle aggregation review: Benefits and vulnerabilities of managing a growing fleet. In *2024 IEEE Texas Power and Energy Conference (TPEC)* (pp. 1–6).

Nguyen, V. L., Tran-Quoc, T., & Bacha, S. (2013). Harmonic distortion mitigation for electric vehicle fast charging systems. In *2013 IEEE Grenoble Conference* (pp. 1–6).

Noel, L., De Rubens, G. Z., Sovacool, B. K., & Kester, J. (2019). Fear and loathing of electric vehicles: The reactionary rhetoric of range anxiety. *Energy Research & Social Science, 48*, 96–107.

Nour, M., Chaves-Ávila, J. P., Magdy, G., & Sánchez-Miralles, Á. (2020). Review of positive and negative impacts of electric vehicles charging on electric power systems. *Energies, 13*(18), 4675.

Nour, M., Ramadan, H., Ali, A., & Farkas, C. (2018). Impacts of plug-in electric vehicles charging on low voltage distribution network. In *2018 International Conference On Innovative Trends In Computer Engineering (ITCE)* (pp. 357–362).

Nutkani, I., Toole, H., Fernando, N., & Andrew, L. P. C (2024). Impact of EV charging on electrical distribution network and mitigating solutions–A review. *IET Smart Grid, 7*(5), 485–502.

Orr, J. A., Emanuel, A. E., & Pileggi, D. J. (2008). Current harmonics, voltage distortion, and powers associated with electric vehicle battery chargers distributed on the residential power system. *IEEE Transactions on Industry Applications*, (4), 727–734.

Our World in Data. (2020). Cars, planes, trains: Where do CO_2 emissions from transport come from? Retrieved from https://ourworldindata.org/co2-emissions-from-transport

Pagani, M., Korosec, W., Chokani, N., & Abhari, R. S. (2019). User behaviour and electric vehicle charging infrastructure: An agent-based model assessment. *Applied Energy, 254*, 113680.

Pal, M., Ali, M. M., & Woo, J. (2006). Exponentiated weibull distribution. *Statistica, 66*(2), 139–147.

Pan, L., Yao, E., & MacKenzie, D. (2019). Modeling EV charging choice considering risk attitudes and attribute non-attendance. *Transportation Research Part C: Emerging Technologies, 102*, 60–72.

Pan, L., Yao, E., Yang, Y., & Zhang, R. (2020). A location model for electric vehicle (EV) public charging stations based on drivers' existing activities. *Sustainable Cities and Society, 59*, 102192.

Papadopoulos, P., Jenkins, N., Cipcigan, L. M., Grau, I., & Zabala, E. (2013). Coordination of the charging of electric vehicles using a multi-agent system. *IEEE Transactions on Smart Grid, 4*(4), 1802–1809.

Pei, M., Huang, Z., Zhang, Z., Wang, K., & Ye, X. (2025). Range anxiety and willingness to pay: Psychological insights for electric vehicle. *Journal of Renewable and Sustainable Energy, 17*(1), 015902.

Philipsen, R., Brell, T., Brost, W., Eickels, T., & Ziefle, M. (2018). Running on empty–Users' charging behavior of electric vehicles versus traditional refueling. *Transportation Research Part F: Traffic Psychology and Behaviour, 59*, 475–492.

Pinthurat, W., & Hredzak, B. (2020). Decentralized frequency control of battery energy storage systems distributed in isolated microgrid. *Energies, 13*(11), 3026.

Pokhrel, S. R., & Hossain, M. B. (2022). Data privacy of wireless charging vehicle to grid (V2G) networks with federated learning. *IEEE Transactions on Vehicular Technology, 71*(8), 9032–9037.

Potoglou, D., Song, R., & Santos, G. (2023). Public charging choices of electric vehicle users: A review and conceptual framework. *Transportation Research Part D: Transport and Environment, 121*, 103824.

Powell, S., Cezar, G. V., Apostolaki-Iosifidou, E., & Rajagopal, R. (2022a). Large-scale scenarios of electric vehicle charging with a data-driven model of control. *Energy, 248*, 123592.

Powell, S., Cezar, G. V., Min, L., Azevedo, I. M., & Rajagopal, R. (2022b). Charging infrastructure access and operation to reduce the grid impacts of deep electric vehicle adoption. *Nature Energy, 7*(10), 932–945.

Powell, S., Kara, E. C., Sevlian, R., Cezar, G. V., Kiliccote, S., & Rajagopal, R. (2020). Controlled workplace charging of electric vehicles: The impact of rate schedules on transformer aging. *Applied Energy, 276*, 115352.

Priyam, T., Ruan, T., & Lv, Q. (2024). Demographic-Based public perception analysis of electric vehicles on online social networks. *Sustainability, 16*(1), 305.

Puech, A., Rigaut, T., Templier, W., & Tournoud, M. (2024). Controlling large electric vehicle charging stations via user behavior modeling and stochastic programming. In *2024 IEEE PES Innovative Smart Grid Technologies Europe (ISGT EUROPE)* (pp. 1–6).

Qian, T., Shi, F., Wang, K., Yang, S., Geng, J., Li, Y., & Wu, Q. (2022). N-1 static security assessment method for power grids with high penetration rate of renewable energy generation. *Electric Power Systems Research, 211*, 108200.

Qin, B., Cai, J., Du, C., Lev, Y., & Guo, C. (2022). Short term forecasting method of charging load based on multilevel discrete wavelet transform and LSTM model. In *Proceedings of the 2022 4th International Academic Exchange Conference on Science and Technology Innovation* (pp. 111–115).

Ramadan, H., Ali, A., Nour, M., & Farkas, C. (2018). Smart charging and discharging of plug-in electric vehicles for peak shaving and valley filling of the grid power. In *2018 Twentieth International Middle East Power Systems Conference (MEPCON)* (pp. 735–739).

Rana, M. M., Alam, S. M., Rafi, F. A., Deb, S. B., Agili, B., He, M., & Ali, M. H. (2025). Comprehensive review on the charging technologies of electric vehicles (EV) and their impact on power grid. *IEEE Access, 13*, 35124–35156.

Renhai, F., Khan, W., Tariq, A., Abbas, M., Yousaf, M. Z., Aziz, A., Bajaj, M., Abdullah, M., & Tuka, M. (2025). Adaptive non-parametric kernel density estimation for under-frequency load shedding with electric vehicles and renewable power uncertainty. *Scientific Reports, 15*(1), 11499.

Richardson, D. B. (2013). Electric vehicles and the electric grid: A review of modeling approaches, Impacts, and renewable energy integration. *Renewable and Sustainable Energy Reviews, 19*, 247–254.

Richardson, P., Flynn, D., & Keane, A. (2011). Optimal charging of electric vehicles in low-voltage distribution systems. *IEEE Transactions on Power Systems, 27*(1), 268–279.

Rizvi, S. A. A., Xin, A., Masood, A., Iqbal, S., Jan, M. U., & Rehman, H. (2018). Electric vehicles and their impacts on integration into power grid: A review. In *2018 2nd IEEE Conference on Energy Internet and Energy System Integration (EI2)* (pp. 1–6).

Rosenblatt, F. (1958). The perceptron: A probabilistic model for information storage and organization in the brain. *Psychological review, 65*(6), 386.

Rubinstein, R. Y., & Kroese, D. P. (2016). *Simulation and the Monte Carlo method.* John Wiley & Sons.

Upstream, Security. (2025). *The hidden cybersecurity risks of the EV charging ecosystem – A critical look. Upstream 2025 Global Automotive & Smart Mobility Cybersecurity Report* https://upstream.auto/cybersecurity-risks-ev-charging-ecosystem/.

Rumelhart, D. E., Hinton, G. E., & Williams, R. J. (1988). Learning internal representations by error propagation. In Neurocomputing: foundations of research (pp. 673–695).

Saadatmandi, S., Chicco, G., Favenza, A., Mozzato, A., Giordano, F., & Arnone, M. (2024). Smart electric vehicle charging for reducing photovoltaic energy curtailment. *Electric Power Systems Research, 230*, 110181.

Sanami, S., Mosalli, H., Yang, Y., Yeh, H. G., Aghdam, A. G. (2025). Demand forecasting for electric vehicle charging stations using multivariate time-series analysis. *arXiv preprint arXiv:2502.16365.*

Saxena, S., Farag, H. E., Hilaire, L. S., & Brookson, A. (2023). A techno-social approach to unlocking vehicle to everything (V2X) integration: A real-world demonstration. *IEEE Access, 11*, 17085–17095.

Sayarshad, H. R. (2025). Securing power grids and charging infrastructure: Cyberattack resilience and vehicle-to-grid integration. *Journal of Transport Geography, 126*, 104231.

Schlee, J., Mousseau, A., Eggebraaten, J., Johnson, B., Hess, H., & Johnson, B. (2009). The effects of plug-in electric vehicles on a small distribution grid. In *41st North American power symposium* (pp. 1–6).

Schmalzl, S., Frey, M., & Gauterin, F. (2025). A spatio-temporal microsimulation framework for charging impact analysis of electric vehicles in residential areas: Sensitivity analysis and benefits of model complexity. *Energies, 18*(13), 3530.

Sevdari, K., Calearo, L., Andersen, P. B., & Marinelli, M. (2022). Ancillary services and electric vehicles: An overview from charging clusters and chargers technology perspectives. *Renewable and Sustainable Energy Reviews, 167*, 112666.

Shafiei, E., Davidsdottir, B., Fazeli, R., Leaver, J., Stefansson, H., & Asgeirsson, E. I. (2018). Macroeconomic effects of fiscal incentives to promote electric vehicles in Iceland: Implications for government and consumer costs. *Energy Policy, 114*, 431–443.

Shahidinejad, S., Filizadeh, S., & Bibeau, E. (2011). Profile of charging load on the grid due to plug-in vehicles. *IEEE Transactions on Smart Grid, 3*(1), 135–141.

Shahriar, S., & Al-Ali, A. R. (2022). Impacts of covid-19 on electric vehicle charging behavior: Data analytics, visualization, and clustering. *Applied System Innovation, 5*(1), 12.

Shanmuganathan, J., Victoire, A. A., Balraj, G., & Victoire, A. (2022). Deep learning LSTM recurrent neural network model for prediction of electric vehicle charging demand. *Sustainability, 14*(16), 10207.

Shariatzadeh, M., Antunes, C. H., & Lopes, M. A. (2024). Charging scheduling in a workplace parking lot: Bi-objective optimization approaches through predictive analytics of electric vehicle users' charging behavior. *Sustainable Energy, Grids and Networks, 39*, 101463.

Shell Recharge Solutions. (2023). EV driver survey report. Retrieved january 25, from https://a.storyblok.com/f/85281/x/f117d74319/sr_ev_driver_survey_2023_uk_final_version.pdf.

Sheng, H., & Xiao, J. (2015). Electric vehicle state of charge estimation: Nonlinear correlation and fuzzy support vector machine. *Journal of Power Sources, 281*, 131–137.

Shepero, M., & Munkhammar, J. (2018). Spatial Markov chain model for electric vehicle charging in cities using geographical information system (GIS) data. *Applied Energy, 231*, 1089–1099.

Singh, S., Saket, R. K., & Khan, B. (2023). A comprehensive state-of-the-art review on reliability assessment and charging methodologies of grid-integrated electric vehicles. *IET Electrical systems in Transportation, 13*(1), e12073.

Sokorai, P., Fleischhacker, A., Lettner, G., & Auer, H. (2018). Stochastic modeling of the charging behavior of electromobility. *World Electric Vehicle Journal, 9*(3), 44.

Sortomme, E., Hindi, M. M., MacPherson, S. J., & Venkata, S. S. (2010). Coordinated charging of plug-in hybrid electric vehicles to minimize distribution system losses. *IEEE transactions on smart grid, 2*(1), 198–205.

Sovacool, B. K., Noel, L., Axsen, J., & Kempton, W. (2018). The neglected social dimensions to a vehicle-to-grid (V2G) transition: A critical and systematic review. *Environmental Research Letters, 13*(1), 013001.

Stacy, E. W. (1962). A generalization of the gamma distribution. *The Annals of Mathematical Statistics, 33*, 1187–1192.

Štogl, O., Miltner, M., Zanocco, C., Traverso, M., & Starý, O. (2024). Electric vehicles as facilitators of grid stability and flexibility: A multidisciplinary overview. *Wiley Interdisciplinary Reviews: Energy and Environment, 13*(5), e536.

Su, J., Lie, T. T., & Zamora, R. (2019). Modelling of large-scale electric vehicles charging demand: A New Zealand case study. *Electric Power Systems Research, 167*, 171–182.

Sun, K., Sarker, M. R., & Ortega-Vazquez, M. A. (2015). Statistical characterization of electric vehicle charging in different locations of the grid. In *2015 IEEE Power & Energy Society General Meeting* (pp. 1–5).

Sun, X., Li, Z., Wang, X., & Li, C. (2019). Technology development of electric vehicles: A review. *Energies, 13*(1), 90.

Sun, X., & Qiu, J. (2021). Hierarchical voltage control strategy in distribution networks considering customized charging navigation of electric vehicles. *IEEE Transactions on Smart Grid, 12*(6), 4752–4764.

Supponen, A., Rautiainen, A., Markkula, J., Mäkinen, A., Järventausta, P., & Repo, S. (2016). Power quality in distribution networks with electric vehicle charging-a research methodology based on field tests and real data. In *2016 eleventh international conference on ecological vehicles and renewable energies (EVER)* (pp. 1–11).

Szinai, J. K., Sheppard, C. J., Abhyankar, N., & Gopal, A. R. (2020). Reduced grid operating costs and renewable energy curtailment with electric vehicle charge management. *Energy Policy, 136*, 111051.

Tang, D., & Wang, P. (2015). Probabilistic modeling of nodal charging demand based on spatial-temporal dynamics of moving electric vehicles. *IEEE Transactions on Smart Grid, 7*(2), 627–636.

Tao, Y., Qiu, J., Lai, S., Sun, X., & Zhao, J. (2022). A data-driven agent-based planning strategy of fast-charging stations for electric vehicles. *IEEE Transactions on Sustainable Energy, 14*(3), 1357–1369.

Tayri, A., & Ma, X. (2025). Grid impacts of electric vehicle charging: A review of challenges and mitigation strategies. *Energies, 18*(14), 3807.

ten Have, S. Y., Gkiotsalitis, K., & Geurs, K. T. (2020). Investigating the future of ultrafast charging: A choice experiment in The Netherlands. *World Electric Vehicle Journal, 11*(4), 70.

Tian, H., Tzelepis, D., & Papadopoulos, P. N. (2021). Electric Vehicle charger static and dynamic modelling for power system studies. *Energies, 14*(7), 1801.

Tian, J., Liu, H., Gan, W., Zhou, Y., Wang, N., & Ma, S. (2025). Short-term electric vehicle charging load forecasting based on TCN-LSTM network with comprehensive similar day identification. *Applied Energy, 381*, 125174.

Tikka, V., Haapaniemi, J., Räisänen, O., & Honkapuro, S. (2022). Convolutional neural networks in estimating the spatial distribution of electric vehicles to support electricity grid planning. *Applied Energy, 328*, 120124.

Tong, X., Ma, Q., Tang, K., Liu, H., & Li, C. (2019). Influence of electric vehicle access mode on the static voltage stability margin and accommodated capacity of the distribution network. *The Journal of Engineering, 2019*(16), 2658–2662.

U.S. Department of Energy. (2024). Impact of electric vehicles on the grid. Congressional report. Retrieved from https://www.energy.gov/sites/default/files/2024-10/Congressional%20Report%20EV%20Grid%20Impacts.pdf

Ul-Haq, A., Cecati, C., Strunz, K., & Abbasi, E. (2015). Impact of electric vehicle charging on voltage unbalance in an urban distribution network. *Intelligent Industrial Systems, 1*(1), 51–60.

Vandael, S., Claessens, B., Ernst, D., Holvoet, T., & Deconinck, G. (2015). Reinforcement learning of heuristic EV fleet charging in a day-ahead electricity market. *IEEE Transactions on Smart Grid, 6*(4), 1795–1805.

Van Fan, Y., Jiang, P., Klemeš, J. J., & Ocłoń, P. (2022). Minimum environmental footprint charging of electric vehicles: A spatiotemporal scenario analysis. *Energy Conversion and Management, 258*, 115532.

Van Kriekinge, G., De Cauwer, C., Sapountzoglou, N., Coosemans, T., & Messagie, M. (2021). Day-ahead forecast of electric vehicle charging demand with deep neural networks. *World Electric Vehicle Journal, 12*(4), 178.

Vapnik, V. (2013). *The nature of statistical learning theory*. Springer Science & Business Media.

Varshney, S., Panda, K. P., Shah, M., Srinivas, B. A., Deshmukh, A., Choudhary, K. K., Bajaj, M., Prokop, L., & Rubanenko, O. (2025). Novel control strategies for electric vehicle charging stations using stochastic modeling and queueing analysis. *Scientific Reports, 15*(1), 21391.

Varshney, S., Srinivas, B. A., Gupta, M., Shah, M., & Bavisi, A. (2024). Stochastic modeling of multiple-server charging stations for electric vehicle networks using feedback strategies: A queueing-theoretic approach. *International Journal of Thermofluids, 24*, 100859.

Virta Global. (2024). Vehicle-to-grid (V2G): Everything you need to know. Retrieved from https://www.virta.global/vehicle-to-grid-v2g

Visakh, A., & Selvan, M. P. (2023). Analysis and mitigation of the impact of electric vehicle charging on service disruption of distribution transformers. *Sustainable Energy, Grids and Networks, 35*, 101096.

Wan, Y., Cao, W., & Wang, L. (2019). A prediction method for EV charging load based on fuzzy inference algorithm. In *2019 Chinese control conference (CCC)* (pp. 2803–2808).

Wang, D., Ge, Y., Cao, J., Lin, Q., & Chen, R. (2023). Charging load forecasting of electric vehicles based on sparrow search algorithm-improved random forest regression model. *The Journal of Engineering, 2023*(6), e12280.

Wang, D., Liao, F., Gao, Z., & Tian, Q. (2022). Analysis of activity duration-related charging behavioral responses of electric vehicle travelers to charging services. *Transport Policy, 123*, 73–81.

Wang, Y., Yao, E., & Pan, L. (2021). Electric vehicle drivers' charging behavior analysis considering heterogeneity and satisfaction. *Journal of Cleaner Production, 286*, 124982.

Wang, Z., Jochem, P., & Fichtner, W. (2020). A scenario-based stochastic optimization model for charging scheduling of electric vehicles under uncertainties of vehicle availability and charging demand. *Journal of Cleaner Production, 254*, 119886.

Wu, Y., Cong, P., & Wang, Y. (2023). Charging load forecasting of electric vehicles based on VMD–SSA–SVR. *IEEE Transactions on Transportation Electrification, 10*(2), 3349–3362.

Xiang, Y., Hu, S., Liu, Y., Zhang, X., & Liu, J. (2019). Electric vehicles in smart grid: A survey on charging load modelling. *IET Smart Grid, 2*(1), 25–33.

Xiao, Y., Qian, B., Ou, J., Lin, W., Hu, H., Lin, X., … Tang, J. (2025). Optimized scheduling of integrated energy systems considering electric vehicles and carbon emission trading. *IEEE Access, 13*, 74147–74162.

Xin, F., Yang, X., Beibei, W., Ruilin, X., Fei, M., & Jianyong, Z. (2024). Research on electric vehicle charging load prediction method based on spectral clustering and deep learning network. *Frontiers in Energy Research, 12*, 1294453.

Xiong, Y., Chu, C. C., Gadh, R., & Wang, B. (2017). Distributed optimal vehicle grid integration strategy with user behavior prediction. In *2017 IEEE Power & Energy Society General Meeting* (pp. 1–5).

Xydas, E., Marmaras, C., Cipcigan, L. M., Jenkins, N., Carroll, S., & Barker, M. (2016). A data-driven approach for characterising the charging demand of electric vehicles: A UK case study. *Applied Energy, 162*, 763–771.

Xydas, E. S., Marmaras, C. E., Cipcigan, L. M., Hassan, A. S., & Jenkins, N. (2013). Forecasting electric vehicle charging demand using support vector machines. In *2013 48th International Universities' Power Engineering Conference (UPEC)* (pp. 1–6).

Yacoub, M. D., Bautista, J. V., & de Rezende Guedes, L. G. (1999). On higher order statistics of the Nakagami-m distribution. *IEEE Transactions on Vehicular Technology, 48*(3), 790–794.

Yan, J., Zhang, J., Liu, Y., Lv, G., Han, S., & Alfonzo, I. E. G. (2020). EV charging load simulation and forecasting considering traffic jam and weather to support the integration of renewables and EVs. *Renewable Energy, 159*, 623–641.

Yan, Q., Lin, H., Li, J., Ai, X., Shi, M., Zhang, M., & Gejirifu, D. (2022). Many-objective charging optimization for electric vehicles considering demand response and multi-uncertainties based on Markov chain and information gap decision theory. *Sustainable Cities and Society, 78*, 103652.

Yang, J., Dong, J., Zhang, Q., Liu, Z., & Wang, W. (2018). An investigation of battery electric vehicle driving and charging behaviors using vehicle usage data collected in Shanghai, China. *Transportation Research Record, 2672*(24), 20–30.

Yang, X., Yun, J., Zhou, S., Lie, T. T., Han, J., Xu, X., Wang, Q., & Ge, Z. (2025a). A spatiotemporal distribution prediction model for electric vehicles charging load in transportation power coupled network. *Scientific Reports, 15*(1), 4022.

Yang, X., Zhang, L., & Han, X. (2025b). Electric vehicle charging load forecasting method based on improved long short-term memory model with particle swarm optimization. *World Electric Vehicle Journal, 16*(3), 150.

Yang, X., Zhuge, C., Shao, C., Huang, Y., Tang, J. H. C. G., Sun, M., Wang, P., & Wang, S. (2022). Characterizing mobility patterns of private electric vehicle users with trajectory data. *Applied Energy, 321*, 119417.

Yu, Z., Yang, C., Wang, Q. (2025). The impact of large-scale EV charging on the real-time operation of distribution systems: A comprehensive review. *arXiv preprint arXiv:2507.21759.*

Yunus, K., De La Parra, H. Z., & Reza, M. (2011). Distribution grid impact of plug-in electric vehicles charging at fast charging stations using stochastic charging model. In *Proceedings of the 2011 14th European Conference on Power Electronics and Applications* (pp. 1–11).

Zainab, F., Naz, K., Mehmood, K. K., Bukhari, S. B. A., Wadood, A., Khalid, H. A., & Park, H. (2023). An optimal joint planning of DGs and electric vehicle charging stations in grid-connected and islanded microgrids. *IET Renewable Power Generation, 17*(7), 1623–1634.

Zamee, M. A., Han, D., Cha, H., & Won, D. (2023). Self-supervised online learning algorithm for electric vehicle charging station demand and event prediction. *Journal of Energy Storage, 71*, 108189.

Zhang, J., Liu, C., & Ge, L. (2022). Short-term load forecasting model of electric vehicle charging load based on MCCNN-TCN. *Energies, 15*(7), 2633.

Zhang, J., Yan, J., Liu, Y., Zhang, H., & Lv, G. (2020). Daily electric vehicle charging load profiles considering demographics of vehicle users. *Applied Energy, 274*, 115063.

Zhang, K., Xu, Y., & Sun, H. (2024a). Bilevel optimal coordination of active distribution network and charging stations considering EV drivers' willingness. *Applied Energy, 360*, 122790.

Zhang, Q. S., & Zhu, S. C. (2018). Visual interpretability for deep learning: A survey. *Frontiers of Information Technology & Electronic Engineering, 19*(1), 27–39.

Zhang, T., Huang, Y., Liao, H., & Liang, Y. (2023). A hybrid electric vehicle load classification and forecasting approach based on GBDT algorithm and temporal convolutional network. *Applied Energy, 351*, 121768.

Zhang, X., Chan, K. W., Li, H., Wang, H., Qiu, J., & Wang, G. (2021). Deep-learning-based probabilistic forecasting of electric vehicle charging load with a novel queuing model. *IEEE Transactions on Cybernetics, 51*(6), 3157–3170.

Zhang, X., Wang, Q., Bi, X., Li, D., Liu, D., Yu, Y., & Tse, C. K. (2024b). Mitigating cascading failure in power grids with deep reinforcement learning-based remedial actions. *Reliability Engineering & System Safety, 250*, 110242.

Zhao, S., Blaabjerg, F., & Wang, H. (2020a). An overview of artificial intelligence applications for power electronics. *IEEE Transactions on Power Electronics, 36*(4), 4633–4658.

Zhao, Y., Chen, X., Liu, P., Nielsen, C. P., & McElroy, M. B. (2025). Future ultrafast charging stations for electric vehicles in China: Charging patterns, grid impacts and solutions, and upgrade costs. *Engineering, 48*, 309–322.

Zhao, Z., Zhang, L., Yang, M., Chai, J., & Li, S. (2020b). Pricing for private charging pile sharing considering EV consumers based on non-cooperative game model. *Journal of Cleaner Production, 254*, 120039.

Zheng, Y., Keith, D. R., Wang, S., Diao, M., & Zhao, J. (2024). Effects of electric vehicle charging stations on the economic vitality of local businesses. *Nature Communications, 15*(1), 7437.

Zhou, D., Guo, Z., Xie, Y., Hu, Y., Jiang, D., Feng, Y., & Liu, D. (2022). Using Bayesian deep learning for electric vehicle charging station load forecasting. *Energies, 15*, 6195.

Zhu, J., Yang, Z., Guo, Y., Zhang, J., & Yang, H. (2019). Short-term load forecasting for electric vehicle charging stations based on deep learning approaches. *Applied Sciences, 9*, 1723.

Demand-side uncertainty and planning for flexibility provision

8

Seyed Pouria Miri[a], *Mohammad Hossein Shaabani*[a],
Hamid Reza Pilehvar Javid[a],
Mohammad Hossein Alizadeh Roknabadi[b], *Hamid Reza Baghaee*[c]
and Gevork B. Gharehpetian[a]
[a] Department of Electrical Engineering, Amirkabir University of Technology, Tehran, Iran, [b] Department of Aerospace Engineering, Amirkabir University of Technology, Tehran, Iran, [c] Department of Electrical Engineering, Tarbiat Modares University, Tehran, Iran

8.1 Introduction

The increasing penetration of renewable energy resources, electrification of demand, and decentralization of power generation have introduced unprecedented levels of uncertainty into modern power systems. Demand side is one of the most critical contributors to this uncertainty, where consumption patterns across residential, commercial, industrial, and agricultural sectors are becoming increasingly volatile and less predictable (Golmohamadi, 2022). Demand-side uncertainty not only complicates short-term operational planning but also has long-term implications for capacity expansion, grid stability, and investment strategies. As such, the ability to effectively plan for and harness flexibility on the demand side has emerged as a central challenge for sustainable energy system development.

Historically, Demand-Side Management (DSM) and Demand Response (DR) programs have been implemented to encourage consumers to modify consumption behavior in response to price signals or system constraints (Nasir et al., 2021). These programs have proven essential for balancing supply and demand, mitigating peak loads, and improving system efficiency. However, the growing complexity of electricity systems, driven by high renewable energy penetration and prosumer participation, requires more advanced approaches that explicitly account for stochasticity in demand profiles (Jabir et al., 2018). Conventional deterministic planning models are increasingly insufficient, as they fail to capture the probabilistic nature of load variations and the interdependencies across sectors. Demand-side uncertainty arises from numerous sources, including occupant behavior, economic cycles, local weather events, and technological adoption such as electric vehicles (EVs) and smart appliances (Mosquera-López et al., 2024). These uncertainties challenge the reliability and economic performance of power systems by introducing fluctuations in load profiles, increasing forecasting errors, and complicating peak demand estimation. Without adequately accounting for this variability, system operators and planners' risk under- or over-investing in infrastructure, leading to inefficiencies and reliability issues.

Stochastic Planning and Modeling for Energy Systems: Methods, Applications, and Developments.
DOI: https://doi.org/10.1016/B978-0-443-45298-7.00006-9

Recent surveys and technical reviews have highlighted that demand-side flexibility must be understood as a multisectoral resource. Flexibility potentials differ significantly across residential, industrial, commercial, and agricultural domains, each influenced by distinct operational constraints, behavioral patterns, and enabling technologies (Bakare et al., 2023). For instance, residential loads offer flexibility primarily through thermostatically controlled appliances and EVs, while industrial facilities may provide large-scale but process-sensitive flexibility. Recognizing these sectoral differences is essential to accurately assess the availability, cost, and reliability of demand-side flexibility. At the same time, digitalization and artificial intelligence (AI) are reshaping DSM and DR paradigms. The integration of advanced metering infrastructure (AMI), Internet of Things (IoT) devices, and data-driven optimization models enables real-time monitoring and intelligent decision-making on the demand side (Aoun et al., 2021; Mousa et al., 2024). Moreover, AI techniques such as machine learning (ML), deep reinforcement learning, and predictive analytics are increasingly applied to improve demand forecasting, detect anomalies, and automate flexibility activation. These innovations allow demand-side resources to actively participate in system balancing, reliability enhancement, and resilience improvement, thereby transforming passive consumers into active contributors of grid stability.

Recent advances in distribution network modeling have highlighted the critical need to account for the relationship between demand uncertainty, load components, and system reconfiguration. Load behavior is strongly influenced by consumer type, voltage variations, and nonlinear consumption dynamics (Mahdavi & Jurado, 2025). Ignoring these aspects may lead to suboptimal flexibility allocation and inefficiencies in planning. A stochastic planning framework that integrates demand-side variability, sector-specific flexibility resources, digital intelligence, and market mechanisms is thus essential for ensuring secure, efficient, and sustainable power system operation. The rise of flexible markets provide an institutional framework for monetizing demand-side flexibility. Local and regional market mechanisms are being designed to incentivize industrial parks, prosumers, and communities to provide demand-side flexibility as a tradable service (Turdybek et al., 2024). Such markets not only enhance grid reliability but also unlock new revenue streams for consumers, encouraging deeper engagement with DSM and DR initiatives. However, effective market participation requires accurate characterization of demand-side variability, robust forecasting techniques, and fair valuation mechanisms for different types of flexible loads.

This chapter addresses these challenges by examining the multifaceted nature of demand-side uncertainty and its implications for flexibility planning. It begins with a detailed analysis of demand-side variability and its systemic impacts in the second section, followed by a discussion on the definition, importance, and quantification of flexibility in Section 8.3. Sections 8.4 and 8.5 explore strategies for enhancing flexibility through generation, transmission, and distribution planning, and assess the role of flexibility markets in enabling effective integration of demand-side resources. The chapter concludes with a forward-looking discussion on research gaps and future directions.

8.2 Demand-side variability

In recent years, greater attention has been directed toward the demand side of power systems, as increasingly variable consumption patterns have been observed across residential, commercial, industrial, and agricultural sectors. The demand side is no longer regarded as a static component of system planning, but rather as a dynamic and uncertain element that must be understood and managed. The variability of demand plays a pivotal role in shaping the flexibility requirements of modern energy systems. With increasing penetration of renewable energy sources, demand-side resources are increasingly leveraged to balance supply-demand mismatches and ensure operational reliability. This section reviews the sources of demand variability, their impacts on power systems, and strategies for managing demand-side uncertainty, drawing exclusively from recent studies.

8.2.1 Sources of variability in demand

Demand-side variability stems from the heterogeneity of consumers across sectors and from external drivers such as weather, climate, and social behaviors. A cross-sectoral survey has shown that flexibility can be unlocked from residential appliances, industrial processes, commercial facilities, and agricultural operations, but each sector exhibits distinct uncertainty profiles (Golmohamadi, 2022).

Residential loads are particularly sensitive to lifestyle patterns and environmental conditions. High-resolution smart meter studies have revealed daily and seasonal fluctuations in household usage, with weather variables such as temperature, rainfall, and sunshine duration significantly influencing hourly demand. For instance, in (Kang & Reiner, 2022), it has been demonstrated that while temperature effects on Irish households are relatively flat across the day, rainfall and sunshine duration exert stronger impacts on behavior, revealing time-dependent sensitivities. Similarly, (Rafiq et al., 2023) have highlighted strong diurnal variations in Dubai's households, reflecting occupancy and appliance use patterns in hot climates. Industrial and commercial loads, while less volatile, also display variability linked to production schedules and facility operations. Industrial demand varies with the flexibility of processes such as cement milling, smelting, and refinery operations, which are constrained by operational and safety requirements (Riaz & Mancarella, 2022). At the distribution grid level, the mix of industrial, residential, and commercial demand components shapes the aggregate variability profile (Mahdavi & Jurado, 2025). Agricultural loads also present variability linked to irrigation schedules, seasonal crop cycles, and livestock activities. Studies have emphasized the coincidence between irrigation demand and peak system load in hot climates, amplifying uncertainty. In greenhouse and livestock farms, heating, cooling, and ventilation systems further introduce variability tied to climatic fluctuations (Golmohamadi, 2022). It should also be mentioned that exogenous factors such as pandemics and climate change exacerbate demand variability. During COVID-19 restrictions, remote work shifted residential load peaks toward midday, altering system-wide consumption patterns (Ku et al., 2022). Long-term climate variability, including changes in temperature and precipitation, introduces further uncertainty in both load and electricity prices.

8.2.2 Impact of variability on power system

Demand variations in power systems are driven by both demand-side dynamics and supply-side uncertainties, and they are becoming increasingly critical as renewable energy penetration grows. Demand variability directly influences system reliability, flexibility requirements, and market operations. Fluctuating consumption can lead to congestion in distribution grids, increased reserve requirements, and challenges for renewable integration. Sudden changes in demand or renewable output cause frequency deviations. This problem is more severe in low-inertia systems, which are dominated by inverter-based renewable resources. Voltage instability and power quality issues are another challenge that comes with variability in power systems. As conventional synchronous generation is replaced by renewables with limited inertia and fault ride-through capability, both transient and small-signal stability challenges rise (Impram et al., 2020). Additionally, variability increases the probability of overloading transmission and distribution assets, accelerating aging, reducing operational lifetimes, and increasing outages (Golmohamadi, 2022).

Variability can also lead to economic challenges. To balance variability, operators must deploy fast-start reserves or spinning reserves, which significantly raises costs. Furthermore, curtailment of renewable generation is another issue that comes with variability in demand and renewable output. This results in lowering system efficiency and undermining decarbonization goals. Variability also forces operators to use additional ancillary services, which in turn increases market expenditures. Another problem that is common when variability is considered is the complication of long-term system design. Without adequate metrics, planners risk either over-investment in flexibility, leading to uneconomic outcomes, or under-investment, causing reliability concerns.

8.2.3 Demand variability management

As variability in the power system, especially demand variability, can lead to many challenges in the operation of the power system, measures must be taken to reduce the impact of uncertain changes in various parameters throughout the system. Many approaches have been proposed to mitigate the impact of demand variability, and they can be categorized into three categories: Demand-side approaches, market-based approaches, and forecasting techniques. These strategies can effectively shape, shift, or stabilize electricity consumption and enhance the flexibility of the power system.

Demand-side approaches include all approaches that directly involve consumers in the process of demand variability management. These strategies play a crucial role in mitigating demand variability by enabling the power system to dynamically adapt to changing consumption patterns. Among these strategies, DR is one of the most prominent mechanisms. This strategy allows consumers to adjust or shift their electricity usage in response to grid conditions or price signals. There are different ways for implementing DR. It can simply be implemented through direct load control, where appliances are remotely controlled by utilities or through price-based programs, such as time-of-use (ToU) tariffs, real-time pricing (RTP), and critical peak pricing (CPP). These mechanisms incentivize consumers to reduce or shift consumption during peak hours (Boopathy et al., 2024). This leads to a flatter load curve and a reduction in the stress on generation and network assets. The integration of new technologies such as IoT further enhances the responsiveness and effectiveness of DR by enabling near real-time coordination between consumers and system operators (Rajesh et al., 2022). Additionally, energy storage systems (ESS) and flexible loads provide additional means to manage demand variability. Storage devices can absorb excess energy during low-demand periods and discharge it when consumption peaks, which gives them the ability to effectively decouple demand from real-time generation. Similarly, flexible loads can be intelligently scheduled to align with grid needs. They can consume energy during off-peak hours, which causes the flattening of the consumption curve. EVs, heating, ventilation, and air conditioning systems, and industrial processes are some of the main flexible loads (Sun et al., 2016). Interconnection of different sectors, such as electricity, transportation, and thermal systems, can also reduce the demand variability. Vehicle-to-grid (V2G) and power-to-heat systems are some examples of these emerging paradigms. Furthermore, the formation of virtual power plants (VPPs) enables coordinated control of distributed flexible resources, providing reliable and scalable demand-side flexibility services.

Market-based mechanisms point to all demand variability management that involves competition. It can manage demand variability management by providing economic incentives. The increasing integration of distributed energy resources (DERs) and flexible loads has prompted the evolution toward more dynamic and decentralized market structures. Price-based mechanisms encourage consumers to adjust their consumption patterns in response to temporal variations in electricity prices. Meanwhile, incentive-based programs, such as capacity payments and ancillary service participation, compensate consumers for providing measurable flexibility services. The introduction of local flexibility markets (LFMs) and peer-to-peer (P2P) trading platforms has further expanded opportunities for distributed flexibility, allowing prosumers, communities, and aggregators to actively participate in grid balancing and congestion management (Forouli et al., 2021).

Data-driven forecasting has emerged as a pivotal tool for mitigating the operational challenges caused by demand variability. With the increase in meter device integration in power systems and the availability of data, using data-driven techniques such as deep learning is possible for forecasting purposes. Accurate short-term and long-term demand forecasts enable system operators, aggregators, and market participants to anticipate fluctuations and allocate flexibility resources efficiently. Traditional statistical methods, such as regression models, have been widely used for load forecasting; however, their capacity to capture nonlinear and stochastic demand patterns is limited. In recent years, ML and AI techniques have significantly advanced forecasting accuracy by learning complex temporal and spatial dependencies in large datasets (Arnob et al., 2023). Despite the promise of data-driven techniques, challenges such as data privacy, model transparency, and generalization across regions must be addressed to fully realize the potential of data-driven forecasting in achieving a flexible power system.

8.3 Distribution system flexibility

Traditional distribution systems were designed for unidirectional power flows and predictable demand profiles. However, the ongoing transformation of power systems, driven by the increasing role of DERs, has brought new challenges for the reliable and economic operation of distribution networks. The variability and uncertainty associated with renewable energy sources, as well as the emergence of flexible demand and distributed storage, have led to more attention to the concept of flexibility in distribution systems as a critical enabler to ensure security of supply. In addition to supporting technical operation, distribution system flexibility plays an important role in shaping future electricity markets. By leveraging flexible resources, distribution system operators (DSOs) can defer grid reinforcements, manage congestion, and provide ancillary services locally. Consequently, understanding the concept of flexibility is essential for advancing both the technical and economic sustainability of modern power systems.

8.3.1 Flexibility definition

Flexibility in power systems is a new concept for which there is no single, universally accepted definition. The main reason for this matter is the wide range of perspectives and different application contexts of flexibility. Some definitions limit this concept to the momentary balance between energy generation and consumption, while others view it more broadly to include applications such as energy storage, demand management, voltage regulation, and power transmission between regions. Nevertheless, most definitions share a common core idea, which is that a flexible power system is able to immediately and economically react to unexpected changes in supply and demand (Gharibvand et al., 2024). In general, flexibility is defined as the ability of a system to adapt to variability and uncertainty in supply and demand in a reliable, economic, and efficient way. Table 8.1 presents a summary of various definitions of flexibility.

Flexibility can be categorized into four main categories based on physical and operational characteristics. Fig. 8.1 compares the time scope and locational scope of the four flexibility categories. This approach provides a more precise understanding of flexibility's specific applications (Gharibvand et al., 2024; Kaushik et al., 2022):

- Energy flexibility: The ability of a power system to balance energy supply and demand over long timeframes (hours to years) on a national or international scale, often by storing surplus energy.
- Power flexibility: The ability of the power system to make instantaneous adjustments (a second to tens of minutes) to generation and consumption to maintain frequency stability.
- Voltage flexibility: The ability of the power system to keep voltage levels within a specific range in a timescale of seconds to minutes to prevent extreme fluctuations.
- Transfer capacity flexibility: The ability of the power system to move energy between different regions to reduce transmission line congestion.

Table 8.1 Different definitions of flexibility in power system.

References	Definition
Peltokeo et al. (2023)	The ability to adapt the planned development of the power system, quickly and at reasonable cost, to any change, foreseen or not, in the conditions which prevailed at the time it was planned.
IEA (2018)	Flexibility expresses the extent to which a power system can modify electricity production or consumption in response to variability, expected or otherwise.
Hussain et al. (2023)	A power system that can manage variability and uncertainty in both generation and demand while maintaining a satisfactory level of reliability at a reasonable cost over different time horizon.
Rahman et al. (2024)	The ability to adapt to dynamic and changing conditions, for example, balancing supply and demand by the hour or minute, or deploying new generation and transmission resources over a period of years.
Hadi et al. (2022)	Flexibility stands for a proficiency that consistently and lucratively handles the net-load forecast errors through different timescales.
IEA (2019)	Power system flexibility is defined as the ability of a power system to reliably and cost-effectively manage the variability and uncertainty of demand and supply across all relevant timescales, from ensuring instantaneous stability of the power system to supporting long-term security of supply.

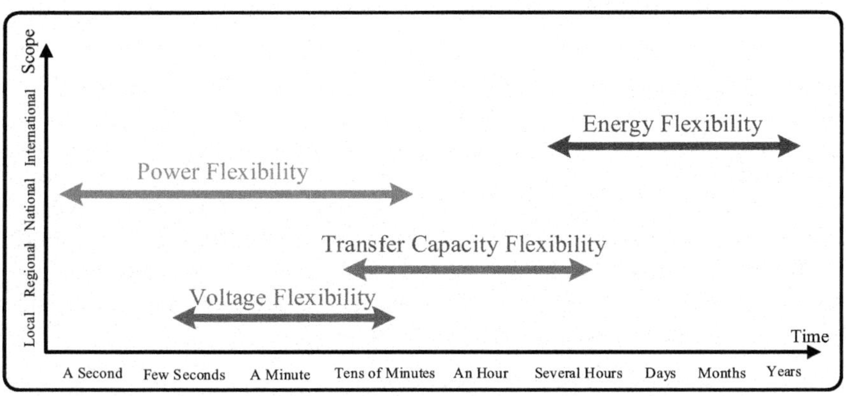

Figure 8.1 Comparison of the four flexibility categories. The horizontal axis shows the time span of flexibility category while vertical axis shows its locational scope.

8.3.2 Importance of flexibility

As the integration of variable renewable energy sources like wind and solar into power systems increases, the concept of flexibility is receiving more attention in recent years. These inherently intermittent and unpredictable sources introduce significant fluctuations and uncertainty into the grid's operation. This poses new challenges for maintaining the real-time balance between supply and demand and ensuring system stability. As a result, flexibility has transformed into a critical necessity for modern and future power systems, enabling them to safely and reliably accommodate the ongoing energy transition.

Variable energy sources, such as wind and solar, have highly variable generation due to their dependence on weather conditions. These constant and unpredictable changes on the supply side require flexible resources that can quickly step in to balance supply and demand. Severe fluctuations can occur in the absence of sufficient flexibility, which in turn leads to grid instability in frequency and voltage. In other words, flexibility can act like a tool for variability and uncertainty management. Furthermore, flexibility can help in ensuring system stability and reliability. Proliferation of variable energy sources is transforming the power system into a stochastic system, and flexibility allows the system to maintain stability despite these fluctuations and prevent widespread blackouts (Yang et al., 2020). At times when renewable energy generation reaches its maximum, the power system may not be able to absorb all the energy produced. In such situations, without sufficient flexibility, grid operators are forced to reduce or curtail the output from these sources. This results in the waste of clean energy and a decrease in the economic return on investments. Flexibility helps the system store this surplus energy or direct it toward flexible loads (Lu et al., 2018). Additionally, flexibility acts as a key enabler for the transition to a low-carbon energy system. By providing the necessary mechanisms to absorb more renewable energy, flexibility paves the way for reducing reliance on fossil fuels and achieving climate goals (Dominković et al., 2020). Fig. 8.2 shows how flexibility can ensure balance in a power system.

8.3.3 Flexibility resources

To provide flexibility in a power system, diverse resources are available, which can be categorized based on different perspectives such as operational time (short-term, mid-term, long-term), scale (small, medium, large), location in the grid (generation, grid, demand), and physical characteristics. Considering flexibility planning, flexibility resources can be divided into five categories to allow for separate planning and development within each category. These categories are generation-side, grid-side, demand-side, ESSs, and policy and regulations.

8.3.3.1 Generation-side

In this category, any power generation unit that can vary its output to balance the supply and demand in the grid is included. This means that everything from conventional power plants to controllable renewable energy sources can be regarded as

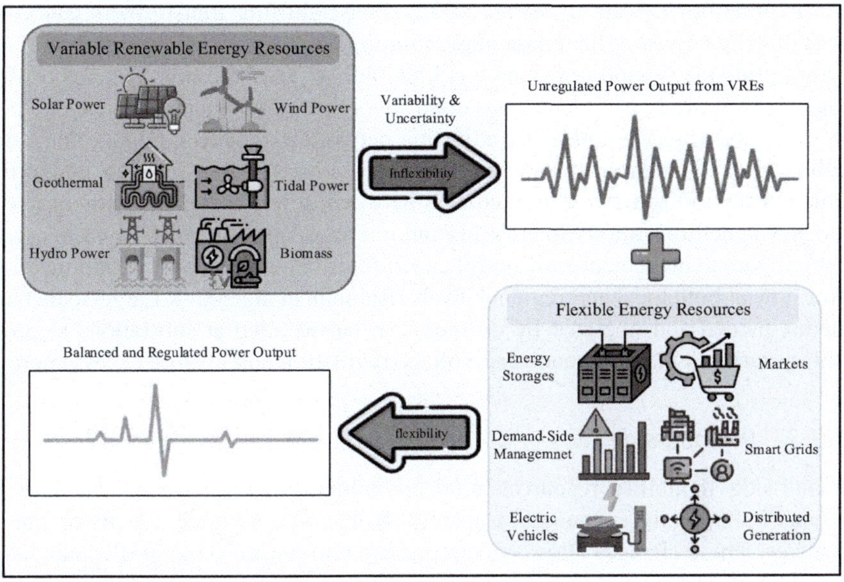

Figure 8.2 The importance of flexibility in providing energy balance. Flexibility sources are used to reduce the variability of renewable energies.

generation-side flexibility resources. It is important for these flexibility resources to have fast ramp-up and ramp-down rates and the ability to start and shut down quickly due to the instantaneous need for power balance in the grid. In flexibility planning, the goal is to optimize the performance of these resources to meet grid fluctuations while maintaining the economic and technical efficiency of the system. Thermal power plants have a limited role in flexibility optimization due to their long startup times and high energy consumption for start-stop cycles. However, with modifications, they can participate in flexibility by offering deep peak load regulation and by using gas-fired units (Babatunde et al., 2020). Hydropower plants can provide flexibility, but their output is affected by seasonal and geographical restrictions. Centralized wind power plants and solar power plants equipped with efficient control mechanisms can provide good flexibility under certain conditions (Kaushik et al., 2022). Furthermore, distributed generation units like micro-turbines and diesel generators can produce energy during times of generation shortage, which improves power and voltage flexibility.

8.3.3.2 Grid-side

Grid-side flexibility resources include the physical equipment and infrastructure of the power system that can be used to manage power flow and enhance grid stability. These flexibility resources influence how power is transmitted and distributed throughout the network. As the penetration of renewable energy sources increases, managing line congestion and voltage fluctuations becomes critical. Grid-side flexibility resources help system operators manage the grid optimally by providing dynamic and fast-acting control capabilities (Kaushik et al., 2022). Phase-shifting transformers can control power flow by adjusting the phase angle to mitigate congestion in transmission lines. Another grid side flexibility resource is Flexible AC Transmission Systems (FACTS) devices, which are power electronics controllers that can dynamically manage power flow and improve grid stability by adjusting network parameters. High-voltage direct current (HVDC) lines also enable long-distance power transmission and provide significant flexibility for power exchange between different regions without needing frequency synchronization. Smart grids and microgrids are advanced systems providing intelligent management and control capabilities that can enable the optimization of power flow at both local and regional levels (Rahman et al., 2024). Large-scale battery systems, owned and operated by utilities, can be installed at substations to provide ancillary services like frequency and voltage regulation and to store excess energy.

8.3.3.3 Demand-side

Demand-side flexibility resources refer to voluntary or mandatory changes in a customer's electricity consumption pattern in response to price signals or financial incentives. These changes allow system operators to optimize the grid's load, leading to a reduction in the need for investment in new infrastructure. Using these flexibility resources, customers can reduce their consumption during peak hours or shift it to off-peak periods. Demand-side flexibility resources are more important in environments with high penetration of variable renewable energy sources, as they can play a vital role in balancing fluctuations in energy production and ensuring stable grid operation (Golmohamadi et al., 2024). DR is a type of demand-side flexibility source, and it refers to a change in electricity use by customers in response to real-time electricity prices or incentive signals from grid operators. The main goal of DR is to reduce demand during periods of high load or grid emergencies (Kaushik et al., 2022). DR is a part of a broader concept called DSM, referring to planned actions to encourage customers to change their electricity consumption profile over time (Li et al., 2021). EVs can act as mobile energy storage resources, and using the V2G capability, they are able to inject power back into the grid during peak hours after they are charged during off-peak hours. VPPs are another demand-side flexibility resource that helps manage the complexities of modern grids. A VPP is a software and control system that combines many small DERs into a single, coordinated unit. In this way, a VPP can provide essential grid services, such as balancing supply and demand, much like a conventional power plant (Wang & Wu, 2021). Electrical devices or processes that can be adjusted or shifted by a power system operator to manage grid conditions are called controllable loads. They are a key tool for DSM as they can be used to balance supply and demand, reduce the need for new power plants, and improve grid stability (Li et al., 2016).

8.3.3.4 Energy storage systems

ESSs are recognized as one of the most important sources of flexibility in modern power systems. These systems have the unique ability to absorb and store electrical energy during one period and then release it in another. ESS can provide various services, including frequency and voltage regulation by responding quickly to small grid fluctuations, and peak shaving by charging during low-cost, off-peak hours and injecting power back into the grid during high-demand periods (Gharibvand et al., 2024). Electromechanical storage systems store electrical energy through chemical reactions or electric fields. Battery storage systems and supercapacitors fall into this category. Batteries are better suited for higher energy applications, while supercapacitors are used to provide very high power for short durations (Detka & Górecki, 2023). Mechanical storage systems convert electrical energy into potential or kinetic energy for storage. Pumped-hydro energy storage is an example of this type of storage, using surplus electricity to pump water to a higher reservoir and releasing it to generate power when needed, making it highly efficient for large-scale, long-duration storage. A flywheel is another example that stores energy as kinetic energy in a rapidly rotating mass. Due to its very fast response time, a flywheel is ideal for providing fast frequency regulation and dynamic stability support for the grid (Mahmoud et al., 2020). Thermal storage and hydrogen storage are other forms of ESSs. The thermal storage method converts electrical energy into heat or cold, which is then stored in an insulated medium. On the other hand, hydrogen storage uses electrolysis to convert surplus electrical energy into storable hydrogen. The stored hydrogen can then be used to produce electricity via fuel cells or gas turbines when needed. This technology has the potential for storing energy for very long periods (Jin et al., 2021).

8.3.3.5 *Policy and regulation*

Flexibility can also be provided by a set of nonphysical resources, including market mechanisms, policies, and regulations. These frameworks create the necessary conditions and incentives for all physical resources from generation to demand to participate in providing flexibility. Market-based instruments provide financial signals to influence the behavior of market participants, and they include real-time electricity pricing, capacity markets, and ancillary service markets. Government policies and industry standards can also be influential in flexibility provision. For example, policies that promote the installation of smart meters can unlock new sources of flexibility. Linking different sectors, such as electricity, heat, and transport, can also be a flexibility enhancement method. Policies that promote a diverse mix of generation sources, including both conventional and a variety of renewables, can increase system flexibility by reducing reliance on any single source. Accurate forecasting of renewable generation and demand is a prerequisite for effective flexibility management. By improving these methods, grid operators can better anticipate supply-demand imbalances and plan for them.

8.3.4 Quantifying flexibility

Flexibility needs to be quantified to evaluate its status in the power system. The accurate quantification of flexibility can help operators and planners in many ways, such as properly planning future investments. Since there is no single, unified definition for flexibility, multiple indices have been introduced to measure it (Shepherd & Mohagheghi, 2024). Flexibility quantification means measuring the ability of a power system to respond to changes and uncertainties in supply and demand. Flexibility must be expressed in figures to assess the adequacy of flexibility sources and compare different flexibility options. Flexibility can be quantified using two methods, whether it can be quantified using electrical parameters or it can be expressed using costs.

Quantification of flexibility based on electrical parameters focuses on evaluating the technical capability of power systems and their components to respond to operational variations. This approach characterizes flexibility through measurable electrical parameters such as voltage, current, power, and energy. Typical parameters include ramping rates, reserve capacities, response times, energy storage capacities, and operational ranges of generation or load. For example, the ramp rate describes how fast a resource can respond to fluctuations, while reserve margins quantify the available power reserves that can be activated to provide flexibility. Advanced formulations may also incorporate spatial and temporal resolution, reactive power capabilities, or dynamic response characteristics, enabling a comprehensive view of frequency and voltage support potential. As an example, the metric expressed in Eq. (8.1) shows how much the peak demand decreases in percentage as flexibility sources are used. In this equation, $Peak_{Flex}$ is the peak demand with flexible loads, $Peak_{Ref}$ is the peak demand without flexible loads, and PR is the peak reduction percentage (Gharibvand et al., 2024).

$$PR = \left(1 - \frac{Peak_{Flex}}{Peak_{Ref}}\right) \times 100 \tag{8.1}$$

Another metric that can be in this category is the self-sufficiency metric, which is illustrated with Eq. (8.2). In this equation, SS shows a numerical evaluation of self-sufficiency, and it expresses the ability of the on-site generation to supply the on-site demand (Li et al., 2021).

$$SS = \frac{Daily\ generation\ directly\ consumed}{Net\ daily\ generation} \tag{8.2}$$

While electrical parameters capture the technical aspects of flexibility, cost-based quantification focuses on its economic evaluation. This approach determines how much it costs to provide a unit of flexibility or to procure it from various sources, such as ESSs, DR, or flexible generation. Flexibility cost quantification bridges the gap between technical capability and market realization, offering an economic perspective that supports efficient pricing, investment prioritization, and policy design for flexibility enhancement. Many metrics have already been proposed that express flexibility in terms of cost. The flexibility index, which is shown in Eq. (8.3) is a cost-based flexibility metric. In this equation, C_{Flex} is the cost when flexibility strategies are used, and C_{Base} is the baseline cost when flexibility is ignored (Li et al., 2021).

$$FI = 1 - \frac{C_{Flex}}{C_{Base}} \tag{8.3}$$

It should also be noted that long-term planning models further integrate investment costs, operational expenditures, and market revenues to assess the economic adequacy of system flexibility under different future scenarios. These flexibility metrics can be defined in various ways based on the needs of the particular study, and it depends on the researchers that how flexibility is integrated in their work.

8.4 Distribution system planning for flexibility

Distribution system planning is undergoing a fundamental transformation driven by the increasing penetration of DERs, EVs, and flexible demand technologies. This shift toward decentralized generation, stochastic renewable outputs, and active consumer participation leads to significant uncertainty and variability in system operation. In this context, adaptation of distribution networks to changing conditions needs the addition of the flexibility concept to the other key planning objectives. Planning for flexibility in distribution systems means the systematic identification, assessment, and allocation of flexible resources to enhance the capability of distribution systems to respond to real-time or forecasted variations in supply and demand. This expands the traditional planning scope beyond conventional reinforcement to include flexible assets such as ESSs, DR programs, and controllable distributed generation. Embedding flexibility considerations into the planning stage can have many benefits, such as deferring costly infrastructure upgrades, enhancing resilience to contingencies, and supporting the integration of renewable energy sources.

8.4.1 Flexibility enhancement using distributed energy resources

DERs have been introduced into the distribution systems as one of the most influential enablers of flexibility. These include small-scale generation units such as photovoltaic (PV) panels, wind turbines, microturbines, and fuel cells that operate close to the point of consumption. ESS also falls into the category of DERs, as they can supply demand in their discharging period. DERs are placed in a decentralized manner, which enhances the adaptability of power systems to local variations in supply and demand. As they have a controllable and modular nature, they can provide fast, localized responses to system disturbances, leading to flexibility and reliability enhancement throughout the distribution systems. Although DERs come with many benefits and flexibility provision capabilities, their planning in the distribution system is challenging.

DERs are important flexibility providers for the distribution system, but renewable-based DERs are sources of variability. PV and wind resources are inherently intermittent and weather-dependent, which adds uncertainty to distribution system operation. This problem can be mitigated by offering controllable active and reactive power support. Another way to reduce the challenges due to the intermittent nature of renewable-based DERs is to integrate them with ESSs. ESSs absorb excess generation during low-demand periods and inject it during peaks, effectively smoothing load curves and mitigating renewable intermittency. The combination of solar or wind systems with storage can lead to dispatchable units, offering both short-term and long-term flexibility. This means co-planning of renewable energy-based DERs and ESSs is mostly needed, which further complicates the planning process (Kalantar-Neyestanaki & Cherkaoui, 2021). To fully exploit the potential of DERs, the introduction of new concepts into the planning process might be needed. VPP is one of these new concepts that has become a promising strategy, combining diverse DERs, flexible loads, and storage systems into a single controllable entity that can participate in energy and ancillary service markets. As shown in Fig. 8.3, VPP acts as an interface between different DERs and the power grid. These new concepts can make the planning process more complicated, and advanced optimization algorithms are needed to find a near-optimal and satisfying planning solution (Maiz et al., 2022).

Distribution system expansion planning is a challenging optimization problem with various continuous and integer variables, linear and nonlinear constraints, and a different number of objectives (Miri & Gharehpetian, 2025). Traditional form of distribution system expansion planning needs a high computational power, and the addition of new assets such as DERs and new concepts like VPPs can further complicate this planning problem. To account for the variability of DERs, stochastic optimization-based approaches are required. As DERs are sources of flexibility, the planning process can quantify flexibility and add it to the objective function so the distribution system can become flexible and cost-effective, simultaneously. It must be noted that DER placement within the network influences congestion management and voltage stability. For this reason, technical constraints must be carefully considered while planning for flexibility enhancement using DERs. Economic evaluation frameworks should also be included to compare the cost-effectiveness of DER-based flexibility against traditional reinforcement investments. Response time of DERs is also an important matter to consider while planning for flexibility. As a flexible distribution system needs to respond to the changes occurring in the demand, it is important to know how fast the available DERs can react to the imbalance and return the distribution system to the normal state. It is vital to ensure that the grid infrastructure is adequate for a certain amount of DER integration. Planners must keep all these challenges in mind while increasing the flexibility of the distribution system, as flexibility enhancement may not always be economically or technically acceptable.

8.4.2 Flexibility provision using demand

DR and DSM have emerged as key enablers of power system flexibility, particularly in systems with high shares of variable renewable energy sources. While traditional flexibility has been supplied by dispatchable generation and network assets, DR and DSM unlock flexibility by focusing on reshaping electricity consumption. The system operator can actively influence demand patterns through control signals, price incentives, and contractual arrangements. This demand-side flexibility can support

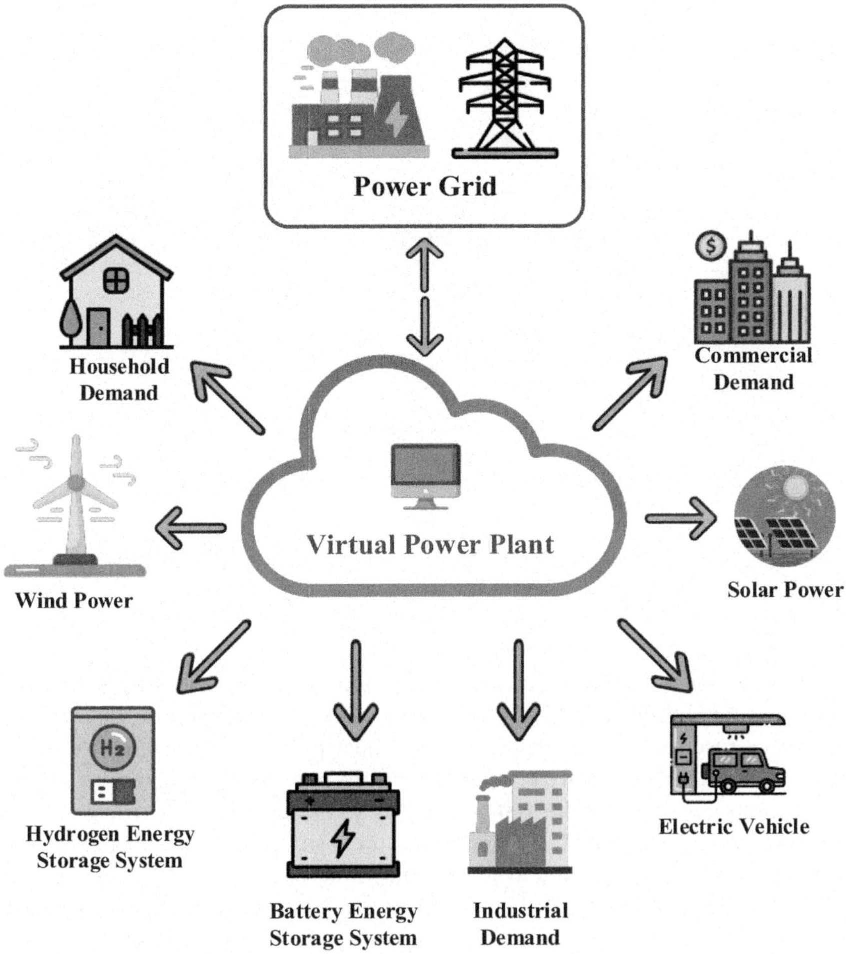

Figure 8.3 VPP role as the connection point of flexible sources and power grid.

balancing, congestion management, and even long-term planning objectives by reducing or deferring costly grid reinforcements. DR programs can be classified into price-based and incentive-based schemes. Price-based DR, such as ToU tariffs, RTP, and CPP, exposes consumers to dynamic prices that reflect system conditions. Consumers can shift consumption from high-price to low-price periods and indirectly provide peak-shaving capabilities. In incentive-based DR, using capacity programs, interruptible load contracts, and ancillary service participation, flexibility volumes can be called upon when needed. These incentive-based programs resemble conventional generation bids and are easier to integrate into system operation (Neves et al., 2015).

DSM offers a broader framework within which DR is embedded, encompassing energy efficiency, load shifting, peak clipping, and valley filling. In the context of

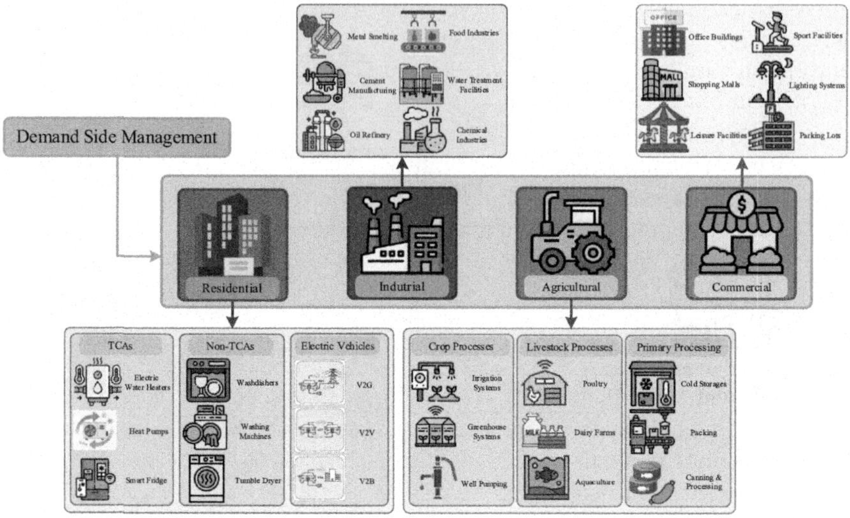

Figure 8.4 Different demand sectors for flexibility provision using DSM. *DSM*, Demand-side management.

flexibility provision, the most relevant DSM actions are those that reshape the temporal load profile without adversely affecting service levels. For instance, Industrial DSM can involve rescheduling batch processes, adjust production rates, or using on-site storage to decouple electricity consumption from real-time production requirements. These actions can enhance the system's ability to mitigate variability and uncertainty in both demand and renewable generation. At the distribution level where a large share of DERs and flexible loads are available; the value of DR- and DSM-based flexibility is particularly pronounced. DR programs can relieve local network constraints, such as thermal overloads and voltage violations, reducing the need for traditional reinforcement. By shifting or curtailing demand in specific feeders or nodes during critical periods, hosting capacity for distributed renewables is increased, providing the possibility of more flexibility enhancement using DERs (Golmohamadi, 2022).

An important dimension of DR and DSM for flexibility provision is the heterogeneity of demand sectors. Residential flexibility is often characterized by a large number of small loads that require aggregation to become visible at the system level. Residential flexibility offers significant potential due to the proliferation of smart appliances, EVs, and heat pumps. Commercial buildings provide flexibility through HVAC systems, lighting, and thermal storage, usually under centralized building management systems. Industrial consumers can offer large and predictable flexibility via process scheduling and on-site generation. Effective planning for flexibility provision must consider these sectoral differences in load profiles and their responsiveness to price or control signals. Fig. 8.4 illustrates the variety of loads that can participate in DSM for flexibility provision. Furthermore, AMI, IoT platforms, and home energy management systems (HEMS), which enable real-time or near-real-time monitoring and control of end-use devices, can increase the value of DR- and DSM-based flexibility.

From a planning standpoint, DR and DSM introduce new decision variables and constraints that must be explicitly modeled when evaluating system expansion and flexibility needs. Planners must estimate the amount, timing, and reliability of demand-side flexibility under different tariff structures, contractual schemes, and technology adoption scenarios. For example, a planner may compare the cost of reinforcing a feeder against deploying DSM programs and automation that unlock sufficient local flexibility to defer reinforcement. Such analyses must account for uncertainties in participation rates, behavioral responses, and regulatory developments. Consumer behavior introduces significant variability in realized flexibility as participation depends on awareness, trust, comfort tolerance, and perceived benefits. Moreover, consumers may increase consumption after DR events, reducing the impact of flexibility gained by DR. Robust planning approaches must therefore consider conservative flexibility estimates or incorporate stochastic/robust optimization techniques to account for these challenges. Demand-side flexibility can complement energy storage, flexible generation, and network reconfiguration. For instance, combining building-level DR with behind-the-meter batteries and rooftop PV can provide more predictable and controllable flexibility profiles and enhance renewable utilization. In planning for flexibility provision, it is therefore important to treat DR and DSM not as isolated measures but as integral components of a multiresource flexibility portfolio that spans generation, network, and demand sides (Zakariazadeh et al., 2014).

8.4.3 Market-based planning for distribution system planning

Market-based planning has emerged as a complement to traditional network reinforcement, enabling DSOs to integrate flexibility in a cost-effective and scalable manner. As the distribution grid transitions from passive to active operation, the role of markets becomes important in valuing and coordinating flexibility sources. Market-based planning incorporates economic signals, uncertainty, and the strategic behavior of participants to determine optimal investments and operations. This shift is the outcome of the growing need to manage uncertainty from renewables, DERs, and prosumers in a dynamic environment.

The main characteristic of market-based flexibility planning is the use of price signals to influence when and where DERs operate. These price signals can originate from wholesale markets, local markets, or dynamic tariffs, each producing different implications for flexibility utilization. As shown by Calvillo et al. (2016), aggregators participating in day-ahead markets can significantly adjust consumption and generation schedules based on hourly prices, which affects both operational decisions and long-term investment strategies. The paper has also indicated that when aggregated DER capacity reaches substantial penetration, aggregators may influence clearing prices, requiring a price-maker model. Ignoring price-makers can lead to misleading planning outcomes, while smaller DER does not change market behavior with their actions. For flexibility planning, this distinction is essential as it determines how economic benefits and grid impacts are evaluated.

LFMs further expand planning options by providing DSOs with alternative, non-network solutions for congestion relief, voltage control, and peak load management.

In (de Lima et al., 2024), it has been highlighted that local energy markets enable DSOs to procure flexibility services from distributed resources instead of just relying on physical reinforcement. Through these markets, DERs can be contracted to alleviate operational constraints, which directly feed into long-term expansion planning by reducing or deferring infrastructure investment. Battery ESS represent one of the most significant enabling technologies in market-based planning. It has been emphasized in (Zhang et al., 2021) that coordinating battery planning with market-clearing mechanisms has enhanced system flexibility and has also maximized trading revenue for distributed participants. When integrated into planning models, ESSs can arbitrage price differences, store excess renewable generation, and participate in both local and wholesale markets. These capabilities alter optimal siting and sizing decisions in the planning process.

The interaction between DSOs, aggregators, prosumers, and market operators adds another layer of complexity to market-based planning. As multiple stakeholders participate in energy markets, coordinated decision-making becomes critical to avoid conflicts and ensure efficient resource utilization. Without coordination, actions in one market may counteract objectives in another. The literature suggests developing multilayered or hierarchical market mechanisms to ensure that local constraints are respected while flexibility resources pursue economic opportunities (de Lima et al., 2024).

By comparing the cost of flexibility provision through markets with the capital cost of new substations, feeders, or reconductoring, planners can identify the most cost-effective solutions. It has been highlighted in (de Lima et al., 2024) that DER and DR often offer superior cost efficiency when supported by market mechanisms that fairly compensate service provision. Additionally, market-based planning supports the integration of decentralized actors, including microgrids, energy communities, and prosumers. These actors not only consume but also produce and trade energy, enabling deeper flexibility provision. This behavioral layer proves that planning models are needed to reflect realistic market interactions.

Market-based planning integrates economic signals and agent behavior to produce more accurate and flexible distribution expansion strategies. By leveraging market-based mechanisms, planners can unlock the full potential of flexibility resources and schedule them in an efficient and competitive way. Market-integrated DERs not only reduce system costs but also improve reliability, support renewable integration, and defer expensive grid reinforcements. Consequently, market-based flexibility planning provides the possibility of managing uncertainty, coordinating decentralized assets, and achieving both economic and operational objectives.

8.4.4 Emerging flexibility sources planning in the distribution system

The transformation of distribution systems is increasingly driven by new categories of distributed resources, enabling DSOs to manage variability and uncertainty in a more efficient way. Among the most prominent developments are EVs equipped with V2G technology, sector-coupled energy systems that shift load between electricity and heat, and community-driven configurations such as microgrids and P2P energy exchanges. Utilizing these resources requires planning approaches that recognize their

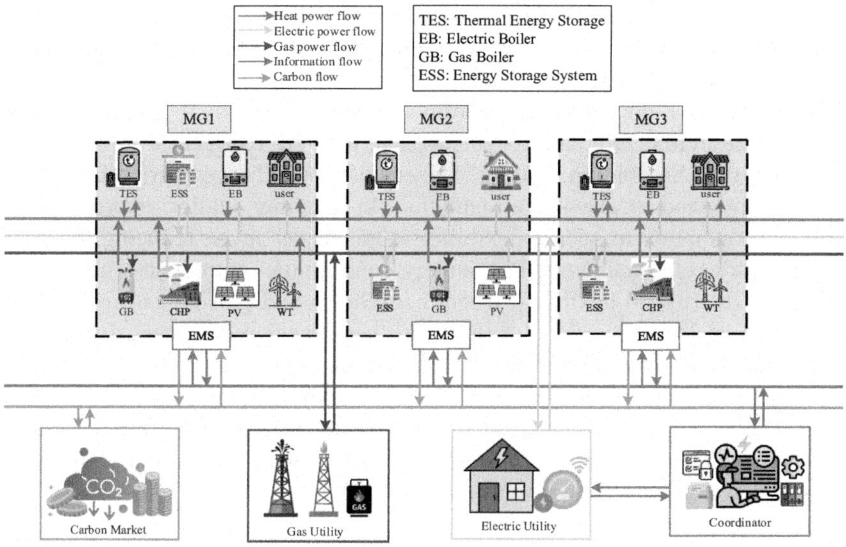

Figure 8.5 Multi-energy system and flexibility provision using power-to-heat concept. The possibility of generating electricity with other energy carriers increases the flexibility level of distribution system.

operational characteristics, uncertainty profiles, and potential interactions with local network constraints.

In V2G technology which is one of the fastest-growing flexibility opportunities within distribution networks, EVs equipped with bidirectional chargers can not only absorb energy during low-demand periods but also inject power back into the grid when needed. This technology can provide peak shaving, local voltage support, and reserve capabilities for the distribution system. As a large number of EVs are expected to be connected to residential and commercial feeders, the flexibility potential of V2G technology can be very high. However, planning the integration of V2G requires explicit modeling of user behavior, arrival and departure times, mobility needs, and battery degradation considerations. Determination of the installation place and capacity of charging infrastructure is another complexity of this planning problem, since clusters of EV chargers can introduce new congestion points or exacerbate voltage drops (Yu et al., 2022). Sector coupling, and particularly power-to-heat conversion, is emerging as a major source of flexibility in distribution systems. Heat pumps, electric boilers, and thermal storage systems can shift large amounts of energy consumption away from peak electrical load periods by utilizing the thermal inertia inherent in buildings or water-based storage. These technologies establish a link between the electricity and heating sectors, enabling multienergy optimization strategies that increase overall system efficiency. An example of a multienergy system is shown in Fig. 8.5 in which electricity and heat can change into each other to supply different demands. The flexibility potential of power-to-heat depends on local climatic conditions, building characteristics, and consumer comfort requirements. Planners must evaluate how shifting heating loads affects feeder loading profiles and the hosting capacity for renewable generation (Pastore, 2023).

Microgrids have the ability to coordinate diverse local energy resources under a single control architecture, and for this reason, they are recognized as key flexibility providers within distribution systems. They can support the main grid through islanding capabilities and local voltage regulation. In planning studies, microgrids enable distribution operators to access flexibility at the feeder level, reducing the need for costly network reinforcements. P2P energy communities introduce a new paradigm for utilizing distributed flexibility, where households and prosumers can trade energy and ancillary services directly. P2P energy exchanges can reduce peak demand locally and enhance the utilization of distributed generation. However, it should be mentioned that they also introduce uncertainty regarding trading behavior. As a result, planners must develop methodologies that incorporate behavioral models, transactive control frameworks, and regulatory constraints to ensure that P2P energy trading enhances distribution system operation (Ponnaganti et al., 2023).

8.5 Flexibility markets

With the rapid increase of renewable energy integration and the addition of their high variability to the power system complexity, traditional approaches of balancing supply and demand, which rely mostly on dispatchable generation, are no longer sufficient to address the challenges made by this variability. To overcome these challenges, the concept of flexibility has emerged, which can be provided through a wide range of sources. The flexibility market has gained increasing attention as a mechanism to efficiently allocate these sources. The main focus of flexibility markets is to provide short-term and localized services that address network constraints like congestion management. Flexible markets can reduce reliance on costly grid reinforcements and increase the integration of DERs by incentivizing participation of different actors. This market-based approach thus represents a promising pathway toward achieving a more adaptive, decentralized, and sustainable power system.

8.5.1 Importance of flexibility markets

Flexibility markets are emerging in the electric power sector that allow the buying and selling of flexibility. This means that the capability to adjust power generation or consumption on short notice to support grid operation can be provided in a market-based environment. These markets typically operate in parallel to traditional wholesale energy markets, and they focus on DERs at the distribution level. In a flexibility market, aggregators or other flexibility service providers can offer demand reductions, increases in local generation, or other grid support services, and DSOs procure these services to manage constraints such as local congestion or voltage issues. Flexibility market enables market-based utilization of distributed flexibility and promotes competition. Such markets unlock DER potential in an economically efficient way by providing a price signal for flexibility, helping DERs to sell services wherever they are most valued (Heinrich et al., 2021). As a result, DSOs are encouraged to use market mechanisms to integrate flexibility in grid planning and operation. This approach empowers prosumers to actively participate in energy markets, turning passive loads into active grid resources (Valarezo et al., 2021). A flexibility market serves as a platform for the exchange of grid support capabilities, enabling DSOs to maintain network reliability by purchasing flexibility from distributed actors with the purpose of avoiding more costly solutions.

One of the main benefits of flexibility markets emergence is the mitigation of the uncertainty of high renewable generation penetration and the variability of demand. By providing flexibility in an efficient way, the flexibility market can utilize flexibility sources to reduce the negative impact of uncertainty and variability. Traditional grid reinforcement strategies alone are no longer efficient nor economically sustainable under high variability and dynamic consumer behavior. Flexibility markets allow system operators to procure demand-side and distributed flexibility services in a cost-effective manner, reducing the need for excessive investment in physical assets. Furthermore, market-based coordination ensures that flexibility is deployed where it is most valuable, improving social welfare and reducing curtailment of renewables. This is important as distribution networks increasingly behave like active resources rather than passive demand sinks. Moreover, flexible markets support the optimal operation of highly distributed systems, addressing congestion, voltage control, and power quality issues at the distribution level. Ongoing European projects show that the adoption of flexibility markets leads to reduced curtailments, improved operational reliability, and creates incentives for investment in distributed storage and controllable assets, further strengthening the energy transition (Rebenaque et al., 2023). The benefits of flexible markets are summarized in Fig. 8.6.

8.5.2 Mechanisms of flexibility markets

Harnessing flexibility resources through market-based mechanisms is viewed as essential for phasing out carbon-intensive generation while preserving system stability and economic efficiency. In Europe, regulators and policymakers therefore promote market-based procurement as a primary way for DSOs to access flexibility services in distribution networks, complementing or replacing purely rule-based or tariff-based measures. Among the different approaches, LFMs have attracted particular attention because they enable a competitive, market-driven exchange of flexibility close to where constraints arise (Heinrich et al., 2021).

A LFM can be understood as a platform or mechanism through which DSOs procure flexibility from connected resources “often via aggregators” in a manner that runs in parallel to, but coordinated with, conventional wholesale energy markets. Through the market, DER owners or aggregators can offer services such as load reduction, load shifting, local generation curtailment, or increased consumption to relieve network constraints or balance local supply and demand (Jin et al., 2020). Fig. 8.7 illustrates the operational structure of an LFM, highlighting the interactions among the main actors “DSO, Balance Responsible Party (BRP), Aggregator, and Prosumers” in the procurement and activation of flexibility services (Olivella-Rosell et al., 2020).

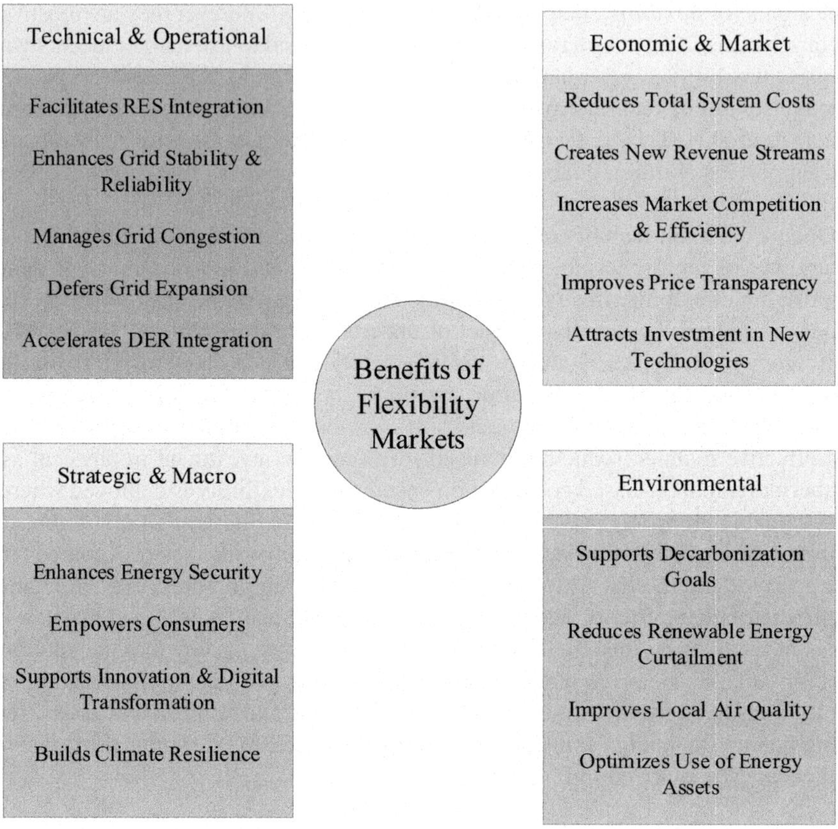

Figure 8.6 Benefits of flexibility markets.

DSOs are encouraged to use market-based solutions to meet flexibility needs and to integrate active customer participation into grid operation (Heinrich et al., 2021). Overall, a flexible market provides a structured environment in which flexibility becomes a tradable commodity or service, thereby enabling the transition from a traditionally passive distribution grid to an active network capable of accommodating high penetration of renewable and distributed resources.

Implementing a flexibility market in practice involves a sequence of coordinated stages that span from initial design to real-time operation, commonly structured into five phases. In the preparation phase, flexibility products and services are defined, participants are registered, and resources undergo technical prequalification; during this stage, the DSO or market operator specifies the types of flexibility needed, such as peak reduction, congestion relief, or voltage support, and sets the technical requirements that providers must meet (Valarezo et al., 2021). The forecasting phase follows, during which DSOs estimate where, when, and how much flexibility will be required by anticipating network constraints, while flexibility providers forecast

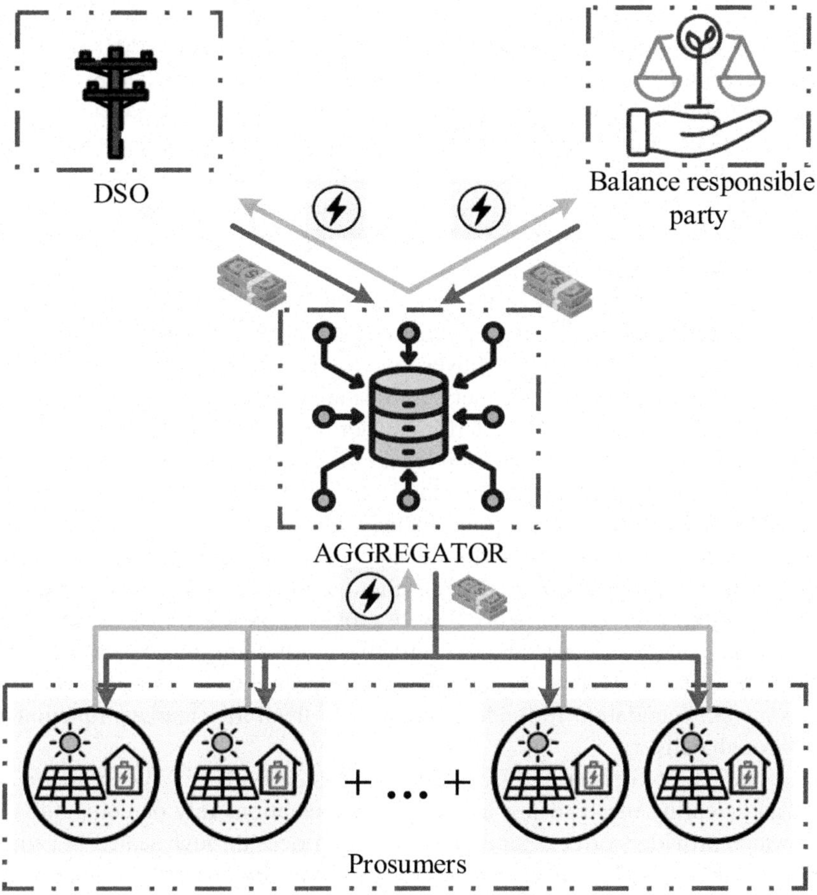

Figure 8.7 The framework of LFM. *LFM,* Local flexibility markets.

their available flexibility based on DER availability, user comfort, and commitments in other markets; the accuracy of these forecasts directly affects procurement efficiency. The process then moves to market operation, where providers submit bids specifying the quantity, location, timing, and price of the flexibility they can deliver, and where the DSO acting as the principal buyer may define required volumes or maximum willingness-to-pay; a clearing algorithm selects winning bids to maximize welfare or minimize procurement cost subject to distribution network constraints, producing a schedule of contracted services and a remuneration structure such as uniform pricing or pay-as-bid (Ma et al., 2024). As real time approaches, the monitoring and activation phase begins, and DSOs continuously assess grid conditions and activate contracted flexibility when operational thresholds are threatened, issuing activation signals instructing providers to adjust consumption or generation accordingly. Coordination with the TSO is essential whenever resources participate in both local and system-level markets to avoid conflicting dispatch decisions (Gerard et al., 2018). Finally, in the measurement and settlement phase, delivered flexibility is verified by comparing metered behavior to a baseline or reference schedule, followed by financial settlement in which providers are compensated for actual delivery and penalties are applied for deviations, making accurate baseline rules crucial to ensure that only genuine flexibility is remunerated (Valarezo et al., 2021). Table 8.2 summarizes the five key operational phases of flexibility market implementation, outlining their objectives, main activities, and the actors responsible for each stage.

These stages are implemented through dedicated market platforms operated by DSOs or independent entities, with structures ranging from one-sided procurement auctions where the DSO is the sole buyer to two-sided exchanges involving multiple buyers such as TSOs or large consumers. European pilots demonstrate both integrated and decentralized models as some platforms coordinate DSO–TSO procurement in a single clearing process, while others run local markets sequentially or in parallel with wholesale and balancing markets. For example, Sweden's CoordiNet project operates a day-ahead DSO flexibility auction prior to the national day-ahead market, whereas the InteGrid project runs a DSO platform independently alongside the TSO's balancing market (Valarezo et al., 2021). Across all configurations, the key actors remain the DSO as the primary buyer, flexibility service providers as sellers, and a market operator overseeing platform operation and settlement. Although responsibilities around qualification, dispatch control, and verification continue to evolve, this framework provides a solid foundation for understanding how flexibility markets function from design through delivery.

The economic performance of flexibility markets depends heavily on their auction formats and clearing mechanisms, as most implementations rely on competitive bidding in which providers offer flexibility at specified prices through sealed-bid auctions that clear based on the DSO's demand. Pricing is typically determined either through pay-as-bid rules, where each accepted bidder is paid its own offer price, or through uniform pricing, in which all accepted providers receive a single market-clearing price derived from the marginal accepted bid (Heinrich et al., 2021); while pay-as-bid offers budget predictability but encourages strategic mark-ups, uniform pricing generally improves efficiency by promoting truthful cost revelation. More advanced mechanisms, such as Vickrey–Clarke–Groves (VCG) auctions have also been explored for flexible markets, as they theoretically induce truthful bidding and maximize social welfare, though they require payments based on each bidder's marginal impact on system cost and may create revenue deficits. In general, market clearing is formulated as an optimization problem that selects winning bids to maximize welfare or minimize DSO procurement cost subject to distribution network constraints. For example, (Ma et al., 2024) model multimicrogrid flexibility trading as a social-welfare maximization problem that incorporates thermal limits and voltage constraints some implementations apply a postclearing adjustment or "collation" phase to ensure that initially cleared schedules remain feasible with detailed power-flow limits before contracts are finalized. This two-step approach helps maintain economic optimality while enforcing the physical realities of distribution networks, prompting advanced platforms to embed

Table 8.2 Key operational phases of a flexibility market.

Phase	Main purpose	Key activities	Responsible actors
Preparation	Establish market framework and ensure technical readiness	• Define flexibility products and services (congestion relief, peak reduction, voltage support) • Register participants and aggregators • Prequalify resources • Set technical and data requirements; standardize product definitions	DSO Market Operator Flexibility Service Providers (FSPs)
Forecasting	Anticipate flexibility needs and resource availability	• DSO forecasts locations and timing of constraints • FSPs forecast available flexibility based on DER capability, user comfort, and other market commitments	DSO FSPs
Market Operation	Procure flexibility through competitive bidding	• Providers submit bids with volume, location, timing, and price • DSO specifies demand or maximum willingness-to-pay • Market clearing algorithm selects winning bids under network constraints • Pricing established (uniform price or pay-as-bid)	Market Operator DSO FSPs

(*continued on next page*)

Table 8.2 Key operational phases of a flexibility market—cont'd

Phase	Main purpose	Key activities	Responsible actors
Monitoring and Activation	Deploy contracted flexibility when network needs arise	• Monitor real-time grid conditions • Activate contracted flexibility when thresholds are reached • Send activation signals • Coordinate with TSO to avoid conflicting activations	DSO TSO FSPs
Measurement and Settlement	Verify delivery and conduct financial settlement	• Compare metered performance to baselines • Confirm delivered flexibility • Apply payments and penalties • Ensure accurate and transparent settlement	Market Operator DSO FSPs

DER, distributed energy resource; *DSO,* distribution system operator; *TSO,* transmission system operators.

power-flow checks, contingency analyses, or iterative clearing to prevent new bottlenecks. Flexibility products themselves may be structured as either capacity reservations or activated energy trades. Ongoing research continues to refine auction rules and pricing designs to balance economic efficiency, incentive compatibility, computational practicality, and regulatory simplicity.

A central challenge in flexibility markets is their coordination with the broader power system and the activation of contracted flexibility. Because DER-based flexibility can provide services to both DSOs and TSOs, coordinated operation is essential to avoid situations in which an activation by one operator exacerbates constraints for the other. Coordination frameworks range from integrated approaches where a shared platform co-clears TSO and DSO need to decentralized approaches in which sequential or information-based coordination is used, such as DSO markets running before TSO balancing markets in some CoordiNet demonstrations (Gerard et al., 2018). In all cases, communication between DSOs and TSOs ensures visibility of resource commitments and prevents conflicting dispatch signals; market rules may also specify priority rights or compensation mechanisms when one operator's actions override the others. Once the market has cleared, the activation process begins as grid conditions approach thresholds that motivate flexibility procurement, prompting the DSO to issue activation signals instructing providers to implement contracted adjustments. Activation may be automated through the market platform or through the DSO's management systems and can occur at day-ahead, intraday, or real-time timescales, depending on the product. The DSO must also ensure that activation does not create new network issues, which may require adjusting volumes or activation patterns; in multimicrogrid configurations, coordinated secondary flexibility exchanges may allow consumption in one microgrid to increase while another decreases under DSO supervision (Ma et al., 2024). After delivery, the provided flexibility is quantified by comparing metered values to baselines or reference schedules, making accurate baseline design essential to market credibility (Valarezo et al., 2021). Settlement then remunerates providers according to the applicable pricing rule and may impose penalties for under-delivery. Together, these coordination and activation processes bridge market outcomes and physical grid operation, underscoring why flexibility markets are typically operated by, or tightly integrated with, DSOs, who must ensure reliable deployment of contracted flexibility. With effective TSO–DSO coordination, these markets allow distributed flexibility to contribute efficiently across both distribution and transmission levels.

8.5.3 Real cases of flexibility markets

The transition of flexibility markets from theoretical concepts to operational reality is a cornerstone of the modern power system's evolution. While the principles of these markets are established in previous sections, their true test lies in practical implementation. This section delves into the experiences of pioneering projects across Europe, which serve as living laboratories. By examining their diverse designs, operational challenges, and achieved outcomes, we can learn critical lessons that are invaluable for regulators, system operators, and market participants worldwide seeking to harness demand-side flexibility.

Table 8.3 Design choices for flexibility market.

Design choice	Description
Market integration	How is the flexibility market sequenced with existing energy markets?
Market operator role	Is the platform operated by a neutral third party?
Remuneration structure	Does the market include payments for reserving capacity, or only for delivered energy?
Product standardization	To what extent are the traded flexibility products standardized?
TSO-DSO coordination	What is the mechanism for coordination between TSOs and DSOs?
Cross-DSO coordination	How do different DSOs cooperate?

DSO, distribution system operator; *TSO*, transmission system operators.

Europe has emerged as a fertile ground for testing flexible markets, with initiatives spanning from fully commercial platforms to large-scale research and demonstration projects. These initiatives can be broadly classified into two complementary categories. The first category is grid-driven market platforms, which are marketplaces where the primary buyers are distribution and transmission system operators (DSOs and TSOs), procuring flexibility to address specific grid constraints like congestion and voltage violations. The second category is aggregator-led platforms focusing on aggregating small-scale DERs to create portfolios large enough to participate either in the grid-driven market or in existing wholesale and balancing markets. A comprehensive review by Valarezo et al. (2021) has highlighted that there is no single "correct" model for the flexibility market, but rather a spectrum of solutions tailored to local regulatory, technical, and market contexts. To make sense of the varied approaches, a structured framework is essential. (Schittekatte & Meeus, 2020) has provided a powerful lens through which to compare four pioneer projects: PicloFlex (United Kingdom), Enera (Germany), GOPACS (Netherlands), and NODES(Norway/Germany). Their analysis revolves around six critical design choices, which are shown in Table 8.3. This framework reveals key insights. For instance, while all four pioneers utilize a third-party market operator to ensure neutrality, their integration strategies differ profoundly. Enera and GOPACS are deeply integrated with intraday markets, while PicloFlex operates as a distinct tendering platform for DSOs. Similarly, the approach to TSO-DSO coordination ranges from the centralized model in GOPACS to more modular approaches in other projects.

Behind every successful flexibility market is a meticulously designed product. A product is not merely energy, and it can be a bundle of technical and commercial attributes that define the service being traded. Heilmann et al. (2020) offer a comprehensive framework for this design process, breaking it down into 23 parameters across technical and trading dimensions. By applying this framework to German pilot projects, they have demonstrated that even markets with similar goals exhibit significant variation in their product specifications, particularly in temporal attributes and technical performance requirements. Temporal attributes define the product's duration, lead time, and activation rules, distinguishing between firm energy blocks and flexible options. This underscores that product design is a strategic tool for matching the specific needs of the grid with the capabilities of different flexibility resources, from industrial loads to residential batteries.

The collective experience from these global pioneers converges on several overarching principles that are crucial for the launch of a flexibility market. The value of flexibility is intrinsically tied to its location within the grid architecture. Effective markets must incorporate locational signals to target constraints efficiently. Another thing to note is that seamless TSO-DSO coordination is a necessity to prevent conflicting actions and to unlock system-wide value, avoiding merely shifting congestion from one voltage level to another. Additionally, well-defined and standardized products, coupled with transparent market rules, are fundamental to attracting a diverse pool of flexibility service providers and building a liquid market. Table 8.4 summarizes five flexibility market projects and compares their objectives, capabilities, network coverage, and operational regions (Rebenaque et al., 2023; Schittekatte & Meeus, 2020).

8.6 Discussion and future directions

This chapter has provided an overview of demand-side uncertainty and the ensuing necessity for flexibility in power systems. While significant strides have been made in understanding and planning for these concepts, several avenues for future research and development remain open. First, regarding the quantification of flexibility, it is suggested that more comprehensive and appropriate metrics be developed for planning purposes. These metrics should holistically integrate all relevant planning parameters, including technical, economic, and reliability constraints, to provide a more accurate representation of system needs and capabilities. Furthermore, the portfolio of potential flexibility resources requires deeper examination. For instance, sector-coupling technologies, such as power-to-heat systems, have been widely studied for their energy conversion efficiency, but their specific potential for providing operational flexibility at the distribution system level is often overlooked. A similar research gap exists for other emerging sources like V2G applications and P2P energy trading, whose impacts on distribution network flexibility are not yet fully characterized or integrated into standard planning models. Concurrently, the mechanisms governing flexibility markets warrant further refinement. More sophisticated market designs are needed to create comprehensive, efficient, and transparent platforms for procuring and dispatching flexibility services. This development must be coupled with a critical analysis of existing market shortcomings. A key challenge to be addressed is the current interdependence between DSOs and TSOs, where a clearer demarcation of roles and responsibilities could enhance market functionality. From a planning perspective, the integration of flexibility should be elevated as a primary objective, rather than merely a constraint or a byproduct. Future planning models should be designed to explicitly maximize system flexibility alongside traditional objectives like cost minimization, thereby enhancing resilience against a broader spectrum of uncertainties. Ultimately,

Table 8.4 Comparison of important flexibility market projects in Europe.

Projects	Operational regions	Network coverage	Strategic objectives	Core capabilities
CoordiNet (2019–2022)	Greece, Spain, Sweden	Cross-border TSO-DSO	Develop and demonstrate coordinated TSO-DSO-consumer flexibility procurement frameworks across multiple European power systems.	Cross-border data sharing, unified flexibility need aggregation, coordinated market clearing, and multi-stakeholder activation protocols.
Enera (2018–2020)	Northwest Germany region	Transmission and distribution systems	Coordinate flexibility trading to resolve grid congestion and minimize renewable energy curtailment through market mechanisms.	Order book management with locational tags, continuous bid matching, market monitoring and verification, settlement processing, and network impact analysis.
InterFlex (2017–2019)	Demonstrations in Czech Republic, France, Germany, Netherlands, and Sweden	Distribution level	Investigate local flexibility potential to relieve grid constraints and enhance DER integration at distribution level.	Pay-as-bid pricing with sanctioning price component, continuous auctions aligned with wholesale markets, availability payments with penalty mechanisms.

(continued on next page)

Piclo Flex (2019–Now)	United Kingdom	Distribution grid level	Streamline and optimize DSO procurement of distributed flexibility services to enhance grid utilization efficiency and minimize infrastructure upgrades.	Digital marketplace for publishing locational flexibility requirements, asset qualification screening, geospatial mapping of flexibility resources, and matching grid needs with technically-suitable distributed assets.
NODES (2018–Now)	Norway and Germany	Transmission and distribution networks	Monetize localized flexibility resources by creating bridges between regional constraints and system-wide balancing markets while optimizing grid operation costs.	Dual-product architecture (LongFlex capacity contracts **and** ShortFlex activation services), seamless integration with intraday and balancing markets, dynamic resource allocation for voltage stabilization and congestion relief.

DSO, distribution system operator; *TSO*, transmission system operators.

a robust framework for distribution network expansion planning that formally incorporates flexibility considerations is essential. Such a framework would guide long-term infrastructure investments in a way that systematically values and leverages flexibility from diverse resources to ensure a secure, reliable, and cost-effective energy transition.

8.7 Conclusion

The increasing penetration of renewable energy resources, along with the decentralization and digitalization of modern power systems, has signified the role of the demand side in ensuring system flexibility and reliability. Flexibility is a concept that measures the capability of a power system to reduce the imbalances occurred as a result of uncertainty and variability of different parameters throughout the system. This chapter has highlighted the origins of demand variability and the challenges it introduces for operational and long-term planning. The discussion has shown that flexibility is no longer a supplementary feature but a necessary requirement for enabling stable, cost-effective, and sustainable power system operation. Flexibility can be provided for the distribution system using different sources such as g DERs, DR, energy storage, and market-based frameworks. A distribution system planner can use these sources to expand the system in a cost-effective and flexible manner. Integration of the flexibility concept in the planning process can lead to a more complicated optimization problem, and careful planning strategies are needed to solve this complex problem efficiently.

In this context, flexibility markets have emerged as a promising mechanism for efficiently allocating distributed flexibility, providing a structured platform where DSOs can procure short-term, localized services such as congestion management and voltage support from aggregators and prosumers. By creating price signals for demand adjustments, distributed generation, and storage utilization, these markets enable flexibility to be traded where it is most valuable, reducing dependence on costly grid reinforcements and mitigating the uncertainties introduced by high renewable penetration. Real-world implementations across Europe, including PicloFlex in the United Kingdom, Enera in Germany, GOPACS in the Netherlands, and NODES in Norway, demonstrate that flexibility markets are not theoretical constructs but operational tools capable of enhancing reliability, lowering curtailment, and incentivizing investment in DERs. These pioneering platforms also reveal that market design must be tailored to local regulatory and technical conditions, and that successful operation relies on clear product definitions, transparent coordination between TSOs and DSOs, and technology-neutral participation rules. Looking ahead, further progress will require refined market architectures, improved quantification of flexibility, and the integration of emerging resources such as V2G, sector-coupling technologies, and P2P trading. As power systems continue their transition toward decentralization and renewable dominance, flexibility markets will play an increasingly central role in supporting adaptive, resilient, and economically efficient grid planning and operation.

AI disclosure

During the preparation of this work the author(s) used Grammarly in order to Correction of grammar mistakes in the text. After using this tool/service, the author(s) reviewed and edited the content as needed and take(s) full responsibility for the content of the publication.

References

Aoun, A., Ghandour, M., Ilinca, A., & Ibrahim, H. (2021). Demand-side management. In *Hybrid renewable energy systems and microgrids* (pp. 463–490). Elsevier. https://doi.org/10.1016/B978-0-12-821724-5.00001-5.

Arnob, S. S., Arefin, A., Saber, A. Y., & Mamun, K. A. (2023). Energy demand forecasting and optimizing electric systems for developing countries. *IEEE Access, 11*, 39751–39775. https://doi.org/10.1109/ACCESS.2023.3250110.

Babatunde, O. M., Munda, J. L., & Hamam, Y. (2020). Power system flexibility: A review. *Energy Reports, 6*, 101–106. https://doi.org/10.1016/j.egyr.2019.11.048.

Bakare, M. S., Abdulkarim, A., Zeeshan, M., & Shuaibu, A. N. (2023). A comprehensive overview on demand side energy management towards smart grids: challenges, solutions, and future direction. *Energy Informatics, 6*, 4. https://doi.org/10.1186/s42162-023-00262-7.

Boopathy, P., Liyanage, M., Deepa, N., Velavali, M., Reddy, S., Maddikunta, P. K. R., Khare, N., Gadekallu, T. R., Hwang, W. J., & Pham, Q. V. (2024). Deep learning for intelligent demand response and smart grids: A comprehensive survey. *Computer Science Review, 51*, 100617. https://doi.org/10.1016/j.cosrev.2024.100617.

Calvillo, C. F., Sánchez-Miralles, A., Villar, J., & Martín, F. (2016). Optimal planning and operation of aggregated distributed energy resources with market participation. *Applied Energy, 182*, 340–357. https://doi.org/10.1016/j.apenergy.2016.08.117.

de Lima, T. D., Lezama, F., Soares, J., Franco, J. F., & Vale, Z. (2024). Modern distribution system expansion planning considering new market designs: Review and future directions. *Renewable and Sustainable Energy Reviews, 202*, 114709. https://doi.org/10.1016/j.rser.2024.114709.

Detka, K., & Górecki, K. (2023). Selected technologies of electrochemical energy storage—A Review. *Energies, 16*, 5034. https://doi.org/10.3390/en16135034.

Dominković, D. F., Junker, R. G., Lindberg, K. B., & Madsen, H. (2020). Implementing flexibility into energy planning models: Soft-linking of a high-level energy planning model and a short-term operational model. *Applied Energy, 260*, 114292. https://doi.org/10.1016/j.apenergy.2019.114292.

Forouli, A., Bakirtzis, E. A., Papazoglou, G., Oureilidis, K., Gkountis, V., Candido, L., Ferrer, E. D., & Biskas, P. (2021). Assessment of demand side flexibility in european electricity markets: A country level review. *Energies, 14*, 2324. https://doi.org/10.3390/en14082324.

Gerard, H., Rivero Puente, E. I., & Six, D. (2018). Coordination between transmission and distribution system operators in the electricity sector: A conceptual framework. *Utilities Policy, 50*, 40–48. https://doi.org/10.1016/j.jup.2017.09.011.

Gharibvand, H., Gharehpetian, G. B., & Anvari-Moghaddam, A. (2024). A survey on microgrid flexibility resources, evaluation metrics and energy storage effects. *Renewable and Sustainable Energy Reviews, 201*, 114632. https://doi.org/10.1016/j.rser.2024.114632.

Golmohamadi, H. (2022). Demand-side flexibility in power systems: A survey of residential, industrial, commercial, and agricultural sectors. *Sustainability, 14*, 7916. https://doi.org/10.3390/su14137916.

Golmohamadi, H. (2022). Demand-side management in industrial sector: A review of heavy industries. *Renewable and Sustainable Energy Reviews, 156*, 111963. https://doi.org/10.1016/j.rser.2021.111963.

Golmohamadi, H., Golestan, S., Sinha, R., & Bak-Jensen, B. (2024). Demand-side flexibility in power systems, structure, opportunities, and objectives: A review for residential sector. *Energies, 17*, 4670. https://doi.org/10.3390/en17184670.

Hadi, M. B., Moeini-Aghtaie, M., Khoshjahan, M., & Dehghanian, P. (2022). A comprehensive review on power system flexibility: Concept, services, and products. *IEEE Access, 10*, 99257–99267. https://doi.org/10.1109/ACCESS.2022.3206428.

Heilmann, E., Klempp, N., & Wetzel, H. (2020). Design of regional flexibility markets for electricity: A product classification framework for and application to German pilot projects. *Utilities Policy, 67*, 101133. https://doi.org/10.1016/j.jup.2020.101133.

Heinrich, C., Ziras, C., Jensen, T. V., Bindner, H. W., & Kazempour, J. (2021). A local flexibility market mechanism with capacity limitation services. *Energy Policy, 156*, 112335. https://doi.org/10.1016/j.enpol.2021.112335.

Hussain, S., Lai, C., & Eicker, U. (2023). Flexibility: Literature review on concepts, modeling, and provision method in smart grid. *Sustain Energy, Grids Networks, 35*, 101113. https://doi.org/10.1016/j.segan.2023.101113.

IEA. (2018). Status of power system transformation 2018. https://www.iea.org/reports/status-of-power-system-transformation-2018

IEA. (2019). Status of power system transformation 2019. https://www.iea.org/reports/status-of-power-system-transformation-2019

Impram, S., Varbak Nese, S., & Oral, B. (2020). Challenges of renewable energy penetration on power system flexibility: A survey. *Energy Strategy Review, 31*, 100539. https://doi.org/10.1016/j.esr.2020.100539.

Jabir, H. J., Teh, J., Ishak, D., & Abunima, H. (2018). Impacts of demand-side management on electrical power systems: A review. *Energies, 11*, 1050. https://doi.org/10.3390/en11051050.

Jin, C., Xiao, J., Hou, J., Wu, X., Zhang, J., & Du, E. (2021). Flexibility improvement evaluation of hydrogen storage based on electricity–hydrogen coupled energy model. *Global Energy Interconnection, 4*, 371–383. https://doi.org/10.1016/j.gloei.2021.09.004.

Jin, X., Wu, Q., & Jia, H. (2020). Local flexibility markets: Literature review on concepts, models and clearing methods. *Applied Energy, 261*, 114387. https://doi.org/10.1016/j.apenergy.2019.114387.

Kalantar-Neyestanaki, M., & Cherkaoui, R. (2021). Coordinating distributed energy resources and utility-scale battery energy storage system for power flexibility provision under uncertainty. *IEEE Transaction on Sustainable Energy, 12*, 1853–1863. https://doi.org/10.1109/TSTE.2021.3068630.

Kang, J., & Reiner, D. M. (2022). What is the effect of weather on household electricity consumption? Empirical evidence from Ireland. *Energy Economics, 111*, 106023. https://doi.org/10.1016/j.eneco.2022.106023.

Kaushik, E., Prakash, V., Mahela, O. P., Khan, B., El-Shahat, A., & Abdelaziz, A. Y. (2022). Comprehensive overview of power system flexibility during the scenario of high penetration of renewable energy in utility grid. *Energies, 15*, 516. https://doi.org/10.3390/en15020516.

Ku, A. L., Qiu, Y., Lou, J., Nock, D., & Xing, B. (2022). Changes in hourly electricity consumption under COVID mandates: A glance to future hourly residential power consumption pattern with remote work in Arizona. *Applied Energy, 310*, 118539. https://doi.org/10.1016/j.apenergy.2022.118539.

Li, B., Shen, J., Wang, X., & Jiang, C. (2016). From controllable loads to generalized demand-side resources: A review on developments of demand-side resources. *Renewable and Sustainable Energy Reviews, 53*, 936–944. https://doi.org/10.1016/j.rser.2015.09.064.

Li, H., Wang, Z., Hong, T., & Piette, M. A. (2021). Energy flexibility of residential buildings: A systematic review of characterization and quantification methods and applications. *Advances in Applied Energy, 3*, 100054. https://doi.org/10.1016/j.adapen.2021.100054.

Lu, Z., Li, H., & Qiao, Y. (2018). Probabilistic flexibility evaluation for power system planning considering its association with renewable power curtailment. *IEEE Transactions on Power Systems, 33*, 3285–3295. https://doi.org/10.1109/TPWRS.2018.2810091.

Ma, Y., Zhang, C., Mi, Z., Zhang, L., & Parisio, A. (2024). Secondary flexibility market mechanism design and response behavior analysis among multi-microgrids with high proportional BTM-RERs. *Applied Energy, 367*, 123345. https://doi.org/10.1016/j.apenergy.2024.123345.

Mahdavi, M., & Jurado, F. (2025). Industrial, residential, and commercial demands relationship with load components in reconfigurable distribution grids. *IEEE Transaction on Industrial Informatics, 21*, 4027–4038. https://doi.org/10.1109/TII.2025.3538120.

Mahmoud, M., Ramadan, M., Olabi, A.-G., Pullen, K., & Naher, S. (2020). A review of mechanical energy storage systems combined with wind and solar applications. *Energy Conversation and Management, 210*, 112670. https://doi.org/10.1016/j.enconman.2020.112670.

Maiz, S., Baringo, L., & García-Bertrand, R. (2022). Expansion planning of a price-maker virtual power plant in energy and reserve markets. *Sustain Energy, Grids Networks, 32*, 100832. https://doi.org/10.1016/j.segan.2022.100832.

Miri, S. P., & Gharehpetian, G. B. (2025). Determination of critical loads best location in distribution expansion planning considering islanded operation mode. *Sustain Energy, Grids Networks, 43*, 101902. https://doi.org/10.1016/j.segan.2025.101902.

Mosquera-López, S., Uribe, J. M., & Joaqui-Barandica, O. (2024). Weather conditions, climate change, and the price of electricity. *Energy Economics, 137*, 107789. https://doi.org/10.1016/j.eneco.2024.107789.

Mousa, H. H. H., Mahmoud, K., & Lehtonen, M. (2024). Recent developments of demand-side management towards flexible DER-rich power systems: A systematic review. *Iet Generation, Transmission and Distribution, 18*, 2259–2300. https://doi.org/10.1049/gtd2.13204.

Nasir, T., Bukhari, S. S. H., Raza, S., Munir, H. M., Abrar, M., Muqeet, H. A. U., Bhatti, K. L., Ro, J. S., & Masroor, R. (2021). Recent challenges and methodologies in smart grid demand side management: State-of-the-Art literature review. *Mathematical Problems in Engineering, 2021*, 1–16. https://doi.org/10.1155/2021/5821301.

Neves, D., Pina, A., & Silva, C. A. (2015). Demand response modeling: A comparison between tools. *Applied Energy, 146*, 288–297. https://doi.org/10.1016/j.apenergy.2015.02.057.

Olivella-Rosell, P., Rullan, F., Lloret-Gallego, P., Prieto-Araujo, E., Ferrer-San-Jose, R., Barja-Martinez, S., Bjarghov, S., Lakshmanan, V., Hentunen, A., Forsström, J., & Ottesen, S. Ø. (2020). Centralised and distributed optimization for aggregated flexibility services provision. *IEEE Transactions on Smart Grid, 11*, 3257–3269. https://doi.org/10.1109/TSG.2019.2962269.

Pastore, L. M. (2023). Combining Power-to-Heat and Power-to-Vehicle strategies to provide system flexibility in smart urban energy districts. *Sustainable Cities Society, 94*, 104548. https://doi.org/10.1016/j.scs.2023.104548.

Peltokeo, S., Summanen, A., Makihannu, T., Asp, T., & Rauhala, T. (2023). Approach to estimate congestion management related flexibility needs during a planned outage in a transmission grid. *CIGRE Science and Engineering*, 30. https://cse.cigre.org/cse-n030/approach-to-estimate-congestion-management-related-flexibility-needs-during-a-planned-outage-in-a-transmission-grid.html.

Ponnaganti, P., Sinha, R., Pillai, J. R., & Bak-Jensen, B. (2023). Flexibility provisions through local energy communities: A review. *Next Energy, 1*, 100022. https://doi.org/10.1016/j.nxener.2023.100022.

Rafiq, H., Manandhar, P., Rodriguez-Ubinas, E., Barbosa, J. D., & Qureshi, O. A. (2023). Analysis of residential electricity consumption patterns utilizing smart-meter data: Dubai as a case study. *Energy and Buildings, 291*, 113103. https://doi.org/10.1016/j.enbuild.2023.113103.

Rahman, M. M., Dadon, S. H., He, M., Giesselmann, M., & Hasan, M. M. (2024). An overview of power system flexibility: High renewable energy penetration scenarios. *Energies, 17*, 6393. https://doi.org/10.3390/en17246393.

Rajesh, P., Shajin, F. H., & Kannayeram, G. (2022). A novel intelligent technique for energy management in smart home using internet of things. *Applied Soft Computing, 128*, 109442. https://doi.org/10.1016/j.asoc.2022.109442.

Rebenaque, O., Schmitt, C., Schumann, K., Dronne, T., & Roques, F. (2023). Success of local flexibility market implementation: A review of current projects. *Utilities Policy, 80*, 101491. https://doi.org/10.1016/j.jup.2023.101491.

Riaz, S., & Mancarella, P. (2022). Modelling and Characterisation of flexibility from distributed energy resources. *IEEE Transactions on Power Systems, 37*, 38–50. https://doi.org/10.1109/TPWRS.2021.3096971.

Schittekatte, T., & Meeus, L. (2020). Flexibility markets: Q&A with project pioneers. *Utilities Policy, 63*, 101017. https://doi.org/10.1016/j.jup.2020.101017.

Shepherd, M. M., & Mohagheghi, S. (2024). Demand flexibility: A review of quantification methods, models, and required data. *Applied Sciences, 14*, 11276. https://doi.org/10.3390/app142311276.

Sun, S., Dong, M., & Liang, B. (2016). Distributed real-time power balancing in renewable-integrated power grids with storage and flexible loads. *IEEE Transactions on Smart Grid, 7*, 2337–2349. https://doi.org/10.1109/TSG.2015.2445794.

Turdybek, B., Tostado-Véliz, M., Mansouri, S. A., Rezaee Jordehi, A., & Jurado, F. (2024). A local electricity market mechanism for flexibility provision in industrial parks involving Heterogenous flexible loads. *Applied Energy, 359*, 122748. https://doi.org/10.1016/j.apenergy.2024.122748.

Valarezo, O., Gómez, T., Chaves-Avila, J. P., Lind, L., Correa, M., Ulrich Ziegler, D., & Escobar, R. (2021). Analysis of new flexibility market models in Europe. *Energies, 14*, 3521. https://doi.org/10.3390/en14123521.

Wang, S., & Wu, W. (2021). Aggregate flexibility of virtual power plants with temporal coupling constraints. *IEEE Transactions on Smart Grid, 12*, 5043–5051. https://doi.org/10.1109/TSG.2021.3106646.

Yang, X., Xu, C., He, H., Yao, W., Wen, J., & Zhang, Y. (2020). Flexibility provisions in active distribution networks with uncertainties. *IEEE Transaction on Sustainable Energy*, 1. https://doi.org/10.1109/TSTE.2020.3012416.

Yu, H., Niu, S., Shang, Y., Shao, Z., Jia, Y., & Jian, L. (2022). Electric vehicles integration and vehicle-to-grid operation in active distribution grids: A comprehensive review on power architectures, grid connection standards and typical applications. *Renewable and Sustainable Energy Reviews, 168*, 112812. https://doi.org/10.1016/j.rser.2022.112812.

Zakariazadeh, A., Jadid, S., & Siano, P. (2014). Stochastic multi-objective operational planning of smart distribution systems considering demand response programs. *Electric Power Systems Research, 111*, 156–168. https://doi.org/10.1016/j.epsr.2014.02.021.

Zhang, C., Qiu, J., Yang, Y., & Zhao, J. (2021). Trading-oriented battery energy storage planning for distribution market. *International Journal of Electrical Power & Energy Systems, 129*, 106848. https://doi.org/10.1016/j.ijepes.2021.106848.

Navigating competition in retailing layer: A risk-averse decision-making model for electricity market retailers

9

Homa Rashidizadeh-Kermani[a,b], *Mostafa Vahedipour-Dahraei*[b], *Sambeet Mishra*[c] and *Miadreza Shafie-khah*[a]
[a] Research and Innovation Division, Nowocert, Dublin, Ireland, [b] Department of electrical and computer engineering, University of Birjand, Birjand, Iran, [c] Department of Electrical engineering, Information Technology and Cybernetics, University of South-Eastern Norway

9.1 Introduction

In recent years, electricity markets have increasingly become complex, involving a huge number of stakeholders. They include electricity retailers, who purchase electricity from wholesale markets and sell it to consumers (Zhou et al., 2024). Retailers' sole responsibility is to achieve maximum anticipated profit while ensuring a stable supply of electricity by taking efficient procurement and pricing decisions regarding energy. Retailers are becoming more susceptible to uncertainties resulting from volatile market conditions and unstable demand patterns. To minimize these risks and maximize procurement decision stability, stochastic and robust optimization techniques have been extensively applied in recent studies do Prado & Qiao (2019). Optimization models, such as bi-level programming, stochastic optimization, and risk measurements like Conditional Value at Risk (CVaR), have been employed in framing retailers' decision problems (Rashidizadeh-Kermani et al., 2021). These mathematical models enable retailers to design bidding and procurement strategies that balance expected profits against potential risks associated with market volatility and consumer trends. These advanced mathematical models facilitate better modeling of uncertainty and enhance the dependability of market participation strategies.

The optimal operation of multicarrier energy systems, integrating electricity and gas networks, is employed in Oraibi et al. (2025), where coordinated management of these networks faces uncertainties such as market prices, renewable generation, and variable loads, which are addressed with a hybrid stochastic-robust optimization framework, simultaneously considering cost minimization, energy conversion technologies, gas storage, and market price fluctuations. One of the key approaches to dealing with uncertainties in energy markets is employing demand response (DR) programs, which enable retailers to actively shift or alter customer consumption patterns based on dynamic market signals. By utilizing robust and stochastic optimization techniques, retailers can come up with risk-averse strategies that safeguard profitability and ensure

Stochastic Planning and Modeling for Energy Systems: Methods, Applications, and Developments.
DOI: https://doi.org/10.1016/B978-0-443-45298-7.00005-7

supply reliability while enhancing operational flexibility in the face of price volatility and demand uncertainty. Besides, DR is an effective way to mitigate procurement risks by allowing real-time modulation of loads, thereby benefiting system stability and cost effectiveness in general (Jordehi et al., 2022). Furthermore, DR renders the retailer more versatile by dynamic load profile modification (do Prado & Qiao, 2020). Cooperative DR among multiple electricity retailers has been explored in Zhang et al. (2023) from the perspective of enhancing the load management strategies' performance further. Within that plan, geographically distributed data centers purchase electricity from different retailers in multiple regions, which enables the use of incentive-based DR programs in a coordinated manner.

In modern electricity markets, industrial loads with flexible demand can reduce their cost or generate additional revenue by adjusting their consumption pattern based on short-term price change (Finhold et al., 2023) or by providing balancing services to the grid (Vahedipour-Dahraie et al., 2022). This flexibility can be maximally utilized in day-ahead (DA) and balancing markets, respectively designed for facilitating short-term position adjustments. To best realize such potential, a systematic optimization framework is proposed for determining the best demand flexibility distribution between such market mechanisms to deliver the maximum economic value for consumers while ensuring grid stability.

Retail competition in deregulated power markets plays a crucial role in driving efficiency, improving customer service, and inducing technological innovation. According to recent studies, retailers increasingly use complex optimization methods and predictive analytics to buy under uncertainty and respond more intelligently to volatile market conditions. Another study in Hossain et al. (2025) highlights how electricity retailers can minimize procurement costs and boost profitability by including stochastic optimization models that account for market price volatilities as well as demand shifts. Moreover, as electricity markets evolve toward decentralized forms, P2P (peer to peer) energy trading models are gaining traction. A work in Abdul-Ghani et al. (2024) explains the new role of retailers as enablers in P2P markets to enable prosumers' local energy trading without threatening the stability of the grid. By adopting this change in paradigm, retailers can come up with innovative business models and establish better customer relations.

In power markets that are deregulated, retailers are under more pressure to remain competitive without incurring imbalance charges based on forecasted versus actual electricity consumption. To address such issues, retailers have increasingly turned to Demand Side Management (DSM) measures and DR programs, which allow them to shift load profiles dynamically and increase their level of flexibility in balancing markets (Rashidizadeh-Kermani et al., 2019). Here, utilizing consumer flexibility via DR can help to partly counteract such charges. For instance, Abate et al. (2021) build a game-theoretical model explaining retailers-consumers interaction underprice uncertainty. Based on their findings, DR programs can help retailers hedge in real-time markets and remain profitable. In a related study, Finhold et al. (2023) propose an optimization model for allocating flexibility between DA and balancing markets, showing that competitive entry into both markets optimizes retailer profits and system stability. Moreover, Yildiz et al. (2017) highlight the role of real-time pricing and distributed control in enabling retailers to effectively implement DR programs while ensuring grid stability. The work in Song et al. (2024) analyzes local electricity market behavior and the role of competitive processes in enhancing efficiency. The study shows that transparent pricing and well-designed market rules can significantly promote healthy competition among retailers and promote demand-side flexibility investments.

Followed by the aforementioned trends, this chapter presents a risk-aware decision-making framework for electricity retailers who operate in both DA and balancing markets under a competitive retailing layer. The CVaR metric is also employed to quantify and regulate the downside risk in order to allow retailers to take into consideration the trade-offs between mean profit and bad-profit scenarios. The main contributions of this paper are presented below:

- A bi-level stochastic optimization model is developed to determine the optimal involvement of the retailer in DA and balancing markets, taking into account the uncertainties.
- A competitive market environment is modeled by incorporating the price responsiveness of flexible loads, which respond to the pricing strategy of competing retailers.
- The CVaR risk measurement metric is added to the problem to model the risk preferences in the decision-making process of the retailer in forming market participation and bidding strategies.

9.2 Stochastic scheduling of risk-aware retailer decision-making problem

In this chapter, a short-term electricity market is modeled, consisting of both the DA market and a real-time energy balancing tool, commonly referred to as the regulating market in Nordpool (Nordic Electricity, n.d.). The retailer is assumed to be a price-taker in both markets and does not possess any market power, and is therefore acting under a competitive retail market. Competition is therefore limited to the retail market and does not encompass strategic bidding or price manipulation across electricity markets. The retailer has a twofold objective:

- To maximize expected revenues from energy trades in the DA market.
- To minimize imbalance costs incurred in the balancing market, subject to satisfying the electricity supply to traditional and responsive loads.

All the markets under consideration are cleared by one auction mechanism. Fig. 9.1 depicts the timeline for DA and balancing market energy trading. From this figure, it can be interpreted that:

- The DA market is cleared on day *d*–1 for delivery of energy on day *d*. Due to the lag involved between clearing of markets and actual delivery, there will always remain real-time versus scheduled mismatches in demand.
- The balancing market covers these imbalances by remaining open until 15 minutes before each delivery hour. It keeps real-time balance by pricing and clearing the deviations away from the DA schedule.

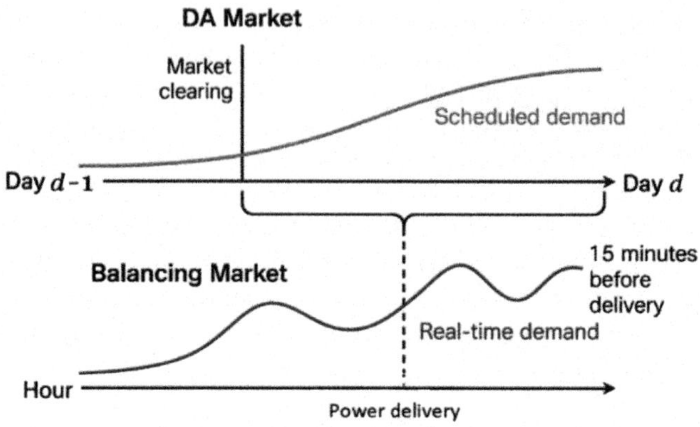

Figure 9.1 Timeline of energy trading in the short-term electricity market. *DA*, Day-ahead.

This chapter presents the retailer's two-stage stochastic bidding strategy in the DA market. In the first stage, the retailer submits bids according to DA and balancing price forecast scenarios, and traditional and responsive load demands to be expected. Once DA prices are realized, the second stage involves determining how much to trade in the balancing market. This is performed via the latest real-time price and demand forecasts. The model particularly addresses:

- Loads' price-responsiveness,
- Price competition between rival retailers,
- Ambiguity regarding the prices charged by said rivals.

In a deregulated and liberalized electricity market consisting of a DA market and a Real-Time balancing market, retailers play a vital role in coordinating distributed energy resources, like responsive loads. Here, the retailer is assumed to be in a competitive market, where he competes with other retailers in acquiring consumers by meeting their energy demands. The retailer would prefer to attain maximum profit from energy trading between the upstream suppliers and the end consumers. On the consumer's side, it is assumed that all loads that respond are equipped with Smart Energy Management Controllers (SEMCs). These devices respond automatically to electricity price signals so that customers can lower their energy bills without any actions taken manually. This system is based on a fast, two-way communication infrastructure between customer premises and the retailer. With this infrastructure, SEMCs can dynamically reroute consumption or even change retailers by price competitiveness with no inconvenience to users. This subjects the retailers to substantial financial risk due to consumer behavior and price fluctuations. Fig. 9.2 illustrates the schematic of this retail energy trading issue.

For the retailer to function effectively under these uncertain and competitive conditions, its pricing strategies and energy trading must be optimally designed in a way that they should maximize expected profit as well as regulates risk. Competitors,

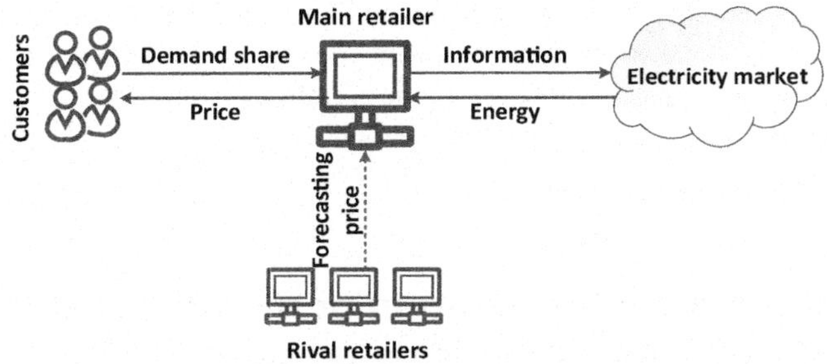

Figure 9.2 Multiple retailers serving the users within a competitive environment.

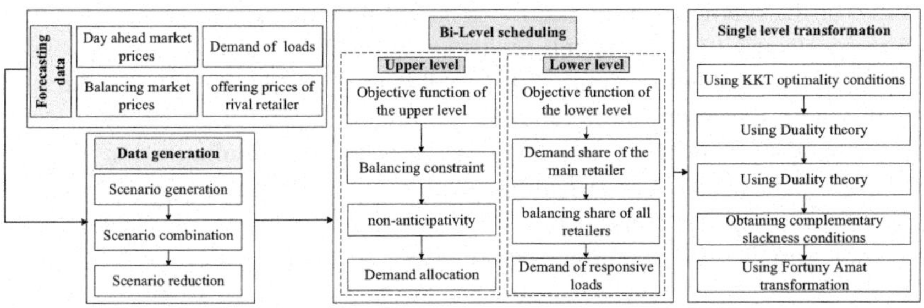

Figure 9.3 The structure of the bi-level stochastic decision-making problem of a retailer in the competition layer. *KKT*, Karush–Kuhn–Tucker.

however, make things complicated, as each retailer must predict the competitors' pricing strategies in a way that they should predict customer switching behavior.

As customer feedback directly influences market share and revenue, it cannot be ignored and results in second-best solutions. To capture this interaction, it is proposed here that a stochastic bi-level optimization framework be used. At the top level, the retailer aims to maximize its expected profit through hourly bids in the DA market and tuning its positions in the Real time (RT) market to clear imbalances. At the lower level, the buyers attempt to minimize their buying prices by choosing between retailers based on prices and consumption schedules. Fig. 9.3 shows how the combination of the two levels reflects the integration of retailer pricing and consumer choice.

When determining prices, retailers are faced with a group of main sources of uncertainty that complicate the world: DA and equilibrating market prices, customer demand, and opposing retailer prices for sale. Pool prices are determined by bids submitted by the market participants and, under the assumption that the retailer is a price taker, independent of retailer action. Also, client demand is not available to the retailer and should never be confused with the actual supply by the retailer. In fact, although aggregate client demand functions are assumed to be independent of the retailer's price, the share supplied by a given retailer is the outcome of clients' individual optimization problems in a choice problem. Finally, the prices charged by other competing retailers are also uncertain and not dependent on the retailer's choice. Retailers therefore have to contend with uncertainty in not just the level of market demand and the market prices, but even in the actions of clients in a competitive market where consumers can switch to other providers if the prices provided are not extremely attractive enough.

9.3 Mathematical formulation of optimization framework in competitive retailing layer

Throughout the chapter, a comprehensive mathematical structure is developed to explain the decision-making of a competitive electricity retailer under uncertainty. The objective of the retailer is to optimally optimize its bidding and energy procurement strategy in electricity markets subject to uncertainty in the behavior of responsive consumers. The problem is framed as a risk-averse stochastic bi-level optimization problem where the upper level captures the retailer's profit maximization objective under uncertainty, and the lower level captures the cost-minimization response of consumers. In the problem, the retailer's (leader) revenue is a function of the amount of energy sold to customers (followers), which is determined by the customers themselves as a response to prices by all other rival retailers.

9.3.1 Upper-level (leader problem): Uncertain retailer scheduling

The issue of high-level optimization attempts to maximize the retailer's expected profit and minimize the potential loss associated with making decisions under uncertainty. For the impact of uncertainty to be captured, the CVaR model is employed, that would be explained in section *E*. This risk-averse approach quantifies economic loss as the expected shortfall at the lower tail of the distribution and thereby enables the retailer to make more robust scheduling decisions under uncertain market behavior and client activity. The profit function of the main retailer is as follows:

$$Maximize \sum_{s \in N_S} \pi_s \sum_{t \in T} \left[-P_{t,s}^{DA} \underset{t,s}{\overset{DA}{\Pr}} + P_{t,s}^{B^+} \underset{t,s}{\overset{B^+}{\Pr}} - P_{t,s}^{B^-} \underset{t,s}{\overset{B^-}{\Pr}} + P_{t,s}^{DR} Pr_t^{DR} + P_{t,s}^{D} Pr_t^{D} \right] \tag{9.1}$$

where $P_{t,s}^{DA}$ and $P_{t,s}^{B+/-}$ are the energy trading in the DA and balancing market, respectively. $P_{t,s}^{DR}$ and $P_{t,s}^{D}$ indicate the total energy of responsive and nonresponsive loads at time t and at scenario s, respectively. $Pr_{t,s}^{DA}$ and $Pr_{t,s}^{B+/-}$ show the electricity price of DA and balancing markets, respectively. Pr_t^{DR} price signals offered to responsive loads and Pr_t^{D} describe the price for selling energy nonresponsive loads π_s is the probability of occurring the scenarios of DA and balancing market prices and the demand of loads.

This objective function incorporates three main components:

- Energy procurement cost from the DA market.

- Balancing market cost incurred due to deviations between DA bids and actual consumption in Real-Time.
- Revenue earned by selling electricity to responsive loads in the competitive layer and to nonresponsive loads in customer side.

The upper-level problem is formulated as a two-stage stochastic program including trading energy in the DA market and deviation compensation in the RT framework. This problem is restricted to the following constraints:

Eq. (9.2) enforces the energy balance condition across all time slots and scenarios, ensuring that supply matches the demand under each possible realization.

$$P^{D}_{t,s} + P^{DR}_{t,s} = P^{DA}_{t,s} - P^{B^+}_{t,s} + P^{B^-}_{t,s} \tag{9.2}$$

The share of energy supplied by the retailer under study at each hour t and scenario ψ is defined in Eq. (9.3). It represents the expected value of the demand served by this retailer, averaged over all possible pricing scenarios of rival retailers. It is worth noting that alternative approaches may also be employed to estimate the demand allocated to the retailer under study.

$$P^{DR}_{t,s} = P^{T_{DR}}_{t,s} \sum_{\psi \in \Psi} \rho^{DR}_{t,\psi} X^{DR}_{R,t,\psi} \tag{9.3}$$

The total load supplied by all retailers is given as $P^{T_{DR}}_{t,s}$, while the share of the under-study retailer is $P^{DR}_{t,s}$. $X^{DR}_{R,t,\psi}$ stands for the percentage of customers supplied by the retailer under study, with the probability of occurrence associated with scenario ψ as $\rho^{DR}_{t,\psi}$.

The nonanticipativity constraint, which enforces that identical DA bids must be submitted across all scenarios with equal DA prices, is formulated as follows

$$P^{DA}_{t,s} = P^{DA}_{t,s'} \tag{9.4}$$

Nonanticipativity implies that if the realizations of the stochastic processes are identical up to a certain stage, then the corresponding decision variables must also take identical values up to that stage (Conejo et al., 2010). Eq. (9.5) imposes upper and lower bounds on the energy traded in the positive and negative balancing markets, reflecting operational market limitations.

$$0 \leq P^{B^+/B^-}_{t,s} \leq \bar{P} \tag{9.5}$$

where $\bar{P}$ denotes the upper limit on the energy that can be exchanged with the network in the balancing market.

Note that here, two sets of scenarios are considered. One for the combination of loads and electricity markets with the scenario of s. The other set of scenarios is considered for the prices offered by rival retailers with scenario ψ. In fact, two separate sets of scenarios are taken.

9.3.2 *Lower level (follower problem): Client decision-making layer*

Here, consumers are classified into responsive and nonresponsive types. Responsive customers can alter their electricity usage in response to price signals, while non-responsive ones follow fixed consumption patterns. The retailer sells electricity to its customers through a Time-Based Rate tariff, while itself purchases energy from wholesale markets. In demand side, responsive clients choose their electricity provider from competing retailers. After each retailer announces its selling price, clients decide which retailer will meet their electricity needs over the planning horizon. These decisions are made under the assumption of perfect information about the retailers' prices, while other factors like pool prices and electricity demand remain uncertain.

The lower-level of the problem explains the energy cost of customers, which should be minimized. Therefore the objective function of this level can be formulated as follows:

$$\underset{X^{DR}_{R,t,\psi},Y^{DR}_{R,R',t,\psi}}{Minimize} \sum_{s\in N_S} \pi_s \hat{P}^{DR} \left[\left(\Pr^{DR}_{R_0,t} X^{DR}_{R_0,t,\psi} + \sum_{\substack{R \in N_R \\ R \neq R_0}} \Pr^{DR}_{R,t,\psi} X^{DR}_{R,t,\psi} \right) + \sum_{R\in N_R} \sum_{\substack{R' \in N_R \\ R \neq R'}} Pr^{DR}_{R,R'} Y^{DR}_{R,R',t,\psi} \right] \tag{9.6}$$

$P^{T_{DR}}_{t,s}$ is the total expected demand of responsive loads. $Pr^{DR}_{R,t,\psi}$ for $R \neq R_0$ stands for the price signals offered by the rival retailers. The share of responsive loads that is served by the under-study retailer is shown by $X^{D}_{R,t,\psi}$. $Y^{D}_{R,R',t,\psi}$ shows the percentage of responsive loads that transferred among the retailers. In the above expression, the expected value of $P^{T_{DR}}_{t,s}$ is obtained as: $\hat{P}^{DR} = \sum_{s\in N_S} \pi_s \sum_{t\in T} P^{T_{DR}}_{t,s}$.

In the above equation, multiple retailers are considered to supply the required energy of responsive loads, in which index R_0 shows the under-study retailer. The first two expressions in Eq. (9.6) express the costs of purchased energy from the under-study retailer and the rivals, respectively. The last term indicates the cost representing the unwillingness of responsive loads to switch among the retailers in order to supply their load. Here, it is assumed that customers can monitor the prices offered by retailers and supply their own loads by switching their load to the most competitive retailer to minimize their energy payments. The tendency of responsive loads to resist switching from retailer *R* to retailer *R'* is modeled by incorporating a switching cost parameter with the price of $Pr^{DR}_{R,R'}$, reflecting the perceived inconvenience involved in the transition process, is modeled with $\sum_{\substack{R' \in N_R \\ R \neq R'}} Pr^{DR}_{R,R'} Y^{DR}_{R,R',t,\psi}$.

The following constraints are considered in the lower-level objective function, with the corresponding dual variables indicated alongside each constraint. Eq. (9.7) specifies the proportion of energy required by responsive loads that is supplied by each retailer:

$$X^{DR}_{R,t,\psi} = X^{init,\ell}_{R,t,\psi} + \sum_{\substack{R \in N_R \\ R' \neq R}} Y^{DR}_{R',R,t,\psi} - \sum_{\substack{R' \in N_R \\ R' \neq R}} Y^{DR}_{R,R',t,\psi} : \varphi_{R,\psi} \tag{9.7}$$

where is the initial proportion of responsive loads supplied by each retailer. The second term indicates the percentage of loads that come from retailers R' to retailer R. The last term shows the percentage of loads that go from retailer R to retailer R', .

Moreover, Eq. (9.8) states that the total demand of responsive loads should be fully met by both the retailer under study and its competitors. In other words, the sum of the demand shares supplied by all retailers equals 100% in each time period t and each scenario ψ as below:

$$X^{DR}_{R,t,\psi} + \sum_{\substack{R \in N_R \\ R \neq R_0}} X^{DR}_{R,t,\psi} = 100\% : \varepsilon_{\psi} \tag{9.8}$$

In the above equations, we have:

$$X^{DR}_{R,t,\psi} \geq 0 : \ \gamma^{X}_{R};\ \forall R \in N_R \tag{9.9}$$

$$Y^{DR}_{R,R\prime,t,\psi} \geq 0, \gamma^{Y}_{R,R'} \forall R,\ R' \in N_R,\ R \neq R'$$

9.3.3 Model of responsive loads

Responsive loads are assumed to actively participate in DR programs by dynamically adjusting their energy consumption in response to both price signals issued by the retailer and their inherent demand elasticity. Demand elasticity quantifies the degree of responsiveness of energy consumption to changes in price signal (Rashidizadeh-Kermani et al., 2019). In pursuit of optimal economic benefit, each consumer can employ a combination of load shifting and load shedding strategies, thereby altering its energy consumption level from P^{int}_{t} to $P^{T_{DR}}_{t,s}$ in period t as formulated below:

$$\boldsymbol{P}^{T_{DR}}_{t,s} = \boldsymbol{P}^{int}_{t} + \Delta \boldsymbol{P}_{t,s} \tag{9.10}$$

The total benefit received by the customers is determined by the following expression:

$$\boldsymbol{S}\left(\boldsymbol{P}^{T_{DR}}_{t,s}\right) = \boldsymbol{B}\left(\boldsymbol{P}^{T_{DR}}_{t,s}\right) - \boldsymbol{P}^{T_{DR}}_{t,s} \boldsymbol{Pr}^{DR}_{t} \tag{9.11}$$

where $S\left(P_{t,s}^{T_{DR}}\right)$ and $B\left(P_{t,s}^{T_{DR}}\right)$ represent the benefit and income of customers at period t and scenario s following the implementation of DR programs. To maximize the benefit of customers, the following criteria must be satisfied:

$$\frac{\partial S\left(P_{t,s}^{T_{DR}}\right)}{\partial P_{t,s}^{T_{DR}}} = \frac{\partial B\left(P_{t,s}^{T_{DR}}\right)}{\partial P_{t,s}^{T_{DR}}} - \Pr_{t}^{DR} = 0 \tag{9.12}$$

In line with the model proposed in Vahedipour-Dahraie et al. (2017), the energy consumption of customers is calculated as follows:

$$\mathbf{P}_{t,s}^{T_{DR}} = \mathbf{P}_{t}^{int} \mathbf{exp} \sum\nolimits_{h \in T} \mathbf{Elas}_{t,h} \ln \left[\frac{\mathbf{Pr}_t^{DR}}{\mathbf{Pr}_t^{DR,int}} + \frac{1}{1 + \mathbf{Elas}_{t,h}^{-1}} \right] \tag{9.13}$$

where $\boldsymbol{Elas}_{t,h}$ indicates the demand elasticity of responsive loads and $\boldsymbol{Pr}_t^{DR,int}$ denotes the average value of $\boldsymbol{Pr}_t^{DR}$.

9.3.4 Conversion of the bi-level optimization problem to the single-level equivalent

One of the key objectives of this chapter is to transform the nonlinear bi-level programming problem formulated in parts A and B into an equivalent single-level mixed-integer linear programming (MILP) form. This conversion enables efficient solution using commercially available optimization solvers, as in Rashidizadeh-Kermani et al. (2020). The steps involved in this transformation are as follows:

- The bi-level programming problem is first transformed into an equivalent single-level nonlinear optimization problem by applying the Karush-Kuhn-Tucker (KKT) optimality conditions to each lower-level Eqs. (9.6)–(9.8). The KKT conditions are applicable here because the lower-level problems are convex with respect to the continuous variables, and the upper-level variables can be treated as fixed parameters from the perspective of the lower-level agents.
- The complementary slackness conditions from the first step, which involve products of Lagrange multipliers and constrained continuous functions, are then linearized. This is done by replacing these bilinear terms with equivalent linear expressions following the approach in Rashidizadeh-Kermani et al. (2022).
- Bilinear products between continuous variables, arising from the previous transformations, are further replaced by equivalent linear expressions based on duality theory as detailed in Carrión et al. (2009).

It should be noted that the single-level nonlinear optimization problem obtained after the first step is a mathematical program with equilibrium constraints (MPEC). From a modeling standpoint, the main contribution of this chapter lies in converting this nonlinear MPEC into a standard MILP problem, which is achieved through the other two steps. The final equivalent single-level MILP formulation of the nonlinear bi-level programming problem is as follows.

The Lagrangian function (L) of the subproblem is determined as in Eq. (9.14). The function encompasses the original objective function of the subproblem and its associated constraints, with each of them being multiplied by its dual variable, called the Lagrange multiplier. The model of the Lagrangian plays a crucial role in restructuring the bi-level problem into a single-level problem under the KKT conditions.

$$
\begin{aligned}
L = \hat{P}^{DR}\left(\Pr^{DR}_{R_0,t} X^{DR}_{R_0,t,\psi} + \sum_{\substack{R \in N_R \\ R \neq R_0}} \Pr^{DR}_{R,t,\psi} X^{DR}_{R,t,\psi} + \sum_{R\in N_R} \sum_{\substack{R' \in N_R \\ R \neq R'}} Pr^{DR}_{R,R'} Y^{DR}_{R,R',t,\psi} \right) \\
+\varphi_{R,\psi}\Big(X^{DR}_{R,t,\psi} - X^{(init,l)}_{(R,t,\psi)} - \sum_{\substack{R \in N_R \\ R' \neq R}} Y^{DR}_{R',R,t,\psi} + \sum_{\substack{R' \in N_R \\ R' \neq R}} Y^{DR}_{R,R',t,\psi}\Big) \\
+\varepsilon_\psi\Big(X^{DR}_{R,t,\psi} + \sum_{\substack{R \in N_R \\ R \neq R_o}} X^{DR}_{R,t,\psi} - 1\Big) \\
- \sum_{R\in N_R} \gamma^X_R X^{DR}_{R,t,\psi} \\
- \sum_{R\in N_R} \sum_{\substack{R' \in N_R \\ R' \neq 0}} Y^{DR}_{R,R',t,\psi}\, \gamma^Y_{R,R'}
\end{aligned}
\tag{9.14}
$$

where $\varphi_{R,\psi}$, ε_ψ, γ^X_R and $\gamma^Y_{R,R'}$ are the Lagrangian multipliers.

In order to derive the necessary optimality conditions for the lower-level problem, both the primal feasibility constraints and the KKT conditions must be considered. These KKT conditions are obtained by formulating the Lagrangian function of the lower-level problem and then setting its partial derivatives with respect to all primal decision variables equal to zero as given in the following constraints. This procedure yields a system of equations that, together with the original feasibility constraints, fully characterizes the necessary conditions for optimality of the lower-level problem.

$$\frac{\partial L}{\partial X^{DR}_{R_0,t,\psi}} = \hat{P}^{DR} Pr^{DR}_{R_0,t} - \varphi_{R_0,\psi} - \varepsilon_\psi - \gamma^X_{R_0} = 0 \tag{9.15}$$

$$\frac{\partial L}{\partial X^{DR}_{R,t,\psi}} = \hat{P}^{DR} Pr^{DR}_{R,t} - \varphi_{R,\psi} - \varepsilon_\psi - \gamma^X_R = 0 : \ \forall R = 1, 2, 3, \ldots \tag{9.16}$$

$$\frac{\partial L}{\partial Y^{DR}_{R,R',t,\psi}} = \hat{P}^{DR} Pr^{DR}_{R,t} - \varphi_{R',\psi} + \varphi_{R,\psi} - \gamma^Y_{R,R'} = 0 \quad \forall R, R' \in N_R, R \neq R' \tag{9.17}$$

$$\gamma^X_R \geq 0 \tag{9.18}$$

$$\gamma^Y_{R,R'} \geq 0 \tag{9.19}$$

$$X^{DR}_{R,t,\psi}\, \gamma^X_R = 0; \ \forall R \in N_R \tag{9.20}$$

$$Y^{DR}_{R,R',t,\psi}\, \gamma^Y_{R,R'} = 0, \forall R, \ R' \in N_R, \ R \neq R' \tag{9.21}$$

where Eqs. (9.20) and (9.21) are the complementary slackness conditions.

The nonlinear complementary slackness conditions, which constitute a fundamental component of the KKT optimality conditions, can be equivalently reformulated as a set of linear constraints by introducing appropriate binary variables along with sufficiently large constants (commonly referred to as big-M parameters). This linearization technique is widely applied in the development of MILP models to substitute nonlinearities and ensure compatibility with standard optimization solvers. From Substituting Eqs. (9.15) and (9.16) in Eqs. (9.18) and (9.20), and from substituting Eq. (9.17) in Eqs. (9.19) and (9.21), we have:

$$\hat{P}^{DR} Pr_{R_0,t}^{DR} - \varphi_{R_0,\psi} - \varepsilon_{\psi} \geq 0 \tag{9.22}$$

$$\hat{P}^{DR} Pr_{R,t}^{DR} - \varphi_{R,\psi} - \varepsilon_{\psi} \geq 0 \tag{9.23}$$

$$\hat{P}^{DR} Pr_{R,t}^{DR} - \varphi_{R',\psi} + \varphi_{R,\psi} \geq 0 \tag{9.24}$$

$$X_{R,t,\psi}^{DR}\left(\hat{P}^{DR_{t,s}^{T_{DR}}} Pr_{R_0,t}^{DR} - \varphi_{R_0,\psi} - \varepsilon_{\psi}\right) = 0 \tag{9.25}$$

$$X_{R,t,\psi}^{DR}\left(\hat{P}^{DR} Pr_{R,t}^{DR} - \varphi_{R,\psi} - \varepsilon_{\psi}\right) = 0 \tag{9.26}$$

$$Y_{R,R',t,\psi'}^{DR} \hat{P}^{DR} Pr_{R,t}^{DR} - \varphi_{R',\psi} + \varphi_{R,\psi}) = 0 \tag{9.27}$$

Note that in complementary slackness conditions, the production of Lagrangian multipliers of $\varphi_{R,\psi}$ and ε_{ψ} with $X_{R,t,\psi}^{DR}$ and $Y_{R,R',t,\psi}^{DR}$ make some nonlinearities that should be replaced with its linear expression. Moreover, the nonlinear term of $P_{t,s}^{DR} Pr_t^{DR}$ can be obtained from using the duality theory as expressed in Carrión et al. (2009). So, the lower-level problem in Eq.(9.6)–(9.9) would be replaced with its dual expression as below:

$$\text{Maximize}\left(X_{R,t,\psi}^{init,\ell} \varphi_{R,\psi} + \varepsilon_{\psi}\right) \tag{9.28}$$

$$\varphi_{R_0,\psi} + \varepsilon_{\psi} \leq \hat{P}^{DR} Pr_{R_0,t}^{DR} \tag{9.29}$$

$$\varphi_{R,\psi} + \varepsilon_{\psi} \leq \hat{P}^{DR} Pr_{R,t}^{DR} \tag{9.30}$$

$$\varphi_{R',\psi} + \varphi_{R,\psi} \leq \hat{P}^{DR} Pr_{R,t}^{DR} \tag{9.31}$$

From the main lower-level problem in Eq. (9.6) and its dual in Eq. (9.28), the following expression would be obtained:

$$P_{t,s}^{DR} Pr_t^{DR} = \frac{P_{t,s}^{DR}}{\hat{P}^{DR}} \sum_{\psi \in \Psi}$$

$$\times \rho_{t,\psi}^{DR}\left(\sum_{\substack{R\in N_R\\ R\neq 0}} -\hat{P}^{DR}Pr_{R,t,\psi}^{DR}X_{R,t,\psi}^{DR} - \sum_{R\in N_R}\sum_{\substack{R'\in N_R\\ R'\neq R}} \hat{P}^{DR}Pr_{R,R'}^{DR}Y_{R,R',t,\psi}^{DR} + \sum_{R\in N_R} X_{R,t,\psi}^{init,\ell}\varphi_{R,\psi} + \varepsilon_{\psi}\right) \tag{9.32}$$

In the above equations, the nonlinear complementary slackness conditions Eqs.(9.25)–(9.27) would be replaced by their corresponding linear statements using big-M formulas as below:

$$\left(\hat{P}^{DR_{t,s}^{T_{DR}}}Pr_{R_0,t}^{DR} - \varphi_{R_0,\psi} - \varepsilon_{\psi}\right) \leq M_1 U_{R_0,t,\psi}^{X} \tag{9.33}$$

$$X_{R_0,t,\psi}^{DR} \leq M_1\left(1 - U_{R_0,t,\psi}^{X}\right) \tag{9.34}$$

$$\left(\hat{P}^{DR}Pr_{R,t}^{DR} - \varphi_{R,\psi} - \varepsilon_{\psi}\right)M_1 U_{R,t,\psi}^{X} \tag{9.35}$$

$$X_{R,t,\psi}^{DR} \leq \left(1 - U_{R,t,\psi}^{X}\right) \tag{9.36}$$

$$(\hat{P}^{DR}Pr_{R,t}^{DR} - \varphi_{R',\psi} + \varphi_{R,\psi}) \leq M_2 U_{R,R',t,\psi}^{Y} \tag{9.37}$$

$$Y_{R,R',t,\psi}^{DR} \leq M_2\left(1 - U_{R,R',t,\psi}^{Y}\right) \tag{9.38}$$

In the above expressions, M_1 and M_2 are selected such that the feasible conditions would be met. Moreover, $U_{R_0,t,\psi}^{X}$ and $U_{R,R',t,\psi}^{Y}$ are binary variables.

9.3.5 Risk-adjusted model

Stochastic programming offers a robust framework for modeling decision-making problems under uncertainty, such as those encountered by retailers in power markets. In the proposed approach, uncertainties related to market prices, client demands, and offering prices by rival retailers are represented as random variables within this framework. To ensure computational tractability, the initial large scenario set is reduced using scenario-reduction techniques (Conejo et al., 2010). Additionally, risk aversion is incorporated by controlling the volatility of the expected profit through the use of CVaR. In a risk-neutral formulation, the objective is solely to maximize the expected profit; however, this can lead to solutions with a significant probability of low or even negative profits (losses). Mathematically, CVaR at a given confidence level α is defined as follows (Rashidizadeh-Kermani et al., 2022):

$$CVaR = \max_{\xi,\eta_s}\left(\xi - \frac{1}{1-\alpha}\sum_{s=1}^{N_s}\pi_s.\eta_s\right) \tag{9.39}$$

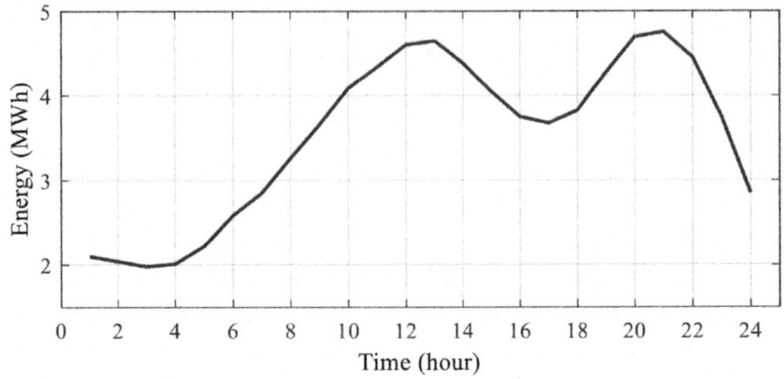

Figure 9.4 Demand profile.

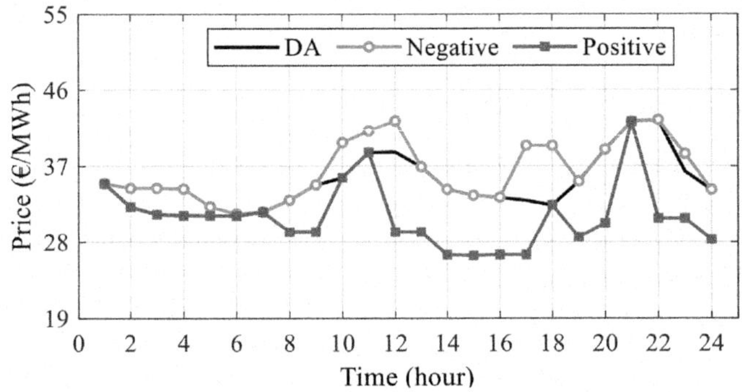

Figure 9.5 Electricity market prices. *DA*, Day-ahead.

Subject to:

$$\eta_s + profit_s - \xi \geq 0; \; \eta_s \geq 0 \tag{9.40}$$

where $profit_s$ stands for the profit in the scenario s,η_s as an auxiliary non-negative variable shows the difference between auxiliary variable ξ and the $profit_s$ when the $profit_s$ is smaller than ξ and π_s is the probability of scenario s.

9.4 Case study and results

9.4.1 Case study

The proposed competitive retail electricity market model is tested using realistic data from the Nordic market, with the forecasted DA and balancing market prices illustrated in Fig. 9.4 and the demand of loads in Fig. 9.5. The scheduling horizon covers one day, divided into 24 hourly intervals, and four competing retailers (Retailer0–Retailer3) are considered to participate in the market.

Table 9.1 Transferring loads among retailers in the competing layer.

Hour	From Retailer0 to Retailer1	From Retailer0 to Retailer2	From Retailer0 to Retailer3	From Retailer1 to Retailer2	From Retailer1 to Retailer3	From Retailer2 to Retailer3
03:00	3.4	-0.2	-9.1	-18.09	-26.9	-8.9
08:00	20.8	1.0	-11.9	-19.83	-32.7	-12.9
12:00	0.2	21.86	-11.5	21.66	-11.7	-33.0
16:00	2.3	21.8	-2.3	19.5	-4.6	-24.1
20:00	-9.2	-8.5	-19.6	0.7	-10.4	-11.1
22:00	19.9	-0.5	-12.4	-20.4	-32.3	-11.9
24:00	17.96	0.9	-8.2	-17.0	-26.1	-9.1

Uncertainties in prices, demand, and rival retailers' offers are modeled using Monte Carlo simulation and a Roulette Wheel mechanism, followed by K-means–based scenario reduction (Arthur & Vassilvitskii, 2007).

Customer demand is divided into responsive and nonresponsive loads, and only responsive loads can choose their preferred retailer.

The optimization is implemented in General algebraic modeling system (GAMS) and solved by the Complex Linear Programming Expert (CPLEX) solver on a PC with an Intel Core i7 at 2.60 GHz processor and 4 GB RAM.

9.4.2 Numerical results

Fig. 9.6 illustrates the hourly DR prices offered by all retailers. As shown, Retailer2 proposes the highest DR prices during peak load hours, while offering the lowest values in off-peak periods, reflecting an aggressive pricing strategy aimed at maximizing profit during high-demand intervals. Retailer1, on the other hand, maintains moderate and relatively stable DR prices throughout the day compared to Retailers2 and 3.

The under-study retailer (Retailer0) adopts a competitive pricing policy, frequently offering lower prices to attract responsive customers and retain its share in the market. This strategic behavior allows the retailer to encourage higher DR participation and consequently manage its procurement cost more effectively. Furthermore, during hours when the DA and balancing market prices are relatively low (e.g., between 14:00 and 16:00, see Fig. 9.3), the examined retailer slightly reduces its DR price. This adjustment reflects a rational decision to minimize incentives for unnecessary load shifting when market conditions are already favorable.

Table 9.1 presents the hourly transfer rates of responsive loads among competing retailers under a 60% DR participation level. Index zero "0" stands for the under-study retailer who is the decision maker in this study, and other numbers refer to the rival retailers. Positive values indicate the correct trend of customers among the retailers, while negative values show the reverse transfer of loads. For example, at hour 3:00, 3.4% of loads were transferred from Retailer0 to Retailer1, while 18.09% of loads come to Retailer2 to Retailer1. The same trend goes for other number of the table.

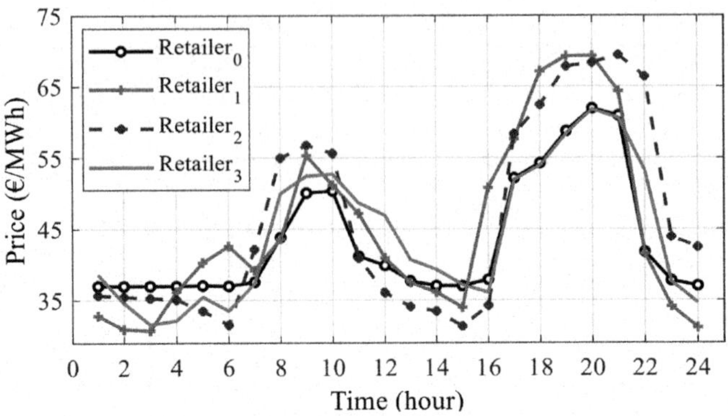

Figure 9.6 Offering price signals by all retailers.

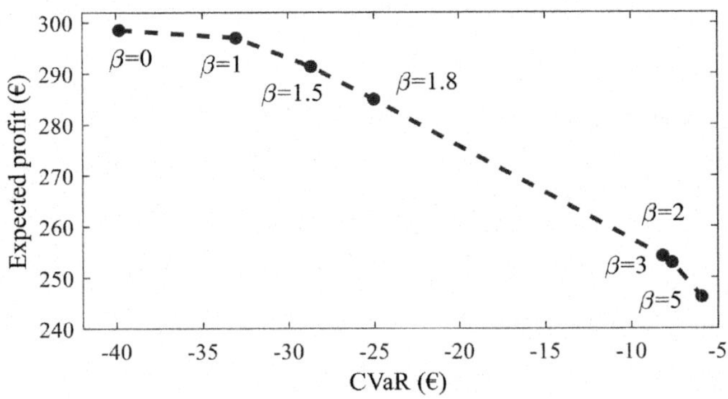

Figure 9.7 The expected profit versus CvaR in multiple β values. *CvaR*, Conditional value at risk.

With this explanation, it can be seen that there are noticeable percentage of loads who choose Retailer0, which indicates its competing offering prices during the day. This brings significant profit for the under-study retailer who competes against rivals in the retail layer to attract a significant portion of responsive loads. In this table, switching activity indicates that all retailers tend to align their prices closer to equilibrium levels, reducing consumers' incentives to switch. Overall, the results validate the capability of the proposed bi-level model in accurately capturing inter-retailer competition and dynamic consumer switching behavior.

The proposed approach is applied to the case study to analyze the risk-constrained decision-making behavior of the retailer under different levels of responsive load participation in DR programs. The expected profit of the under-study retailer versus the CVaR for various values of the risk aversion parameter is illustrated in Fig. 9.7. In all simulations, a 95% confidence level is considered for CVaR calculation.

As observed, with increasing values of β, the expected profit of the retailer gradually decreases, while the CVaR improves (i.e., becomes less negative), indicating a lower exposure to financial risk. This trend shows that a more risk-averse retailer tends to adopt conservative bidding and pricing strategies, prioritizing profit stability over higher average returns. When the retailer operates under a risk-neutral condition (β close to zero), it achieves the maximum expected profit but is more exposed to potential losses in unfavorable market scenarios. Conversely, at higher β values, the retailer sacrifices a portion of its expected profit to significantly reduce the likelihood of financial losses, leading to a more stable and risk-averse operational strategy.

Table 9.2 presents the optimal offered energy determined by each retailer under different levels of DR participation and risk aversion coefficient (β). As can be observed, for a fixed level of DR participation, the total quantity of power offered by the retailer in the DA market decreases as β increases. This indicates that with higher risk aversion, the retailer becomes more conservative, preferring to reduce exposure to market uncertainty.

The reduction in DA trading is accompanied by an increase in participation in the positive balancing market, as this market is relatively less volatile. Hence, risk-averse retailers tend to allocate a greater share of their energy to the balancing market to hedge against profit variability, even though it may reduce their expected average profit.

Furthermore, as β increases, the energy traded in the negative balancing market decreases, indicating that the retailer aims to minimize costs associated with energy shortages or imbalance penalties. At high β values, the retailer's strategy becomes nearly constant, reflecting a stable risk-averse behavior where the offered power in each market segment remains fixed to ensure predictable profit levels.

9.5 Conclusion

Despite substantial progress in DSM, risk modeling, and market-based pricing of electricity, work on formulating an integrated, risk-averse decision-making system for electricity retailers in a competitive market with flexible loads is in its infancy. This chapter addresses this research gap by introducing a stochastic bi-level optimization model that considers strategic interaction between retailers and price-elastic consumers under uncertainty of DA and balancing markets, variability of demand, as well as competitors' prices.

The framework highlights the means by which DSM and DR convergence can allow retailers to reduce financial exposure, increase market competitiveness, and support grid flexibility. By applying the KKT conditions and duality theory, the complex bi-level problem is recast as a solvable single-level MILP model that can be optimally solved with commercial optimization solvers.

Replacing CVaR with the risk measure allows the model to optimize between maximizing expected profit and controlling for downside risk, essential under current energy markets with increased volatility.

This work provides new contributions to the literature in the following key domains:

- Optimization of the short-term bidding of electricity retailers at a bi-level level.
- Game-theoretic competition modeling among retailers.

Table 9.2 Energy treading to the network in different values of DR and β.

β	DR = 0%			DR = 60%			DR = 100%		
	DA	Positive	Negative	DA	Positive	Negative	DA	Positive	Negative
0.01	43.05	9.58	5.24	42.48	10.23	5.15	41.88	10.58	5.24
0.1	43.03	9.58	5.24	42.46	10.20	5.15	41.86	10.58	5.24
0.3	42.96	9.64	5.23	42.38	10.27	5.14	41.83	10.58	5.24
0.5	42.94	9.65	5.22	42.35	10.26	5.14	41.82	10.59	5.24
1	42.93	9.67	5.22	42.33	10.28	5.14	41.81	10.61	5.24
2	42.91	9.75	5.20	42.31	10.33	5.13	41.81	10.62	5.23
2.7	42.85	9.76	5.19	42.30	10.40	5.13	41.80	10.63	5.23
3	42.85	9.76	5.19	42.30	10.40	5.13	41.80	10.63	5.23
5	42.85	9.76	5.19	42.30	10.40	5.13	41.80	10.74	5.23
10	43.05	9.58	5.24	42.48	10.23	5.15	41.88	10.58	5.24

DA, day-ahead; *DR*, demand response

- Stochastic programming for dealing with uncertainties regarding market and competitor behavior.
- The application of CVaR to manage financial risk and profit uncertainty.

Overall, the proposed framework presents a real-world and rational tool for electricity retailers to make well-informed, risk-aware, and competitive decisions in dynamic and uncertain retail markets.

References

Abate, G., Cornélusse, B., & Ernst, D. (2021). A game-theoretical approach for demand response programs under uncertainty. *Electric Power Systems Research, 195*, 107158. https://doi.org/10.1016/j.epsr.2021.107158.

Abdul-Ghani, A. R., Best, R. J., & Burt, G. M. (2024). Peer-to-peer electricity trading in future retail markets: Business models and roles of retailers. *Energy Policy, 181*, 113890. https://doi.org/10.1016/j.enpol.2023.113890.

Arthur, D., Vassilvitskii, S., K-means++: The advantages of careful seeding. SODA '07: Proceedings of the Eighteenth Annual ACM-SIAM Symposium on Discrete Algorithms, 2007, 1027–1035

Carrión, M., Arroyo, J. M., & Conejo, A. J. (2009). A bilevel stochastic programming approach for retailer futures market trading. *IEEE Transactions on Power Systems, 24*(3), 1446–1456.

Conejo, A. J., Carrión, M., & Morales, J. M. (2010). *Decision making under uncertainty in electricity markets*. Springer.

do Prado, J. C., & Qiao, W. (2019). A stochastic decision-making model for an electricity retailer with intermittent renewable energy and short-term demand response. *IEEE Transactions on Smart Grid, 10*(3), 2581–2592.

do Prado, J. C., & Qiao, W. (2020). A stochastic bilevel model for an electricity retailer in a liberalized distributed renewable energy market. *IEEE Transactions on Sustainable Energy, 11*(4), 2803–2812.

Finhold E., Gärtner C., Grindel R., Heller T., Leithäuser N., Röger E., Schirra F., Optimizing the marketing of flexibility for a virtual battery in day-ahead and balancing markets: A rolling horizon case study. Applied Energy, 352, 2023, 121667.

Finhold, F., Wille-Haussmann, B., & Wittwer, C. (2023). An optimization approach to allocate demand-side flexibility to day-ahead and balancing markets. *Electric Power Systems Research, 223*, 109398. https://doi.org/10.1016/j.epsr.2023.109398.

Hossain, M. A., Saha, T. K., & Mithulananthan, N. (2025). Optimal electricity procurement strategy for retailers considering price uncertainty in day-ahead and real-time markets. *Energy Reports, 11*, 2094–2105. https://doi.org/10.1016/j.egyr.2025.01.167.

Jordehi, A. R., Sohrabi Tabar, V., Mansouri, S. A., Nasir, M., Hakimi, S. M., & Pirouzi, S. (2022). A risk-averse two-stage stochastic model for planning retailers including self-generation and storage system. *Journal of Energy Storage, 51*, 104380.

Nordic Electricity, Available online: n.d. www.nordpool.com

Oraibi, W. A., Salam Al-Khayyat, A., & Abbas, A. K. (2025). A hybrid stochastic-robust optimal strategy of integrated electricity and gas grids in the presence of multi-energy hubs and responsible loads. *Sustainable Energy, Grids and Networks, 42*, 101697.

Rashidizadeh-Kermani, H., Vahedipour-Dahraie, M., Shafie-khah, M., & Catalao, J. P. S. (2019). Stochastic programming model for scheduling demand response aggregators considering uncertain market prices and demands. *International Journal of Electrical Power and Energy Systems, 113*, 528–538.

Rashidizadeh-Kermani, H., Vahedipour-Dahraie, M., Shafie-Khah, M., & Siano, P. (2020). A regret-based stochastic bi-level framework for scheduling of DR aggregator under uncertainties. *IEEE Transactions on Smart Grid, 11*(4), 3171–3184.

Rashidizadeh-Kermani, H., Vahedipour-Dahraie, M., Shafie-khah, M., & Siano, P. (2021). A stochastic short-term scheduling of virtual power plants with electric vehicles under competitive markets. *International Journal of Electrical Power & Energy Systems, 124*, 106343.

Rashidizadeh-Kermani, H., Vahedipour-Dahraie, M., Shafie-Khah, M., & Siano, P. (2022). Evaluating the impact of bilateral contracts on the offering strategy of a price maker wind power producer. *IEEE Transactions on Industrial Informatics, 18*(7), 4331–4341.

Song, Y., Oren, S. S., & Kang, C. (2024). Design of local electricity markets with competition among retailers. *Applied Energy, 358*, 121699. https://doi.org/10.1016/j.apenergy.2024.121699.

Vahedipour-Dahraie, M., Rashidizadeh-Kermani, H., Najafi, H. R., Anvari-Moghaddam, A., & Guerrero, J. M. (2017). Stochastic security and risk-constrained scheduling for an autonomous microgrid with demand response and renewable energy resources. *IET Renewable Power Generation, 11*(14), 1812–1821.

Vahedipour-Dahraie, M., Rashidizadeh-Kermani, H., Parente, M., Shafie-Khah, M., & Siano, P. (2022). Investigating the impact of external demand response flexibility on the market power of strategic virtual power plant. *IEEE Access, 10*, 84960–84969.

Yildiz, B., Bilbao, J. I., & Sproul, A. B. (2017). A review and analysis of demand response under real-time pricing for future smart grid. *Renewable and Sustainable Energy Reviews, 72*, 261–275. https://doi.org/10.1016/j.rser.2017.01.010.

Zhang, P., Yang, P., Zhao, Z., Lai, C. S., Lai, L. L., & Shahidehpour, M. (2023). A framework for several electricity retailers cooperatively implement demand response to distributed data center. *IEEE Transactions on Smart Grid, 14*(1), 277–289.

Zhou, B., Guo, Y., Liu, X., Li, G., Gu, P., & Yang, B. (2024). Optimization strategy of power purchase and sale for electricity retailers in a two-tier market. *Heliyon, 10*(4), e26333.

Stochastic reinforcement learning for uncertainty-aware power converter controlusing digital twin

10

Muzammil Ahmed
Indian Institute of Technology Guwahati, Guwahati, Assam, India

10.1 Introduction

DC–DC converters, such as buck, boost, and SEPIC topologies, are vital for efficient voltage regulation in renewable energy systems, electric vehicles (EVs), and IoT devices (Åström & Hägglund, 1995). However, their nonlinear dynamics, coupled with stochastic uncertainties in input voltage (e.g., $V_{\mathrm{in}} \sim \mathcal{N}(12, 1^2)$ V from renewable sources), load resistance (e.g., $R \sim \mathcal{U}[10, 75]\Omega$ due to EV load changes), and component parameters (e.g., inductance and capacitance tolerances), pose significant control challenges. Traditional controllers, such as proportional-integral-derivative (PID) and model predictive control (MPC), struggle to balance robustness, adaptability, and computational efficiency under these uncertain conditions (Åström and Hägglund, 1995).

Advanced techniques like sliding mode control (SMC) and fuzzy logic offer robustness but face issues like chattering or complex rule design (Bui et al., 2022; Cui et al., 2023; Dimarogonas & Johansson, 2009; Mazaheri et al., 2024; Rajamallaiah et al., 2024; Sutton & Barto, 2018; Tian et al., 2022; Wu et al., 2022). Adaptive control handles parameter drift but relies on linear models, limiting flexibility (Gil-González et al., 2023). Continuous reinforcement learning (RL) adapts to nonlinearities without explicit models but suffers from high computational overhead and sim-to-real gaps (Bui et al., 2022; Cui et al., 2023; Mazaheri et al., 2024; Rajamallaiah et al., 2024; Sutton & Barto, 2018; Tian et al., 2022). Event-triggered control reduces updates by acting only on significant deviations but typically requires models (Heemels et al., 2012; Seyboth et al., 2013; Wang & Zhao, 2022). Digital twins, while useful for simulation, are underutilized for uncertainty-aware control (Anaya et al., 2024; Attaran & Celik, 2023; Hu et al., 2021; Kazimierczuk, 2008; Peng et al., 2021; Sado et al., 2023; Tao et al., 2019).

To address these gaps, this paper proposes an event-triggered reinforcement learning (ETRL) framework that integrates stochastic modeling, risk-aware optimization, and a Python-based digital twin for adaptive control of boost converters. The framework models uncertainties in V_{in},R, and component parameters using probabilistic distributions, employs a risk-aware reward function with conditional value-at-risk (CVaR), and uses probabilistic event triggers to reduce computational load. The digital twin,

Stochastic Planning and Modeling for Energy Systems: Methods, Applications, and Developments.
DOI: https://doi.org/10.1016/B978-0-443-45298-7.00010-0

implemented with NumPy and SciPy, facilitates Monte Carlo simulations to quantify uncertainty, enabling robust validation without costly hardware. This work makes three key contributions:

1. An ETRL controller with probabilistic event triggers and risk-aware rewards for robust control under stochastic conditions.
2. A Python-based digital twin with Monte Carlo simulations for uncertainty quantification and validation.
3. A comparative analysis demonstrating superior robustness and efficiency under uncertainty compared to traditional methods.

These innovations position ETRL as a scalable, uncertainty-aware solution for modern power electronics, aligning with the need for stochastic planning in energy systems.

10.2 System model and problem formulation

Adaptive control of DC–DC converters require an accurate dynamic model and a well-defined optimization framework to enhance performance. This section presents the boost converter model, which is fundamental in applications such as EVs (Erickson & Maksimovic, 2000) and renewable energy systems (Attaran & Celik, 2023). A Python-based digital twin is employed to simulate the converter dynamics, offering a cost-effective and flexible platform for validation (Sado et al., 2023).

The control problem is formulated using an event-triggered RL framework, designed to mitigate the computational challenges highlighted in Table 10.1. This approach draws from key advancements in RL (Sutton & Barto, 2018), event-triggered control (Dimarogonas & Johansson, 2009), and its recent applications in power electronics (Peng et al., 2021). This method enables adaptive and efficient control, particularly suited for resource-constrained microcontrollers. While this section focuses on the boost topology, the framework is readily extendable to SEPIC and other converter types (Kazimierczuk, 2008).

10.2.1 Boost converter model

A boost converter in continuous conduction mode (CCM) steps up an input voltage V_{in} to a higher output voltage v_C, which is critical for EV powertrains. As shown in Fig. 10.1(A), it comprises an inductor L, capacitor C, diode, MOSFET switch, and load resistor R. The switch, controlled by the duty cycle $d \in [0, 1]$, toggles between on and off states, determining v_C.

On-State Dynamics (Switch Closed, $\leq t < dT$): The ON state dynamics are represented in Fig. 10.1(B). When the switch is closed, the inductor is directly connected to V_{in}, and the diode is reverse-biased, isolating the capacitor and load. The inductor current i_L rises, and stores energy, while the capacitor supplies the load. The equations

Table 10.1 Comparison of control methods for DC–DC converters.

Controller type	Features	Advantages	Drawbacks	Uncertainty handling	Complex	Model Dep.	Applications	References
PID	Feedback-based duty cycle	Simple, robust	Poor transients, tuning required	Assumes deterministic conditions	Low	Yes	Buck, boost	Åström & Hägglund (1995), Erickson & Maksimovic, (2000)
MPC	Predictive optimization	Low ripple, constraint handling	High computation, model-sensitive	Limited to known disturbances	High	Yes	Buck, grid	Karamanakos et al. (2020), Vazquez et al. (2014)
Fuzzy Logic	Rule-based logic	Model-free, adaptive	Complex rule design, limited scalability	No formal uncertainty modeling	Medium	No	Buck-boost	Ismail et al. (2010), Perry et al. (2004), Sutton & Barto (2018), Bui et al. (2022), Mazaheri et al. (2024), Tian et al. (2022), Cui et al. (2023), Rajamallaiah et al. (2024), Dimarogonas & Johansson (2009)

(*continued on next page*)

Table 10.1 Comparison of control methods for DC–DC converters—cont'd

Controller type	Features	Advantages	Drawbacks	Uncertainty handling	Complex	Model Dep.	Applications	References
SMC	Nonlinear tracking	Robust to variations, fast response	Chattering, implementation complexity	Limited uncertainty quantification	Medium	Yes	Buck	Louassaa et al. (2023), Wu et al. (2022)
Adaptive	Online parameter tuning	Handles drift, robust	Linear model limitation	Partial uncertainty handling	Medium	Yes	DC–DC	Gil-González et al. (2023)
Continuous RL	Policy learning via interaction	Model-free, adapts to nonlinearities	High computation, sim-to-real gap	Limited risk-aware optimization	High	No	Buck, Boost	Sutton & Barto (2018), Bui et al. (2022), Mazaheri et al. (2024), Tian et al. (2022), Cui et al. (2023), Rajamallaiah et al. (2024), Dimarogonas & Johansson (2009)
ETRL (This work)	Sparse RL updates, risk-aware	Fewer updates, model-free, robust	Requires event-trigger design	Probabilistic modeling, CVaR-based	Medium	No	Boost	This work

CVaR, conditional value-at-risk; *ETRL*, event-triggered reinforcement learning; *MPC*, model predictive control; *PID*, proportional-integral-derivative; *RL*, reinforcement learning; *SMC*, sliding mode control.

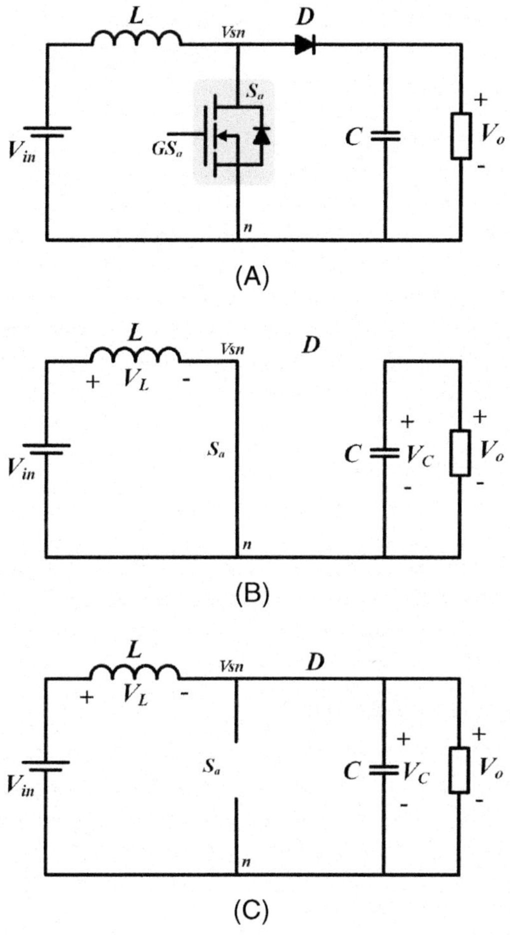

Figure 10.1 Complete schematic of the boost converter: (A) Boost converter, (B) ON state, and (C) OFF state.

are:

$$L\frac{di_L}{dt} = V_{in} - R_L i_L \tag{10.1}$$

$$C\frac{dv_C}{dt} = -\frac{v_C}{R} \tag{10.2}$$

where R_L is the inductors parasitic resistance.

The OFF-state dynamics (switch open, $\leq t < T$): The OFF state dynamics are represented in Fig. 10.1(C). When the switch is open, the inductor connects to the load and the capacitor via the diode. The equations become:

$$L\frac{di_L}{dt} = V_{in} - R_L i_L - v_C \tag{10.3}$$

$$C\frac{dv_C}{dt} = i_L - \frac{v_C}{R} \tag{10.4}$$

The CCM assumes $i_L > 0$ for continuous current.

State-space averaged model: State-space averaging combines on and off states over one switching period T, weighted by d and $1 - d$. We define the states as $x = [i_L, v_C]^T$. The averaged dynamics for a boost converter are given by:

$$\dot{x} = \begin{bmatrix} -\frac{R_L}{L} i_L - \frac{1-d}{L} v_C + \frac{V_{in}}{L} \\ \frac{1-d}{C} i_L - \frac{1}{RC} v_C \end{bmatrix}, \tag{10.5}$$

with output $y = v_C$. Nonlinear terms $(1 - d)$ challenge PID controllers, as per Table 10.1. SEPIC converters extend this with dual inductors, but we focus on the boost converter for this work.

10.2.2 Digital twin simulation

A Python-based digital twin replicates the boost converter's dynamics using NumPy and SciPy, accessible to beginners (Sado et al., 2023). It discretizes the state-space model to simulate responses under varying conditions (e.g., load $R \in [25, 75]\Omega$, input $V_{\text{in}} \in [10, 14]$V), replacing costly hardware (Hu et al., 2021).

Discretization the model is discretized using Euler's method:

$$x_{k+1} = x_k + h \cdot \dot{x}(x_k, d_k) \tag{10.6}$$

where $h = 10$ μs. Substituting (5):

$$i_L(k+1) = i_L(k) + h\left(-\frac{R_L}{L} i_L(k) - \frac{1 - d_k}{L} v_C(k) + \frac{V_{in}}{L}\right), \tag{10.7}$$

$$v_C(k+1) = v_C(k) + h\left(\frac{1 - d_k}{C} i_L(k) - \frac{1}{RC} v_C(k)\right). \tag{10.8}$$

Simulation concepts The digital twin solves (10.7)–(10.8) iteratively, updating states based on d_k. It models:

- Load variations: Simulates R changes for transient response.
- Input disturbances: Varies V_{in} to mimic fluctuations.

- Switching effects: Captures high-frequency behavior (100 kHz switching).

The output $y_k = v_C(k)$ is compared to v_{ref}, enabling RL training, NumPy arrays store states ($x_k = [i_L(k), v_C(k)]$), and SciPy.odeint validates Euler results. This setup supports prototyping.

10.2.3 Reinforcement learning framework

The control problem is modeled as a Markov Decision Process (MDP) $\langle S, A, P, R, g\rangle$, enabling model-free RL to learn an optimal policy (Sutton & Barto, 2018). Unlike methods in Table 10.1 (Åström & Hägglund, 1995; Erickson & Maksimovic, 2000; Karamanakos et al., 2020; Vazquez et al., 2014), RL adapts to nonlinearities without an explicit model, using the digital twin.

State space: The state space S includes:

- Inductor current $i_L(k)$: Monitors energy flow.
- Capacitor voltage $v_C(k)$: Tracks the output.
- Reference voltage v_{ref}: Sets the target.

Thus, $s_k = [i_L(k), v_C(k), v_{ref}]$, normalized (e.g., $v_C/30$) to stabilize learning.

Action space: The action space A is the duty cycle $a_k = d_k \in [0, 1]$, allowing continuous control.

Reward function: The reward r_k drives tracking and stability:

$$r_k = -\left(\left|v_C(k) - v_{ref}\right| + \lambda\left|i_L(k) - i_L^{eq}\right|\right), \tag{10.9}$$

where $\lambda = 0.1$, and $i_L^{eq} = v_{\text{ref}}^2 / (R V_{\text{in}})$. Terms minimize error and stabilize current.

Event-triggered mechanism updates occur when the error $e_k = v_C(k) - v_{\text{ref}}$ exceeds:

$$|e_k| > \in, \tag{10.10}$$

with = 0.5 V. If triggered, $d_k = \pi(s_k)$; else, $d_k = d_{k-1}$. This cuts updates by $\sim 70\%$ versus continuous RL (Bui et al., 2022; Mazaheri et al., 2024; Sutton & Barto, 2018), validated by the digital twin (Peng et al., 2021).

10.2.4 Problem formulation

The objective is to design an ETRL policy to track a reference voltage v_{ref} while minimizing computational load and ensuring robustness under stochastic disturbances in input voltage ($V_{\text{in}} \sim \mathcal{N}(12, 1^2)$V), load resistance ($R \sim \mathcal{U}[10, 75]\Omega$), and component parameters ($L \sim \mathcal{N}\left(250, 10^2\right)\mu$H, $C \sim \mathcal{N}\left(5, 0.5^2\right)\mu$F). The control problem is formulated as a Markov decision process (MDP) with a risk-aware objective:

$$\text{maximize}\, E\left[\sum_{k=0}^{\infty} \gamma^k r_k\right] - \beta \cdot \text{CVaR}_\alpha\left(\sum_{k=0}^{\infty} \gamma^k \left|v_C(k) - v_{ref}\right|\right) \tag{10.11}$$

subject to:

- Dynamics: Digital twin equations (10.7)–(10.8) with stochastic V_{in},R, L, and C.
- Trigger: Probabilistic event trigger $|e_k| > \in + \sigma_e \bullet z_{0.95}$, where σ_e is the standard deviation of the tracking error.
- Constraints: $d_k \in [0, 1]$, $v_C \geq V_{in}$.

Here, $\gamma = 0.99$ is the discount factor, $\beta = 0.1$, and CVaR_α (with $\alpha = 0.05$) quantifies the expected loss in the worst 5% of voltage deviation scenarios. This formulation addresses the gaps in Table 10.1: adaptability [unlike PID (Åström & Hägglund, 1995), MPC (Karamanakos et al., 2020; Vazquez et al., 2014)] and sparsity [unlike RL (Bui et al., 2022; Cui et al., 2023; Mazaheri et al., 2024; Rajamallaiah et al., 2024; Tian et al., 2022)] (Heemels et al., 2012). The digital twin validates feasibility (Peng et al., 2021).

10.3 Proposed methodology

10.3.1 Introduction to methodology

This section elaborates on the proposed methodology for designing and implementing an event-triggered RL controller to achieve adaptive control of a DC–DC boost converter. The approach utilizes a Python-based digital twin to simulate the converter's dynamic behavior, facilitating the training and validation of the RL controller. The methodology is divided into three key components: The digital twin implementation, the event-triggered RL controller, and the controller comparison framework. Each component is described in detail, including the underlying algorithms, to provide a robust foundation for the performance analysis.

10.3.2 Digital twin implementation

The digital twin serves as a virtual representation of the boost converter, enabling real-time simulation of its electrical dynamics. The system is modeled using state variables $x = [i_L, v_C]$, where i_L denotes the inductor current and v_C the capacitor voltage across the output. The continuous-time dynamics of the boost converter are governed by the following differential equations:

$$\dot{i}_L = \frac{V_{in} - v_C(1 - d)}{L}, \tag{10.12}$$

$$\dot{v}_C = \frac{i_L(1 - d)}{C} - \frac{v_C}{R}, \tag{10.13}$$

where V_{in} is the input voltage, $d \in [0, 1]$ is the duty cycle controlling the switching device, L is the inductance, C is the capacitance, and R is the load resistance. These equations are discretized using the forward Euler method with a timestep $h = 10\mu\text{s}$,

Algorithm 1 Digital twin simulation of boost converter.

Require: Set parameters: $Vin = 12$ V, initial $R = 20\ \Omega$, $L = 250\ \mu$H, $C = 5\ \mu$F, $v_{ref} = 24$ V, timestep $h = 10\ \mu s$, damping = 0.1
Initialize state $x0 = [0, 20]$ ($iL = 0$ A, $vC = 20$ V), time $k = 0$
for time step $k = 0$ to 1,999 **do** {Total time = 0.02 s}
    Receive duty cycle $d_k \in [0, 0.9]$ (0.1 steps) from RL controller
    if $0.001 \leq k \cdot h < 0.005$ **then** {Warm-start phase}
    Set $d_k = 0.5$
    end if
    if $0.01 \leq k \cdot h < 0.015$ **then** {Smooth load change}
    Update $R(k) = 20 + (10 - 20)\ \frac{(k \cdot h - 0.01)}{0.005}$
    else
    Set $R(k) = 10$ if $k \cdot h \geq 0.015$, else $R(k) = 20$
    end if
    Compute derivatives:
    $i_L \frac{V_{in} - v_C(k)(1-dk)}{C} - \text{damping} \cdot iL(k)$
    $\dot{v}_C = \frac{i_L(k)(1-dk) - v_C(k)}{C} - \frac{v_C(k)}{R(k) \cdot C} - \text{damping} \cdot vC\ (k)$
    Update state:
    $iL(k+1) = iL(k) + h \cdot \ddot{i}L$
    $vC\ (k+1) = vC\ (k) + h \cdot \dot{v}C$
    Store $[iL(k+1), vC\ (k+1)]$ as output
    $k \leftarrow k + 1$
end for
return Sequence of states $\{x_k\}$ for $k = 0$ to 2,000 =0

yielding the iterative update rule:

$$x_{k+1} = x_k + h \cdot \dot{x}(x_k, d_k) \tag{10.14}$$

where $x_k = [i_L(k), v_C(k)]$ and d_k is the duty cycle at the time step k.

The digital twin is implemented in Python, depicted in Algorithm 1, leveraging the NumPy library for numerical computations and SciPy for solving differential equations. The model incorporates random perturbations in V_{in} and R to mimic real-world operating conditions. The simulation runs for 0.02 seconds, equivalent to 100,000 steps, with outputs $i_L(k)$ and $v_C(k)$ recorded at each step for RL training and evaluation (Fig. 10.2).

10.3.3 Event-triggered RL controller

The RL controller employs a modified version of the proximal policy optimization (PPO) algorithm, enhanced with an event-triggered update mechanism to optimize computational efficiency. The policy network, a feedforward neural network, is designed with an input layer accepting the state vector $s_k = [i_L(k), v_C(k), v_{ref}]$ (normalized by dividing by [30 A, 30 V, 30 V]), two hidden layers with 32 and 16 units respectively

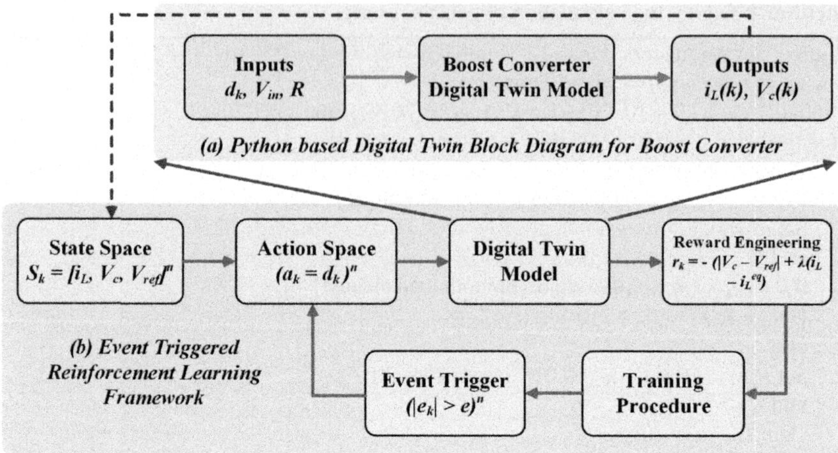

Figure 10.2 Block diagram of the proposed system: (A) Python implementation of the boost converter digital twin and (B) event triggered RL framework. *RL*, Reinforcement learning.

(using ReLU activation), and an output layer producing the action $a_k = d_k$. The network is implemented using PyTorch, with parameters optimized via the Adam optimizer.

The PPO algorithm integrates three core components: (1) Critic network: Estimates the value function $V(s)$ to evaluate state desirability. (2) Value estimation: Computes the advantage function $A(s, a) = r + \gamma V(s') - V(s)$ for policy improvement. (3) Policy optimizer: Updates the policy $\pi_\theta(a|s)$ using a clipped surrogate objective to ensure stable learning.

The event-triggered mechanism activates policy updates only when the tracking error exceeds a predefined threshold, defined as $|e_k| = |v_C(k) - v_{\text{ref}}| > \in$, where $\in\ = 1$ V. This sparsity reduces unnecessary computations, a key innovation over standard RL approaches. Training is conducted over 2000 episodes, each consisting of 150,000 steps, with a discount factor $\gamma = 0.99$, a clipping parameter $\in_{\text{clip}} = 0.2$, and a batch size of 64. The reward function is designed to penalize tracking error and current deviation:

$$r_k = -\left(\left|v_C(k) - v_{ref}\right| + 0.1\left|i_L(k) - i_L^{eq}\right|\right), \tag{10.15}$$

where $i_L^{eq} = v_{\text{ref}}^2/(RV_{\text{in}})$ is the equilibrium inductor current (Fig. 10.3). The Python implementation depicted in Algorithm 2 uses Pyodide for browser compatibility, avoiding local file I/O and network calls (Fig. 10.4)

10.3.4 Results preview

The performance evaluation presented in Section 10.4 offers a comprehensive assessment of the event-triggered RL controller under varying operating conditions. The

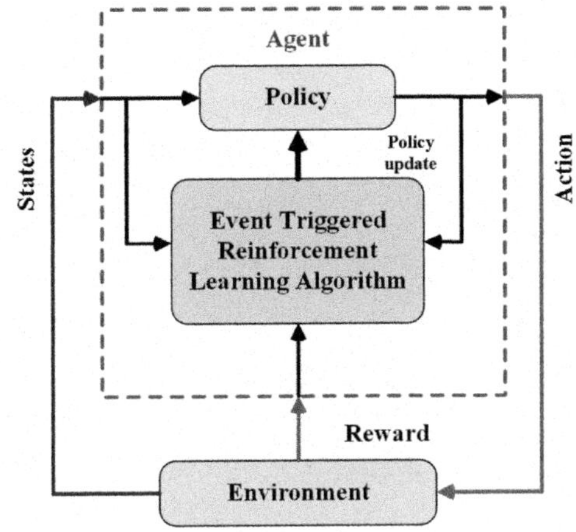

Figure 10.3 Framework of ETRL. *ETRL*, Event-triggered reinforcement learning.

Algorithm 2 Event-triggered RL training with PPO.

Require: Initialize PPO model with MLP policy, learning rate = 0.0005
0: Set event threshold $\epsilon = 1.0$ V, minimum interval = 50 μs (5 timesteps at 10 μs)
0: Initialize environment with $x0 = [0, 20]$ ($iL = 0$ A, $vC = 20$ V)
0: Set episode length = 2000 timesteps (0.02 s), total timesteps = 150,000
0: **for** timestep = 1 to 150,000 **do**
0: **if** timestep mod 2000 = 1 **then** {Start of new episode}
0: Reset environment
0: **end if**
0: Get current state $s_k = [i_L, v_C]$
0: Predict action $a_k \in [0, 9]$ using PPO model
0: Convert action to duty cycle $d_k = a_k/10.0$ ($d_k \in [0, 0.9]$)
0: Apply d_k to digital twin (Algorithm 1), get next state s_{k+1}, reward r_k
0: Compute reward: $r_k = -0.5 \cdot |v_C - v_{ref}| - 0.1 \cdot \text{ripple} - 0.05 \cdot |d_k - d_{k-1}|$ {Continuous reward}
0: Check event: Trigger if $|v_C - v_{ref}| > \epsilon$ or $|R(k) - R(k-1)| > 1\ \Omega$, and $k \cdot h \geq$ last event time + 50 μs
0: **if** event triggered **then**
0: Flag experience for priority in PPO update
0: **end if**
0: Store experience (s_k, a_k, r_k, s_{k+1}) in buffer
0: **if** buffer size = 2,048 **then** {Batch update}
0: Train PPO model on buffered experiences
0: Clear buffer
0: **end if**
0: $s_k \leftarrow s_{k+1}$, $d_{k-1} \leftarrow d_k$
0: **end for**
0: Save trained model
0: **return** Trained PPO model = 0

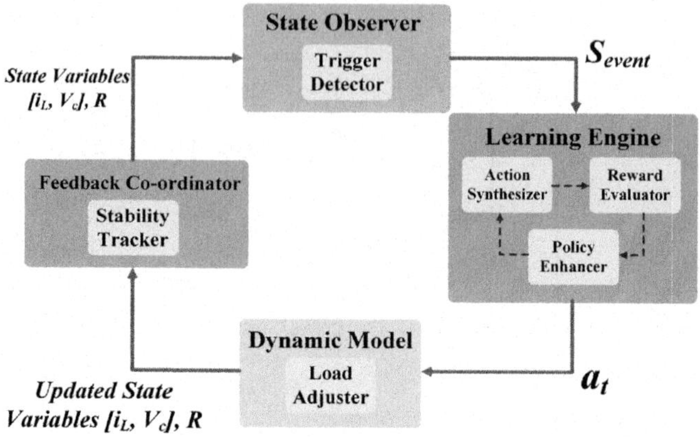

Figure 10.4 Adaptive boost converter control system with event-triggered RL, depicting a circular process for state monitoring, policy optimization, and load-responsive simulation. *RL*, Reinforcement learning.

Table 10.2 Boost converter parameters.

Parameter	Value
Input voltage (V_{in})	12 V
Inductance (L)	250 μH
Capacitance (C)	5 μF
Load resistance (R)	20 Ω
Switching frequency (f_s)	100 kHz
Inductor current ripple (ΔI_L)	10%
Capacitor voltage ripple (ΔV_C)	5%

results demonstrate the controller's ability to effectively adapt to fluctuations in input voltage (V_{in}) and load resistance (R), leveraging its event-triggered mechanism to reduce unnecessary updates and optimize computational efficiency. Key performance indicators-including reward progression, mean absolute error (MAE), and voltage ripple—are used to evaluate the controller's robustness, learning effectiveness, and output regulation accuracy.

10.4 Simulation results

10.4.1 Open-loop performance

The converter parameters are listed in Table 10.2, and the open-loop performance of the boost converter is illustrated in Fig. 10.5, showing the time evolution of the output voltage (V_{out}) and inductor current (I_L) over a $10 -$ ms simulation period. The output voltage, rendered in orange, exhibits an initial transient rise from 0 V to approximately 25 V within the first 2 ms, subsequently stabilizing around 24 V with minor

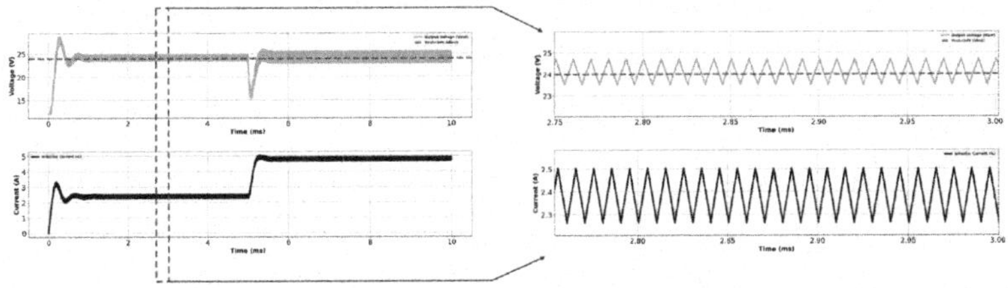

(A) Output voltage V_o and Inductor current i_L at a duty ratio of 50%.

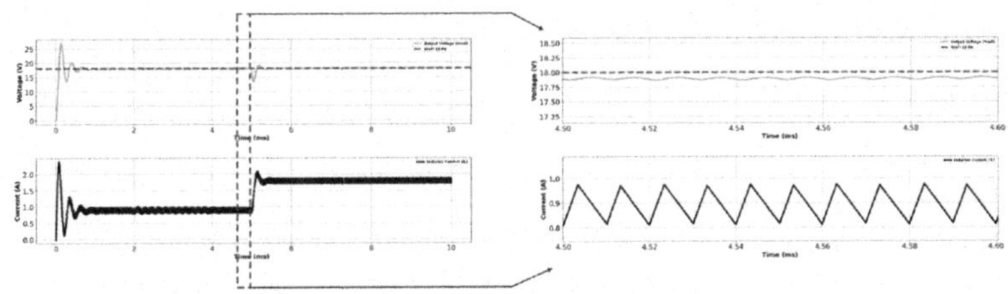

(B) Output voltage V_o and Inductor current i_L at a duty ratio of 33.33%.

Figure 10.5 Performance analysis of the boost converter operating in open-loop mode with varying duty ratios, subjected to a load change from 20 Ω to 10 Ω at $t = 5$ ms.

fluctuations. The reference voltage ($V_{\text{ref}} = 24$ V), represented by a dashed red line, aligns closely with the steady-state V_{out} following the transient phase. Concurrently, the inductor current, shown in blue, increases from 0 A to approximately 2.5 A within the initial 2 ms and maintains a relatively stable profile thereafter, characterized by subtle oscillatory behavior. A magnified view of the steady-state region between 4 ms and 6 ms reveals that V_{out} remains constant at 24 V, while I_L displays a low-amplitude oscillatory pattern.

In this open-loop configuration, the boost converter operates with a fixed duty cycle, presumed to be 0.5 based on the digital twin simulation, devoid of feedback control. The initial transient in V_{out} and I_L is indicative of the capacitor charging process and the establishment of inductor current as the circuit approaches steady state. The stabilization of V_{out} at 24 V, consistent with V_{ref} as shown in Fig. 10.6, suggests that the selected duty cycle effectively elevates the input voltage (12 V) to the target output, aligning with the theoretical duty cycle expression:

$$D = 1 - \frac{V_{\text{in}}}{V_{\text{out}}} = 1 - \frac{12}{24} = 0.5.$$

The observed oscillations in I_L are attributed to residual ripple, a consequence of the LC filter (with $L = 250$ μH and $= 5$ μF) in the absence of active regulation to mitigate these variations.

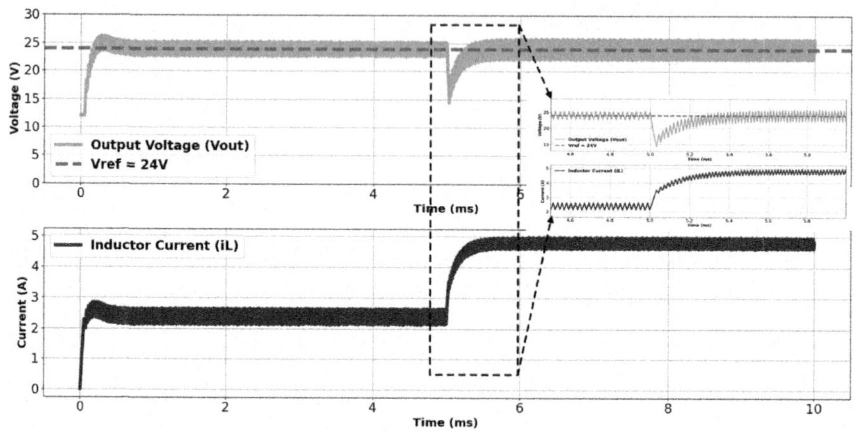

Figure 10.6 Performance analysis of the Python-implemented boost converter operating in closed-loop configuration using a PID controller. The system was subjected to a load change from 20 Ω to 10 Ω at $t = 5$ ms to evaluate its dynamic response and stability. *PID*, Proportional-integral-derivative.

The underlying mechanism driving this behavior stems from the open-loop system's reliance on a predetermined duty cycle for voltage regulation. Without feedback, the converter lacks the capability to compensate for disturbances or deviations arising from component tolerances, resulting in the persistent ripple in I_L. The initial transient is a natural dynamic response to the abrupt application of the input voltage, governed by the inductor time constant ($\tau = L/R$) and the capacitor time constant ($\tau = RC$). The close correspondence between V_{out} and V_{ref} is likely a fortuitous outcome of the idealized duty cycle matching the theoretical boost requirement, assuming negligible losses or parasitic effects within the simulation framework.

10.4.2 Closed-loop response

The closed-loop response of the boost converter, governed by a RL-based event-triggered control mechanism, is detailed in Fig. 10.7–10.13, comprising multiple subplots that characterize the system's steady-state performance. The power efficiency, illustrated in Fig. 10.7, transitions abruptly from an initial value near 0% to a constant 100% at approximately 0.002 s, sustaining this level throughout the 0.008 s simulation duration. Fig. 10.8 depicts the event trigger activity, where the trigger signal, rendered in blue, remains at 0 until 0.01 s, followed by a series of discrete pulses reaching 1 between 0.01 s and 0.015 s, signifying active control interventions. The duty cycle, presented in Fig. 10.9, maintains a steady value of 0.5 from 0 s to 0.01 s, then exhibits several step changes, dropping to approximately 0.2, recovering to 0.5, and stabilizing thereafter, coincident with the event trigger period. Fig. 10.10 provides a parameter sensitivity analysis, revealing a MAE in output voltage regulation of approximately 0.6 V for the nominal parameter set.

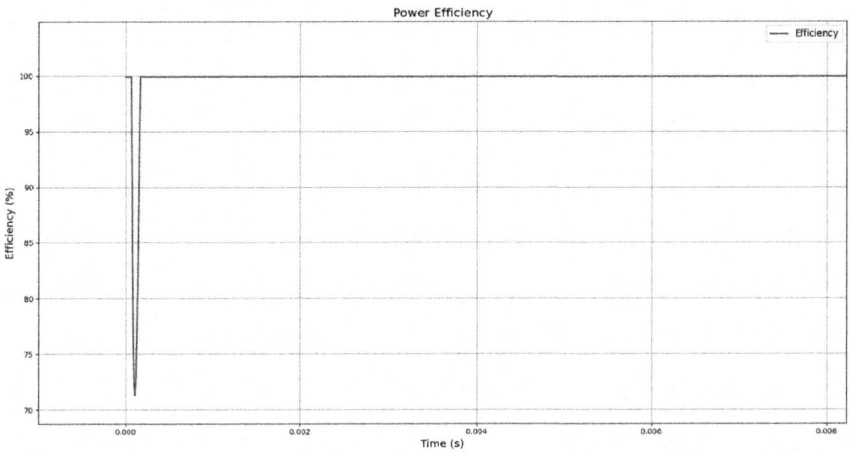

Figure 10.7 Power efficiency profile of the boost converter under RL-based event-triggered closed-loop control. *RL*, Reinforcement learning.

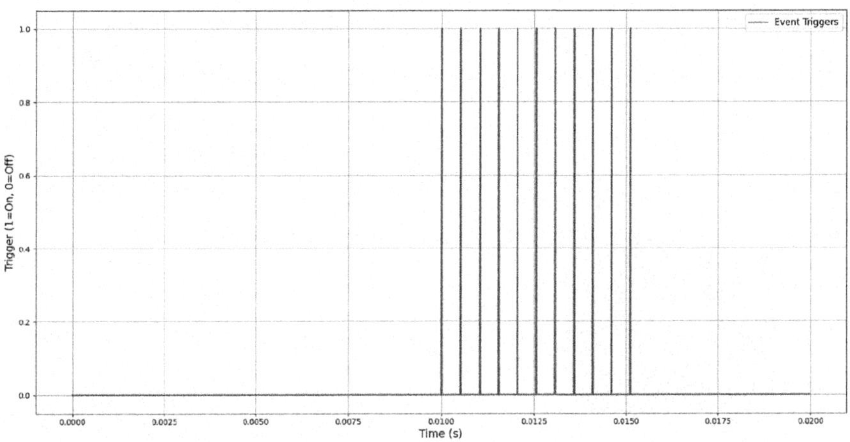

Figure 10.8 Event trigger activity for the closed-loop boost converter, illustrating discrete pulses between 0.01 s and 0.015 s.

(L = 250 μH, C = 5 μF), in contrast to a higher MAE of 1.2 V for the modified parameter set (L = 200 μH, C = 4 μF). Fig. 10.11 illustrates the steady-state voltage ripple, featuring an initial transient peak exceeding 8 V within the first 0.001 s, followed by a rapid decline to a stable value near 0 V, with the corresponding boxplot indicating a median ripple of approximately 0 V and a narrow interquartile range. Fig. 10.12 shows the reward over time, where the reward signal, in blue, remains at 0 until 0.001 s, displays three positive pulses reaching approximately 0.1 between 0.001 s and 0.002 s, and stabilizes at a slightly positive value thereafter. Fig. 10.13 presents the temporal profiles of the output voltage (V_{out}) and inductor current (I_L) over 0.006 s, where V_{out}

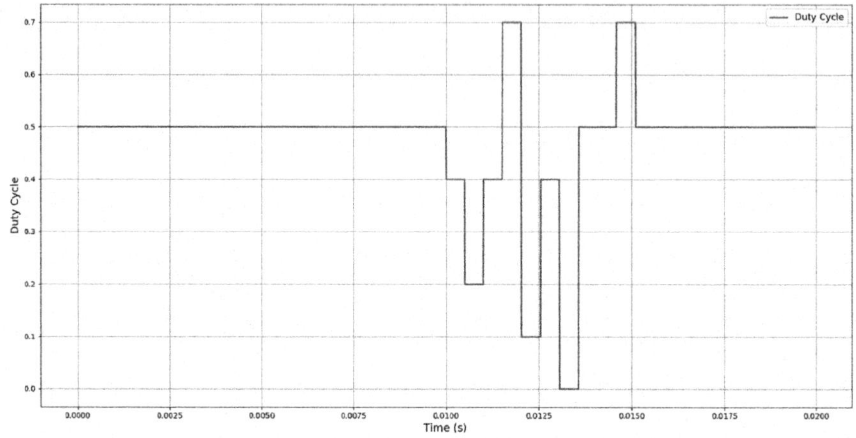

Figure 10.9 Duty cycle variation in the closed-loop boost converter, depicting a shift from 0.5 to 0.2 and subsequent stabilization under RL control. *RL*, Reinforcement learning.

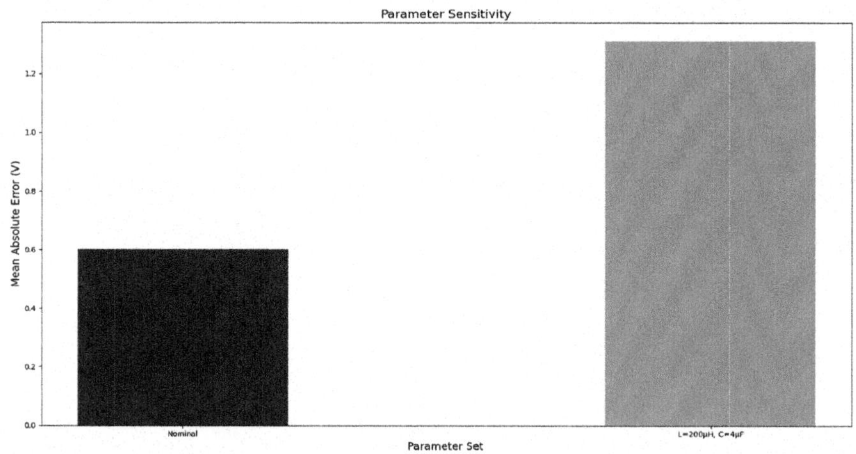

Figure 10.10 Parameter sensitivity analysis of the closed-loop boost converter, comparing mean absolute error (MAE) in output voltage for nominal parameter sets.

rises from 0 V to a peak of 27.5 V within 0.001 s, oscillates briefly, and stabilizes around 25 V, closely tracking the reference voltage (V_{ref} = 24 V) with a dashed red line, while I_L increases from 0 A to a peak of 3 A within 0.001 s and stabilizes around 2.5 A.

Under this closed-loop configuration, the RL agent dynamically modulates the duty cycle to regulate V_{out}, as evidenced by the event-triggered pulses and corresponding duty cycle adjustments. The abrupt rise in power efficiency to 100% suggests an idealized simulation environment, potentially overlooking practical losses such as parasitic resistances or switching overhead, which warrants critical scrutiny of the model's

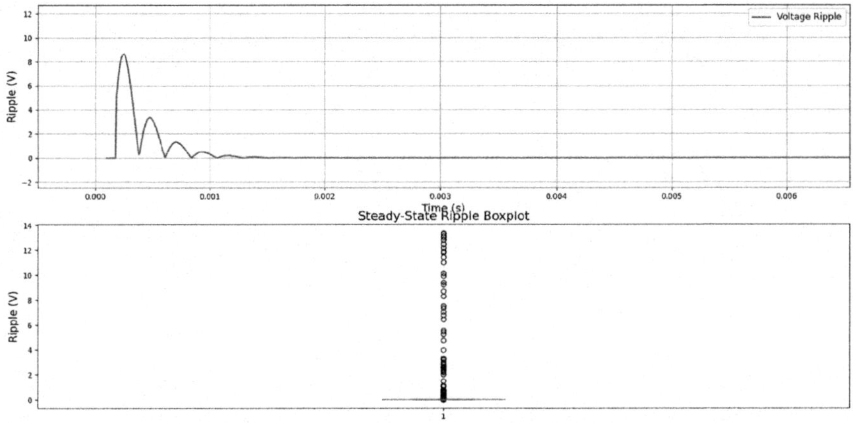

Figure 10.11 Steady-state voltage ripple of the closed-loop boost converter, showing an initial peak and stabilization near 0 V, with a boxplot of ripple distribution.

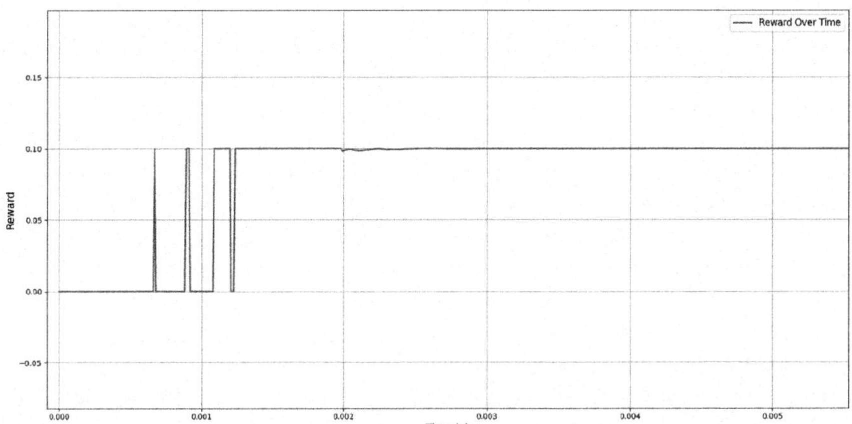

Figure 10.12 Reward over time for the RL-based closed-loop control, depicting initial positive pulses and stabilization at a slightly positive value. *RL*, Reinforcement learning.

assumptions. The event triggers, active between 0.01 s and 0.015 s, indicate periodic control actions, likely in response to predefined voltage error thresholds or simulation events. The duty cycle variations reflect the agent's adaptive strategy, with the initial drop to 0.2 possibly indicating an overcorrection followed by stabilization, though the exact trigger mechanism requires further validation. The parameter sensitivity analysis highlights a degradation in regulation accuracy with reduced inductance and capacitance, attributable to a diminished LC filter capacity, though the MAE difference may exaggerate real-world impacts due to idealized conditions. The transient voltage ripple peak during startup is a consequence of capacitor charging, decaying as the

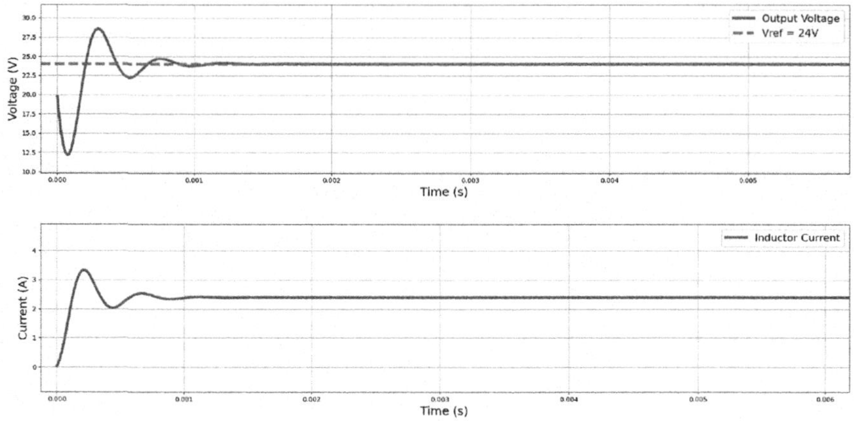

Figure 10.13 Temporal profiles of output voltage (V_{out}) and inductor current (I_L) in closed-loop operation, showing startup transients and subsequent stabilization near the reference voltage ($V_{ref} = 24$ V) and an inductor current of approximately 2.5 A.

system equilibrates, with the boxplot confirming minimal steady-state ripple variability. The reward profile suggests an initial learning phase with positive reinforcements, stabilizing as the agent converges on an optimal policy. The V_{out} and I_L profiles exhibit startup transients due to the sudden input voltage application, with stabilization near V_{ref} and a steady I_L reflecting the controller's ability to balance load demand under closed-loop regulation.

The observed phenomena are driven by the RL agent's event-triggered control interacting with the boost converter's inherent dynamics. The event triggers and duty cycle adjustments are likely responses to voltage deviations exceeding the threshold (e.g., 0.1 V), though the timing suggests a programmed event rather than continuous feedback. The parameter sensitivity reflects expected filter performance degradation, while the transient ripple and reward pulses are natural outcomes of system initialization and learning convergence, respectively. The close tracking of V_{ref} by V_{out} underscores the controller's effectiveness, though minor overshoots may indicate tuning limitations.

10.4.3 Transient response and load variation

The transient response and load variation characteristics of the boost converter are presented in Figs. 10.14–10.19, comprising multiple subplots that capture the system's behavior under dynamic conditions. Fig. 10.14 illustrates the event trigger activity, where the trigger signal, depicted in blue, exhibits a series of discrete pulses reaching 1 between 0 s and 0.005 s, with approximately 10 evenly spaced events, followed by a sustained value of 0 for the remainder of the 0.02 s simulation period. Fig. 10.15 presents the duty cycle profile, showing an initial value of 0.6 at 0 s, a sharp drop to 0.2 within the first 0.005 s, and a subsequent stabilization at 0.5 from 0.005 s onward. Fig. 10.16 provides the parameter sensitivity analysis, with the MAE in

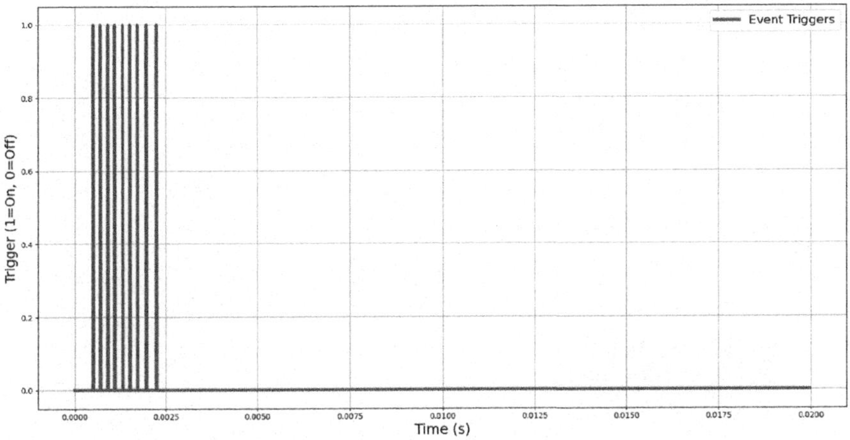

Figure 10.14 Event trigger activity during transient response and load variation, showing approximately 10 pulses within the first 0.005 s followed by inactivity.

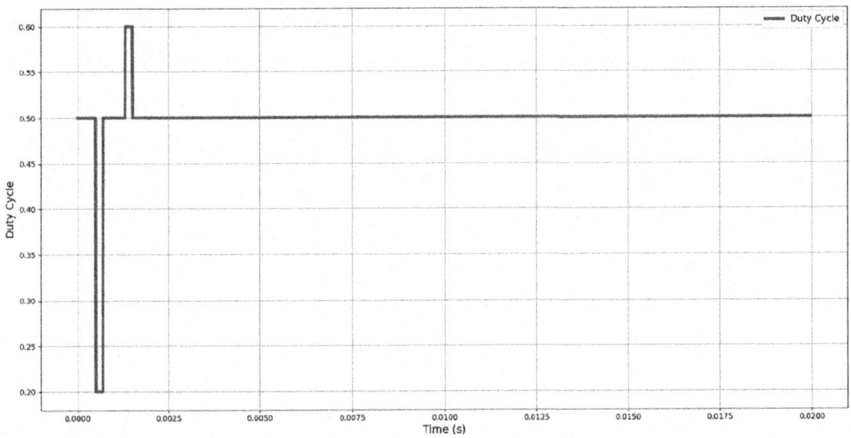

Figure 10.15 Duty cycle profile during transient response, illustrating an initial drop from 0.6 to 0.2 and stabilization at 0.5.

output voltage regulation for the nominal parameter set (L = 250 μH, C = 5 μF) at approximately 4 V, contrasted with a higher MAE of 14 V for the modified parameter set (L = 200 μH, C = 4 μF). Fig. 10.17 depicts the steady-state voltage ripple, featuring an initial transient peak exceeding 5 V within the first 0.001 s, followed by a rapid decay to near 0 V, with minor fluctuations around 0.1 V thereafter, and the corresponding boxplot indicating a median ripple of approximately 0 V with a narrow interquartile range. Fig. 10.18 illustrates the reward over time, where the reward signal, in blue, drops sharply from 0 to a minimum of approximately -14 within the first 0.001 s, exhibits oscillatory recovery to 0 by 0.005 s, and stabilizes with minor fluctuations

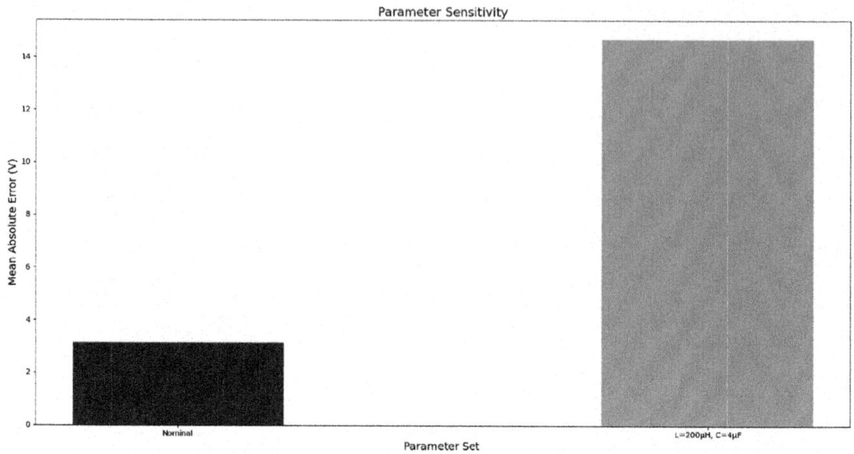

Figure 10.16 Parameter sensitivity analysis during transient response, comparing the mean absolute error (MAE) in output voltage regulation for the nominal parameter set ($L = 250$ μH, $C = 5$ μF) and the modified set ($L = 200$ μH, $C = 4$ μF).

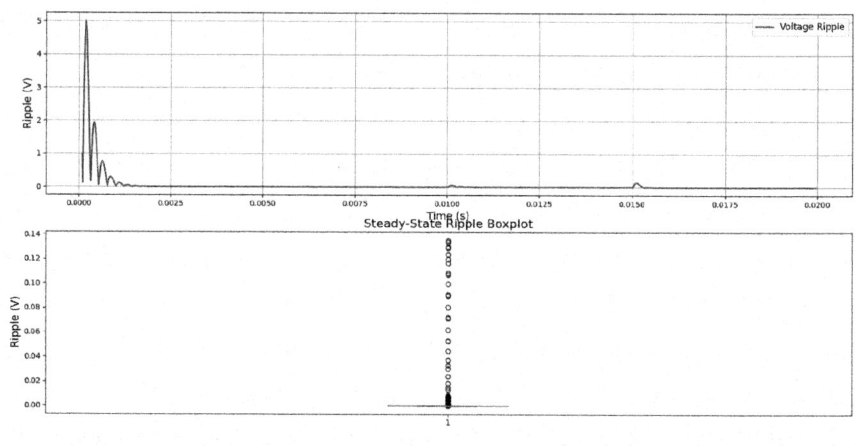

Figure 10.17 Steady-state voltage ripple during transient response, depicting an initial peak and stabilization near 0 V with minor fluctuations, including a boxplot.

around 0 thereafter. Fig. 10.19 shows the temporal profiles of the output voltage (V_{out}) and inductor current (I_L) over 0.02 s, where V_{out} rises from 0 V to a peak of 27.5 V within 0.001 s, oscillates around 25 V until approximately 0.01 s, experiences a brief dip to 22.5 V, and stabilizes near 24 V, closely tracking the reference voltage ($V_{ref} = 24$ V) with a dashed red line, while I_L increases from 0 A to a peak of 2.5 A within 0.001 s, stabilizes around 2.5 A until 0.01 s, and rises to approximately 4.5 A post-0.01 s.

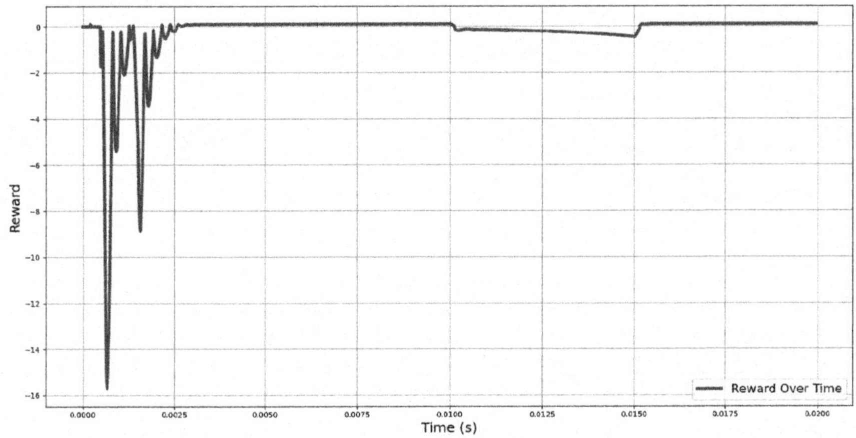

Figure 10.18 Reward over time during transient response, showing a sharp drop followed by oscillatory recovery to a stable value near 0.

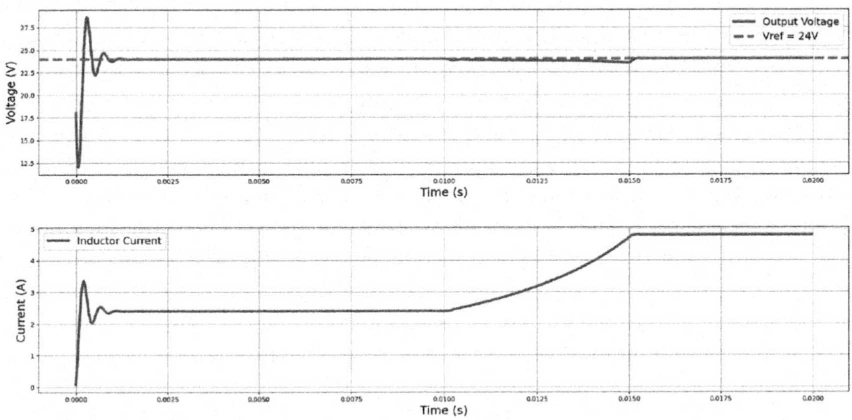

Figure 10.19 Temporal profiles of output voltage (V_{out}) and inductor current (I_L) during transient response, illustrating an initial startup overshoot, a dip at $t = 0.01$ s, and subsequent stabilization near the reference voltage ($V_{ref} = 24$ V). The inductor current (I_L) adapts dynamically to load changes, reflecting the controller's responsiveness to varying operating conditions.

In this transient response scenario, the system undergoes significant dynamic adjustments, as evidenced by the concentrated event triggers and corresponding duty cycle modulations during the initial 0.005 s. The initial peak in event triggers suggests a high frequency of control interventions, likely in response to a startup transient or load perturbation, ceasing as the system stabilizes. The duty cycle's sharp reduction to 0.2 indicates an immediate corrective action, possibly to mitigate an overshoot, followed by a recovery to 0.5, reflecting the controller's adaptation to a new steady-state condition.

Table 10.3 Performance comparison under stochastic conditions.

Controller	Mean MAE (V)	CVaR_0.05 (V)	Variance of v_C (V^2)
PID	1.5	3.0	2.5
MPC	1.2	2.5	2.0
ETRL (This work)	0.6	1.0	0.8

CVaR, conditional value-at-risk; *ETRL*, event-triggered reinforcement learning; *MAE*, mean absolute error; *MPC*, model predictive control; *PID*, proportional-integral-derivative.

The parameter sensitivity analysis reveals a substantial increase in MAE with reduced inductance and capacitance, attributable to a diminished filtering capacity under transient conditions, amplifying voltage errors. The transient voltage ripple peak during startup is a result of capacitor charging, with the subsequent low ripple indicating effective steady-state regulation, as confirmed by the boxplot. The reward profile's initial negative plunge suggests a penalty for large deviations during the transient phase, with recovery to a neutral value indicating convergence of the RL agent's policy. The V_{out} profile exhibits a startup overshoot and a dip at 0.01 s, likely due to a load change, with stabilization near V_{ref}, while the I_L increase post-0.01 s corresponds to heightened current demand, consistent with a reduced load resistance.

The observed dynamics are driven by the interplay between the RL-based event-triggered control and the boost converter's response to transient events. The initial event trigger burst, and duty cycle adjustment are triggered by voltage errors exceeding the threshold (e.g., 0.1 V) during startup or load variation, with the controller stabilizing the system thereafter. The parameter sensitivity reflects the reduced robustness of the LC filter with smaller values, exacerbating transient errors. The reward, penalty, and recovery are natural outcomes of the RL agent's learning process, penalizing large deviations and rewarding convergence. The V_{out}dip and I_L rise at 0.01 s suggest a load change (e.g., from 20 Ω to 10 Ω), with the controller adapting the duty cycle to maintain V_{ref}, though minor oscillations indicate potential underdamping.

10.4.4 Robustness under stochastic conditions

To evaluate the ETRL controller's robustness under stochastic operating conditions, Monte Carlo simulations (1000 trials) were conducted with probabilistic parameters: input voltage $V_{in} \sim \mathcal{N}(12, 1^2)$V, load resistance $R \sim \mathcal{U}[10, 75]\Omega$, inductance $L \sim \mathcal{N}\left(250, 10^2\right)\mu$H, and capacitance $C \sim \mathcal{N}\left(5, 0.5^2\right)\mu$F. The performance of ETRL is compared against PID and MPC controllers, with results summarized in Table 10.3. The ETRL controller achieves a MAE of 0.6 V, a ($CVaR_{0.05}$) of 1.0 V, and a variance of v_C of 0.8 V^2, outperforming PID (MAE: 1.5 V, CVaR: 3.0 V, Var: 2.5 V^2) and MPC (MAE: 1.2 V, CVaR: 2.5 V, Var: 2.0 V^2). Fig. 10.20 shows the probability density of the output voltage v_C for each controller, with ETRL exhibiting the tightest clustering around the reference voltage $v_{ref} = 24$ V, indicating superior robustness. The probabilistic event-trigger mechanism reduces control updates by approximately 75% compared to continuous RL, enhancing computational efficiency under uncertainty.

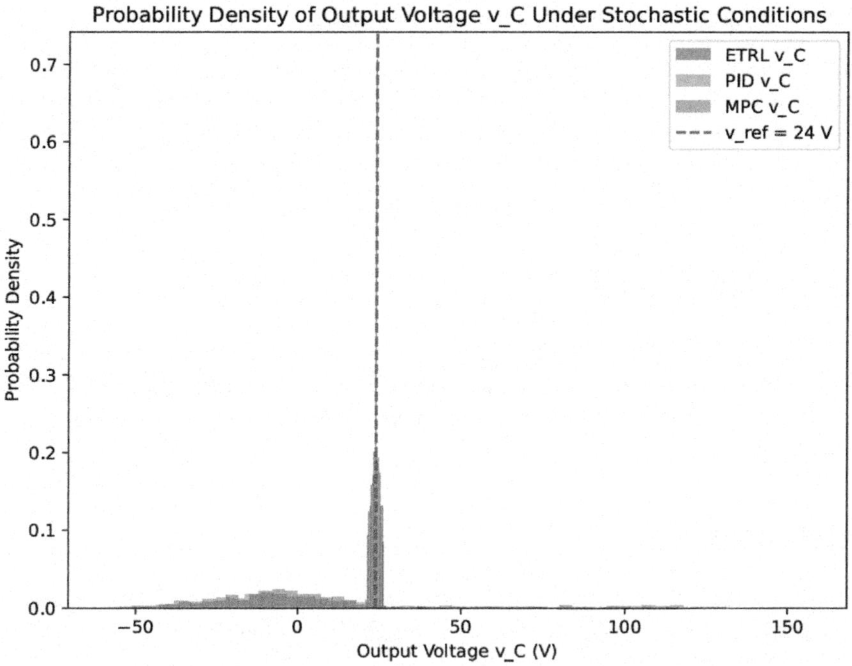

Figure 10.20 Probability density of output voltage v_C under stochastic conditions ($V_{\text{in}} \sim \mathcal{N}(12, 1^2)$ V, $R \sim \mathcal{U}[10, 75]\Omega$) for ETRL, PID, and MPC, computed from 1000 Monte Carlo trials. ETRL shows the tightest clustering around $v_{\text{ref}} = 24$ V. *ETRL*, Event-triggered reinforcement learning; *MPC*, model predictive control; *PID*, proportional-integral-derivative.

10.5 Conclusion

This work presents an ETRL framework integrated with a Python-based digital twin for adaptive control of DC–DC boost converters, with a focus on stochastic modeling and uncertainty management. By incorporating probabilistic models for input voltage ($V_{\text{in}} \sim \mathcal{N}(12, 1^2)$V), load resistance ($R \sim \mathcal{U}[10, 75]\Omega$), and component parameters, a risk-aware reward function with CVaR, and a probabilistic event-trigger mechanism, the ETRL framework ensures robust voltage regulation under uncertain conditions. Monte Carlo simulations (1000 trials) demonstrate ETRL's superior performance, achieving a mean MAE of 0.6 V, $\text{CVaR}_{0.05}$ of 1.0 V, and variance of 0.8 V^2, compared to PID (MAE: 1.5 V, CVaR: 3.0 V) and MPC (MAE: 1.2 V, CVaR: 2.5 V), as shown in Table 10.3 and Fig. 10.20. The probabilistic event trigger reduces control updates by 75%, enhancing computational efficiency. The digital twin enables cost-effective uncertainty quantification, making ETRL a scalable, uncertainty-aware solution for next-generation power electronics, aligned with stochastic planning in energy systems.

10.5.1 Practical considerations and limitations

Practical implementation on resource-constrained microcontrollers may face challenges due to the computational demands of Monte Carlo simulations and PPO training. Precomputing statistical models offline and validating via hardware-in-the-loop (HIL) testing can address these issues. Future work will explore real-world deployment to bridge the sim-to-real gap.

Acknowledgments

The authors would like to express their sincere gratitude to all those who contributed to the completion of this work. Special thanks to Farnaz Tamanna Chowdhury for her constant motivation and unwavering encouragement throughout the course of this research.

References

Anaya, V., Alberti, E., & Scivoletto, G. (2024). A manufacturing digital twin framework. In J. Soldatos (Ed.), *Artificial intelligence in manufacturing* (pp. 167–186). Springer. doi:10.1007/978-3-031-46452-210.

Åström, K. J., & Hägglund, T. (1995). *PID controllers: Theory, design, and tuning* (2nd ed.). Instrument Society of America Research.

Attaran, M., & Celik, B. G. (2023). Digital twin: Benefits, use cases, challenges, and opportunities. *Decision Analytics Journal, 6*, 100165. doi:10.1016/j.dajour.2023.100165.

Bui, V.-H., Chang, F., Su, W., Wang, M., Murphey, Y. L., Da Silva, F. L., Huang, C., Xue, L., & Glatt, R. (2022). Deep neural network-based surrogate model for optimal component sizing of power converters using deep reinforcement learning. *IEEE Access, 10*, 78702–78712. doi:10.1109/ACCESS.2022.3194267.

Cui, C., Yang, T., Dai, Y., Zhang, C., & Xu, Q. (2023). Implementation of transferring reinforcement learning for DC-DC buck converter control via duty ratio mapping. *IEEE Transactions on Industrial Electronics, 70*(6), 6141–6150. doi:10.1109/TIE.2022.3192676.

Dimarogonas, D. V., & Johansson, K. H. (2009). Event-triggered control for multi-agent systems. In *Proceedings of the 48h IEEE Conference on Decision and Control (CDC)* (pp. 7131–7136). doi:10.1109/CDC.2009.5399776.

Erickson, R. W., & Maksimovic, D. (2000). *Fundamentals of power electronics* (2nd ed.)). Kluwer Academic Publishers.

Gil-González, W., Riffo, S., Montoya, O. D., Restrepo, C., & Hernández, J. C. (2023). Adaptive voltage control for second-order DC-DC converters supplying an unknown constant power load: A generalized PBC plus damping injection design. *IEEE Access, 11*, 47390–47409. doi:10.1109/ACCESS.2023.3275083.

Heemels, W. P. M. H., Johansson, K. H., & Tabuada, P. (2012). An introduction to event-triggered and self-triggered control. In *2012 IEEE 51st IEEE Conference on Decision and Control (CDC)* (pp. 3270–3285). doi:10.1109/CDC.2012.6425820.

Hu, W., Zhang, T., Deng, X., Liu, Z., & Tan, J. (2021). Digital twin: A State-of-the-Art review of its enabling technologies, applications and challenges. *Journal of Intelligent Manufacturing and Special Equipment, 2*(1)), 1–34.

Ismail, N. F. N., Musirin, I., Baharom, R., & Johari, D. (2010). Fuzzy logic controller on DC/DC boost converter. In *2010 IEEE International Conference on Power and Energy* (pp. 661–666). doi:10.1109/PECON.2010.5697663.

Karamanakos, P., Liegmann, E., Geyer, T., & Kennel, R. (2020). Model predictive control of power electronic systems: Methods, results, and challenges. *IEEE Open Journal of Industry Applications, 1*, 95–114. doi:10.1109/OJIA.2020.3020184.

Kazimierczuk, M. K. (2008). *Pulse-width modulated DC-DC power converters*. Wiley. doi:10.1002/9780470694640.

Louassaa, K., Chouder, A., & Rus-Casas, C. (2023). Robust nonsingular terminal sliding mode control of a buck converter feeding a constant power load. *Electronics, 12*(3)), 728. doi:10.3390/electronics 12030728.

Mazaheri, N., Santamargarita, D., Bueno, E., Pizarro, D., & Cobreces, S. (2024). A deep reinforcement learning approach to DC-DC power electronic converter control with practical considerations. *Energies, 17*(14), 3578. doi:10.3390/en17143578.

Peng, Y., Zhao, S., & Wang, H. (2021). A digital twin based estimation method for health indicators of DC-DC converters. *IEEE Transactions on Power Electronics, 36*(2), 2105–2118. doi:10.1109/TPEL.2020.3009600.

Perry, A. G., Feng, G., Liu, Y.-F., & Sen, P. C. (2004). A new design method for PI-like fuzzy logic controllers for DC-to-DC converters. In *2004 IEEE 35th Annual Power Electronics Specialists Conference: 5* (pp. 3751–3757). doi:10.1109/PESC.2004.1355138.

Rajamallaiah, A., Karri, S. P. K., & Shankar, Y. R. (2024). Deep reinforcement learning based control strategy for voltage regulation of DC-DC buck converter feeding CPLs in DC microgrid. *IEEE Access, 12*, 17419–17430. doi:10.1109/ACCESS.2024.3358412.

Sado, K., Hannum, J., & Booth, K. (2023). Digital twin modeling of power electronic converters. In *2023 IEEE Electric Ship Technologies Symposium (ESTS)* (pp. 86–90). doi:10.1109/ESTS56571.2023.10220465.

Seyboth, G. S., Dimarogonas, D. V., & Johansson, K. H. (2013). Event-based broadcasting for multi-agent average consensus. *Automatica, 49*(1), 245–252.

Sutton, R. S., & Barto, A. G. (2018). *Reinforcement learning: An introduction* (2nd ed.). MIT Press.

Tao, F., Zhang, H., Liu, A., & Nee, A. Y. C. (2019). Digital twin in industry: State-of-the-Art. *IEEE Transactions on industrial informatics, 15*(4), 2405–2415. doi:10.1109/TII.2018.2873186.

Tian, F., Cobaleda, D. B., & Martinez, W. (2022). Deep reinforcement learning for DC-DC converter parameters optimization. In *2022 IEEE 31st International Symposium on Industrial Electronics (ISIE)* (pp. 325–330). doi:10.1109/ISIE51582.2022.9831660.

Vazquez, S., León Galván, J. I., García Franquelo, L., Rodríguez, J., Young, H. A., Márquez Alcaide, A., & Zanchetta, P. (2014). Model predictive control: A review of its applications in power electronics. *IEEE Industrial Electronics Magazine, 8*(1), 16–31. doi:10.1109/MIE.2013.2290138.

Wang, Y., & Zhao, J. (2022). Periodic event-triggered sliding mode control for switched uncertain T-S fuzzy systems with a logistic adaptive event-triggering scheme. *IEEE Transactions on Fuzzy Systems, 30*(10), 4115–4126. doi:10.1109/TFUZZ.2022.3141749.

Wu, L., Liu, J., Vazquez, S., & Mazumder, S. K. (2022). Sliding mode control in power converters and drives: A review. *IEEE/CAA Journal of Automatica Sinica, 9*(3), 392–406. doi:10.1109/JAS.2021.1004380.

Planning for distributed energy resources and microgrids—Scope: Stochastic siting, sizing, and control of DER clusters in diverse contexts

11

Hamid HassanzadehFard[a] *and Seyed Mehdi Hakimi*[b,c]
[a]Department of Electrical Engineering, Mi.C., Islamic Azad University, Miyaneh, Iran, [b]Department of Electrical Engineering, Dam.C., Islamic Azad University, Damavand, Iran, [c]Renewable Energy Research Center, Dam.C., Islamic Azad University, Damavand, Iran

11.1 Introduction

Nowadays, electrical demand has been growing rapidly due to economic growth and a higher standard of living. Consequently, traditional power plants face several environmental and technical challenges to keep pace with the growing demand (Nezhadpashaki et al., 2020). Renewable energy sources (RESs) are the predominant forms of power generation units that have received increasing interest for their potential in electrifying various regions worldwide (Gharib & Mohammadbeigi, 2025). Moreover, RESs must be optimally connected and operated to increase the efficiency of the system. The optimization of RESs across various systems has been explored in multiple studies.

Sifou et al. (2025) has presented a modified Harris Hawks optimizer for designing a hybrid sustainable and renewable energy system in order to minimize the cost while the reliability constraint is satisfied. Ba-swaimi et al. (2025) has used Homer software for optimization of a micro-grid (MG) to reduce the cost of energy (COE). Güven (2025) has proposed a novel hybrid salp swarm Kepler optimization algorithm for optimum designing of a hybrid system (HS). Increasing the renewable energy fraction (REF), minimizing COE and environmental emissions, and reducing grid dependency are the primary objectives of this work.

Tooryan et al. (2020b) has used particle swarm optimization (PSO) to determine the optimal capacity and energy management (EM) of an MG. Minimizing the total cost (TC), environmental pollution (EP), and fuel consumption are the objectives of this work. In Hakimi et al. (2011), has suggested an optimization methodology to the problem of sizing in a hybrid power system such that the residential demand is satisfied. The main purpose of (Dargahi et al., 2024; HassanzadehFard et al., 2023a; Tooryan et al., 2012) is to achieve the optimum planning and EM of an MG considering reliability indices.

Stochastic Planning and Modeling for Energy Systems: Methods, Applications, and Developments.
DOI: https://doi.org/10.1016/B978-0-443-45298-7.00008-2

Simulated annealing algorithm has been suggested in (Gharib & Mohammadbeigi, 2025) to find the optimal design of MG where the total annual costs (TAC) and environmental emissions are considered as the objective function (OF). Non-dominated sorting genetic algorithm (GA) has been used in (Gönül et al., 2024) to determine the best capacity of hybrid MG. An optimal EM strategy has been proposed in (HassanzadehFard et al., 2021) for optimum design and EM of an HS. A multiobjective optimization (MOO) approach has been implemented in (Nkalo et al., 2024) to optimize a HS configuration in order to reduce the COE. PSO has been applied in (HassanzadehFard et al., 2015b) for achieving the optimum configuration of a hybrid MG system. A cost-effective methodology has been proposed in (HassanzadehFard et al., 2023b) to achieve the optimal design of an HS considering methane production from livestock manure. A PSO-based optimization solution has been implemented in (Tooryan et al., 2020a) for optimization and EM of distributed generations (DGs) for a residential microgrid (MG), while the TCs are minimized.

The aforementioned studies focus on the optimum design of various HSs, and they only achieve the best capacity of DGs. Additionally, different optimization approaches have been suggested to determine the best sizing and siting of RESs in distribution networks (DNs). In (Patel et al., 2025), Multiobjective PSO has been utilized to obtain the optimum capacity and placement of DGs and shunt capacitors in a DN to minimize the power losses (PLs) and improve the voltage profile. A two-stage approach has been suggested in Ghani & Farjah (2025) to achieve the best capacity and site of DGs and charging stations in the distribution grid. Minimizing greenhouse gas emissions, total harmonic distortion (THD), and PLs are the objectives of this work. The MOO problem has been formulated in Guindi & Kamel (2024) to achieve the best capacity and location of RESs and electric vehicle charging stations in DN for minimizing the PLs and improving the voltage profile. Water flow optimization (WFO) methodology has been used in Jahed et al. (2024) for optimization of DGs in DN in order to reduce the economic losses and annual active/reactive PLs.

Sellami et al. (2022) has presented an improved MOPSO algorithm to find the best location and capacity of DGs in DN. The main objectives are minimizing the PLs and improving the voltage profile. In Hadidian-Moghaddam et al. (2018), the ant lion optimization approach has been applied for optimization of DG in DN to reduce the cost of purchased energy from the upstream network, DGs' application cost, DS losses, voltage deviation (VD), and improve the reliability. The purpose of Khenissi et al. (2024) is to achieve the best placement and capacity of a photovoltaic (PV) in redial distribution system for the minimization of PLs and enhancement of the voltage profile. A novel techno-economic approach has been introduced in Chreim et al. (2024) to find the best capacity and location of battery energy storage systems (BESSs) in residential communities with the aim of TC minimization.

Multiobjective grey wolf optimization (MOGWO) has been applied in Bendriss et al. (2024) for optimization of DGs in DN. Minimizing the PLs, reducing VD, and enhancing the voltage stability index (VSI) are the main purposes of the mentioned work. Alao et al. (2024) has presented a methodology based on PSO to find the best capacity and location of DGs for seasonal demands in DN to reduce PLs, VD, and improve the VSI. An optimization methodology has been developed in Apribowo et al. (2024) in order to identify the best location and capacity of a battery energy storage system, while demand response is considered. The equilibrium optimizer (EO) approach has been used in Hashem et al. (2021) to detect the best capacity and location of wind turbines (WTs) and superconducting magnetic energy storage (SMESs) in DN. Minimizing PLs and VD, and enhancing VSI are the objectives of this study. Al-Ammar et al. (2021) has proposed an artificial bee colony (ABC) approach to achieve the best capacity and location of DGs in DN in order to reduce energy cost, power loss, and average voltage drop.

GA has been implemented in Nezhadpashaki et al. (2020) to find the best location and capacity of DG in DN in order to reduce the total active PLs. Shuffled frog leap and binary particle swarm optimization methodologies have been suggested in Hassan et al. (2020) to achieve the best location and capacity of the DGs, which can significantly minimize PL and improve VSI. Essallah et al. (2019) has proposed a methodology for optimization of DG in DN to enhance the voltage stability and reduce the total PLs. Whale optimization methodology has been suggested in Wong et al. (2019) for optimum allocation of BESS in DN to reduce the PLs. Quantum particle optimization algorithm (QPSO) has been utilized in Dashtaki et al. (2024) for optimization of DGs in the DN considering an electricity market. Huy et al. (2020) has applied the differential evolution methodology to optimally integrate DGs into the DN to minimize PLs. A QPSO-based approach has been applied in Hakimi et al. (2019) for optimization of RESs, where the stochastic behavior of the RESs has been considered in Hasankhani & Hakimi (2021). HassanzadehFard & Jalilian (2018a) has implemented the PSO approach to attain the best sitting and sizing of RESs in DN in order to reduce THD, PLs, the TCs of DGs and EPs.

According to the aforementioned studies, a vital contributing factor in DNs is achieving the best location and sizing of RESs in order to reduce the PLs and improve the voltage profile. Therefore the primary purpose of this chapter is to achieve the best capacity and placement of RESs in the DNs. Therefore the contributions of the chapter can be listed as follows:

- Stochastic optimization of RESs in DN using different meta-heuristic algorithms.
- Optimum planning and EM of DN for improving the voltage profile and minimizing the PLs.
- Achieving the optimum electricity exchanged between the MG and electrical network (EN) while TC is minimized.

In the following Section 11.2, the modeling of the applied DGs is briefly described. Section 11.3 demonstrates the mathematical expression of RESs optimization in DN. The uncertainty analysis is provided in Section 11.4, which is followed by the DN optimization approach in Section 11.5. The numerical results are evaluated in Section 11.6, and Section 11.7 concludes the chapter.

11.2 System modeling

This section defines the methodology to model the key system components. Various DGs are connected directly to the DN to satisfy the loads in the system. The modeling approach encompasses the wind turbine (WT), PV, fuel cell (FC), BESS, and grid interconnection.

11.2.1 Photovoltaic

The electricity generated by PVs (P^t_{PV}) varies with changes in ambient temperature (T_c) and solar irradiance (G^t), as follows (Dashtaki et al., 2024).

$$P^t_{PV} = P^{rated}_{PV} \times \frac{G^t}{G_{ref}} \times \left[\beta \times \left(T_c + \left(G^t \times 0.0256\right) - T_{ref}\right) + 1\right] \tag{11.1}$$

in which:

$$\begin{cases} \beta = -3.7 \times 10^{-3}\left(\frac{1}{^\circ C}\right) \\ T_{ref} = 25^{oC} \\ G_{ref}\left(\frac{W}{m^2}\right) = 1000 \\ t \in \{1, 2, 3, ...T^{\max}\} \end{cases} \tag{11.2}$$

where P^{rated}_{PV} is the rated power of PV at standard conditions.

11.2.2 Wind turbine

The electricity generated by WTs (P^t_{WT}) can be achieved by the equations below (Dargahi et al., 2024):

$$P^t_{WT} = \begin{cases} 0 & V^t < V_{ci}, V^t > V_{co} \\ P^{\max}_{WT} \times \left(\frac{V^t - V_{ci}}{V_{rated} - V_{ci}}\right)^3 & V_{ci} \le V^t < V_{rated} \\ P_{rated} & V_{rated} \le V^t \le V_{co} \end{cases} \tag{11.3}$$

where $P^{\max}_{WT}$ represents the maximum power of WT, $V^t$is wind velocity, V_{rated} is nominal wind velocity, V_{co} is cut out wind velocity, and V_{ci} is cut in wind velocity.

11.2.3 Fuel cell

In this study, FC units—another type of DG—are applied in the proposed system to provide the required demand. The amount of consumed fuel of the FC ($Fuel^t_{FC}$) can be obtained according to its generated power (P^t_{FC}) as follows (HassanzadehFard et al., 2020; Shahrabi et al., 2021):

$$Fuel^t_{FC} = \lambda_{FC} \times P^t_{FC} \tag{11.4}$$

in which λ_{FC} is the fuel consumption required to produce 1 kWh of energy by FCs.

11.2.4 Battery energy storage system

BESS is used in HSs to store excess electricity generated by RESs (HassanzadehFard & Jalilian, 2018b; HassanzadehFard et al., 2011a). The SOC level in the BESS is expressed as:

$$SOC^{t}_{Battery} = \begin{cases} SOC^{t-1}_{Battery} + \left(P^{t-1}_{Renewable} - \frac{P^{t-1}_{Demand}}{\eta_{inverter}}\right) \times \eta^{Charging}_{Battery} & Charging\,mode \\ SOC^{t-1}_{Battery} - \left(\frac{\frac{P^{t-1}_{Demand}}{\eta_{inverter}} - P^{t-1}_{Renewable}}{\eta^{Discharging}_{Battery}}\right) & Discharging\,mode \end{cases} \tag{11.5}$$

$\eta_{inverter}$ is the inverter efficiency, $P^{t-1}_{Renewable}$ is the electricity generated by RESs, and P^{t-1}_{Demand} is the electrical demand. $\eta^{Charging}_{Battery}$ and $\eta^{Discharging}_{Battery}$ are the battery charging/discharging efficiency.

11.2.5 Uncertainty analysis

In the present paper, the Beta distribution function is used to model uncertainties in the generated electricity of the PVs as follows (Hung et al., 2014):

$$f_b(r) = \begin{cases} \frac{\Gamma(\alpha+\beta)}{\Gamma(\alpha)\Gamma(\beta)} r^{\alpha-1}(1-r)^{\beta-1} & 0 \le r \le 1, \alpha \ge 0, \beta \ge 0 \\ 0 & otherwise \end{cases} \tag{11.6}$$

where $f_b(r)$ is the Beta distribution, r is the solar irradiance (kW/m^2), α and β are the parameters of $f_b(r)$, which are calculated by the average μ and standard deviation δof solar radiation r:

$$\alpha = \frac{\mu \times \beta}{1-\mu} \tag{11.7}$$

$$\beta = (1-\mu) \times \left(\frac{\mu \times (1+\mu)}{\delta^2} - 1\right) \tag{11.8}$$

Wind velocity fluctuations and forecasting complexities of the generated electricity of the WT result in high uncertainties (Hassanzadehfard et al., 2011b). Therefore, modeling uncertainty in the electricity generation of WTs is necessary to address the optimal planning and EM of the EH. In our chapter, a Rayleigh distribution function, which is derived from a Weibull distribution, is applied to formulate the uncertainties of wind power as follows:

$$f_w(v) = \frac{k}{c}\left(\frac{v}{c}\right)^{k-1} \times \exp\left(-\left(\frac{v}{c}\right)^{k}\right) \tag{11.9}$$

where k and c denote the shape parameter and scale parameter (m/s), respectively.

11.3 Optimization problem formulation

In this chapter, the main problem is determining the optimum capacity and location of DGs in the DN to reduce the TC of the MG and satisfy the constraints. The reduction of PLs, the TC of DGs, and the penalty for EPs are considered as the economic factors. Load balance constraint, voltage and power flow limits, and generated power of DGs are also considered as the main technical constraints. The mentioned factors are expressed as:

11.3.1 Proposed objective function

Identifying the optimum capacity and location of RESs in MG with the minimization of total planning and operational costs and the improvement of the voltage profile presents the primary objectives of the MG's operator. The determination of the appropriate capacity and location for each of the DG choices is made by assessing environmental, economic, and technological factors. Thus, the proposed OF is formulated as follows:

$$OF = \sum_{s=1}^{S} \psi(s) \left[\begin{array}{l} \sum\limits_{i \in DG} \left[C^i_{Cap}(s) + C^i_{Rep}(s) + C^i_{O\&M}(s) \right] + \\ \sum\limits_{t=1}^{T^{mac}} \left[C^t_{Fuel}(s) + C^t_{Pur}(s) - C^t_{Sell}(s) + C^t_{Loss}(s) + C^t_{Em}(s) \right] \end{array} \right] \tag{11.10}$$

where $\psi(s)$ denotes the probability of scenario s, and S is the number of scenarios. $C^i_{Cap}(s)$, $C^i_{Rep}(s)$ and $C^i_{O\&M}(s)$are the capital cost (CC), replacement cost (RC), and operation and maintenance (OM) costs of each DG, respectively. $C^t_{Fuel}(s)$, $C^t_{Pur}(s)$, and $C^t_{Sell}(s)$ are the costs of purchased fuel, electricity purchased from the EN, and electricity sold to the EN, respectively.

In the proposed system, the MG operator is responsible for satisfying the required demands and PLs. Optimum installation and EM of RESs is driven by cost reduction.

11.3.1.1 Cost model of renewable energy sources

The cost of RESs includes various items, such as capital/replacement/OM costs, which are formulated as follows:

$$C^i_{Total} = C^i_{Cap} + C^i_{Rep} + C^i_{O\&M} \tag{11.11}$$

in which,

$$C^i_{Cap} = CC^i_{DG} \times N^i_{DG} \tag{11.12}$$

$$C^i_{Rep} = k^i_{DG} \times RC^i_{DG} \times N^i_{DG} \tag{11.13}$$

$$C^{i}_{O\&M} = CRF^{ir}_{R} \times OM^{i}_{DG} \times N^{i}_{DG} \tag{11.14}$$

$$CRF^{ir}_{R} = \left[\frac{(1+ir)^{R}}{(1+ir)^{R}-1}\right] \times ir \tag{11.15}$$

$$k^{i}_{DG} = \sum_{n=1}^{Y^{i}_{DG}} \frac{1}{(1+ir)^{n\times L^{i}_{DG}}} \tag{11.16}$$

$$Y^{i}_{DG} = \begin{cases} \left[\frac{R}{L^{i}_{DG}}\right] - 1 & \text{if } R \text{ is dividable to } L^{i}_{DG} \\ \left[\frac{R}{L^{i}_{DG}}\right] & \text{if } R \text{ is not dividable to } L^{i}_{DG} \end{cases} \tag{11.17}$$

where CC^{i}_{DG}, RC^{i}_{DG} and OM^{i}_{DG} denote the CC, RC, and OM costs for each applied DG, respectively. Y^{i}_{DG} denotes how many times each DG is replaced over the lifetime of the project (R). L^{i}_{DG} represents the project lifetime and the lifetime of each DG.

11.3.1.2 Cost model of power losses

Another objective is to reduce the TC of power loss (C^{Total}_{Loss}) and is illustrated as follows:

$$C^{Total}_{Loss} = \left\{ \sum_{t=1}^{T^{\max}} \sum_{L=1}^{m} \left(P^{t}_{Loss_L}(r_L, i_L) \right) \right\} \times \frac{\mu_{Loss}}{CRF^{ir}_{R}} \tag{11.18}$$

where, $P^{t}_{Loss_L}(r_L, i_L)$ denotes the power loss of the Lth branch, m is the number of branches, and μ_{Loss} is the penalty factor for the PL.

11.3.1.3 Cost model of environmental pollutions

During the combustion of fuel in DG for generating electrical power, various types of EPs are released. In this chapter, the penalty for EPs (C^{Total}_{Em}) is considered as follows:

$$C^{Total}_{Em} = \left\{ \sum_{e=1}^{E} \sum_{t=1}^{T^{\max}} \left(E^{t}_{e} \right) \right\} \times \frac{\mu_{Emission}}{CRF^{ir}_{R}} \tag{11.19}$$

in which, E^{t}_{e} refers to the amount of $e$the greenhouse gas emission at each hour, and $\mu_{Emission}$ is the penalty factor for EPs.

11.3.1.4 Cost model of consumed fuel

In the proposed MG, FCs consume fuel to generate electricity in the DN. The cost of the consumed fuel (C_{Fuel}^{Total}) is formulated as below:

$$C_{Fuel}^{Total} = \left\{ \sum_{t=1}^{T^{\max}} Fuel_{FC}^{t} \right\} \times \frac{\mu_{Fuel}}{CRF_{R}^{ir}} \tag{11.20}$$

in which, $\mu_{Fuel}$is the cost of natural gas.

11.3.1.5 Cost model of purchased electrical power from the electrical network

In the proposed system, the electricity can be purchased from the EN to provide the required demands in DN. The cost of purchased electricity from the EN can be mathematically represented as follows:

$$C_{Pur}^{Total} = \left\{ \sum_{t=1}^{T^{\max}} P_{Pur}^{t} \right\} \times \frac{\mu_{Pur}}{CRF_{R}^{ir}} \tag{11.21}$$

where μ_{Pur}refers to the cost of purchased energy, and $P_{Pur}^{t}$is the purchased electricity from EN at each hour.

11.3.1.6 Cost model of sold energy to the electrical network

The cost of selling power to the EN is expressed as:

$$C_{Sell}^{Total} = \left\{ \sum_{t=1}^{T^{\max}} P_{Sell}^{t} \right\} \times \frac{\mu_{Sell}}{CRF_{R}^{ir}} \tag{11.22}$$

where μ_{Sell} represents the cost of sold electricity, and P_{Sell}^{t} is the electricity sold to the EN.

11.3.2 System constraints

In addressing the DG allocation problem, the proposed optimization approach adheres to all inequality and equality constraints, as follows:

11.3.2.1 Load balance constraint

The produced electrical power by applied RESs and purchased energy from the EN must supply the required demand (P^t_{Demand}) and PLs (P^t_{Loss}) in the DN. The load balance is considered as:

$$\sum_{i \in DG} P^t_i \pm P^t_{EN} = P^t_{Demand} + P^t_{Loss} \tag{11.23}$$

where P^t_{EN} and P^t_i are the purchased electricity from EN, and the power produced by each DG, respectively,

11.3.2.2 Available energy in battery

The available energy in BESS is limited based on the minimum ($SOC^{\min}$) and maximum ($SOC^{\max}$) allowable ranges as (HassanzadehFard et al., 2015a):

$$SOC^{\min} \leq SOC^t \leq SOC^{\max} \tag{11.24}$$

in which:

$$SOC^{\min} = SOC^{\max} \times (1 - DOD) \tag{11.25}$$

DOD is the maximum acceptable depth of BESS discharge.

11.3.2.3 Voltage limits

The voltage magnitude of all buses ($|V^t_i|$) must be kept within an acceptable interval (HassanzadehFard & Jalilian, 2021):

$$V^{\min} \leq |V^t_i| \leq V^{\max} \tag{11.26}$$

$V^{\min}$ and $V^{\max}$ are the minimum and maximum acceptable limits of voltage.

11.3.2.4 Distributed generation technical constraint

Each installed DG in the MG produces power ($P^t_{DG,i}$) between its lowest ($P^{\min}_{DG,i}$) and highest ($P^{\max}_{DG,i}$) generation capability as:

$$P^{\min}_{DG,i} \leq P^t_{DG,i} \leq P^{\max}_{DG,i} \tag{11.27}$$

11.3.2.5 Power flow constraint

The current through each line (I_{ij}^{t}) must be constrained by a maximum limit ($I_{ij}^{\max}$), as follows:

$$I_{ij}^{t} \leq I_{ij}^{\max} \tag{11.28}$$

11.3.2.6 Exchanged electricity constraints

The MG can exchange electricity by EN, considering the following constraints:

$$P_{Pur}^{\min} \leq P_{Pur}^{t} \leq P_{Pur}^{\max} \tag{11.29}$$

$$P_{Sell}^{\min} \leq P_{Sell}^{t} \leq P_{Sell}^{\max} \tag{11.30}$$

where$P_{Sell}^{\min}$, $P_{Sell}^{\max}$, $P_{Pur}^{\min}$, and $P_{Pur}^{\max}$, denote the minimum and maximum limits for sold/purchased electricity to/from the EN.

11.4 Distribution system optimization methodology

Different meta-heuristic optimization algorithms are applied in order to determine the optimum location and size of various RESs in MG. The proposed approach enables comparison, energy purchasing from the EN, installation of different types of DGs, as well as the penalty for EPs, PLs, fuel consumption, etc. In the proposed system, the MG operator has the following options to provide the loads:

- Installing various types of RESs in the MG.
- Purchasing the required electricity from the EN.

The main objectives of the developed methodology are to reduce the TC of DN. As mentioned before, the proposed minimization includes:

- Economic data of various RESs.
- Costs of PLs.
- Costs of purchased electricity from the EN.
- Costs of EPs.
- Costs of purchased fuel.

In this chapter, different optimization algorithms are implemented to solve the suggested optimization problem while satisfying the system constraints. Fig. 11.1 illustrates the flowchart of the proposed approach.

11.5 Simulation and numerical results

The proposed approach for achieving the optimum location and sizing of RESs has been applied in the MATLAB software. To demonstrate the effectiveness of the

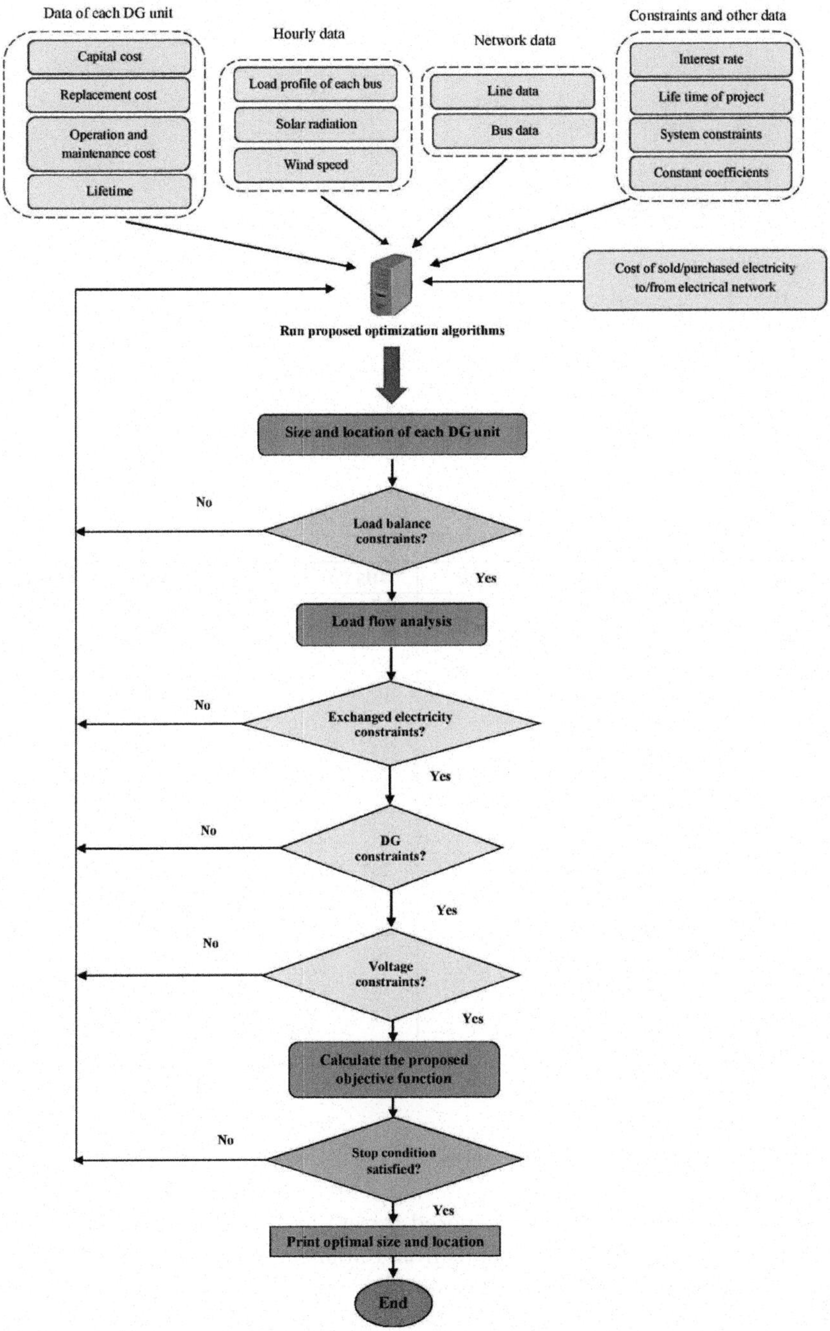

Figure 11.1 Flowchart of the proposed approach.

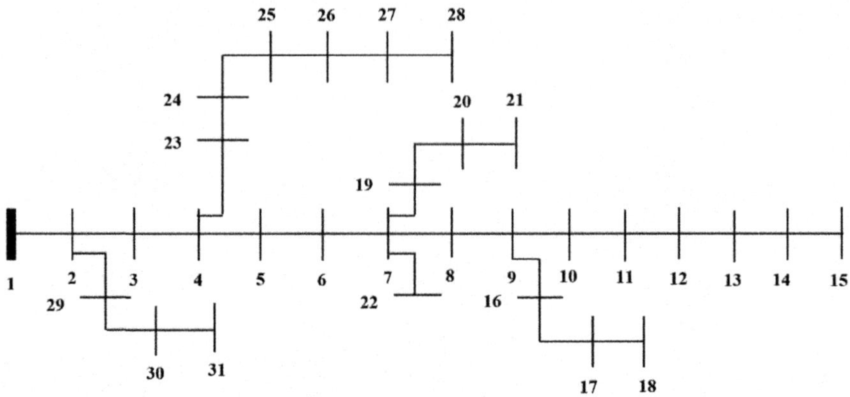

Figure 11.2 Schematic diagram of the proposed MG. *MG*, Microgrid.

Table 11.1 The peak of demands for each bus (p.u).

Bus No.	Peak of demand	Bus No.	Peak of demand
1	0	17	0.025
2	0.032	18	0.026
3	0.026	19	0.028
4	0.079	20	0.042
5	0.038	21	0.014
6	0.046	22	0.016
7	0.052	23	0.018
8	0.034	24	0.018
9	0.056	25	0.022
10	0.066	26	0.024
11	0.042	27	0.025
12	0.024	28	0.022
13	0.038	29	0.038
14	0.021	30	0.033
15	0.035	31	0.015
16	0.045		

suggested method, the modified version of the 31-bus DN is selected, as shown in Fig. 11.2.

The total active power of the proposed system is considered as equal to 20 MW. The line data of the proposed system is from (Grainger & Civanlar, 1985). Table 11.1 depicts the peak of demand for each bus.

The hourly wind speed, solar irradiation, and electrical demand are shown in Figs. 11.3–11.5.

The losses in the proposed DN amount to 1810 kW. The voltage magnitude of different buses without DG installation is illustrated in Fig. 11.6.

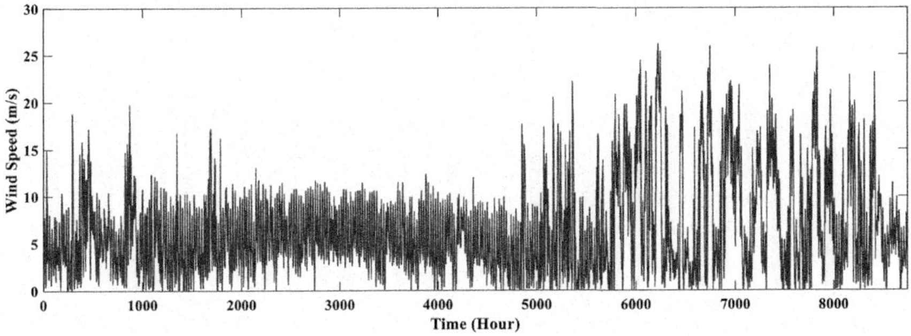

Figure 11.3 Wind speed.

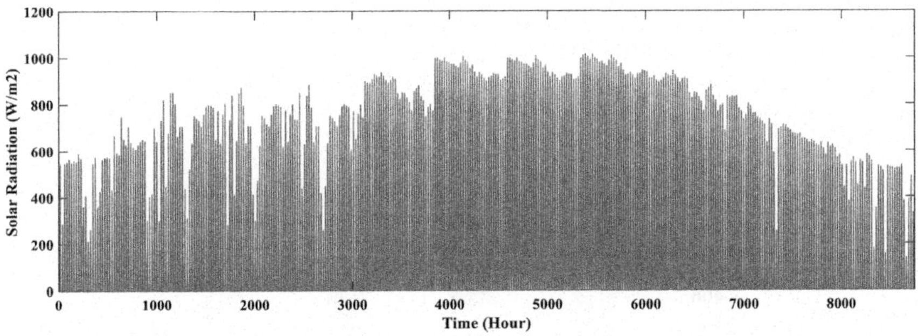

Figure 11.4 Solar radiation.

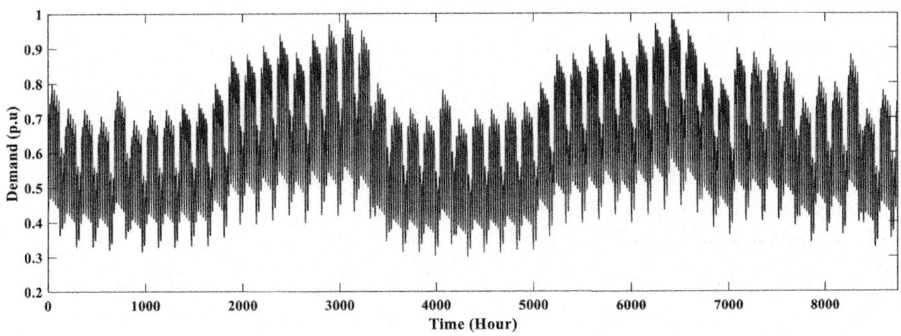

Figure 11.5 Electrical demands.

Fig. 11.6 clearly shows that voltage magnitudes at certain buses fall outside the safe interval (i.e., [0.95–1.05]). From this figure, the minimum magnitude of voltage is observed at bus 15, which is equal to 0.848 p.u., which is not within the acceptable limit. As mentioned, different optimization methodologies are used to obtain the best

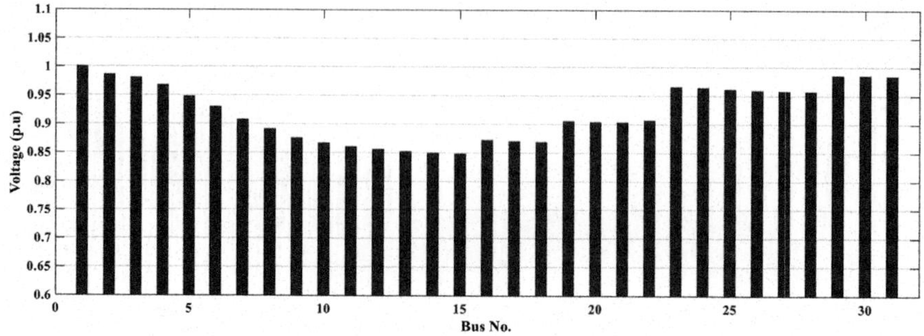

Figure 11.6 The voltage profile before DG installation. *DG,* Distributed generation.

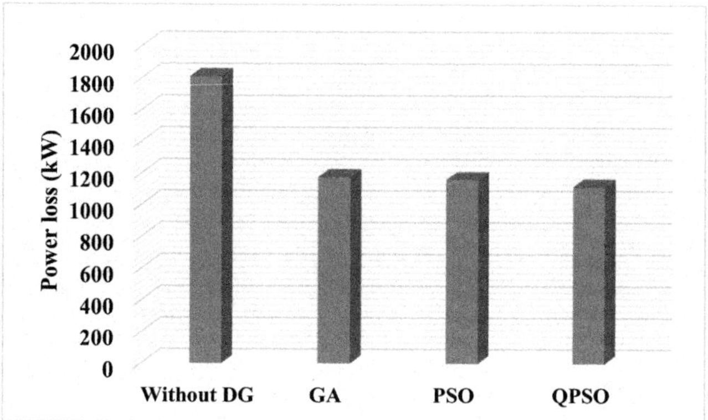

Figure 11.7 The PLs for different algorithms. *PLs*, Power losses.

locations and capacities of various RESs in the system. The optimum location and size of RESs are illustrated in Table 11.2.

Figs. 11.7 and 11.8 depict a comparison between the PLs and COE after the optimum sizing and siting of RESs applying different methodologies.

After installing the DGs in the MG, the PLs are significantly reduced. Also, it is clear that the QPSO, in comparison with GA and PSO, in achieving the best location and capacity of RESs, reduced the PLs of the DN.

The results show that QPSO's global search capability is significantly superior to other methodologies. Consequently, only the results of QPSO have been evaluated in the remaining paragraphs. The voltage profile for the peak load condition after optimization of RESs is illustrated in Fig. 11.9.

This figure clearly shows that the voltage profile was improved within the acceptable limit by the optimum installation of RESs. Before the optimization of RESs, the voltage magnitude was unacceptable in some buses. As shown in Fig. 11.9, the voltage profile was improved to an acceptable limit by optimum installation of RESs.

Table 11.2 The simulation results.

Algorithm	PV		WT		FC		BESS		Loss (kW)
	Size	Location	Size	Location	Size	Location	Size	Location	
GA	5025	20	4673	6	9500	13	8674	8	1177
PSO	5175	18	4558	6	9363	13	8750	9	1160
QPSO	4950	28	4500	8	9500	15	8500	23	1115

BESS, battery energy storage system; *FC*, fuel cell; *GA*, genetic algorithm; *PSO*, particle swarm optimization; *PV*, photovoltaic; *QPSO*, quantum particle optimization algorithm; *WT*, wind turbine.

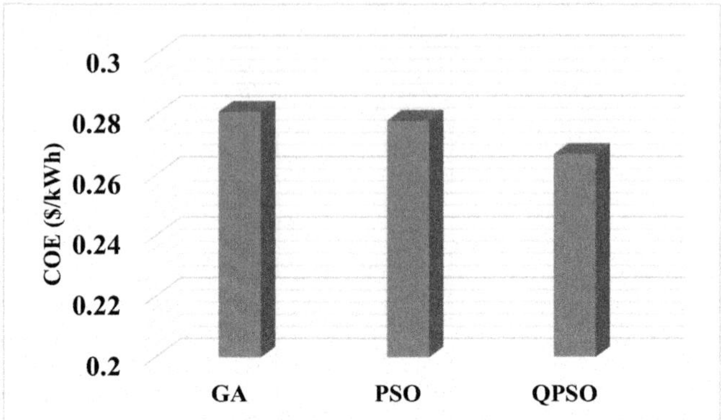

Figure 11.8 The COE for different algorithms. *COE,* Cost of energy.

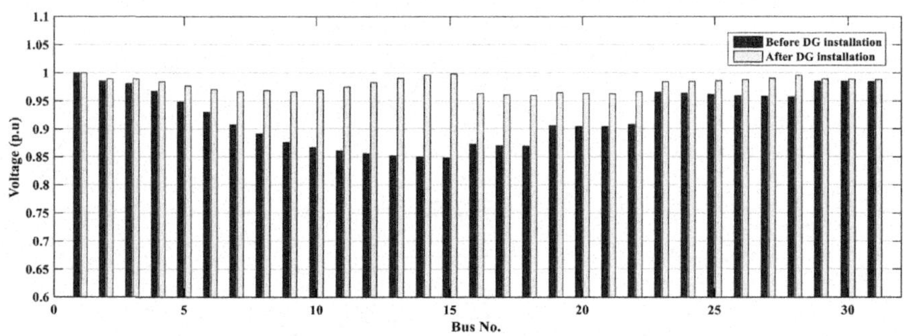

Figure 11.9 The voltage profile after RESs installation. *RESs,* Renewable energy resources.

To verify the efficiency of the suggested approach, four typical days are considered as follows:

- Scenario#1: windy-sunny.
- Scenario#2: windy-cloudy.
- Scenario#3: calm-sunny.
- Scenario#4: calm-cloudy.

The voltage profile for the mentioned days is illustrated in Fig. 11.10. In addition, the maximum and minimum magnitude of voltage is illustrated in Table 11.3.

Different typical days are considered in this study, including windy-sunny days, windy-cloudy days, calm-sunny days, and calm-cloudy days. The minimum magnitude of voltage is equal to (1) windy-sunny day: 0.9620 p.u, (2) windy-cloudy day: 0.9599 p.u, (3) calm-sunny day: 0.9522 p.u, and (4) calm-cloudy day: 0.95 p.u. The amount of generated energy by different components to satisfy demands is shown in Fig. 11.11.

From Fig. 11.11, the amount of energy produced by FCs is dominating. Because the FC can produce electricity all the times during a year. It is important to note that the remaining demand is met by the EN at the final step. Moreover, the environmental

Table 11.3 The maximum/minimum magnitude of voltage for different scenarios.

Scenario#1		**Scenario#2**		**Scenario#3**		**Scenario#4**	
V_{min}	V_{max}	V_{min}	V_{max}	V_{min}	V_{max}	V_{min}	V_{max}
0.9620	1	0.9599	1	0.9522	1	0.95	1

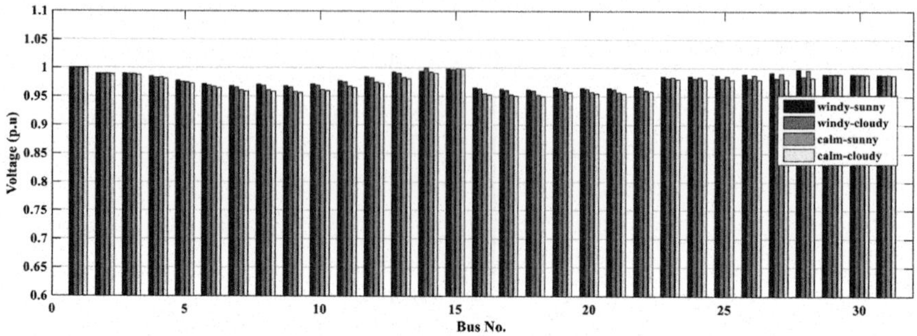

Figure 11.10 The voltage profile for different typical days.

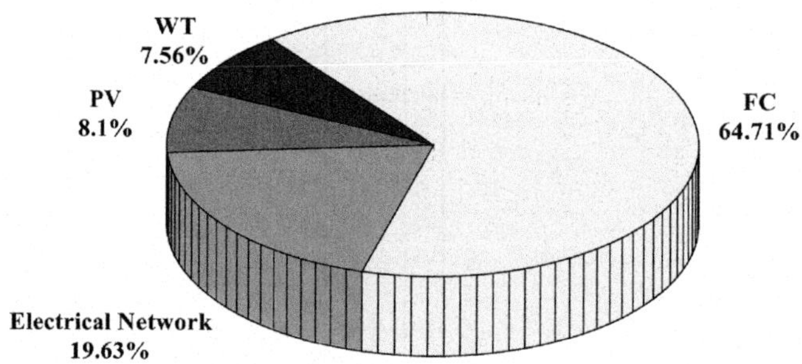

Figure 11.11 The amount of generated energy by different components.

Table 11.4 EPs for different scenarios.

Scenario	NO_X	SO_2	CO_2	C
Before DG installation (tons)	105.888	141.44	68,380	18,649.36
After optimization (tons)	20.7858	27.7646	13,423	3660.87

EPs, environmental pollution; *DG*, distributed generation.

concerns are also considered in the OF. To depict the efficiency of the suggested methodology, a comparison based on different types of EPs for different scenarios is illustrated in Table 11.4.

In the first scenario, the amounts of NO_X, SO_2, CO_2, and C are equal to 105.888 tons, 141.44 tons, 68,380 tons, and 18,649.36 tons, respectively. After optimal installation of RESs in DN, the greenhouse gas emissions have decreased by approximately 80.37%. Therefore it can be concluded that by applying RESs in DN and reducing the use of fossil fuels in power plants, the EPs would be remarkably reduced.

11.6 Conclusion

In the present chapter, an optimization approach is utilized for stochastic optimum sizing and siting of RESs in the MG. To achieve the main purpose of the chapter, different optimization approaches, including GA, PSO, and QPSO, are suggested to solve the proposed optimization problem. The developed methodology is tested on a 31-bus DN to reduce the PLs, cost of RESs, EPs, and improve the voltage profile. The results confirmed that the voltage profile improved and PLs were reduced by installing the optimum RESs. After the installation of RESs, EPs remarkably reduced by more than 80%. It can be concluded that CO_2 emissions were reduced from 68,380 tons to 13,423 tons when the RESs are optimally installed in the DS. Simulation results approved that the QPSO had the best solution and optimal results. Furthermore, QPSO showed a high capability in determining the optimal solutions.

References

Al-Ammar, E. A., Farzana, K., Waqar, A., Aamir, M., Saifullah, A. U. H., Zahid, M., & Batool, M. (2021). ABC algorithm based optimal sizing and placement of DGs in distribution networks considering multiple objectives. *Ain Shams Engineering Journal, 12*(1), 697–708.

Alao, M. A., Popoola, O. M., & Ayodele, T. R. (2024). An improved particle swarm optimisation algorithm for optimum placement and sizing of biogas-fueled distributed generators for seasonal loads in a radial distribution network. *Energy Reports, 12*, 1531–1550.

Apribowo, C. H. B., Hadi, S. P., Wijaya, F. D., Setyonegoro, M. I. B., & Sarjiya (2024). Optimal sizing and placement of battery energy storage system for maximum variable renewable energy penetration considering demand response flexibility: A case in Lombok power system, Indonesia. *Energy Conversion and Management: X, 23*, 100620.

Ba-swaimi, S., Verayiah, R., Ramachandaramurthy, V. K., ALAhmad, A. K., & Padmanaban, S. (2025). Optimal configuration and sizing of integrated hybrid renewable energy systems for sustainable power supply in healthcare buildings. *Results in Engineering, 26*, 104800.

Bendriss, B., Sayah, S., & Hamouda, A. (2024). Efficient multi-objective optimization approach for solving optimal DG placement and sizing problem in distribution systems. *Journal of Engineering Research, 13*(4), 3295–3307.

Chreim, B., Esseghir, M., & Merghem-Boulahia, L. (2024). SPLANDID—Optimal sizing, placement, and management of centralized and distributed shared battery energy storage systems in residential communities: Application to smart grids. *Sustainable Cities and Society, 113*, 105694.

Dargahi, V., HassanzadehFard, H., Tooryan, F., & Tourian, F. (2024). Reliable cost-efficient integration of pumped hydro storage in islanded hybrid microgrid for optimum decarbonization. *Energy, 312*, 133615.

Dashtaki, A. A., Hakimi, S. M., Farahani, E. S., & HassanzadehFard, H. (2024). Simultaneous optimal location and sizing of DGs in distribution system considering different types of MGs in an electricity market. *Multiscale and multidisciplinary modeling, experiments and design, 7*, 2443–2459.

Essallah, S., Khedher, A., & Bouallegue, A. (2019). Integration of distributed generation in electrical grid: Optimal placement and sizing under different load conditions. *Computers & Electrical Engineering, 79*, 106461.

Ghani, R., & Farjah, E. (2025). A two-stage approach for determination of optimal size and site of distributed generations and charging stations in the distribution system, considering load growth and the electric vehicle penetration. *Energy, 319*, 134953.

Gharib, M. R., & Mohammadbeigi, A. (2025). Optimal sizing, techno-economic, and environmental assessment of hybrid renewable energy systems. *Energy Strategy Reviews, 59*, 101721.

Gönül, Ö, Duman, A. C., & Güler, Ö (2024). Multi-objective optimal sizing and techno-economic analysis of on- and off-grid hybrid renewable energy systems for EV charging stations. *Sustainable Cities and Society, 115*, 105846.

Grainger, J. J., & Civanlar, S. (1985). Volt/var control on distribution systems with lateral branches using shunt capacitors and voltage regulators part III: The numerical results. *IEEE Transactions on Power Apparatus and Systems, 104*(11), 3278–3283.

Guindi, M., & Kamel, R. M. (2024). Optimal location and sizing of renewable distributed generations and electric vehicle charging stations. *Renewable Energy, 235*, 121272.

Güven, A. F. (2025). A novel hybrid Salp Swarm Kepler optimization for optimal sizing and energy management of renewable microgrids with EV integration. *Energy, 334*, 137696.

Hadidian-Moghaddam, M. J., Arabi-Nowdeh, S., Bigdeli, M., & Azizian, D. (2018). A multi-objective optimal sizing and siting of distributed generation using ant lion optimization technique. *Ain Shams Engineering Journal, 9*(4), 2101–2109.

Hakimi, S. M., Bagheritabar, H., Hasankhani, A., Shafie-khah, M., Lotfi, M., & Catalão, J. P. (2019). Planning of smart microgrids with high renewable penetration considering electricity market conditions. In *2019 IEEE International Conference on Environment and Electrical Engineering and 2019 IEEE Industrial and Commercial Power Systems Europe (EEEIC/I&CPS Europe)* (pp. 1–5). doi:10.1109/EEEIC.2019.8783333.

Hakimi, S. M., Moghaddas-Tafreshi, S. M., & HassanzadehFard, H. (2011). Optimal sizing of reliable hybrid renewable energy system considered various load types. *Journal of Renewable Sustainable Energy, 3*, 062701.

Hasankhani, A., & Hakimi, S. M. (2021). Stochastic energy management of smart microgrid with intermittent renewable energy resources", in electricity market. *Energy, 219*, 119668.

Hashem, M., Abdel-Salam, M., El-Mohandes, M. T., Nayel, M., & Ebeed, M. (2021). Optimal placement and sizing of wind turbine generators and superconducting magnetic energy storages in a distribution system. *Journal of Energy Storage, 38*, 102497.

Hassan, A. S., Sun, Y., & Wang, Z. (2020). Multi-objective for optimal placement and sizing DG units in reducing loss of power and enhancing voltage profile using BPSO-SLFA. *Energy Reports, 6*, 1581–1589.

HassanzadehFard, H., Bahreyni, S. A., Dashti, R., & Shayanfar, H. A. (2015a). Evaluation of reliability parameters in micro-grid. *Iranian Journal of Electrical and Electronic Engineering, 11*(2), 127–136.

HassanzadehFard, H., Dargahi, V., & Tooryan, F. (2023a). Energy systems decarbonization: Design optimization of a commercial building MG system considering high penetration of renewable energies. In*: Energy systems transition: Digitalization, decarbonization, decentralization and democratization*. Springer, Cham https://doi.org/10.1007/978-3-031-22186-6_1.

HassanzadehFard, H., & Jalilian, A. (2018a). Optimal sizing and location of renewable energy based dg units in distribution systems considering load growth. *International Journal of Electrical Power & Energy systems, 101*, 356–370.

HassanzadehFard, H., & Jalilian, A. (2018b). Optimal sizing and siting of renewable energy resources in distribution systems considering time varying electrical/heating/cooling loads using PSO algorithm. *International Journal of Green Energy, 15*(2), 113–128.

HassanzadehFard, H., & Jalilian, A. (2021). Optimization of DG units in distribution systems for voltage sag minimization considering various load types. *Iranian Journal of Science and Technology, Transactions of Electrical Engineering, 45*, 685–699.

HassanzadehFard, H., Moghaddas-Tafreshi, S. M., & Hakimi, S. M. (2011a). Effect of energy storage systems on optimal sizing of islanded micro-grid considering interruptible loads. *3rd International Youth Conference on Energetics.*

Hassanzadehfard, H., Moghaddas-Tafreshi, S. M., & Hakimi, S. M. (2011b). Optimal sizing of an islanded micro-grid for an area in North-West Iran using particle swarm optimization based on reliability concept. *World Renewable Energy Congress-Sweden.*

HassanzadehFard, H., Moghaddas-Tafreshi, S. M., & Hakimi, S. M. (2015b). Optimization of grid-connected microgrid consisting of PV/FC/UC with considered frequency control. *Turkish Journal of Electrical Engineering & Computer Sciences, 23*, 1–16.

HassanzadehFard, H., Tooryan, F., Collins, E. R., Jin, S., & Ramezani, B. (2020). Design and optimum energy management of a hybrid renewable energy system based on efficient various hydrogen production. *International Journal of Hydrogen Energy, 45*(55), 30113–30128.

HassanzadehFard, H., Tooryan, F., & Dargahi, V. (2023b). Standalone hybrid system energy management optimization for remote village considering methane production from livestock manure. *International Journal of Hydrogen Energy, 48*(29), 10778–10796.

HassanzadehFard, H., Tooryan, F., Dargahi, V., & Jin, S. (2021). A cost-efficient sizing of grid-tied hybrid renewable energy system with different types of demands. *Sustainable Cities and Society, 73*, 103080.

Hung, D. Q., Mithulananthan, N., & Lee, K. Y. (2014). Determining PV penetration for distribution systems with time-varying load models. *IEEE Transactions on Power Systems, 29*(6), 3048–3057.

Huy, P. D., Ramachandaramurthy, V. K., Yong, J. Y., Tan, K. M., & Ekanayake, J. B. (2020). Optimal placement, sizing and power factor of distributed generation: A comprehensive study spanning from the planning stage to the operation stage. *Energy, 195*, 117011.

Jahed, Y. G., Mousavi, S. Y. M., & Golestan, S. (2024). Optimal sizing and siting of distributed generation systems incorporating reactive power tariffs via water flow optimization. *Electric Power Systems Research, 231*, 110278.

Khenissi, I., Guesmi, T., Alshammari, B. M., Alqunun, K., Almalaq, A., Alturki, M., & Neji, R. (2024). A hybrid chaotic bat algorithm for optimal placement and sizing of DG units in radial distribution networks. *Energy Reports, 12*, 1723–1741.

Nezhadpashaki, M. A., Karbalaei, F., & Abbasi, S. (2020). Optimal placement and sizing of distributed generation with small signal stability constraint. *Sustainable Energy, Grids and Networks, 23*, 100380.

Nkalo, U. K., Inya, O. O., Obi, P. I., Bola, A. U., & Ewean, D. I. (2024). A modified multi-objective particle swarm optimization (M-MOPSO) for optimal sizing of a solar–wind–battery hybrid renewable energy system. *Solar Compass, 12*, 100082.

Patel, P., Patil, N., Mohsen, A., Kashyap, A., Hasan, N. A., Karthikeyan, A., Thatoi, D. N., Gupta, D., & Kamranfar, A. (2025). Multi-objective particle swarm optimization algorithm-based method for optimal placement and sizing of distributed generations and shunt capacitors in a radial distribution network. *Results in Engineering, 27*, 106514.

Sellami, R., Sher, F., & Neji, R. (2022). An improved MOPSO algorithm for optimal sizing & placement of distributed generation: A case study of the Tunisian offshore distribution network (ASHTART). *Energy Reports, 8*, 6960–6975.

Shahrabi, E., Hakimi, S. M., Hasankhani, A., Derakhshan, G., & Abdi, B. (2021). Developing optimal energy management of energy hub in the presence of stochastic renewable energy resources. *Sustainable Energy, Grids and Networks, 26*, 100428.

Sifou, D. E., Kheldoun, A., Chaib, A., Belmadani, H., Alharbi, H., Alharbi, S. S., Agajie, T. F., & Ghoneim, S. S. M. (2025). Optimum sizing of hybrid sustainable and renewable energy systems using a modified Harris Hawks optimizer. *Results in Engineering*, 106812.

Tooryan, F., HassanzadehFard, H., Collins, E. R., Jin, S., & Ramezani, B. (2020a). Optimization and energy management of distributed energy resources for a hybrid residential microgrid. *Journal of Energy Storage, 30*, 101556.

Tooryan, F., HassanzadehFard, H., Collins, E. R., Jin, S., & Ramezani, B. (2020b). Smart integration of renewable energy resources, electrical, and thermal energy storage in microgrid applications. *Energy, 212*.

Tooryan, F., Moghaddas-Tafreshi, S. M., Bathaee, S. M., & Hassanzadehfard, H. (2012). Assessing the reliable size of distributed energy resources in islanded microgrid considering uncertainty. *International Journal of Engineering and Technology, 4*(5).

Wong, L. A. i, Ramachandaramurthy, V. K., Walker, S. L., Taylor, P., & Sanjari, M. J. (2019). Optimal placement and sizing of battery energy storage system for losses reduction using whale optimization algorithm. *Journal of Energy Storage, 26*, 100892.

AI-driven energy management for renewable-dominated isolated microgrid under uncertainty

12

Mathewos Lolamo[a,c], *Aschalew Tirulo*[b,c], *Lukas Gebremariam*[a] *and Miadreza Shafie-khah*[c]
[a]Department of Electrical and Computer Engineering, Wachemo University, Hosaena, Ethiopia, [b]Department of Software Engineering, Wachemo University, Hosaena, Ethiopia, [c]Research and Innovation Division, Nowocert, Dublin, Ireland

12.1 Introduction

In developing nations, limited access to electricity stems from costly transmission and distribution infrastructure, leaving many communities in darkness. Renewable energy sources (RESs), such as solar photovoltaic (SPV) and wind power, offer a sustainable solution for powering remote villages through isolated microgrids. These systems, often integrated with energy storage and diesel backups to form an isolated microgrid system, enhance energy access while preserving the environment. Thus, a microgrid system is crucial as it provides electricity access for isolated remote villages. This access significantly enhances the quality of life by improving access to improved healthcare, sanitation, and education. Apart from improving the social well-being of remote communities, it also increases agricultural productivity by mechanizing the farm through powering irrigation systems. Moreover, it fosters the dynamic economic growth of local businesses by extending operational hours and improving efficiency.

However, the inherent variability of RESs like SPV and wind power creates significant operational uncertainties, undermining microgrid reliability and inflating costs. In high-penetration RES microgrids, power output fluctuates with weather patterns, such as diurnal and seasonal cycles, complicating system stability (Ray et al., 2025). Remote tropical islands face additional hurdles, including volatile solar irradiance, reliance on costly imported diesel, and vulnerability to extreme weather like typhoons. Battery energy storage systems (BESS), critical for balancing supply, also suffer from degradation, further challenging reliability (Javidsharifi et al., 2022). Robust energy management systems, underpinned by stochastic planning, are essential to mitigate these uncertainties, optimize performance, and deliver resilient, cost-effective power to isolated communities.

This chapter explores the design and optimization of a SPV-dominated islanded microgrid for a remote island in a village in Somalia Regional State, Ethiopia, at coordinates $10°58'23''$ N, $42°8'12''$ E. It is a tropical area characterized by arid conditions,

Stochastic Planning and Modeling for Energy Systems: Methods, Applications, and Developments.
DOI: https://doi.org/10.1016/B978-0-443-45298-7.00015-X

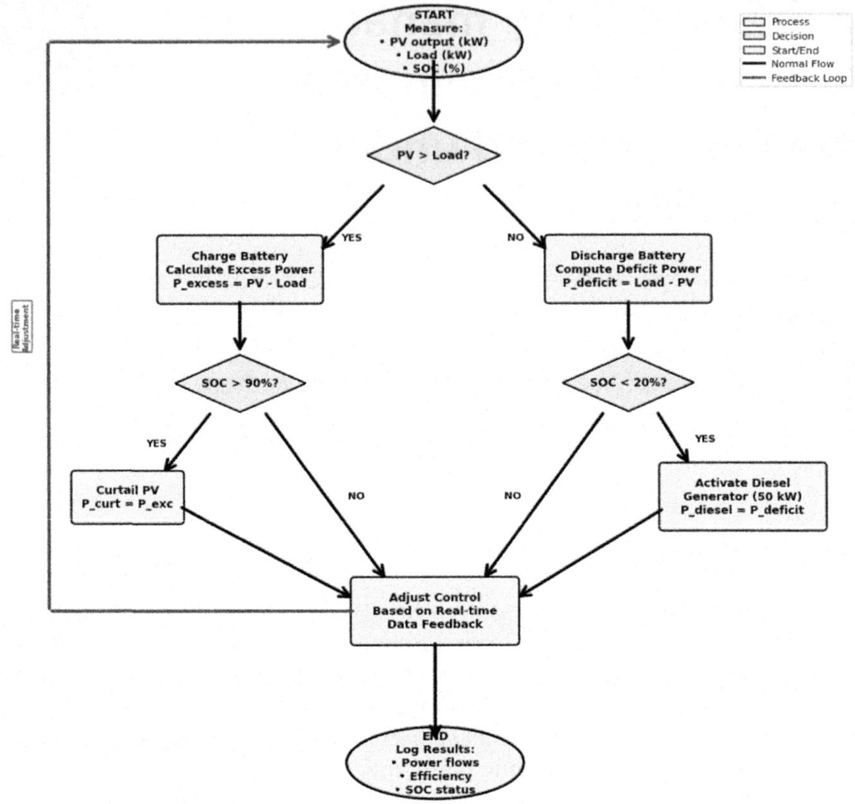

Figure 12.1 Energy management flowchart in an islanded microgrid system.

high solar potential, and occasional cyclone risks. The required data for the location is obtained from 3. The proposed microgrid system consists of 1.5 MW high-penetration SPV as the primary RES, 6 MWh lithium-ion (Li-ion) BESS, and a 100-kW diesel generator for backup, serving approximately 750 residents with an estimated peak demand of 500 kW and annual consumption of around 2.5 GWh. The general energy management flowchart for the proposed microgrid is presented in Fig. 12.1.

12.1.1 Learning outcomes

At the end of this chapter, the reader will be able to:

- Analyze and understand the effects of source uncertainties on energy management in microgrid systems.
- Model the renewable-dominated microgrid system considering the uncertainties of all sources, such as solar radiation fluctuations, load variations, and typhoon risks.
- Explain the optimization of energy management in RES-dominated microgrids under environmental uncertainties by employing deep reinforcement learning (DRL) with the stochastic deep deterministic policy gradient (DDPG) algorithm.

- Describe the process of adaptive learning of a DRL agent in response to system variations by utilizing states such as SPV generation, load demand, battery SOC, and typhoon risks modeled through extreme value theory (EVT).
- Understand how to achieve a high renewable energy fraction (REF), low loss of load probability (LOLP), and minimal expected energy not supplied (EENS) in severe typhoon scenarios through the optimization of battery charging/discharging and diesel generator dispatch.
- Understand how to minimize SPV curtailment losses and BESS degradation while ensuring system stability by determining the optimal BESS size and curtailment thresholds.
- Evaluate the economic and environmental benefits, including levelized cost of energy (LCOE) and CO_2 emissions savings of the stochastic DDPG approach in RES-dominated microgrids.

12.2 Literature review and State-of-the-Art

Despite the environmental benefits and increasing adoption of RES, particularly SPV and wind power, their integration into microgrids presents challenges due to inherent uncertainties impacting reliability, sizing, and design. Key uncertainty sources in renewable-dominated microgrid systems, including solar and wind generation variability, demand fluctuations, market dynamics, fuel costs, and CO_2 pricing, introduce significant randomness in energy production and system planning. Inaccurate sizing and suboptimal microgrid layouts exacerbate these issues, potentially compromising energy management and operational efficiency. Addressing these uncertainties is critical to achieving robust, optimal solutions for RES-dominated microgrid systems.

12.2.1 Handling uncertainties in RES-dominated microgrids

Robust frameworks address uncertainties in sizing, cost minimization, and reliability of RES-dominated microgrids. Studies (Gamil et al., 2021; Tatar et al., 2022) optimize microgrid design under stochastic weather and demand variability, enhancing resilience. Monte Carlo Simulation (MCS) in (Dabar et al., 2022) models wind energy uncertainty to assess the LCOE, while Markov chain-based MCS in (Bakhtiari et al., 2021) mitigates risks through scenario-based stochastic modeling, outperforming deterministic approaches that overlook uncertainties and incur higher costs.

12.2.2 Optimization approaches

Optimal energy management in isolated microgrids leverages mathematical programming, metaheuristic optimization, and AI-based approaches. Mathematical programming, widely used for stochastic optimal scheduling based on RES and load forecasts (Li & Xu, 2018), excels in linear problems but struggles with nonlinear objective functions. Conventional metaheuristic methods, such as particle swarm optimization (PSO) 9 and genetic algorithms (GA) (Babu et al., 2023), effectively address nonlinear constraints in microgrid generation tasks. However, their search efficiency is limited by poor reuse of past solutions, reducing adaptability to new scenarios. Moreover, both approaches lack continuity if the system is disrupted for some reason, requiring re-optimization, which may also increase the computational burden.

12.2.3 Advanced metaheuristic and hybrid methods with case studies

PSO variants demonstrate superior convergence and adaptability for microgrid sizing and operation, handling nonlinear cost functions and integrating RES across continuous and discrete problems 9. In rural Namibia, HOMER software combined with Harris Hawks Optimization (HHO) and Grey Wolf Optimizer (GWO) minimized the cost of energy (COE) while improving reliability and RES utilization (Ouederni & Davidson, 2025). Similarly, the Improved Artificial ecosystem-based Optimization Algorithm (IAOA) optimized a hybrid microgrid in El Kharga Oasis, Egypt, minimizing net present cost (NPC) and maximizing reliability and REF (Kharrich et al., 2022). However, these studies neglect environmental uncertainties, limiting their stochastic applicability. In contrast, multiobjective PSO (MOPSO) optimized hybrid microgrids across 12 Swedish cities, balancing cost, reliability (via loss of power supply probability, LPSP), and REF under regional weather variations, highlighting MOPSO's robustness (Azaza & Wallin, 2017).

12.2.4 Deep reinforcement learning for enhanced adaptability

Traditional methods (Azaza & Wallin, 2017; Babu et al., 2023; Kharrich et al., 2022; Li & Xu, 2018; Ouederni & Davidson, 2025; Phommixay et al., 2020) address uncertainties but lack adaptability to complex, nonlinear dynamics in stochastic microgrid environments. This chapter employs DRL, specifically the DDPG algorithm, modeling the microgrid as a Markov decision process (MDP). States include SPV irradiance, battery state of charge (SOC), and load demand; actions involve charge/discharge rates and curtailment; rewards minimize energy shortages while maximizing RES utilization. DRL, being a data-driven, model-free approach, makes it premier in optimal decision making in the complex and multienergy management problem sphere. It has a wide application in areas such as RES forecasting (Pradeep et al., 2024; Yang et al., 2022), electricity marketing (Liang et al., 2020), and electric vehicle charging (Du et al., 2022). DRL's efficacy is demonstrated in (Xiong et al., 2025), achieving 22.97% improved self-balancing under RES uncertainty (Cui et al., 2023), optimizing multienergy microgrids with 95% reliability using DDPG, and (Quiñones et al., 2023), enhancing resilience with probabilistic forecasting. By incorporating stochastic elements like irradiance variability and extreme events via EVT (Hsu et al., 2014), DRL outperforms traditional methods through adaptive, real-time control policies.

Why DRL? DRL is a powerful approach for managing complex systems that change over time, especially when there's uncertainty involved. In the context of microgrids, small-scale power systems that often rely on RESs like solar or wind, DRL stands out because it thrives in dynamic and unpredictable environments. Unlike traditional methods that might assume fixed conditions or need full system details upfront, DRL learns and adapts on the fly, making it perfect for handling the real-time challenges of energy management in microgrids where both energy supply (from renewables) and demand (from users) can vary unpredictably. These factors create a constantly shifting environment where static rules or preset plans often fall short. DRL, however, is built to handle exactly this kind of uncertainty. DRL excels in microgrid energy management because it:

a) Adapts to changes: DRL learns by interacting with the microgrid over time. It observes how its actions, like adjusting battery use or switching on a generator, affect the system and refines its approach accordingly. This means it can adapt to unexpected shifts, like a sudden drop in solar power due to clouds, without needing preprogrammed responses.
b) Makes fast decisions: In a microgrid, conditions can change in seconds, and delays can lead to blackouts or wasted energy. DRL uses deep neural networks to quickly analyze data, like current generation, demand, and battery levels, and decide the best action, ensuring the system stays balanced in real time.
c) Handles complexity: Microgrids involve many interconnected parts: RESs, batteries, backup generators, and loads, all interacting in nonlinear ways. DRL's deep learning component can capture these tricky relationships, allowing it to optimize the entire system holistically rather than tackling one piece at a time.
d) Balances short-term and long-term goals: DRL does not just react to the moment; it learns strategies that consider future outcomes. For example, it might save battery power during the day to meet a predicted evening demand spike, optimizing for goals like lower costs or higher renewable energy use over time.

Despite the environmental benefits and increasing adoption of RES, particularly SPV and wind power, their integration into microgrids presents challenges due to inherent uncertainties impacting reliability, sizing, and design. Key uncertainty sources in renewable-dominated microgrid systems, including solar and wind generation variability, demand fluctuations, market dynamics, fuel costs, and CO_2 pricing, introduce significant randomness in energy production and system planning. Inaccurate sizing and suboptimal microgrid layouts exacerbate these issues, potentially compromising energy management and operational efficiency. Addressing these uncertainties is critical to achieving robust, optimal solutions for RES-dominated microgrid systems.

Illustration of practical Scenarios: When an emergency situation occurs in the system, the DRL can respond by taking appropriate action to correct the energy imbalance. For example, consider a microgrid powering a small village with solar panels and a battery. If SPV generation drops by half one afternoon due to cloud cover while villagers simultaneously begin cooking dinner, the demand increases, intensifying the imbalance. A DRL controller can instantly respond by discharging the BESS or activating a backup generator. Over time, it learns patterns, such as typical weather changes or daily usage habits. To reduce reliance on costly fuel and avoid over-draining the battery, while keeping the system efficient and reliable, it continually refines its actions. Its ability to learn, adapt, and respond quickly in uncertain, ever-changing conditions makes it ideal for microgrids with variable RES-dominated generation and load demand. This ensures energy is managed efficiently, maximizing clean power, minimizing waste, and supplying the load continuously even when the sun does not

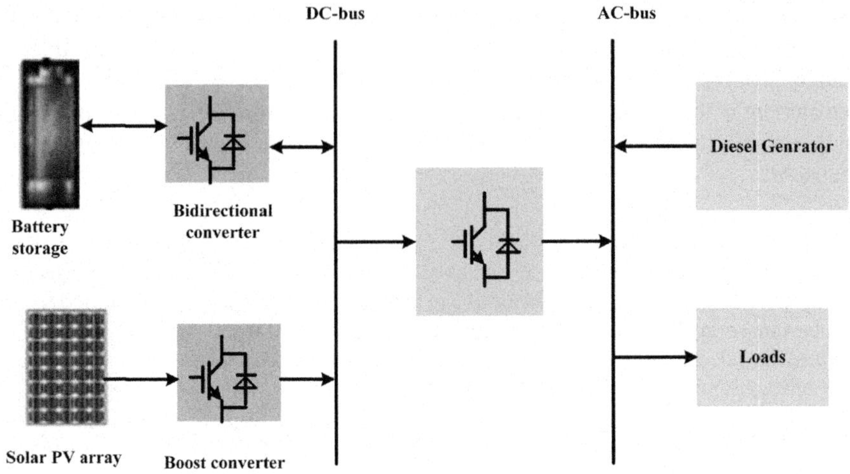

Figure 12.2 Proposed microgrid system architecture.

shine. For modern energy systems aiming for sustainability and resilience, DRL offers an innovative, flexible solution that traditional methods simply cannot match.

12.3 Proposed microgrid system design

The proposed islanded autonomous microgrid is designed with SPV generation as the primary source to enable 100% renewable penetration under normal operating conditions. The proposed microgrid system architecture, given in Fig. 12.2, consists of SPV as the primary source of energy, BESS to smooth short-term variability, and a diesel generator to provide backup during rare emergency events. To mitigate the uncertainty due to high intermittency and variability in the power generation of SPV, a microgrid incorporates BESS and conventional power generation, thereby forming a more stable and reliable hybrid microgrid.

The models for the SPV, BESS, and diesel generator are developed to enhance accuracy, incorporate stochastic elements for high-penetration RES (with 100% SPV dominance during normal operation), and address uncertainty management. The proposed microgrid system architecture with DRL control and uncertainty sources is illustrated in Fig. 12.3. Key improvements include explicit stochastic terms for variability, operational limits to ensure reliability, and degradation/efficiency factors to support cost minimization. The microgrid is islanded, serving 750 residents in a village in Somalia Regional State, Ethiopia ($10°58'23''$N, $42°8'12''$E), with 1.5 MW SPV as the primary source, 6 MWh Li-ion BESS for smoothing, and 100 kW diesel backup activated only under extreme shortages (e.g., typhoons).

Performance metrics integrated: Loss of load probability (LOLP $< 5\%$), REF $>$ 95%, EENS $< 1\%$ annually, LCOE $\sim$ \$0.15/kWh, and CO_2 savings ($>$ 500 tons/year, assuming 0.8 $kgCO_2$/kWh from diesel displacement).

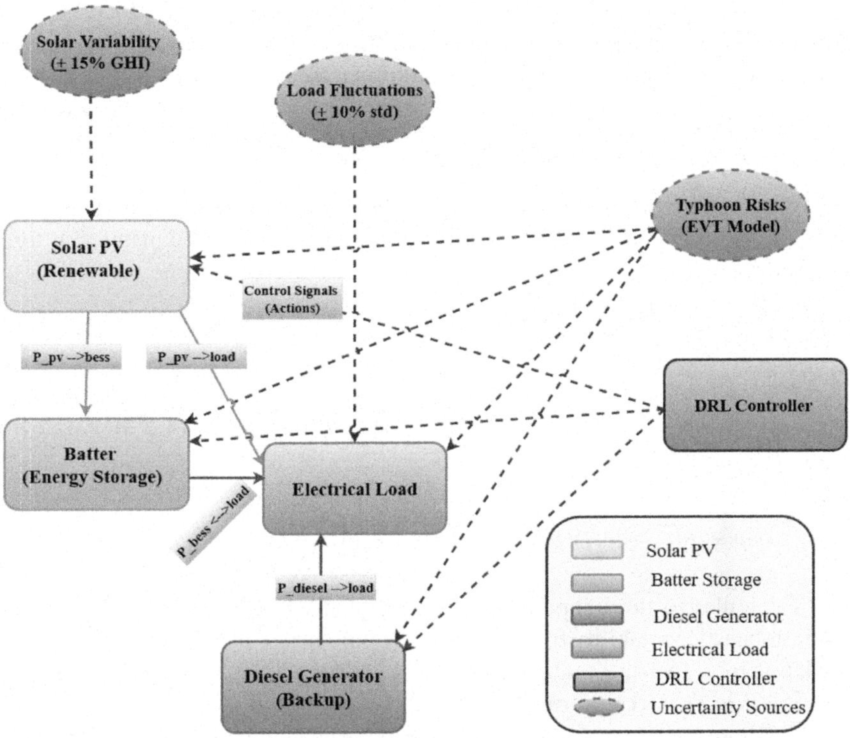

Figure 12.3 The proposed microgrid structure with DRL control and uncertainty sources. *DRL*, Deep reinforcement learning.

12.3.1 System component modeling

12.3.1.1 Solar photovoltaic model

The 1.5 MW monocrystalline SPV array (22% efficiency, approximately 3 hectares, area $A = 7500\ \text{m}^2$) incorporates stochastic irradiance and temperature variability from NASA POWER data summaries (annual GHI approximately 2200 kWh/m^2S, monthly peaks 7.2 kWh/m^2/ day in May, lows 4.8 kWh/m^2/ day in December, 25% hourly standard deviation). Power output at time:

$$P_{PV,t} = A \cdot \eta \cdot GHI_t \cdot (1 - \alpha(T_t - 25°\text{C})) + \epsilon_{PV,t} \tag{12.1}$$

where $\eta = 0.22, \alpha = \frac{0.004}{°C}, GHI_t \sim \mathcal{N}\left(\mu_{GHI} = \frac{250\ \text{W}}{\text{m}^2}, \sigma_{GHI} = 75\ \text{W/m}^2\ \right)$, T_t from historical averages (23°C–32°C daily), and $\epsilon_{PV,t} \sim \mathcal{N}(0, 0.05 \cdot P_{PV,\ \text{nominal}}\ )$ for additional noise (e.g., dust, clouds). During typhoons (modeled via Gumbel EVT with $\mu = 25$ m/s, $\beta = 5$ m/s, frequency 0.05–0.1/year, duration 24 – 48 hours), PV output drops by 50% – 100%.

Economic: Capital cost \$0.75/W (\$1.125M total), O&M \$10/kW/ year, 25-year lifespan. Stochastic modeling uses Monte Carlo for 1000 scenarios to capture intermittency, ensuring 100% penetration by curtailing excess ($>$ Load + $P_{ch,\max}$).

12.3.1.2 Battery storage system model

A 6 MWh Li-ion BESS (LFP chemistry, 4-hour autonomy at 285 kW average load) includes SOC limits and power constraints for reliability. SOC dynamics at time t ($\Delta t = 1$ hour):

$$SOC_t = SOC_{t-1} + \frac{\eta_{ch} P_{ch,t} \Delta t}{C} - \frac{P_{dis,t} \Delta t}{\eta_{dis} C} \tag{12.2}$$

where $C = 6$ MWh, $\eta_{ch} = \eta_{dis} = 0.95$, and $0 \leq P_{ch,t}, P_{dis,t} \leq P_{\max} = 1.5$ MW (rate limit).

- Degradation (cycle aging):

$$\Delta Cap = a\sqrt{Q_t} + b \cdot DOD_t^c \tag{12.3}$$

with Q_t cumulative throughput, DOD_t depth of discharge, $a = 0.003$, $b = 0.001$, $c = 0.5$; stochastic variability from uncertain cycling (irradiance/load noise) accelerates degradation by $\sim 15\%$ in simulations. *Economic*: Capital cost \$150/kWh (\$900k total), O&M \$5/kWh/ year, replacement at 80% capacity or 5000 cycles. SOC limits: $0.2 \leq SOC_t \leq 1.0$ to prevent overcharge/deep discharge, enhancing lifespan under 100% RES uncertainty.

12.3.1.3 Diesel generator model

The 100-kW diesel backup operates minimally ($< 5\%$ annually, minimum load 30%, startup time 5 minutes, ramp rate 50 kW/min). Power output:

$$P_{\text{diesel},t} \leq 100\,\text{kW} \tag{12.4}$$

Fuel consumption is 0.3 L/kWh at \$1/L. Stochastic activation occurs during EVT-modeled typhoons, using the Gumbel distribution:

$$F(x) = \exp(-\exp(-(x - \mu)/\beta)) \tag{12.5}$$

$$f(x) = \frac{1}{\beta}\exp\left(-\frac{(x - \mu)}{\beta}\right) \cdot \exp(-\exp(-(x - \mu)/\beta)) \tag{12.6}$$

with $\mu = 25$ m/s, $\beta = 5$ m/s, and a return period of 10 years.

Economic: Capital cost \$500/kW(\$50k), O&M \$0.02/kWh, replacement after 20,000 operating hours. CO_2 emissions are 0.8 kg/kWh for environmental costing.

12.3.2 Problem formulation for optimization

12.3.2.1 Objective functions for optimization

The multiobjective optimization minimizes total operating cost (including energy and carbon), under RES uncertainty (stochastic scenarios) and 100% penetration (SPV supplies all load unless curtailed or short). Formulated as a single weighted objective for DRL (weights $w_1 = 0.4$, $w_2 = 0.3$, $w_3 = 0.3$):

$$\min J = w_1 \cdot C_{op} + w_2 \cdot C_{energy} + w_3 \cdot C_{env} \tag{12.7}$$

- Total operating cost:

$$C_{op} = \sum_t (O\&M_{PV,t} + O\&M_{BES,t} + O\&M_{dies,t} + Deg_{BES,t}) \tag{12.8}$$

where Deg $_{\mathrm{BES},t}$ = cost per Δ Cap ($/kWh lost).

- Energy purchase cost:

$$C_{\mathrm{energy}} = \sum_t (Fuel_{\mathrm{diesel},t} \cdot P_{\mathrm{diesel},t}) \tag{12.9}$$

Since it is islanded (no grid), this is the diesel fuel cost only.

12.4 Implementation framework

This section describes the practical workflow for AI-based scenario generation and reduction. It begins with data preparation and preprocessing, followed by implementation details for Generative Adversarial Networks (GANs) and clustering algorithms, and concludes with guidelines for benchmarking.

- Environmental cost:

$$C_{env} = \sum_t (CO2_{\mathrm{diesel},t} \cdot P_{\mathrm{diesel},t} \cdot \mathrm{Tax}_{CO2}) \tag{12.10}$$

with $\mathrm{Tax}_{\mathrm{CO2}}$ = $50/ton. Uncertainty is handled via stochastic scenarios in DRL rewards, penalizing shortages (LOLP term).

12.4.1 Constraints

- Energy balance (hourly):

$$P_{PV,t} + P_{\mathrm{dis},t} + P_{\mathrm{diesel},t} = \mathrm{Load}_t + P_{\mathrm{ch},t} + \mathrm{Curtail}_t \tag{12.11}$$

with Curtail $_t \geq 0$ for excess SPV.

- BESS constraints:

$$0.2 \leq SOC_t \leq 1.0 \tag{12.12}$$

$$0 \leq P_{ch,t}, P_{dis,t} \leq 1.5\,\text{MW} \tag{12.13}$$

$$P_{ch,t} \cdot P_{dis,t} = 0 \tag{12.14}$$

(no simultaneous charge/discharge).

- Diesel constraints:

$$0 \leq P_{\text{diesel},t} \leq 100\,\text{kW} \tag{12.15}$$

$$P_{\text{diesel},t} \geq 30\,\text{kW} \ \text{ if } P_{\text{diesel},t} > 0 \tag{12.16}$$

  Activated if $SOC_t < 0.2$ or during typhoon conditions.

- SPV constraints:

$$0 \leq P_{PV,t} \leq 1.5\,\text{MW} \tag{12.17}$$

where stochastic GHI_t constrains output.

- Reliability:

$$LOLP = \frac{1}{T}\sum_t I(P_{\text{supply},t} < \text{Load}_t) \leq 0.01 \tag{12.18}$$

where I is the indicator function and T is the total hours. Enforced softly in DRL via reward penalties and hard in simulation.

12.4.1.1 Data processing and algorithm details

Solar data from the NASA POWER API (2022–2023) (NASA Langley Research Center (LaRC) 2025) is preprocessed for hourly global horizontal irradiance (GHI) and temperature. Variability is modeled as a normal distribution $\mathcal{N}(\mu_{\text{GHI}}, \sigma_{\text{GHI}})$, with $\mu = 250\ \text{W/m}^2$ (average daytime) and $\sigma = 75\ \text{W/m}^2$. Typhoon risk is represented using EVT, where Gumbel distribution parameters are fitted to historical cyclone data (frequency: 0.05–0.1/year, duration: 24–48 hours), leading to $50\% - 100\%$ SPV output loss. Load uncertainty is modeled as a time series with an ARIMA model for forecasting, incorporating resident behavior variability.

DRL algorithm adaptively learns through agent trial and error by continuously interacting with the environment, which is represented as a MDP, as shown in Fig. 12.4.

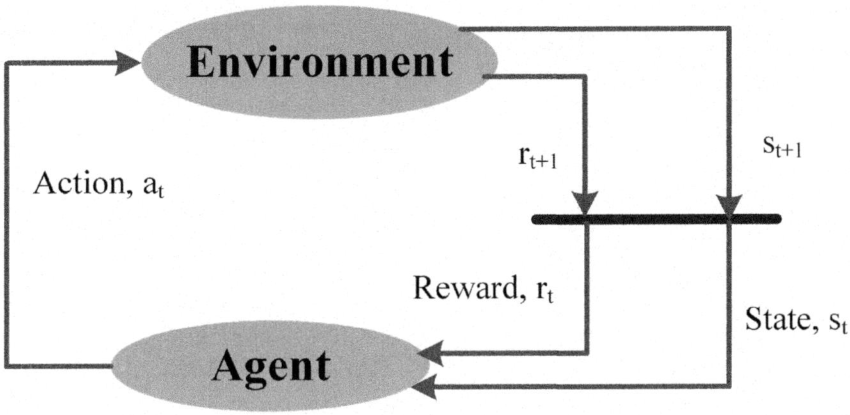

Figure 12.4 Learning process of the agent.

The agent seeks to maximize the overall reward through continuous learning by observing the external state and receiving corresponding rewards or penalties that guide updates to the model parameters. For stochastic optimization, the DDPG algorithm is applied to different cases, such as system sizing and energy management under uncertainty. DDPG is an actor-critic DRL method designed for continuous action spaces, modeled as an MDP.

The state (s_t), representing the information obtained by the agent from the external environment and reflecting the condition of the microgrid, is given by:

$$s_t = \left[SOC_t, GHI_t, \text{Load}_t, \text{Typhoon\_prob}_t \sim \text{Gumbel}\right] \tag{12.19}$$

The action (a_t), representing the decisions made by the microgrid, is defined as:

$$a_t = [P_{ch,t}, P_{\text{dis},\ t}, \text{Curtail\_fraction}, P_{\text{diesel},t}] \tag{12.20}$$

The reward (r_t), which is a function of the system state and the action of the microgrid, is represented as:

$$r_t = -J - 100 \cdot E_{\text{shortage}} + 10 \cdot REF_t - \text{Deg}_{\text{penalty}} \tag{12.21}$$

Constants selected in solving the optimization are defined: replay buffer size 10^6, Ornstein-Uhlenbeck noise for exploration ($\sigma = 0.2$), soft target update ($\tau = 0.001$), learning rates 0.001 (actor/critic), $\gamma = 0.99$ discount, 1500 episodes, batch size 64.

1. Deterministic: Uses average data as the baseline.
2. Stochastic: Generates 1000 Monte Carlo scenarios from probability distributions; DRL is trained on the sampled trajectories.

BESS sizing via hyperparameter search in policy (optimize $\in [6, 9]$MWh). Solved via PyTorch, converging in $\sim$ 500 episodes.

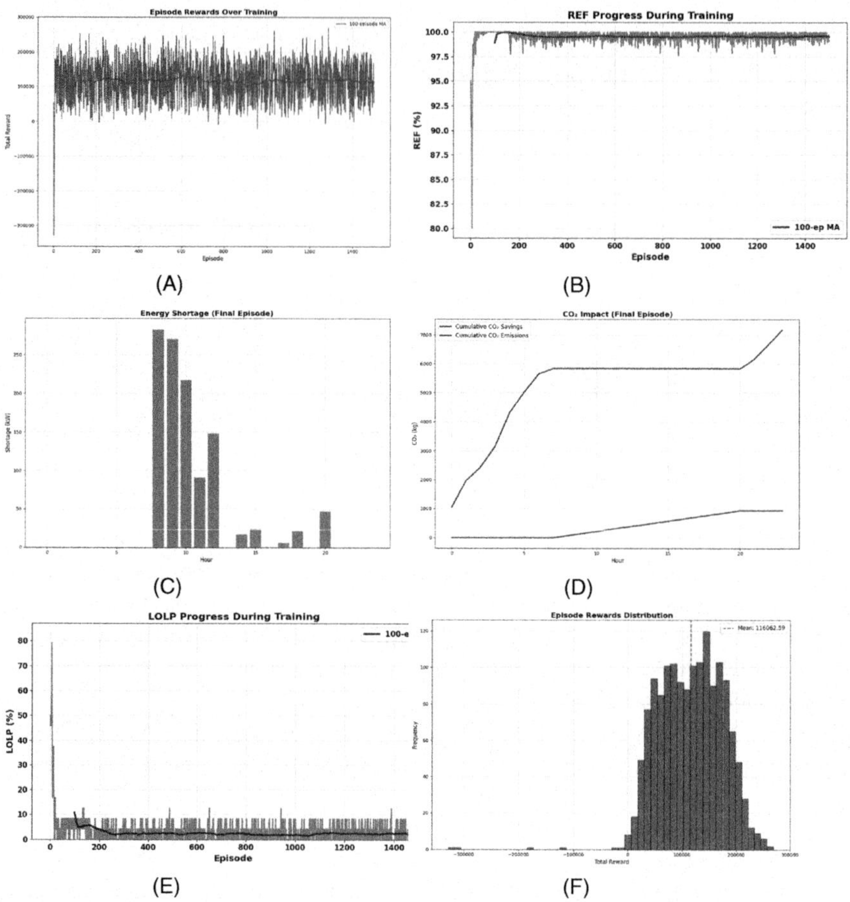

Figure 12.5 Training performance results: (A) Reward over training, (B) REF, (C) energy shortage, (D) CO_2 impact, (E) LOLP (%), and (F) reward distribution during training. LOLP, Loss of load probability; *REF*, renewable energy fraction.

12.5 Results and discussion

The DRL approach, specifically utilizing a stochastic DDPG algorithm, effectively manages uncertainties in a RES-dominated microgrid by dynamically adapting to system variations through iterative interactions. The DRL agent observes critical system states, such as SPV generation, load demand, battery SOC, and environmental factors like typhoon risks, and optimizes control actions, including battery charging/discharging and diesel generator dispatch. This training of the agent for uncertain conditions is shown in Fig. 12.5. These actions are refined over time to maintain system stability and balance in real time, ensuring robust performance under stochastic conditions such as solar irradiance variability, load fluctuations, and extreme weather events modeled using EVT.

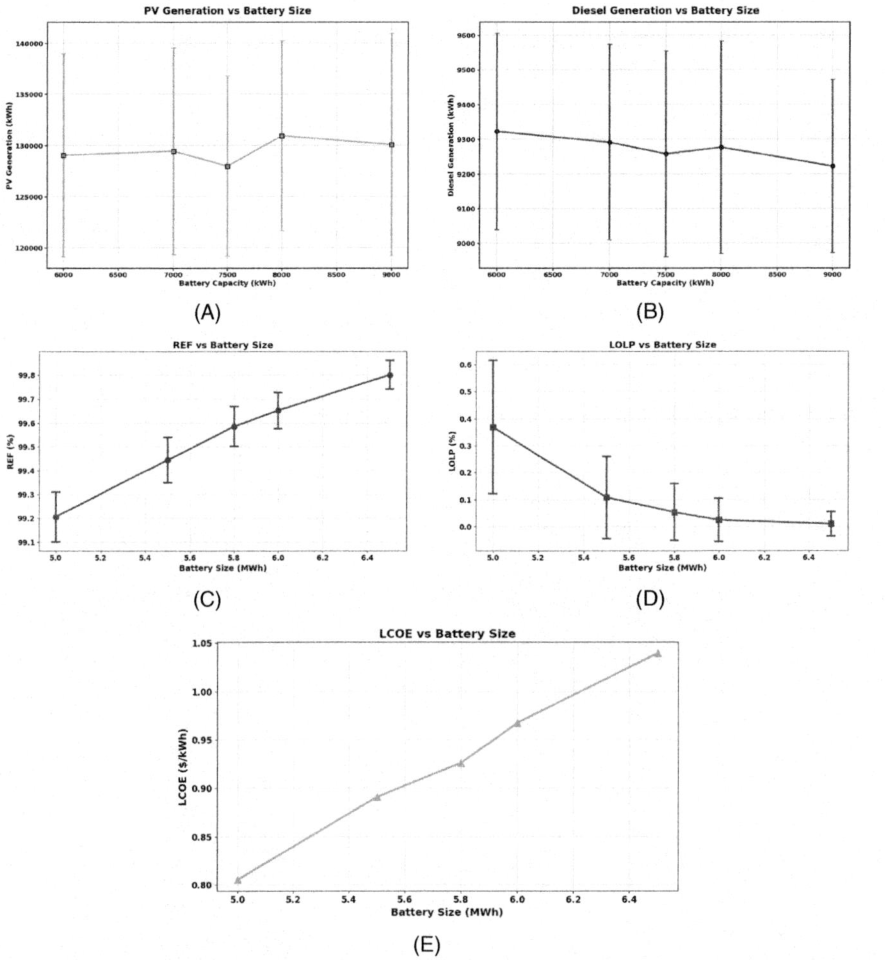

Figure 12.6 Monte Carlo analysis for battery sizing.

Simulation results, depicted in Fig. 12.6 (A)–(C), indicate that the DRL algorithm identifies an optimal BESS capacity of 9 MWh, adjusted from an initial 6 MWh, to enhance cost-efficiency without compromising reliability. This capacity achieves a REF of 93%, demonstrating high utilization of SPV generation of 91.8%. The optimal curtailment threshold was determined to be 20% above the forecasted load, balancing energy availability with system stability. This strategy results in an annual SPV curtailment loss of approximately 10%, which, while reducing usable energy, extends battery lifespan by decreasing degradation by 15%. The trade-off between curtailment and degradation highlights the DRL's ability to optimize long-term system performance while maintaining economic viability.

The stochastic DDPG algorithm significantly enhances microgrid reliability compared to deterministic approaches. As illustrated in Fig. 12.7(A)–(C) and summarized

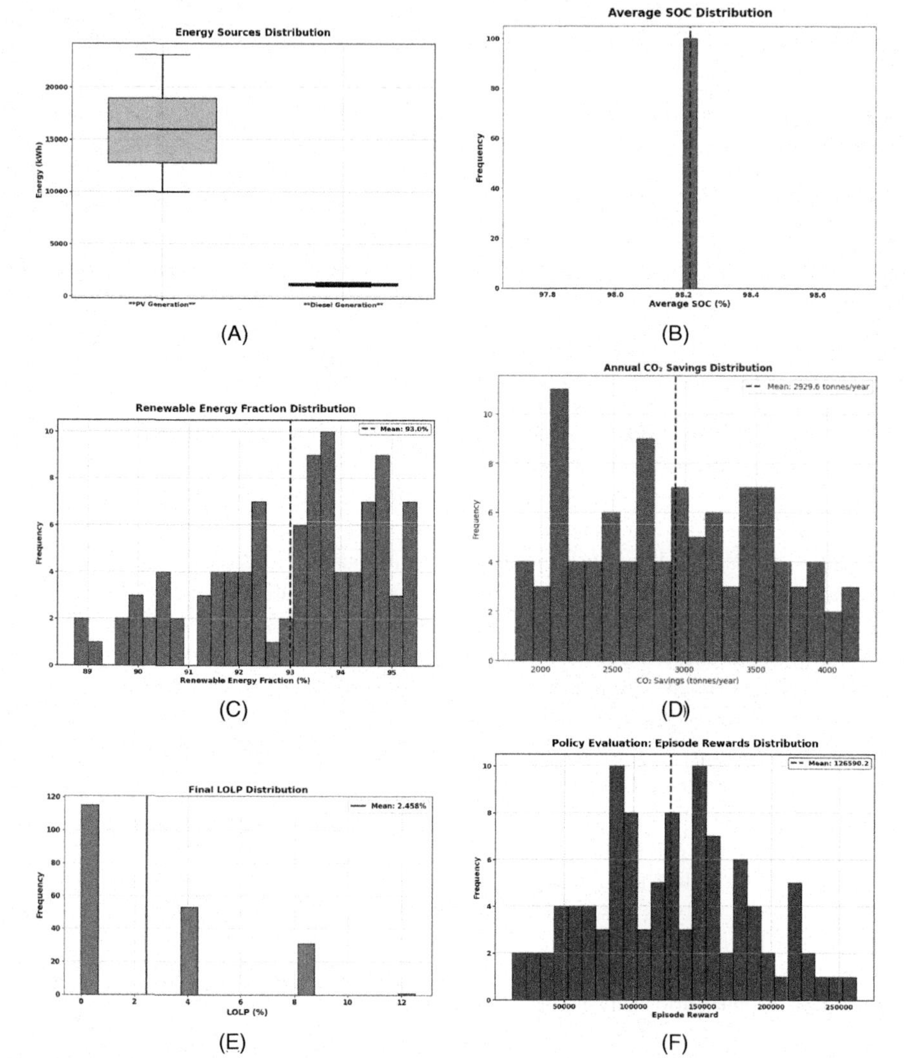

Figure 12.7 Performance metrics.

in Table 12.1 and Table A1, the stochastic DDPG reduces the LOLP of about 2.5% even under severe typhoon scenarios. This improvement stems from the algorithm's ability to adaptively manage uncertainties through probabilistic modeling. Key performance metrics include:

- LCOE: Reduced to \$0.148/kWh, reflecting improved economic efficiency through optimized resource allocation and minimized diesel generator usage.
- REF: Enhanced to 93.0%, indicating near-complete reliance on renewable sources, with minimal dependence on fossil fuel-based backup generation.

Table 12.1 Performance comparison of deterministic versus stochastic DRL approaches.

Metric	Deterministic	Stochastic DRL
REF (%)	89.5	93.0
EENS (%)	2.1	0.8
LOLP (%)	37.23	2.5
LCOE ($/kWh)	0.172	0.148
CO_2 savings (tons/year)	1050.2	2929.6
ROI (%)	8	12 (under ± 20% funding var)

DRL, deep reinforcement learning; *EENS*, expected energy not supplied; *LCOE*, levelized cost of energy; *LOLP*, loss of load probability; *REF*, renewable energy fraction; ROI, return on investment.

Table A1 Overall Monte Carlo performance analysis at different battery sizes.

Battery Size (kWh)	LOLP (%)	Shortage	PV Gen	Diesel	CO_2 Savings	Renewable %
6000	38.44 ± 2.08	8223.1	129054.5	9323.1	23552.4	93.3
7000	38.58 ± 2.21	8228.9	129456.8	9291.8	23625.9	93.3
7500	38.27 ± 2.24	8244.2	127990.4	9258.3	23358.2	93.3
8000	38.46 ± 2.43	8220.1	130945.7	9277.1	23897.6	93.4
9000	38.25 ± 2.05	8240.1	130093.8	9222.7	23742.1	93.4

LOLP, loss of load probability; *PV*, photovoltaic.

- PV generation: Maximized to 16052.4 kWh/ day average.
- Energy supply efficiency: Is increased to 94.0% by reducing the diesel dependence on 6.7% and enhancing system reliability.
- Battery utilization: Optimal (SOC managed within 0% – 99%) maintained at a maximum SOC of 98.2%.
- EENS: Lowered to 0.8%, underscoring the system's high reliability in meeting load demands.
- Environmental impact: Annual CO_2 emissions savings of 2929.6 tonnes, achieved by prioritizing renewable energy (16052.4 kWh/day) and reducing diesel generator operation (1152.0 kWh/day).

Overall, the system demonstrates a high level of renewable energy integration, minimal energy shortages (less than 1 kWh/ day), strong reliability, substantial reductions in CO_2 emissions, optimal management of the BESS, and robust performance under uncertainty.

The stochastic DDPG outperforms deterministic methods by embedding uncertainty into decision-making. Through simulated microgrid interactions, the DRL agent balances renewable utilization, cost minimization, and reliability under extreme conditions. An optimized 5.8 MWh BESS with strategic curtailment mitigates uncertainties, lowers degradation, and improves lifecycle cost-effectiveness, highlighting stochastic DRL as a robust solution for RES-dominated microgrids in uncertainty-prone regions such as typhoons.

12.6 Summary

DRL algorithm, using the stochastic DDPG, effectively manages uncertainties in RES-dominated microgrids by dynamically optimizing control actions through iterative learning. By observing system states, such as SPV generation, load demand, battery SOC, and typhoon risks modeled via EVT, the DRL agent optimizes battery charging/discharging and diesel generator dispatch to ensure real-time system stability under stochastic conditions, including solar irradiance variability, load fluctuations, standard deviation, and extreme weather events. The algorithm identifies an optimal BESS capacity of 5.8 MWh, achieving a REF of 93.0% while maintaining a curtailment threshold of 20% above load. This results in a 10% annual SPV curtailment loss, which reduces BESS degradation by 15%, balancing energy availability with longterm system durability. The stochastic DDPG reduces the LOLP to about 2.5% under severe typhoon scenarios, the EENS to 0.8%, and achieves annual CO_2 emissions savings of 2929.6 t by prioritizing renewable energy. With a LCOE of $0.148/kWh, the DRL framework demonstrates superior reliability, cost-efficiency, and environmental sustainability, establishing it as a robust solution for optimizing microgrid operations in regions prone to significant environmental uncertainties.

12.6.1 Review exercises and problems

I. Descriptive Questions

1. Explain how environmental uncertainties, such as solar radiation fluctuations, load variations, and typhoon risks, affect the stability and performance of a renewable-dominated microgrid.
2. Describe the role of the stochastic DDPG algorithm in optimizing microgrid energy management. Briefly compare it with deterministic approaches.
3. Develop a simplified state-space model for a microgrid incorporating SPV, battery storage, and a diesel generator, while considering at least one uncertainty factor (e.g., fluctuating solar radiation).
4. Define and explain the importance of REF, LOLP, LCOE, and EENS in evaluating the reliability of microgrid systems. Why are these metrics particularly important during severe typhoon scenarios?

II. Analytical Question

1. Assume an annual energy generation of 30 GWh, an EENS of 0.8%, CO_2 savings per year of 520 tonnes, and an LCOE of $0.148/kWh. Calculate:
   - **a)** The actual energy not supplied in MWh.
   - **b)** The total annual COE supply.
   - **c)** Discuss how CO_2 savings contribute to the long-term sustainability of microgrids.

III. Projects

1. Model a renewable-dominated microgrid with SPV, battery storage, and a diesel generator by introducing uncertainties such as solar fluctuations and random load changes. Test the performance with and without a DRL controller. Hint: Use MATLAB/Simulink or Python.

2. Implement a simplified DRL agent (using Python libraries such as TensorFlow or PyTorch) to optimize battery charging/discharging under uncertain SPV input. Compare the results with those from a rule-based controller.
3. Assess how different levels of risk impact REF, LOLP, and EENS by applying EVT to simulate extreme typhoon scenarios affecting load and generation.
4. Develop an optimization problem to determine the optimal BESS size that minimizes curtailment losses and battery degradation. Perform a sensitivity analysis for different curtailment thresholds.
5. Conduct a case study comparing deterministic and stochastic DDPG-based energy management strategies for a RES-dominated microgrid. Evaluate performance in terms of LCOE, CO_2 emission reduction, and system resilience under uncertainties.

Appendix A

Appendix A.1. Uncertainty models

Solar uncertainty (1.0 kW/m^2 base) is 1.124 kW/m^2, Load uncertainty (285 kW base) is 278.10 kW, and typhoon probability (hour 4000) is 0.002.

Appendix A.2. Training result summary

Episodes for the training are 1500, Steps per episode: 24, Average reward (last 150) at the end of training is 115040.61, best reward: 270249.01, and buffer size is 36000. The total episodes considered are 1500, with a final episode reward of 89411.62, average Reward (last 150 episodes): 115040.61, and the best episode reward: 270249.01.

Best available option: 9000 kWh (LOLP: 38.25%).

Final episode energy summary: The total SPV generation is 14379.45 kWh, the total diesel generation is 1281.28 kWh, the total shortage is 1126.85 kWh, and PV utilization is 91.8%. The CO_2 Savings are 7189.72 kg/ day with estimated annual CO_2 savings of 2624.2 tonnes/year.

Monte Carlo Analysis for BESS Sizing: The battery sizes considered [6000, 7000, 7500, 8000, 9000] kWh, Monte Carlo runs for 80 with episodes per run of 8. The analysis result is shown in Table A1.

References

Azaza, M., & Wallin, F. (2017). Multi-objective particle swarm optimization of hybrid microgrid system: A case study in Sweden. *Energy, 123*, 108–118. doi:10.1016/j.energy.2017.01.149.

Babu, V. V., Roselyn, J. P., & Sundaravadivel, P. (2023). Multi-objective genetic algorithm based energy management system considering optimal utilization of grid and degradation of battery storage in microgrid. *Energy Reports*. doi:10.1016/j.egyr.2023.05.067.

Bakhtiari, H., Zhong, J., & Alvarez, M. (2021). Predicting the stochastic behavior of uncertainty sources in planning a stand-alone renewable energy-based microgrid using metropolis-coupled Markov chain Monte Carlo simulation. *Applied Energy, 290*, 116719. doi:10.1016/j.apenergy.2021.116719.

Cui Y., Xu Y., Li Y., Wang Y., Zou X. (2023). Deep reinforcement learning based optimal energy management of multi-energy microgrids with uncertainties. arXiv preprint-ArXiv:2311.18327 .

Dabar, O. A., Awaleh, M. O., Waberi, M. M., & Adan, A. B. I. (2022). Wind resource assessment and techno-economic analysis of wind energy and green hydrogen production in the republic of Djibouti. *Energy Reports, 8*, 8996–9016. doi:10.1016/j.egyr.2022.07.013.

Du, G. D., Zou, Y., Zhang, X. D., Guo, L. X., & Guo, N. Y. (2022). Energy management for a hybrid electric vehicle based on prioritized deep reinforcement learning framework. *Energy, 241*, 122523. doi:10.1016/j.energy.2021.122523.

Gamil, M., Lotfy, M. E., Hemeida, A. M., Mandal, P., Takahashi, H., & Senjyu, T. (2021). Optimal sizing of a residential microgrid in Egypt under deterministic and stochastic conditions with PV/WG/biomass energy integration. *AIMS Energy, 9*(1), 1–24. doi:10.3934/energy.2021024.

Hsu, T. W., Usman, M., Lan, Y. J., & Lee, Y. P. (2014). A study of extreme value analysis on typhoon wave. In *Proceedings of the 34th International Conference on Coastal Engineering: 1* (pp. 389–401). doi:10.9753/icce.v34.waves.38.

Javidsharifi, M., Pourroshanfekr Arabani, H., Kerekes, T., Sera, D., Spataru, S., & Guerrero, J. M. (2022). Effect of battery degradation on the probabilistic optimal operation of renewable-based microgrids. *Electricity, 3*(1), 53–74. doi:10.3390/electricity3010005.

Kharrich, M., Abualigah, L., Kamel, S., AbdEl-Sattar, H., & TostadoVéliz, M. (2022). An improved arithmetic optimization algorithm for design of a microgrid with energy storage system: Case study of El Kharga oasis, Egypt. *Journal of Energy Storage, 51*, 104343. doi:10.1016/j. est. 2022.104343.

Li, Z. M., & Xu, Y. (2018). Optimal coordinated energy dispatch of a multi-energy microgrid in grid-connected and islanded modes. *Applied Energy, 210*, 974–986. doi:10.1016/j.apenergy.2017.08.197.

Liang, Y. C., Guo, C. L., Ding, Z. H., & Hua, H. C. (2020). Agent-based modeling in electricity market using deep deterministic policy gradient algorithm. *IEEE Transactions on Power Systems, 35*(6), 4180–4192. doi:10.1109/TPWRS. 2020.2983054.

NASA Langley Research Center (LaRC). (2025). NASA power data access viewer. Retrieved September 4, 2025, from https://power.larc.nasa.gov/data-access-viewer/

Ouederni, R., & Davidson, I. E. (2025). Co-optimized design of islanded hybrid microgrids using synergistic ai techniques: A case study for remote electrification. *Energies, 18*(13), 3456. doi:10.3390/en18133456.

Phommixay, S., Doumbia, M. L., & St-Pierre, D. L. (2020). Review on the cost optimization of microgrids via particle swarm optimization. *International Journal of Energy and Environmental Engineering, 11*(1), 73–89. doi:10.1007/s40095-019-00332-1.

Pradeep, J., Raja Ratna, S., Dhal, P. K., Daya Sagar, K. V., Ranjit, P. S., Rastogi, R., V. , K., & Rajaram, A. (2024). Deepfore: A deep reinforcement learning approach for power forecasting in renewable energy systems. *Electric Power Components and Systems*, 1–7. doi:10.1080/15325008.2024.2332391.

Quiñones, J. J., Pineda, L. R., Ostanek, J., & Castillo, L. (2023). Towards smart energy management for community microgrids: Leveraging deep learning in probabilistic forecasting of renewable energy sources. *Energy Conversion and Management, 300*, 117440. doi:10.1016/j.enconman.2023.117440.

Ray, S., Ali, A. M., Eshchanov, T., & Khudoynazarov, E. (2025). Optimal energy management of the smart microgrid considering uncertainty of renewable energy sources and demand response programs. *Operations Research Forum, 6*(2), 1–24. doi:10.1007/s43069-025-00450-z.

Tatar, S. M., Akulker, H., Sildir, H., & Aydin, E. (2022). Optimal design and operation of integrated microgrids under intermittent renewable energy sources coupled with green hydrogen and demand scenarios. *International Journal of Hydrogen Energy, 47*, 27848–27865. doi:10.1016/j.ijhydene.2022.06.130.

Xiong, B., Zhang, L., Yang, Y., Hu, F., & Fang, F. (2025). Deep reinforcement learning for optimal microgrid energy management with renewable energy and electric vehicle integration. *Applied Soft Computing, 176*, 113180. doi:10.1016/j.asoc.2025.113180.

Yang, R., Liu, H., Nikitas, N., Duan, Z., Li, Y. F., & Li, Y. (2022). Short-term wind speed forecasting using deep reinforcement learning with improved multiple error correction approach. *Energy, 239*, 122128. doi:10.1016/j.energy.2021.122128.

Microgrid and power network state estimation with the open-source tool GridCal (aPAC)

13

Mostafa Eidiani[a], Kumars Rouzbehi[b], Saeed Tavakkoli[c], Atiye Ghasemzadeh[c] and Mohsen Masoumi[c]
[a]Consultant for Equipment Research & QC Office, Khorasan Regional Electric Company, Mashhad, Razavi Khorasan, Iran, [b]Department of Systems Engineering and Automatic Control, University of Seville, Sevilla, Spain, [c]Azarakhsh Energy Maham Shargh Engineering Company, Mashhad, Razavi Khorasan, Iran

13.1 Introduction

The power industry, as a fundamental pillar of economic and social development, has always faced complex technical and operational challenges. Accurate power system analysis and optimization play a crucial role in ensuring grid stability and reliability (Eidiani et al., 2002, 2008, 2011). With the continuous expansion and growing complexity of power networks, the demand for advanced simulation and analysis tools has become more critical than ever.

The history of power system analysis software dates back to the 1960s and 1970s, when the first computational tools were developed for load flow and stability studies (Eidiani & Heidari, 2023). Over time, these tools have evolved from basic calculation programs such as load flow (Auer et al., 2019) into sophisticated systems capable of performing a wide range of analyses—from state estimation (SE) (Eidiani, 2023c; Eidiani et al., 2022c; Eidiani & Zeynal, 2022) to dynamic simulations (Eidiani, 2021a; Eidiani & Zeynal, 2014). Commercial software such as DIgSILENT PowerFactory (Eidiani, 2023a; Eidiani et al., 2023; PowerFactory Technical Documentation 2021) and PSS®E (Siemens Power Technologies International., 2020) have since become industry standards.

However, over the past two decades, the rise of open-source software (OSS) has introduced a paradigm shift in this field. Tools such as MATPOWER (Zimmerman & Murillo-Sánchez, 2017), OpenDSS (Dugan, 2012), PyPSA (Brown et al., 2018), PowerSAS (Liu et al., 2023), and VeraGrid (Peñate Vera, 2024) have gained attention from researchers and industry professionals due to their code transparency, customization flexibility, and cost efficiency. These solutions demonstrate that open-source alternatives can effectively meet power system analysis needs.

In many countries, restricted access to advanced commercial software has intensified the need for a comprehensive, systematic evaluation of open-source capabilities.

Stochastic Planning and Modeling for Energy Systems: Methods, Applications, and Developments.
DOI: https://doi.org/10.1016/B978-0-443-45298-7.00019-7

This chapter systematically reviews over 50 reputable studies and research theses, providing a comparative analysis of commercial and open-source power system software (Hay & Ferguson, 2021; Johnson, 2022; Khanh et al., 2025; Makaa, 2024; Nawaz et al., 2023; Shidling, 2021; Su et al., 2024; van Ouwerkerk, 2022; van Ouwerkerk et al., 2022). The goal is to offer practical insights for optimizing tool selection and guiding indigenous software development.

The significance of this study becomes evident when considering the power industry's two main challenges: The demand for higher accuracy and efficiency in grid analysis, contrasted with budgetary and technological constraints. By examining the strengths and limitations of both commercial and open-source solutions, this chapter serves as a practical guide for researchers and power industry professionals.

Developing native software based on open-source frameworks can reduce technological dependency while enabling tailored solutions for region-specific grid challenges. Through detailed comparative analysis, this chapter establishes a scientific foundation for informed decision-making. The findings serve as a roadmap for researchers, policymakers, and utility managers in selecting and developing power system analysis tools.

The study proceeds with a comprehensive evaluation of SE capabilities. It begins with an analysis of professional software, followed by an assessment of open-source solutions and a comparative analysis of both approaches. Section 13.3 covers the Fundamentals of State Estimation, Section 13.4 details the State Estimation Methodology with VeraGrid, and Section 13.5 presents the Case Studies (using DIgSILENT and Azarakhsh power analysis calculator [aPAC]). Finally, the conclusions and references are provided.

13.2 Comparison of open-source and professional software in power system analysis

13.2.1 Fundamental concepts of open-source software

OSS is defined by its publicly accessible source code, which users can freely study, modify, and redistribute. This inherent transparency enables two critical advantages: (Eidiani et al., 2011) Complete visibility into computational methodologies, and (Eidiani, 2008) unlimited customization potential for specific research or operational needs. These characteristics stand in direct contrast to proprietary solutions such as DIgSILENT, which maintain closed-source architectures and require commercial licensing for legal use with unauthorized versions constitute copyright infringement.

Within power system engineering, OSS has gained particular significance in academic and research contexts, where its adaptability supports innovative algorithm development and reproducible studies. The collaborative nature of open-source projects further accelerates technological advancements while removing financial barriers inherent in commercial software ecosystems.

13.2.2 Key advantages of open-source software

OSS demonstrates clear advantages over proprietary solutions for power system analysis, as this chapter will substantiate. Unlike closed-source proprietary systems, open-source tools provide complete transparency of algorithms, enabling verification of results, customization for specific needs, and faster identification of security vulnerabilities through community review. The collaborative development model allows continuous improvements and adoption of new standards like IEC 61850, while eliminating licensing costs make these solutions particularly valuable for budget-constrained projects. These characteristics make OSS ideal for applications requiring reproducibility, customization, and cost efficiency.

While proprietary software like DIgSILENT may offer more polished interfaces and dedicated technical support, its closed nature and high costs present substantial limitations. Restricted access to source code and expensive licensing models make proprietary solutions less suitable for research and small-scale applications. Therefore, while both approaches have value depending on project requirements, open-source tools are generally superior in academic and research contexts where transparency, adaptability, and affordability are prioritized.

13.2.3 Comparison of professional power system software

First, a general overview and comparison of the software tools is provided. Then, the focus is specifically placed on their SE modules.

13.2.3.1 Comprehensive comparison

PowerFactory is a powerful tool suitable for analyzing systems ranging from large transmission grids to local distribution networks and renewable energy sources such as wind and solar farms(Eidiani & Ghavami, 2022; Eidiani et al., 2022a; Ghardashi et al., 2022). It is especially good for complex simulations like harmonics and power grid stability, and it lets you automate tasks with scripting. PSS®E (Siemens) is widely used for transmission planning, with strong features for load flow, short circuits, and stability studies. It works well with other Siemens tools, making it a go-to for large power companies. ETAP is user-friendly and packed with features for designing and analyzing electrical systems, including safety studies like arc flash analysis. It is widely adopted in industrial and commercial projects. CYME focuses on distribution networks, helping utilities model things like power flow and protection systems, and it works well with mapping software (GIS). Neplan is a flexible tool that can handle planning, optimization, and reliability checks, and you can customize it with scripting. Each tool has its strengths: PowerFactory and PSS®E for transmission, ETAP and CYME for distribution/industry, and Neplan if you need a more adaptable solution.

13.2.3.2 Comparison of the state estimation module of professional power system software

This section evaluates SE capabilities across five industry-standard tools.

13.2.3.2.1 Algorithms and technical capabilities

All platforms employ the weighted least squares (WLS) method. PowerFactory distinguishes itself through dynamic state estimation (DSE) and phasor measurement unit (PMU) integration. PSS®E excels in robust outlier detection, while ETAP and CYME specialize in real-time distribution network analysis. Neplan provides unique features for integrating renewable energy and energy storage (Zeynal & Eidiani, 2014).

13.2.3.2.2 Accuracy and performance

PowerFactory and PSS®E demonstrate superior accuracy, with PowerFactory leading in PMU support and anomaly detection. ETAP and CYME show optimal performance for distribution networks, while Neplan performs best in hybrid system applications.

13.2.3.2.3 Integration and application

For transmission systems and advanced research, PowerFactory and PSS®E remain preferred. ETAP and CYME better serve distribution network needs, while Neplan specializes in distributed generation (Parhamfar et al., 2024) scenarios. Selection should align with specific project requirements and system characteristics.

13.2.3.2.4 Implementation requirements and limitations

Each software presents distinct implementation considerations. PowerFactory requires significant computational resources for full functionality, while PSS®E demands extensive training for optimal use. ETAP and CYME offer more accessible deployment but with reduced transmission-level capabilities. Neplan's renewable energy modules require additional configuration for complex scenarios. Common limitations include reliance on accurate network models and potential convergence issues in large-scale systems. Hardware requirements vary substantially, with PowerFactory and PSS®E needing high-performance workstations for advanced simulations, whereas ETAP and CYME operate efficiently on standard hardware. These factors critically influence software selection based on available infrastructure and technical expertise.

13.2.3.3 Final comparison summary

While all five software packages are capable of SE performance, DIgSILENT PowerFactory holds distinct advantages. Its DSE with PMU integration surpasses competitors in accuracy for large-scale networks. The software's robust bad data detection and multilevel validation algorithms outperform standard WLS implementations in other tools. PowerFactory uniquely combines transmission-level precision with distribution network adaptability, offering broader applicability than specialized alternatives. These technical strengths, coupled with real-time monitoring capabilities, position it among the most comprehensive solutions for modern grid challenges.

The following section presents an equivalent comparative analysis of open-source alternatives.

13.2.4 Comparison of open-source software tools

As in the previous section, first, a general overview and comparison of the software tools is provided. Then, the focus is specifically placed on their SE modules.

13.2.4.1 Comprehensive comparison

The evaluation of open-source power system analysis tools reveals distinct strengths and limitations across platforms. OpenSE serves as a lightweight Python option for academic research despite its documentation limitations, while VeraGrid offers broader industrial applicability with dynamic estimation capabilities and an intuitive interface. PowerWorld provides high accuracy comparable to industry standards through its C++/Python implementation, though its open-source version has functional constraints.

MATPOWER remains the educational standard with robust documentation, but carries MATLAB dependency challenges. Pandapower (Thurner et al., 2018) excels in distribution system analytics with Python-based machine learning (ML) integration, whereas InterPSS delivers smart grid versatility through Java at the cost of an outdated interface and programming requirements.

The selection criteria for open-source tools can be prioritized as follows:

- Academic research: OpenSE or MATPOWER
- Industrial applications: VeraGrid or PowerWorld
- Smart analytics: Pandapower or InterPSS

with considerations for project requirements, user expertise, and available computational resources.

13.2.4.2 Comparison of state estimation modules

SE serves as a critical function in power system management and analysis, enabling optimal network state determination using measurement data such as voltage, active/reactive power, and current values. This section provides a comparative evaluation of leading OSS solutions specifically for SE applications.

13.2.4.2.1 OpenSE

This Python-based tool employs WLS for SE in both transmission and distribution networks, with PMU support. While lightweight and highly customizable for research, its limited documentation and small user base present adoption challenges.

13.2.4.2.2 VeraGrid

This Python-based power system analysis tool features WLS and LAV estimation algorithms with an intuitive GUI. While excelling at large-scale transmission/distribution network analysis and dynamic estimation, its computational demands may challenge resource-limited users.

13.2.4.2.3 PowerWorld simulator open-source edition

This hybrid C++/Python tool employs WLS and fast algorithms for load shedding for dynamic and static analysis. While offering industry-standard accuracy and interface quality, its open-source version has reduced functionality compared to the commercial edition.

13.2.4.2.4 MATPOWER

This MATLAB-based tool (Python-compatible via Octave) offers WLS and PMU-based SE, renowned for its thorough documentation and academic adoption. While a power system analysis benchmark, its dependency on MATLAB remains a limitation, although Octave offers a partial alternative.

13.2.4.2.5 Pandapower

This Python library excels in distribution network analysis, integrating SCADA/PMU data for precise SE. Its seamless Pandas integration enables advanced data processing and ML applications, though transmission network capabilities are more limited.

13.2.4.2.6 InterPSS

This Java-based framework offers WLS and heuristic algorithms for smart grid applications, featuring platform independence through the JVM. While flexible for researchers, its outdated interface and reliance on Java may pose challenges for nontechnical users.

13.2.4.2.7 PYPOWER

This Python-based MATPOWER port enables basic power analysis (load flow, short circuit) with NumPy/SciPy integration. Ideal for education and small-scale research, it offers license-free MATPOWER compatibility but lacks advanced features like dynamic estimation.

13.2.4.2.8 PSAT

This MATLAB-based toolbox performs comprehensive power system analyses from load flow to dynamic simulations, featuring a GUI and custom model support (Milano, 2005). While valuable for academic research, its MATLAB dependency and limited scalability for large networks constrain industrial applications.

13.2.5 Final comparison between professional and open-source software

Our comprehensive analysis reveals a nuanced reality in power system tools—optimal solutions depend on specific operational contexts. DIgSILENT PowerFactory maintains its position as the industry benchmark for transmission networks, delivering the precision required by large utilities, while specialized tools like ETAP/CYME serve distribution needs, and Neplan supports renewable integration.

The open-source ecosystem demonstrates remarkable diversity:

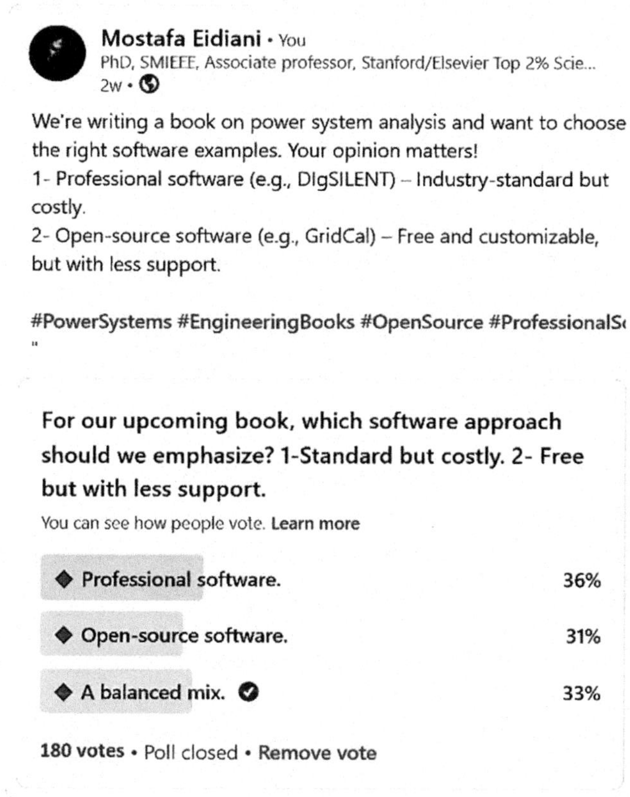

Figure 13.1 LinkedIn survey results. This chart presents the final results of a LinkedIn survey comparing industrial software and open-source software.

- Academic-focused solutions (OpenSE, MATPOWER) enabling research.
- Hybrid tools (VeraGrid, Pandapower) blending traditional analysis with data science.
- Future-oriented platforms (InterPSS) for smart grid development.

Three key trends emerge from both our analysis and authoritative references (Fig. 13.1) and the LinkedIn industry survey (Fig. 13.2):

- Functional convergence: Open-source tools are rapidly matching proprietary capabilities.
- Economic shifts: Budget pressures favor accessible solutions without compromising quality.
- Innovation pathways: Collaborative development accelerates feature advancement.

While professional software remains essential for mission-critical operations today, the industry trajectory clearly points toward open-source solutions gaining prominence. This transition reflects both technological evolution (bridging functionality gaps) and strategic imperatives (achieving cost efficiency and customization). The coming decade is expected to witness open-source tools move from complementary options to primary

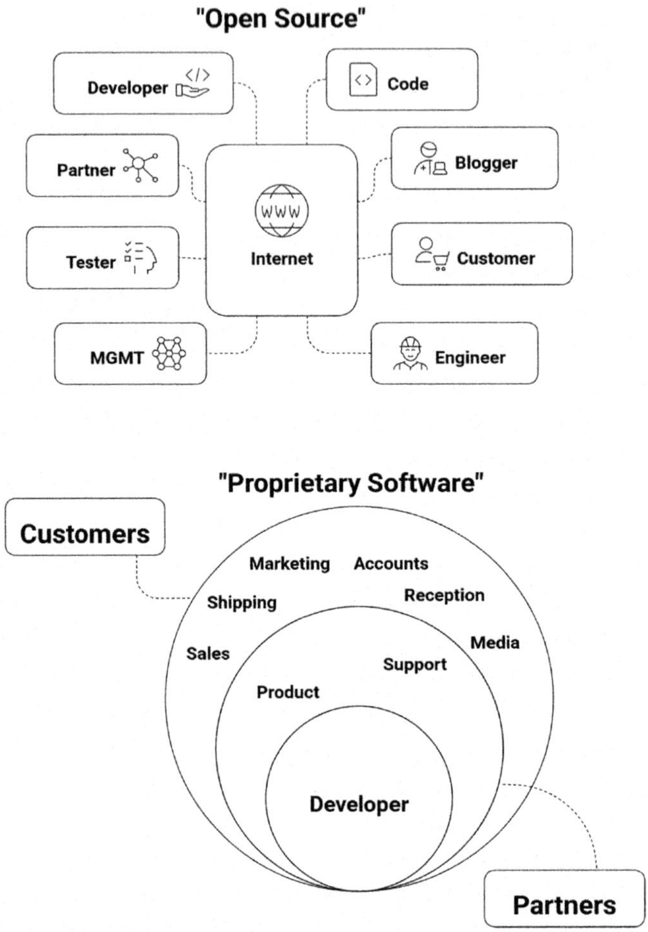

Figure 13.2 A comparative analysis of open source versus proprietary software. This figure provides a straightforward comparison of open-source and proprietary software.

solutions, particularly for research institutions and forward-looking utilities embracing digital transformation.

13.3 Fundamentals of state estimation

Power system SE determines the most probable bus voltage magnitudes and angles using incomplete or redundant measurements, such as voltage magnitudes, active power (P), and reactive power (Q) flows. Despite measurement errors, missing data, or redundancies, the estimator computes optimal state variables to enable critical control center functions such as load dispatch, economic operation, and reliability assessment. This process reconstructs the system's true state from imperfect measurements and forms the foundation of real-time grid monitoring.

The following are the popular methods of SE in power systems.

13.3.1 Weighted least squares method

Among various estimation techniques, the WLS method has emerged as an industry standard due to its mathematical simplicity, computational efficiency, and robustness in handling measurement uncertainties. The WLS method operates by minimizing the weighted sum of squared differences between the actual measurements and their corresponding calculated values, where the weights are inversely proportional to the measurement variances. This approach effectively accounts for varying measurement accuracies, assigning higher importance to more reliable data points. The solution is typically obtained through an iterative process using the Newton–Raphson method, which linearizes the nonlinear power system equations around an operating point. While WLS offers numerous advantages, including straightforward implementation and good statistical properties, it can also be sensitive to bad data. This necessitates the use of additional detection and identification schemes. These characteristics make WLS particularly suitable for real-time applications in energy management systems, where it forms the foundation for various advanced grid analytics and control functions.

The mathematical formulation of this method is summarized below.

13.3.1.1 Mathematical model

The state vector typically includes voltage magnitudes (V) and voltage angles (δ) for all buses, except that the slack bus angle is fixed.

$$x = [\delta_2, \delta_3, \ \ldots, \ \delta_N, V_1, V_2, \ \ldots, V_N]^T \tag{13.1}$$

Here, N is the number of buses, and the slack bus has a fixed angle ($\delta_1 = 0$) (Eidiani et al., 2011).

Measurements include line power flows, bus power injections, and voltage magnitudes.

$$z = h(x) + e \tag{13.2}$$

(z) is a measurement vector (e.g., P_{ij}, Q_{ij}, P_i, Q_i, V_i), $h(x)$ is a nonlinear function relating states to measurements, and (e) is measurement error (assumed Gaussian with zero mean and covariance matrix R) (Eidiani, 2008).

13.3.1.2 WLS formulation

The goal is to minimize the weighted sum of squared errors:

$$J(x) = \sum_{i=1}^{m} \frac{(z_i - h_i(x))^2}{R_{ii}} = [z - h(x)]^T R^{-1} [z - h(x)] \tag{13.3}$$

(m) is the number of measurements, and (R) is the diagonal covariance matrix of measurement errors (σ_i^2) (Eidiani, 2002).

Since $h(x)$ is nonlinear, the optimization problem is solved iteratively as:

a. Linearize around the current estimate (x_k):

$$\Delta z^k = H(x^k)\Delta x^k + e \tag{13.4}$$

where ($\Delta z^k = z - h(x^k)$) is the residual vector, and ($H(x^k) = \frac{\partial h(x)}{\partial x}\Big|_{x=x^k}$) is the Jacobian matrix (Eidiani & Rouzbehi, 2024).

a. Update the state vector:

$$\Delta x^k = (H^T R^{-1} H)^{-1} H^T R^{-1} \Delta z^k,\ x^{k+1} = x^k + \Delta x^k \tag{13.5}$$

b. Terminate when the changes in the state vector or the residuals fall below a predefined tolerance $\Delta x^k < \varepsilon$, ($e.g.$, $\varepsilon = 10^{-4}$)

13.3.1.3 Advantages and limitations

The WLS method offers several advantages, including simplicity of implementation, numerical stability, and the ability to assign weights to measurements according to their accuracy (σ_i). However, it has notable limitations: The method is sensitive to bad data, necessitating additional detection techniques such as the Largest Normalized Residual Test, and it requires computationally intensive recalculations of the Jacobian matrix at each iteration.

13.3.2 Robust estimation and extended Kalman filter

13.3.2.1 Robust estimation (M-Estimators)

Robust estimation methods enhance traditional WLS approaches for power system SE by addressing their vulnerability to measurement errors. While WLS works effectively with accurate data, its reliance on squared errors makes it highly sensitive to outliers from instrument failures or communication problems. These robust techniques modify the error-weighting approach to reduce the impact of abnormal measurements while preserving valid data. By maintaining the core WLS framework but with more balanced error handling, they provide system operators with reliable state estimates even when some input data is flawed. This approach combines the WLS's computational efficiency with improved resilience, making it practical for real-world grid monitoring where perfect data quality cannot be guaranteed.

13.3.2.1.1 Mathematical formulation

In this method, the quadratic objective function of WLS is replaced by a robust cost function (e.g., Huber, Cauchy).

$$J(x) = \sum_{i=1}^{m} \rho\left(\frac{z_i - h_i(x)}{\sigma_i}\right)^2 \tag{13.6}$$

where ρ is a symmetric function that penalizes outliers less severely than WLS (e.g., Huber function) (Eidiani & Rouzbehi, 2025):

$$\rho(r) = \begin{cases} \frac{r^2}{2}, & if|r| \le c \\ c|r| - \frac{c^2}{t}, & \text{otherwise} \end{cases} \tag{13.7}$$

where (c) is the tuning parameter and (r) is the normalized error ($r = \frac{(z_i - h_i(x))}{\sigma_i}$) (Nawaz et al., 2023).

13.3.2.1.2 Advantages and limitations

The Iteratively Reweighted Least Squares (IRLS) method is commonly employed to solve nonlinear optimization problems, offering advantages such as resilience against bad data (tolerating up to 50% contamination) and computational simplicity compared to extended Kalman filter (EKF), though it has limitations including the need to tune robustness parameters (e.g., threshold c) and potentially slower convergence than WLS for clean datasets.

13.3.2.2 Extended Kalman filter

The EKF is designed to address nonlinear dynamics and time-varying states in power systems, enabling accurate tracking of transient stability and load variations. By linearizing the system model around the current operating point at each time step, the EKF effectively estimates system states in real-time despite nonlinearities. This capability makes EKF particularly valuable for applications requiring DSE, such as monitoring generator angles during faults or following rapid load changes.

13.3.2.2.1 Mathematical formulation

In this section, the formulas are summarized in four parts: State-Space Model, Prediction Step, Update Step, and Iteration.

$$x_{k+1} = f(x_k) + \omega_k \tag{13.8}$$

(Process model) (Eidiani, 2021b)

$$z_k = h(x_k) + v_k \tag{13.9}$$

(Measurement model) (Eidiani, 2022)where (ω_k,v_k) are process and measurement noise, respectively. The EKF linearizes the nonlinear system equations around the current state estimate using a first-order Taylor series expansion. This creates a time-varying linear approximation of the system, allowing the Kalman filter equations to be applied.

Based on the system model and the previous state estimate, the EKF predicts the next state and its associated uncertainty.

The EKF uses new sensor measurements to refine the state estimate. It combines the predicted state with the measurement data, taking into account the uncertainties of both.

The prediction and update steps are repeated iteratively as new measurements become available, continually refining the state estimate.

13.3.2.2.2 Advantages and limitations

The EKF tracks dynamic states like generator angles during faults, which allows handling process noise, but requires Jacobian updates (computationally expensive) and suffers from linearization errors. Compared to WLS (static estimation) and Robust Estimation (noise handling), EKF uniquely handles time-varying systems but demands more resources. It's ideal for real-time PMU applications where dynamic tracking outweighs computational costs, while WLS suits static analysis, and Robust Estimation manages noisy offline data. Each method serves distinct estimation needs based on system dynamics and data quality.

13.3.3 Advanced algorithms, incorporating phasor measurement units, machine learning hybrids, and uncertainty quantification

In the following, four important issues in SE are examined.

13.3.3.1 Phasor measurement units in state estimation

PMUs provide synchronized, high-speed voltage and current phasor measurements, significantly improving SE accuracy compared to traditional SCADA systems. By offering time-stamped data with GPS synchronization, PMUs enable dynamic state tracking and better detection of grid anomalies. However, their widespread adoption is limited by high costs, communication infrastructure requirements, and the need for careful placement to ensure observability. Despite these challenges, PMUs are crucial for real-time monitoring in wide-area power systems, particularly in applications like fault detection, oscillation monitoring, and renewable energy integration.

13.3.3.2 Machine learning hybrids in state estimation

ML enhances traditional SE by improving bad data detection, topology identification, and parameter calibration. Hybrid approaches combine physics-based models with ML techniques (e.g., neural networks for anomaly detection or reinforcement learning for adaptive filtering) to process noisy or incomplete measurements. While ML improves robustness and computational efficiency, challenges include the need for extensive training data, model interpretability, and integration with legacy systems. Future advancements will focus on explainable AI and real-time learning to ensure reliability in dynamic grid conditions.

13.3.3.3 Uncertainty quantification in state estimation

Uncertainty quantification (UQ) addresses measurement errors, forecast inaccuracies, and model mismatches in SE, providing probabilistic confidence intervals for decision-making. Techniques like Bayesian inference, Monte Carlo methods, and interval analysis help quantify uncertainties from renewable generation fluctuations, PMU noise, or cyber threats. UQ improves grid resilience by enabling risk-aware control strategies, though it increases computational complexity. As power systems grow more stochastic with distributed energy resources, UQ will become essential for robust SE and adaptive grid management.

13.3.3.4 Advanced algorithms in state estimation

Recent advancements in SE algorithms focus on overcoming nonlinearities, computational bottlenecks, and real-time constraints inherent in large-scale power systems. Distributed optimization techniques, such as alternating direction method of multipliers (ADMM), enable decentralized SE for multiarea grids, while robust estimators like Huber M-estimation improve resilience against bad data and cyber attacks. Graph neural networks (GNNs) are emerging as powerful tools for topology-aware SE, leveraging the inherent graph structure of power networks. Additionally, real-time Kalman filter variants and compressed sensing techniques are being adapted for high-speed dynamic estimation in systems with high PMU penetration. These cutting-edge approaches collectively push the boundaries of accuracy, scalability, and computational efficiency, though challenges remain in balancing model complexity with practical implementation requirements for mission-critical grid operations.

13.4 State estimation with VeraGrid: Methodology

SE in VeraGrid is implemented based on the general mathematical formulation of power system SE (outlined in Section 13.3), where the WLS method iteratively minimizes measurement residuals to obtain the best estimate of system states. VeraGrid translates this formulation into a modular software architecture, ensuring that each step—from measurement acquisition to Jacobian construction and iterative updates—is encapsulated within well-defined layers. The implementation details the WLS solution process with Levenberg–Marquardt (LM) damping, from initialization to convergence criteria. Practical considerations such as island detection, observability checks, and results integration are also addressed. The complete software implementation details follow below.

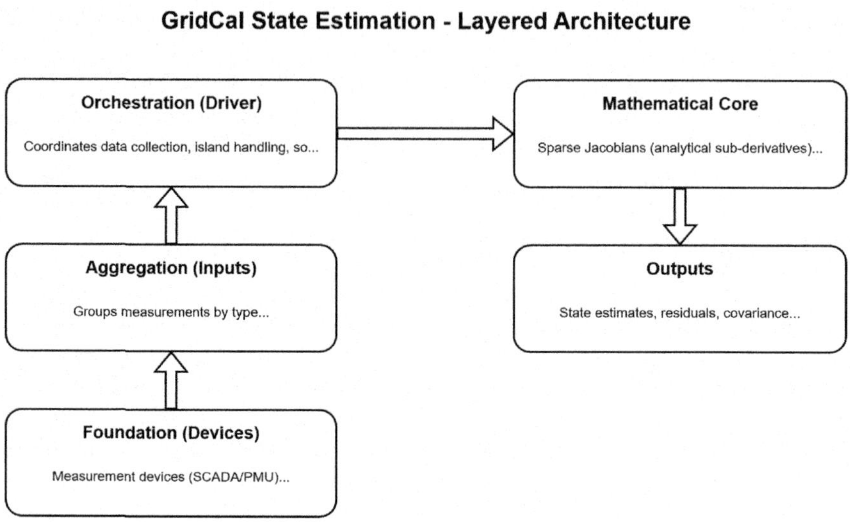

Figure 13.3 VeraGrid layered architecture. This figure illustrates a layered software architecture for power grid state estimation, showing data flow from device-level inputs through aggregation, mathematical computation, and orchestration to outputs.

13.4.1 Architectural overview

VeraGrid implements SE as a modular driver that follows the same simulation scaffold as power flow and other analyses. The module is organized in five cooperating layers according to the following diagram (See Fig. 13.3).

- Foundation (devices): Measurement device classes represent SCADA/PMU readings and carry three core attributes: The measured value, its standard deviation (uncertainty σ), and a link to the network element (bus or branch).
- Aggregation (inputs): A dedicated input container groups measurements by type and exposes uniform access to values and uncertainties.
- Orchestration (driver): The SE driver coordinates data collection, island handling, solution, and results packaging.
- Mathematical core: Sparse Jacobians are assembled from analytical derivatives and solved in a WLS framework with LM damping.
- Output: The final results of the SE—such as estimated voltages, angles, flows, and residuals—are packaged and presented in a consistent format, enabling integration with visualization tools, reporting modules, or downstream applications.

This separation ensures clear data flow, strong type safety, and easy extensibility, allowing new sensors or algorithms to be introduced without disrupting existing components. See Fig. 13.3. This figure illustrates a layered software architecture for power grid SE, showing data flow from device-level inputs through aggregation, mathematical computation, and orchestration to outputs.

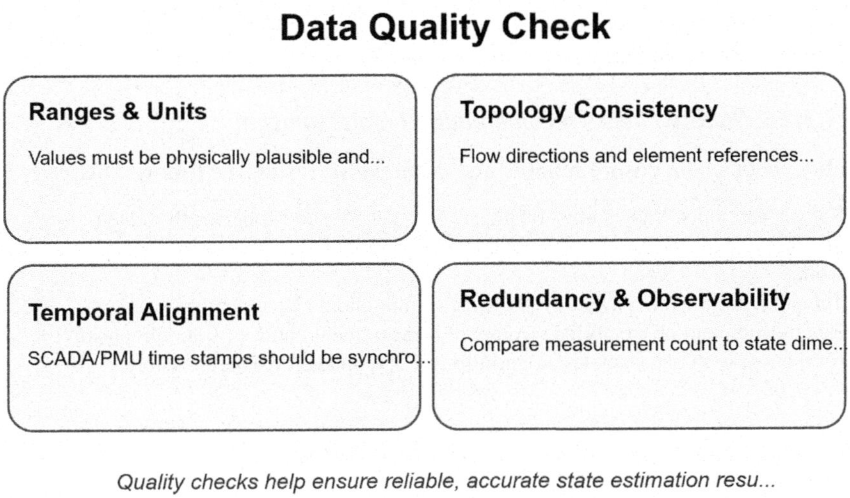

Figure 13.4 Data quality checks improved. This figure outlines the data quality checks performed to ensure reliable and accurate state estimation results.

13.4.2 Measurement model and supported types

SE consumes a stacked vector of measurements along with a matching vector of standard deviations (σ). The supported measurement types and their native locations/units are:

- Bus injections: Active P_i and reactive Q_i (MW, MVAr; at buses).
- Voltage magnitude: V_m (p.u., at buses).
- Branch quantities: Active and reactive "from-end" power flows *Pf, Qf* (MW, MVAr; on branches) and branch current magnitude $|If|$ (A).

The sign conventions are as follows:*Pf, Of* > 0 denote flow from the "from" bus toward the "to" bus; *Pi* > 0 denotes net generation at a bus.

Typical uncertainty guidelines (for selecting σ and consequently weights) are as follows: $Vm \approx 0.1\%–0.5\%$ (PMU $\approx 0.1\%$, standard VT $\approx 0.5\%$); $P, Q \approx 1–3\%$ (digital $\approx 1\%$, analog $\approx 2\%–3\%$); $|I| \approx 1\%–2\%$ (digital CT $\approx 1\%$, analog $\approx 2\%$).

Before solving, all values are converted to per unit, and the diagonal weight matrix is formed as $W = \mathrm{diag}(1/\sigma^2)$. This ensures that higher-quality sensors have greater influence on the estimate.

13.4.3 Data ingestion and preparation

At run time, the driver collects the available sensors, maps them to buses/branches, and stacks them in a fixed documented order [| *Pi* | *Qf* | *Qi* | | *If* | | *Vm*] with a σ vector in the same order. The lists grow dynamically, without zero-padding. If a weak prior is needed at an unmetered point, a pseudo-measurement can be added with a conservative value and a large σ, ensuring that it contributes negligible weight. Fig. 13.4 illustrates the four essential preprocessing validation checks—data quality, topology, temporal alignment, and redundancy—required for reliable power system SE.

13.4.3.1 Quality checks recommended before solution

Quality checks help ensure reliable and accurate SE results, as follows: See Fig. 13.4:

- Ranges and units: Values must be physically plausible and consistently scaled.
- Topology consistency: Flow directions and element references must match the active network.
- Temporal alignment: SCADA/PMU time stamps should be synchronized.
- Redundancy and observability: Compare measurement count to state dimension ($\approx 2n - 1$ for n buses with one slack angle); aim for $>1 \times$ redundancy where feasible.

13.4.4 Measurement equations and Jacobian

The state vector comprises bus angles and voltage magnitudes. For a given state, predicted measurements $h(x)$ are computed for the stacked set above. The Jacobian $H = \partial h/\partial x$ is assembled as a sparse block matrix from analytical component Jacobians for:

- bus complex-power injections,
- branch complex-power flows (from-end; to-end if used),
- branch current magnitudes, and
- voltage magnitudes.

Only the rows corresponding to available measurements are retained, keeping the linear systems compact and well-conditioned.

13.4.5 Solver and convergence control

VeraGrid solves the WLS problem using LM damping:

1. Initialization. Flat start $V = 1\angle 0$ unless a warm start is supplied; measurements and σ are in per unit; W is built from σ.
2. Iterative update. At each iteration, the solver computes residuals $r = z - h(x)$, forms the damped normal equations $(H^{\mathrm{T}}WH + \lambda I)\Delta x = H^{\mathrm{T}}Wr$; updates the state, and adapts λ by decreasing it when progress is good and increasing it when not, to avoid oscillations and improve robustness.
3. Stopping criteria. Maximum iteration cap or sufficiently small state change (typical tolerances 10^{-6}–10^{-4} pu).

   The solver returns a convergence report (status, iteration count, final residual norm, elapsed time).

13.4.6 Islands, observability, and resilience

Disconnected islands are detected and solved independently, improving robustness and enabling parallel execution on large systems. If the normal matrix is singular/ill-conditioned (insufficient information), the driver flags nonconvergence rather than

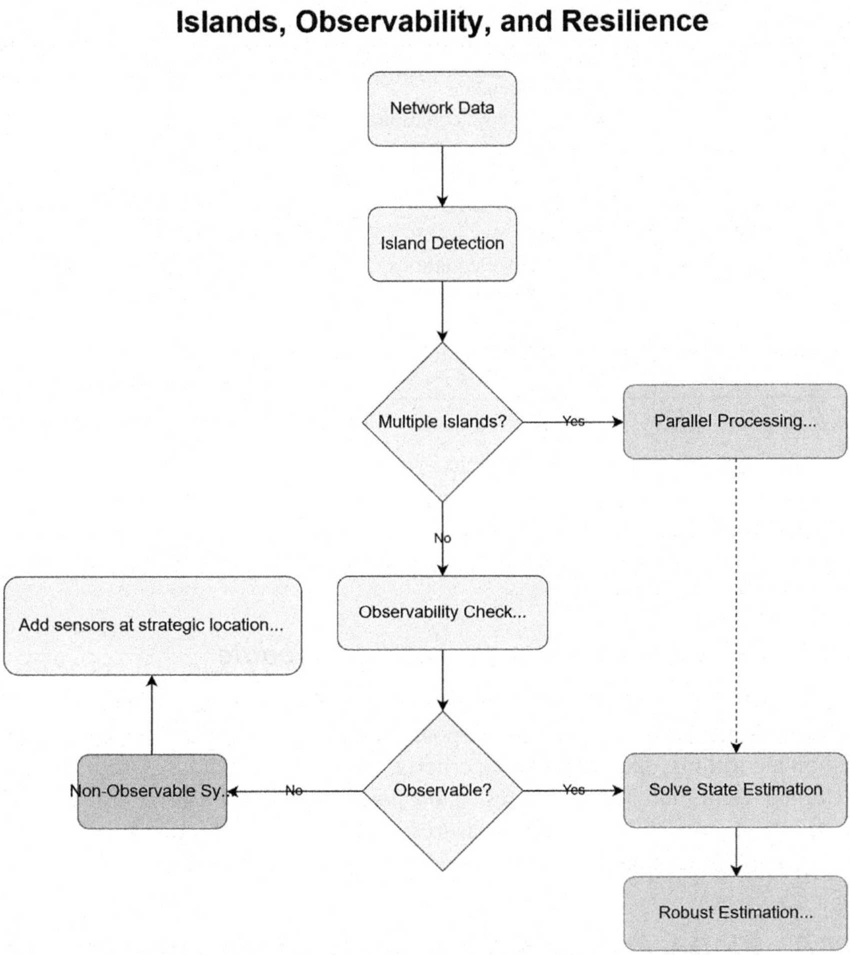

Figure 13.5 Islands observability resilience. This figure depicts a decision flow for ensuring power system observability by detecting islands, performing checks, and applying solutions to enable robust, parallel state estimation.

producing an artificial solution. Adding sensors, introducing weak pseudo-measurements (large σ), or revisiting placements are the recommended remedies. See Fig. 13.5. This figure depicts a decision flow for ensuring power system observability by detecting islands, performing checks, and applying solutions to enable robust, parallel SE.

13.4.7 Results packaging and postprocessing

SE results are returned in the same data structure as VeraGrid's power-flow module, ensuring seamless compatibility with plotting and downstream tools. Reported quantities

Table 13.1 Comparison between PowerFlow and state estimation in power systems.

Aspect	PowerFlow	State estimation
Purpose	Find a steady-state consistent with equations	Infer the state from noisy measurements
Inputs	Loads, generator set-points	SCADA/PMU measurements (+σ)
Unknowns	Bus voltages, some injections	Bus voltage magnitudes and angles
Constraint	Power-balance equations	Weighted residual minimization
Algorithm	Newton–Raphson/fast-decoupled	Levenberg–Marquardt WLS
Convergence target	Power mismatch → 0	Measurement residuals → minimum

PMU, phasor measurement unit; *WLS*, weighted least squares.

include bus voltages/angles and P, Q, branch flows at both ends, current magnitudes, thermal loading, and system losses. Postprocessing uses the estimated state to compute these electrical quantities, so users can compare SE and power-flow outcomes directly.

13.4.8 Relationship to the power-flow module

Table 13.1 compares the key differences between Power Flow and SE in power system analysis, covering aspects such as purpose, inputs, unknowns, governing equations, solution algorithms, and convergence criteria.

Unified results and data structures enable power flow to validate SE through power-balance checks, while SE provides measured initial conditions for power flow studies—both operating on the same network model for consistent analysis.

13.4.9 SCADA–Phasor measurement unit fusion and uncertainty configuration

Heterogeneous data are fused naturally through weights. Assign a smaller σ to PMU phasors than to legacy SCADA to reflect higher accuracy and synchronization; PMU will then dominate overlapping quantities. Enforce time alignment must be enforced during preprocessing to avoid spurious residuals. If cross-correlations are unknown but suspected, use conservative σ to avoid overconfidence.

13.4.10 Bad-data screening (recommended operational loop)

In operational environments, the accuracy of SE can be significantly affected by erroneous or corrupted measurements. A practical approach to addressing this issue involves an iterative bad-data screening process that is applied externally to the WLS solver. The process begins by running the SE and calculating the normalized residuals for all measurements. The largest residual is then examined, and if it exceeds a

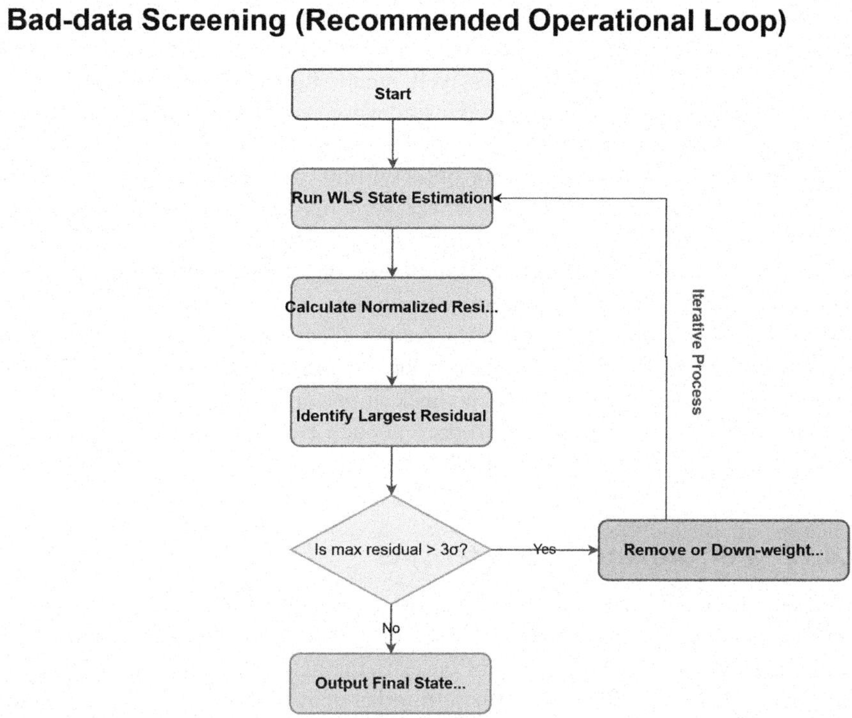

Figure 13.6 Bad data screening loop. This figure illustrates an iterative bad-data screening loop that identifies and removes outliers based on normalized residuals to ensure robust state estimation.

predefined statistical threshold—typically three times the standard deviation (3σ)—the corresponding measurement is considered a potential outlier. Instead of discarding large portions of data, only the suspect measurement is removed or assigned a lower weight, thereby reducing its influence on the solution. The SE is then executed again with the updated measurement set, and the residuals are recalculated. This cycle continues until all normalized residuals fall within acceptable limits. By following this procedure, the WLS method retains its mathematical simplicity while achieving a high degree of resilience against the influence of bad data, ensuring that the final estimated state closely reflects the true operating conditions of the network. See Fig. 13.6. This figure illustrates an iterative bad-data screening loop that identifies and removes outliers based on normalized residuals to ensure robust SE.

13.4.11 Diagnostics and common issues

During the practical application of SE, several recurring issues may arise that require systematic diagnosis. One common problem is nonconvergence, which often stems from insufficient measurement redundancy or suboptimal sensor placement. In such cases, it is advisable to review the network's measurement configuration, verify that all units and sign conventions are correct, and ensure that breaker statuses are accurately represented. In some situations, excessively small standard deviation values (σ) can over-weight certain measurements, making convergence more difficult; relaxing these values can improve solver stability.

Another frequent challenge is poor observability, which occurs when parts of the network lack sufficient measurement coverage. This situation can be addressed by confirming that each electrical island contains an adequate number of sensors, with priority given to voltage magnitude measurements and strategically placed power flow or injection measurements that maximize coverage.

Numerical issues can also impair accuracy. These may occur if measurement variances (σ) are set to zero or negative values, if magnitudes are improperly scaled, or if units are inconsistent. A thorough check of per-unit conversion bases, kV levels, and data scaling factors is essential to prevent such errors and to maintain numerical stability throughout the estimation process.

13.4.12 Performance characteristics

The computational performance of the SE process in VeraGrid is influenced by several structural and operational factors. The assembly of the Jacobian matrix grows approximately in proportion to the number of buses in the network, while the computational cost of matrix factorization scales with the number of available measurements. Memory usage increases linearly with both system size and the total number of sensors, making the method efficient in terms of storage requirements for networks of varying scales.

In terms of runtime, small systems with fewer than one hundred buses typically solve in less than a second, while medium-sized networks containing between one hundred and one thousand buses complete in a matter of seconds. For large-scale systems exceeding one thousand buses, runtimes may extend to tens of seconds, depending on hardware capabilities and the complexity of the measurement set.

Scalability is influenced by both measurement density and network topology. Higher measurement density generally leads to improved estimation accuracy but increases computation time. Conversely, networks with well-connected topologies and realistically assigned measurement variances (σ) tend to converge more quickly, striking a balance between precision and computational efficiency.

13.4.13 From VeraGrid to aPAC: Extending open-source state estimation for operational readiness

The methodological framework outlined in Sections 13.4.1–13.4.12 demonstrates that VeraGrid provides a robust analytical foundation for SE, with a clear modular architecture, comprehensive measurement handling, and mathematically rigorous solvers. These capabilities make it a valuable platform for research, education, and proof-of-concept studies. However, when the objective shifts from academic experimentation to continuous operation within a control center environment, a different set of priorities emerges. Real-time integration with SCADA/EMS infrastructure, compliance with industry communication protocols, optimization for uninterrupted execution, and alignment with operator-oriented workflows become critical factors.

In practice, deploying VeraGrid in such operational contexts reveals several gaps that are not inherent flaws but rather natural consequences of its research-oriented design philosophy. Native support for real-time telemetry exchange via protocols such as ICCP/TASE.2 or IEC 60870-5-104 is absent, requiring significant customization for field deployment. While its weighted least-squares SE performs well on static datasets, maintaining stability and accuracy under conditions of high measurement noise, rapid topology changes, or incomplete data demands additional mechanisms for adaptive solver tuning and bad-data resilience. Furthermore, while VeraGrid's analytical capabilities are robust and suitable for engineering-level analysis, the SE module operates exclusively through its API and does not natively offer visual reporting features, such as operational dashboards, multilingual displays, or utility-specific performance indicators. These features are essential for real-time control room operations and are typically required for effective network monitoring and decision-making.

To bridge this gap between analytical strength and operational applicability, our team at Azarakhsh Energy Moham Shargh has developed the aPAC platform. aPAC inherits VeraGrid's transparent and extensible computational core, preserving the advantages of open-source flexibility, while systematically augmenting it with the capabilities necessary for 24/7 real-world deployment. The system incorporates native SCADA/EMS data interfaces fully compliant with utility communication standards, automated and operator-tunable bad-data detection loops, integrated real-time contingency screening, and a high-availability execution environment designed to ensure continuity under stringent operational constraints. In addition, aPAC delivers reporting and visualization modules tailored to the language, regulatory environment, and decision-making practices of regional grid operators.

In this way, aPAC represents not a departure from VeraGrid but rather its evolution into a production-grade solution. By retaining the proven analytical methodology described in the preceding subsections and embedding it within an operationally mature software environment, aPAC provides utilities with a SE tool that combines the transparency and adaptability of open-source development with the robustness, integration capability, and situational awareness essential for modern power system operation.

The next section applies these concepts in practice by introducing aPAC's SE capabilities through a detailed case study on the IEEE 14-Bus test network. This demonstration not only illustrates the complete execution of the estimation process within aPAC but also highlights how its extended features, such as native SCADA/EMS data handling, enhanced bad-data detection, and customized reporting, contribute to both accuracy and operational readiness. By comparing the results with established benchmarks, Section 13.5 underscores aPAC's ability to merge the analytical rigor of VeraGrid with the reliability, integration, and situational awareness demanded in modern power system control environments.

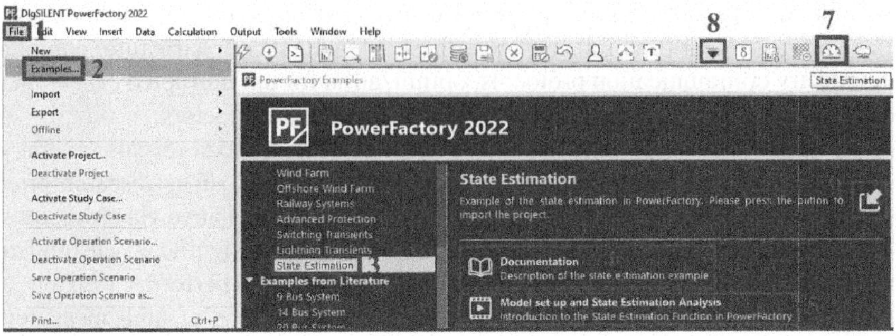

Figure 13.7 DIgSILENT PowerFactory state estimation. This figure illustrates how to execute the demo example within the state estimation section.

13.5 Case studies (DIgSILENT and aPAC)

13.5.1 DIgSILENT PowerFactory

This section demonstrates the capabilities of the SE module in professional software, specifically DIgSILENT PowerFactory. This demonstration uses the 2022 version of the software, which does not differ significantly from earlier releases. For users with access to the software, Fig. 13.7 illustrates how to execute the demo example within the SE section.

From the File (item 1) and Examples (item 2) menus, users can access a sample article (item 4) and a training video (item 5) relevant to this section. Clicking on the example (item 6) loads and executes the SE network file. To run the SE, click on icon 7 in the Toolbox section (item 8).

After executing the example (item 6 in Fig. 13.7), the interface opens the view shown in Fig. 13.8, which displays a 20-bus distribution network with complete SE information. Three scenarios are predefined for this network, as shown in section (item 1) of Fig. 13.8. To activate a scenario (Study Case), right-click on it and select the "Active" option. Section (Eidiani, 2008) of Fig. 13.8 shows the network graph, while section (Eidiani, 2002) provides the textual output generated by the software.

For this example, three scenarios are considered:

A. Undisturbed (Safe) measurement: In this scenario, all meters are in service, and each reports the same value as the load-flow solution. The estimated bus and line values closely match the measured values. All buses are observable, and all measurements are valid.

B. Disturbed measurement: In this scenario, assume all meters are in the circuit, but some measurements are affected by noise. The output information for the estimated buses and lines remains close to the measured values. However, the software identifies that some measurements are incorrect.

C. Unobservable network: In this scenario, some meters are not functioning, and some operational meters are sending incorrect values. The software detects all invalid measurements. Although a load flow cannot be executed due to insufficient information, the SE module can still perform the estimation and distinguish correct from incorrect measurements.

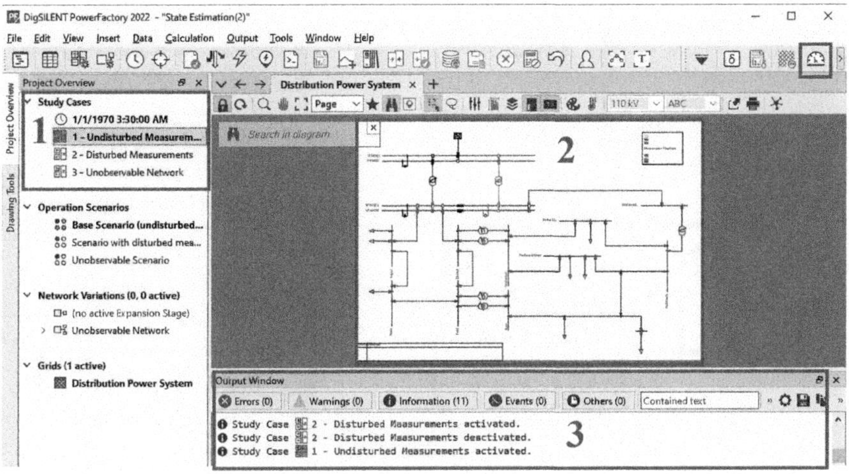

Figure 13.8 Demo example of DIgSILENT PowerFactory state estimation. This figure demonstrates how to run a state estimation demo example in DIgSILENT PowerFactory.

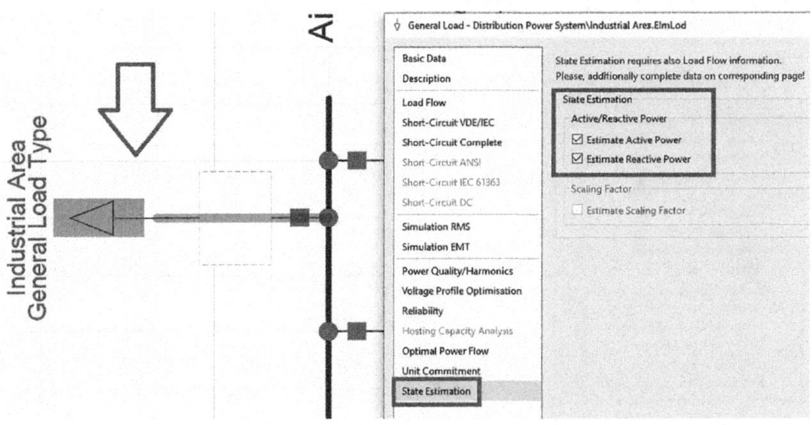

Figure 13.9 Load state estimation output. In this figure, users can enable or disable estimation of active and reactive load power by double-clicking on a load and accessing the state estimation subwindow.

Note that the load-flow output may include active and reactive load power, as well as transformer tap positions. For example, as shown in Fig. 13.9, users can enable or disable estimation of active and reactive load power by double-clicking on a load and accessing the SE subwindow. Furthermore, by right-clicking on any entry (item 1 in Fig. 13.10) within the New Devices section, a new measuring device (item 3) can be added. Alternatively, the Show Ext. the measurements section (item 4, Fig. 13.10) provides a list of all installed measurements. The three scenarios are examined in sequence.

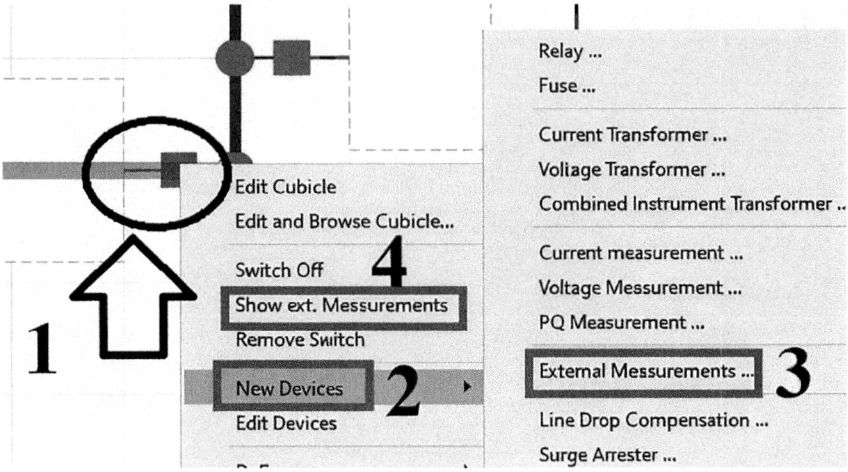

Figure 13.10 New measurements and old measurements. This figure shows how to add new measurements and old measurements.

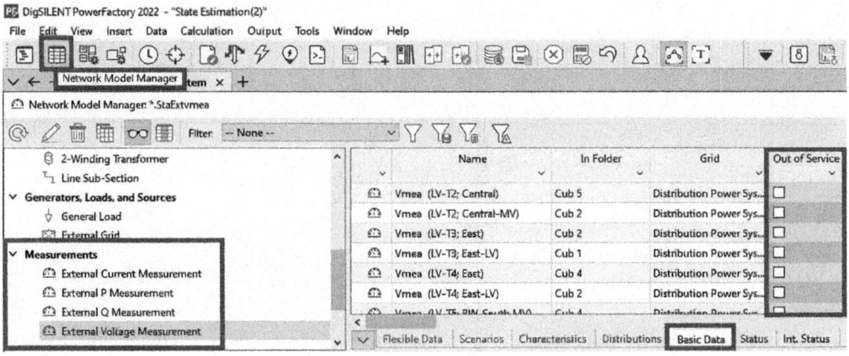

Figure 13.11 All measurements are in service (active). This figure shows that all measurements are available and compatible with the load-balancing solution.

13.5.1.1 Undisturbed measurements

The three scenarios differ primarily in terms of data availability and quality. In the first scenario (Fig. 13.11), all measurements are available and consistent with the load-flow solution. After running a load flow (Fig. 13.12, panel A), executing SE (Fig. 13.12, panel B) produces load estimates nearly identical to the original values. To test estimator performance, the active power load at Feeder B was modified from 2 MW to 0.5 MW (Fig. 13.12, panel C). Despite this change, the state estimator correctly

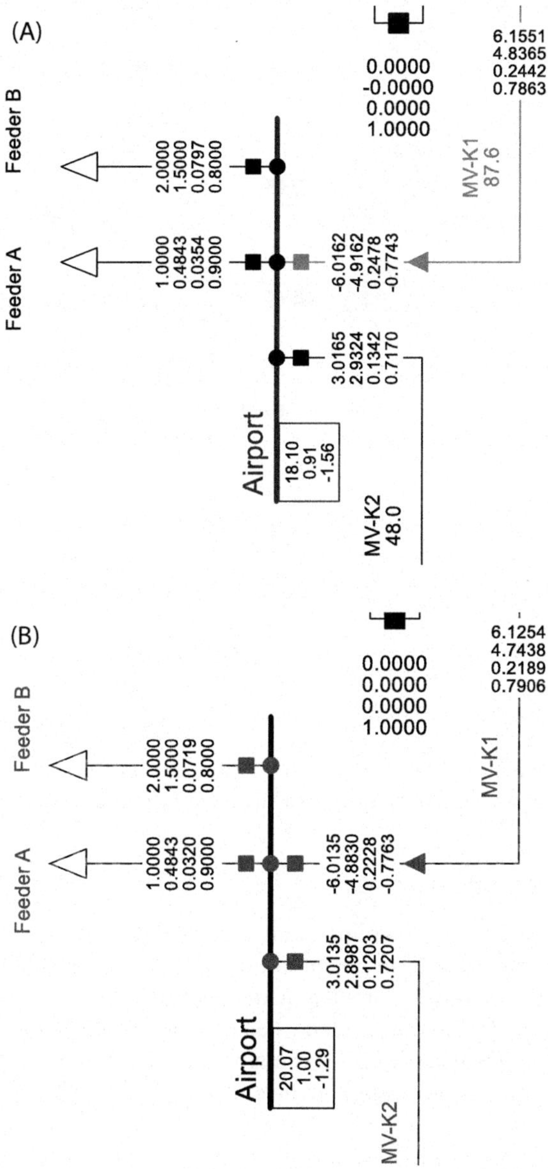

Figure 13.12 (A) Load flow amount in base case, (B) state estimation in base case and load change 12-C, (C) load flow amount with load change of feeder B. These figures show that after running a load flow, executing state estimation produces load estimates that are nearly identical to the original values.

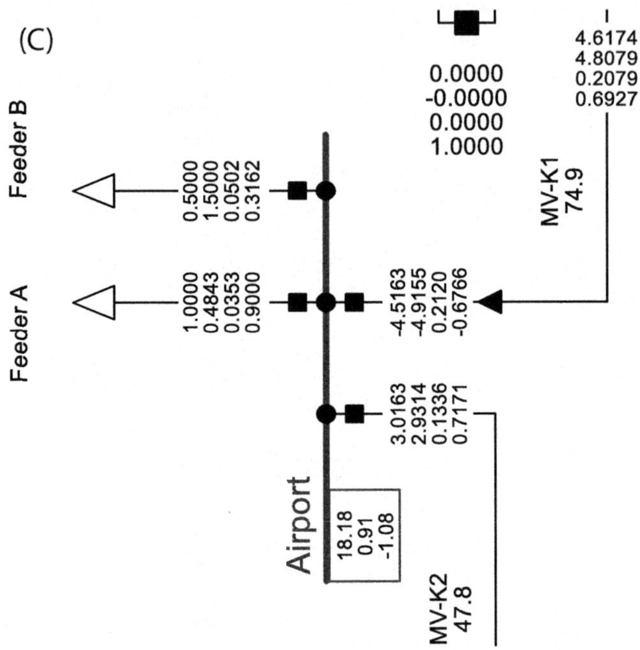

Figure 13.12, cont'd.

recovered the original 2 MW value, demonstrating its robustness in rejecting errors and maintaining accuracy when sufficient measurement redundancy exists.

13.5.1.2 Disturbed measurements

In this scenario, all voltage measurements are unavailable. The remaining active power, reactive power, and current measurements are still available but are affected by elevated noise. Although the load-flow output remains unchanged, the SE result (Fig. 13.13, panel A) shows slight deviations, as seen when compared to the base case in Fig. 13.12, panel B. This illustrates that while SE remains functional under noisy measurement conditions, its accuracy is moderately reduced.

13.5.1.3 Unobservable network

In this scenario, all voltage and reactive power measurements, along with more than half of the active power and current measurements, are unavailable. The remaining measurements are affected by significant noise. As a result, although the load-flow output remains constant, the SE result (Fig. 13.12, panel B) shows significant deviation, which is evident when compared to Fig. 13.12, panel B, and Fig. 13.13, panel A. This highlights the critical role of complete and accurate measurements in ensuring reliable SE, particularly in unobservable networks.

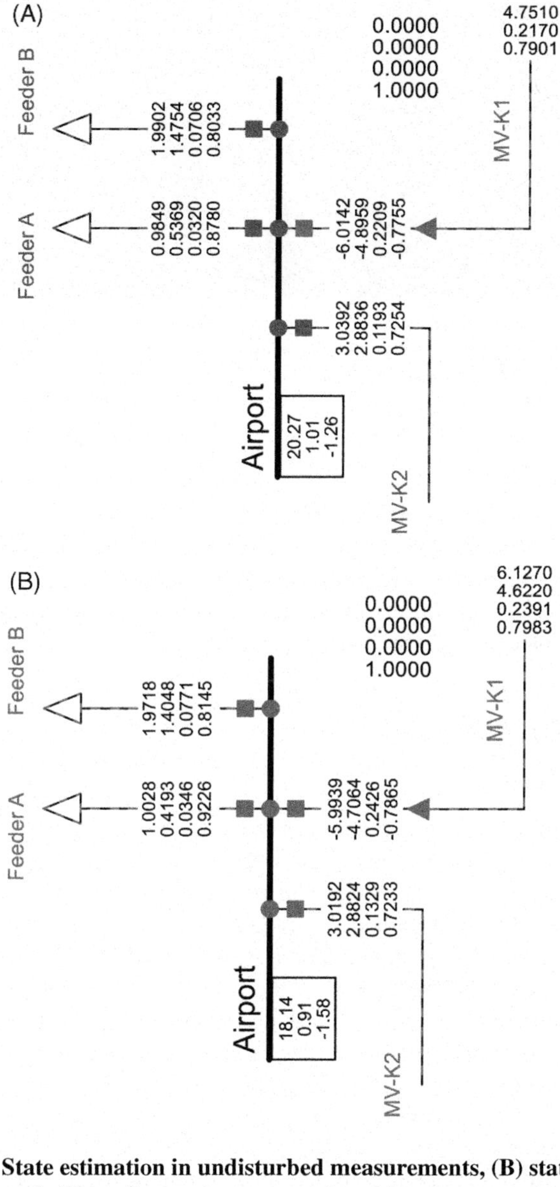

Figure 13.13 (A) State estimation in undisturbed measurements, (B) state estimation in unobservable network. These figures demonstrate that although state estimation remains functional with noisy measurement data, its accuracy is moderately compromised.

13.5.1.4 Conclusion

This section evaluated the SE module of DIgSILENT PowerFactory 2022 using a 20-bus demo network. Three scenarios were tested: Undisturbed, disturbed (with

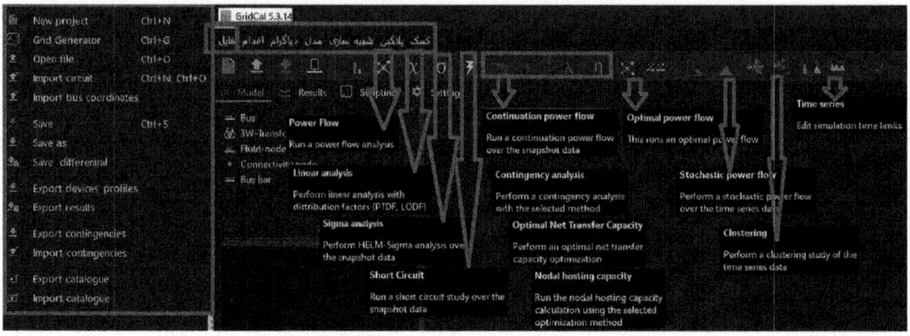

Figure 13.14 Illustrates the graphical structure of the software's main screen under windows. This figure illustrates the graphical layout of the software's main screen in the Windows environment.

noise), and unobservable (with missing and erroneous data) measurements. The results demonstrated the module's robustness, showing high accuracy with good data and its ability to detect errors and provide estimates even under severely compromised conditions, where traditional load flow would fail. The software's color-coded output effectively visualized node observability and measurement validity for the user.

The next section examines the performance of OSS in SE analysis.

13.5.2 GridCal (VeraGrid)

13.5.2.1 Introduction

VeraGrid's origins trace back to 2016 with the initial release of an open-source project developed by Santiago Peñate. This Python-based software package was groundbreaking due to its unique combination of powerful computational algorithms and advanced visualization capabilities. Designed from the outset to provide comprehensive power system analysis at users' fingertips, it laid a strong foundation for future development.

The project entered a new phase with the formation of eRoots in 2022, marking a shift toward focused development. eRoots was established to tackle critical challenges in modern power systems, particularly those arising in renewable-rich grids, such as short-circuit analysis and dynamic stability. This strategic focus led to rapid capability expansion, and by 2025, VeraGrid had evolved into a comprehensive platform offering enhanced steady-state analysis and sophisticated dynamic and stability modules.

Today, VeraGrid stands as a complete professional-grade solution built on open-source principles. The platform offers extensive simulation capabilities, including power flow, optimal power flow, and dynamic RMS simulations, with seamless interoperability with industry-standard formats like PSS/E and CIM/CGMES. With unlimited visualization tools for creating detailed maps and schematics, VeraGrid fulfills its original vision of providing engineers with powerful, intuitive tools for designing and securing modern power systems.

Fig. 13.14. The graphical structure of the software's main screen under Windows.

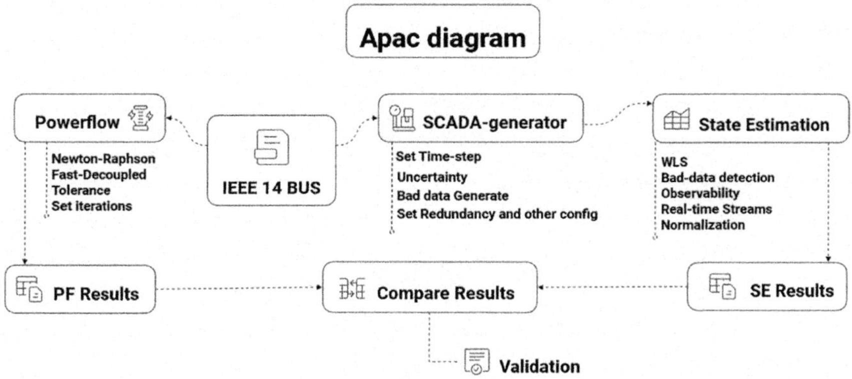

Figure 13.15 aPAC diagram. This figure provides a publication-ready overview of aPAC's core capabilities and their integration into the end-to-end workflow, emphasizing the conceptual process rather than low-level implementation details. *aPAC*, Azarakhsh power analysis calculator.

From Fig. 13.14, the following points can be observed:

- A key advantage of the software is that its interface icons can be easily customized to support different languages, such as Persian and English.
- The main page provides direct access to essential power system analysis functions, such as Power-flow or Load-flow, Optimal power flow (Eidiani & Mahnani, 2023; Eidiani et al., 2022; Zeynal et al., 2014a; 2014b), Short circuit, Continuation power flow (Eidiani & Yazdanpanah, 2011), Contingency analysis, Transfer capability (ATC or FCTTC) (Eidiani, 2023b; Eidiani et al., 2002; Eidiani et al., 2008; Eidiani et al., 2010), and Stochastic power flow.
- The SE analysis feature is not available on the main page.

The following section details how to access and use the SE module.

13.5.2.2 Case study (IEEE14-Bus in aPAC) and result

This section provides a concise, publication-ready overview of aPAC's core capabilities and their integration into the end-to-end workflow, emphasizing on the conceptual process rather than low-level implementation details (Fig. 13.15). The case study demonstrates the preparation and validation of the IEEE 14-bus benchmark, serving as a foundation for applying SE algorithms to a real test network within aPAC.

13.5.2.3 Loading the IEEE 14-Bus benchmark in aPAC

The packaged case *"IEEE14.zip.gridcal"* is opened in aPAC as the starting point. The IEEE 14-bus test system specifications are assumed to comply with established global standards, and the analysis continues using a preserved working copy. Upon ingestion, aPAC (1) validates references and unit consistency, (2) converts all quantities to per-unit, (3) constructs the network graph and detects islands, (4) assigns Slack, PV, and PQ

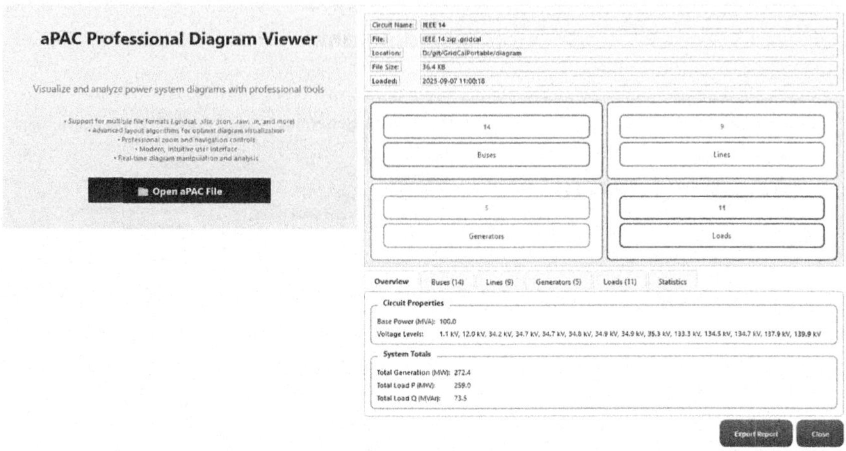

Figure 13.16 Load package file and get information. This figure illustrates the load package file and get information.

bus roles with operational bounds, and (5) assembles the numerical structures (Ybus, Yf, Yt) along with an initial state vector.

Verification criteria include confirming the presence of 14 buses and approximately 20 branches/transformers, ensuring exactly one slack bus, proper PV and PQ generator designation, the absence of unintended islands, and consistent base MVA and kV values. If anomalies are detected, missing references or unit bases must be corrected, or file paths unblocked in the case of an empty diagram. Finally, a baseline power-flow is executed to establish the reference operating point, which serves as the foundation for subsequent SCADA data generation and SE studies. Figs. 13.16 and 13.17 illustrate the IEEE-14 bus test system loaded in aPAC and its corresponding single-line diagram.

13.5.2.4 Generating SCADA-realtime measurements in aPAC

Once a validated power-flow operating point is available, aPAC can generate synthetic SCADA-style measurements for voltages, injections, and branch quantities. These synthetic measurements drive SE and visualization, while configurable options allow users to control their scope, credibility, and diagnostic value.

The generation process is governed by several core options. The timestep can be set either as a single snapshot—suitable for focused studies, or as a time series to evaluate trends and robustness over longer intervals. Uncertainty (σ/weights) reflects the relative trust assigned to measurement channels; higher-quality instruments are assigned lower variance and therefore exert greater influence on the estimation process. Noise is then applied in accordance with the declared σ, transforming ideal physical quantities into realistic readings that resemble field measurements.

To safeguard realism and robustness, aPAC applies multiple quality controls. Validation checks that ranges, units, and references are internally consistent, enabling quick

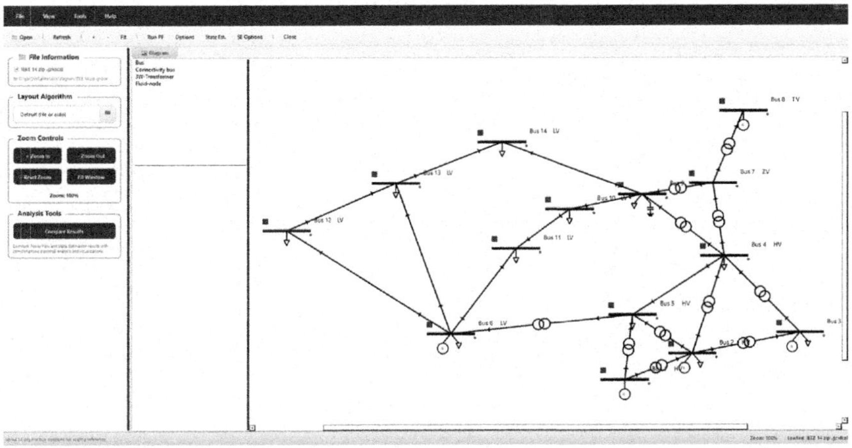

Figure 13.17 Drawing network diagram from package file. This figure illustrates the IEEE-14 bus test system loaded in aPAC and its corresponding single-line diagram. *aPAC*, Azarakhsh power analysis calculator.

detection of data issues. Bad-data simulation deliberately introduces gross errors or dropouts, allowing resilience testing and chi-square screening to identify implausible snapshots for further diagnostics. Redundancy analysis quantifies measurement coverage relative to system unknowns, highlighting where additional sensors would provide the greatest improvement in observability. The generated dataset is exported as an organized and well-labeled Excel file containing IDs, timestamps, values, uncertainties, and status flags. This structured output provides a clean, auditable bridge to the state estimator, reducing or eliminating the need for additional preprocessing.

In summary, the timestep option defines the temporal scope of the experiment; uncertainty and noise encode instrument trust; validation safeguards data quality; bad-data tools test estimator resilience; redundancy analysis guides sensor placement; and structured export ensures a reproducible workflow from scenario definition to SE. Figs. 13.18–13.20 illustrate the SCADA-realtime measurement generation workflow in aPAC, covering file selection and noise configuration, quality validation framework, and the resulting measurement statistics.

13.5.2.5 Baseline power-flow in aPAC

The purpose of the baseline power-flow analysis is to establish a deterministic operating point from the assembled network. This reference point serves as the anchor for subsequent SE and visualization, ensuring consistency and accuracy in subsequent analyses.

The inputs for the power-flow analysis consist solely of the internal network model, which includes bus type and voltage targets, branches and transformers impedances and ratings, generators active power setpoints and reactive power limits, and load demand

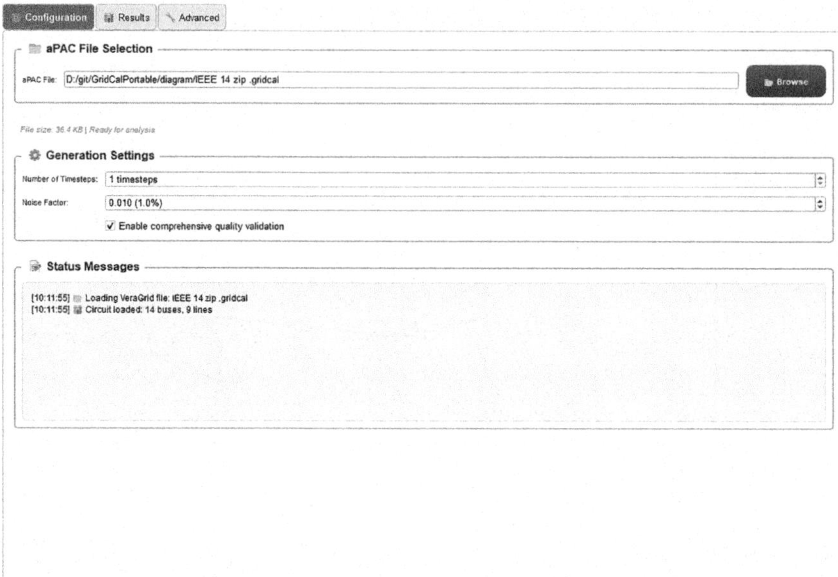

Figure 13.18 Load package file for measurements generate. This figure illustrates the workflow for generating SCADA-real-time measurements in aPAC, with this section highlighting the "Load Package File For Measurement Generation" step. *aPAC*, Azarakhsh power analysis calculator.

specifications. aPAC normalizes all inputs to per-unit values and provides a sensible initial guess to initiate the calculation. aPAC offers several solvers to compute the power-flow, each suitable for different study needs. The Newton–Raphson method is a general-purpose solver that quickly converges to a solution when the system is close to the operating point. The Fast-Decoupled method is more efficient for large transmission systems, while linear DC and simplified AC solvers are useful for screening studies and warm starts. For particularly challenging points, trust-region and holomorphic variants are available to improve convergence. The choice of solver depends on the specific study goals and the numerical behavior of the system.

Settings for the power-flow analysis include tolerance, which specifies the desired accuracy of the solution mismatch, and maximum iterations, which define the computational effort cap. Additionally, optional settings for enforcing Q-limit constraints and modest tap adjustments are available, ensuring that operational bounds are respected while aiding the convergence of the solution.

The outputs from the power-flow analysis include key parameters for both buses and branches. For buses, the output consists of voltage magnitudes and angles, net active and reactive power injections, postlimit type, and voltage-violation flags. For branches and transformers, the outputs include from/to active and reactive power flows, currents, apparent power, losses, and loading relative to thermal limits. These outputs help identify areas of stress within the system and confirm the balance of power.

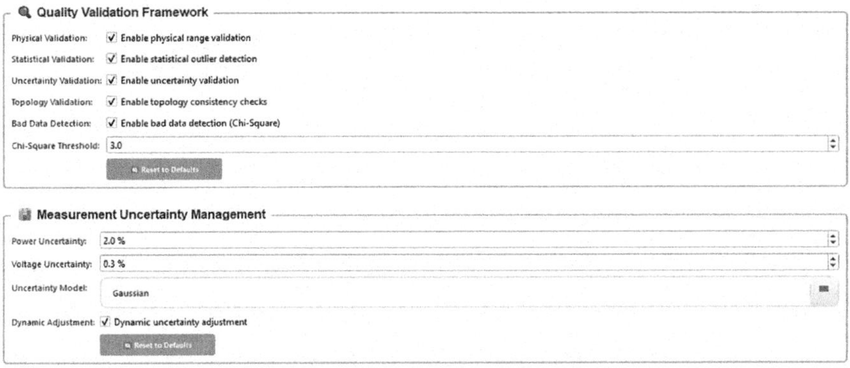

Figure 13.19 Advanced features config. This figure illustrates the workflow for generating SCADA real-time measurements in aPAC, with this section highlighting the "Advanced Features Configuration" step. *aPAC*, Azarakhsh power analysis calculator.

Finally, the results can be exported as ordered tables containing stable IDs, units, and per-unit bases. These exports are valuable for reporting and for direct comparison with SE results. While differences between power-flow and SE outputs are expected due to measurement noise, significant deviations may indicate underlying modeling or measurement issues, requiring further investigation. The baseline power-flow in aPAC is demonstrated in Figs. 13.21–13.23, which present solver parameter configuration, the single-line diagram of the network, and a tabular summary of the analysis results.

13.5.2.6 State estimation with weighted least-squares in aPAC

SE employs a WLS estimator to reconcile noisy and sometimes inconsistent measurements with the physical network model, producing a best-effort snapshot of the system state for operations, analysis, and visualization. In aPAC, this process is streamlined, emphasizing practical use rather than algorithmic depth. The estimator relies on two key inputs: A validated network model, including bus roles, device limits, and branch impedances and ratings, and a measurement set, typically generated via SCADA workflows or imported from supervisory systems in tabular form, including values, uncertainties, statuses, and timestamps.

The WLS approach weights measurements according to their specified confidence, ensuring that more accurate channels exert a stronger influence on the solution. It also applies normalization to align disparate units and scales, preventing distortion and ensuring that weights reflect confidence rather than raw magnitude. Built-in bad-data detection evaluates the overall fit for plausibility, flagging suspicious channels through residual-based tests when inconsistencies arise and guiding users on when to trust the estimate and where further investigation is required.

Redundancy analysis assesses coverage and robustness, quantifying the level of independent information available and identifying weak spots where additional measurements could improve observability. This distinction prevents mistaking solvability

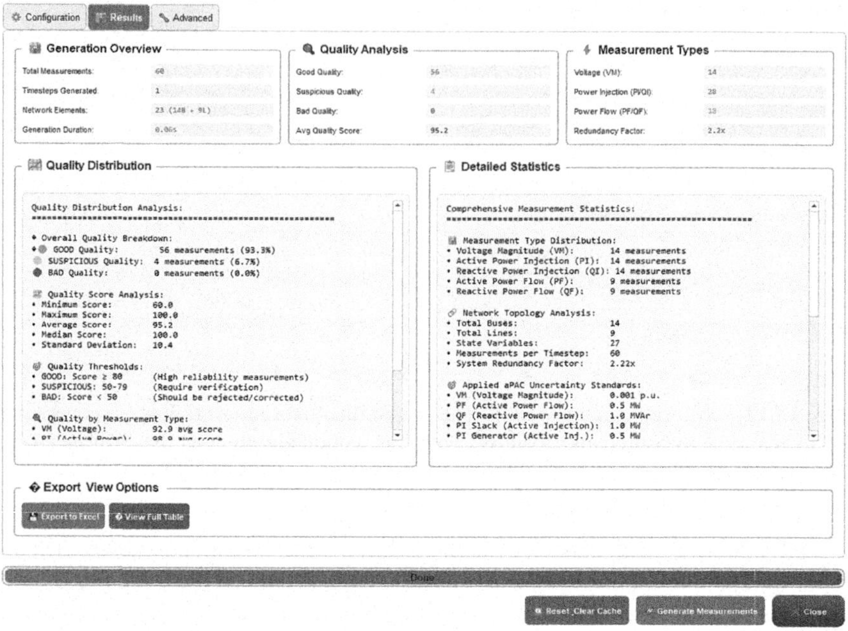

Figure 13.20 Show results summary and export measurement file. This figure illustrates the workflow for generating SCADA real-time measurements in aPAC, with this section highlighting the "View Results Summary and Export Measurement File" step. *aPAC*, Azarakhsh power analysis calculator.

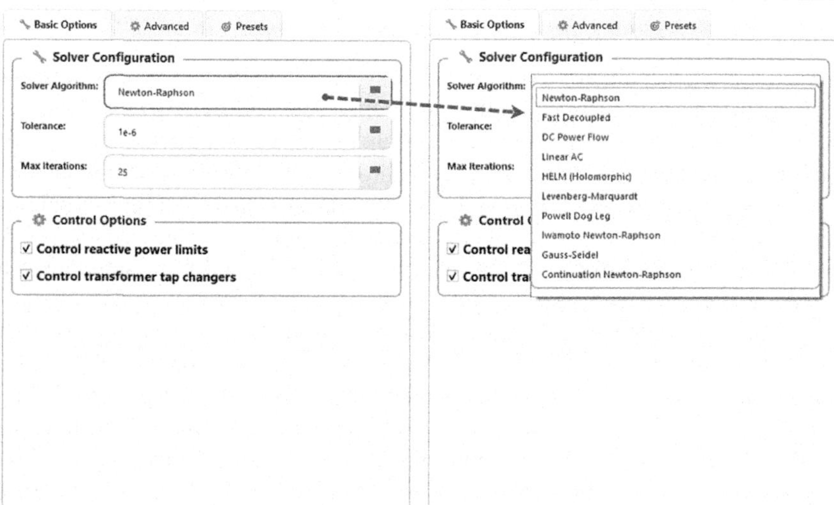

Figure 13.21 Advanced configuration for PowerFlow solver. This figure demonstrates the baseline power-flow in aPAC, presenting the advanced configuration options for the power-flow solver. *aPAC*, Azarakhsh power analysis calculator.

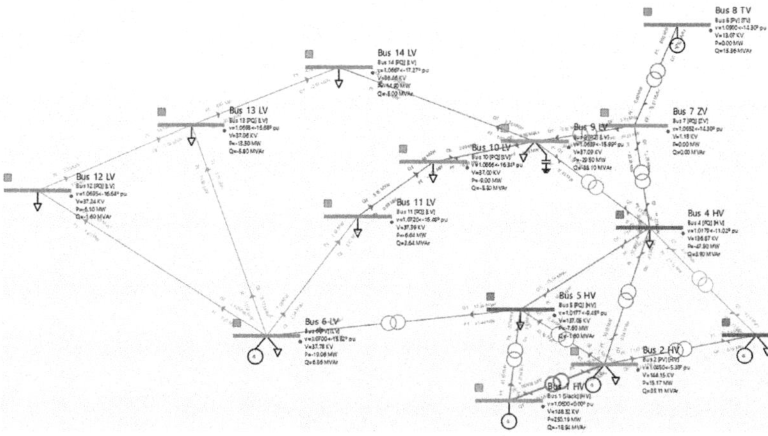

Figure 13.22 Show PowerFlow results on network diagram. This figure demonstrates the baseline power-flow in aPAC and shows the power-flow results displayed on a network diagram. *aPAC*, Azarakhsh power analysis calculator.

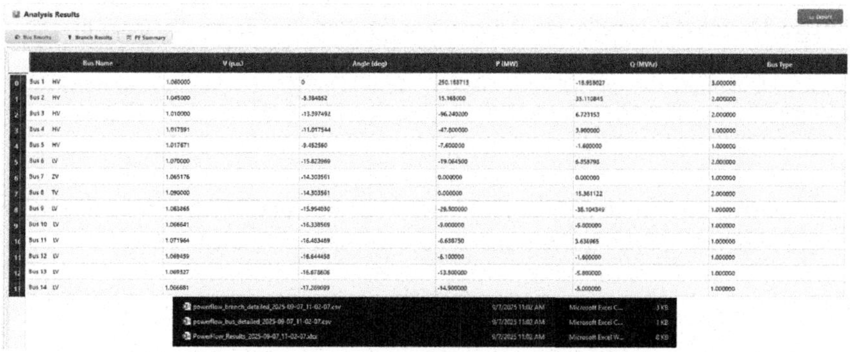
Analysis Results

	Bus Name	V (p.u.)	Angle (deg)	P (MW)	Q (MVAr)	Bus Type
0	Bus 1 HV	1.060000	0	250.168715	-18.958027	3.000000
1	Bus 2 HV	1.045000	-5.384852	15.165000	35.110845	2.000000
2	Bus 3 HV	1.010000	-13.397492	-96.240200	6.723153	2.000000
3	Bus 4 HV	1.017591	-11.017544	-47.800000	3.900000	1.000000
4	Bus 5 HV	1.017671	-9.452560	-7.600000	-1.600000	1.000000
5	Bus 6 LV	1.070000	-15.822969	-19.064500	6.858798	2.000000
6	Bus 7 ZV	1.065176	-14.303561	0.000000	0.000000	1.000000
7	Bus 8 TV	1.090000	-14.303561	0.000000	15.361122	2.000000
8	Bus 9 LV	1.063265	-15.994030	-29.500000	-38.104349	1.000000
9	Bus 10 LV	1.066641	-16.338569	-9.000000	-5.800000	1.000000
10	Bus 11 LV	1.071964	-16.483469	-6.638750	3.636965	1.000000
11	Bus 12 LV	1.069459	-16.644458	-6.100000	-1.600000	1.000000
12	Bus 13 LV	1.069327	-16.678606	-13.500000	-5.800000	1.000000
13	Bus 14 LV	1.066681	-17.269099	-14.900000	-5.000000	1.000000

Figure 13.23 Export PowerFlow results. This figure demonstrates the baseline power-flow in aPAC and presents the "Export PowerFlow Results" functionality. *aPAC*, Azarakhsh power analysis calculator.

for reliability. aPAC supports both static file imports for on-demand estimation and real-time streams, aligning measurements within small windows through consistent validation, normalization, and quality checks, thereby enabling seamless lab-to-live transitions.

Results include voltage magnitudes and angles at buses, residuals and flags for inconsistent channels, and flows and currents at branches and transformers, with any overloads noted. These outcomes can be exported in tables consistent with power-flow formats, enabling side-by-side comparisons that highlight meaningful differences. Unlike deterministic power-flow, the state estimate reflects real-world uncertainties, offering a credible and traceable representation for learning, interpretation, and future

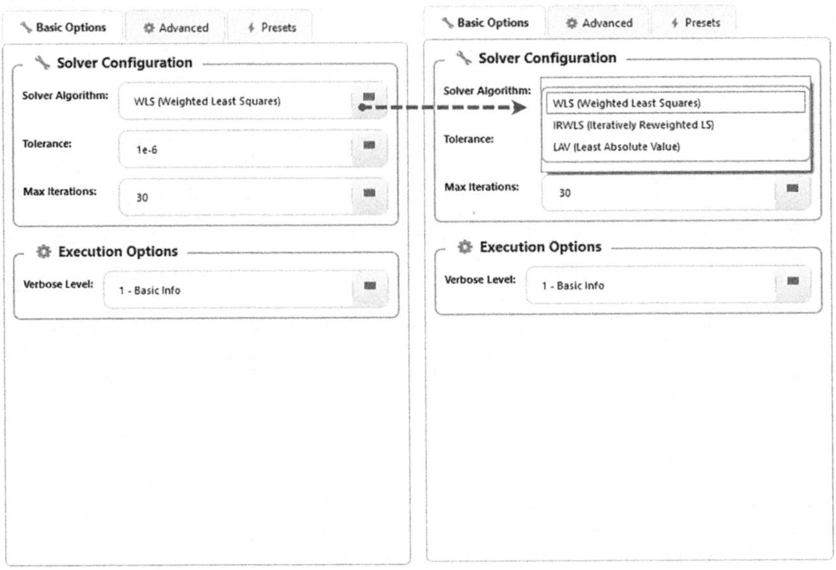

Figure 13.24 Advanced configuration for state estimation. This figure illustrates the state estimation process in aPAC using the weighted least squares (WLS) method, highlighting the "Advanced Configuration for State Estimation" section. *aPAC*, Azarakhsh power analysis calculator.

measurement improvements. Figs 13.24 and 13.25 illustrate SE in aPAC using the WLS approach, highlighting solver setup and the resulting estimation summary.

13.5.2.7 Comparing power-flow and state estimation in aPAC

This section compares the deterministic power-flow solution with the measurement-driven state estimate, highlighting practical differences. Although both approaches rely on the same network snapshot and per-unit basis, their philosophies diverge. Power-flow enforces the physical model strictly, producing a consistent solution without accounting for uncertainty. By contrast, SE incorporates noisy, weighted measurements, reconciling them with the network model to deliver the most probable operating point. As a result, discrepancies between the two highlight the influence of measurement noise, redundancy, and bad data handling.

Fig. 13.26 illustrates this comparison, showing how the state estimate deviates slightly from the deterministic solution while remaining statistically consistent with the available evidence.

Inputs Used for the Comparison

Power-flow relies solely on the internal network model, including bus types, device settings, and branch impedances. SE, by contrast, uses the same model in conjunction with a measurement set—whether SCADA-generated or file-based—that includes

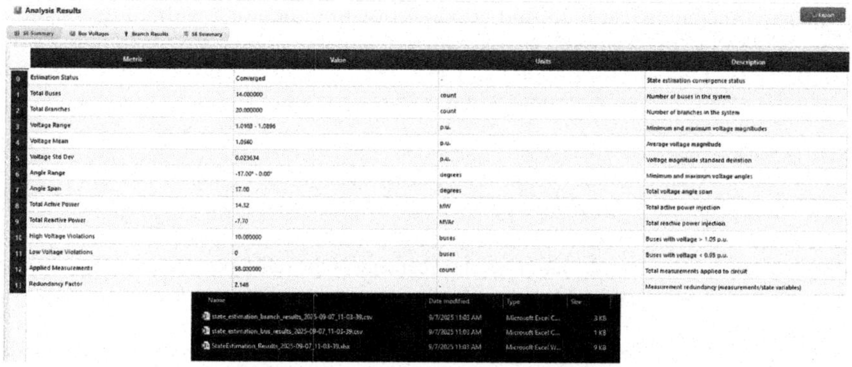

Figure 13.25 Export state estimation results. This figure illustrates the state estimation process in aPAC using the weighted least squares (WLS) method and highlights the "Export State Estimation Results" functionality.

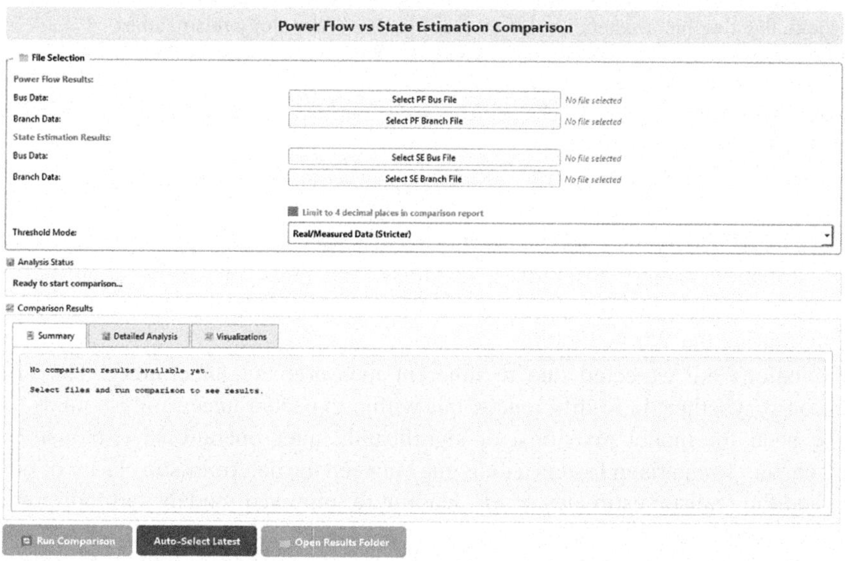

Figure 13.26 Import PowerFlow and stateestimation results for comparison. This figure illustrates how the state estimate slightly deviates from the deterministic solution while maintaining statistical consistency with the available evidence.

identifiers, values, uncertainties, statuses, and timestamps. To ensure direct comparability, aPAC normalizes and aligns all units across both approaches.

Outputs Brought to a Common Footing

Both analyses produce voltage magnitudes and angles at buses, as well as flows, currents, apparent power, losses, and loading indicators for branches and transformers.

SE further contributes residuals and goodness-of-fit indicators, with all results aligned side by side to facilitate direct comparison.

Methods for Comparing Results and Their Meaning

- Pointwise difference: Deltas (e.g., ΔVm, ΔVa) quantify deviations from the power-flow solution. Small, scattered differences indicate measurement noise, while systematic shifts suggest potential model biases.
- Relative error: Expresses deviations as a percentage of the reference values, aiding interpretation across different scales. Absolute differences are preferable near zero to avoid misleading percentages.
- Confidence-weighted comparison: Normalizes deviations by measurement uncertainty (σ), highlighting unexpected differences in accordance with WLS principles.
- Aggregate statistics: Mean absolute error, RMSE, histograms, and zonal averages help distinguish between random discrepancies and systematic skew.
- Constraint-level comparison: Focuses on operational impacts such as voltage violations or overloads, signaling conditions that may require corrective action.
- Topology-aware checks: Ensures comparisons are performed within the same electrical island, flagging inconsistencies due to data filtering or topology mismatches.
- Consistency tests: Verifies that net power injections align with system losses, revealing potential measurement bias or over-fitting.
- Time-series comparisons: Tracks deviations over multiple timesteps, distinguishing persistent biases from isolated outliers.

Exports for Documentation and Audit aPAC exports a table with power-flow and state-estimation values, differences, residuals, and flags, preserving identifiers and units for reproducibility.

Interpreting the Whole Picture

Deviations are expected due to inherent measurement uncertainty. The critical question is whether these differences fall within expected uncertainty bounds, point to the need for model revisions, or significantly alter operational conclusions. In practice, this comparison fosters a dialogue between the deterministic clarity of power-flow and the realistic estimates of SE, leading to improved models and more reliable measurements.

Fig. 13.27 compares bus voltage magnitudes obtained from power-flow and SE, showing close alignment across all buses with only minor deviations.

Fig. 13.28 illustrates the comparison of active (Pf) and reactive (Qf) branch power flows between power-flow and SE, demonstrating strong consistency between the two methods.

Fig. 13.29 presents the comparison of bus active power injections (Pi) and reactive power injections (Qi), confirming high correlation and minimal discrepancies between power-flow and SE.

Fig. 13.30 summarizes the statistical evaluation of power-flow and SE, reporting root mean square errors and correlation coefficients that indicate overall high accuracy and agreement.

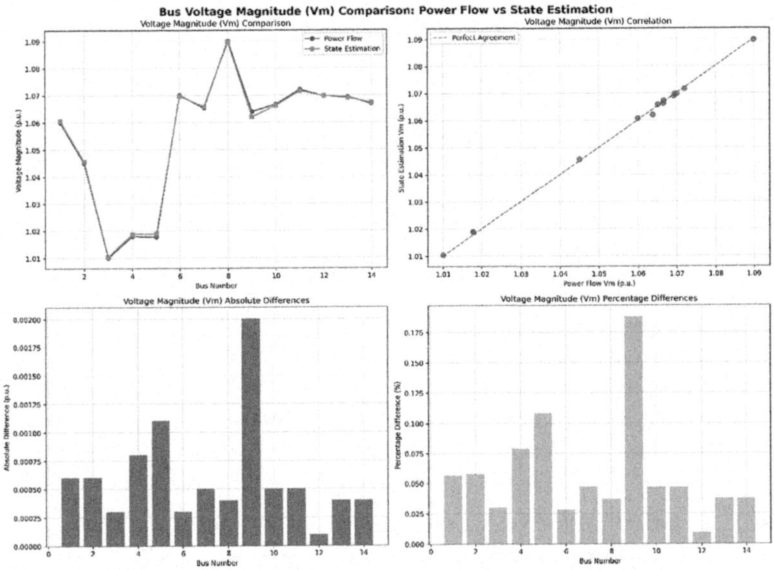

Figure 13.27 Bus voltage magnitude comparison. This figure compares the bus voltage magnitudes obtained from power-flow analysis and state estimation, demonstrating close alignment across all buses with only minor deviations.

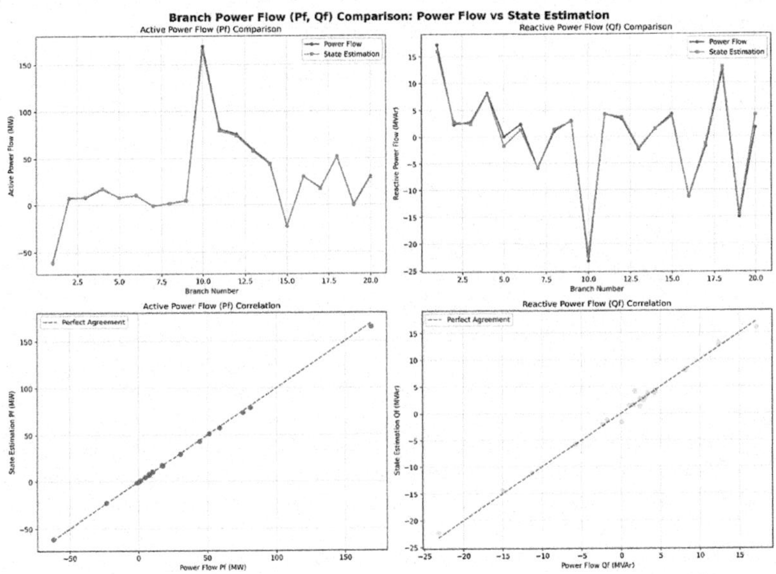

Figure 13.28 Branch PowerFlow Pf and Qf Comparison. This figure compares the active (Pf) and reactive (Qf) branch power flows obtained from power-flow analysis and state estimation, demonstrating strong consistency between the two methods.

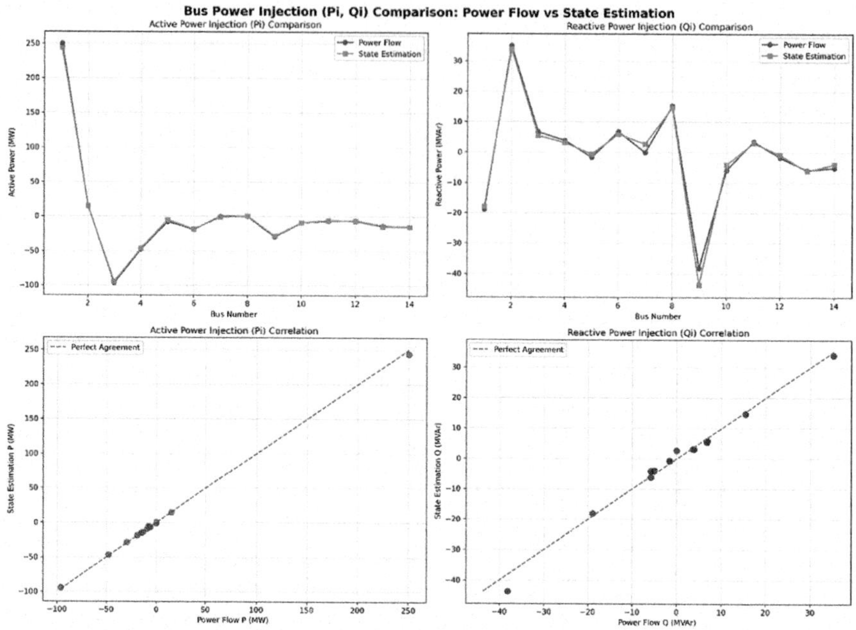

Figure 13.29 Branch PowerFlow Pf and Qf Comparison. This figure presents a comparison of bus active (Pi) and reactive (Qi) power injections, confirming a high correlation and minimal discrepancies between power-flow analysis and state estimation.

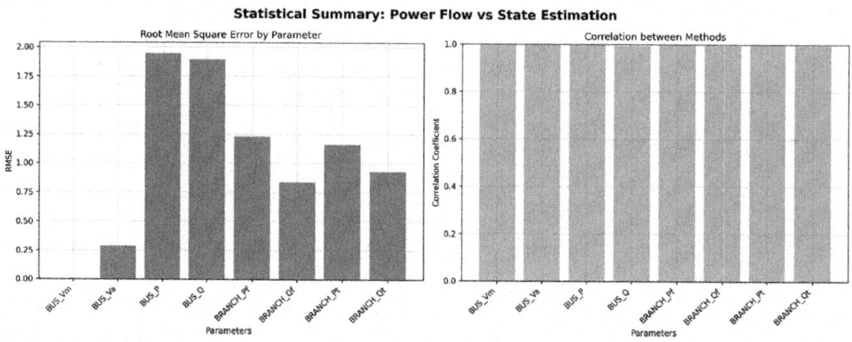

Figure 13.30 Statistical summary PowerFlow and state estimation. This figure summarizes the statistical evaluation of power-flow and state estimation, reporting root mean square errors (RMSE) and correlation coefficients that indicate high overall accuracy and strong agreement between the methods.

13.6 Conclusion and future directions

This case study demonstrated the application of SE using aPAC, which is built upon the robust computational foundation of VeraGrid. The IEEE 14-bus test network served as a valuable benchmark, enabling validation of aPAC's core capabilities in synthesizing SCADA-style measurements, performing SE, and comparing results with traditional power-flow analyses. The key finding is that aPAC's SE provides a more realistic snapshot of the power system by accounting for uncertainties and measurement noise—an essential feature for enhancing real-time decision-making. By integrating bad-data detection and redundancy analysis, aPAC ensures that results remain reliable even in the presence of erroneous or missing data, thereby improving the overall observability of power networks.

Nevertheless, several limitations were identified. While aPAC performed effectively with standard measurement sets, the accuracy and stability of the SE process are still influenced by data quality, particularly when measurements are sparse or heavily corrupted by noise. Moreover, the computational performance of aPAC, though efficient for medium-sized networks, may face scalability challenges when applied to large-scale systems with thousands of buses. These findings highlight the need for further optimization of the solver and Jacobian matrix assembly, especially in the context of real-time applications.

In terms of potential enhancements, VeraGrid offers several promising avenues for advancing the accuracy and scalability of its SE module. One important direction is the adoption of adaptive algorithms that can dynamically adjust to varying data quality and real-time system conditions. Whereas static approaches may struggle under rapidly changing network configurations or in the presence of significant measurement noise, DSE with continuous updates could directly address these challenges. Another critical gap lies in real-time integration with SCADA/EMS environments, since VeraGrid's current architecture does not natively support telemetry protocols such as ICCP/TASE.2 or IEC 60870-5-104, which are indispensable for continuous data exchange in operational contexts.

Beyond integration, improvements to VeraGrid's mathematical core could enable more effective handling of incomplete or inconsistent datasets. While the platform performs well with ideal measurement sets, real-world conditions frequently involve missing or unreliable data. Strengthening its bad-data resilience and incorporating more sophisticated methods for dealing with such imperfections would significantly broaden its operational applicability. Likewise, the solver and Jacobian assembly, though efficient for small- and medium-scale networks, may become bottlenecks for very large systems. Leveraging distributed or cloud-based computing architectures would allow VeraGrid to process larger datasets efficiently, making it more suitable for high-demand, real-time applications. Finally, expanding its analytics portfolio to include ML techniques for predictive maintenance, fault detection, and contingency analysis could transform VeraGrid into a proactive decision-support platform, extending its role from analytical tool to operational enabler in modern power system management.

At the same time, aPAC represents the natural operational extension of VeraGrid, and its future development should focus on augmenting VeraGrid's analytical strengths with the functionalities required for continuous utility deployment. Priority areas for expansion include strengthening native SCADA/EMS interfaces with robust, field-proven telemetry modules, developing operator-focused visualization dashboards that are multilingual and compliant with local regulatory requirements, and integrating real-time contingency screening into the operational workflow. Expanding aPAC's analytical scope to include voltage stability monitoring, dynamic simulations, and renewable variability forecasting would transform it from a SE platform into a comprehensive operational analysis environment.

Another key direction for aPAC is enhancing its resilience and high-availability features to ensure reliable operation in 24/7 control room environments. This includes implementing automated bad-data loops, redundant servers, and seamless failover mechanisms to guarantee uninterrupted performance. To further support operators, aPAC could incorporate training modules and simulators, bridging the gap between computational analysis and human decision-making. Finally, as a platform handling critical infrastructure, aPAC must continue to evolve with a strong emphasis on cyber-security, embedding encryption, access control, and audit trails as standard features.

In conclusion, while VeraGrid provides a solid analytical foundation for SE research and education, its future lies in refining algorithms, scalability, and adaptability to real-world data. aPAC, by contrast, carries this foundation into the operational domain, where real-time integration, visualization, resilience, and security are central priorities. Together, VeraGrid and aPAC define a complementary trajectory: VeraGrid advancing the methodological frontier, and aPAC ensuring that these innovations can be trusted and applied in the mission-critical environments of modern power system operation.

References

Auer, S., Liße, J., Mandha, S. R. & Horn, C. (2019). Power-Flow-Constrained Asset Optimization for Off-Grid Power Systems Using Selected Open-Source Frameworks. In 4th International Hybrid Power Systems Workshop. Crete, Greece.

Brown, T., Hörsch, J., & Schlachtberger, D. (2018). PyPSA: Python for power system analysis. *Journal of Open Research Software, 6*(1). doi:10.5334/jors.188.

Dugan, R. C. (2012). *Reference Guide The Open Distribution System Simulator (OpenDSS)*. Palo Alto, CA, USA: Electric Power Research Institute (EPRI).

Eidiani, M. (2002). Atc evaluation by CTSA and POMP, two new methods for direct analysis of transient stability. *Proceedings of the IEEE Power Engineering Society Transmission and Distribution Conference, 3*, 1524–1529. doi:10.1109/TDC.2002.1176824.

DIgSILENT GmbH. (2021). PowerFactory technical documentation. DIgSILENT GmbH. https://www.digsilent.de/en/faq-powerfactory.html.

Eidiani, M., 2008, 2008 IEEE 2nd International Power and Energy Conference, doi: 10.1109/PECON.2008.4762525 494 496. Assessment of voltage stability with the new NRS.

Eidiani, M. (2021a). A reliable and efficient holomorphic approach to evaluate dynamic available transfer capability. *International Transactions on Electrical Energy Systems, 31*(11). doi:10.1002/2050-7038.13031.

Eidiani, M. (2021b). An efficient differential equation load flow method to assess dynamic available transfer capability with wind farms. *IET Renewable Power Generation, 15*(16), 3843–3855. doi:10.1049/rpg2.12299.

Eidiani, M. (2022). A new load flow method to assess the static available transfer capability. *Journal of Electrical Engineering & Technology, 17*(5), 2693–2701. doi:10.1007/s42835-022-01105-3.

Eidiani, M. (2023a). *Modeling renewable energy resources using DIgSILENT PowerFactory software power systems operation with 100% renewable energy sources* (pp. 165–202). Elsevier. doi:10.1016/B978-0-443-15578-9.00013-3.

Eidiani, M. (2023b). A new hybrid method to assess available transfer capability in AC–DC networks using the wind power plant interconnection. *IEEE Systems Journal, 17*(1), 1375–1382. doi:10.1109/jsyst.2022.3181099.

Eidiani, M. (2023c). A rapid state estimation method for calculating transmission capacity despite cybersecurity concerns. *IET Generation, Transmission & Distribution, 17*(20), 4480–4488. doi:10.1049/gtd2.12747.

Eidiani, M., Asadi, S. M., Faroji, S. A., Velayati, M. H., & Yazdanpanah, D. (2008). Minimum distance, a quick and simple method of determining the static ATC. In *2008 IEEE 2nd International Power and Energy Conference* (pp. 490–493). doi:10.1109/PECON.2008.4762524 490 493.

Eidiani, M., & Ghavami, A. (2022). Iran New network design for simultaneous use of electric vehicles, photovoltaic generators, wind farms and energy storage. In *2022 9th Iranian Conference on Renewable Energy and Distributed Generation (ICREDG)* (pp. 1–5). doi:10.1109/ICREDG54199.2022.9804534.

Eidiani, M., & Heidari, V. (2023). *Fundamentals of power systems analysis 1: Problems and solutions* (pp. 1–215). CRC Press. doi:10.1201/9781003394433.

Eidiani, M., & Mahnani, S. (2023). Optimally and independent planned microgrid with solar-wind and biogas hybrid renewable systems by HOMER. *Majlesi Journal of Electrical Engineering, 17*(2), 153–158. doi:10.30486/mjee.2023.1974843.1021.

Eidiani, M., & Rouzbehi, K. (2024). *Advanced topics in power systems analysis: Problems, methods, and solutions* (pp. 1–120). CRC Press. doi:10.1201/9781003506751.

Eidiani, M., & Rouzbehi, K. (2025). *Fundamentals of power system transformers: Modeling, analysis, and operation* (pp. 1–113). CRC Press. doi:10.1201/9781003537564.

Eidiani, M., Shanechi, M. H. M., & Vaahedi, E. (2002). Fast and accurate method for computing FCTTC (first contingency total transfer capability). In *Proceedings International Conference on Power System Technology: 2*. doi:10.1109/ICPST.2002.1047595.

Eidiani, M., & Yazdanpanah, D. (2011). Minimum distance, a quick and simple method of determining the static ATC. *Journal of Electrical Engineering, 11*(2), 16. Available from http://www.jee.ro/articles/WN1286616566W4cb035f61e1ac.pdf .

Eidiani, M., & Zeynal, H. (2014). New approach using structure-based modeling for the simulation of real power/frequency dynamics in deregulated power systems. *Turkish Journal of Electrical Engineering and Computer Sciences, 22*(5), 1130–1146. doi:10.3906/elk-1208-90.

Eidiani, M., & Zeynal, H. (2022). Determination of online DATC with uncertainty and state estimation. In *2022 9th Iranian Conference on Renewable Energy and Distributed Generation (ICREDG)* (pp. 1–6). doi:10.1109/ICREDG54199.2022.9804581.

Eidiani, M., Zeynal, H., Ghavami, A., & Zakaria, Z. (2022a). Comparative analysis of monofacial and bifacial photovoltaic modules for practical grid-connected solar power plant using PVsyst. In *2022 IEEE International Conference on Power and Energy (PECon)* (pp. 499–504). doi:10.1109/PECon54459.2022.9988872.

Eidiani, M., Zeynal, H., & Shaaban, M. (2022b). A detailed study on prevailing atc methods for optimal solution development. In *2022 IEEE International Conference on Power and Energy (PECon)* (pp. 299–303). doi:10.1109/PECon54459.2022.9988775.

Eidiani, M., Zeynal, H., Zadeh, A. K., Mansoorzadeh, S., & Nor, K. M. (2011). Voltage stability assessment: An approach with expanded Newton Raphson-Sydel. In *2011 5th International Power Engineering and Optimization Conference* (pp. 31–35). doi:10.1109/PEOCO.2011.5970424 31 35.

Eidiani, M., Zeynal, H., Zadeh, A. K., & Nor, K. M. (2010). Exact and efficient approach in static assessment of available transfer capability (ATC). In *2010 IEEE International Conference on Power and Energy* (pp. 189–194). doi:10.1109/PECON.2010.5697580 189 194.

Eidiani, M., Zeynal, H., & Zakaria, Z. (2022c). Development of online dynamic ATC Calculation integrating state estimation. In *2022 IEEE International Conference in Power Engineering Application (ICPEA)* (pp. 1–5). doi:10.1109/ICPEA53519.2022.9744694.

Eidiani, M., Zeynal, H., & Zakaria, Z. (2023). A comprehensive study on the renewable energy integration using DIgSILENT. In *2023 IEEE 3rd International Conference in Power Engineering Applications (ICPEA)* (pp. 197–201). doi:10.1109/ICPEA56918.2023.10093153.

Ghardashi, G., Gandomkar, M., Majidi, S., Eidiani, M., & Dadfar, S. (2022). Accuracy and speed improvement of microgrid islanding detection based on PV using frequency-reactive power feedback method. In *2022 International Conference on Protection and Automation of Power Systems (IPAPS): 16* (pp. 1–8). doi:10.1109/IPAPS55380.2022.9763190.

Hay, S., Ferguson, A., A review of power system modelling platforms and capabilities. IET, (2021), Available from: https://www.theiet.org/media/9413/3-a-review-of-power-system-modelling-platforms-and-capabilities.pdf.

Johnson, B. (2022). Free simulation tools for power quality and grid-integration of renewable energies cases. *International Journal of Electrical Power & Energy Systems, 45*(1), 78–89.

Khanh, N. B., Ngo, P. L., Giap, L. N., An, T. N. T., Tien, T. B., Vinh, T. T., Nghia, L. V., & Dat, T. T. (2025). A review of open-source energy system modeling tools. *International Journal of Advances in Applied Sciences, 14*(2), 469–480. doi:10.11591/ijaas.v14.i2.pp469-480.

Liu, J., Yao, R., Qiu, F., Liu, Y., & Sun, K. (2023). PowerSAS.m—An open-source power system simulation toolbox based on semi-analytical solution technologies. *IEEE Open Access Journal of Power and Energy, 10*, 222–232. doi:10.1109/oajpe.2023.3245040.

Makaa, B., Top 10 software tools for power systems engineers: A comprehensive guide. EEEK, (2024) Available from: https://www.eeekenya.com/top-10-software-tools-for-power-systems-engineers-a-comprehensive-guide/.

Milano, F. (2005). An open source power system analysis toolbox. *IEEE Transactions on Power Systems, 20*(3), 1199–1206. doi:10.1109/TPWRS.2005.851911.

Nawaz, S., Bhardwaj, E., Sharma, D. R., & Parmar, D. (2023). Comparison of different simulation tools for load flow analysis in power systems. *International Journal of Advanced Engineering, Management and Science, 9*(5), 088–093. doi:10.22161/ijaems.95.12.

Ouwerkerk, J., Hainsch, K., Candas, S., Muschner, C., Buchholz, S., Günther, S., Huyskens, H., Berendes, S., Löffler, K., Bußar, C., Tardasti, F., Köckritz, L., & Bramstoft, R. (2022). Comparing open source power system models - A case study focusing on fundamental modeling parameters for the German energy transition. *Renewable and Sustainable Energy Reviews, 161*, 112331. doi:10.1016/j.rser.2022.112331.

Parhamfar, M., Eidiani, M., & Abtahi, M. (2024). Distributed energy storage system: Case study. In *Distributed energy storage systems for digital power systems* (pp. 395–422). Elsevier. doi:10.1016/B978-0-443-22013-5.00013-7.

Peñate Vera, S. (2024). GridCal: A cross-platform power systems software written in Python with user interface, used in academia and industry [Computer software]. GitHub. https://github.com/SanPen/GridCal

Shidling, C., Top 25 simulation software in power system tools for engineers. CS Electrical & Electronics, (2021) Available from: https://cselectricalandelectronics.com/top-25-simulation-softwares-in-power-system-tools-for-engineer/.

Siemens Power Technologies International. (2020). *PSS/E Program Operation Manual.* Schenectady, NY: Siemens PTI.

Su, T., Peng, J., Selim, A., Zhao, J., & Tan, J. (2024). A survey of open-source power system dynamic simulators with grid-forming inverter for machine learning applications. *2025 IEEE Power & Energy Society General Meeting (PESGM).* doi:10.48550/arXiv.2412.08065.

Thurner, L., Scheidler, A., Schafer, F., Menke, J. H., Dollichon, J., Meier, F., Meinecke, S., & Braun, M. (2018). Pandapower—An open-source python tool for convenient modeling, analysis, and optimization of electric power systems. *IEEE Transactions on Power Systems, 33*(6), 6510–6521. doi:10.1109/TPWRS.2018.2829021.

van Ouwerkerk, J. (2022). Comparing open-source power system models -A case study focusing on fundamental modeling parameters for the German energy transition. *Renewable and Sustainable Energy Reviews, 161*, 112331.

Zeynal, H., & Eidiani, M. (2014). Hydrothermal scheduling flexibility enhancement with pumped-storage units. In *22nd Iranian Conference on Electrical Engineering (ICEE)* (pp. 820–825). doi:10.1109/IranianCEE.2014.6999649.

Zeynal, H., Jiazhen, Y., Azzopardi, B., & Eidiani, M. (2014a). Flexible economic load dispatch integrating electric vehicles. In *2014 IEEE 8th International Power Engineering and Optimization Conference (PEOCO)* (pp. 520–525). doi:10.1109/PEOCO.2014.6814484 520 525.

Zeynal, H., Jiazhen, Y., Azzopardi, B., & Eidiani, M. (2014b). Impact of electric vehicle's integration into the economic VAr dispatch algorithm. In *2014 IEEE Innovative Smart Grid Technologies-Asia (ISGT ASIA)* (pp. 780–785). doi:10.1109/ISGT-Asia.2014.6873892.

Zimmerman, R. D., & Murillo-Sánchez, C. E. (2017). MATPOWER: Steady-state operations, planning, and analysis tools for power systems research and education. *IEEE Transactions on Power Systems, 32*, 4649–4659. doi:10.1109/TPWRS.2017.2705104.

AI-driven scenario generation and reduction for renewable-rich energy systems: RNN-WGAN synthesis and deep clustering

14

Fatemeh Aminizade[a], *Mahmoud Mollayousefi Zadeh*[b], *Farzaneh Mohammadi*[c], *Mohammad Hossin Alizadeh Roknabadi*[d] *and Seyed Mohammad Taghi Bathaee*[b]

[a]Department of Mathematics & Computer Science, Amirkabir University of Technology, Tehran, Iran, [b]Department of Electrical Engineering, K. N. Toosi University of Technology, Tehran, Iran, [c]Department of Information Technology Management, Tarbiat Modares University, Tehran, Iran, [d]Department of Aerospace Engineering, Amirkabir University of Technology, Tehran, Iran

14.1 Introduction

In response to the growing uncertainty introduced by renewable energy sources, there is a clear and accelerating research trend toward artificial intelligence (AI) and data-driven methods that leverage large datasets and advanced architectures to produce scenarios that are more realistic, diverse, and computationally efficient than traditional techniques. This chapter presents the formulations and underlying concepts of cutting-edge models, most notably generative adversarial networks (GANs) and deep clustering frameworks, alongside other emerging machine learning (ML) techniques for scenario generation and reduction. To concretely illustrate their potential, a real-world dataset is used to evaluate these methods and compare their performance with conventional approaches. Specifically, a recurrent neural network-based Wasserstein GAN (RNN-WGAN) is employed to generate a large set of high-fidelity 48-hour wind power profiles, which are rigorously benchmarked, and deep clustering is then applied to compress this large synthetic set into a small, representative, and computationally tractable collection of scenarios for downstream stochastic optimization. All developed codes are provided on GitHub as supplementary material, enabling readers to replicate the results, expand their knowledge, and build new solutions for stochastic planning and modeling in energy systems.

Stochastic Planning and Modeling for Energy Systems: Methods, Applications, and Developments.
DOI: https://doi.org/10.1016/B978-0-443-45298-7.00013-6

14.2 Generative adversarial networks

14.2.1 Generative adversarial networks overview

GANs, first proposed by Goodfellow et al. (2014), represent a groundbreaking class of deep learning architectures for generating synthetic data. A GAN consists of two neural networks trained in a competitive, zero-sum game. One network, the generator (G), aims to produce data indistinguishable from a real dataset, while the other, the discriminator (D), strives to accurately differentiate between real and synthetic samples. This adversarial dynamic establishes a feedback loop wherein the generator improves by continuously generating more realistic samples, and the discriminator refines its ability to distinguish between real and fake data, driving both networks toward optimal performance.

A GAN's architecture comprises two core components:

- Generator (G): Parameterized by θ_G, this network takes a random noise vector $\mathbf{z}$ (typically sampled from a Gaussian distribution) and transforms it into a synthetic data sample $\mathbf{x}_G = G(\mathbf{z})$. The generator is typically a deep neural network that up samples the input noise into data mimicking the target distribution.
- Discriminator (D): Parameterized by θ_D, this network takes a data sample $\mathbf{x}$ (either real or synthetic) and outputs a scalar probability $D(\mathbf{x})$ indicating the likelihood that the sample is real. The discriminator is typically a neural network designed for binary classification.

The adversarial training process involves alternating updates to both networks, with the ideal goal of reaching a Nash equilibrium where the generated data distribution is statistically indistinguishable from the real data distribution.

The training of a standard GAN is formalized as a minimax optimization problem. The discriminator aims to maximize the objective by correctly classifying real and fake samples, while the generator aims to minimize it by producing samples that fool the discriminator. The objective function is defined as:

$$\min_{\theta_G} \max_{\theta_D} L(D, G) = \mathbb{E}_{\mathbf{x}\sim\mathbb{P}_{\text{data}}}\left[logD(x)\right] + \mathbb{E}_{\mathbf{z}\sim\mathbb{P}_{\mathbf{z}}}\left[log(1 - D(G(z)))\right] \tag{14.1}$$

In this adversarial game, the generator G seeks to minimize the objective while the discriminator D simultaneously attempts to maximize it, forming a competitive "tug-of-war." The first term, $\mathbb{E}_{\mathbf{x}\sim\mathbb{P}_{\text{data}}}\left[\log D(\mathbf{x})\right]$, represents the discriminator's ability to correctly classify real data samples, and the discriminator strives to make this value as large as possible. Conversely, the second term, $\mathbb{E}_{\mathbf{z}\sim\mathbb{P}_{\mathbf{z}}}\left[\log(1 - D(G(\mathbf{z})))\right]$, penalizes the discriminator for incorrectly classifying generated (fake) samples as real. The generator attempts to minimize this penalty by producing synthetic samples that are indistinguishable from real data. Ideally, this adversarial process converges to an equilibrium at which the discriminator outputs 0.5 for all inputs, indicating that it can no longer distinguish between real and generated samples. The general GAN architecture is demonstrated in Fig. 14.1.

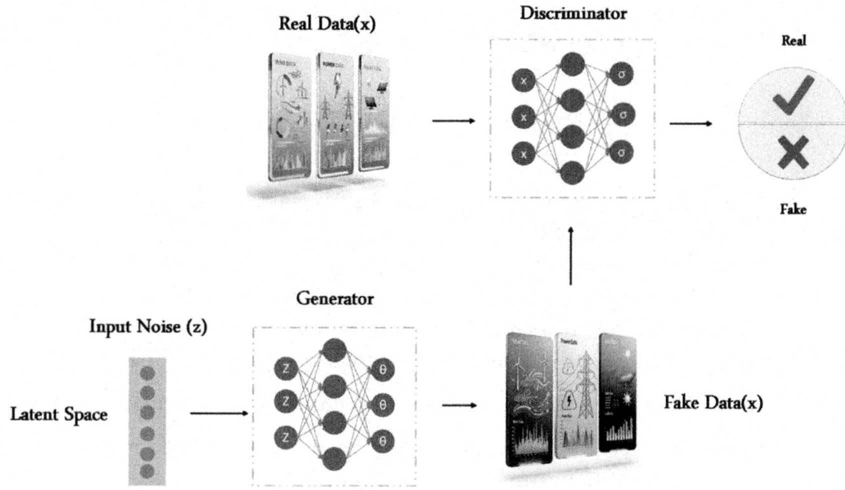

Figure 14.1 GAN model architecture. *GAN,* Generative adversarial network.

14.2.2 A taxonomy of generative adversarial networks architectures for energy systems

While the foundational GAN concept is powerful (Goodfellow et al., 2014), its direct application to the complex, time-dependent data found in energy systems is hindered by significant challenges. Traditional scenario generation methods, such as Monte Carlo (MC) simulation or Copula functions, often rely on strict statistical assumptions that may not hold in practice (Ma et al., 2023; Tiancheng et al., 2022). GANs offer a superior data-driven alternative, yet the original architecture is notoriously unstable and prone to issues like mode collapse, where the generator fails to capture the full diversity of the real data (Fekri et al., 2020; Yilmaz et al., 2024). Consequently, a suite of specialized GAN architectures is now available to enhance stability, model temporal dynamics, and enable controllable scenario generation for energy applications.

Addressing training instability: Wasserstein GAN with gradient penalty (WGAN-GP)

A primary challenge in early GANs is the vanishing gradient problem, where the discriminator improves too quickly and stops providing a useful learning signal to the generator. The breakthrough solution is the WGAN, which replaces the discriminator with a "critic" that scores the realism of an input rather than making a binary classification (Fekri et al., 2020). This approach optimizes the Wasserstein distance, providing smoother and more reliable gradients that greatly improve training stability.

A key refinement, regarded as the State-of-the-Art for stable and high-fidelity generation, is the WGAN-GP (Kaneva et al., 2025; Zhang et al., 2024). Instead of the problematic "weight clipping" used in the original WGAN, this variant applies a soft constraint on the critic's gradients to enforce the 1-Lipschitz condition. To further improve performance, especially on smaller datasets, a "consistency term" can be added to the loss function to enforce this property more robustly (Jiang et al., 2018). The objective for the critic in WGAN-GP is to maximize:

$$\mathcal{L}_D = \mathbb{E}_{\hat{\mathbf{x}}\sim\mathbb{P}_g}[D(\hat{\mathbf{x}})] - \mathbb{E}_{\mathbf{x}\sim\mathbb{P}_r}[D(x)] - \lambda\mathbb{E}_{\hat{\mathbf{x}}\sim\mathbb{P}_{\hat{\mathbf{x}}}}\left[\left(\nabla_{\hat{\mathbf{x}}}D(\hat{\mathbf{x}})_2 - 1\right)^2\right] \tag{14.2}$$

where the final term is the gradient penalty that regularizes the critic by penalizing deviations of its gradient norm from 1.

The loss function consists of two main components. The first is the critic's scoring gap, $\mathbb{E}_{\tilde{\mathbf{x}}\sim\mathbb{P}_g}[D(\tilde{\mathbf{x}})] - \mathbb{E}_{\mathbf{x}\sim\mathbb{P}_r}[D(\mathbf{x})]$, which encourages the critic to maximize the difference in realism scores between generated (fake) and real samples—a larger gap indicating a more effective critic. The second component is the stability term, $-\lambda\mathbb{E}_{\hat{\mathbf{x}}\sim\mathbb{P}_{\hat{\mathbf{x}}}}\left[\left(\nabla_{\hat{\mathbf{x}}}D(\hat{\mathbf{x}})_2 - 1\right)^2\right]$, which serves as a regularization mechanism to enforce smooth gradients during training. By penalizing the critic whenever the gradient norm deviates from 1, this term prevents the emergence of erratic gradients that could destabilize training. The hyperparameter λ controls the strength of this enforcement, balancing stability with the critic's ability to distinguish real from generated data.

Mastering temporal dynamics: Sequential architectures

Energy data are fundamentally time-series, yet standard GANs treat each data point as independent, failing to capture crucial temporal correlations. To address this limitation, several advanced architectures have been developed.

- Recurrent GANs (R-GANs) for sequential patterns: To model the sequential nature of energy data, R-GANs incorporate RNNs, particularly long short-term memory (LSTM) units, into the generator and discriminator (Fekri et al., 2020). By maintaining a hidden state, these models can learn long-range temporal dependencies, preserving the characteristic patterns and autocorrelations found in real-world load profiles and energy consumption data (Yilmaz & Korn, 2022).
- Convolutional GANs (Conv1D-GANs) for local patterns: An alternative, and often more stable, approach involves using 1D convolutional GANs (Conv1D-GANs). Inspired by the success of deep convolutional GANs (DCGANs) (Chen et al., 2018), this architecture uses 1D convolutional layers to identify local temporal patterns, such as morning ramp-ups or evening peaks, within a sliding window. These models are often more computationally efficient and stable than R-GANs, performing particularly well on smaller datasets where recurrent models might overfit (Kaneva et al., 2025).
- Transformer-based GANs for long-range dependencies: The current State-of-the-Art for capturing complex, long-range temporal dependencies is the Transformer architecture. The Transformer-WGAN-GP (TWGAN-GP) model integrates the Transformer's powerful self-attention mechanism into the generator (Gu et al., 2024). This allows the model to dynamically weigh the importance of all points in the time series relative to each other, giving it a superior ability to model long-distance correlations compared to both CNNs and RNNs. The TWGAN-GP has demonstrated superior performance over standard WGAN-GP and traditional methods for renewable energy scenario generation.

Enabling controllable generation: Conditional GANs (CGANs)

For practical energy applications, generating random, realistic scenarios is often insufficient. Instead, operators need scenarios that are conditional on specific external factors. The CGAN architecture achieves this by feeding auxiliary information, $\mathbf{y}$ (a label or condition) to both the generator and the discriminator (Chen et al., 2018). The generator must learn to produce outputs that correspond to the given condition, while the discriminator must learn to judge if a sample is real and if it matches its condition. The minimax objective is modified as:

$$\min_{G} \max_{D} L(D, G) = \mathbb{E}_{\mathbf{x} \sim \mathbb{P}_{\text{data}}} \left[logD(x \mid y) \right] + \mathbb{E}_{\mathbf{z} \sim \mathbb{P}_{\mathbf{z}}} \left[log(1 - D(G(z \mid y))) \right] \tag{14.3}$$

This loss function operates like the standard GAN, but with an added piece of information, $\mathbf{y}$. In a CGAN framework, a guiding label $\mathbf{y}$ (such as "weekday," "sunny," or the digit "7") serves as an explicit conditional input that informs both the generator and the discriminator about the desired characteristics of the data. The conditional discriminator $D(\mathbf{x}|\mathbf{y})$ evaluates samples by addressing a two-part question: Whether the sample $\mathbf{x}$ appears realistic and whether it correctly corresponds to the given label $\mathbf{y}$. Simultaneously, the conditional generator $G(\mathbf{z}|\mathbf{y})$ is tasked with producing samples from random noise $\mathbf{z}$ that are not only realistic but also aligned with the specified condition provided by $\mathbf{y}$. This conditioning mechanism enables controlled and targeted sample generation while ensuring label-consistent discrimination.

This framework is extremely powerful for energy systems, enabling the generation of scenarios conditioned on weather events, time of day, season, or specific operational states like high ramp events (Chen et al., 2018).

While more complex architectures incorporating mechanisms like self-attention (e.g., SAGAN) exist, recent studies suggest a crucial trade-off. For some energy time-series like electricity prices, simpler and well-regularized models such as WGAN-GP often provide a more favorable balance between synthetic data quality and computational cost (Yilmaz et al., 2024). This architectural landscape is summarized in Fig. 14.2.

14.2.3 Challenges and future directions for GANs

This section has charted the evolution of GANs, from their foundational min-max concept to the specialized architectures essential for modern energy system analysis. While the original GAN framework is powerful, its direct application is hindered by training instability and an inability to model the complex temporal dependencies inherent in energy data. Classic challenges such as mode collapse remain a primary concern (Chen et al., 2018). Consequently, architectures such as the WGAN-GP, CGANs, and sequence-aware models using RNNs and Transformers are necessary tools for generating realistic and actionable scenarios. The key challenge remains the synthesis of these advanced techniques for instance, developing a conditional Transformer-based WGAN-GP that combines stability, controllability, and mastery of long-range temporal dependencies.

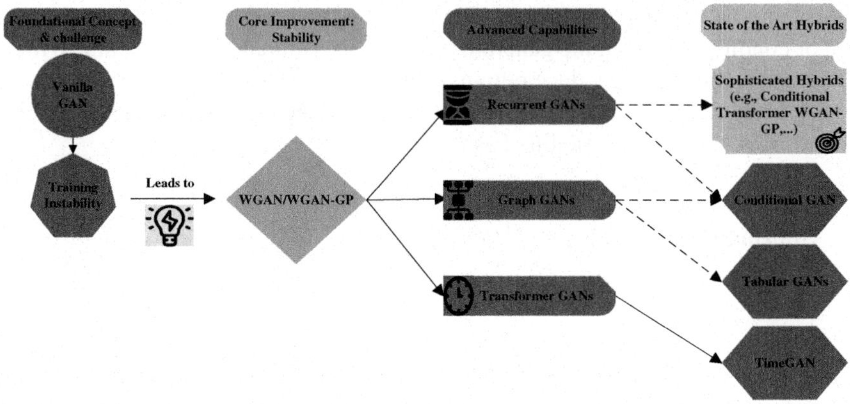

Figure 14.2 A hierarchical taxonomy of GAN architectures for energy systems, illustrating the progression from the foundational GAN to specialized models addressing training instability (WGAN-GP), temporal dynamics (R-GAN, Conv1D-GAN, Transformer-GAN), and conditional generation (CGAN). *Conv1D-GAN*, Convolutional GAN; *GAN*, generative adversarial network; *R-GAN*, recurrent GAN; *WGAN-GP*, Wasserstein GAN.

14.3 Deep clustering

14.3.1 Overview of deep clustering

Generative models such as GANs can produce vast and diverse sets of synthetic scenarios. However, their direct use in downstream stochastic optimization is often computationally prohibitive. This creates a critical bottleneck: Although thousands of plausible futures may be generated, only a tractable subset can be optimized. Scenario reduction addresses this challenge by selecting a small but representative subset that preserves the essential statistical and temporal properties of the original large set. Achieving this balance is particularly vital when dealing with time-series scenarios in energy systems.

A deep clustering model is typically composed of two main parts:

- A deep neural network (encoder): Parameterized by θ_e, this network, often the encoder part of an autoencoder, maps a high-dimensional input scenario $\mathbf{x}$ into a low-dimensional latent representation $\mathbf{z} = f_{\theta_e}(\mathbf{x})$. The core goal is to learn a mapping where similar scenarios are projected close to one another in the latent space.
- A clustering loss function ($\mathcal{L}_c$): This objective function operates on the latent representations $\mathbf{z}$ to encourage the formation of distinct, well-separated clusters. This loss is minimized jointly with the network's representation learning objective.

The key innovation of deep clustering is the joint optimization of both the encoder parameters θ_e and the cluster assignments. This creates a powerful feedback loop: The feature representation is progressively improved to be more "cluster-friendly," while the cluster assignments are refined based on this improved representation. The general architecture is shown in Fig. 14.3.

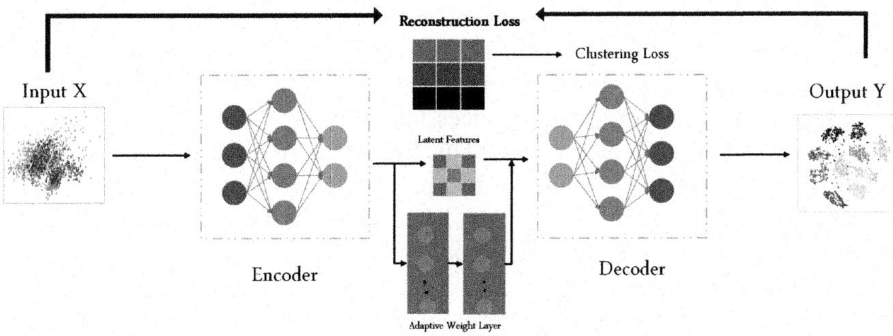

Figure 14.3 The architecture of deep embedded clustering.

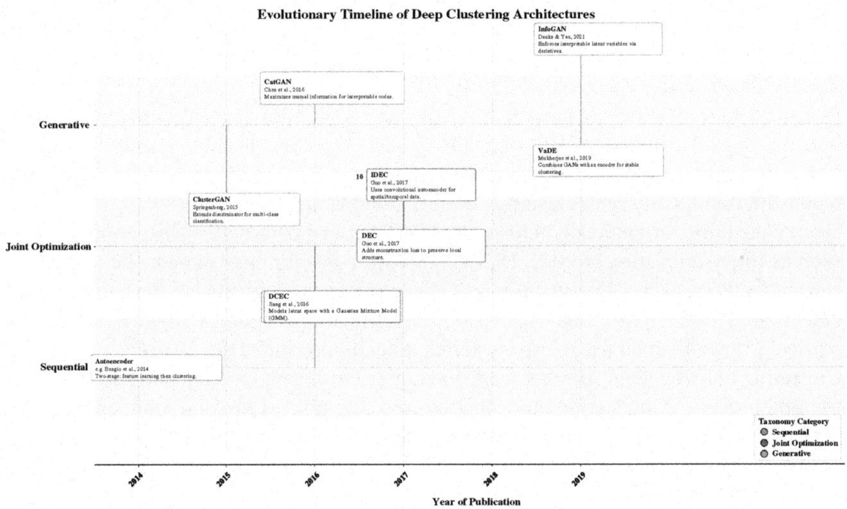

Figure 14.4 Taxonomy of deep clustering architectures for scenario reduction.

14.3.2 A taxonomy of deep clustering architectures

Deep clustering methods have evolved from sequential pipelines to end-to-end joint optimization frameworks and generative approaches. The principal motivation is to overcome the inefficiency of decoupled methods, where feature learning and clustering are performed independently (Ren et al., 2024). For energy scenarios, which are both nonlinear and temporally structured, joint optimization is particularly advantageous. Broadly, deep clustering architectures can be classified into three categories, as shown in Fig. 14.4.

Sequential (two-stage) approaches

The most straightforward application follows a two-stage paradigm: Representation learning followed by clustering. While computationally simple, this approach is limited because the learned features are not optimized with clustering objectives in mind. Autoencoder + K-Means: A deep autoencoder is first trained to minimize reconstruction loss, thereby learning a compressed representation of the input scenarios (Ren et al., 2024). Once trained, the encoder produces latent vectors, which are clustered using traditional algorithms such as k-means. This hybrid approach has attracted significant attention in recent years, as it combines neural representation learning with established clustering algorithms (Ren et al., 2024). However, its sequential nature makes it inherently suboptimal compared to end-to-end models.

Joint optimization approaches

To overcome these limitations, modern frameworks jointly optimize feature learning and cluster assignment in an end-to-end fashion.

- Deep embedded clustering (DEC) is a seminal joint optimization method. It begins by pretraining an autoencoder and then fine-tunes the encoder to simultaneously refine the latent space and optimize cluster assignments. DEC minimizes the Kullback-Leibler divergence between a soft cluster assignment and an auxiliary target distribution. A limitation, however, is that fine-tuning may distort the embedding space, reducing the representativeness of intermediate features (Xie et al., 2016).
- Improved DEC (IDEC) addresses this issue by incorporating local structure preservation into the objective function. By maintaining reconstruction loss alongside clustering loss, IDEC prevents degradation of the embedding space. The final optimization thus balances reconstruction fidelity with clustering quality (Guo et al., 2017).
- Deep convolutional embedded clustering (DCEC): For data with spatio-temporal patterns, such as time-series load profiles, DCEC employs convolutional autoencoders to extract localized features. DCEC jointly optimizes clustering and reconstruction objectives, producing embeddings that capture both spatial correlations and temporal dynamics. This makes it particularly well-suited for energy system applications (Guo et al., 2017).
- Clustering autoencoders: Recent work integrates clustering directly into the autoencoder's training process. A clustering autoencoder uses a combined loss function, allowing feature learning and clustering to occur simultaneously. For example, clustering can be achieved via online algorithms such as Sinkhorn-Knopp, reducing computational cost while maintaining accuracy (Xie et al., 2016).

Generative architectures for clustering

Generative models aim to learn the joint distribution $p(x, y)$ and synthesize realistic samples, which can also be leveraged for clustering. These methods fall into two categories: VAE-based and GAN-based.

- Variational autoencoder (VAE)-based clustering: VAEs typically assume a unit Gaussian prior, which is unsuitable for clustering as it enforces a unimodal latent distribution (Jiang et al., 2016). To address this, Gaussian mixture priors are introduced.
- Variational deep embedding (VaDE): Extends VAEs by modeling the latent space with a Gaussian mixture model, enabling multiclass clustering (Jiang et al., 2016).
- BasisDeVAE: Introduces derivative-based decoder structures to enforce interpretability and domain-informed constraints in latent-variable modeling (Jiang et al., 2016).
- GAN-based clustering: While GANs excel at generation, their standard formulation does not enforce cluster structures. Several variants address this limitation:
- CatGAN: Extends the discriminator's objective to multiclass classification (Springenberg, 2015).

- InfoGAN: Maximizes mutual information between latent codes and outputs, yielding interpretable features (Chen et al., 2016).
- ClusterGAN: Combines discrete latent variables with continuous codes, supported by an encoder for inverse mapping, thereby preserving cluster geometry (Mukherjee et al., 2019).

14.3.3 Challenges and future directions for deep clustering

Deep clustering serves as a critical bridge between large-scale scenario generation and tractable stochastic optimization. By mitigating the curse of dimensionality and jointly learning latent representations with clustering objectives, it enables high-quality scenario reduction. Future challenges include improving the interpretability of learned features and devising principled methods for selecting the number of clusters (K). A promising direction lies in integrating clustering directly into generative architectures, enabling tailored scenario sets optimized for specific downstream applications.

14.4 Scenario generation and reduction

14.4.1 The role of scenarios in stochastic programming

Stochastic programming models are indispensable tools for decision-making under uncertainty. A common and practical approach to solving such models involves approximating a continuous probability distribution with a discrete one composed of a finite number of scenarios. This process is bifurcated into two critical stages: Scenario generation, where a large, comprehensive set of potential futures is created, and scenario reduction, where this set is distilled into a computationally manageable subset.

The computational complexity of stochastic programs scales unfavorably with the number of scenarios. While a large initial set is vital for an accurate approximation of the underlying stochastic process, the resulting optimization problem is often intractable. This trade-off necessitates the use of scenario reduction techniques, which aim to select a small but representative subset of scenarios that preserves the essential statistical properties of the original set, thereby making the optimization problem solvable (Römisch, 2009). The quality of the final solution is therefore critically dependent on how effectively this reduced set represents the original uncertainty.

14.4.2 A taxonomy of scenario generation architectures

Scenario generation methods aim to create a rich set of scenarios that accurately reflects the possible future states of uncertain variables. The techniques documented in the literature can be broadly classified into two paradigms: Explicit methods, rooted in statistical and probabilistic theory, and implicit methods, which are data-driven and leverage deep learning architectures (Zheng et al., 2025). A hierarchical chart of these methods is shown in Fig. 14.5.

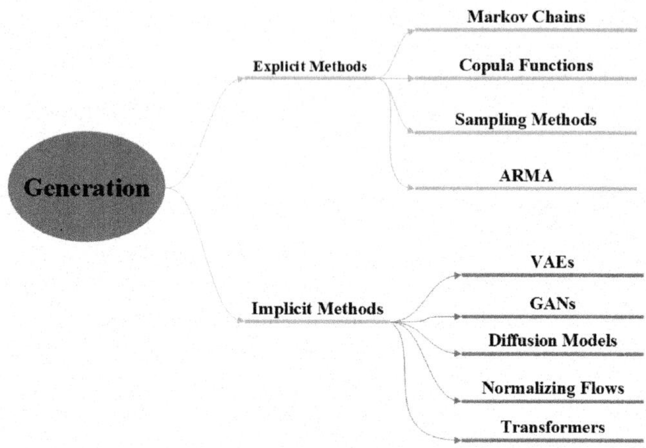

Figure 14.5 The hierarchical chart of the scenario generation architectures.

Scenario generation aims to construct a realistic and representative set of future time series $\{s^1,\ldots, s^N\}$ for renewable generation (e.g., wind, solar). Let $S = \{s^1,\ldots, s^N\}$ be generated using one or more of the scenario generation methods. Each scenario is assigned equal or empirical probability:

$$p_n = \frac{1}{N}, \quad \text{or based on likelihood estimation or clustering} \tag{14.4}$$

Statistical and probabilistic (explicit) methods

Explicit methods employ precise mathematical formulations to construct scenario models, particularly for renewable energy generation (Zheng et al., 2025). The core principle is to assume that key parameters, such as wind speed or solar irradiance, follow specific probability distributions, whose parameters are then fitted using historical data. While these methods offer high interpretability, they may struggle to capture highly complex, nonlinear system dynamics (Zheng et al., 2025).

- Markov chains: A stochastic process that models the next state of a system based solely on its current state. It is effective for capturing temporal correlations but has limitations in modeling spatial correlations across multiple sites (Zheng et al., 2025).
- Copula functions: A powerful statistical tool for flexibly representing nonlinear dependencies between multiple correlated variables. By separating the marginal distributions from their dependence structure, Copulas can generate realistic multivariate scenarios that preserve complex interdependencies (Le-Anh et al., 2023; Zheng et al., 2025).
- Sampling methods: These methods are used to discretize continuous probability models. (MC) Sampling involves repeated random sampling from a distribution, while Latin hypercube sampling is an advanced variant that ensures samples uniformly cover the entire variable space, often achieving higher accuracy with fewer samples (Zheng et al., 2025).
- Autoregressive moving average (ARMA): A widely used time-series model that captures trends and fluctuations by assuming the current value depends on past values and past

prediction errors. It is well-suited for simulating the dynamics of weather-related variables (de Heij et al., 2021; Zheng et al., 2025).

Deep learning (implicit) methods

With the advancement of deep learning, implicit generative models have become a primary tool for scenario generation. These models utilize the nonlinear mapping capabilities of neural networks to automatically learn intricate distributional features from data, without relying on predefined functional forms (Zheng et al., 2025).

- VAEs: A deep generative model comprising an encoder, which maps data to a latent space, and a decoder, which reconstructs data from it. By sampling from the learned latent distribution, VAEs can generate diverse and realistic new scenarios with a relatively stable training process (Wang et al., 2022; Zheng et al., 2025).
- GANs: GANs feature a generator and a discriminator trained in an adversarial process. The generator learns to produce synthetic scenarios so realistic that the discriminator cannot distinguish them from real data. This competitive training allows GANs to capture the complex patterns inherent in energy data (Dumas et al., 2022; Zheng et al., 2025).
- Diffusion models: A powerful new class of generative models that operate via a two-step process: A forward process systematically adds noise to data, and a reverse process learns to denoise it, starting from pure noise to generate a clean sample. This approach can produce higher-quality samples than GANs, albeit at a higher computational cost (Zheng et al., 2025).
- Normalizing flows (NFs): NFs learn a sequence of invertible transformations to map a simple base distribution to a complex target data distribution. Unlike VAEs or GANs, NFs can be trained by maximizing the exact data likelihood, offering a more direct and precise method for learning the distribution (Dumas et al., 2022).
- Transformer-based models: Originally from natural language processing, Transformers use a "self-attention" mechanism to weigh the importance of different points in a time series. This enables them to capture complex long-range dependencies, making them highly effective for probabilistic forecasting and scenario generation (Rellier et al., 2024).

14.4.3 A taxonomy of scenario reduction techniques

Once a large scenario set is generated, reduction techniques are applied to select a smaller, representative subset. The objective is to mitigate the computational burden for downstream optimization while preserving the salient stochastic properties of the original set. A taxonomy is shown in Fig. 14.6.

In stochastic optimization for energy systems with high renewable penetration, the uncertainty of renewable generation is typically represented by a set of time-series scenarios $\{s^1,\dots, s^N\}$ with associated probabilities $\{p_1,\dots, p_N\}$. Large scenario sets improve accuracy but drastically increase the computational burden in scheduling problems (e.g., unit commitment, economic dispatch). Scenario reduction aims to find a smaller set $C=\{c^1,\dots, c^K\}$, with $K \ll N$, that preserves the essential statistical and temporal features of the original set while maintaining the quality of downstream decisions.

Traditional scenario reduction relies on probability metrics (e.g., Kantorovich–Rubinstein/ Wasserstein distance) and clustering-based methods (e.g., *k*-means, *k*-medoids, forward/ backward selection). With the advent of data-driven methods, ML

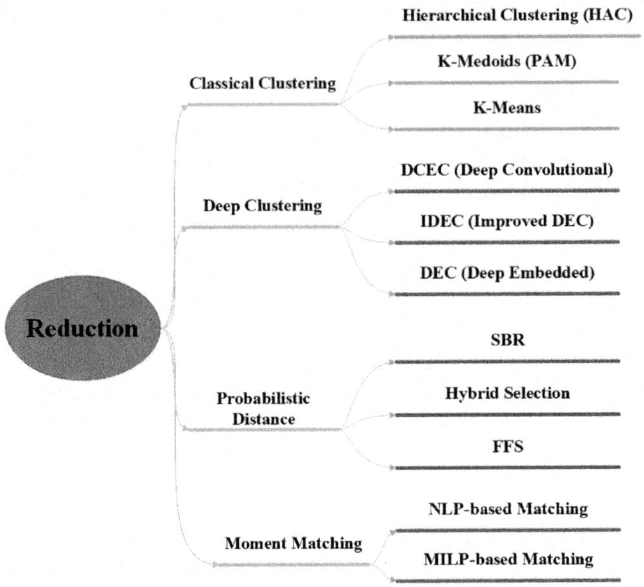

Figure 14.6 The hierarchical chart of the scenario reduction architectures.

and AI techniques can extract nonlinear features from large historical datasets for more effective reduction.

Probabilistic distance-based methods

This class of methods selects a scenario subset by minimizing a probability distance metric between the original and reduced distributions. A common metric is the Kantorovich distance (Heitsch & Römisch, 2003).

- Simultaneous backward reduction (SBR): This algorithm begins with the full set and iteratively eliminates the scenario whose removal causes the smallest increase in probability distance, continuing until the target size is reached (Heitsch & Römisch, 2003).
- Fast forward selection (FFS): In contrast, this algorithm starts with an empty set and iteratively adds the scenario that results in the largest decrease in probability distance. It is typically faster than SBR but may yield a slightly less optimal approximation (Heitsch & Römisch, 2003).

Moment matching methods

The goal of moment matching is to construct a reduced set whose statistical moments (e.g., mean, covariance, skewness) closely match those of the original distribution. This is often framed as an optimization problem to minimize the deviation between the moments of the two sets (Høyland & Wallace, 2001).

Clustering-based methods

Clustering is an intuitive approach that groups scenarios based on a similarity measure. A single representative scenario is then chosen from each cluster (e.g., the centroid or medoid) to form the reduced set. These methods are widely applied in areas like residential demand response modeling (Räsänen et al., 2010).

- Classical clustering: Algorithms such as K-means, K-medoids, and hierarchical agglomerative clustering (HAC) are applied directly to the high-dimensional scenario data. However, their performance can degrade in high-dimensional spaces due to the "curse of dimensionality."
- Deep clustering: To overcome the limitations of classical methods, deep clustering integrates deep neural networks with clustering objectives. A common architecture employs a deep autoencoder to learn a compressed, low-dimensional latent representation of the scenarios. A traditional clustering algorithm is then applied to this information-rich latent space, leading to more meaningful and robust clusters by capturing complex nonlinear patterns (Danks & Yau, 2021).

14.4.4 Challenges and future directions for scenario modeling

A central challenge in scenario-based stochastic programming is the trade-off between the accuracy of the stochastic representation and computational tractability. Aggressive reduction may discard crucial information, such as extreme events, leading to suboptimal decisions. For explicit methods, this involves developing more flexible models to capture complex, unknown distributions (Zheng et al., 2025). For implicit methods, challenges include mitigating the training instability of GANs, the "blurriness" of VAE outputs, and the computational cost of diffusion models (Zheng et al., 2025). A key frontier is the integration of generation and reduction into end-to-end frameworks tailored for specific downstream optimization tasks.

14.5 Case study

To render the application of the previously discussed techniques concrete, this section presents a complete, two-stage worked example. The complete source code, data, and models used in this case study are publicly available in a GitHub repository, which has been archived for long-term preservation (Amini, 2025). First, a sophisticated RNN-WGAN is used to generate a large set of high-fidelity 48-hour wind power profiles, and its performance is rigorously benchmarked. Second, deep clustering is applied to reduce this large set into a small, representative, and computationally tractable collection of scenarios suitable for downstream optimization tasks.

14.5.1 Scenario generation: High-fidelity wind power profiles with RNN-WGAN-GP

14.5.1.1 Problem statement and model benchmarking

The objective is to generate a statistically robust set of 48-hour wind power scenarios using data from a wind turbine SCADA system (Erisen, 2020). To demonstrate the efficacy of modern neural architectures, a sequence-aware Wasserstein GAN with gradient penalty, built upon a RNN-WGAN-GP, is trained. Its performance is critically compared against three widely used statistical benchmarks: An autoregressive integrated moving average (ARIMA) model, a Gaussian Copula, and a Gaussian MC simulation. The evaluation framework is designed to be comprehensive, assessing distributional fidelity, temporal dynamics, and covariance structure. The detailed training procedure for the RNN-WGAN-GP is outlined in Algorithm 14.1.

```
Algorithm 14.1 Training Procedure for RNN-WGAN-GP for Wind Scenario
Synthesis
Require: Real data loader $\mathcal{D}$ of time-series sequences,
total epochs T, critic steps $n_{c}=5$,
learning rate $\eta$, gradient penalty coefficient $\lambda=10$,
latent dimension $d_{z}$.
Initialize RNN-based Critic $D_{\theta_{D}}$ and RNN-based Generator
$G_{\theta_{G}}$ parameters.
for $t \leftarrow 1$ to $T$ do
for $i \leftarrow 1$ to $n_{c}$ do
Sample a mini-batch of real sequences $\left\{\mathbf{x}^(j)\right\}_
{j=1}^{m} \sim \mathcal{D}$.
Sample a mini-batch of latent vectors $\left\{\mathbf{z}^(j)\right\}
_{j=1}^{m} \sim \mathcal{N}\left(0, I_{d_{z}}\right)$.
Generate a mini-batch of fake sequences $\tilde{\mathbf{x}}^(j)
\leftarrow G_{\theta_{G}}\left(\mathbf{z}^(j)\right)$.
Sample $\epsilon \sim U[0,1]$ and form interpolated sequences
$\hat{\mathbf{x}} \leftarrow \epsilon \mathbf{x}+(1-\epsilon)
\tilde{\mathbf{x}}$.
Compute the gradient penalty $p_{g} \leftarrow\left(\left\|\nabla_
{\hat{\mathbf{x}}} D_{\theta_{D}}(\hat{\mathbf{x}})\right\|_{2}-
1\right)^{2}$.
Compute critic loss: $\mathcal{L}_{D} \leftarrow
\mathbb{E}\left[D_{\theta_{D}}(\tilde{\mathbf{x}})\right]-
\mathbb{E}\left[D_{\theta_{D}}(\mathbf{x})\right]+\lambda \cdot
\mathbb{E}\left[p_{g}\right]$.
Update critic parameters: $\theta_{D} \leftarrow \operator-
name{Adam}\left(\nabla_{\theta_{D}} \mathcal{L}_{D}, \theta_{D},
\eta\right)$.
end for
Sample a mini-batch of latent vectors $\left\{\mathbf{z}^(j)\right\}_
{j=1}^{m}$.
Compute generator loss: $\mathcal{L}_{G} \leftarrow-
\mathbb{E}\left[D_{\theta_{D}}\left(G_{\theta_{G}}(\mathbf{z})\right)
\right]$.
Update generator parameters: $\theta_{G} \leftarrow \operator-
name{Adam}\left(\nabla_{\theta_{G}} \mathcal{L}_{G}, \theta_{G},
\eta\right)$.
end for
```

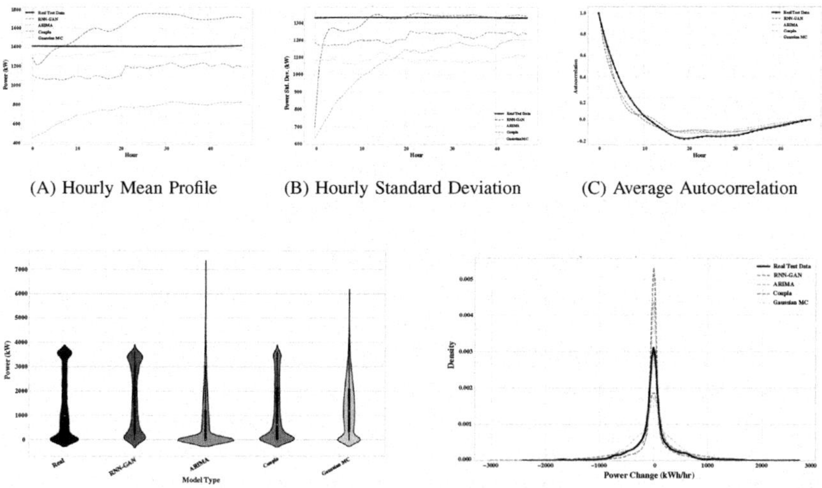

(A) Hourly Mean Profile (B) Hourly Standard Deviation (C) Average Autocorrelation

(D) Overall Value Distribution (E) Ramp Rate Distribution

Figure 14.7 Comprehensive evaluation of scenario generation models.

Table 14.1 Quantitative performance metrics (lower is better).

Model	Wasserstein-1 (1W)	ACF MAE	Cumulative P-Norm	Energy Score (ES)	CRPS/SP	Avg. DTV
RNN-WGAN	72.89	0.008	6.27×10^6	112.17	1.14×10^6	1566.08
ARIMA	226.47	0.002	1.43×10^7	182.26	1.38×10^6	1724.85
Copula	233.95	0.014	2.08×10^7	238.12	1.48×10^6	1707.19
Gaussian MC	295.51	0.008	1.42×10^7	88.59	1.34×10^6	1.30×10^4

ACF, autocorrelation function; *ARIMA*, autoregressive integrated moving average; *CRPS*, continuous ranked probability score; *DTV*, discrete total variation; *MAE*, mean absolute error; *MC*, Monte Carlo; *RNN-WGAN*, recurrent neural network-based Wasserstein GAN; *SP*, spherical score.

14.5.1.2 Results and comprehensive analysis

The comparative evaluation, summarized in Fig. 14.7, provides conclusive evidence of the RNN-WGAN's superior performance across all evaluated criteria. The analysis is presented by dissecting the key statistical properties that define a realistic time-series scenario.

Distributional fidelity. A fundamental test of a generative model is its ability to replicate the marginal distribution of the underlying data. Panel D of Fig. 14.7 shows the violin plots of power values, where the RNN-WGAN distribution is a near-perfect match for the bimodal, non-Gaussian profile of the real test data. This visual finding is quantitatively validated by the Wasserstein distance (1W) metric in Table 14.1, a standard for evaluating generative models in energy applications (Li et al., 2021). The RNN-WGAN achieves a score of 72.89, which is over three times lower than the next best model (ARIMA at 226.47), indicating a statistically superior match to the true data distribution. Furthermore, the hourly mean profile in Panel A shows that only the RNN-WGAN correctly captures the central tendency of the power output across the 48-hour horizon.

Temporal dynamics and correlation. Beyond static distributions, a successful model must capture the intricate hour-to-hour relationships. The average autocorrelation function (ACF), mean absolute error (MAE) in Panel C illustrates the "memory" of the process, a key validation technique for generated wind series (Dong et al., 2022). The RNN-WGAN's ACF curve aligns precisely with that of the real data. More critically, the ramp-rate distribution in Panel E, which is vital for power systems stability analysis, is captured with exceptional accuracy by the RNN-WGAN. It correctly models the sharp peak around zero and the heavy tails, a focus of specialized generative models designed explicitly to capture such events (Yoon et al., 2019).

Variability and covariance structure. The final component of realism is the model's ability to represent the correct amount of uncertainty and the complex covariance structure. Panel B shows that the RNN-WGAN is the only model that accurately traces the hourly standard deviation of the real data. This is powerfully confirmed by the cumulative P-norm (a proxy for the covariance frobenius norm) in Table 14.1. Comparing the full covariance matrix is a robust method for validating the joint temporal structure of generated wind scenarios (Koshiyama et al., 2021). The RNN-WGAN's score of 6.27×10^6 is over twice as effective as its competitors, demonstrating that it has learned not just simple correlations, but the entire high-dimensional dependency structure of the time series. Other key metrics in the table include the energy score (ES), continuous ranked probability score (CRPS), spherical score (SP), and average discrete total variation (DTV), which provide a comprehensive evaluation of the model's performance in capturing uncertainty and variability.

In summary, the inherent ability of the recurrent architecture to process information sequentially, combined with the distributional precision of the WGAN-GP training framework, results in a model that is unequivocally superior for this task. The statistical benchmarks, while useful, prove insufficient for capturing the nonlinear, non-Gaussian, and temporally complex nature of wind power data.

14.5.2 Scenario reduction: Deep clustering of synthetic wind profiles

Having generated a high-fidelity set of 5000 synthetic wind scenarios using the validated RNN-WGAN, the pipeline proceeds to the reduction stage. Deep clustering is now applied to distill this large set into a small, computationally tractable, and representative collection.

Problem statement and model architecture

The objective is to reduce a set of $N = 5,000$ synthetic 48-hour wind profiles into a representative set of $K = 5$ scenarios. A DEC model built upon a convolutional autoencoder is employed for its ability to learn temporal features from the 1D time-series data. The process follows two main stages:

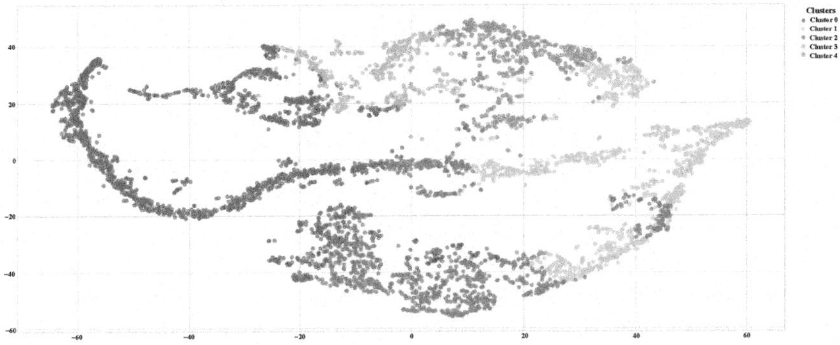

(A) t-SNE visualization of 5,000 scenarios into 5 distinct clusters.

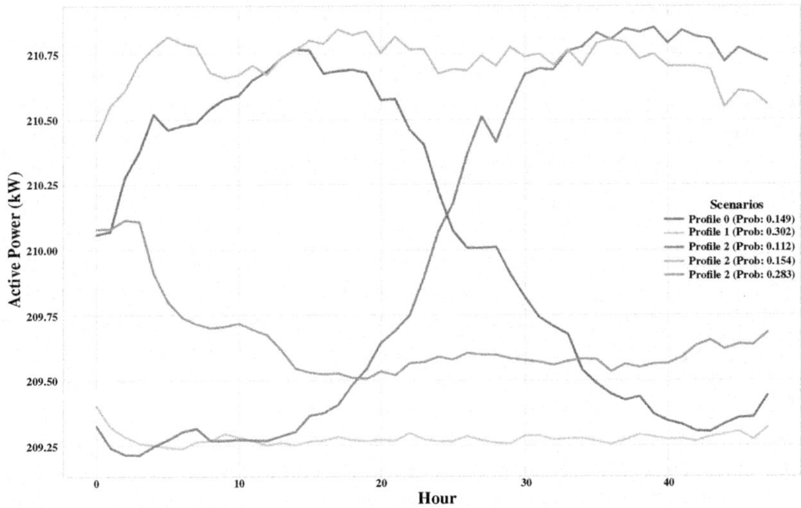

(B) The 5 representative profiles (decoded centroids) with their probabilities.

Figure 14.8 Evaluation of the DEC scenario reduction into 5 clusters. The left panel visualizes the cluster structure in the latent space, while the right panel shows the resulting archetypal wind profiles. *DEC*, Deep embedded clustering.

1. Autoencoder pretraining: The convolutional autoencoder is first trained to learn a robust, low-dimensional representation of the wind profiles.
2. DEC fine-tuning: The model is then fine-tuned by jointly optimizing the clustering loss (Kullback-Leibler divergence) and the reconstruction loss.

Evaluation: Analysis of clustering results

The success of the scenario reduction is evaluated through a multifaceted analysis of the final clustering results, shown in Fig. 14.8. The evaluation considers the intrinsic structure of the learned feature space, the physical interpretation of the final representative profiles, and the quantitative fidelity of the reduced set.

A note on data scaling. It is critical to note that the results in Fig. 14.8 are presented on the GAN's native normalized scale (approx. [-1, 1]), not in physical units (kW). While the near-zero quantitative metrics in Panel C confirm an excellent statistical match on this normalized scale, a complete engineering analysis requires un-scaling the results to their physical domain to interpret errors in megawatts and assess economic impact. The following qualitative analysis, however, remains valid as it focuses on the shape and relative importance of the discovered patterns.

Intrinsic evaluation: Latent space visualization. Panel A of Fig. 14.8 displays the t-SNE projection of the 5000 scenarios in the learned latent space. T-SNE (t-distributed stochastic neighbor embedding) is a data visualization technique used to see the structure of high-dimensional data (like our 48-point wind profiles) in a low-dimensional map, typically a 2D plot. Its main goal is to arrange the data points so that similar scenarios appear close together and dissimilar scenarios appear far apart. The visualization confirms the model's effectiveness, as the scenarios are organized into well-defined, colored clusters with clear separation. The nonrandom, arch-like structure of the manifold suggests that the autoencoder has learned a meaningful continuum of wind behavior, allowing the DEC algorithm to identify distinct operational zones.

Analysis of representative profiles. The five representative profiles, decoded from the cluster centroids, reveal distinct and physically meaningful archetypes of wind turbine operation over a 48-hour horizon. Unlike previous normalized results, these profiles are presented directly in physical units of active power (kW), allowing for a direct interpretation of their operational and economic implications. The model has autonomously discovered the following patterns, with their corresponding probabilities of occurrence:

- Profile 1—The low-wind state (Prob: 30.2%): This profile represents the most common operational mode, characterized by sustained low power output. Its high probability of occurrence makes this archetype fundamental for assessing baseline power deficits and planning for necessary reserve capacity.
- Profile 4—The production trough (Prob: 28.3%): As the second most frequent scenario, this profile captures a significant dip in power generation followed by a slow, sustained recovery. This pattern is crucial for understanding intra-day variability and the challenges of managing mid-range, fluctuating wind resources.
- Profile 3—The high-production state (Prob: 15.4%): This archetype illustrates periods of optimal wind conditions with consistently high-power output, likely near the turbine's rated capacity. It serves as the key scenario for forecasting maximum potential energy yield.
- Profile 0—The major ramp-down event (Prob: 14.9%): This profile is defined by a sharp, significant decrease from high production to a lower operational state. Such events are critical for system reliability studies, as they represent a sudden loss of generation that the grid must absorb.
- Profile 2—The major ramp-up event (Prob: 11.2%): As the rarest archetype, this profile captures a steep and substantial increase in power from a very low initial state. Such significant ramping events are of paramount importance for grid stability analysis, particularly in assessing the grid's ability to handle rapid injections of renewable power.

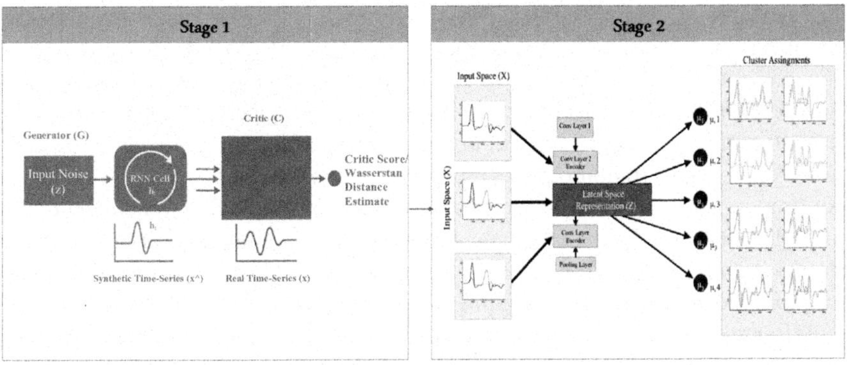

Figure 14.9 The efficient pipeline for generative scenario analysis, detailing the two primary stages from raw data processing to the final representative scenario set.

14.5.3 Conclusion: An efficient generation and reduction pipeline

In conclusion, this case study has demonstrated a complete, end-to-end pipeline for converting raw, high-volume time-series data into a small, statistically robust, and computationally tractable set of representative scenarios. The workflow, depicted in Fig. 14.9, begins by leveraging a State-of-the-Art generative model (RNN-WGAN-GP) to produce a large, high-fidelity synthetic dataset. This is followed by a sophisticated deep clustering technique (DEC) that intelligently distills this dataset into a handful of archetypal scenarios. This two-stage process ensures that the final scenarios are not only realistic and diverse but also suitable for robust decision-making in downstream applications such as stochastic optimization and risk analysis.

14.6 Research gaps and future trends

Scenario generation and reduction techniques simulate unknown and uncertain future trajectories under various conditions for uncertain variables, enabling the development of distributionally robust optimization models. Based on the published literature, we can claim that, in the context of these techniques, AI and data-driven methods have become functional solutions for modeling uncertainty in the energy sector. However, there are some research gaps and future trends in this domain that require further explanation. The idea of identifying research gaps and future directions has been challenging for many researchers; therefore in this section, we first elaborate on research barriers and then reveal future trends.

14.6.1 Overarching challenges in generative models

Various GAN strategies have been developed to address the primary issues with GAN training methods, including gradient vanishing, convergence, as well as mode collapse. GAN models struggle with some challenges that make them infeasible for application or inefficient as per the intended needs (Saxena & Cao, 2021). Attempts are consistently employed to correct these challenges. Diminishing gradients slow or even terminate learning, and mode collapse happens when a generator provides different inputs to one control output class, generating nondiversified examples (Sharma et al., 2024). Incompatible, repetitive, and inconsistent generator and discriminator training lead to nonconvergence (Chaudhary et al., 2025). Some recent publications highlight these GAN challenges, reporting diminishing gradients, which prevent or slow down learning, and mode collapse, which occurs when the generator produces closely related instances by assigning different inputs to the identical class at the output (Barsha & Eberle, 2025; Cobbinah et al., 2025). Nonconvergence is triggered by training the generator and discriminator via approaches that are inconsistent, repetitive, and unstable. These challenges in the realm of AI and data-driven methods in scenario generation and reduction immediately impact the quality, diversity, and reliability of generated scenarios, contributing to energy systems challenges in representing unusual events, ensuring robust optimization, and creating representative datasets for decision-making. One of the most common challenges with one limitation of GANs is that they often cannot settle on an outcome (Sampath et al., 2021). During GAN training, it is observed that theoretically, it is ensured to converge with direct modification of the density functions. However, in practice, we modify the sample generation function generator (He et al., 2022). It results in variations as a consequence. At this stage in the operation, the model may be trained for a prolonged period of time, which may generate a wide range of categories of samples. The GAN model operates effectively indeed in case the generator and discriminator achieve equilibrium if the discriminator loss is 0.5 (Nguyen et al., 2017). To improve GAN convergence, other regularization methods are also implemented. This was achieved by adding noise to discriminator inputs, as reported in (Liu & Hsieh, 2019), and by constraining discriminator weights, as suggested by (Li et al., 2021). The output from the discriminator approximates to zero because it has been trained to its maximum capability, which triggers the convergence rate to a reduction. The intensity of the generator is erroneous when the discriminator is not adequately trained.

When employing gradient descent in order to train artificial neural networks for generation or even classification, the vanishing gradient issue arises (Durgadevi, 2021). The generator is enhanced with discriminator feedback. The generator will stop training if the discriminator is very strong (Che et al., 2020). The variables of neural networks are obtained employing back-propagation approaches. These processes were using a chain rule that moves layer by layer to evaluate the gain of the loss function with respect to the network variables (Pan et al., 2020). The chain rule begins at the final layer and moves backwards until it attains the initial layer (Hitaj et al., 2017). This enables the derivatives of the initial layers to be calculated.

When the generator produces the identical visuals or a minuscule portion of the training visuals, mode-collapse occurs. An effective generator has to generate many training images in all categories (Zhou et al., 2023). When the discriminator cannot detect fake information, mode collapse happens, and the generator persists in generating the same information to deceive it. Therefore mode collapse is a frequent cause for failure when a generator unit only delivers defined categories of actual samples and has little diversity among the data (Cobbinah et al., 2025). As a result, enhancing the variety of the generated samples remains a challenging task for scholars. This is a common problem with GANs (Sampath et al., 2021). This issue restricts the GAN's application in many computer vision domains.

14.6.2 Future research directions

So, what is next? There are several future research directions to extend in the applications of AI in the evolving landscape of energy systems. In this chapter, the dynamic topic modeling was employed to present the dominant, emerging, and maturing themes that are shaping this given field. Thus, a robust bibliometric methodology was used, centered on the BERTopic modeling framework. To conduct this, a comprehensive corpus of 632 scientific articles from 1992 to 2025 was compiled. A targeted dataset was curated from the Web of Science (WoS) database by executing the following topic search (TS) query. The query is structured around three core conceptual pillars: (1) A comprehensive set of terms for AI and optimization, (2) keywords for scenario analysis, and (3) a vast vocabulary for energy systems. This process yielded a specialized corpus of 632 documents. The following search query (Table 14.2) was used to classify documents.

Then, the titles and abstracts of these documents were subjected to a standard natural language preprocessing pipeline, which included converting text to lowercase, removing punctuation and numerical digits, eliminating common English stopwords, and applying lemmatization to reduce words to their base forms. The core of our analysis utilized BERTopic, a State-of-the-Art topic modeling technique. Unlike traditional models like latent dirichlet allocation (LDA), BERTopic leverages transformer-based embeddings to capture the deep semantic meaning of the text. It then uses a density-based clustering algorithm to identify coherent topics and a class-based TF-IDF procedure to generate human-interpretable topic representations. Crucially, by associating each document with its publication year, we enabled BERTopic's dynamic topic modeling capability. This allows us to track the frequency and prominence of each identified research topic over time, revealing key intellectual trends and shifts in focus, which are visualized in the accompanying trends chart. Fig. 14.10 shows the prospective future research avenues on generative AI in the scenario generation context.

According to Fig. 14.10, our analysis reveals a significant paradigm shift from isolated, predictive modeling of system components toward the use of advanced generative AI for holistic, high-fidelity scenario simulation, paving the way for the development of resilient, decarbonized energy futures. Our analysis identifies the most powerful emerging trend as "Topic 1: Generative Adversarial Deep Load." This topic, which has seen explosive growth in just the last few years, centers on the use of sophisticated deep learning models like GANs. This signals a pivotal move away from simply forecasting a likely future to generating thousands of highly realistic, synthetic scenarios for load profiles, wind patterns, and solar output. This new capability allows for much more rigorous stress-testing of grid infrastructure against a wider range of possibilities. The largest and most persistent research area is "Topic 0: Load Demand

Table 14.2 Search query applied in WoS database.

Topic Search (TS)
TS = (("artificial intelligence" OR "AI" OR "machine learning" OR "ML" OR "deep learning" OR "DL" OR "data-driven" OR "data based" OR "predictive model*" OR "predictive analytics" OR "neural network*" OR "ANN" OR "convolutional neural network*" OR "CNN" OR "recurrent neural network*" OR "RNN" OR "long short-term memory" OR "LSTM" OR "transformer model*" OR "GAN" OR "generative adversarial network*" OR "support vector machine*" OR "SVM" OR "extreme learning machine*" OR "reinforcement learning" OR "RL" OR "Bayesian network*" OR "ensemble learning" OR "gradient boosting" OR "XGBoost" OR "random forest*" OR "decision tree*" OR "meta-learning" OR "transfer learning" OR "fuzzy logic" OR "digital twin" OR "optimization algorithm*" OR "metaheuristic*" OR "genetic algorithm*" OR "swarm intelligence" OR "particle swarm optimization" OR "PSO" OR "ant colony optimization" OR "ACO" OR "evolutionary computation") AND ("scenario generation" OR "scenario planning" OR "scenario analysis" OR "scenario modeling" OR "scenario simulation" OR "future scenario*" OR "scenario reduction" OR "scenario optimization" OR "scenario forecasting" OR "scenario assessment") AND ("energy system*" OR "energy sector" OR "electric power" OR "power grid*" OR "power system*" OR "smart grid*" OR "microgrid*" OR "renewable energy" OR "clean energy" OR "wind energy" OR "wind power" OR "wind farm*" OR "solar energy" OR "solar power" OR "solar farm*" OR "photovoltaic*" OR "PV system*" OR "hydropower" OR "biomass energy" OR "energy forecasting" OR "energy optimization" OR "energy planning" OR "load forecasting" OR "demand response" OR "energy storage" OR "distributed generation"))

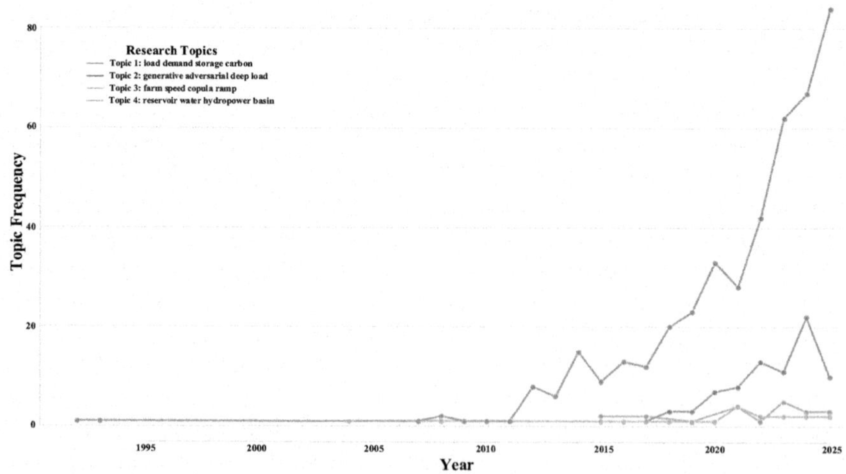

Figure 14.10 Future research in generative AI for scenario generation.

Storage Carbon." As the foundational topic, its continued growth underscores that the core challenge is no longer optimizing individual components in isolation but achieving holistic system integration. This research addresses the complex interplay between managing electricity demand, integrating energy storage, and meeting decarbonization targets. As renewable penetration grows, the complexity of this systemic challenge intensifies, keeping it at the forefront of the research agenda. Topics focused on specific renewable sources, such as "Topic 2: Farm Speed Copula Ramp" (wind) and "Topic 3: Reservoir Water Hydropower," represent more mature research streams. While consistently present, their growth has stabilized. This indicates that the initial challenge of modeling the uncertainty of a single renewable source is now a well-established field. The research frontier is shifting from how to model wind to understanding how wind variability impacts the entire grid, a question that ties directly back into the core challenge of system integration.

References

Amini, F., Generative-energy-scenarios: Source code for scenario generation and reduction. (2025). https://github.com/fateme-amini/Generative-Energy-Scenarios

Barsha, F. L., & Eberle, W. (2025). An in-depth review and analysis of mode collapse in generative adversarial networks. *Machine Learning, 114*(6), 141.

Chaudhary, A., Rastogi, R., Mattoo, A., Kumar, P., Kumari, T., & Dubey, D. (2025). Generative adversarial networks (GANs). *Generative AI: Disruptive technologies for innovative applications* (pp. 29–55). Springer Nature. doi:10.1002/9781394302932.ch2.

Che, T., Zhang, R., Sohl-Dickstein, J., Larochelle, H., Paull, L., Cao, Y., & Bengio, Y. (2020). Your GAN is secretly an energy-based model and you should use discriminator driven latent sampling. *Advances in Neural Information Processing Systems, 33*, 12275–12287.

Chen, X., Duan, Y., Houthooft, R., Schulman, J., Sutskever, I., & Abbeel, P. (2016). InfoGAN: Interpretable representation learning by information maximizing generative adversarial nets. In *Advances in neural information processing systems 29 (NIPS 2016)* (pp. 2172–2180).

Chen, Y., Wang, Y., Kirschen, D., & Zhang, B. (2018). Model-free renewable scenario generation using generative adversarial networks. *IEEE Transactions on Power Systems, 33*(3), 3265–3275.

Cobbinah, M., Nunoo-Mensah, H., Adjei, P. E., Acheampong, F. A., Acquah, I., Tchao, E. T., Adu-Manu, K. S., de-la Hema, D., & Abaidoo, E. (2025). Diversity in stable GANs: A systematic review of mode collapse mitigation strategies. *Engineering Reports, 7*(6), e12967.

Danks, D., & Yau, C. (2021). Basisdevae: Interpretable simultaneous dimensionality reduction and feature-level clustering with derivative-based variational autoencoders. In *International conference on machine learning* (pp. 2410–2420).

de Heij, P. M., de Vletter, P. N. B., & van der Sluis, L. (2021). Generating electricity price forecasting scenarios using an autoregressive model. In *2021 International conference on probabilistic methods applied to power systems (PMAPS)* (pp. 1–6).

Dong, W., Chen, X., & Yang, Q. (2022). Data-driven scenario generation of renewable energy production based on controllable generative adversarial networks with interpretability. *Applied Energy, 308*, 118387.

Dumas, J., Wehenkel, A., Lanaspeze, D., Cornélusse, B., & Sutera, A. (2022). A deep generative model for probabilistic energy forecasting in power systems: Normalizing flows. *Applied Energy, 305*, 117871.

Durgadevi, M. (2021). Generative adversarial network (GAN): A general review on different variants of GAN and applications. In *2021 6th International conference on communication and electronics systems (ICCES)* (pp. 1–8).

Erisen, B. (2020). Wind turbine SCADA dataset. Retrieved September 16, 2025, from https://www.kaggle.com/datasets/berkerisen/wind-turbine-scada-dataset

Fekri, M. N., Ghosh, A. M., & Grolinger, K. (2020). Generating energy data for machine learning with recurrent generative adversarial networks. *Energies, 13*(1), 130.

Goodfellow, I., Pouget-Abadie, J., Mirza, M., Xu, B., Warde-Farley, D., Ozair, S., Courville, A., & Bengio, Y. (2014). Generative adversarial nets. In *Advances in neural information processing systems 27 (NIPS 2014)* (pp. 2672–2680).

Gu, L., Xu, J., Ke, D., Deng, Y., Hua, X., & Yu, Y. (2024). Short-term output scenario generation of renewable energy using transformer-Wasserstein generative adversarial nets-gradient penalty. *Sustainability, 16*(24), 10936.

Guo, X., Gao, L., Liu, X., & Yin, J. (2017). Improved deep embedded clustering with local structure preservation. In *Proceedings of the twenty-sixth international joint conference on artificial intelligence (IJCAI-17)* (pp. 1753–1759).

Guo, X., Liu, X., Zhu, E., & Yin, J. (2017). Deep convolutional embedded clustering. In *International conference on neural information processing (ICONIP)* (pp. 1–10).

He, Y. L., Li, X. Y., Ma, J. H., Lu, S., & Zhu, Q. X. (2022). A novel virtual sample generation method based on a modified conditional Wasserstein GAN to address the small sample size problem in soft sensing. *Journal of Process Control, 113*, 18–28.

Heitsch, H., & Römisch, W. (2003). Scenario reduction algorithms in stochastic programming. *Computational Optimization and Applications, 24*(2), 187–206.

Hitaj, B., Ateniese, G., & Perez-Cruz, F. (2017). Deep models under the GAN: Information leakage from collaborative deep learning. In *Proceedings of the 2017 ACM SIGSAC Conference on computer and communications security* (pp. 603–618).

Høyland, K., & Wallace, S. W. (2001). Generating scenario trees for multistage decision problems. *Management Science, 47*(2), 295–307.

Jiang, C., Mao, Y., Chai, Y., Yu, M., & Tao, S. (2018). Scenario generation for wind power using improved generative adversarial networks. *IEEE Access, 6*, 62193–62203.

Jiang, Z., Zheng, Y., Tan, H., Tang, B., Zhou, H. (2016). Variational deep embedding: An unsupervised and generative approach to clustering. arXiv preprint arXiv:1611.05148.

Kaneva, T., Valova, I., Gabrovska-Evstatieva, K., & Evstatiev, B. (2025). A data-driven approach for generating synthetic load profiles with GANs. *Applied Sciences, 15*(14), 7835.

Koshiyama, A., Firoozye, N., & Treleaven, P. (2021). Generative adversarial networks for financial trading strategies fine-tuning and combination. *Quantitative Finance, 21*(5), 797–813.

Le-Anh, T., Tran-Quoc, T., & Le, V.-H. (2023). Stochastic scenario generation method based on copula and k-means for optimal microgrid energy management. *Applied Sciences, 13*(7), 4137.

Li, S., Wu, J.-Q., Xiao, X., Chao, F., Mao, X.-Y., & Ji, R. (2021). Revisiting discriminator in GAN compression: A generator-discriminator cooperative compression scheme. *Advances in Neural Information Processing Systems, 34*, 28560–28572.

Liu, X., & Hsieh, C.-J. (2019). Rob-GAN: Generator, discriminator, and adversarial attacker. In *Proceedings of the IEEE/CVF Conference on computer vision and pattern recognition (CVPR)* (pp. 11234–11243).

Ma, X., Liu, Y., Yan, J., & Wang, H. (2023). A wgan-gp-based scenarios generation method for wind and solar power complementary study. *Energies, 16*(7), 3114.

Mukherjee, S., Asnani, H., Lin, E., & Kannan, S. (2019). ClusterGAN: Latent space clustering in generative adversarial networks. *Proceedings of the AAAI Conference on Artificial Intelligence, 33*, 4610–4617.

Nguyen, T. D., Le, T., Vu, H., & Phung, D. (2017). Dual discriminator generative adversarial nets. *Advances in Neural Information Processing Systems, 30.*

Pan, Z.-Y., Yu, W.-J., Wang, B.-L., Xie, H.-T., Sheng, V. S., Lei, J.-J., & Kwong, S. (2020). Loss functions of generative adversarial networks (GANs): Opportunities and challenges. *IEEE Transactions on Emerging Topics in Computational Intelligence, 4*(4), 500–522.

Räsänen, T., Voukantsis, D., Niska, H., Karatzas, K., & Kolehmainen, M. (2010). Data-based method for creating electricity use load profiles using large amount of customer-specific hourly measured electricity use data. *Applied Energy, 87*(11), 3538–3545.

Rellier, D., Ferreira, P.M.d.C., Fontane, F. (2024). Transformer-based models for probabilistic time series forecasting with explanatory variables. arXiv:2407.13945.

Ren, Y., Pu, J., Yang, Z., Xu, J., Li, G., Pu, X., Yu, P. S., & He, L. (2024). Deep clustering: A comprehensive survey. *IEEE Transactions On Neural Networks and Learning Systems, 36*(4), 5858–5878.

Römisch, W. (2009). Scenario reduction techniques in stochastic programming. *Stochastic algorithms: Foundations and applications* (pp. 1–14). Springer. doi:10.1007/978-3-642-04944-6_1.

Sampath, V., Maurtua, I., Aguilar Martin, J. J., & Gutierrez, A. (2021). A survey on generative adversarial networks for imbalance problems in computer vision tasks. *Journal of Big Data, 8*(1), 1–49.

Saxena, D., & Cao, J. (2021). Generative adversarial networks (GANs) challenges, solutions, and future directions. *ACM Computing Surveys (CSUR), 54*(3). doi:10.1145/3446374.

Sharma, P., Kumar, M., Sharma, H. K., & Biju, S. M. (2024). Generative adversarial networks (GANs): Introduction, taxonomy, variants, limitations, and applications. *Multimedia Tools and Applications, 83*(41), 88811–88858.

Springenberg, J. T. (2015). Unsupervised and semi-supervised learning with categorical generative adversarial networks. arXiv:1511.06390.

Tiancheng, S., Gejirifu, D., Hao, C., Wenzhang, G., Yalin, Z., & Long, Q. (2022). Scenario generation for wind power using generative adversarial networks. *Journal of Physics: Conference Series, 2320*, 012019.

Wang, C., Tindemans, S. H., Palensky, P. (2022). Generating contextual load profiles using a conditional variational autoencoder. arXiv:2209.04056.

Xie, J., Girshick, R., & Farhadi, A. (2016). Unsupervised deep embedding for clustering analysis. In *Proceedings of the 33rd International conference on machine learning: 48* (pp. 478–487).

Yilmaz, B., & Korn, R. (2022). Synthetic demand data generation for individual electricity consumers: Generative adversarial networks (GANs). *Energy and AI, 9*, 100161.

Yilmaz, B., Laudagé, C., Korn, R., & Desmettre, S. (2024). Electricity GANs: Generative adversarial networks for electricity price scenario generation. *Commodities, 3*(3), 254–280.

Yoon, J., Jarrett, D., & Van der Schaar, M. (2019). Time-series generative adversarial networks. *Advances in Neural Information Processing Systems, 32*, 1–11.

Zhang, X., Li, D., & Fu, X. (2024). A novel Wasserstein generative adversarial network for stochastic wind power output scenario generation. *IET Renewable Power Generation, 18*, 37313742.

Zheng, K., Sun, Z., Song, Y., Zhang, C., Zhang, C., Chang, F., Yang, D., & Fu, X. (2025). Stochastic scenario generation methods for uncertainty in wind and photovoltaic power outputs: A comprehensive review. *Energies, 18*(3), 503.

Zhou, T., Li, Q., Lu, H., Cheng, Q., & Zhang, X. (2023). Gan review: Models and medical image fusion applications. *Information Fusion, 91*, 134–148.

Intelligent energy management for renewable energy communities and microgrids: Models, algorithms, and practical constraints

15

Danial Zendehdel, Gianluca Ferro, Enrico De Santis and Antonello Rizzi
Department of Information Engineering, Electronics and Telecommunications, Sapienza University of Rome, Italy

Case study/activity

Objective

To conceptualize and structure a hierarchical, uncertainty-aware energy management system (EMS) framework for renewable energy communities (RECs) and microgrids (MGs). The chapter's goal is to position EMS as a central enabler for the integration of renewable sources, energy storage, and peer-to-peer energy sharing under uncertainty.

Scope

This work is a theoretical and methodological synthesis, not a site-based or numerical case study. It reviews existing EMS architectures, identifies key design layers (forecasting, optimization, control), and proposes a generalized framework that integrates reinforcement learning (RL) and blockchain-based peer-to-peer trading concepts. The focus is conceptual rather than empirical.

Audience

Academics, graduate students, and professionals researching smart grids, EMS design, or renewable integration. It is also valuable to policymakers or developers interested in future EMS structures for RECs.

Expected results and deliverables

Since no case study was conducted, results here refer to conceptual outcomes, such as:

A hierarchical EMS diagram emphasizing the tertiary (scheduling/dispatch) layer.
Mathematical formulations representing typical EMS objectives and constraints.

Stochastic Planning and Modeling for Energy Systems: Methods, Applications, and Developments.
DOI: https://doi.org/10.1016/B978-0-443-45298-7.00020-3

Integration of uncertainty modeling, state of charge (SoC)/state of health (SoH) equations, and degradation cost terms.

Comparative summary tables of EMS strategies from the literature. These are deliverables of understanding, not numerical results.

Safety considerations

No physical implementation. The "safety" aspect is reflected in operational constraints (battery SoC/SoH limits, grid balance equations) discussed in the framework.

Actions taken/workflow/tools used/simulations and analyses

Comprehensive literature review on EMS and RL under uncertainty.

Conceptual structuring of EMS layers and control hierarchies.

Development of representative mathematical formulations.

Preparation of figures and tables summarizing State-of-the-Art studies.

Challenges and solutions

Challenge: Integrating diverse control and learning paradigms into one coherent EMS framework. Solution: A hierarchical structure combining optimization-based and RL-based layers.

Challenge: Representing uncertainty rigorously without a numerical case. Solution: Using stochastic formulations and qualitative performance summaries.

Results

Conceptual results include:

A structured EMS framework adaptable to different community energy settings.

A notation and mathematical foundation for future empirical work.

Comparative analysis showing how emerging AI and blockchain tools complement traditional EMSs.

Learning and knowledge outcomes

Understanding of EMS hierarchies and uncertainty management.

Clarity on the transition from rule-based and mixed-integer linear programming (MILP) EMS to RL-based autonomous EMS.

Basis for future implementation or simulation of the proposed framework.

15.1 Introduction

The worldwide exponential growth of electricity consumption in buildings and industries is causing global warming through the emission of greenhouse gas, and this trend continues with ongoing contributions arising from unsustainable energy use

(Inter governmental Panel on Climate Change, 2023). Various policies have been enforced to mitigate climate change, while renewable energy shows potential to overcome this phenomenon. The steps toward transition to more sustainable, cost-effective, and decentralized systems are happening through the use of renewable energy sources (RES), like solar, wind, hydro, and, in some cases, energy storage in batteries and hydrogen systems (Ochoa-Correa et al., 2025). The deployment of RES has been enhanced in response to global regulations. The Renewable Energy Directive (RED II, Directive (EU) 2018/2001) encourages the adoption of renewable energy as a central goal of the Union's energy strategy (European Commission Joint Research Centre, 2024). The updated directive raises the binding target from 32% to at least a 42.5% share for renewables in EU energy usage, aiming to achieve 45% (European Environment Agency, 2025; REN21, 2021). In addition, RESs have priority grid access through regulations and low operating costs, which make them affordable. This growth helped solar photovoltaic (PV) and wind power become the most affordable sources of electricity. The statistics show that PV generation will become the first RES up to 2029 as solar PV generation increased by a record 320 TWh (up 25%) in 2023, reaching over 1600 TWh.

Despite these advances, RES are inherently intermittent and stochastic, creating challenges for a reliable energy supply. Another key enabler of sustainability and decentralization in modern power systems is the deployment of energy storage systems (ESS). Residential consumers may adopt their consumption based on the energy production; however, this could not be feasible in terms of hospitals or other similar places with work commitment (Cucchiella et al., 2016). Recent innovations in battery technologies, particularly in lithium-ion and emerging chemistries, have significantly enhanced performance, longevity, and efficiency, while also reducing costs. These advancements have made ESS more affordable and scalable, thereby facilitating their integration with RES and strengthening the flexibility, stability, and reliability of decentralized energy systems (Zakeri & Syri, 2015). ESS is the key to lowering the peak demand cost by storing excess energy generated during off-peak periods and releasing it during peak demand periods, further reducing the need to draw power from the grid (National Energy Installers, 2024). The regulation on lithium batteries (European Union, 2023) supports the integration of renewable energy sources into the grid. This helps stabilize the energy supply and enhances the reliability of renewable energy systems, making energy systems more sustainable and resilient. ESS has also provided a solution to the intermittency issue of RES, while changing the role of citizens, as one can store the energy surplus and sell it to the main grid or another household at another time.

While ESS is central to reliability, batteries alone cannot capture the full complexity of modern decentralized systems. They are best understood as a component of distributed energy resources (DERs), a broader category that includes small-scale, geographically distributed generation, storage units, and controllable loads. Unlike centralized fossil-fuel power plants, DERs offer localized flexibility and can be deployed in diverse topologies, including grid-connected, islanded, or hybrid configurations. By leveraging DERs, consumers and prosumers can not only cover their own demand but also participate in energy exchange, thus reshaping the traditional producer–consumer relationship. The aggregation and coordination of DERs naturally lead to the concept of the MG. An MG is a low or medium-voltage network that integrates distributed generation, storage, and loads, with the primary objective of optimizing renewable utilization while minimizing reliance on fossil-fuel resources (Saleem et al., 2024). MGs can operate in grid-connected mode, synchronizing with the utility network, or in islanded mode, supplying local demand independently for limited periods. Typical MG compositions include solar PV, wind turbines, biomass units, and battery storage, allowing smart grids to enhance resilience and reduce greenhouse gas emissions.

The EU has introduced the concept of the REC (Capillo et al., 2024). RECs are legally recognized entities that empower citizens, local authorities, and small enterprises to collectively generate, store, and consume renewable energy. Their purpose is not merely financial gain but the provision of environmental, economic, and social benefits to members and the wider community (European Commission Joint Research Centre, 2024). Through RECs, the role of the citizen evolves from passive consumer to active stakeholder in the energy transition, reinforcing the decentralization of modern power systems. While this chapter focuses broadly on decentralized systems, islanded MGs offer an illustrative stress-test for EMS design. Their limited interconnection and low rotational inertia make them acutely sensitive to the stochasticity of RES, highlighting the need for robust forecasting, storage coordination, and inverter-aware control. Lessons from islands, e.g., Nusa Penida and the Canary Islands, demonstrate how grid-forming capabilities and carefully scheduled battery energy storage system (BESS) can sustain frequency and voltage quality under large renewable shares (Ochoa-Correa et al., 2025).

15.2 Energy management systems

The optimization of energy flows in smart grids and RECs is carried out through specific systems called "EMS." An EMS, as described in (Javed et al., 2021), is an automated intelligent system that manages the energy flows among the network elements by controlling the production, storage, distribution, and consumption of energy. This section first outlines the primary objectives of an EMS, emphasizing how they may differ depending on the type of MG and how they often come into conflict with one another. It then introduces the hierarchical control architecture, which enables the coordination of different functions across distinct time scales. Subsequently, the most widely adopted optimization methods are examined, ranging from rule-based strategies to classical mathematical formulations and metaheuristic techniques, up to machine learning (ML) approaches, with particular emphasis on their application in complex and uncertain scenarios. The chapter concludes by addressing the main open challenges, including scalability, data quality, and privacy, as well as the role of uncertainty associated with generation and demand.

15.2.1 Energy management systems objectives

The roles of an EMS depend on the type of MG. In a grid-connected system, they typically include demand response, local self-consumption, and cost minimization while accounting for market dynamics; by contrast, in an isolated MG, the focus shifts toward unit commitment, economic dispatch, and load control management (Tasmant et al., 2025). More broadly, EMSs aim to reduce operational costs (e.g., fuel consumption and generator cycling), mitigate environmental impacts and emissions, and ensure system reliability by enhancing grid resilience, maintaining stability, and reducing supply deficiencies (Esparza et al., 2025).

Since EMSs coordinate multiple MGs components, their objectives are often heterogeneous or even conflicting. For instance, maximizing renewable energy utilization and minimizing grid dependence may conflict with limiting battery degradation caused by excessive cycling. Similar trade-offs arise in energy communities, where the self-consumption interests of individual units may not align with community-wide priorities such as maximizing benefits from energy sharing.

15.2.2 Hierarchical control in energy management systems

Given the complexity that characterizes the smart grid environment, it is evident that the various control functions within a MG are not only heterogeneous in nature but also operate across very different temporal scales. To efficiently manage all requirements in a MG, hierarchical control architectures have been developed, in which distinct levels are defined to handle tasks that must be executed within specific time frames. Communication among levels occurs through the exchange of variables, which are essential to ensure that the different tasks are fulfilled simultaneously. For instance, this approach guarantees power balance within the MG while simultaneously determining the optimal power flows required for optimal power dispatching (Yamashita et al., 2020). Hierarchical control is typically structured into three levels: Primary, secondary, and tertiary control (Hasheminasab et al., 2024). The primary level operates on very short time scales, in the order of a few seconds, with the goal of mitigating the most significant frequency deviations. The secondary level acts over a longer time horizon, gradually restoring the system to a stable equilibrium condition. Finally, the tertiary control concerns production scheduling, regulating generator power output, and defining economic decisions to minimize generation costs while simultaneously ensuring demand satisfaction and compliance with operational constraints. Fig. 15.1 illustrates the hierarchical framework along with the main tasks associated with each of the three layers. Considering a household that can generate RES through PV solar energy, the harvested solar energy is denoted $P_{PV,DC,\,i}$ of the i-th rooftop solar energy system of capacity $P_{STC,i}$in terms of the irradiance G radiated to the surface of the solar panels can be estimated through:

$$P_{PV,DC,i} = P_{STC,i} \times G \times (1 + K(T_c - T_{STC})), \qquad (15.1)$$

where T_c and T_{STC}, are the actual and standard test condition cell temperatures of the solar panel ($T_{STC} = 25$), and k is the temperature coefficient, respectively (Saleem et al., 2024). If the efficiency of the associated inverter is η_i the converted AC power from

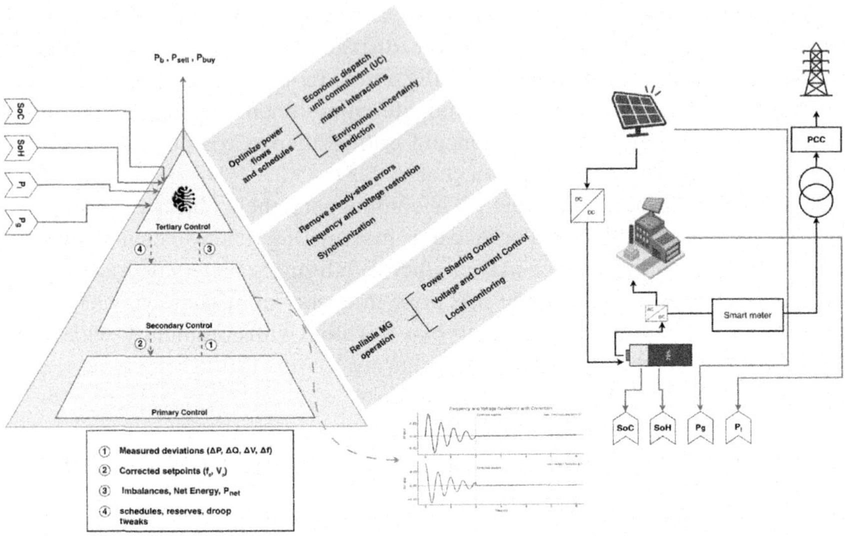

Figure 15.1 Hierarchical control architecture of a microgrid. Hierarchical control architecture of a microgrid. Primary control ensures local reliability through power sharing and voltage/current regulation. Secondary control corrects steady-state errors by restoring frequency and voltage, as illustrated in the inset plots. Tertiary control optimizes dispatch, market interactions, and scheduling under uncertainties.

the harvested solar power can be written as,

$$P_{PV,AC,i} = \eta_i P_{PV,DC,i} \quad . \tag{15.2}$$

Total harvested AC power from the rooftop solar energy systems in the MG is given by:

$$P_{PV,AC,T} = \sum_{i=1}^{N} P_{PV,AC,i} \tag{15.3}$$

Furthermore, PV energy generation is significantly influenced by meteorological conditions, such as cloud cover or sunny summer days. Consequently, numerous research studies have focused on weather forecasting, aiming to integrate supplemental information into the tertiary level of control for enhanced scheduling performance. At the tertiary layer, additional uncertainties are considered, such as day-ahead generation and demand for scheduling energy with respect to constraints of the EMS and ESS. The present work focuses on the tertiary level, as it encompasses the stochastic planning and control of DERs in MGs.

At the tertiary level, the EMS optimization problem can be expressed compactly as a stochastic program:

$$\min_{P_t^{\mathrm{grid}},P_t^{\mathrm{batt}}} \mathbb{E}_{\xi_t}\left[\sum_{t\in\mathcal{T}}\left(c_t^{\mathrm{grid}}P_t^{\mathrm{grid}} + c^{\mathrm{deg}}\Delta\mathrm{SoH}_t + c^{\mathrm{pen}}s_t\right)\right] \tag{15.4}$$

$$P_t^{\mathrm{grid}} = P_t^{\mathrm{load}}(\xi_t) - P_t^{\mathrm{PV}}(\xi_t) - P_t^{\mathrm{batt}} - s_t$$

$$\mathrm{SoC}_{t+1} = \mathrm{SoC}_t + \frac{\eta_{ch}P_t^{batt,+} - P_t^{batt,-}/\eta_{dis}}{E_{nom}}\Delta t$$

$$\mathrm{SoC}_{\mathrm{min}} \le \mathrm{SoC}_t \le \mathrm{SoC}_{\mathrm{max}}$$

$$-P_{\mathrm{max}}^{\mathrm{batt,dis}} \le P_t^{\mathrm{batt}} \le P_{\mathrm{max}}^{\mathrm{batt,ch}}$$

$$\Delta\mathrm{SoH}_t = f_{\mathrm{deg}}\left(P_t^{\mathrm{batt}}, T_t, \mathrm{SoC}_t\right)$$

In Eq. 15.4, $\mathbb{E}_{\xi_t}[\cdot]$ denotes the expectation operator over the random variable ξ_t, which represents the stochastic disturbances affecting renewable generation $P_t^{\mathrm{PV}}(\xi_t)$ and load demand $P_t^{\mathrm{load}}(\xi_t)$. The decision variables are the grid power exchangeP_t^{grid} and the battery power P_t^{batt}, both determined at each time step $t \in \tau$. The objective function minimizes the expected total cost, which includes the energy cost from or to the grid, $c_t^{\mathrm{grid}}P_t^{\mathrm{grid}}$, and the degradation cost of the battery, proportional to the incremental loss of SoH, $\Delta\mathrm{SoH}_t$ weighted by the unit cost c^{deg}. The penalty term $c^{\mathrm{pen}}s_t$ that accounts for constraint relaxation through the non-negative slack variable s_t. The degradation model $f_{\mathrm{deg}}(\bullet)$ quantifies the incremental aging of the battery as a function of power throughput, temperature, and SoC.

15.2.3 Energy management systems optimization methods

Multiple optimization techniques have been investigated in the scientific literature on smart grids. In the context of smart grids, EMS plays a pivotal role in coordinating DERs, ensuring operational reliability, and optimizing economic and environmental objectives. Over the years, a wide variety of optimization methods have been developed to address the challenges posed by the intrinsic variability of renewable generation, the stochastic nature of load demand, and the technological constraints of ESS. These approaches can be broadly classified into different families, each characterized by specific strengths and limitations. At the most fundamental level, rule-based strategies provide simple and computationally efficient solutions, although often at the expense of optimality. More sophisticated alternatives rely on classical mathematical programming methods, such as linear and nonlinear programming, which enable the rigorous formulation of EMS optimization problems but suffer from scalability and efficiency issues in nonlinear or highly complex scenarios. To overcome such drawbacks, metaheuristic techniques have been extensively explored, exploiting stochastic

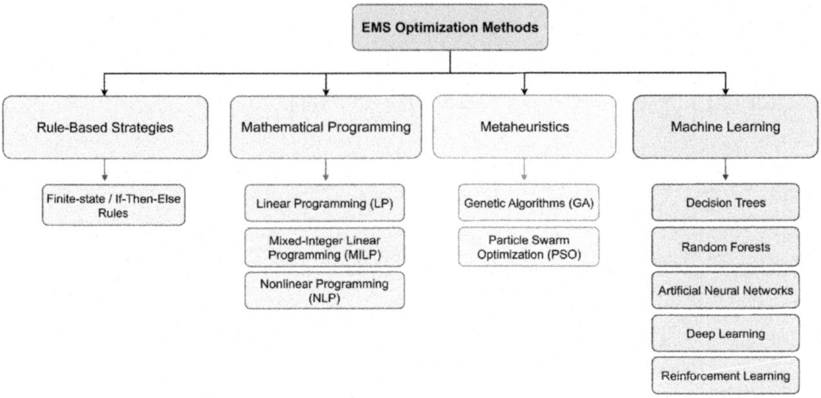

Figure 15.2 Fundamental EMS optimization methods. Fundamental EMS optimization methods. *EMS*, Energy management system.

search processes inspired by natural phenomena to provide near-optimal solutions even in nonconvex and multiobjective contexts.

More recently, the rise of large-scale data availability and advanced computational resources has fostered the adoption of machine learning (ML)-based methods, which offer new paradigms for learning decision-making strategies directly from data. Within this category, a variety of models have been applied, ranging from decision trees and ensemble methods to artificial neural networks (ANNs), deep learning (DL) architectures, and RL frameworks. Fig. 15.2 illustrates the main optimization methods described in this work.

To provide a comparative overview of recent applications, Table 15.1 summarizes representative EMS case studies reported in the literature, highlighting their optimization strategies and performance metrics. The following subsections provide an overview of these optimization methods, highlighting their operational principles, applications in MGs, and relative advantages and limitations.

15.2.3.1 Rule-based approaches

Basic control strategies, often referred to as rule-based methods, operate without explicitly optimizing an objective function. Instead, the EMS behavior is determined by predefined rules and constraints (typically implemented using if-then-else logic) based on the current system state. For example, (Restrepo et al., 2021) proposed a rule-based energy management system (RBEMS) for a Canadian MG, structured around a finite-state machine to coordinate DERs. Their work compared this approach with an optimization-based energy management system (OBEMS) formulated as a linear programming unit commitment model. The RBEMS required less computational effort and was easier to implement; however, it delivered sub-optimal resource allocation and resulted in higher long-term operational costs compared with the OBEMS.

Table 15.1 Summary of selected EMS case studies illustrating different optimization paradigms and key performance metrics.

Case study description	EMS strategy	Main assumptions and performance results
Isolated hybrid microgrid with a 4 MW PV plant and three 2 MW diesel gensets (Azuara-Grande et al., 2021).	Rule-based, decentralized and MILP opt.	Annual dynamic simulation minimizing total marginal cost (fuel + wear + BESS degradation). Achieved 40% reduction in PV curtailment, 1.5% reduction in diesel generation, 8% higher battery throughput.
Microgrid with renewable generators (PV, wind) and 24 kWh Li-ion BESS (De Santis et al., 2017).	HGA + FLC versus fuzzy-GA.	Discrete 15-min timestep without reactive power control. Maximized yearly accounting profit, increasing by 67% and reducing costs by 25%. Improved interpretability and computation time.
A microgrid including RES, gas microturbine, and BESS under grid-connected operation (Li & Liu, 2024).	CLS-PSO	CLS-PSO optimization reduced operational cost by 24% and energy conversion cost by 31%. Combined PV + storage achieved an overall cost decrease of 29% relative to baseline.
Electrical grid with wind turbines, PV plant, and large-scale storage (Ruddick et al., 2025).	TreeC (decision tree EMS).	CMA-ES simulation-based optimization without forecasting. Controlled real/reactive power for all assets. Achieved performance comparable to MPC with perfect forecasts, outperforming RL-based EMS (SAC).
Grid-connected hybrid system (PV, wind, fuel cell, microturbine, BESS) (Vaish et al., 2024).	Ensemble learning	Minimize total system cost and battery charging cost while sizing seasonal battery capacity. Generation cost reduced 15% versus PSO and 10% versus ANN; charging cost reduced 25%. 12% smaller optimal capacity needed; RES curtailment reduction 10%.

(*continued on next page*)

Table 15.1 Summary of selected EMS case studies illustrating different optimization paradigms and key performance metrics—cont'd

Case study description	EMS strategy	Main assumptions and performance results
Laboratory-scale hybrid AC/DC microgrid integrating PV, wind turbine, and battery energy storage system. Experiments conducted in both grid-connected and stand-alone modes (Kang et al., 2021).	ANN-based.	SOC, distributed generation, and grid power used as ANN inputs. Improved power-management efficiency and reduced complexity versus conventional EMS.
Islanded AC microgrid with hybrid PV/BESS DGs (Srikanth et al., 2025).	CNN-based via E-COA.	CNN maps irradiance and voltage to control frequency. Reduced MAE, MSE, and RMSE by up to 80% versus baseline CNN and hybrid metaheuristics. Faster convergence and lower uncertainty.

ANN, artificial neural network; *BESS*, battery energy storage system; *CNN*, convolutional neural network; *COA*, coati optimization; *DDPG*, deep deterministic policy gradient; *EMS*, energy management system; *MILP*, mixed integer linear programming; *MPC*, model predictive control; *PSO*, particle swarm optimization; *PV*, photovoltaic; *RES*, renewable energy source; HGA, hybrid genetic algorithm; FLC, fuzzy logic controller; GA, genetic algorithm; CLS, constrained least squares; CMA-ES, covariance matrix adaptation evolution strategy; SAC, soft actor-critic; MAE, mean absolute error; MSE, mean squared error; RMSE, root mean squared error; DGs, distributed generators.

15.2.3.2 Classical methods

Classical approaches, currently employed mostly for benchmarking purposes, consist of formulating mathematical problems that incorporate the constraints of the energy system, whose optimal solutions are obtained through dedicated solvers. Depending on the nature of the problem, different variants of these methods have been developed. Among them, MILP stands out, as it involves both integer and continuous variables. A MILP-based EMS for the management of a BESS was developed (Azuara-Grande et al., 2021), where optimization was carried out using forecasted data on energy production and consumption, while considering multiple objectives, including operational generation costs and battery degradation. In the numerous experiments reported in the literature, systems based on linear programming have proven effective for linear problems, ensuring the existence of optimal solutions. However, such systems tend to be computationally demanding and unsuitable for problems that are inherently nonlinear. Although certain models attempt to overcome the limitations of linearity, such as those relying on nonlinear programming, issues of low efficiency and limited capability in handling complex dynamics persist. Indeed, nonlinear programming is highly dependent on initial conditions and may converge to local minima, often requiring multiple runs to obtain satisfactory solutions, as highlighted in (Jamii et al., 2022).

15.2.3.3 *Metaheuristic techniques*

Within the domain of optimization-based methodologies, an important branch in energy management is represented by Metaheuristic methods, which integrate fundamental heuristic techniques to enable a more effective exploration of the search space. Specifically, a metaheuristic is defined as an iterative process that guides and modifies heuristic procedures in order to facilitate the generation of optimal or near-optimal (Voß et al., 2012). As the term suggests, these methodologies rely on heuristic tools for solution search, complemented by higher-level strategies that support the pursuit of optimal outcomes (Rezk et al., 2023). Metaheuristic methods have been extensively applied in smart grid contexts, particularly for energy flow optimization, scheduling, and planning problems (De Santis et al., 2013, 2017a).

These models were developed to overcome the limitations of classical approaches based on mathematical programming, such as poor performance in nonconvex problems and the inability to effectively address multiobjective formulations. By relying on stochastic search, metaheuristic models mitigate the risk of converging to local minima, since the generation of multiple solutions enhances exploration capabilities. Among the main metaheuristic search methods, particular relevance is attributed to those belonging to the fields of evolutionary computation and swarm intelligence (SI). Evolutionary computation models include methods that simulate the evolution of population members, each representing a potential solution to the optimization problem. Within this field, genetic algorithms stand out, as they mimic natural selection and have proven successful in tackling problems with complex objective functions (Capillo et al., 2023). For instance, in Golshannavaz et al. (2016), GAs was employed to solve the stochastic scheduling problem of energy and reserves in reconfigurable MGs, enabling the management of uncertainties associated with load and renewable generation while achieving a reduction in daily operational costs.

SI methods are inspired by the dynamics of self-organized groups of biological organisms that interact both with each other and with their surrounding environment. Interest in such behaviors has led to the emulation of natural interactions observed in various species, including dragonfly swarming, bacterial growth, bird flocking, and ant colonies. Among the most prominent SI approaches is particle swarm optimization (PSO), which draws inspiration from bird flocking behavior. PSO is modeled after the dynamics of bird flocks, where the group follows the leader positioned closest to the food source. This behavior is translated into algorithms for solving optimization problems, where the flock of birds is represented by a swarm of particles, and each particle corresponds to a potential solution. The search space is explored by the swarm, which collectively identifies the optimal solution to the problem (Shami et al., 2022). Compared to genetic algorithms, PSO does not require specific encodings, which makes its implementation more straightforward. Furthermore, the effectiveness and speed of such methods make them particularly suitable for contexts where maintaining low computational costs is fundamental. An exemplary case of PSO effectiveness is reported in (Li & Liu, 2024), where the algorithm is applied to optimize the energy management of a MG comprising PV generation, a battery storage system, and a gas microturbine. In this context, PSO, enhanced with chaotic local search, demonstrated a remarkable ability to avoid premature convergence to local minima and to identify globally optimal solutions. The experimental results highlighted a reduction in daily operational costs and primary energy conversion expenses. Metaheuristic methods, however, present several drawbacks, in that the large number of objective function evaluations makes them computationally more demanding than classical mathematical approaches, and parameter tuning can be challenging, potentially leading to a drastic increase in convergence time (Papadimitrakis et al., 2021).

15.2.3.4 Machine learning methods

As the complexity of the optimization problem increases, several challenges may arise, including the presence of nonlinearities, multiobjective functions, and extremely large search spaces. In this context, ML methods become pivotal, as they enable the management of the intrinsic variability of environments such as smart grids through the exploitation of large datasets. An algorithm is defined as ML-based when its operation relies exclusively on data, without an explicit solution to the problem at hand; instead, various patterns can be identified. Unlike a design problem, in a ML problem, there is no exact mapping between input and output, meaning that the algorithm is not explicitly programmable (Dahrouj et al., 2021). Consequently, ML algorithms are applied to extract patterns from datasets by employing mathematical, statistical, optimization, and knowledge discovery techniques (Azevedo et al., 2024). The vast majority of ML approaches can be categorized into three main areas: (1) Supervised learning, (2) unsupervised learning, and (3) RL. In supervised learning methods, the objective is to estimate the relationship between input data and target variables known a priori (ground truth). During training, the model generates predictions that are compared against the targets, and the discrepancy between the two is quantified by a loss function. Through optimization algorithms, the model parameters are updated to minimize this loss and thereby improve predictive performance. In unsupervised learning methods, by contrast, no target variables are available, in that the algorithm seeks to uncover hidden structures within the data, such as recurring patterns or relationships among samples, which may lead to clustering or dimensionality reduction. A RL problem is based on mapping situations (states) to actions with the goal of maximizing a numerical reward. The agent must discover the most effective actions through trial and error. The outcome of an action affects not only the immediate reward but also future rewards. The task of RL is therefore to solve a sequential decision-making problem, which entails selecting a consistent sequence of actions (policy). In RL, unlike in supervised learning, the agent must learn solely from its own experience, since in most interactive problems it is not possible to provide data that encompasses all possible situations, nor labels indicating correct or incorrect behaviors. Furthermore, in contrast to unsupervised learning, the objective is not to uncover hidden structures, but rather to maximize a reward signal (Sutton et al., 1998). Despite the numerous applications and the wide range of algorithms validated in the scientific literature, ML methods also present inherent limitations and challenges. Chief among these is the strong dependence on large, high-quality datasets, which are essential for enabling ML agents to cope with the significant variability of energy environments. Equally critical are the challenges posed by the high computational costs associated with complex models such as deep neural networks (DNN), along with the substantial training and inference times, which may prove particularly demanding in real-time contexts such as MGs.

15.2.4 Machine learning for energy management in microgrids

Given the wide range of applications of ML-based EMS in the field of MGs, the following outlines the current trends related to the development of ML methods for energy management.

1. Decision Trees and Random Forests (RFs). A decision tree is a hierarchical structure that iteratively splits input data into homogeneous subsets by making binary decisions at each node based on a splitting function (Breiman et al., 2017). Decision trees are known for their interpretability, as the sequence of decisions leading to a prediction can be easily understood and visualized. This makes them a highly useful tool in contexts where justifying the model's decisions is important. In Ruddick et al. (2025), the authors propose TreeC, which leverages a metaheuristic optimization algorithm (CMA-ES) to directly construct a decision tree that dictates EMS actions based on historical data. The approach achieves, in the management of a simulated power grid, performance comparable to that of an MPC with perfect forecasts and advanced RL methods. Furthermore, in the case of a residential heating system with a heat pump, it delivers results similar to MPC and significantly superior to RL, thereby demonstrating the feasibility of developing interpretable systems without compromising efficiency. However, decision trees can suffer from overfitting when constructed without restrictions on depth or minimum leaf size. To mitigate this problem, techniques such as pruning are applied, which remove branches of the tree that do not significantly contribute to the model's predictive ability. RF is an ensemble learning method that employs a large number of decision trees. In this approach, the process begins by defining the number of trees to be included in the forest. For each individual tree, a model is constructed using decision trees. The available features are then subjected to random permutations, which are evaluated and compared, and the outcomes of all possible prediction combinations are subsequently provided (Forootan et al., 2022). In Vaish et al. (2024), RF has been applied to optimize DERs within a MG, focusing on seasonal battery sizing and cost minimization. Results showed that RF outperforms PSO and ANNs in reducing total system and battery charging costs while ensuring optimal operation.
2. ANNs. ANNs are computational models inspired by the structure of the biological nervous system. They consist of interconnected processing units, called neurons, which work collectively to address complex problems (Abdolrasol et al., 2021). Each neuron typically applies a weighted sum of its inputs followed by a nonlinear activation function, allowing the network to approximate complex nonlinear mappings between input and output spaces (Hornik et al., 1989). ANNs are particularly effective in situations were identifying patterns or extracting trends is challenging. An energy management application of ANN is presented in Kang et al. (2021), where the authors propose a centralized control strategy for hybrid AC/DC MGs in which the interlinking converter acts as the main controller and an ANN, trained on data from distributed generation, load demand, SoC, and grid power exchange, selects the optimal operating mode with low error. This model simplified EMS operation and enabled efficient management of charging, discharging, grid interaction, and islanded conditions.
3. DL Models. DL is a subfield of ML based on ANNs with multiple layers, which enables the automatic learning of hierarchical feature representations from Goodfellow et al. (2016).

Recent studies highlight how the increasing penetration of renewable sources, such as wind and solar, together with the active involvement of consumers, make smart grids more uncertain and complex to manage. Traditional methods, mainly based on physical modeling and numerical analysis, often struggle to cope with uncertainty and partial observability, limiting their applicability in future grid scenarios. At the same time, the deployment of monitoring infrastructures such as advanced metering infrastructures and wide area monitoring systems generates massive amounts of data, enabling the use of data-driven approaches. In this context, DL has been recognized as one of the keys enabling technologies for the future development and success of smart grids (Zhang et al., 2018). Recent research has further demonstrated the potential of DL combined with optimization strategies. For instance, in (Srikanth et al., 2025) is proposed a convolutional neural network (CNN)-based EMS for PV/BESS MGs, where the CNN weights are tuned using an enhanced coati optimization (E-COA) algorithm. The system enables near real-time state-of-charge balancing and significantly improves forecasting accuracy. Simulation results showed that the E-COA-CNN model achieved lower prediction errors compared to conventional methods, highlighting its superior performance in managing hybrid PV-battery systems efficiently.

4 RL

Modern power systems are complex in nature and include many different uncertainties, which make it hard for a linear optimization algorithm to generalize. Model-free RL algorithms are able to mitigate these issues by sampling instead of depending on a full mathematical model (Li et al., 2020). The main difference between RL and other optimization methods is the fact that RL methods do not require a collection and labeled dataset, which makes them a suitable approach for real-world applications. RL and deep reinforcement learning (DRL) methodologies are grounded in the framework of the Markov decision process (MDP) or its variants, like the partially observable MDP. An MDP is described by the tuple (S, A, P, R, γ). Here, S signifies the set of all potential states within the environment, while A represents the available actions the agent can undertake. The transition probabilities are defined by the distribution $P(s'|s, a)$, indicating the chance of transitioning from state $s \in S$ to state $s' \in S$upon executing action $a \in A$. The reward function $R : S \times A \rightarrow \mathbb{R}$ provides immediate scalar feedback for each state-action pair. The discount factor $\gamma \in [0, 1]$ signifies how much the agent values immediate rewards over future ones, with higher values indicating a greater focus on long-term gains (Odonkor & Lewis, 2019; Sutton et al., 1998). Moreover, the agent's actions are guided by the policy $\pi(a|s)$, which defines the likelihood of executing an action a given the state s. The aim of RL is to find an optimal policy π^*that enhances the cumulative expected reward over time. Notably, the Q-function is a tool to assess the policy.

$$Q^{\pi}(s, a) = \mathbf{E}_{\pi}\left[\sum_{t=0}^{\infty} \gamma^{t} r(s_t, a_t) | s_0 = s, a_0 = a\right] \tag{15.5}$$

Thus, Eq. 15.5 signifies the expected return when beginning at state s, performing action a, and subsequently adhering to policy π. Following the success of deep Q-network (DQN) (Mnih et al., 2013), different researchers considered DQN as an approach to employ EMS. (Salaorni et al., 2025) investigated the advantage of DQN in a weather-aware framework, in which they investigated the impact of day-ahead weather information on dispatching DERs. Kahraman & Yang, 2022 demonstrated that while DQN can effectively manage the battery's SoC within operational limits, its performance in minimizing operational costs is limited due to the discretization of the action space. The DQN has several limitations, in that it cannot handle continuous action spaces and experiences instability as dimensionality increases. DQN assigns scores to actions and selects the highest-scoring action, problematic in continuous or large action spaces. It also has memory inefficiencies and long training times due to the experience replay technique and is restricted to off-policy methods like Q-iterations (Mnih et al., 2016). On the other hand, deep deterministic policy gradient (DDPG) and twin delayed deep deterministic policy gradient (TD3) exhibit a greater ability to leverage electricity price fluctuations and optimize battery usage to reduce costs. The approach proposed in (Odonkor & Lewis, 2019) employed the DDPG algorithm to optimize the charge and discharge policies of a shared battery system within a building cluster by removing the need for action discretization and allowing for continuous decision-making. Although their design focuses on minimizing the demand in peak hours, this work can be enhanced by considering battery degradation; to this end Table 15.2, summarizes some of the major works based on RL.

15.2.5 Challenges in energy management systems

EMSs in MGs face a series of inherent challenges and limitations that complicate their implementation and optimization. The uncertainty and variability of renewable energy RES and load demand constitute a fundamental difficulty. Inaccurate forecasting of these stochastic factors, combined with energy price fluctuations, can result in operational inefficiencies, increased costs, and system instability (Esparza et al., 2025).

Scalability represents another critical challenge. In fact, optimization techniques become increasingly difficult to apply as MGs expand in size and complexity (Ojo et al., 2025). Centralized control systems, while capable of providing globally optimal solutions, encounter issues with stability and scalability, as processing large volumes of data at a single point becomes inefficient. Similarly, distributed EMS may face computational complexity and communication overhead as the systems grow. Data scarcity and quality also represent a significant concern. Artificial intelligence and ML algorithms require accurate and high-quality data to ensure reliable forecasting and optimal management. The lack or ambiguity in data is a major issue, as historical records may not adequately capture future scenarios or uncommon events. Moreover, optimized EMS are often insufficiently validated under diverse climatic conditions and real-world datasets. The inconsistency of experimental settings across studies further complicates the selection of effective ML algorithms for generalizable applications, underscoring the need for common evaluation criteria and data sharing (Elsisi et al., 2023).

Another critical issue in EMSs concerns data privacy, particularly due to the growing reliance on digital communication networks and the massive collection of information from smart meters. These devices can disclose personal details regarding users' energy consumption patterns, making robust privacy protection indispensable. Regulatory

Table 15.2 Summary of RL and DRL methodologies for EMS in microgrids and smart homes.

Method/algorithm	States	Actions	Agent type and scope	Objective(s)
Single-agent methods				
Prioritized experience replay SAC, (Bao & Xu, 2023).	Operation cost, RES generation, load, SoC, market prices.	Control of DGs and ESS; continuous actions.	Single-agent, single MG.	Minimize cost, improve stability and economy.
DDPG with adapted replay buffer and safety layer (da Silva André et al., 2022).	Battery SoE, grid conditions, lookahead constraints.	Charge/discharge decisions (continuous).	Single-agent, single BESS in active distribution grid.	Cost minimization while satisfying grid and battery constraints.
Imitation learning with MILP expert; DNN policy approximation (Gao et al., 2021).	MG conditions, ESS status, market prices.	ESS scheduling (continuous, multiaction).	Single-agent, multiple ESSs in a MG.	Economic operation, minimize cost, handle intermittency.
Comparison of DQN, DDPG, TD3 for HEMS (Kahraman & Yang, 2022).	Day-ahead prices, net load, battery SoC.	Grid/storage scheduling; DDPG/TD3 (continuous), DQN (discrete).	Single-agent, single HEMS (prosumer).	Minimize cost, keep SoC within limits.
Double deep Q-learning (DDQL) (Bui et al., 2019).	Not explicitly detailed (ESS state).	ESS management.	Single-agent, single MG.	Manage ESS under uncertainty.
Model-free RL with DDPG; BESS scheduling as MDP (Selim et al., 2023).	BESS SoC, grid tariff price.	Battery/grid set-points (continuous, 5-min intervals).	Single-agent, residential BESS.	Bill minimization, extend battery life, optimize energy use.

(continued on next page)

MultiAgent Methods				
SAC with asymmetric actor-critic, (Guiducci et al., 2023).	Real-time data, shared consumption and generation.	BESS charging/discharging ([-1,1]).	Centralized REC control (3–9 entities).	Maximize social welfare, incentivize self-consumption.
Entropy-based collective multiagent DRL (Yang et al., 2019).	Not detailed (EVs and ESS scheduling).	EV and ESS scheduling.	Multiagent, large-scale households.	Reduce operating cost, peak load.
MultiAgent PPO (MAPPO); centralized training, independent execution (Lee et al., 2020).	Not detailed (demand response).	Appliance scheduling.	Multiagent, residential MG.	Demand response.
Multiagent DRL (Cao et al., 2021).	PV inverter states.	Coordinated PV inverter control.	Multiagent, distribution network.	Voltage regulation.

BESS, battery energy storage system; *DDPG*, deep deterministic policy gradient; *DNN*, deep neural networks; *DRL*, deep reinforcement learning; *DQN*, deep Q-network; *EMS*, energy management system; *ESS*, energy storage system; *EV*, electric vehicle; *HEMS*, home energy management system; *MG*, microgrid; *MILP*, mixed integer linear programming; *REC*, renewable energy community; *RL*, reinforcement learning; *RES*, renewable energy source; *SoC*, state of charge; *SoE*, state of energy; *TD3*, twin delayed deep deterministic policy gradient.

frameworks play a crucial role, as data privacy legislation can strongly influence the design and operational strategies of EMS (Lee et al., 2024). To address this challenge, federated learning (FL) has emerged as a promising ML approach specifically designed to handle data privacy concerns in EMS. FL is a paradigm that enables collaborative and distributed training of ML models across multiple nodes, such as MGs or substations, without requiring the exchange of raw and sensitive data with a central (Joshi et al., 2023). Instead of sharing the entire consumption history or real-time measurements, FL involves transmitting only model parameters or encrypted updates to a central aggregation server, thereby contributing to the preservation of both privacy and data security.

15.3 Handling uncertainty in microgrids

Uncertainty is a critical factor to consider in smart grids, as it can undermine both network stability and resilience. The main sources of uncertainty can be categorized as environmental, economic, and technical (Pourbehzadi et al., 2019).

Environmental sources of uncertainty are primarily associated with weather conditions, which drive the stochastic behavior of renewable energy generation, as in the case of solar and wind power. Moreover, regardless of the nature of the MG (residential or industrial), electricity consumption by MG users represents one of the major sources of (Cabrera-Tobar et al., 2022). Load profiles are characterized by temporal variability, challenges in achieving accurate forecasts, and a strong dependence on external factors such as weather conditions, consumption habits, and production activities. Hence, it is fundamental to implement reliable forecasting algorithms for both production and consumption in order to ensure the optimal scheduling of DERs. Among technical factors, key subjects are the occurrence of faults in energy infrastructures (Ma et al., 2022) and the aging of ESS. ESSs introduce uncertainty into the MG environment, since the State of Chaict and strongly influence power flow management in MGs. Accurately estimating the SoC and SOH remains a challenging task, and recent studies emphasize advancements in real-time estimation methods of such variables. In the subsequent sections, the main estimation and forecasting methods of key variables required for the stochastic planning of energy resources in MGs under uncertainty are examined in detail.

15.3.1 Load and production forecasting

Forecasting of load and energy generation in MGs is of paramount importance, as the outcome of such predictions represents a key element for an EMS to effectively carry out tasks such as demand response, unit commitment, and economic dispatching (Verdone et al., 2025). Grid operators also require highly accurate forecasts to manage the energy balance and cope with unexpected load peaks. From the perspective of energy utilities and retailers, demand forecasting is fundamental in planning power generation and supporting energy trading with the main grid. An estimation of future energy demand and generation also provides highly valuable insights for the management of DERs within MGs and their efficient utilization. This includes the intelligent use of storage systems to mitigate degradation, the effective exploitation of renewable energy production, and the appropriate siting and sizing of DERs during the design phase of MGs. Forecasting in this context is closely related to the task of anomaly detection in MGs, since in both cases the objective is to provide future information on energy usage to enable real-time diagnostics of power systems.

Renewable sources such as PV panels and wind turbines exhibit high variability in energy production, as it depends on several meteorological and environmental factors, including solar irradiation, temperature, humidity, and wind speed. On the consumer side, energy usage behavior also depends on multiple factors of both environmental and social nature. Seasonal patterns at different scales can be observed in such energy series, for instance, due to climatic variations as well as holidays and events that shape social life in urban areas. Although tools are available to assist in detecting these patterns, the high heterogeneity in consumption behaviors and energy usage characterizing MGs in residential and industrial contexts still makes forecasting a challenging task. For this purpose, weather forecasts often represent highly valuable information to be integrated into predictive models; however, the frequent inaccuracy of such data further complicates the forecasting task. To address these challenges, tools such as smart meters have been introduced in smart grids, enabling the collection of detailed data on local networks. As a result, in recent years, it has become possible to build extensive datasets for training ML and DL models, alongside the development of advanced data preprocessing and feature extraction techniques. Furthermore, the spread of IoT technologies has provided a significant boost to communication among local network nodes, allowing the aggregation of MG data and delivering real-time feedback for model training (Tasmant et al., 2025).

The following discussion will provide an overview of current methodologies applied in the field of energy forecasting, along with the factors that motivate the selection of specific approaches and techniques. After describing the main models, the discussion will focus on the metrics used for forecast evaluation, as well as on the current challenges and limitations of the presented algorithms.

15.3.1.1 Handling time series data

The recent acceleration in the development of real-time meters and sensing technologies for smart grids has laid the foundation for the construction of extensive datasets, whose informational content can be effectively exploited for load and generation forecasting. For this reason, advanced techniques of data preprocessing, feature selection, and extraction are essential to ensure accurate and reliable predictions. Among the most prominent preprocessing techniques for forecasting are data decomposition methods (Devaraj et al., 2021).

Time series such as electrical load and PV generation are, in most cases, characterized by nonstationarity and heteroscedasticity, due to statistical properties such as medium and long-term trends as well as seasonality at multiple time scales. A widely employed tool to address these issues is empirical mode decomposition (EMD), an adaptive, data-driven technique designed for the analysis of nonstationary and nonlinear signals. Its distinct feature lies in not relying on predetermined basis functions, as in the case of Fourier or wavelet transforms, but rather in extracting elementary oscillatory components directly from the signal (Rilling et al., 2003). In Yin et al. (2024), EMD was introduced to decompose high-frequency sub-hourly load series into stationary subsequences, thereby simplifying the modeling of long-range correlations. When integrated into the hybrid EMDHM model, EMD enabled the isolation of trends, periodicities, and fluctuations, yielding a reduction in forecasting RMSE of 26.8%.

Recent advances have highlighted the importance of multiseasonal trend decomposition as a preliminary step in forecasting models. In this context, Bandara et al. (2025) introduced the multiple seasonal-trend decomposition using loess (MSTL) algorithm, an extension of seasonal-trend decomposition using loess (STL) that enables the decomposition of time series characterized by multiple seasonalities. The approach iteratively applies the STL procedure to each identified seasonal cycle, separately estimating the seasonal components, the trend, and the residual. The authors evaluate MSTL on simulated series (with different combinations of deterministic and stochastic seasonalities) as well as on a real hourly electricity demand series from the state of Victoria (Australia), comparing it with State-of-the-Art methods such as STR, TBATS, and Prophet. The results demonstrate that MSTL provides competitive and often superior estimates, particularly in capturing seasonal and trend components, while requiring substantially lower computational cost compared to other approaches, thereby proving its effectiveness and scalability.

Historical data on consumption and generation alone may be insufficient to address the complexity of energy management; many studies indeed showed the importance of exploiting exogenous information such as weather forecasts and additional data collected from sensors within the MGs. In this regard, the work in Luo et al. (2024) proposes a short-term load forecasting method based on a stacking integration algorithm, which combines a CNN with a BiLSTM-Attention mechanism and XGBoost. A central aspect of the approach is the use of weather forecasts. Specifically, maximum temperature, minimum temperature, and type of weather conditions are included as input variables together with the day type. Since meteorological information is not directly interpretable by the machine, the authors transform it into numerical form, for example, by representing the weather condition as a vector distinguishing between sunny and cloudy days. Experimental results show that the integration of weather forecasts significantly enhances prediction accuracy. Furthermore, the sensitivity analysis on temperature confirms that the model maintains robust performance under moderate variations, while more extreme changes may still negatively affect forecasting precision.

15.3.1.2 Selecting the appropriate model

Selecting a model for energy forecasting tasks is not a straightforward procedure, as no universally standardized options exist that are suitable for the wide range of scenarios encountered in environments. Several factors influence the choice of one model over

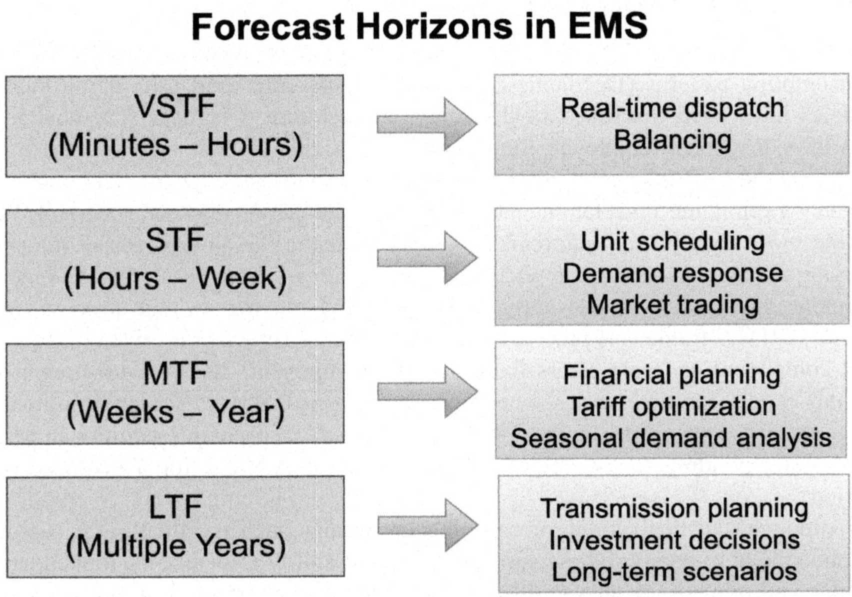

Figure 15.3 Forecast horizons. Forecast horizons and associated smart grid tasks.

another; among the most relevant are the type of data to be processed, the available computational resources, and the forecasting horizon (Wazirali et al., 2023).

Depending on the time horizon, forecasting tasks can be classified into very short-term (VSTF), short-term (STF), medium-term (MTF), and long-term (LTF) (Chandrasekaran & Paramasivan, 2025). VSTF spans from minutes to hours and is carried out to support decision-making for dispatching schedules and balancing generation and demand. STF covers the interval from one hour to one week and is critical for tasks such as demand response, unit dispatching, and energy trading. MTF refers to forecasts up to one year ahead and is applied in financial planning, tariff formulation, and seasonal demand forecasting. Finally, LTF concerns predictions over a time span of several years and is employed for planning future transmission networks and capital investments. Fig. 15.3 shows a summary of forecast horizons and corresponding typical tasks.

Furthermore, it is essential to account for the varying performance that a forecasting model may achieve across different time horizons in order to select the most robust and accurate model for the required task. Since many short-term and ultra-short-term applications demand real-time predictions, inference time becomes a crucial factor in model selection. For long-term forecasts, on the other hand, it is important that the chosen methodology is robust against overfitting and reliable in terms of generalizing long-term trends (Sangrody et al., 2018).

15.3.1.3 *Methodologies for load forecasting*

Given the wide range of applications and the complex dynamics that characterize consumption patterns (De Santis et al., 2017), numerous load forecasting methodologies have been discussed in the scientific literature. Classical statistical methods have long been employed for load prediction due to their relative ease of implementation and interpretability. Among these, autoregressive models stand out, as they exploit the correlation between future and past values of a variable. One of the most widely used autoregressive approaches is the autoregressive integrated moving average (ARIMA), applicable to nonstationary time series; it combines an autoregressive process (AR), integration (I), and a moving average process (MA). In the ARIMA framework, the current value depends on past observations from the AR component and past errors from the MA component. A well-known extension of this model for time series with pronounced seasonality is seasonal autoregressive integrated moving average (SARIMA), which incorporates additional parameters corresponding to the seasonal counterparts of ARIMA for a given seasonal period.

Although statistical models present clear limitations, such as difficulties in capturing nonlinearities in complex time series and limited ability to detect interdependencies among multiple exogenous variables, several State-of-the-Art applications still employ these techniques, either in combination with data decomposition methods or within hybrid algorithms that integrate statistical models with ML or DL approaches. In this regard, the work in (Baesmat et al., 2025) introduces a novel hybrid methodology for load forecasting, termed SP-RF-ARIMA, which combines the predictive capabilities of RF and ARIMA while incorporating variable selection through the minimum redundancy maximum relevance and maximum synergy method. Tested on real-world data from Duke Energy Ohio and Kentucky, the method first forecasts consumption, then analyzes the residuals and integrates them into the final estimation, thereby improving accuracy. The results demonstrated a reduction of over 40% in both MAE and RMSE compared to other approaches. To overcome the limitations of statistical models, numerous ML algorithms have been introduced for load forecasting tasks, as they handle nonlinearities more effectively and can be trained on larger datasets. A widely adopted ML approach in this context is extreme gradient boosting (XGB), valued for its ability to process large-scale data and achieve high performance in tasks such as classification and regression. At its core, XGB relies on gradient boosted decision trees, where each model is trained by using the residual errors of the previous model as new targets (Harikrishnan et al., 2025). A successful application of XGB is presented in (Banik & Biswas, 2024), where a forecasting model is built on a stacked ensemble of RF and XGBoost, combined through logistic regression. Applied to data from Agartala, the model achieved very high accuracy, with an R^2 of 0.98 for electricity consumption and solar generation and 0.99 for wind generation, significantly outperforming both the individual algorithms and other benchmark models. Another widely applied ML methodology is support vector regression (SVR), an extension of support vector machines (SVM) to regression problems, where the goal is to estimate a function that approximates observed data within a certain tolerance, balancing prediction accuracy with model complexity. Introduced by Drucker et al. (1996), this technique enables the construction of both linear and nonlinear models through kernel functions, demonstrating strong generalization capabilities even in high-dimensional spaces. A recent application of SVR is presented in Jia et al. (2025), where a kernel-free SVR model is proposed based on a reduced quadratic surface, enhanced with variable and historical data weighting techniques. The aim is to improve the accuracy and robustness of load forecasting while simultaneously reducing computational costs. Tests conducted on benchmark datasets and real-world data, including GEFCOM2012 and ISO New England, showed superior performance compared to widely used statistical methods and traditional kernel-based variants.

DL models, owing to their ability to automatically extract features from large-scale data, have in many cases proven more suitable for energy forecasting tasks than their "shallow" ML counterparts. The application of DL techniques can enhance MG operations by offering advantages such as the ability to capture complex patterns and the capability to estimate uncertainty, features particularly valuable for short-term decision support (Wazirali et al., 2023). Among the most established DL methodologies are long short-term memory (LSTM) networks, a class of recurrent networks that address the well-known vanishing gradient problem of traditional recurrent neural networks (RNNs). At the core of the LSTM architecture are memory cells, designed to retain information over extended periods. These cells feature a distinctive structure that allows them to critically learn and discard information depending on the context of the input sequence (Zaboli et al., 2024). Among the effective applications of LSTMs is the study in (Majeed et al., 2025), where an LSTM model leverages historical consumption data, renewable generation, and meteorological variables, successfully capturing complex temporal dependencies and reducing forecasting errors compared to alternative approaches. The results demonstrate high accuracy (in terms of RMSE, MAE, and MAPE), fast computation, and good stability, highlighting the superiority of LSTM over GRU, RNN, and CNN. DL techniques based on attention mechanisms are gaining increasing relevance in load forecasting, as they can enhance both interpretability and learning performance (Wang et al., 2024). The attention mechanism is a sequential encoder–decoder model designed to decompose complex tasks into smaller subtasks, focusing attention on specific network components, thereby proving to be a powerful tool for improving DL models (Chandrasekaran & Paramasivan, 2025). The potential of this approach is illustrated in (Cavus & Allahham, 2025), where ST-CALNet is introduced, an innovative hybrid model that integrates CNN, LSTM, and attention mechanisms to improve short-term electricity demand forecasting in smart grids with high renewable penetration. Tested on real-world MG data, the model outperformed traditional methods and other neural networks in terms of MAE and RMSE, while also ensuring greater interpretability.

15.3.1.4 Forecasting solar and wind generation

Forecasting of solar and wind power generation is essential for the efficient and reliable management of MGs. The inherently variable and unpredictable nature of these resources, which are highly dependent on meteorological conditions, makes forecasting a challenging task. Forecasting methodologies can be broadly categorized into three groups: (1) Physical models, (2) statistical models, and (3) computational intelligence-based models (ML and DL). Physical models, such as those relying on numerical weather prediction, are based on meteorological parameters and physical equations; however, they are computationally expensive and less suitable for short-term forecasting (Aslam et al., 2021).

In addition to the aforementioned ARIMA and SARIMA, another widely employed statistical model is the Kalman filter. By exploiting the principles of linear system dynamics, this algorithm optimally estimates the system states based on observed inputs and outputs, ensuring accurate forecasts even under uncertain data conditions (Zaboli et al., 2024). The main limitation of this approach lies in its difficulty in capturing nonlinear relationships within the data or analyzing long-term dependencies. To overcome these constraints, the Kalman filter is often integrated into hybrid models.

Among ML approaches, SVR models have shown significantly lower error metrics for PV solar forecasting compared to traditional linear regression models, thereby improving demand–supply balance and reducing costs (Singh et al., 2024). DL models can extract robust spatio-temporal features for various applications in power systems. A wide range of studies employ hybrid architectures that integrate DNN, RNN, or LSTM-based networks, since these techniques are effective in capturing temporal correlations within data, albeit with varying levels of accuracy. An interesting example of a hybrid algorithm is presented in Xue et al. (2024), where the authors introduce a PV power forecasting model based on a multiscale CNN-LSTM network, enriched with spatio-temporal feature fusion and a two-stage cascaded configuration. The proposed model reduced RMSE by approximately 30% compared to standard CNN, LSTM, and CNN-LSTM models, with the cascaded version achieving further accuracy improvements. In Belletreche et al. (2024), the effectiveness of hybrid approaches for wind power forecasting is demonstrated. Specifically, the study employs a hybrid DL model (Conv-DA-LSTM-GSA) that integrates CNN, a two-layer LSTM, and an attention mechanism, optimized through grid search and applied to meteorological data from the Adrar wind farm. The model achieved superior performance, recording a 0.97% reduction in RMSE compared to the nonoptimized version and showing more substantial improvements as a function of wind speed, with reductions of up to 22.9% under strong wind conditions.

Regardless of the inherent advantages and limitations of the aforementioned algorithms, as previously noted, it is always necessary to assess whether a given model is suitable for the forecasting horizon of interest. Although no standard rules exist for model selection based on the time horizon, the main approaches can generally be categorized according to short-, medium-, or long-term forecasting needs. Owing to their low computational requirements and reduced risk of overfitting, statistical models may represent a reasonable choice for real-time applications that do not demand extremely high accuracy. Conversely, ML models are often effective for long-term forecasting, as they provide a reasonable trade-off between computational complexity and predictive accuracy (Wang et al., 2024). DL models, on the other hand, rank among the most effective approaches for short-term forecasting, given their ability to handle a large number of features and extensive historical datasets, as well as to capture highly complex temporal dynamics (Aurangzeb et al., 2021).

15.3.1.5 Limitations and challenges in energy forecasting

Load and energy production forecasting is affected by numerous challenges and limitations, both intrinsic to the forecasting process and related to State-of-the-Art approaches. The main difficulties stem from the intermittency, variability, and unpredictability of DERs, particularly nondispatchable ones such as PV and wind, which reduce grid flexibility and complicate their integration into power systems. A further significant limitation lies in data availability and quality. The increasing penetration of behind-the-meter (BTM) DERs in residential and commercial buildings makes load forecasting even more complex, as it requires dedicated procedures to analyze datasets for different DERs, which are often not publicly accessible (Zaboli et al., 2024). In addition, ML models frequently overlook socioeconomic factors and meteorological conditions, undermining predictive accuracy (Devaraj et al., 2021). Customer privacy concerns also play a non-negligible role, restricting data sharing and accessibility, thus hindering the development of more robust forecasting systems (Aslam et al., 2021). Further challenges are posed by the high computational demands of DL models and their limited interpretability, which has led to the growing research interest in explainable AI (Wang et al., 2024).

15.4 Composing battery energy storage system and renewable energy source

An essential component of contemporary ESS is comprised of lithium-ion batteries (LIB). A survey highlighted that improvements in LIB production have significantly reduced their costs and increased accessibility, revealing a 40% decrease in global average turnkey ESS prices, dropping from 2023 down to US$165/kWh in 2024. This makes LIBs a crucial element in EMS (Sui & Song, 2020).

LIBs are now the primary power source for clean energy vehicles and are extensively utilized in electric vehicles (EVs) and ESS. This widespread adoption is attributed to their exceptional characteristics such as high energy and power density, long cycle life and extended shelf life, low self-discharge rate (Yang et al., 2023), lightweight and compact form factor, fast charging and discharging capabilities, high coulombic efficiency, and scalability from small modules to grid-scale deployments (Dini et al., 2024; Jiang et al., 2021). The benefits of coordinated PV–BESS scheduling, peak shaving, ramp-rate limiting, and voltage support are particularly consequential in islanded systems. Here, BESS not only arbitrages energy but also supplies synthetic inertia and fast frequency response through grid-forming control, which the EMS must value explicitly in the objective or constraints (Ochoa-Correa et al., 2025).

15.4.1 State of charge estimation in lithium-ion batteries

LIBs are affected by various uncertainties stemming from their complex chemical nature, operational conditions, and inherent degradation processes. Two key indicators of LIB performance are the SoC and the SoH. The SoC reflects the amount of energy currently stored in the battery, while the SoH characterizes the degree of aging and remaining performance over the battery's lifecycle (Enrico et al., 2025; Paschero et al., 2016; Xing et al., 2014). Accurate estimation of these parameters is essential for an EMS to maximize energy utilization, ensure safe operation, and prolong battery lifespan.

A straightforward approach to estimate SoC is the Coulomb counting method, which measures the current entering and leaving the battery and integrates these measurements over time to compute the transferred charge (Dini et al., 2024):

$$SoC(t) = SoC(t_0) + \frac{\int (I_b - I_{loss})\, dt}{C_{rated}} \tag{15.6}$$

In Eq. 15.6, $SoC(t_0)$ represents the initial SoC at time t_0, while I_b and I_{loss} denote the battery current and current losses, respectively (Dini et al., 2024). Despite its simplicity, Coulomb counting is highly sensitive to initial SoC settings and cumulative measurement errors, which may propagate over long operation periods (Demirci et al., 2024; Paschero et al., 2016). Moreover, extreme SoC levels, such as over-charging or over-discharging, accelerate degradation and shorten battery lifetime. For this reason, advanced controllers typically constrain SoC within safe operating ranges, e.g., 20%–80%, depending on battery type and manufacturer specifications (Mussi et al., 2024).

Since SoC cannot be directly measured, it is often inferred from other measurable quantities such as current, voltage, and temperature (Luzi et al., 2018). In this context, the relationship between open circuit voltage (OCV) and SoC has been widely employed. OCV-SOC curves are usually derived through controlled laboratory procedures in which batteries are charged and discharged at low current rates and then rested at different SoC levels to measure stable OCV values (Chen et al., 2019). These curves are then compiled into lookup tables or mathematical functions that map OCV to SoC. However, their practical application is limited. For example, in real-world systems such as EVs, batteries rarely rest long enough to achieve stable OCV readings, making direct lookups impractical. Furthermore, OCV-SOC curves shift with temperature and aging, while certain chemistries (e.g., $LiFePO_4$) exhibit flat OCV regions, where even small voltage errors can result in large SoC estimation inaccuracies (Xing et al., 2014).

To address these limitations, researchers have proposed a variety of advanced nonlinear approaches, Table 15.3 that embed the OCV–SOC relationship into dynamic models or adaptively update it during operation. These methods integrate model-based estimation, state observers (e.g., extended and unscented Kalman filters), and data-driven ML techniques to cope with nonlinearities, uncertainties, and real-time variability. Such innovations enhance SoC estimation accuracy by: (1) Explicitly capturing the nonlinear OCV–SOC dependence, (2) dynamically adapting to external

Table 15.3 Comparison of advanced OCV–SoC based SoC estimation methods.

Paper	Main innovation	Reported performance / outcome
Paschero et al. (2016)	Nonlinear circuit model with quasi-stationary voltage (V_{qst}) as OCV; multicell EKF using round-robin scheme.	Mean error halved compared to TTC model; multicell SoC tracking feasible with error < 5% in slow dynamics, ~ 15% in aggressive cycles.
Luzi et al. (2018)	Gray-box design with dedicated NNs for instantaneous, dynamic, and quasi-stationary components; preserves SoC influence in V_{out}.	Outperformed black-box NNs (RBF, WNN, ELM); robust SoC estimation on NASA dataset with lower MSE.
Luzi et al. (2019)	White-box ECM enhanced by NNs for resistors/capacitors; FLNN for OCV–SoC curve.	Best trade-off of accuracy and computational cost; superior generalization under EV driving cycles.
Chen et al. (2019)	OCV curve reconstructed online using "good points" and RLS with parameter backtracking; local update possible without full cycles.	Max SoC error < 3%; jitters in OCV identification reduced by ~ 80%; practical for EV real-time BMS.
Xing et al. (2014)	Explicit OCV–SoC–T lookup table integrated with UKF; addresses flat regions of $LiFePO_4$.	SOC RMSE smaller than single OCV–SoC models; robust from 0°C–50°C and insensitive to initial SoC guess.

ECM, equivalent circuit models; *EV*, electric vehicle; *OCV*, open circuit voltage; *SoC*, state of charge; EKF, extended kalman filter; NNs, neural networks; RBF, radial basis function; WNN, wavelet neural network; ELM, extreme learning machine; TTC, total transfer capability; RLS, recursive least squares.

factors such as temperature and aging, and (2) enabling real-time responsiveness in dynamic operating conditions.

It is well known that the OCV–SoC curve of Li-ion batteries is not only nonlinear but also temperature dependent: Variations in ambient temperature alter diffusion processes and internal resistance, producing shifts in the voltage response for the same SoC value. As a consequence, SoC estimation based on a fixed OCV–SoC regression may lead to biases when the battery operates under conditions different from those used for model identification. This effect is more critical in the flat regions of the OCV curve, where even small voltage deviations translate into significant SoC errors. In our experiments (Zendehdel et al., 2024), the ECM was trained using data at 25°C and then tested also at 0°C and 45°C. The 25°C dataset provided the baseline regression, while the discharging results at 25°C and 45°C in highlight how the same model reacts under different thermal conditions. Although the average errors remain below 10%, local deviations appear more pronounced at 45°C, especially during steep current transients. These results, Fig. 15.4, confirm that temperature can directly affect SoC estimation accuracy, reinforcing the need to consider temperature effects in future extensions of the model.

15.4.2 State of health

SoH is a health prognosis instrument and represents the condition of a battery compared to its initial state (BOL), reflecting the battery's capability to store energy and supply power (Luzi et al., 2018; von Bülow & Meisen, 2023). The SoH can be interpreted in many ways, like the ratio of current and initial chargeable capacity of the battery or the maximum releasable energy during fully charging and discharging in the present and initial states, respectively (Su et al., 2024).

$$SoH_c = \frac{C}{C_{init}} \times 100\%, \tag{15.7}$$

$$SoH_E = \frac{E}{E_{init}} \times 100\%. \tag{15.8}$$

The battery degradation is mostly caused by complex and irreversible electrochemical reactions between the anode and cathode. This nonlinear process is affected by many factors, such as the chemical material of the battery, which leads to loss of active materials (Su et al., 2024), ambient temperature, number of full charging/discharging. Empirically, it is shown that battery degradation is largely due to calendar and cyclic aging (Mussi et al., 2024). Calendar aging is driven primarily by time, temperature, and storage SoC, whereas cycle aging depends heavily on depth-of-discharge, charge/discharge rate, and cycling frequency (Birkl et al., 2017; Kehrer et al., 2020). Many early RES-BESS integration studies overlooked degradation, focusing solely on energy balances and economics. However, more recent works integrate degradation models, often based on rain flow cycle counting or semi-empirical aging laws, directly into EMS optimization (Schimpe et al., 2018; Teixeira et al., 2021). While this shift

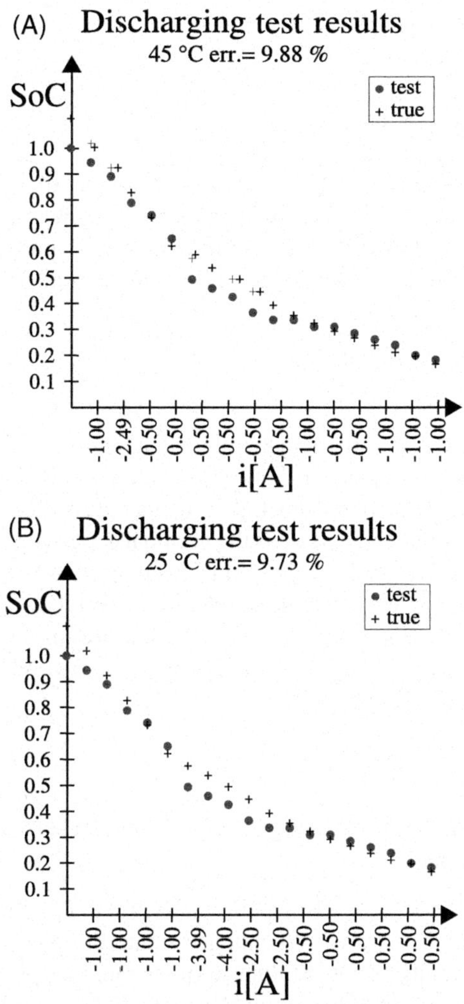

Figure 15.4 Comparison of discharging experiment results at 25°C (*left*) and 45°C (*right*). Temperature shifts in the OCV curve and internal resistance lead to larger local SoC deviations at higher temperature. *OCV*, Open circuit voltage; *SoC*, state of charge.

marks a significant advancement, degradation models remain approximate and context-dependent, raising concerns about generalizability across chemistries and operating conditions.

The degradation model incorporates calendar aging as its foundational element, leading to an expression for time-dependent degradation as follows: Where T_{total} denotes the total elapsed time in hours, and $\mathrm{Life}_{cal}^{80\%}$ indicates the calendar lifespan until the battery diminishes to 80% of its original capacity [for instance, 15 years for

LFP according to Hesse et al. (2017)].

$$\text{aging}_{\text{cal}} = \frac{T_{\text{total}}}{\text{Life}_{\text{cal}}^{80\%} \cdot 8760} \tag{15.9}$$

This equation describes the intrinsic deterioration over time, independent of usage cycles, which is combined with cyclic aging to determine the overall decline in the SoH. The cyclic component of degradation is defined by the battery's energy throughput, encompassing both charge and discharge cycles (Fuchs et al., 2012).

$$\text{aging}_{\text{cyc}} = 0.5 \times \frac{\sum \left|P_t^{\text{batt}}\right| \Delta t}{\text{Life}_{\text{cyc}}^{80\%} \times C_b}. \tag{15.10}$$

Cyclic aging is calculated mathematically as: Where $\text{Life}_{\text{cyc}}^{80\%}$ refers to the number of complete cycles required for the battery to reduce to 80% of its initial capacity, P_t^{batt} represents the battery power at any time t (using the absolute value for throughput), Δt signifies the time increment, and C_b is the nominal capacity of the battery. The factor 0.5 adjusts for the average half-cycle inclusion, employing terminal power without efficiency alterations to reflect full aging implications. Total aging is calculated through a superposition method: Where $\text{aging}_{\text{cyc}}$ directly measures the cyclic aging impact. The SoH is expressed as $SoH_t = \max(1 - 0.2 \bullet \text{aging}_{\text{tot}}, \alpha_{\text{replace}})$, with $\alpha_{\text{replace}} = 0.7$ representing modern warranty standards.

$$\text{aging}_{\text{tot}} = \text{aging}_{\text{cal}} + \text{aging}_{\text{cyc}}. \tag{15.11}$$

15.4.3 Coupling battery energy storage system with renewable generation

The integration of BESS with RES, particularly solar PV and wind, has been extensively investigated as a strategy to enhance reliability and self-sufficiency at both residential and community levels. Studies demonstrate that properly sized BESS can absorb excess PV generation during midday peaks and redistribute it during evening demand hours, thereby alleviating reverse power flows and mitigating voltage rise issues in low-voltage feeders (Hesse et al., 2017; Prakasam et al., 2022). In wind-integrated systems, BESS reduces curtailment during high wind periods and smooths short-term fluctuations, thus improving forecast accuracy and market participation (Tan et al., 2017).

Although these benefits are well-documented, challenges persist. Many optimization-based frameworks still assume ideal battery performance, neglecting degradation and thermal constraints, which leads to overly optimistic projections of economic viability (Vetter et al., 2005; Wang et al., 2011). Furthermore, most grid-integration studies focus on energy arbitrage and load-shifting while overlooking the provision of ancillary services such as voltage support, inertial response, and synthetic frequency regulation that inverter-interfaced BESS can deliver (Denholm et al., 2019; Mathew & Balachandran, 2021).

15.4.4 Operational strategies and control architectures

Multiple operational strategies have been proposed to coordinate BESS with RES. Model predictive control (MPC) frameworks enable predictive scheduling that accounts for forecasted generation and demand, offering optimal dispatch under uncertainty. However, MPC solutions are computationally expensive and require accurate forecasts. Alternatively, heuristic and rule-based methods are computationally simpler but often result in suboptimal outcomes, particularly under volatile price and weather conditions (Palma et al., 2024).

Recent advances in RL have shown promising results for BESS-RES coordination by directly learning policies from interaction with stochastic environments. For instance, (Gao et al., 2021) employed imitation learning to accelerate RL convergence using expert solutions from MILP formulations, while (Sage et al., 2023) demonstrated long-horizon dispatch improvements by leveraging imperfect demonstrations. Despite these advancements, RL approaches are often criticized for their lack of interpretability and the difficulty of explicitly handling degradation costs (Eichelbeck et al., 2022). Safe RL variants attempt to overcome this limitation by embedding feasibility checks or risk-sensitive constraints (Jeong et al., 2024; Yang et al., 2021), though real-world deployment remains limited.

15.4.5 Research gaps and perspectives

Despite considerable progress, several gaps persist in the integration of BESS with RES:

- Most studies adopt simplified equivalent circuit models (ECM) for real-time feasibility, whereas high-fidelity physics-based models are rarely incorporated due to computational burden, limiting accurate degradation prediction (Nejad et al., 2016).
- While RL and data-driven EMS approaches show adaptability, their lack of explainability and hardware validation hinders industrial adoption.
- Multiobjective formulations that jointly optimize economics, degradation, and ancillary services remain underdeveloped, especially under decentralized peer-to-peer trading paradigms.
- Experimental validations on community MGs are scarce, and most field deployments are limited to pilot projects, leaving uncertainties about scalability and long-term reliability.

In summary, coupling BESS with RES delivers significant technical, economic, and environmental benefits, yet the path toward scalable, degradation-aware, and interpretable control architectures remains open. A comprehensive framework integrating physics-based models, advanced estimation of SoC/SoH/RUL, and safe RL-based dispatch strategies is essential for bridging the gap between theoretical potential and practical deployment.

15.5 Conclusions and outlook

This chapter has examined EMS as the coordinating layer of decentralized, renewable-rich power systems, with emphasis on the tertiary level where scheduling decisions balance economics, reliability, and sustainability under uncertainty. Within the hierarchical control framework, we highlighted how forecasting quality and explicit uncertainty quantification directly affect dispatch performance, especially as PV and wind variability interact with battery dynamics. Degradation-aware operation was shown to be essential: Accurate SoC/SoH estimation, temperature compensation, and throughput-based aging proxies must be embedded in the scheduler to avoid overstating economic value and accelerating asset degradation.

From a methodological perspective, no single optimization family is universally dominant. Mathematical programming provides a solid reference when system physics and constraints are tractable, while metaheuristics offer flexibility in nonconvex, multiobjective problems. Data-driven approaches, including ML and RL, are increasingly attractive for handling high-dimensional features and sequential decision-making under uncertainty.

Looking forward, three priorities stand out:

- Reduced-order electrochemical–thermal surrogates: Develop models with quantified error bounds to bridge the gap between fidelity and tractability in long-horizon scheduling.
- Reliable, interpretable, and privacy-aware energy management: Design EMS that generalize across diverse contexts and datasets, preserve user privacy, and explicitly manage uncertainty through advanced forecasting.
- Battery-aware scheduling: Integrate battery operation into scheduling frameworks to minimize overall operational costs while preserving lithium-ion battery longevity.

The integration of uncertainty handling and battery optimization into EMS will enable the development of robust, efficient, and scalable energy management frameworks. Such systems will support reliable operation under system constraints, deliver measurable economic benefits, and extend asset life, thereby fostering sustainable and resilient energy communities in the transition toward decentralized power systems.

Acknowledgments

This study was carried out within the MOST – Sustainable Mobility Center and received funding from the European Union Next-GenerationEU (PIANO NAZIONALE DI RIPRESA E RESILIENZA (PNRR) – MISSIONE 4 COMPONENTE 2, INVESTIMENTO 1.4 – D.D. 1033 17/06/2022, CN00000023). This manuscript reflects only the authors' views and opinions, neither the European Union nor the European Commission can be considered responsible for them.

AI disclosure

During the preparation of this work, the author(s) used OpenAI in order to better format the sentences in English. After using this tool/service, the author(s) reviewed and edited the content as needed and take(s) full responsibility for the content of the publication.

List of acronyms and notations

Acronym		Notations		
ANN	Artificial neural network	Symbol	Description	Unit
ARIMA	Autoregressive integrated moving average	$P_{PV,AC,\ i}$	AC power from the *i*-th PV system	kW
BESS	Battery energy storage system	η_i	Inverter efficiency	–
CNN	Convolutional neural network	SoC_t	Battery state of charge at time *t*	%
DERs	Distributed energy resources	SoH_t	Battery state of health at time *t*	%
DL	Deep learning	c^t_{grid}	Grid electricity price at time *t*	€/kWh
DRL	Deep reinforcement learning	P^t_{batt}	Battery power at time *t* (positive = discharge)	kW
EMS	Energy management system	γ	Discount factor in reinforcement learning	–
EMD	Empirical mode decomposition	E_{batt}	Battery nominal energy capacity	kWh
ESS	Energy storage system	P^t_{grid}	Power exchanged with the grid at time *t*	kW
HEMS	Home energy management system	η_{ch}, η_{dis}	Charge/discharge efficiency	–
LIB	Lithium-ion battery	θ_t	Battery degradation proxy (throughput or rainflow)	kWh-eq
LSTM	Long short-term memory	$c_{deg}, c_{curt}, \rho_k$	Weights for degradation, curtailment, and soft constraints	–
MDP	Markov decision process	$\Pi_{\mathcal{A}}(\cdot)$	Safety-layer projection onto feasible action set	–

(*continued on next page*)

—cont'd

Acronym		**Notations**		
MG	Microgrid	μ_θ, Q_w	Actor and critic neural networks in DDPG	–
MILP	Mixed integer linear programming	Δt	Time step	min
ML	Machine learning	λ_t^{buy}, λ_t^{sell}	Electricity buy/sell prices	€/kWh
MPC	Model predictive control	E_{init}	Initial energy stored in the battery	kWh
MTF	Medium-term forecasting	V, I	Voltage and Current	V, A
OBEMS	Optimization-based energy management system	P, Q	Active and Reactive Power	kW, kVAr
OCV	Open circuit voltage	H	Overall efficiency	–
PCC	Point of common coupling			
PSO	Particle swarm optimization			
PV	Photovoltaic			
REC	Renewable energy community			
RES	Renewable energy source			
RF	Random forest			
RL	Reinforcement learning			
RNN	Recurrent neural network			
SARIMA	Seasonal autoregressive integrated moving average			
SI	Swarm intelligence			
SoC	State of charge			

(*continued on next page*)

Acronym		**Notations**		
SoE	State of energy			
SoH	State of health			
STF	Short-term forecasting			
STL	Seasonal-trend decomposition using Loess			
TBATS	Trigonometric, box-cox transformation, ARMA errors, trend and seasonal components			
TD3	Twin delayed deep deterministic policy gradient			
VSTF	Very short-term forecasting			
XGB	Extreme gradient boosting			

References

Abdolrasol, M. G. M., Hussain, S. M. S., Ustun, T. S., Sarker, M. R., Hannan, M. A., Mohamed, R., Ali, J. A., Mekhilef, S., & Milad, A. (2021). Artificial neural networks based optimization techniques: A review. *Electronics, 10*(21), 2689.

Aslam, S., Herodotou, H., Mohsin, S. M., Javaid, N., Ashraf, N., & Aslam, S. (2021). A survey on deep learning methods for power load and renewable energy forecasting in smart microgrids. *Renewable and Sustainable Energy Reviews, 144*, 110992.

Aurangzeb, K., Alhussein, M., Javaid, K., & Haider, S. I. (2021). A pyramid-CNN based deep learning model for power load forecasting of similar-profile energy customers based on clustering. *IEEE Access, 9*, 14992–15003.

Azevedo, B. F., Rocha, A. M. A. C., & Pereira, A. I. (2024). Hybrid approaches to optimization and machine learning methods: A systematic literature review. *Machine Learning, 113*(7), 4055–4097.

Azuara-Grande, L. S., Arnaltes, S., Alonso-Martinez, J., & Luis Rodriguez-Amenedo, J. (2021). Comparison of two energy management system strategies for real-time operation of isolated hybrid microgrids. *Energies, 14*(20), 6770.

Baesmat, K. H., Shokoohi, F., & Farrokhi, Z. (2025). SP-RF-ARIMA: A sparse random forest and ARIMA hybrid model for electric load forecasting. *Global Energy Interconnection, 8*(3), 486–496.

Bandara, K., Hyndman, R. J., & Bergmeir, C. (2025). MSTL: A seasonal-trend decomposition algorithm for time series with multiple seasonal patterns. *International Journal of Operational Research, 52*(1), 79–98.

Banik, R., & Biswas, A. (2024). Enhanced renewable power and load forecasting using RF-XGBoost stacked ensemble. *Electrical Engineering, 106*(4), 4947–4967. doi:10.1007/s00202-024-02273-3.

Bao, G., & Xu, R. (2023). A data-driven energy management strategy based on deep reinforcement learning for microgrid systems. *Cognitive Computation, 15*(2), 739–750.

Belletreche, M., Bailek, N., Abotaleb, M., Bouchouicha, K., Zerouali, B., Guermoui, M., Kuriqi, A., Alharbi, A. H., Sami Khafaga, D., & El-Shimy, M. (2024). Hybrid attention-based deep neural networks for short-term wind power forecasting using meteorological data in desert regions. *Scientific Reports, 14*(1), 21842.

Birkl, C. R., Roberts, M. R., McTurk, E., Bruce, P. G., & Howey, D. A. (2017). Degradation diagnostics for lithium ion cells. *Journal of Power Sources, 341*, 373–386. doi:10.1016/j.jpowsour.2016.12.011.

Breiman, L., Friedman, J., Olshen, R. A., & Stone, C. J. (2017). *Classification and regression trees*. Chapman and Hall/CRC.

Bui, V.-H., Hussain, A., & Kim, H.-M. (2019). Double deep Q -learning-based distributed operation of battery energy storage system considering uncertainties. *IEEE Transactions on Smart Grid, 11*(1), 457–469.

Cabrera-Tobar, A., Massi Pavan, A., Petrone, G., & Spagnuolo, G. (2022). A review of the optimization and control techniques in the presence of uncertainties for the energy management of microgrids. *Energies, 15*(23), 9114.

Cao, D., Zhao, J., Hu, W., Ding, F., Huang, Q., & Chen, Z. (2021). Attention enabled multi-agent DRL for decentralized volt-VAR control of active distribution system using PV inverters and SVCs. *IEEE transactions on sustainable energy, 12*(3), 1582–1592.

Capillo, A., De Santis, E., Mascioli, F. M. F., & Rizzi, A. (2023). On the performance of multi-objective evolutionary algorithms for energy management in microgrids. In *International joint conference on computational intelligence* (pp. 3–15).

Capillo, A., De Santis, E., Mascioli, F. M. F., Rizzi, A. (2024). An Online hierarchical energy management system for energy communities, complying with the current technical legislation framework. arXiv preprint arXiv:2402.01688.

Cavus, M., & Allahham, A. (2025). Spatio-temporal attention-based deep learning for smart grid demand prediction. *Electronics, 14*(13), 2514.

Chandrasekaran, R., & Kumar Paramasivan, S. (2025). Advances in deep learning techniques for short-term energy load forecasting applications: A review. *Archives of Computational Methods in Engineering, 32*(2), 663–692. doi:10.1007/s11831-024-10155-x.

Chen, X., Lei, H., Xiong, R., Shen, W., & Yang, R. (2019). A novel approach to reconstruct open circuit voltage for state of charge estimation of lithium ion batteries in electric vehicles. *Applied Energy, 255*, 113758.

Cucchiella, F., D'adamo, I., & Gastaldi, M. (2016). Photovoltaic energy systems with battery storage for residential areas: An economic analysis. *Journal of Cleaner Production, 131*, 460–474.

Dahrouj, H., Alghamdi, R., Alwazani, H., Bahanshal, S., Ahmad, A. A., Faisal, A., Shalabi, R., Alhadrami, R., Subasi, A., & Al-Nory, M. T. (2021). An overview of machine learning-based techniques for solving optimization problems in communications and signal processing. *IEEE Access, 9*, 74908–74938.

da Silva André, J., Stai, E., Stanojev, O., & Hug, G. (2022). Battery control with lookahead constraints in distribution grids using reinforcement learning. *Electric Power Systems Research, 211*, 108551.

Demirci, O., Taskin, S., Schaltz, E., & Demirci, B. A. (2024). Review of battery state estimation methods for electric vehicles-Part I: SOC estimation. *Journal of energy storage, 87*, 111435.

Denholm, P., Sun, Y., & Mai, T. (2019). The value of energy storage for grid applications. *Nature Communications, 10*(1), 1–9. doi:10.1038/s41467-019-09729-7.

De Santis, E., Rizzi, A., & Sadeghian, A. (2017a). Hierarchical genetic optimization of a fuzzy logic system for energy flows management in microgrids. *Applied Soft Computing, 60*, 135–149.

De Santis, E., Rizzi, A., Sadeghiany, A., Mascioli, F., & Massimo, F. (2013). Genetic optimization of a fuzzy control system for energy flow management in micro-grids. In *Joint IFSA world congress and NAFIPS annual meeting (IFSA/NAFIPS)* (pp. 418–423). doi:10.1109/IFSA-NAFIPS.2013.6608437.

De Santis, E., Sadeghian, A., & Rizzi, A. (2017b). A smoothing technique for the multifractal analysis of a medium voltage feeders electric current. *International Journal of Bifurcation and Chaos, 27*(14), 1750211.

Devaraj, J., Madurai Elavarasan, R., Shafiullah, G. M., Jamal, T., & Khan, I. (2021). A holistic review on energy forecasting using big data and deep learning models. *International Journal of Energy Research, 45*(9), 13489–13530.

Dini, P., Colicelli, A., & Saponara, S. (2024). Review on modeling and SOC/SOH estimation of batteries for automotive applications. *Batteries, 10*(1), 34.

Drucker, H., Burges, C. J., Kaufman, L., Smola, A., & Vapnik, V. (1996). Support vector regression machines. *Advances in Neural Information Processing Systems, 9*.

Eichelbeck, D., Schenke, J., & Ulbig, A. (2022). Contingency-aware economic dispatch using reinforcement learning with safety layers. *Electric Power Systems Research, 208*, 107833. doi:10.1016/j.epsr.2022.107833.

Elsisi, M., Amer, M., & Su, C.-L. (2023). A comprehensive review of machine learning and IoT solutions for demand side energy management, conservation, and resilient operation. *Energy, 281*, 128256.

Enrico, De. S., Vanessa, P., Massimiliano, L., & Antonello, R. (2025). Degradation mechanisms and differential curve modeling for non-invasive diagnostics of lithium cells: An overview. *Renewable and Sustainable Energy Reviews, 211*, 115349. doi:10.1016/j.rser.2025.115349.

Esparza, A., Blondin, M., & Trovão, J. P. F. (2025). A review of optimization strategies for energy management in microgrids. *Energies, 18*(13), 3245.

European Commission Joint Research Centre. (2024). Renewable energy recast to 2030 (RED II). https://joint-research-centre.ec.europa.eu/welcome-jec-website/reference-regulatory-framework/renewable-energy-recast-2030-red-ii_en

European Environment Agency. (2025). Share of energy consumption from renewable sources in Europe. https://www.eea.europa.eu/en/analysis/indicators/share-of-energy-consumption-from

European Union. (2023). Regulation (EU) 2023/1542 of the European Parliament and of the Council. https://eur-lex.europa.eu/legal-content/EN/TXT/PDF/?uri=CELEX:32023R1542

Forootan, M. M., Larki, I., Zahedi, R., & Ahmadi, A. (2022). Machine learning and deep learning in energy systems: A review. *Sustainability, 14*(8), 4832.

Fuchs, G., Lunz, B., Leuthold, M., & Sauer, D. U. (2012). Technology overview on electricity storage. *ISEA, Aachen, Juni, 26*, 12.

Gao, S., Xiang, C., Yu, M., Tan, K. T., & Lee, T. H. (2021). Online optimal power scheduling of a microgrid via imitation learning. *IEEE Transactions on Smart Grid, 13*(2), 861–876.

Golshannavaz, S., Afsharnia, S., & Siano, P. (2016). A comprehensive stochastic energy management system in reconfigurable microgrids. *International Journal of Energy Research, 40*(11), 1518–1531.

Goodfellow, I., Bengio, Y., & Courville, A. (2016). *Deep learning*. The MIT Press.

Guiducci, L., Palma, G., Stentati, M., Rizzo, A., & Paoletti, S. (2023). A Reinforcement learning approach to the management of renewable energy communities. In *2023 12th Mediterranean conference on embedded computing (MECO), Budva, Montenegro* (pp. 1–8). doi:10.1109/MECO58584.2023.10154979.

Harikrishnan, G. R., & Sreedharan, S. (2025). Advanced short-term load forecasting for residential demand response: An XGBoost-ANN ensemble approach. *Electric Power Systems Research, 242*, 111476.

Hasheminasab, S., Alzayed, M., & Chaoui, H. (2024). A review of control techniques for inverter-based distributed energy resources applications. *Energies, 17*(12), 2940.

Hesse, H.-C., Schimpe, M., Kucevic, D., & Jossen, A. (2017). Economic potential of PV battery systems compared to conventional battery systems. *Applied Energy, 189*, 310–319. doi:10.1016/j.apenergy.2016.12.153.

Hornik, K., Stinchcombe, M., & White, H. (1989). Multilayer feedforward networks are universal approximators. *Neural networks, 2*(5), 359–366.

Inter governmental Panel on Climate Change. 2023, Climate change 2023: Synthesis Report. Summary for policymakers. https://www.ipcc.ch/report/ar6/syr/downloads/report/IPCC_AR6_SYR_SPM.pdf

Jamii, J., Trabelsi, M., Mansouri, M., Mimouni, M. F., & Shatanawi, W. (2022). Non-linear programming-based energy management for a wind farm coupled with pumped hydro storage system. *Sustainability, 14*(18), 11287.

Javed, H., Muqeet, H. A., Shehzad, M., Jamil, M., Khan, A. A., & Guerrero, J. M. (2021). Optimal energy management of a campus microgrid considering financial and economic analysis with demand response strategies. *Energies, 14*(24), 8501.

Jeong, H., Kim, S., & Lee, Y.-J. (2024). Time-varying constraint-aware reinforcement learning for safe battery dispatch. *IEEE Transactions on Smart Grid, 15*(3), 1776–1789. doi:10.1109/TSG.2023.3332290.

Jia, Y., Zhou, S., Wang, Y., Lin, F., & Gao, Z. (2025). A quadratic ν-support vector regression approach for load forecasting. *Complex & Intelligent Systems, 11*(1), 123. doi:10.1007/s40747-024-01730-7.

Jiang, B. o, Dai, H., & Wei, X. (2021). A cell-to-pack state estimation extension method based on a multilayer difference model for series-connected battery packs. *IEEE Transactions on Transportation Electrification, 8*(2), 2037–2049.

Joshi, A., Capezza, S., Alhaji, A., & Chow, M.-Y. (2023). Survey on AI and machine learning techniques for microgrid energy management systems. *IEEE/CAA Journal of Automatica Sinica, 10*(7), 1513–1529.

Kahraman, A., Yang, G. 2022, Home energy management system based on deep reinforcement learning algorithms. IEEE Access, 9, 153108–153115

Kang, K.-M., Choi, B.-Y., Lee, H., An, C.-G., Kim, T.-G., Lee, Y.-S., Kim, M., Yi, J., & Won, C.-Y. (2021). Energy management method of hybrid AC/DC microgrid using artificial neural network. *Electronics, 10*(16), 1939.

Kehrer, D., Nuhic, A., Schuster, S. F., & Jossen, A. (2020). Calendar aging of lithium-ion batteries: A review. *Journal of Energy Storage, 32*, 101859. doi:10.1016/j.est.2020.101859.

Lee, J., Wang, W., & Niyato, D. (2020). Demand-side scheduling based on multi-agent deep actor-critic learning for smart grids. In *2020 IEEE international conference on communications, control, and computing technologies for smart grids (SmartGridComm)* (pp. 1–6).

Lee, S., Seon, J., Hwang, B., Kim, S., Sun, Y., & Kim, J. (2024). Recent trends and issues of energy management systems using machine learning. *Energies, 17*(3), 624.

Li, P., & Liu, Q. (2024). Optimizing energy management strategies for microgrids through chaotic local search and particle swarm optimization techniques. *Heliyon, 10*(17).

Li, Y., Tang, H., Lv, K., Wang, K., & Wang, G. (2020). Optimization of dynamic dispatch for multiarea integrated energy system based on hierarchical learning method. *IEEE Access, 8*, 72485–72497.

Luo, S., Wang, B., Gao, Q., Wang, Y., & Pang, X. (2024). Stacking integration algorithm based on CNN-BiLSTM-Attention with XGBoost for short-term electricity load forecasting. *Energy Reports, 12*, 2676–2689.

Luzi, M., Mascioli, F. M. F., Paschero, M., & Rizzi, A. (2019). A white-box equivalent neural network circuit model for SoC estimation of electrochemical cells. *IEEE Transactions On Neural Networks and Learning Systems, 31*(2), 371–382.

Luzi, M., Paschero, M., Rizzi, A., Maiorino, E., & Mascioli, F. M. F. (2018). A novel neural networks ensemble approach for modeling electrochemical cells. *IEEE Transactions on Neural Networks and Learning Systems, 30*(2), 343–354.

Ma, G., Li, J., & Zhang, X.-P. (2022). A review on optimal energy management of multimicrogrid system considering uncertainties. *IEEE Access, 10*, 77081–77098.

Majeed, M. A., Phichaisawat, S., Asghar, F., & Hussan, U. (2025). Data-driven optimized load forecasting: An LSTM based RNN approach for smart grids. *IEEE Access*.

Mathew, B., & Balachandran, P. (2021). Ancillary services from battery energy storage systems: A comprehensive review. *Renewable and Sustainable Energy Reviews, 146*, 111203. doi:10.1016/j.rser.2021.111203.

Mnih, V., Badia, A. P., Mirza, M., Graves, A., Lillicrap, T., Harley, T., Silver, D., & Kavukcuoglu, K. (2016). Asynchronous methods for deep reinforcement learning. *Proceedings of Machine Learning Research, 48*, 1928–1937.

Mnih, V., Kavukcuoglu, K., Silver, D., Graves, A., Antonoglou, I., Wierstra, D., & Riedmiller, M. (2013). Playing Atari with deep reinforcement learning. arXiv preprint arXiv:1312.5602. ,

Mussi, M., Pellegrino, L., Pindaro, O. F., Restelli, M., & Trovò, F. (2024). A Reinforcement learning controller optimizing costs and battery state of health in smart grids. *Journal of Energy Storage, 82*, 110572.

National Energy Installers. (2024). Peak demand opportunities. https://nationalenergyinstallers.com/newowhee/2024/07/NEI-Peak-Demand-Opportunities.pdf

Nejad, S. A., Gladwin, D. T., & Stone, D. A. (2016). Review and comparison of modeling techniques for lithium-ion batteries. *Renewable and Sustainable Energy Reviews, 60*, 132–147. doi:10.1016/j.rser.2016.01.060.

Ochoa-Correa, D., Arévalo, P., & Martinez, S. (2025). Pathways to 100% renewable energy in island systems: A systematic review of challenges, solutions strategies, and success cases. *Technologies, 13*(5), 180.

Odonkor, P., & Lewis, K. (2019). Automated design of energy efficient control strategies for building clusters using reinforcement learning. *Journal of Mechanical Design, 141*(2), 021704.

Ojo, K. E., Saha, A. K., & Srivastava, V. M. (2025). Microgrids' control strategies and real-time monitoring systems: A comprehensive review. *Energies, 18*(13), 3576.

Palma, G., Guiducci, L., Stentati, M., Rizzo, A., & Paoletti, S. (2024). Reinforcement learning for energy community management: A European-scale study. *Energies, 17*(5), 1249.

Papadimitrakis, M., Giamarelos, N., Stogiannos, M., Zois, E. N., Livanos, N. A.-I., & Alexandridis, A. (2021). Metaheuristic search in smart grid: A review with emphasis on planning, scheduling and power flow optimization applications. *Renewable and Sustainable Energy Reviews, 145*, 111072.

Paschero, M., Storti, G. L., Rizzi, A., Mascioli, F. M. F., & Rizzoni, G. (2016). A novel mechanical analogy-based battery model for SoC estimation using a multicell EKF. *IEEE Transactions on Sustainable Energy, 7*(4), 1695–1702.

Pourbehzadi, M., Niknam, T., Aghaei, J., Mokryani, G., Shafie-khah, M., & Catalão, J. P. S. (2019). Optimal operation of hybrid AC/DC microgrids under uncertainty of renewable energy resources: A comprehensive review. *International Journal of Electrical Power & Energy Systems, 109*, 139–159.

Prakasam, P., Krishnasamy, V., & Rajasekar, N. (2022). A review on battery energy storage systems for grid applications. *Renewable and Sustainable Energy Reviews, 162*, 112403. doi:10.1016/j.rser.2022.112403.

REN21. (2021). Renewables in cities. 2021 global status report. https://www.ren21.net/wp-content/uploads/2019/05/REC_2021_full-report_en.pdf

Restrepo, M., Cañizares, C. A., Simpson-Porco, J. W., Su, P., & Taruc, J. (2021). Optimization- and rule-based energy management systems at the Canadian renewable energy laboratory microgrid facility. *Applied Energy, 290*, 116760.

Rezk, H., Olabi, A. G., Sayed, E. T., & Wilberforce, T. (2023). Role of metaheuristics in optimizing microgrids operating and management issues: A comprehensive review. *Sustainability, 15*(6), 4982.

Rilling, G., Flandrin, P., & Goncalves, P. (2003). On empirical mode decomposition and its algorithms. *IEEE-EURASIP workshop on nonlinear signal and image processing NSIP-03*.

Ruddick, J., Camargo, L. R., Putratama, M. A., Messagie, M., & Coosemans, T. (2025). TreeC: A method to generate interpretable energy management systems using a metaheuristic algorithm. *Knowledge-Based Systems, 309*, 112756.

Sage, M., Staniszewski, M., & Zhao, Y. F. (2023). Economic battery storage dispatch with deep reinforcement learning from rule-based demonstrations. In *2023 International conference on control, automation and diagnosis (ICCAD)* (pp. 1–6). doi:10.1109/ICCAD57653.2023.10152299.

Salaorni, D., Bianchi, F., Trovò, F., & Restelli, M. (2025). A reinforcement learning approach for optimal control in microgrids. In *2025 International joint conference on neural networks (IJCNN)* (pp. 1–8). doi:10.1109/IJCNN64981.2025.11227643.

Saleem, M. I., Saha, S., Izhar, U., & Ang, L. (2024). Optimized energy management of a solar battery microgrid: An economic approach towards voltage stability. *Journal of Energy Storage, 90*, 111876.

Sangrody, H., Zhou, N., Tutun, S., Khorramdel, B., Motalleb, M., & Sarailoo, M. (2018). Long term forecasting using machine learning methods. In *2018 IEEE power and energy conference at illinois (PECI)* (pp. 1–5). doi:10.1109/PECI.2018.8334980.

Schimpe, M., Hesse, H.-C., Müller, S., Kucevic, D., & Jossen, A. (2018). Comprehensive modeling of temperature-dependent degradation mechanisms in lithium-ion batteries. *Journal of Power Sources, 397*, 157–169. doi:10.1016/j.jpowsour.2018.07.061.

Selim, A., Mo, H., Pota, H., & Dong, D. (2023). Optimal scheduling of battery energy storage systems using a reinforcement learning-based approach. *IFAC-PapersOnLine, 56*(2), 11741–11747.

Shami, T. M., El-Saleh, A. A., Alswaitti, M., Al-Tashi, Q., Summakieh, M. A., & Mirjalili, S. (2022). Particle swarm optimization: A comprehensive survey. *IEEE Access, 10*, 10031–10061.

Singh, A. R., Kumar, R. S., Bajaj, M., Khadse, C. B., & Zaitsev, I. (2024). Machine learning-based energy management and power forecasting in grid-connected microgrids with multiple distributed energy sources. *Scientific Reports, 14*(1), 19207.

Srikanth, D., Durga Sukumar, G., & Sobhan, P. V. S. (2025). A convolutional neural network based energy management system for photovoltaic/battery systems in microgrid using enhanced coati optimization approach. *Journal of Energy Storage, 119*, 116252.

Su, L., Xu, Y., & Dong, Z. (2024). State-of-health estimation of lithium-ion batteries: A comprehensive literature review from cell to pack levels. *Energy Conversion and Economics, 5*(4), 224–242.

Sui, Y. u, & Song, S. (2020). A multi-agent reinforcement learning framework for lithium-ion battery scheduling problems. *Energies, 13*(8), 1982.

Sutton, R. S., & Barto, A. G. (1998). *Reinforcement learning: An introduction*: 1. MIT press Cambridge.

Tan, X., Yang, P., & Xu, C. (2017). Battery energy storage system (BESS): A review of applications, architectures, and challenges. *Renewable and Sustainable Energy Reviews, 36*, 82–110. doi:10.1016/j.rser.2014.04.007.

Tasmant, H., Bossoufi, B., Alaoui, C., & Siano, P. (2025). A review of machine learning and IoT-based energy management systems for AC microgrids. *Computers and Electrical Engineering, 127*, 110563.

Teixeira, A. L., Narayan, A., & Devaraj, A. (2021). Alexandre Oudalov, Degradation-aware optimization of lithium-ion battery energy management. *Applied Energy, 302*, 117520. doi:10.1016/j.apenergy.2021.117520.

Vaish, J., Tiwari, A. K., & Kaimal, S. (2024). Multi-objective optimization of distributed energy resources based microgrid using random forest model. *Bulletin of Electrical Engineering and Informatics, 13*(1), 67–75.

Verdone, A., Panella, M., De Santis, E., & Rizzi, A. (2025). A review of solar and wind energy forecasting: From single-site to multi-site paradigm. *Applied Energy, 392*, 126016.

Vetter, J. ü, Novak, P., Wagner, M. R., Veit, C., Möller, K.-C., Besenhard, J. üO., Winter, M., Wohlfahrt-Mehrens, M., Vogler, C., & Hammouche, A. (2005). Ageing mechanisms in lithium-ion batteries. *Journal of Power Sources, 147*(1-2), 269–281. doi:10.1016/j.jpowsour.2005.01.006.

von Bülow, F., & Meisen, T. (2023). A review on methods for state of health forecasting of lithium-ion batteries applicable in real-world operational conditions. *Journal of Energy Storage, 57*, 105978.

Voß, S., Martello, S., Osman, I. H., & Roucairol, C. (2012). *Meta-Heuristics: Advances and trends in local search paradigms for optimization*. Springer https://books.google.it/books?id=6CzoBwAAQBAJ.

Wang, J., Liu, P., Hicks-Garner, J., Sherman, E., Soukiazian, S., Verbrugge, M., Tataria, H., Musser, J., & Finamore, P. (2011). Cycle-life model for graphite-LiFePO4 cells. *Journal of Power Sources, 196*(8), 3942–3948.

Wang, X., Wang, H., Bhandari, B., & Cheng, L. (2024). AI-empowered methods for smart energy consumption: A review of load forecasting, anomaly detection and demand response. *International Journal of Precision Engineering and Manufacturing-Green Technology, 11*(3), 963–993.

Wazirali, R., Yaghoubi, E., Abujazar, M. S. S., Ahmad, R., & Vakili, A. H. (2023). State-of-the-art review on energy and load forecasting in microgrids using artificial neural networks, machine learning, and deep learning techniques. *Electric power systems research, 225*, 109792.

Xing, Y., He, W., Pecht, M., & Tsui, K. L. (2014). State of charge estimation of lithium-ion batteries using the open-circuit voltage at various ambient temperatures. *Applied energy, 113*, 106–115.

Xue, H., Ma, J., Zhang, J., Jin, P., Wu, J., & Du, F. (2024). Power forecasting for photovoltaic microgrid based on multiscale CNN-LSTM network models. *Energies, 17*(16), 3877.

Yamashita, D. Y., Vechiu, I., & Gaubert, J.-P. (2020). A review of hierarchical control for building microgrids. *Renewable and Sustainable Energy Reviews, 118*, 109523.

Yang, Y., Hao, J., Zheng, Y., & Yu, C. (2019). Large-scale home energy management using entropy-based collective multiagent deep reinforcement learning framework. In *Proceedings of the 28th International Joint Conference on Artificial Intelligence (IJCAI'19)* (pp. 630–636). AAAI Press. doi:10.5555/3367032.3367122.

Yang, Y., He, J., Chen, C., & Wei, J. (2023). Balancing awareness fast charging control for lithium-ion battery pack using deep reinforcement learning. *IEEE Transactions on Industrial Electronics, 71*(4), 3718–3727.

Yang, Z., Chen, X. i, Zhou, M., Zhao, J., Zhang, S., & Chen, C. (2021). WCSAC: Worst-case soft actor critic for safety-constrained reinforcement learning. In *Proceedings of the AAAI Conference on Artificial Intelligence: 35* (pp. 10639–10646).

Yin, C., Wei, N., Wu, J., Ruan, C., Luo, X. i, & Zeng, F. (2024). An empirical mode decomposition-based hybrid model for sub-hourly load forecasting. *Energies, 17*(2), 307.

Zaboli, A., Kasimalla, S. R., Park, K., Hong, Y., & Hong, J. (2024). A comprehensive review of behind-the-meter distributed energy resources load forecasting: Models, challenges, and emerging technologies. *Energies, 17*(11), 2534.

Zakeri, B., & Syri, S. (2015). Electrical energy storage systems: A comparative life cycle cost analysis. *Renewable and Sustainable Energy Reviews, 42*(2015), 569–596.

Zendehdel, D., Capillo, A., De Santis, E., & Rizzi, A. (2024). An extended battery equivalent circuit model for an energy community real time EMS. In *2024 International Joint Conference on Neural Networks (IJCNN)* (pp. 1–9).

Zhang, D., Han, X., & Deng, C. (2018). Review on the research and practice of deep learning and reinforcement learning in smart grids. *CSEE Journal of Power and Energy Systems, 4*(3), 362–370.

DER clusters in diverse contexts: Stochastic siting, sizing, and control for distributed energy resources and microgrids planning

16

Uday Chandra Sikkam, Buddhadeva Sahoo and Chandan Kumar Shiva
Department of Electrical and Electronics Engineering, SR University, Warangal, Telangana, India

Case study/activity

Objective

The case study is determined to illustrate the advantages of stochastic optimization in improving the siting, sizing, and control of the distributed energy resources (DERs) in the microgrids under uncertainty. It explores the possibilities of having cost-effective, reliable, and resilient energy systems with probabilistic and AI-assisted ones that respond to renewable and demand variability.

Scope

The paper is about stochastic models of DER planning and operation, where Monte Carlo simulation is used with scenario-based optimization and AI-based control.

Limitations: The analysis had drawbacks of simulation limits, accessibility of data, and computational sources.

Workflow literature review—formulation of stochastic models—simulation and analysis—validation and discussion.

Budget: The project is a research-based simulative study, which does not involve considerable experimental or fiscal costs, except for the normal computing infrastructure.

Audience

Researchers, graduate students, and practitioners in the field of power systems, renewable energy planning, and smart grid design will find this case study useful. It will also be useful to policymakers and industry-scale engineers who are engaged in the energy transition, the implementation of microgrids, and the AI-supported energy systems.

Stochastic Planning and Modeling for Energy Systems: Methods, Applications, and Developments.
DOI: https://doi.org/10.1016/B978-0-443-45298-7.00016-1

The study can be followed by the reader familiar with power systems, optimization, and control theory.

Rationale

It was performed as part of a critical lack of knowledge in the conventional deterministic planning of DERs, which does not reflect the uncertainty in renewable generation and load demand. The major issue was the absence of strong structures to deal with stochastic variability and provide stability, reliability, and cost-effectiveness of the system. The purpose of the experiment was to test the probabilistic method and its combination with AI and digital technologies to have smarter planning.

Expected results and deliverables

The paper provides a complete stochastic planning and control framework of a DER-based microgrids and demonstrates that uncertainty may be used to make more intelligent decisions.

Reader benefits:

- Knowing the distinction between deterministic and stochastic structure.
- Knowledge of advanced optimization methods, including Monte Carlo, chance-constrained, and AI-assisted optimization.
- Understanding the issues of implementation in real life and digital integration.

Safety considerations

Since only simulation and analytical modeling are involved in the case study, there are no physical safety risks. However, readers should be focused on data integrity/cybersecurity and model validation aspects when taking part in digital twin (DT)/IoT-based stochastic frameworks in the real environment.

Actions taken/workflow/tools used/simulations and analyses

Step-by-step process:

- Identification of the problem and literature analysis.
- Formulation of deterministic baseline models. Stochastic optimization of siting and sizing models based on probabilistic variables.
- Implementation of Monte Carlo simulation and multiobjective algorithms (NSGA-II, PSO).
- Generalization of adaptive control based on AI (reinforcement learning [RL], stochastic model predictive control [SMPC]). Performance analysis in changing conditions.
- Assessment of issues and suggestions for future research directions.

Tools used: Applied mathematics—Mathematical software such as MATLAB, Python, optimization programs/software—e.g., Pyomo, TensorFlow for RL modules.

Flowchart: Problem definition, data modeling, stochastic optimization, AI-based control, evaluation, future insights.

Challenges and solutions

Challenges:

- Great count of calculations in extensive stochastic modeling.
- Little actual time flexibility of models.
- The lack of data and the problem of interoperability in digital integration.

Solutions

- Hybrid optimization and applied scenario reduction to simplify the scenario.
- Online learning introduced AI-assisted adaptive control.
- Federated learning and blockchain integration proposal to exchange data in a safe manner.

Results

The results proved that stochastic and AI-based models offer more resistant and cost-effective solutions as compared to deterministic approaches. They enhanced the reliability, lowered the operation cost, and augmented adaptability in uncertainty of renewable.

Expectedness: They were based on theoretical predictions and enforced by meta theta-simulation of multiple scenarios.

Learning and knowledge outcomes

Stochastic intelligence: It is the process of converting uncertainty to actionable insight.

The synchronization of DTs and IoT allows real-time flexibility.

What could have been done better: Extension of the analysis to hybrid AC/DC systems and field data validation.

What must not be changed? The integrated stochastic-AI methodology with the multiobjective structure as they become a base for realistic DER planning.

Recommendations for future research:

- Federated and quantum inspired optimization.
- Market mechanisms based on blockchain.
- On-the-fly co-simulation environments between AI, DT, and stochastic control.

DER clusters in diverse contexts: Stochastic siting, sizing, and control for DERs and microgrids planning

16.1 Introduction

16.1.1 Background and motivation

The world is changing toward renewable energy, whose decentralized sources (photovoltaics, wind turbines, and energy storage systems) are gaining more traction. DERs

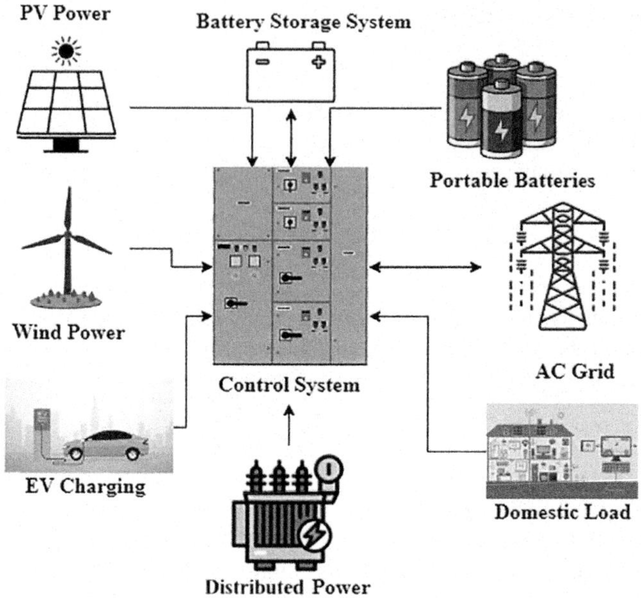

Figure 16.1 Overview of DER components and uncertainty sources in modern microgrids. *DER*, Distributed energy resource.

have emerged as important facilitators of smart grids because they enhance reliability, energy independence, and decarbonization (Kundur et al., 2009). By 2030, the global electricity demand will require more than 45% of its power through distributed renewables, and microgrids and virtual power plants (VPPs) will be the main integration sites (International Energy Agency, 2024).

Centralized generation models are no longer able to make flexibility compromises and high transmission loss, but DERs allow the local generation and adaptive control, which leads to better voltage control, less congestion, and the overall system resilience (Vasquez et al., 2010). Nevertheless, the intermittent and stochastic characteristics of the renewable production create complexities in the planning and operations.

Solar photovoltaic (PV) production is a variable that is sensitive to the amount of solar irradiance, temperature, and the weather; wind energy is a variable that is sensitive to the speed of wind; and consumer demand is a variable sensitive to socioeconomic factors (Sun et al., 2025). The uncertainties may cause an imbalance in supply and demand, poor quality of power, and reliability. Fig. 16.1 Predicts the DER components and uncertainty sources in modern microgrids. The overall structure of DER components and the associated uncertainty sources is illustrated in Fig. 16.1.

16.1.2 Deterministic versus stochastic planning paradigms

Historically, DER planning has been based on deterministic planning models where load is fixed, generation is known, and pricing is predictable. In the real world, these

Table 16.1 Comparison of deterministic and stochastic planning approaches.

Parameter	Deterministic approach	Stochastic approach	References
Input data	Fixed values	Probability distributions	Domenech et al. (2021)
Output	Single-point solution	Probabilistic or expected outcome	Bakhtiari et al. (2021)
Uncertainty handling	Ignored or simplified	Explicitly modeled	Noori et al. (2024)
Robustness	Low	High	Noori et al. (2024)
Computation	Low complexity	High complexity	Bakhtiari et al. (2021)

simplifications can give suboptimal or infeasible results (Domenech et al., 2021). To address this shortcoming, stochastic optimization uses randomly varying variables as a way of representing the uncertainties, as well as allowing one to examine a large number of possible outcomes rather than a single estimate. Probably, the optimal processing methods that describe the probabilistic renewable and load aspects are Monte Carlo simulation, probability density modeling, and scenario-based decomposition (Bakhtiari et al., 2021). Deterministic models are less computationally complicated, but they are not robust to variability. Conversely, the stochastic methods are more complicated but provide more reliable and cost-effective solutions, particularly in systems where renewable penetration is high and where the energy markets are volatile (Noori et al., 2024). A comparison of deterministic and stochastic planning methodologies is summarized in Table 16.1

16.1.3 Importance of stochastic siting, sizing, and control of distributed energy resources

Critical features of DER planning are siting and sizing. Ideal siting choices location to cause minimal losses, deviations in voltage, and congestion, whereas size, or capacity, fixes levels of cost versus reliability (Grisales-Noreña et al., 2022). These decisions should be made with uncertainty, taking into account variations in renewable generation, load increase, and equipment reliability, because otherwise, the concerns may result in increased costs, over-sizing, or unreliable system behavior (Amjady et al., 2017).

Stochastic models, such as chance-constrained optimization, two-stage stochastic programming, and robust optimization, directly include risk tolerance in the planning process whereby operational constraints are met with the specified degree of certainty and unneeded conservatism is avoided (Zhang et al., 2024). In addition, the adaptive control strategies are needed to operate DER clusters in microgrids in a coordinated manner. Predictive control is based on AI that can be used to schedule the system in real-time and increase its stability (depending on the operating conditions) (Hasani et al., 2024).

16.1.4 Multiobjective nature of distributed energy resource planning

DER planning has to balance various contradicting goals, including the reduction of cost, maximization of the use of renewable, enhancement of reliability, and minimization of emissions. Multiobjective evolutionary algorithms (MOEAs) such as non-dominated sorting genetic algorithm II (NSGA-II), multi-objective particle swarm optimization (MOPSO), and SPEA2 are often used in stochastic optimization to provide Pareto-optimal solutions (or solutions that trade off these conflicting objectives) (Moura & de Almeida, 2010). The added uncertainty can be used in such formulations to evaluate the strength of the solution under different operating conditions, whereas preference-based techniques with weights or fuzzy membership functions can be used to adjust the results to the priorities of the decision-maker (Moura & de Almeida, 2010).

Eq. (16.1): Generalized stochastic multiobjective optimization

$$\min_{x} |\mathbb{E}[f_i(x, \xi)], i = 1, \ldots, m \tag{16.1}$$

subject to

$$g(x, \xi) \leq 0, h(x, \xi) = 0$$

where x represents decision variables (location, capacity, control set-points), and ξ denotes random parameters (generation, demand, market prices).

16.2 Deterministic and Stochastic DER Planning Frameworks

Planning of DERs must be done with special consideration of the lack of uncertainties in renewable generation, demand patterns, and network conditions. Even though deterministic approaches have long been used to serve the expansion planning, the growing use of variables and decentralized resources has placed stochastic models in a better and stronger position. This part discusses the two frameworks, their mathematical basis, and how they apply in the real world of DER implementation.

16.2.1 Overview of deterministic planning

Deterministic planning will make assumptions regarding some of the most important inputs, like the load demand, the energy prices, and the generation output. These models generally minimize economic cost or loss of the system with the help of linear, nonlinear, and mixed-integer optimization methodology (El-Khattam et al., 2004).

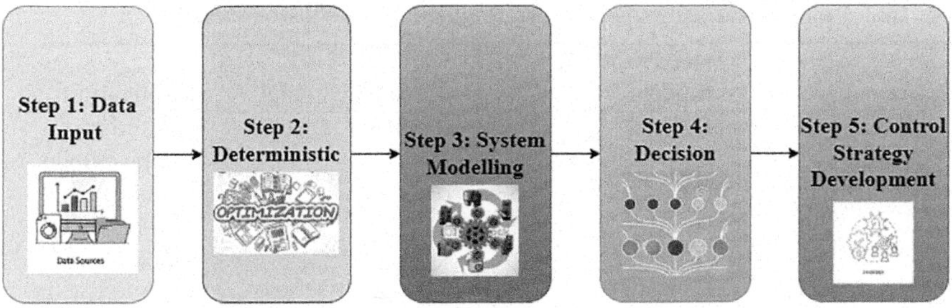

Figure 16.2 Deterministic planning workflow assuming fixed load and generation profiles.

16.2.1.1 Mathematical formulation

A generic deterministic DER planning problem can be expressed as:

$$\min_{x} C_{\text{total}} = \sum_{i=1}^{N} \left(C_{\text{inv},i} + C_{\text{op},i} \right) \tag{16.2}$$

subject to:

$$P_{\text{gen},i} + P_{\text{grid}} = P_{\text{load}}, V_{\min} \leq V_i \leq V_{\max} \tag{16.3}$$

where $C_{\text{inv},i}$ and $C_{\text{op},i}$ represent investment and operational costs, $P_{\text{gen},i}$ is local generation, and V_i is the bus voltage magnitude.

Deterministic frameworks are computationally efficient and well-suited for small-scale planning, often providing closed-form or near-optimal solutions. They may, however, cause overconfidence in decision-making and plants or operations that cannot be most optimal through their disregard of variability, which are uncovered to real-world variability (Al-Saffar et al., 2021). The deterministic planning workflow used in this study is presented in Fig. 16.2.

16.2.2 Limitations of deterministic planning

Deterministic models may be conceptually easy to use, but they do not model system behavior in the face of renewable uncertainty. They have weaknesses such as a low degree of robustness to predict deviations, reliance on one operating scenario, the possible mispricing of the investment cost, and the inability to adapt to changing distributed architecture. Such inadequacies have resulted in the greater use of stochastic and probabilistic models, which directly include variability (Reddy et al., 2017). A comparative assessment of deterministic and stochastic models is presented in Table 16.2.

Table 16.2 Comparative assessment of deterministic and stochastic models.

Parameter	Deterministic model	Stochastic model	Implication
Uncertainty	Ignored	Explicitly modeled	Reliability improvement
Data requirement	Low	High	Requires probabilistic datasets
Computational cost	Low	High	Trade-off with accuracy
Solution type	Point estimate	Probability distribution	Risk quantification
Suitability	Small-scale, static	Large-scale, dynamic	Realistic operation

16.2.3 *Stochastic planning paradigm*

In contrast to deterministic frameworks, the stochastic planning models define the uncertainties, e.g., wind speed, solar irradiance, load, and market prices, as random variables whose probability distributions are known or estimated (Massucco et al., 2020). The method aims at achieving the best expected performance and making systems viable as well as maintainable relative to uncertainty and acceptable risk.

A general stochastic optimization problem is expressed as:

$$\min_{x} \mathbb{E}_{\xi}[C(x, \xi)] \tag{16.4}$$

subject to:

$$g(x, \xi) \leq 0, h(x, \xi) = 0$$

where x denotes decision variables and ξ the vector of random parameters (generation, demand, prices, etc.) (Li et al., 2015).

The stochastic approaches are computationally expensive and yet offer statistically strong and economically sound solutions. They produce probability distributions of key performance indicators, including cost, losses, and renewable curtailment, to assist in risk-based planning and decision-making.

16.2.4 *Classification of stochastic approaches*

Stochastic planning methods are generally classified as scenario-based, probabilistic, robust, or hybrid AI-assisted approaches, depending on how uncertainty is modeled and propagated.

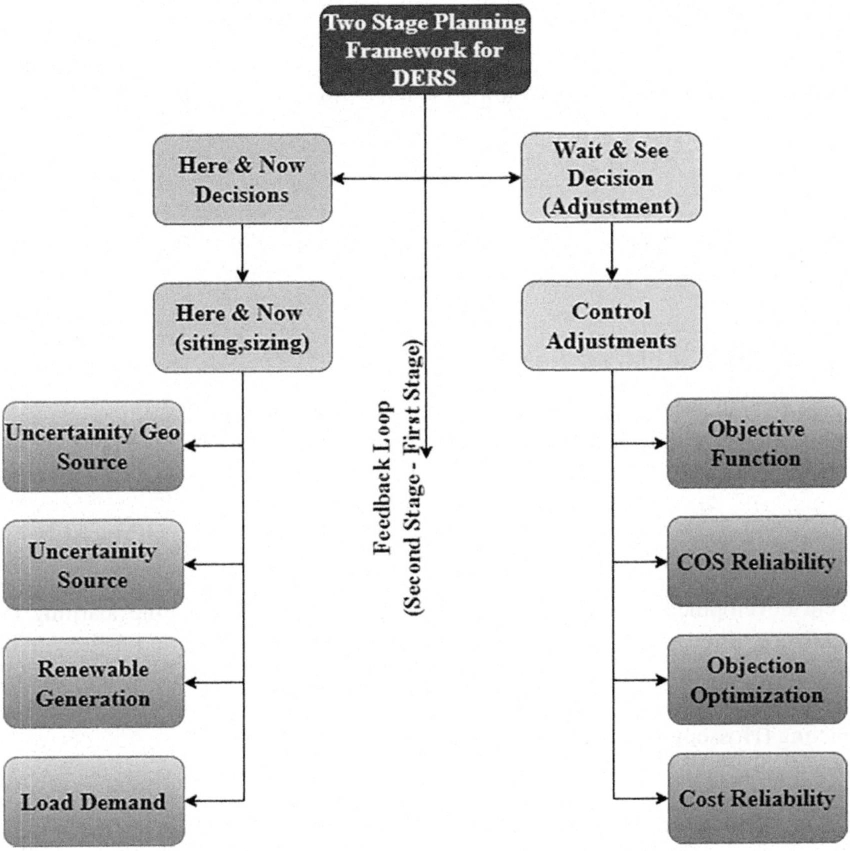

Figure 16.3 Structure of two-stage scenario-based stochastic programming for DER planning. *DER*, Distributed energy resource.

16.2.4.1 Scenario-based stochastic programming

Scenario-based stochastic programming (SBSP) is a form of uncertainty in the sense that it consists of a number of discrete scenarios, each of which indicates a potential occurrence of variables like renewable generation or market prices, and has an associated probability p_s. The objective is to minimize the expected total cost:

$$\min_{x} \sum_{s=1}^{S} p_s C(x_s) \tag{16.5}$$

The two-stage version separates first-stage (investment) and second-stage (operation) decisions, allowing adaptive responses to realized uncertainties (Osifeko et al., 2025). The structure of two-stage SBSP is shown in Fig. 16.3.

16.2.4.2 Probabilistic and chance-constrained optimization

The system constraints are expressed probabilistically, ensuring feasibility with a specified confidence level. For example, voltage limits are imposed as:

$$P(V_i \in [V_{\min}, V_{\max}]) \geq 1- \in \tag{16.6}$$

where $\in$ is the acceptable violation probability. Such methods balance reliability and cost through probabilistic safety margins (Zhou et al., 2020).

16.2.4.3 Robust optimization

In contrast to the use of probability distributions, robust optimization aims at worst-case conditions on identified sets of uncertainty and provides conservative but guaranteed solutions when statistical information is scarce or unavailable (Zhang et al., 2018).

16.2.4.4 Hybrid AI-assisted stochastic models

Recent developments combine stochastic optimization and machine learning (ML) to enhance forecasting, approximate probability distributions, and better solution convergence. Examples of approaches include deep reinforcement learning (DRL) to adaptive-decision making and Gaussian process regression (GPR) to probabilistic-modeling (Hosseini et al., 2021).

16.2.5 Comparative evaluation of optimization algorithms

A wide variety of algorithms have been applied for stochastic DER planning. Classical mathematical techniques (e.g., linear programming, nonlinear programming, and mixed-integer formulations) are being increasingly replaced by metaheuristic and hybrid algorithms for handling nonlinear, multiobjective, and nonconvex problems Table 16.3.

More and more, hybrid methods, where metaheuristics are applied to stochastic simulation (e.g., GA + Monte Carlo or NSGA-II + DRL), are increasingly preferred since they can address nonlinear constraints and multiobjective uncertainty more efficiently (Al-Saffar et al., 2021; Güven et al., 2023).

Deterministic and stochastic models are at one extreme and the other extreme of the planning; deterministic methods have been applied in preliminary studies or small-scale studies with few variables, and stochastic optimization is necessary as renewable penetration and system complexity increase. Moving ahead, hybrid models that combine AI-based prediction, stochastic programming, and robust control are the most promising options as they can be used to make adaptive decisions to meet short-term uncertainty in operations and long-term resilience to investments. Table 16.3 lists commonly used optimization algorithms for stochastic DER planning and their typical applications.

Table 16.3 Optimization algorithms used in stochastic DER planning.

Algorithm	Type	Feature	Typical application	References
MILP/MINLP	Classical	Exact solution, small scale	Grid-connected DERs	Domenech et al. (2021)
Genetic algorithm (GA)	Metaheuristic	Global search capability	DER siting/sizing	Noori et al. (2024)
Particle swarm optimization (PSO)	Swarm intelligence	Fast convergence	Multiobjective optimization	Grisales-Noreña et al. (2022)
NSGA-II/MOPSO	Evolutionary	Pareto-front generation	Multiobjective planning	Güven et al. (2023)
Deep RL	AI-based	Adaptive learning	Real-time decision-making	Al-Saffar et al. (2021)

DER, distributed energy resource; *RL*, reinforcement learning; MILP, mixed-integer linear programming; MINLP, mixed integer nonlinear programming.

16.3 Stochastic siting of distributed energy resources

Derivation of DERs has a lot of impact on system performance, reliability and cost in general. Poor location may result in overloading of the feeder, voltage variations, and quality of power. The conventional deterministic techniques do not consider uncertainties in renewable production, load dynamics, and network realities. Conversely, stochastic siting models are stochastic models that use probabilistic models and scenario analysis to have strong performance at varying operating conditions (Wu et al., 2020).

16.3.1 Concept and objectives of stochastic siting

Stochastic siting aims at optimizing the placement of the DERs to reduce the cost of the system, losses, and voltage changes due to uncertainty. It directly includes randomness in renewable availability, load growth, and component reliability in the objective and constraint formulations (Chihota et al., 2022).

The general objective function for stochastic siting is expressed as:

$$\min_{x} |\mathbb{E}_{\xi}\left[w_1 C_{\text{loss}}(x, \xi) + w_2 C_{\text{voltage}}(x, \xi) + w_3 C_{\text{investment}}(x, \xi)\right] \tag{16.7}$$

subject to operational, technical, and probabilistic constraints:

$$g(x, \xi) \leq 0, h(x, \xi) = 0 \tag{16.8}$$

where x represents the candidate site selection vector, and ξ encapsulates random parameters such as solar irradiance, wind speed, and demand variations.

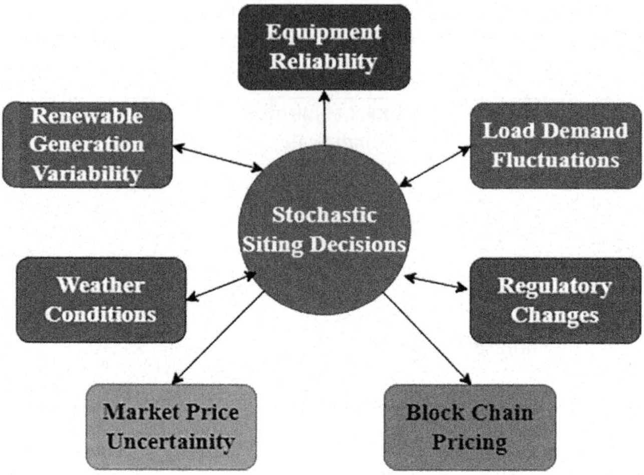

Figure 16.4 Classification of uncertainty sources influencing stochastic siting decisions.

The objective weights w_1,w_2,w_3 allow planners to balance between system efficiency, reliability, and economic performance.

16.3.2 Uncertainty modeling for siting decisions

Accurate representation of uncertainty is fundamental to stochastic siting. Typical sources of uncertainty include:

1. Renewable variability—Solar irradiance and wind speed are modeled using Beta and Weibull probability distributions, respectively (Manohar et al., 2020).
2. Load fluctuations—Daily and seasonal demand variations are represented using normal or log-normal distributions (Dab et al., 2024).
3. Equipment availability—Discrete random variables with reliability indices or failure probabilities capture component outages (Karafotis et al., 2022).
4. Market price volatility—Modeled using historical price distributions or autoregressive stochastic processes (Tehrani et al., 2022).

The major uncertainty sources affecting DER siting are categorized in Fig. 16.4. The joint probability distribution function (PDF) for all uncertainties is given by:

$$f(\xi) = f_{\text{solar}}(G) \cdot f_{\text{wind}}(v) \cdot f_{\text{load}}(P_L) \cdot f_{\text{price}}(\pi) \tag{16.9}$$

where G, v, P_L, and π represent irradiance, wind velocity, load demand, and market price, respectively.

Monte Carlo simulation is often used to generate random samples from these PDFs, which are then used to evaluate performance indices under thousands of possible realizations (Farh et al., 2024).

16.3.3 Formulation of the stochastic siting problem

The multiobjective stochastic siting problem typically aims to minimize expected losses, investment cost, and voltage deviation while maximizing reliability.

Objective function:

$$\min_{x} |\mathbb{E}_{\xi}\left[w_1 P_{\text{loss}}(x, \xi) + w_2 V_{\text{dev}}(x, \xi) + w_3 C_{\text{inv}}(x, \xi)\right] \tag{16.10}$$

Constraints:

1. Power balance:

$$\sum_{i} P_{\text{DER},i}(\xi) + P_{\text{grid}}(\xi) = P_{\text{load}}(\xi)$$

2. Voltage limits:

$$V_{\min} \leq V_i(\xi) \leq V_{\max}$$

3. Thermal limits:

$$| I_l(\xi) | \leq I_{l,\max}$$

4. Siting binary constraint:

$$x_i \in \{0, 1\} \forall i$$

where $x_i = 1$ indicates installation of a DER at the node i, and $x_i = 0$ otherwise (Saber et al., 2024).

16.3.4 Techniques for stochastic siting optimization

16.3.4.1 Monte Carlo-based evaluation

Monte Carlo Simulation (MCS) is widely applied for stochastic evaluation by generating numerous random realizations of uncertain parameters and statistically assessing performance metrics such as expected cost and reliability (F. Samadi Gazijahani et al., 2018).

$$\mathbb{E}[C(x, \xi)] \approx \frac{1}{N} \sum_{k=1}^{N} C(x, \xi_k) \tag{16.11}$$

While accurate, it becomes computationally intensive for large networks with multiple DER types.

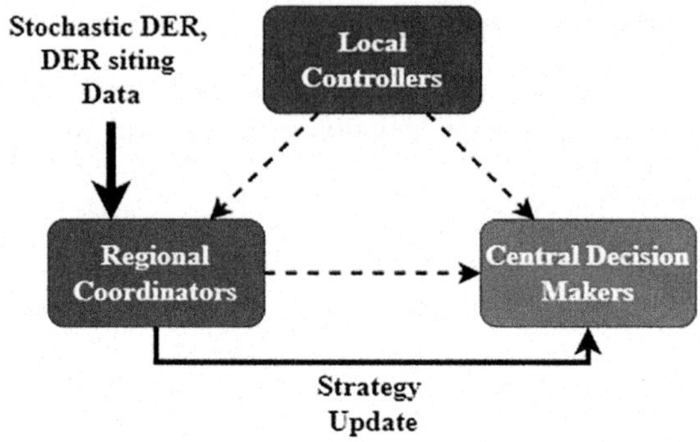

Figure 16.5 Multiagent decision framework for stochastic DER siting. *DER*, Distributed energy resource.

16.3.4.2 Scenario reduction techniques

To improve tractability, algorithms such as forward selection and backward deletion identify a smaller set of representative scenarios with minimal loss of accuracy (Hu et al., 2025).

16.3.4.3 Multiobjective evolutionary algorithms

MOEAs, including NSGA-II, MOPSO, and SPEA2, optimize siting decisions across conflicting objectives by encoding candidate solutions and evaluating fitness under stochastic conditions, enabling robust trade-off analysis (Seifpour et al., 2024).

16.3.4.4 Game-theoretic and agent-based siting

These approaches treat DERs as strategic agents that interact competitively or cooperatively for network resources, using equilibrium models to maximize system-wide benefits under uncertainty (Andoni et al., 2021). The multiagent decision-making process for stochastic DER siting is depicted in Fig. 16.5.

16.3.5 Spatial and network considerations

Siting decisions depend strongly on the network's topology and spatial characteristics. Stochastic frameworks, therefore incorporate both geographical and electrical constraints:

1. Geographical clustering—Spatial correlation models capture regional similarities in solar and wind resource availability (Zhang et al., 2024).
2. Electrical connectivity—Power-flow sensitivity factors (e.g., loss and voltage sensitivity) help identify nodes that provide maximum operational benefit.
3. Network partitioning—Large distribution systems are split into zones to simplify and scale stochastic optimization.

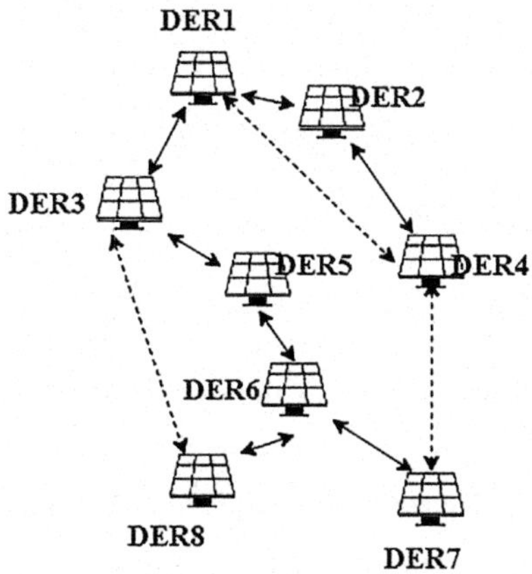

Figure 16.6 Geographic correlation modeling in stochastic DER siting. *DER*, Distributed energy resource.

The geographic correlation modeling used for regional resource analysis is shown in Fig. 16.6.

16.4 Stochastic sizing of distributed energy resources

Whereas siting is the whereabouts of DERs to be traversed, sizing is the extent of capacity the resources should possess in order to sustain the technical, economical and reliability in the presence of uncertainty. Because of the stochastic behavior of renewable output and load, deterministic sizing tends to oversize or undersize with unneeded investment or reduce reliability. Stochastic sizing models rather strike a balance between cost and performance in a variety of potential operating conditions (Wu et al., 2020).

16.4.1 Significance of stochastic sizing

In the context of power-system planning, optimal sizing is a method that guarantees that both distributed resources, including PV systems, wind turbines, and battery storage business efficiently but at minimal lifecycle costs. Nonetheless, renewable output is also uncertain because of the variability of the environment, and output in storage reduces with cycling and aging effects (Jiménez-Vargas et al., 2023). Stochastic sizing deals with these issues by introducing the idea of probabilistic models of solar irradiance, wind speed, and load profiles into the sizing optimization process so that a capacity decision based on stochastic models will be reliable and cost-effective given changing conditions.

The overarching objective is expressed as:

$$\min_{x} \mathbb{E}_{\xi}\left[C_{\text{inv}}(x) + C_{\text{op}}(x, \xi) + C_{\text{penalty}}(x, \xi)\right] \tag{16.12}$$

subject to probabilistic technical constraints:

$$P(V_i \in [V_{\min}, V_{\max}]) \geq 1 - \epsilon_V, P(P_{\text{loss}} \leq P_{\max}) \geq 1 - \epsilon_P \tag{16.13}$$

where x represents DER capacities and ξ the set of uncertain variables.

16.4.2 Stochastic optimization models for sizing

The stochastic optimization methods of DER sizing are classified into a few broad categories, namely, two-stage stochastic programming, chance-constrained optimization, robust optimization, and hybrid metaheuristic techniques.

16.4.2.1 Two-stage stochastic sizing model

This framework distinguishes between first-stage investment decisions (DER capacities) and second-stage operational responses (dispatch under uncertain conditions). The general structure is:

$$\min_{x} C_{\text{inv}}(x) + \mathbb{E}_{\xi}[Q(x, \xi)] \tag{16.14}$$

subject to:

$$Ax \leq b, T(x, \xi) \in \Omega$$

where $Q(x, \xi)$ is the second-stage cost function for each scenario.

Two-stage models are computationally efficient and can be used when the network is medium-sized, but they can fail to capture sequential temporal dependencies (Mohan et al., 2015).

16.4.2.2 Chance-constrained sizing

In this formulation, the constraints are met with a specific probability, whereby it is less conservative as compared to the robust optimization and more reliable as compared to deterministic techniques. For example:

$$P\left(P_{\text{grid}} \leq P_{\max}\right) \geq 0.95 \tag{16.15}$$

This is particularly useful when the uncertainty of renewable generation is overpowering the system, and flexibility is required without too much safety margin (Soltani et al., 2020).

Table 16.4 Classification of stochastic sizing approaches.

Method	Risk modeling	Strength	Limitation	References
Two-stage stochastic	Scenario-based	Balances cost and reliability	Limited temporal detail	Mohan et al. (2015)
Chance-constrained	Probability-based	Risk-adjustable design	Needs accurate PDFs	Soltani et al. (2020)
Robust	Set-based	Guaranteed feasibility	Conservative solution	Xie et al. (2022)
Hybrid	Combined	Flexible and resilient	High complexity	Nourollahi et al. (2021)

PDF, probability distribution function.

16.4.2.3 Robust and hybrid optimization

Robust optimization makes use of worst-case costs instead of probability distributions in reducing uncertainty sets. It is desired in cases where there are few or unreliable statistical data (Xie et al., 2022). Hybrid techniques between robust and stochastic frameworks, where probabilistic sampling is used to capture short-term variations and uncertainty sets are used to capture long-term factors, are used to provide both flexibility and resilience (Nourollahi et al., 2021). A comparison of stochastic sizing approaches and their strengths is given in Table 16.4.

16.4.3 Objective functions and constraints

Stochastic DER sizing models can integrate multiple objectives such as economic, technical, and environmental performance. Common objective functions include:

1 Economic objective*:*

$$\min |\mathbb{E}_{\xi}\left[C_{inv} + C_{op} - R_{energy}\right] \tag{16.16}$$

where R_{energy} represents expected revenue from energy sales or cost savings (Shufian et al., 2022).

2 Reliability objective*:*

$$\max |P_{\text{reliab}} = P(EENS < EENS_{\max}) \tag{16.17}$$

ensuring supply-demand balance under uncertainty.

3 Environmental objective:

$$\min \mathbb{E}_{\xi}[E_{\text{CO2}}]$$

which accounts for emission variability under stochastic generation (Gao et al., 2023).

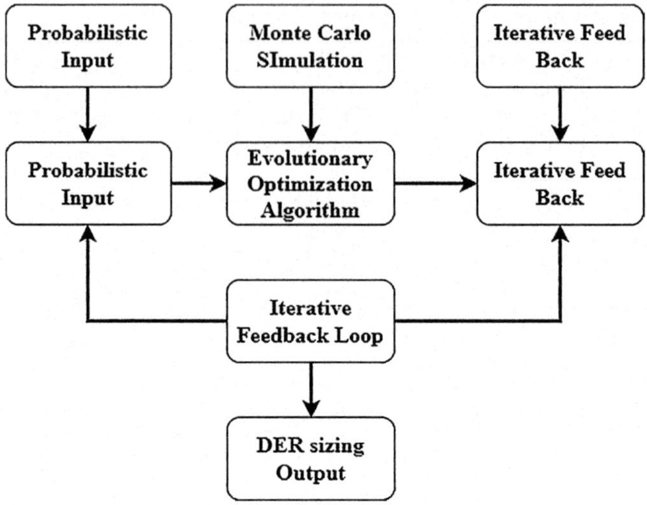

Figure 16.7 Monte Carlo-assisted evolutionary optimization loop for stochastic DER sizing. *DER*, Distributed energy resource.

4 Multiobjective hybrid model:

$$\min_{x} | \mathbb{E}_{\xi}[w_1 C_{\text{tot}} + w_2 L_{\text{index}} + w_3 E_{\text{CO2}}] \tag{16.18}$$

balancing cost (C_{tot}), losses (L_{index}), and emissions (E_{CO2}) through weighted summation or Pareto-front search (Moura & de Almeida, 2010).

16.4.4 Methods of solving stochastic sizing problems

16.4.4.1 Scenario-based decomposition

SBD divides complex stochastic optimization into multiple smaller subproblems, each corresponding to a scenario. Parallel computation techniques or decomposition algorithms (e.g., Benders decomposition) are often used for scalability (Zhang et al., 2023).

16.4.4.2 Monte Carlo—enhanced metaheuristics

Hybrid methods couple metaheuristics like genetic algorithms (GA), particle swarm optimization (PSO), and differential evolution (DE) with Monte Carlo sampling to evaluate uncertainty within each iteration (Jafari Siahroodi et al., 2022). The Monte Carlo–enhanced evolutionary optimization loop is illustrated in Fig. 16.7.

Table 16.5 Common algorithms in stochastic sizing applications.

Algorithm	Approach	Key feature	Application	References
Benders Decomposition	Scenario-based	Decomposes complex stochastic models	Multiperiod optimization	Zhang et al. (2023)
GA + MCS	Hybrid	Adaptive population evaluation	PV–wind–battery hybrid	Jafari Siahroodi et al. (2022)
NSGA-II	Multiobjective	Pareto-front generation	Techno-economic optimization	Moura & de Almeida (2010)
Deep RL	AI-based	Adaptive decision-making	Real-time capacity control	Feriani et al. (2021)

GA, genetic algorithms; *RL*, reinforcement learning.

16.4.4.3 AI-enabled predictive sizing

Recent developments integrate ML and RL to predict optimal capacity levels dynamically based on evolving weather, market, and demand data. RL agents can continuously adapt sizing decisions in response to real-time uncertainty feedback (Feriani et al., 2021). Table 16.5 summarizes the key algorithms used in stochastic DER sizing and their applications.

Stochastic sizing assists in efficient and robust planning of DER capacity using uncertainty to provide a full spectrum of uncertainties in the decision-making process. Hybrid stochastic-AI approaches are more flexible and more accurate in their predictions, whereas risk-aware expressions ensure the right balance between cost and reliability. In the future, the multilayered optimization structures, which combine RL and DTs, will enable scalable, adaptive, and self-learning distributed energy networks, which give continuous feedback between real and simulated systems.

16.5 Control Strategies for DERs Under Stochastic Conditions

16.5.1 Objectives and challenges of distributed energy resource control

The main objectives of DER control are to:

1. Maintain power balance between generation and load under fluctuating renewable output.
2. Regulate voltage and frequency within permissible limits.
3. Optimize economic dispatch while accounting for stochastic uncertainty.
4. Ensure system resilience under communication, component, or cyber disturbances.

However, several challenges complicate these goals in stochastic systems:

- Rapid fluctuations in renewable power cause real-time mismatches.
- Data latency and packet losses affect coordination in decentralized grids.
- Control tuning parameters may become suboptimal as system dynamics evolve (Zhang et al., 2019).

16.5.2 Hierarchical control architecture

The control of DER clusters in microgrids typically follows a three-tier hierarchical structure comprising primary, secondary, and tertiary control layers (Vasquez et al., 2010). Each layer operates at different time scales and decision hierarchies.

16.5.2.1 Primary control—fast local regulation

This layer is used to stabilize local voltages and frequencies based on droop control laws. Droop coefficients are actively adjusted to indicate stochastic changes in the generation or load.

$$f = f_0 - K_P(P - P_0), V = V_0 - K_Q(Q - Q_0) \tag{16.19}$$

where K_P and K_Q are active and reactive power droop constants (Bhatt et al., 2021).

The adaptive droop tuning systems with a fuzzy or neural algorithm improve the resilience in stochastic operation.

16.5.2.2 Secondary control—system restoration

Secondary control rectifies the deviations that are added by primary control by communicating centralized or distributed algorithms.

A consensus-based secondary control law can be represented as:

$$\dot{x}_i = -\sum_{j \in N_i} a_{ij}\left(x_i - x_j\right) + b_i(x_i - x_{\text{ref}}) \tag{16.20}$$

where a_{ij} are adjacency weights among DER nodes (Espina et al., 2023).

16.5.2.3 Economic and power flow optimization tertiary control

On top of this, tertiary control has control over the economic dispatch, power trading, and grid-interactive operations. It is based on optimization of prediction and decision-making that is market-conscious (Hou et al., 2018). The hierarchical control structure for DER microgrids is presented in Fig. 16.8.

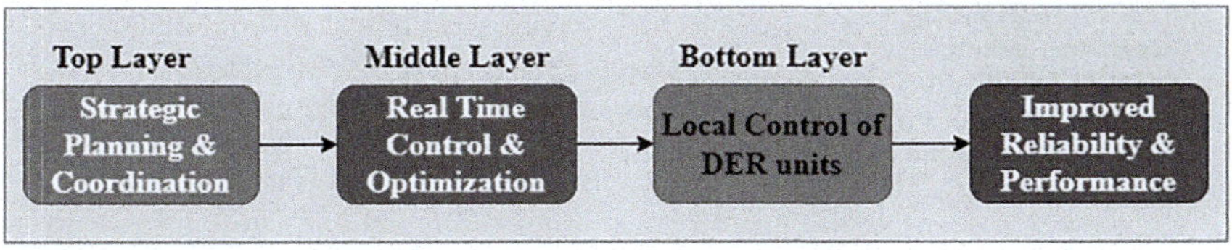

Figure 16.8 Three-layer hierarchical control structure for stochastic DER microgrids. *DER*, Distributed energy resource.

Table 16.6 Comparison of control architectures under stochastic conditions.

Control type	Features	Strengths	Limitations	References
Centralized	Single controller with full data visibility	Global optimization	Communication bottlenecks	Zarco-Soto et al. (2021)
Decentralized	Local controllers with minimal communication	Scalable and fault-tolerant	Suboptimal global coordination	Li et al. (2015)
Distributed	Peer-to-peer consensus-based control	Balances autonomy and coordination	Sensitive to network latency	Baros and Ilić (2017)

16.5.3 Centralized versus decentralized control under uncertainty

The control architecture of stochastic microgrids can follow either centralized, decentralized, or hybrid (distributed) configurations Table 16.6.

Centralized architectures perform well for small systems but lack scalability. Decentralized control enhances resilience but may lead to suboptimal global performance. Distributed approaches, using multiagent consensus algorithms, provide a balance by combining local autonomy with global synchronization (Baros and Ilić, 2017).

16.5.4 Stochastic model predictive control

The capability to deal with multivariable optimization and constraints in real time has made model predictive control (MPC) one of the most potent control methods in the DER systems. In stochastic settings, SMPC builds on top of MPC by incorporating probabilistic predictions and uncertainty archetypes into the control horizon (Hu et al., 2025).

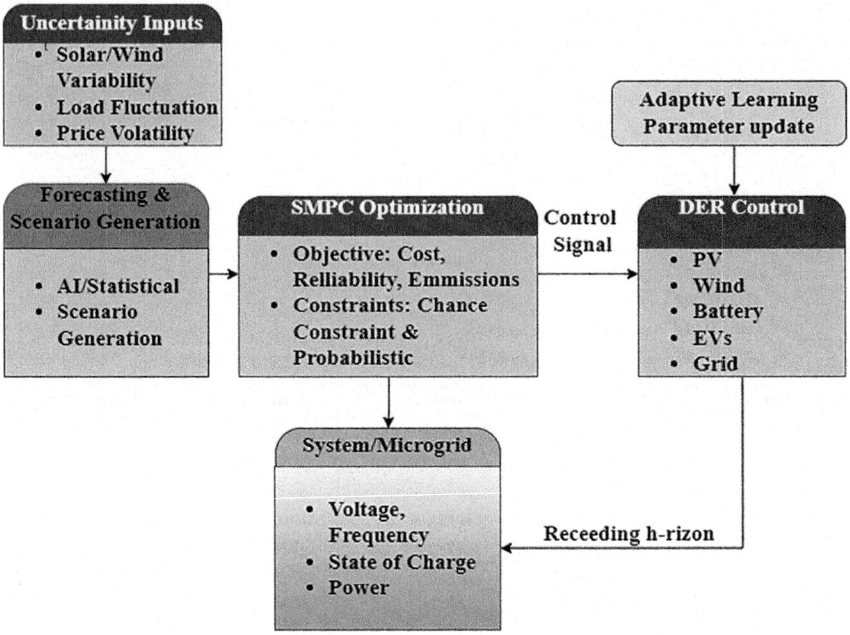

Figure 16.9 SMPC-based adaptive DER control structure. *DER*, Distributed energy resource; *SMPC*, stochastic model predictive control.

Mathematical formulation

$$\min_{u(0),..,u(N-1)} \mathbb{E}\left[\sum_{k=0}^{N-1} \ell(x_k, u_k, \xi_k) + V_f(x_N)\right] \tag{16.21}$$

subject to:

$$x_{k+1} = f(x_k, u_k, \xi_k), P(g(x_k, u_k, \xi_k) \leq 0) \geq 1- \in \tag{16.22}$$

where ξ_k represents random disturbances and ε the allowable constraint violation probability.

SMPC controllers predict future states over a horizon N, re-optimizing decisions at every time step based on updated forecasts. The SMPC-based adaptive DER control system is modeled in Fig. 16.9.

16.5.5 AI-based distributed control

ML techniques increasingly support DER coordination:

RL

- Learns optimal policies from operational feedback.

- Handles nonlinearity and unknown dynamics.
- Suitable for plug-and-play DER participation (Hu et al., 2025).

Deep multiagent RL (MARL)

- Enables DER clusters to negotiate power flows and share information efficiently.
- Achieves near-global optimality without centralized command (Al-Saffar et al., 2021).

Forecast-aware controllers

- AI-based forecasts guide proactive power scheduling.
- Improves resilience and operation costs under stochastic disturbances.

These methods support evolving microgrids with highly variable conditions and prosumer participation.

Centralized architectures are good with small systems but cannot be scaled. Decentralized control has a good level of resilience but can give rise to weak global performance. In response to these two opposing solutions, distributed methods with multiagent consensus algorithms offer a compromise between local agency and global coordination (Baros and Ilić, 2017). Hierarchical control provides stability whereas SMPC and AI-based distributed strategies provide flexibility in uncertainty. With the increasing DER penetration, the next-generation microgrid control will be characterized by hybrid architectures with predictive optimization and multiagent learning.

16.6 Challenges, research gaps, and future directions

In spite of impressive improvements in stochastic planning, control and digital integration of DERs, there remain a number of challenges that do not allow large-scale adoption and reliable operation. These issues are as a result of data availability constraints, scalability in computing, assurances in cyberspace security, and policy noninteroperability. All these are important barriers to be overcome in order to achieve the best in intelligent and adaptive energy systems.

16.6.1 Technical challenges

16.6.1.1 High computational complexity

Stochastic siting and sizing models are multidimensional uncertainty, and usually demand very high numbers of scenarios to be generated (thousands of runs). The result is the exponentially increasing computation time and memory consumption, especially in the case of high-resolution temporal data (Roberts et al., 2021). Even though complexity reduction methods (e.g., K-means clustering, backward deletion) reduce loss, the methods may compromise the fidelity. The necessity of real-time stochastic optimization that is compatible with DT and IoT data streams is also one of the research bottlenecks.

16.6.1.2 *Limited real-time adaptability*

The majority of the existing stochastic models are off-the-shelf and use past information. Nonetheless, renewable generation and market prices are nonstationary and are not fixed, but they change with time. Shifting to online adaptive models, where probability distributions and decision parameters are updated dynamically, is a continuing problem (Garg et al., 2013).

16.6.1.3 *Lack of interoperability in communication protocols*

The IoT devices and standards of communication of heterogeneous services, like the Message Queuing Telemetry Transport (MQTT) or Constrained Application Protocol (CoAP), and Open Platform Communications – Unified Architecture (OPC-UA), are not always seamlessly interoperable. This lack of consistency constrains data integration between platforms of DT-IoT-cyber-physical systems (CPS) and real-time control synchronization (Wen et al., 2025).

16.6.2 Data and modeling challenges

16.6.2.1 *Data scarcity and privacy*

Stochastic modeling involves extensive detailed data of renewable output, load demand, and market prices. Nonetheless, its availability of data is not high owing to ownership challenges and privacy concerns among the utilities and the customers (Habibi et al., 2022). This gap could be overcome by federated data governance models and privacy-preserving ML (PPML).

16.6.2.2 *Model generalization*

Stochastic models constructed using AI and conditioned to particular geographic or climatic circumstances are prone to fail in generalizing to new areas. Transfer learning and domain adaptation models are needed to make sure that it has scalability and cross-regional robustness (Lv et al., 2022).

16.6.2.3 *Validation of digital twin fidelity*

DTs in stochastic energy systems should work efficiently in terms of reflecting physical conditions with few lags or errors. Nonetheless, the large-scale, distributed DT networks find this real-time fidelity difficult to maintain technically because of latency, noise, and synchronization errors (Liu et al., 2024).

16.6.3 Cybersecurity and reliability challenges

Stochastic DER systems have become more vulnerable to cyber-attacks, which are capable of affecting workflows, destroying probabilistic models, or altering control signals due to increasing digitalization. The possible attacks are a data injection, spoofing, man-in-the-middle attacks, and malware propagation (Taveras Cruz et al., 2025).

The use of blockchain-based authentication, artificial intelligence-based anomaly detection, and quantum-resistant encryption is being investigated, and its interplay with stochastic optimization structures is still at its very beginning.

Reliability issues also continue to exist, that provide a constant communication between network congestion or partial system failure. CPS architecture self-healing and fault-tolerant can be integrated in order to increase the reliability in stochastic and adversarial conditions (Piardi et al., 2020).

16.6.4 Economic and policy barriers

16.6.4.1 Lack of unified market mechanisms

The existing electricity markets are not currently stochastic, DER-friendly. DERs are not frequently considered in dynamic pricing, carbon credits, and ancillary service mechanisms because of their changing output maps. Lack of risk-adjusted models of compensation turns away investors (Pressmair et al., 2021).

16.6.4.2 Policy inconsistency

Regulatory frameworks are diverse among countries and even regions, and they serve to deter standardization. Data sharing, cybersecurity compliance, as well as interoperability certification of DER technologies require uniform policies (IEC, 2024).

16.6.4.3 Investment risk and uncertainty

Stochastic planning models are viewed by investors as being highly risky because of the probability of their results. Economic returns include such financial tools as green insurance and performance-based guarantees that can be used to encourage adoption and dispel uncertainty (Xiao et al., 2023).

16.6.5 Future research directions

In order to address the specified issues, new areas of research comprise:

1. AI-DT co-simulation platforms: Building coupled simulation platforms in which stochastic models are continuously synchronized with the DT in real time (Addo et al., 2025).
2. Optimization hybrid stochastic-RL: A method to train RL agents to vary probability distributions according to experimental data (Khan et al., 2024).
3. Quantum-inspired and parallel computation: Using quantum algorithms and high-performance computing to speed up stochastic optimization and scenario assessment (Yeboah et al., 2025).
4. Privacy-preserving federated energy learning (FEL): The joint training of AI on utilities without raw data exchange, with confidentiality, and enhanced generalization (Li et al., 2023).

5. Blockchain-integrated market design: Development of transparent and decentralized trading systems of DERs, based on blockchain to document and certify stochastic energy transactions (Salmani et al., 2022).

All of these directions in the future are aimed at the goal of having intelligent, secure, and responsive energy ecosystems that cannot be confined due to uncertainty but can learn it.

16.7 Conclusions

The shift to decentralized and sustainable power systems is altering the traditional principles of energy production, functioning, and regulation. Photovoltaic systems, wind turbines, energy storage units, and electric vehicles are the building blocks of this change, combined to form the category of DERs. But their unpredictable and changing character dictates the need to develop new ways of planning that transcend deterministic assumptions.

The chapter has given a detailed discussion of stochastic planning and operation structures of DER systems, including the problem of siting, sizing, and control in the face of uncertainty. A combination of probabilistic, AI-assisted, and digital technologies is a framework for the next generation of adaptable and resilient microgrids that will react smart to the changing conditions.

16.7.1 Summary of key contributions

16.7.1.1 Stochastic planning foundations

The chapter has started by comparing the deterministic models with the stochastic models, and it has been seen that the latter models are superior to the former models in their nature of modeling the variability of renewable resources, load demand, and market dynamics. The random variables and the probabilistic factor assure an optimal solution to the problem, along with viability to the real-world variations.

16.7.1.2 Siting and sizing frameworks

Stochastic optimal siting and sizing of DERs based on uncertainty in renewable availability, load dynamics, and component reliability was shown in one unified stochastic formulation. Multiobjective optimization and Monte Carlo simulations indicated that they can be used to represent trade-offs between cost, reliability, and emissions. The evolutionary algorithms cross-fertilized with the stochastic modeling became a strong tool for complex multiscenario systems.

16.7.1.3 Control strategies under uncertainty

The section about the issue of control has highlighted the significance of hierarchical, distributed, and AI-based control models. Adaptive droop control and SMPC to

RL-based decision-making: Both these methods offer the real-time flexibility they ensure that a system is stable in a fluctuating environment and market.

16.7.1.4 Digital Integration with digital twin, Internet of Things, and cyber-physical systems

Among the significant contributions of the chapter, we can outline the synergy of DTs, Internet of Things (IoT), and CPS in stochastic DER environments. Combined, these technologies allow acquiring data in real time, modeling virtually, and providing predictive analytics and linking the physical and digital energy facility.

16.7.1.5 Challenges and future outlook

This chapter has indicated that there is a scarcity of data, interoperability issues, cyber-security, and fractured issues in policy. This was augmented with future directions that focused on AI-DT co-simulation, federated learning, quantum-inspired optimization, and blockchain-enabled energy markets as one of the avenues toward autonomous and safe distributed energy systems.

16.7.2 Strategic insights

The study conducted within the framework of this chapter implies that four primary principles will characterize the further development of DER planning and microgrid operation, namely:

16.7.2.1 Stochastic intelligence

Uncertainty needs to be learned, and not simply suppressed. Stochastic optimization based on data analysis reinforced by learning will convert the uncertainty into actionable intelligence.

16.7.2.2 Digital synchronization

The predictive decision-making and constant optimization in physical energy systems (both micro and macro-scale) will be possible due to real-time mirroring of the physical systems with DTs.

16.7.2.3 Resilient and secure operation

It will need cyber-physical resilience and transparency via blockchain to protect distributed networks from failures and other malicious attacks, as well as to guarantee trust among participants.

16.7.2.4 Integrated market ecosystems

The energy system will become a decentralized market in which prosumers, utilities, and aggregators can freely communicate and trade using autonomous AI-based trading and optimization systems.

16.7.3 Vision for the next decade

In 2035, it is anticipated that the distributed energy systems will develop into self-learning, interoperability ecosystems where AI, edge computing, and DTs will work to handle stochasticity in the real time.

Microgrids will be self-sufficient entities, as they will make energy deals, optimize local generation, and make contributions to the overall grid operations based on flexibility and resilience. Stochastic optimization, cyber-physical intelligence, and policy harmonization are the three components according to which the next-generation grid, which is not only technically efficient but also economically viable and environmentally sustainable, will be developed.

16.7.4 Concluding remarks

The introduction of a sustainable energy future relies on the magnitude of uncertainty that can be modeled, handled, and turned to advantage. The stochastic approaches enable the energy planners and policymakers to make informed, flexible, and future-ready decisions. The combination of these techniques with digital and AI technologies would radically change DERs to be a passive asset, a moving component of a self-regulating grid.

Finally, intelligence in the paradox of humanizing the world through stochastic intelligence, digital synchronization, and secure cyber-physical integration will open the door to a robust, decarbonizing, and intelligent energy ecosystem, one of the enablers toward reaching global net-zero objectives and setting the process of clean energy transition into high gear.

References

Addo, K., Kabeya, M., & Ojo, E. E. (2025). AI-Powered Digital Twin Co-Simulation Framework for Climate-Adaptive Renewable Energy Grids. *Energies, 18*(21), 5593.

Al-Saffar, M., & Musilek, P. (2021). Distributed optimization for distribution grids with stochastic DER using multi-agent deep reinforcement learning. *IEEE access, 9*, 63059–63072.

Amjady, N., Attarha, A., Dehghan, S., & Conejo, A. J. (2017). Adaptive robust expansion planning for a distribution network with DERs. *IEEE Transactions on Power Systems, 33*(2), 1698–1715.

Andoni, M., Robu, V., Couraud, B., Früh, W. G., Norbu, S., & Flynn, D. (2021). Analysis of strategic renewable energy, grid and storage capacity investments via Stackelberg-cournot modelling. *IEEE Access, 9*, 37752–37771.

Bakhtiari, Hamed, Zhong, Jin, & Alvarez, Manuel (2021). Predicting the stochastic behavior of uncertainty sources in planning a stand-alone renewable energy-based microgrid using Metropolis–coupled Markov chain Monte Carlo simulation. *Applied Energy, 290*, 116719.

Baros, S., & Ilić, M. D (2017). A consensus approach to real-time distributed control of energy storage systems in wind farms. *IEEE Transactions on Smart Grid, 10*(1), 613–625.

Bhatt, N., Sondhi, R., & Arora, S. (2021). Droop control strategies for microgrid: A review. In *Advances in Renewable Energy and Electric Vehicles: Select Proceedings of AREEV 2020* (pp. 149–162).

Chihota, M. J., Bekker, B., & Gaunt, T. (2022). A stochastic analytic-probabilistic approach to distributed generation hosting capacity evaluation of active feeders. *International Journal of Electrical Power & Energy Systems, 136*, 107598.

Dab, K., Nagarsheth, S. H., Amara, F., Henao, N., Agbossou, K., Dubé, Y., & Sansregret, S. (2024). Uncertainty quantification in load forecasting for smart grids using non-parametric statistics. *IEEE Access, 12*, 138000–138017.

Domenech, Carmen Bas, Riaz, Shariq, & Mancarella, Pierluigi (2021). Distribution system planning under uncertainty: A comparative analysis of decision-making approaches. In *2021 IEEE Madrid PowerTech*. IEEE.

El-Khattam, W., & Salama, M. M. (2004). Distributed generation technologies, definitions and benefits. *Electric power systems research, 71*(2), 119–128.

Espina, E., Cárdenas-Dobson, R. J., Simpson-Porco, J. W., Kazerani, M., & Sáez, D. (2023). A consensus-based distributed secondary control optimization strategy for hybrid microgrids. *IEEE Transactions on Smart Grid, 14*(6), 4242–4255.

Farh, H. M. H., Al-Shamma'a, A. A., Alaql, F., Omotoso, H. O., Alfraidi, W., & Mohamed, M. A. (2024). Optimization and uncertainty analysis of hybrid energy systems using Monte Carlo simulation integrated with genetic algorithm. *Computers and Electrical Engineering, 120*, 109833.

Feriani, A., & Hossain, E. (2021). Single and multi-agent deep reinforcement learning for AI-enabled wireless networks: A tutorial. *IEEE Communications Surveys & Tutorials, 23*(2), 1226–1252.

Güven, A. F., Yörükeren, N. U. R. A. N., Tag-Eldin, E., & Samy, M. M. (2023). Multi-objective optimization of an islanded green energy system utilizing sophisticated hybrid metaheuristic approach. *IEEE Access, 11*, 103044–103068.

Gao, J., & Huang, H. (2023). Stochastic optimization for energy economics and renewable sources management: A case study of solar energy in digital twin. *Solar Energy, 262*, 111865.

Garg, V., Jayram, T. S., & Narayanaswamy, B. (2013, June). Online optimization with dynamic temporal uncertainty: Incorporating short term predictions for renewable integration in intelligent energy systems. In *Proceedings of the AAAI Conference on Artificial Intelligence: 27* (pp. 1291–1297).

Gazijahani, F. Samadi, & Salehi, J. (July 2018). Optimal Bilevel Model for Stochastic Risk-Based Planning of Microgrids Under Uncertainty. *IEEE Transactions on Industrial Informatics, 14*(7), 3054–3064.

Grisales-Noreña, L. F., Restrepo-Cuestas, B. J., Cortés-Caicedo, B., Montano, J., Rosales-Muñoz, A. A., & Rivera, M. (2022). Optimal location and sizing of distributed generators and energy storage systems in microgrids: A review. *Energies, 16*(1), 106.

Habibi, M., Vahidinasab, V., & Sepasian, M. S. (2022). A privacy-preserving approach to day-ahead TSO-DSO coordinated stochastic scheduling for energy and reserve. *IET Generation, Transmission & Distribution, 16*(1), 163–180.

Hasani, A., Heydari, H., & Golsorkhi, M. S. (2024). Enhancing microgrid performance with AI-based predictive control: Establishing an intelligent distributed control system. *IET Generation, Transmission & Distribution, 18*(15), 2499–2508.

Hosseini, M. M., & Parvania, M. (2021). Artificial intelligence for resilience enhancement of power distribution systems. *The Electricity Journal, 34*(1), 106880.

Hou, X., Sun, Y., Lu, J., Zhang, X., Koh, L. H., Su, M., & Guerrero, J. M. (2018). Distributed hierarchical control of AC microgrid operating in grid-connected, islanded and their transition modes. *IEEE Access, 6*, 77388–77401.

Hu, J., Yan, P., & Tan, G. (2025). A two-layer optimal scheduling method for microgrids based on adaptive stochastic model predictive control. *Measurement Science and Technology, 36*(2), 026208.

Hu, J., & Li, H. (2019). A new clustering approach for scenario reduction in multi-stochastic variable programming. *IEEE Transactions on Power Systems, 34*(5), 3813–3825.

International Energy Agency. (2024). *Renewables 2024*. Global Outlook.

Jafari Siahroodi, H., Mojallali, H., & Mohtavipour, S. S. (2022). A new stochastic multi-objective framework for the reactive power market considering plug-in electric vehicles using a novel metaheuristic approach. *Neural Computing and Applications, 34*(14), 11937–11975.

Jiménez-Vargas, I., Rey, J. M., & Osma-Pinto, G. (2023). Sizing of hybrid microgrids considering life cycle assessment. *Renewable Energy, 202*, 554–565.

Karafotis, P. A., Evangelopoulos, V. A., & Georgilakis, P. S. (2022). Reliability-oriented reconfiguration of power distribution systems considering load and RES production scenarios. *IEEE Transactions on Power Delivery, 37*(6), 4668–4678.

Khan, N. H., Wang, Y., Jamal, R., Iqbal, S., Elbarbary, Z. M. S., Alshammari, N. F., & Jurado, F. (2024). A novel modified artificial rabbit optimization for stochastic energy management of a grid-connected microgrid: A case study in China. *Energy Reports, 11*, 5436–5455.

Kundur, P. (2009). Power System Stability and Control TMH 8th reprint.

Li, Q., Chen, F., Chen, M., Guerrero, J. M., & Abbott, D. (2015). Agent-based decentralized control method for islanded microgrids. *IEEE Transactions on Smart Grid, 7*(2), 637–649.

Li, C., & Grossmann, I. E. (2021). A review of stochastic programming methods for optimization of process systems under uncertainty. *Frontiers in Chemical Engineering, 2*, 622241.

Li, Y., He, S., Li, Y., Shi, Y., & Zeng, Z. (2023). Federated multiagent deep reinforcement learning approach via physics-informed reward for multimicrogrid energy management. *IEEE Transactions on Neural Networks and Learning Systems, 35*(5), 5902–5914.

Liu, W., Fu, Y., Guo, Y., Wang, F. L., Sun, W., & Zhang, Y. (2024). Two-timescale synchronization and migration for digital twin networks: A multi-agent deep reinforcement learning approach. *IEEE Transactions on Wireless Communications, 23*(11), 17294–17309.

Lopes, J. A. P., Hatziargyriou, N., & Mutale, J. (2011a). Integrating distributed generation into electric power systems. *Proceedings of the IEEE, 99*(1), 28–39.

Lv, L., Wu, Z., Zhang, L., Gupta, B. B., & Tian, Z. (2022). An edge-AI based forecasting approach for improving smart microgrid efficiency. *IEEE Transactions on Industrial Informatics, 18*(11), 7946–7954.

Manohar, M., Koley, E., & Ghosh, S. (2020). Stochastic weather modeling-based protection scheme for hybrid PV–wind system with immunity against solar irradiance and wind speed. *IEEE Systems Journal, 14*(3), 3430–3439.

Massucco, S., Mosaico, G., Pongiglione, P., Saviozzi, M., & Silvestro, F. (2020, August). Probabilistic planning for distribution networks including optimal der regulation and storage allocation. In *2020 IEEE Power & Energy Society General Meeting (PESGM)* (pp. 1–5). IEEE.

Mohan, V., Singh, J. G., & Ongsakul, W. (2015). An efficient two stage stochastic optimal energy and reserve management in a microgrid. *Applied energy, 160*, 28–38.

Moura, P. S., & de Almeida, A. T. (2010). Multi-objective optimization of a mixed renewable system with demand-side management. *Renewable and Sustainable Energy Reviews, 14*(5), 1461–1468.

Noori, A., Tavassoli, B., & Fereidunian, A. (2024). Evaluation of Prosumer Networks for Peak Load Management in Iran: A Distributed Contextual Stochastic Optimization Approach. arXiv:2409.00493.

Nourollahi, R., Salyani, P., Zare, K., & Mohammadi-Ivatloo, B. (2021). Resiliency-oriented optimal scheduling of microgrids in the presence of demand response programs using a hybrid stochastic-robust optimization approach. *International Journal of Electrical Power & Energy Systems, 128*, 106723.

Osifeko, M., & Munda, J. (2025). Scenario-Based Stochastic Optimization for Renewable Integration Under Forecast Uncertainty: A South African Power System Case Study. *Processes, 13*(8), 2560.

Piardi, L., Leitão, P., & de Oliveira, A. S. (2020, July). Fault-tolerance in cyber-physical systems: Literature review and challenges. In *2020 IEEE 18th International Conference on Industrial Informatics (INDIN): 1* (pp. 29–34). IEEE.

Pressmair, G., Kapassa, E., Casado-Mansilla, D., Borges, C. E., & Themistocleous, M. (2021). Overcoming barriers for the adoption of Local Energy and Flexibility Markets: A user-centric and hybrid model. *Journal of Cleaner Production, 317*, 128323.

Reddy, S. S., Sandeep, V., & Jung, C. M. (2017). Review of stochastic optimization methods for smart grid. *Frontiers in Energy, 11*(2), 197–209.

Roberts, C., Ngo, S. T., Milesi, A., Scaglione, A., Peisert, S., & Arnold, D. (2021, May). Deep reinforcement learning for mitigating cyber-physical der voltage unbalance attacks. In *2021 American Control Conference (ACC)* (pp. 2861–2867). IEEE.

Saber, H., Ranjbar, H., & Mazaheri, H. (2024, February). A stochastic multi-objective planning framework for distributed energy resources as an alternative to transmission expansion. In *2024 IEEE Texas Power and Energy Conference (TPEC)* (pp. 1–6). IEEE.

Salmani, H., Rezazade, A., & Sedighizadeh, M. (2022). Stochastic peer to peer energy trading among charging station of electric vehicles based on blockchain mechanism. *IET Smart Cities, 4*(2), 110–126.

Seifpour, M., Asghari, S. A., & Ghobaei-Arani, M. (2024). A stochastic multi-objective optimization method for railways scheduling: a NSGA-II-based hybrid approach: M. Seifpour et al. *The Journal of Supercomputing, 80*(2), 2128–2163.

Shen, X., Luo, Z., Xiong, J., Liu, H., Lv, X., Tan, T., & Dai, Y. (2021). Optimal hybrid energy storage system planning of community multi-energy system based on two-stage stochastic programming. *IEEE Access, 9*, 61035–61047.

Shufian, A., Hoque, M. J. A. M., Kabir, S., & Mohammad, N. (2022). Optimized performance and economic assessment for hybrid island microgrid system considering uncertainties. *Technology and Economics of Smart Grids and Sustainable Energy, 7*(1), 32.

Soltani, N. Y., & Nasiri, A. (2020). Chance-constrained optimization of energy storage capacity for microgrids. *IEEE Transactions on Smart Grid, 11*(4), 2760–2770.

Su, T., Zhao, J., Pei, Y., Yao, Y., & Ding, F. (2024). Analytic neural network gaussian process enabled chance-constrained voltage regulation for active distribution systems with PVs, batteries and EVs. *IEEE Transactions on Power Systems, 40*(3), 2037–2049.

Sun, H. Y., Mu, H. R., Sun, J., Han, H. G., & Qiao, J. F. (2025). Output Feedback Synthesis for Networked Control Systems With Packet Dropouts and Multiple Probability Sampling Periods: The Stochastic Communication Protocol Case. *IEEE Transactions on Cybernetics.*

Taveras Cruz, A. J., Aybar-Mejía, M., Colon-González, C. G., Mariano-Hernández, D., Hernandez, J. C., Andrade-Rengifo, F., & Hernández-Callejo, L (2025). Cybersecurity in MAS-Based Adaptive Protection for Microgrids—A Review. *Electronics, 14*(18), 3663.

Tehrani, S., Juan, J., & Caro, E. (2022). Electricity spot price modeling and forecasting in European markets. *Energies, 15*(16), 5980.

Vasquez, J. C., Guerrero, J. M., Miret, J., Castilla, M., & De Vicuna, L. G. (2010). Hierarchical control of intelligent microgrids. *IEEE Industrial Electronics Magazine, 4*(4), 23–29.

Wen, H., Huang, G., Liu, Y., Chen, W., Xu, G., & Lei, B. (2025, October). Toward Seamless IoT Interoperability for Energy and Mobility Integration: A Systemic Perspective. In *2025 25th Conference of the Electric Power Supply Industry (CEPSI)* (pp. 935–940). IEEE.

Wu, D., Ma, X., Huang, S., Fu, T., & Balducci, P. (2020). Stochastic optimal sizing of distributed energy resources for a cost-effective and resilient Microgrid. *Energy, 198*, 117284.

Xiao, D., Chen, H., Cai, W., Wei, C., & Zhao, Z. (2023). Integrated risk measurement and control for stochastic energy trading of a wind storage system in electricity markets. *Protection and Control of Modern Power Systems, 8*(4), 1–11.

Xie, R., Wei, W., Shahidehpour, M., Wu, Q., & Mei, S. (2022). Sizing renewable generation and energy storage in stand-alone microgrids considering distributionally robust shortfall risk. *IEEE Transactions on Power Systems, 37*(5), 4054–4066.

Yeboah, S. J., Nunoo, S., Attachie, J. C., Ansere, J. A., & Gyamfi, E. (2025). Quantum-Enhanced PPO for Demand Follower Model in Stochastic Industrial Smart Grid Environment. *IEEE Access, 14*, 966–982.

Zarco-Soto, F. J., Zarco-Periñán, P. J., & Martinez-Ramos, J. L. (2021). Centralized control of distribution networks with high penetration of renewable energies. *Energies, 14*(14), 4283.

Zhang, X., Xia, P., Peng, F., Xiao, M., Zhao, T., Fu, A., … Sun, X. (2024). Multiple spatial-temporal scales assessment of solar and wind resources potential integrating geospatial-technology-correlation indicators: A case study of Hunan Province. *Energy, 304*, 132036.

Zhang, C., Xu, Y., & Dong, Z. Y. (2018). Probability-weighted robust optimization for distributed generation planning in microgrids. *IEEE Transactions on Power Systems, 33*(6), 7042–7051.

Zhang, Y., Wang, J., & Li, Z. (2019). Uncertainty modeling of distributed energy resources: Techniques and challenges. *Current Sustainable/Renewable Energy Reports, 6*(2), 42–51.

Zhang, W., Wang, K., Jacquillat, A., & Wang, S. (2023). Optimized scenario reduction: Solving large-scale stochastic programs with quality guarantees. *INFORMS Journal on Computing, 35*(4), 886–908.

Zhou, Q., Shahidehpour, M., Paaso, A., Bahramirad, S., Alabdulwahab, A., & Abusorrah, A. (2020). Distributed control and communication strategies in networked microgrids. *IEEE Communications Surveys & Tutorials, 22*(4), 2586–2633.

Zhang, Y., & Wang, J. (2022). Fuzzy multi-objective optimization in DER planning. *Renewable Energy, 189*, 1450–1463.

Stochastic modeling for energy storage and hydrogen systems in hybrid electric platforms

17

Nazanin Pourmoradi[a], *Mohammadtaghi Ameli*[a] *and Hossein Ameli*[b]
[a]Department of Electrical Engineering, Shahid Beheshti University, Tehran, Iran, [b]EEE department, Control and Power Group, Imperial College, London, United Kingdom

17.1 Introduction

Electrochemical energy storage has become a cornerstone of contemporary electric powertrains and grid-interfaced systems, with batteries and supercapacitors (SCs) increasingly embedded in hybrid electric platforms. In these applications, model choice is pivotal: Reduced-order, prevalidated representations enable credible architecture exploration, control synthesis, and fast simulation without the computational burden of full electrochemical formulations. Within battery management system (BMS) and energy management system (EMS), ECMs are widely preferred because they retain essential transient dynamics while remaining identifiable from laboratory procedures such as hybrid pulse power characterization (HPPC) (Nkembi et al., 2024). Their parameters can be expressed as functions of operating conditions: State of charge (SoC), state of health (SoH), current direction, and temperature; thereby elevating prediction fidelity in realistic duty cycles (Chen & Rincon-Mora, 2006). In contrast, high-fidelity electrochemical models (Partial differential equations [PDE]-based) often impose prohibitive computational cost for real-time or large-scale system design (Saidani et al., 2017). The practical consequence for hybrid platforms is clear: ECMs provide a tractable backbone for control-oriented analysis, SoC/SoH estimation, and architecture-level co-design (Nkembi et al., 2024).

Parallel to batteries, SCs serve as high-power buffers that absorb sub-second power surges and mitigate battery current peaks (Naseri et al., 2022). A spectrum of terminal models, from basic resistor–capacitor (RC) cores and Randall/first-order variants to Debye, multibranch (e.g., Zubieta/Rafik), ladder, and transmission-line models, has been developed to capture the distributed electrochemical and thermal phenomena governing short- and long-timescale behavior. For control and power-electronics studies, RC families remain attractive due to ordinary differential equation (ODE)-level simplicity and parameter identifiability, while refined variants incorporate voltage-, frequency-, and temperature-dependent elements to maintain accuracy across operating bands up to the kilohertz range. These SC models support tasks ranging from efficiency estimation and loss partitioning to health monitoring and frequency-domain role allocation within hybrid energy storage systems (HESS) (Jiya et al., 2018; Naseri et al., 2022).

Stochastic Planning and Modeling for Energy Systems: Methods, Applications, and Developments.
DOI: https://doi.org/10.1016/B978-0-443-45298-7.00018-5

Hydrogen FCs, particularly proton exchange membrane fuel cells (PEMFCs), complement electrochemical storage by providing relatively uniform baseload power when integrated with HESS on the vehicle DC bus. Stack-level modeling, coupled with appropriate high-gain, often isolated DC–DC interfaces (e.g., dual-active-bridge or inductor–inductor–capacitor (LLC)), enables safe, efficient power transfer from low-voltage stacks to traction inverters (Wang et al., 2022). However, the efficacy and durability of any multisource hybrid electric vehicle (HEV) architecture, series, parallel, series–parallel, or complex, are ultimately constrained by converter selection and control. Bidirectional DC–DC converters mediate energy between sources and bus, enforce current and voltage safety, and shape transient responses (Zimmermann et al., 2016). Nonisolated high-gain families (interleaved, quadratic, coupled-inductor, switched-capacitor/inductor) and isolated families (phase-shifted Dual active bridge [DAB], current-fed with active clamp) each offer distinct trade-offs in gain, switching loss, device stress, and control complexity; selection hinges on required voltage ratio, isolation needs, power density, and cost. Multiinput topologies can reduce system mass and cost by aggregating sources while preserving deep discharge of SC banks for effective energy utilization (Mamun et al., 2022).

Crucially, design and operation proceed under pervasive uncertainty. Driving profiles, gradients, ambient and component temperatures, auxiliary loads, and parameter drift (e.g., internal resistance, polarization time constants) jointly determine the stochastic load spectrum experienced by HESS and power electronics. A scenario-based, uncertainty-quantified (UQ) workflow is therefore essential (Li et al., 2021). Representative natural cycles are complemented with data-augmented profiles generated by random sampling over velocity-transition probability surfaces; together, these furnish broad spectral coverage for EMS calibration and stress testing. Performance indicators, battery peak/RMS current, SoC occupancy distributions for battery and SC, converter losses, thermal margins, and aging proxies are evaluated distributionally rather than at single points. Risk-aware formulations then introduce chance constraints (e.g., respecting SoC/temperature windows with high probability) and tail-focused measures (e.g., conditional value-at-risk [CvaR] of peak current or equivalent aging cost), aligning online EMS tuning with offline sizing to avoid designs that are efficient on average yet fragile in the tails (Li et al., 2023; Pal et al., 2021).

Within this framework, role assignment becomes principled: SCs shoulder fast transients; batteries supply the slow energy component; and smoother sources (e.g., fuel cells) provide baseload. Converter-level controls enforce bidirectionality, limit recovery currents during regenerative braking, and maintain a clean DC link. Co-design of sizing, EMS policy, and reliability constraints closes the loop between component physics and system-level objectives, improving peak-current containment, thermal behavior, and effective aging. Finally, when on-board renewables (e.g., rooftop PV) are considered, their integration is assessed realistically; primarily to offset auxiliary loads and reduce peak charging, while life-cycle considerations (materials, recyclability, safety) inform policies that connect instantaneous performance to sustainability outcomes over the value chain. The overarching contribution is a stochastic, distribution-aware methodology that elevates hybrid electric platform design from mean-centric evaluation to defensible, tail-robust decision-making.

17.2 Stochastic models of energy-storage components

In recent years, electrochemical energy storage, namely batteries and SCs, has seen growing penetration in electric power generation and distribution infrastructures as well as in automotive applications. Within this context, abstract or reduced-order models assume heightened importance; they streamline design workflows while enabling credible simulation of architectures that integrate storage devices. Crucially, such models must be both accurate and prevalidated; otherwise, design outcomes and performance assessments lack reliability.

17.3 Li-ion batteries: ECMs, thermal coupling, and stochastic aging/SoH

Optimal battery operation requires a BMS that continuously monitors and acts upon key indicators, including SoC, SoH, temperature, and terminal voltage (Urooj & Nasir, 2024). Depending on the target phenomena and required fidelity, battery modeling approaches span a spectrum: (1) Empirical or analytical formulations that capture behavior with compact mathematical relations; (2) detailed electrochemical models; and (3) equivalent-circuit representations with a small set of lumped parameters.

In practice, when designing energy management controls and BMSs, most designers turn to equivalent circuits because they are simple, few-parameter, and largely linear, and can reproduce the transient responses of the battery with good accuracy with the same few resistors and capacitors. Another advantage of this approach is that it is not dependent on the specific cell chemistry and can be implemented on a variety of battery technologies (Mama et al., 2025). The parameters of such circuits also come directly from practical experiments under a range of real-world conditions, such as temperature variations, current magnitude/direction, charge level, and even aging effects.

In contrast, electrochemical models, despite their high accuracy, open us to the world of nonlinear PDEs and many unknown parameters; as a result, they require heavy computational resources, and a simulation can take hours to days to run (Shi et al., 2024). For this reason, for system applications and real-time design, equivalent circuits are usually a more practical choice.

In power electronics studies, batteries are often represented using equivalent-circuit models. Numerous variants have been proposed in the literature, and Fig. 17.1 depicts a representative configuration.

Fig. 17.1(A) shows the simplest equivalent circuit framework for a battery, the R_{int} model, which is the basis for many subsequent models (Hussein, 2021). The idea is simple: A state-dependent voltage source to represent V_{OC} that varies with SoC and temperature, and a series resistor (R_i) that sums up the ohmic losses in the internal paths from the junctions to the electrode and electrolyte (Einhorn et al., 2010). This model is suitable for describing the steady state, but it does not cover the transient responses and dynamic behavior of the cell; it is precisely this weakness that motivates the move to higher-order models (with RC branches) (Saidani et al., 2017). To compensate for R_{int}'s weakness in capturing transient behavior, we usually resort to the Thevenin

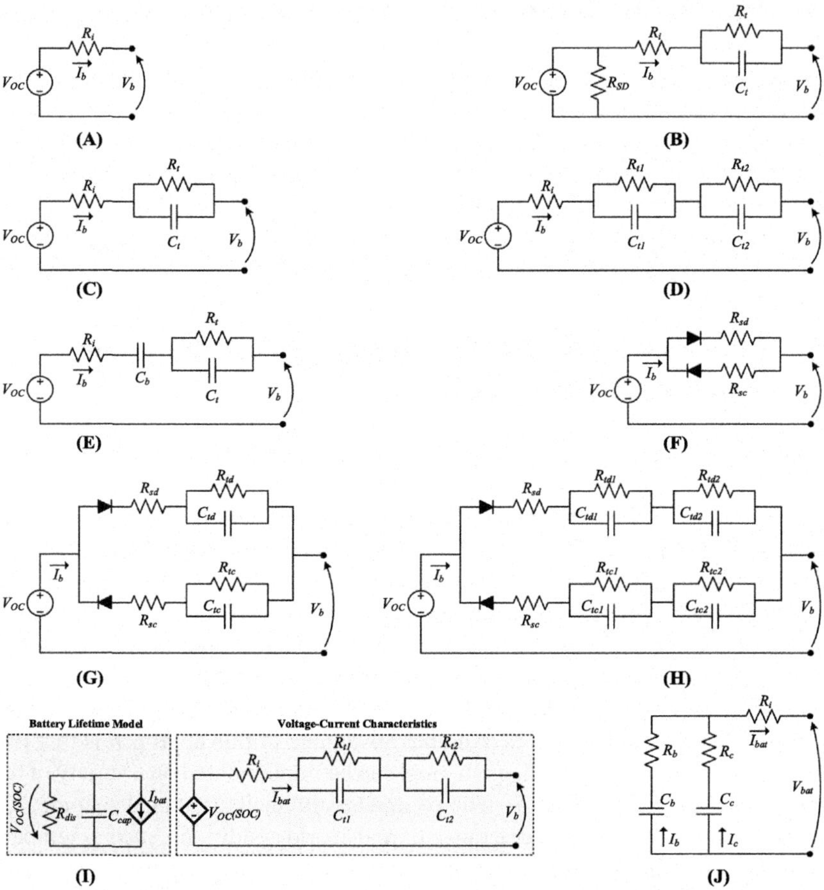

Figure 17.1 Common battery equivalent-circuit models: (A) the simple Rint model without dynamic behavior; (B) the first-order leaky model; (C) the simplified first-order leaky model; (D) the second-order model; (E) the PNGV model; (F) the modified Rint model with separate charging and discharging paths; (G) the first-order Thevenin model with separate charging and discharging paths; (H) the second-order model with separate charging and discharging paths; (I) the runtime model, in which the battery capacity/SoC dynamics and transient behavior are represented by two separate circuits; and (j) the NREL reference RC model, which has been widely used in automotive simulations [16].
From Nkembi, A. A., Simonazzi, M., Santoro, D., Cova, P., & Delmonte, N. (2024). Comprehensive review of energy storage systems characteristics and models for automotive applications. *Batteries, 10*(3). https://doi.org/10.3390/batteries10030088.

model [Fig. 17.1(B)]: a parallel RC branch is added to capture the instantaneous charge and discharge responses, and a leakage resistor is placed in parallel with the voltage source to represent the self-discharge of the cell in the no-load state (Sun & Shu, 2011). In many applications, this leakage resistor has a negligible effect and can be omitted [Fig. 17.1(C)] (Lee et al., 2008). If more accuracy is needed, we can achieve a second-order/bipolar (DP) model by adding a second RC: One RC captures the fast response and the other the slow behavior (Tjandra et al., 2014). Of course, increasing the order and adding more RCs, even up to the fifth order; improves accuracy, but also increases computational cost; so, a trade-off between accuracy and complexity must be made (Tran et al., 2021).

The Partnership for a New Generation of Vehicles (PNGV) model [Fig. 17.1(E)] is a family of equivalent circuit models developed for automotive applications and still widely used today (Johnson, 2002). Structurally, it is similar to the model in Fig. 17.1(C), but with one important difference: The addition of a large capacitor C_b to capture better the open-circuit voltage (V_{OC}) variations that occur in practice due to the accumulation of charge current over time during charging and discharging (Zhou et al., 2021). If more accuracy is required, we can increase the order of the PNGV model by adding more parallel RC branches to more realistically reflect the transient behavior of the cell during the charging/discharging process (Saldaña et al., 2019).

The National renewable energy laboratory (NREL)/SAFT model [Fig. 17.1(J)] is the same framework that is widely used in ADVISOR, a MATLAB/Simulink automotive simulation tool (Johnson, 2002). The idea is simple and efficient:

- Two capacitors, C_b (bulk) to represent the main storage capacity and C_c (surface) to obtain fast charge/discharge responses, the latter being much smaller than the former.
- The three resistors also play the role of different layers of losses/phenomena: R as the terminal/internal resistance, R_b to represent surface diffusion/scattering phenomena, and R_c to penetrate the internal layers (He et al., 2011; Zhang et al., 2016).

In this arrangement, the SoC can be obtained directly from the voltage across C_b; therefore, this model is both accurate for transient simulation and useful for state estimation and EMS design without entering the complexity of electrochemical models.

To make these battery equivalent circuits behave more realistically, it is sufficient not to keep R and C constant but to define them as dependent on the operating conditions, i.e., they vary with the SoC, SoH, charge/discharge direction, and temperature. This significantly increases the accuracy of the model's prediction. The relationships for the most used model are given below (Chen & Rincon-Mora, 2006):

$$V_{OC}(SoC) = -1.031*e^{-35*SoC}+3.685+0.2156*SoC-0.1178*SoC^2+0.3201*SoC^3 \quad (17.1)$$

$$R_i(SoC) = 0.1562 * e^{-24.37*SoC} + 0.07446 \quad (17.2)$$

$$R_{t1}(SoC) = 0.3208 * e^{-29.14*SoC} + 0.04669 \quad (17.3)$$

$$C_{t1}(SoC) = -752.9 * e^{-13.51*SoC} + 703.6 \quad (17.4)$$

$$R_{t2}(SoC) = 6.603 * e^{-155.2*SoC} + 0.04984 \quad (17.5)$$

$$C_{t2}(SoC) = -6056 * e^{-27.12*SoC} + 4475 \quad (17.6)$$

Table 17.1 Calculated parameters of a single-cell battery model across different SoC levels.

SoC (%)	V_{OC} (V)	R_i (Ω)	R_{t1} (Ω)	C_{t1} (F)	τ_{t1} (s)	R_{t2} (Ω)	C_{t2} (F)	τ_{t2} (s)
10	3.6746	0.088115	0.064096	508.613	32.6	0.049841	4072.9	203.0
20	3.7250	0.075654	0.047634	653.102	31.1	0.049840	4448.3	221.7
50	3.8034	0.074461	0.046690	702.723	32.8	0.049840	4475.0	223.0
80	3.9460	0.074460	0.046690	703.585	32.9	0.049840	4475.0	223.0
100	4.1029	0.074460	0.046690	703.599	32.9	0.049840	4475.0	223.0

Soc, State of charge.

To make the battery equivalent circuit model behave more realistically, the parameters R and C were not considered constant and were calculated as a function of SoC according to the presented equations. This method allows the model to reproduce the fundamental changes in no-load voltage, internal resistance, and loading dynamics under different conditions, an issue that is of particular importance in electric vehicle applications.

For numerical investigation, the model parameters were calculated at several practical points of the charge range (10%, 20%, 50%, 80%, and 100%). The results are given in Table 17.1 and show that the battery discharge voltage increases with increasing SoC, from around 3.67 V at 10% charge to more than 4.10 V at 100%. This behavior is quite consistent with the typical characteristic of lithium-ion cells, and the relatively steep slope of the curve in the low SoC region is the region where the voltage sensitivity to charge changes is high.

On the other hand, the battery series resistance decreases rapidly with increasing SoC and approaches a nearly constant value after about 50%. This, in practice, means that the battery heats up less and can handle higher currents at higher states of charge. The resistance of the short transient branch also behaves similarly, and the equivalent capacitance of this branch increases to about 700 Farads. The product of R and C in this branch, the time constant τ, is in the range of 31–33 seconds, indicating that this branch is responsible for reproducing the medium-term voltage dynamics.

The long transient branch is inherently slower, with a time constant of around 220–230 seconds, representing slower phenomena such as ion diffusion or thermal stability. The only caveat to using the equations is that at very low SoCs (especially near zero), the output of some functions becomes unrealistic and produces negative capacitances. This is primarily a consequence of the exponential fit and can be managed in practice by assuming a minimum SoC (e.g., 5 or 10%) or by imposing a limit on the values.

Overall, the findings suggest that the SoC model not only provides a more accurate representation of voltage variations but can also realistically simulate transient behavior and internal resistance. This ability is of great importance in investigating the performance of electric vehicles under conditions such as acceleration, regenerative braking, and rapid load fluctuations.

To extract the parameters of the equivalent circuit models from the actual data, a laboratory test is usually performed. The most common option is the HPPC test: A series of controlled current pulses applied to the battery to measure its transient responses. In this experimental method, the battery is first discharged. Then, a charge

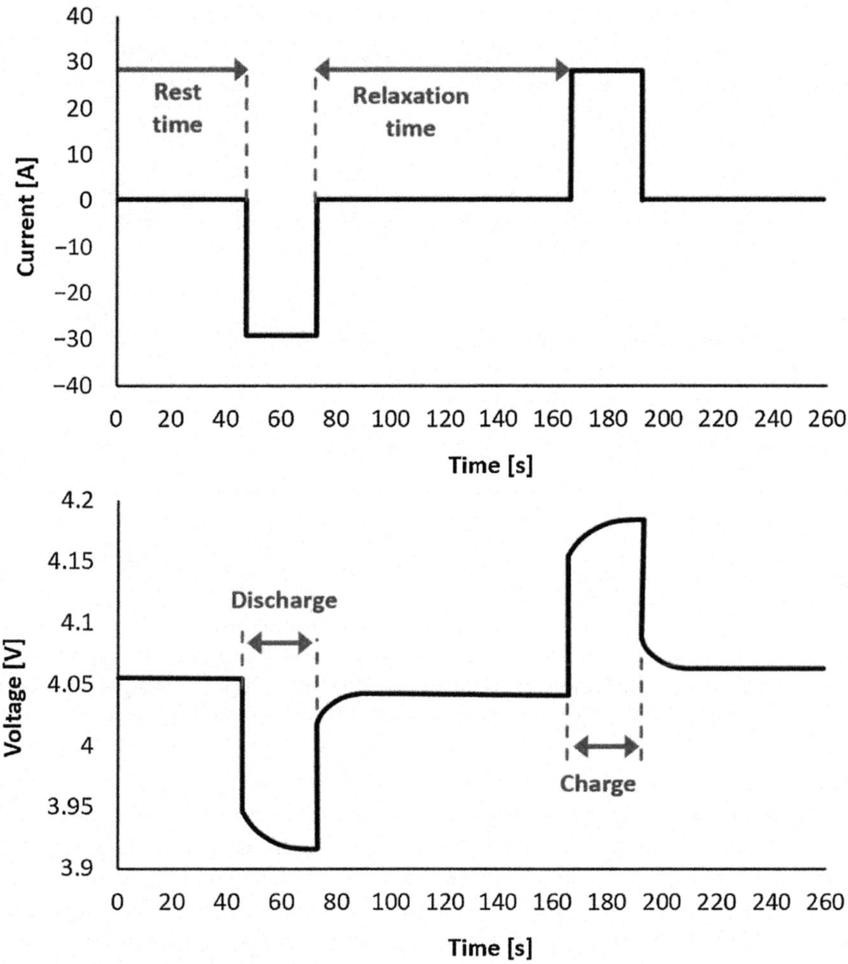

Figure 17.2 HPPC test sample template used to extract/tune battery model parameters. *HPPC*, Hybrid pulse power characterization.
From Nkembi, A. A., Simonazzi, M., Santoro, D., Cova, P., & Delmonte, N. (2024). Comprehensive review of energy storage systems characteristics and models for automotive applications. *Batteries, 10*(3). https://doi.org/10.3390/batteries10030088.

cycle according to a specific schedule is performed. During these steps, the voltage, current, and temperature are simultaneously measured and stored to determine the system's transient behavior (Thanagasundram et al., 2012); an example of how the current and voltage change during this test is shown in Fig. 17.2.

To verify the accuracy of the extracted results, the desired parameters were applied to the proposed model in the Cadence environment to simulate the battery voltage response against the same pulsed discharge currents that were used to extract the

Table 17.2 Accuracy of model parameter extraction for lithium-ion polymer batteries.

Puls discharge current (mA)	Max error voltage (mV)	Runtime error[a] (%)
80	15	0.039
160	17	0.118
320	18	0.020
640	21	0.029

From Chen, M., & Rincon-Mora, G. A. (2006). Accurate electrical battery model capable of predicting runtime and I–V performance. IEEE Transactions on Energy Conversion, 21 (2), 504–511.
[a] Runtime Error = [Runtime (Sim) – Runtime (Exp)] / Runtime (Exp).

Table 17.3 Model validation results (lithium-ion polymer batteries).

Load profiles	Max error voltage (mV)	Runtime error (%)
Continuous discharge	15	0.395
Pulse charge	30	0.133
4-Step discharge	20	0.338

From Chen, M., & Rincon-Mora, G. A. (2006). Accurate electrical battery model capable of predicting runtime and I–V performance. IEEE Transactions on Energy Conversion, 21 (2), 504–511.

parameters. Table 17.2 shows the voltage error and simulation execution time for each discharge current level, and Fig. 17.2 presents the simulation results along with the experimental data for discharge currents of 80, 320, and 640 mA. The proposed model is able to accurately reproduce the voltage of lithium-ion polymer batteries with an error of less than 21 mV and a runtime of less than 0.12%. This close agreement indicates the high accuracy of the parameter extraction process.

To validate the model extracted from lithium-ion polymer batteries, three different charge profiles, including continuous discharge, pulse discharge, and four-stage cyclic discharge, were applied to the battery. In the first case, the battery was discharged with a constant current of 80 mA. Fig. 17.3(A) shows the simulation results compared with the experimental data, with a maximum voltage difference of 15 mV and a runtime error of 0.395%. In the second case, the battery was subjected to a pulse charge profile with a constant current of 80 mA and then a constant voltage of 4.1 V. As shown in Fig. 17.3(B), the simulation results generally agree well with the experimental data. However, there is a slight difference during the transition period from constant current to constant voltage. This difference is because the charging circuit modeled in Cadence has slightly different characteristics from the actual circuit implemented on the prototype PCB. In the third step, the battery was discharged with a four-step cyclic load profile consisting of currents of 0, 400, 160, and 640 mA; the results are shown in Fig. 17.3(C). The simulations are in good agreement with the experimental data, with a maximum voltage error of 20 mV and a runtime error of 0.338%. These results indirectly confirm the assumptions regarding the small effect of the number of cycles and self-discharge (Table 17.3).

The close agreement between the simulation results and experimental data for NiMH and Li-ion polymer batteries shows that the proposed electrical model is able to predict the steady-state and transient voltage responses with high accuracy and keep

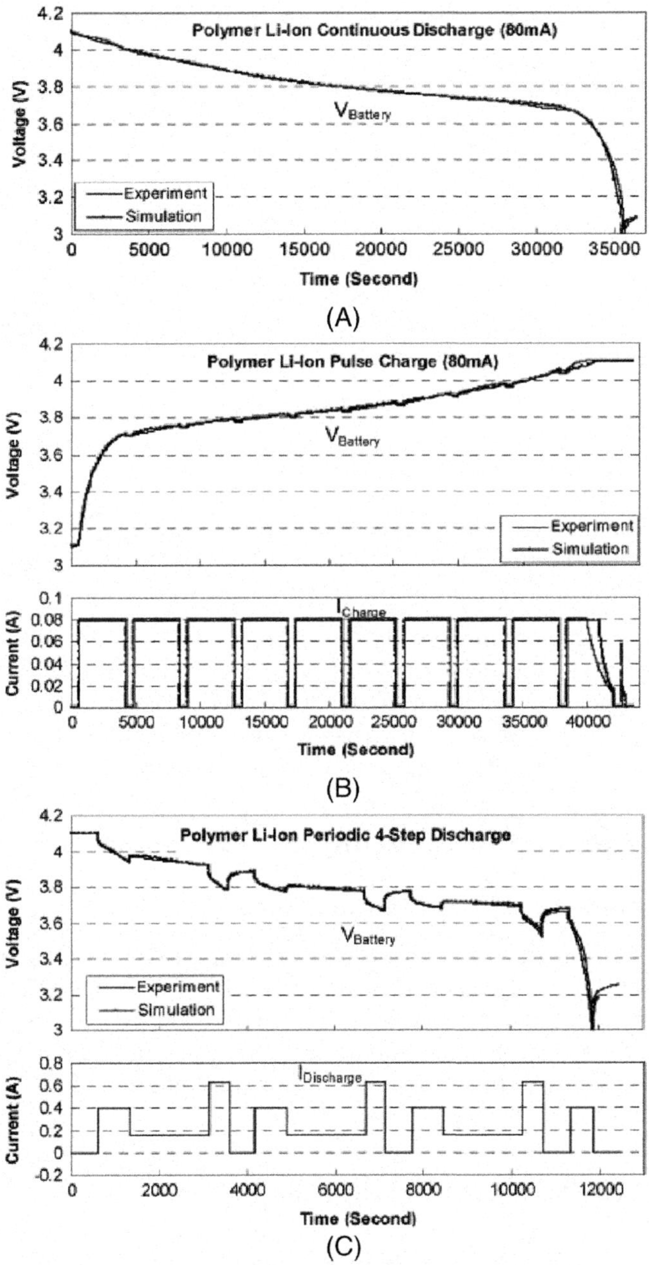

Figure 17.3 Comparison of simulation results with experimental data for lithium-ion polymer batteries: (A) continuous discharge with a current of 80 mA, (B) pulse discharge with a current of 80 mA, and (C) four-stage cyclic discharge.
From Chen, M., & Rincon-Mora, G. A. (2006). Accurate electrical battery model capable of predicting runtime and I–V performance. *IEEE Transactions on Energy Conversion, 21* (2), 504–511.

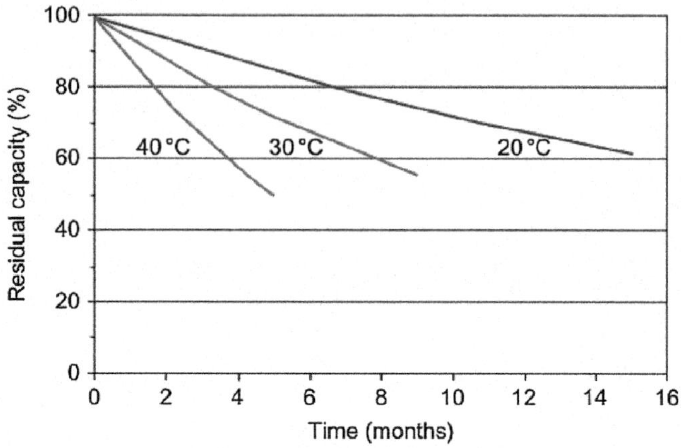

Figure 17.4 Battery capacity at different temperatures.
From Garche, J., Moseley, P. T., & Karden, E. (2015). Lead-acid batteries for hybrid electric vehicles and battery electric vehicles. In Advances in battery technologies for electric vehicles (pp. 75–101). Elsevier. https://doi.org/10.1016/B978-1-78242-377-5.00005-4.

the simulation execution time optimal. Furthermore, the model is fully implemented in the Cadence simulator, an industry-standard platform, allowing for circuit, block, or system-level simulation and its integration with other circuits and systems. It is worth noting that the model can be easily generalized to different types of batteries (e.g., lead-acid, NiCd, Li-ion) and even alternative power sources (e.g., fuel cells).

17.4 Key factors in reducing battery capacity

17.4.1 Operation at high temperature

One of the most important factors in battery performance is proper control of voltage and charging. At elevated temperatures, internal chemical reactions intensify compared to cooler conditions, and the self-discharge rate rises. If low-current self-discharge continues, charging becomes ineffective, and the electrodes can suffer sulfate-like passivation. As shown in Fig. 17.4, higher operating temperature leads to a noticeable drop in available capacity (Garche et al., 2015).

17.5 Excessive depth of discharge

To extend service life, deep discharges should be limited; discharging more deeply causes a sharp decline in cycle life and capacity. Fig. 17.5 (room temperature) illustrates that the deeper a battery is discharged, the fewer cycles it survives and the faster its capacity falls. For example, at ~30% depth of discharge (DoD), the cycle life can be ~1200 cycles before capacity falls to ~60%, while at ~50% DoD, the life drops to ~550 cycles before reaching ~60% capacity (CANBAT, 2025).

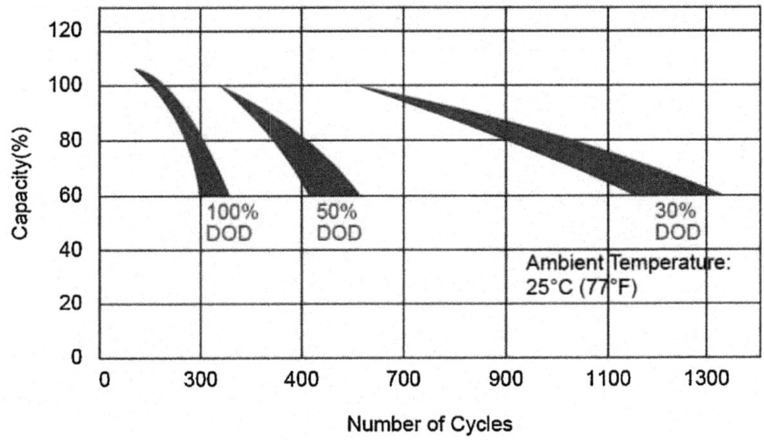

Figure 17.5 Capacity of CANBAT absorbent glass mat (AGM) battery at room temperature at different DoD percentages. *DoD*, Depth of discharge.
From CANBAT. (2025). *AGM vs gel batteries.*
https://canbat.com/guide-agm-batteries-gel-batteries.

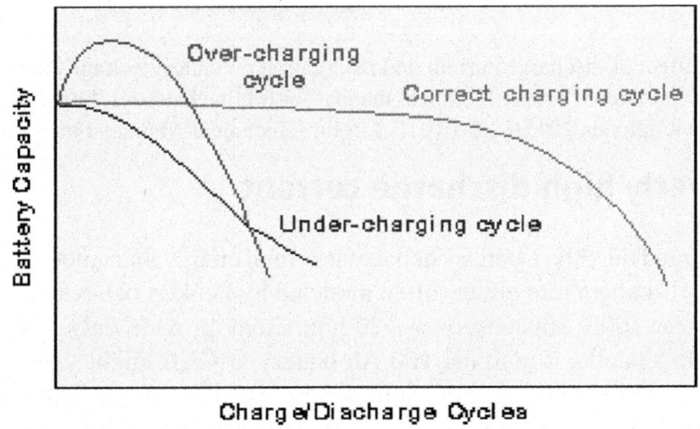

Figure 17.6 Effect of improper charging on battery capacity.
From PV Education. (2025). Characteristics of lead acid batteries. https://www.pveducation.org/pvcdrom/lead-acid-batteries/characteristics-of-lead-acid-batteries.

17.6 Improper charging

Continuous overcharging or chronic undercharging both damage a battery and shorten its life. Overcharge drives gas evolution and electrolyte loss through pressure-relief paths, reducing capacity over time. Persistent undercharge forms a resistive surface layer on the positive plate that blocks full acceptance of charge; active-material shedding may follow, further cutting capacity and life. Fig. 17.6 shows how incorrect charging practices degrade capacity (PV Education, 2025).

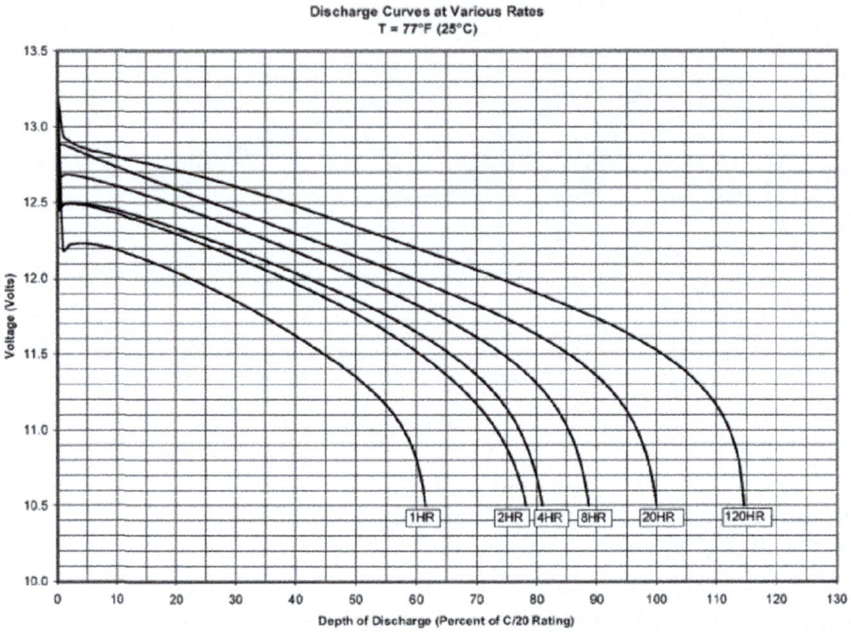

Figure 17.7 Effect of discharge current and its duration on battery voltage and capacity. From C. B. Corporation. (2019). Technical manual for lifeline batteries. https://lifelinebatteries.com/wp-content/uploads/2015/12/6-0101F-Lifeline-Technical-Manual-Final-5-06-19.pdf.

17.7 Overly high discharge current

When discharged quickly, a battery delivers less total energy than under slow discharge. This is the well-known rate effect (often modeled by Peukert behavior): A battery that can deliver near 100% capacity over ~20 hours may provide only ~60% if emptied in ~1 hour. Practically, a nominal 100 Ah battery at C/20 might yield only ~30 Ah usable in a one-hour, high-current event when accounting for ~50% DoD and rate-related losses. Fig. 17.7 illustrates how discharge current and duration affect voltage and available capacity (C. B. Corporation. (2019).

17.8 Lack of capacity equalization in series-connected packs

Due to unequal internal cells and differing cut-off thresholds, the charge/discharge capacity of a series pack is limited by the weakest cell, which reaches its cut-off first (Fig. 17.8). Charging stops when any single cell hits its upper cut-off; thus, total pack capacity (C_T) is set by the cell with the smallest chargeable capacity (C_C) if no balancing strategy exists. Likewise, during discharge, the pack must stop when any cell hits its lower cut-off, so CT is determined by the smallest releasable capacity (C_R).

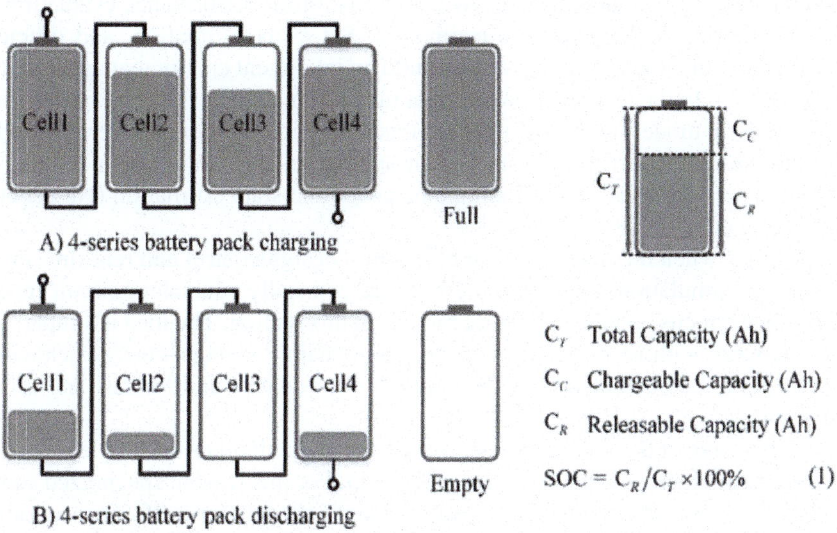

Figure 17.8 Schematic of a series-connected battery pack during charging and discharging, and the definition of its capacity.

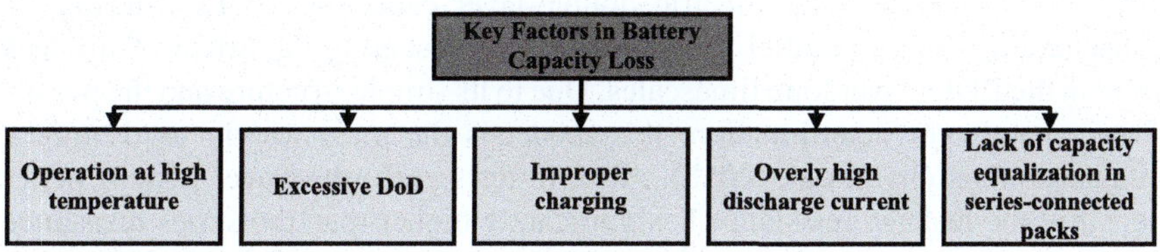

Figure 17.9 The main drivers of capacity loss in batteries.

This "weakest-cell" effect reduces usable capacity and overall performance, and by the "barrel effect," the pack's service life is constrained by the shortest-lived cell.

Fig. 17.9 summarizes the main drivers of capacity loss in batteries. In short: (1) high temperature raises self-discharge and accelerates fade; (2) overcharge or undercharge degrades materials and reduces capacity; (3) deep discharges (especially beyond ~50% DoD) shorten life; (4) high discharge currents cause rate-dependent losses; and (5) cell imbalance in series packs limits usable capacity and life to that of the weakest cell. Managing temperature, charge profile, DoD, discharge rate, and cell balancing helps preserve battery or lithium-ion battery capacity and cycle life.

17.9 Supercapacitors: RC/fractional models, electro-thermal behavior, and transient roles

With the increasing use of SCs in electric vehicles and propulsion systems, having an accurate terminal model that represents their dynamic behavior has become a design requirement. Such a model provides us with tools: From performance evaluation and health monitoring to internal loss estimation, SoC/SoH estimation, and efficiency calculation, all of which are necessary to build an intelligent energy management (Niu & Yang, 2011). In the technical literature, a range of models for SC can be seen from electrochemical models and equivalent circuits to fractional-order and thermal models; alongside these, data-driven and intelligent approaches such as neural networks and fuzzy logic are also used to approximate behavior and state estimation (Naseri et al., 2022; Zhang et al., 2018).

In practice, when it comes to power electronics design/control and real-time energy management simulation, engineers' best choice is usually equivalent circuit models; that is why we focus on precisely these models in this section. The idea is simple: With an RC network with identified parameters, the dynamic voltage–current behavior of the SC can be reconstructed to be both suitable for control and for fast and reliable simulation.

The advantage of this approach is its simplicity: Since it relies on simple ODEs, it is both easy to implement and computationally light. For this reason, specialized sources usually refer to several reference equivalent circuit models for a SC, an overview of which can be seen in Fig. 17.10.

Fig. 17.10(A) shows the classical equivalent circuit model, which consists of only three components: A capacitor (C), an equivalent series resistance (R_s) representing the ohmic losses in the electric double-layer capacitor (EDLC) during charging/discharging, and a parallel resistance (R_p) representing the effect of the leakage current in the EDLC over long timescales. Due to its simple structure and the possibility of easier parameter determination, this model is the most widely used among the available options (Jiya et al., 2018). Also, in tests performed over a short period of time, since the leakage resistance is significantly higher than the series resistance, its effect is usually ignored; as a result, the circuit design is simplified to a single series branch comprising a resistor and a capacitor (Jiya et al., 2018).

In Fig. 17.10(B), the first-order/Randall model has one important change from the basic equivalent circuit: The addition of an ESL, or equivalent series inductor L, to account for both the inductance of the capacitor leads and the current loops within the capacitor (Cultura & Salameh, 2015; Helseth, 2019). This component becomes dominant in the high-frequency bands, generally above 100 kHz, and its absence can make the model response unrealistic in those bands.

The model in Fig. 17.10(C) shows another approach: Using a Debye polarization cell for the SC. This version modifies the classical circuit by adding a faradic capacitance in series with a parallel resistance, here called the charge transfer resistance (Kai et al., 2017; Nelms et al., 2001). The result is a model that, in addition to the terminal behavior, also represents part of the charge transfer kinetics and polarization phenomena in the EDLC; thus, it is advantageous for more detailed analyses in specific frequency/operational ranges.

In the configuration of Fig. 17.10(D), we have a series R_s that contains all the "front-of-line" resistive components from the current collector contacts to the electrode and electrolyte, and is responsible for the instantaneous voltage drop (Devillers et al., 2014). The simplicity and easy extraction of features have made this model a very popular

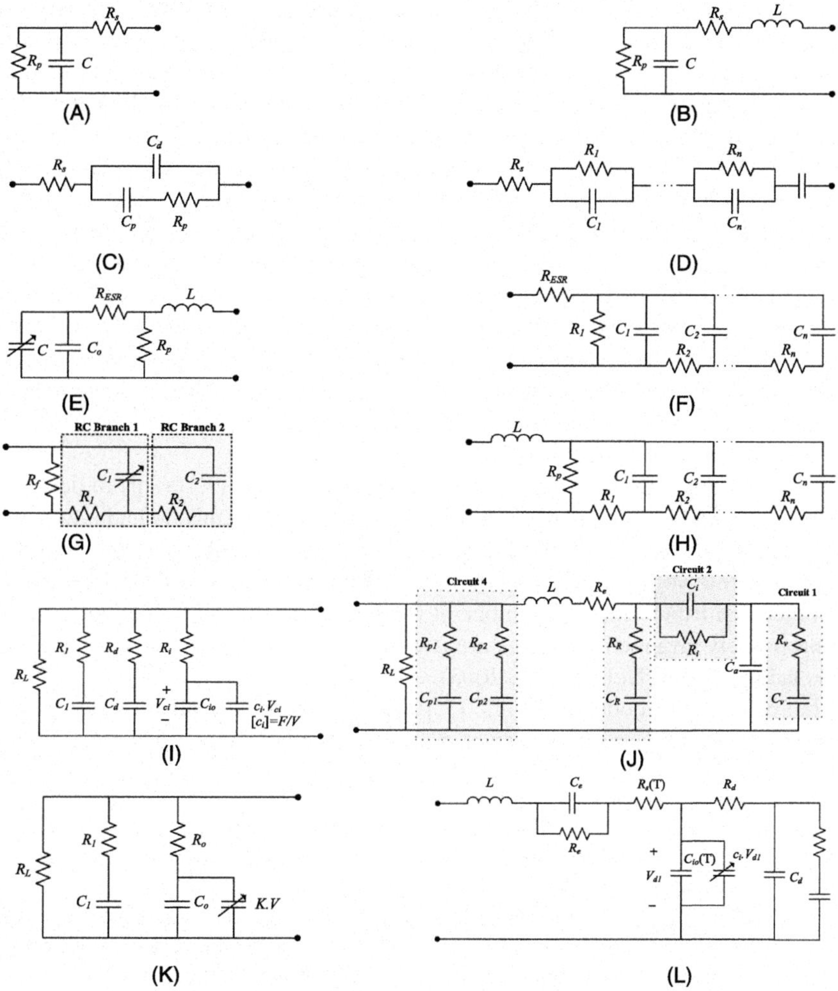

Figure 17.10 Terminal models of supercapacitor (SC): (A) Basic core CECM with C, series resistance and parallel leakage; (B) First order adding an RC for main transients; (C) Debay capturing the effect of polarization/charge transfer; (D) Standard multibranch RC correct order dynamics for response time spectrum; (E) Modified CECM with higher accuracy over a wider operating range; (F) RC chain ladder to approximate distributed electrode behavior; (G) Two branches separating fast/slow charging and discharging dynamics; (H) TLM continuum model of resistance/capacitance in electrode depth; (I) Zubita experimental framework with measurable parameters; (J) Rafik experimental variant with better transient fit; (K) Faranda light simplification for low-cost simulation; (L) High-frequency model with temperature dependence introducing the variation of parameters with temperature in HF bands. *CECM*, Cauer equivalent circuit model; *HF*, high frequency; *TLM*, transmission line model.

choice in analysis (Jiya et al., 2018). Also, when the experiment is carried out for only a short time, the leakage resistance is sufficiently small that its contribution to the series resistance can be neglected. As a result, the circuit is reduced to a simpler structure, namely a series combination of a resistor and a capacitor (Jiya et al., 2018). Such an arrangement shines when we want to capture high-frequency behavior: Speedy charge/discharge cycles or highly oscillatory power pulses that push simpler models (Naseri et al., 2022).

The model presented in Fig. 17.10(E), proposed in (Cultura & Salameh, 2015), is an evolved version of the conventional equivalent circuits for SCs and seems to be more suitable for slow discharge conditions. In this model, the total cell capacitance is divided into two parts: A fixed value and a part that changes with the terminal voltage, and their relationship is given by the equation $C(V_C) = C_0 + KV_C$ Other circuit components, including the equivalent series resistance, long-term leakage resistance, and inductance of the connection paths, are used without changes compared to the classical interpretations (Cultura & Salameh, 2015).

In (Nelms et al., 2003) used a ladder arrangement [Fig. 17.10(F)]) to describe the behavior of the SC, which partly replicates the logic of the equivalent classical model. The results show that in slow discharge scenarios, the classical model usually provides a more accurate voltage prediction than the ladder version. The main reason for this difference is attributed to the number of rungs used in the ladder network; the more inappropriately arranged the rungs are (too few or too many), the lower the accuracy of the ladder model (Nelms et al., 2003).

The bifurcated model of Fig. 17.10(G) provides a realistic picture of the charge/discharge dynamics by separating the SC behavior into a fast path and a slow path.

- In the fast branch (R_1, C_1), C_1 is voltage-dependent and is modeled as $C_1 = C_0 + KV_1$, as in Fig. 17.10(E); it is this branch that shapes the multisecond responses and determines the instantaneous behavior of the SC (Cabrane et al., 2016).
- The slow branch (R_2, C_2) is responsible for the redistribution of internal energy after the completion of charge/discharge; a phenomenon that occurs on a minute scale and affects the final equilibrium voltage (Argyrou et al., 2018; Vural et al., 2009).
- The resistance R_f is also for long-term leakage in standby mode and can usually be ignored if the cycles are fast.

To increase accuracy, especially when dealing with high-frequency bands, it is better to add the inductive effect of the connection paths to the model to make the high-frequency response more realistic.

The transmission line model (TLM) in Fig. 17.10(H) can be considered as a more advanced version of the bifurcated model. This framework explicitly incorporates the distributed behavior of the SC, which comes from the nonlinear capacitances and electrolyte resistances in the porous electrodes, which allows both charge redistribution and transient and long-term dynamics to be reconstructed with better accuracy (Naseri et al., 2022; Zhang et al., 2018). Its disadvantage is that both the identification of components (due to the large number of RC branches) becomes more difficult, and the computational time increases. On the other hand, the parameters are generally extracted by EIS/impedance spectroscopy, a method that is not well representative for high currents and may hide part of the real behavior (Du, 2009).

The well-known Zubieta model [Fig. 17.10(I)] (Zubieta & Bonert, 2000) covers three different time periods of SC behavior by placing three RC branches together and also introduces self-discharge with a considerable parallel leakage resistance (Şahin et al., 2021). Specifically:

- The third branch is designed for the first few seconds and takes on the details of the start-up with a voltage-dependent capacity.
- The second branch (R_d, C_d), or delayed branch, covers the response immediately after onset from a few seconds to about 10 minutes.
- The first branch (R_1, C_1) represents long-term behavior (more than 10 minutes) (Seim, 2011).

This arrangement represents the physical effects inherent in the EDL, from ion diffusion to charge redistribution in porous electrodes, in a terminal form, and is efficient for transient analysis and SC condition monitoring.

The Rafik model [Fig. 17.10(J)] (Rafik et al., 2007) is a logical continuation of Zubieta's work, except that it explicitly accounts for parameter variation with frequency, voltage, and temperature. In addition to the three basic components L for the high-frequency responses, $R_e$for the parasitic losses of the connections, and C_a as the principal capacitance, it has four subcircuits, each of which covers a part of the physical reality of the EDLC:

- Subcircuit 1: With R_V and C_V (voltage-dependent capacitance that increases linearly with voltage), it corrects the voltage behavior in low-frequency bands.
- Subcircuit 2: Considers the temperature dependence of the electrolyte resistance at low frequencies.
- Subcircuit 3: Effective for intermediate frequencies and effectively increases the total adequate capacity.
- Subcircuit 4: This is a version of the same mid/long-term branch as Zubita, and models temperature-dependent leakage and internal charge redistribution (self-discharge).

The result of these components is a rich and realistic model. However, the price is heavy parameter testing: To fine-tune the parameters, multiple EIS tests must be performed, which is time-consuming and not always available (Navarro et al., 2020).

The Faranda model [Fig. 17.10(K)] (New, 2004) is essentially a simplified version of the Zubieta model that eliminates one RC branch, thereby reducing the number of experimental parameters and measurements required, thereby reducing the risk of error. Like Zubieta, the first branch (with two distinct capacitors) is responsible for the primary energy storage, while the second branch captures the medium and long-term behavior of the EDLC (Jiya et al., 2018). Another advantage of this approach is the explicit parameter-estimation method proposed by Faranda himself, based on experimental measurements. Since the complexity of the parameters in the Zubieta model makes it challenging to implement, this simplified version of Faranda usually provides sufficient accuracy with good agreement with experimental data (Şahin et al., 2021).

As shown in Fig. 17.10(L), a model has been presented in (Berrueta et al., 2014) that is able to reproduce the transient behavior of a SC with reasonable accuracy up to frequencies of about 1 kHz. In addition to considering the self-discharge phenomenon, this structure also includes internal processes related to charge transfer and thermal effects in the modeling (Berrueta et al., 2014). Architecturally, this framework is close to the Rafik model, but its parameterization is more difficult and multistage, which has been reported as a significant drawback of this model (Berrueta et al., 2019).

17.10 Proton-exchange membrane (PEM) fuel cells and hydrogen in HESS: Equivalent-circuit modeling and uncertainties

In recent years, the use of hydrogen as a fuel for automotive propulsion systems has been increasingly pursued due to its abundance, cleanliness, and high energy density (~120 MJ/kg; approximately three times that of gasoline/diesel). However, one of the main obstacles to the widespread adoption of hydrogen vehicles is the challenge of in-vehicle storage of flammable hydrogen (Cheng et al., 2024). Among the options, compressed storage is the most promising on-vehicle solution: Gaseous hydrogen is stored in high-pressure tanks, which are typically classified into five categories, Type I–V (depending on liner and composite material/weight, pressure tolerance, hydrogen penetration resistance, and cost). Liquid and cryo-compressed technologies are also under development. For hydrogen to become "feasible," the energy density of the storage system must be increased, costs reduced, and interoperability with the vehicle system improved; a comparative assessment of storage options in terms of density, pressure, temperature, and cost has been reported (Gómez & Santos, 2023; Shin & Ha, 2023).

The hydrogen in the tank is then fed to the FC and produces low-voltage DC power; this power is transferred to the bus and drive via high-gain power electronics converters. Fig. 17.11 shows the working mechanism of a fuel cell. At the anode, a catalyst splits hydrogen into protons and electrons. The protons pass through an electrolyte, while the electrons flow through an external circuit, creating a direct electric current. At the cathode, these protons react with oxygen to form the final product, water (Bethoux, 2020). The main reactions of this process are as follows:

$$Anode : 2H_{2(g)} \rightarrow 4H^{+} + 4e^{-}$$

$$Cathode : O_{2(g)} + 4H^{+} + 4e^{-} \rightarrow 2H_2O_{(1)}$$

$$Overall : 2H_{2(g)} + O_{2(g)} \rightarrow 2H_2O_{(1)} + heat$$

Since the voltage of a single cell is usually less than 1 V, cells must be stacked in series to achieve practical voltage levels. Fuel cells are classified based on the type of electrolyte, which determines the operating temperature/performance characteristics (increasing temperature increases fuel impurity tolerance but also increases material/technological requirements and startup time). In automotive applications, PEMFC is an important option due to its lower operating temperature and faster startup.

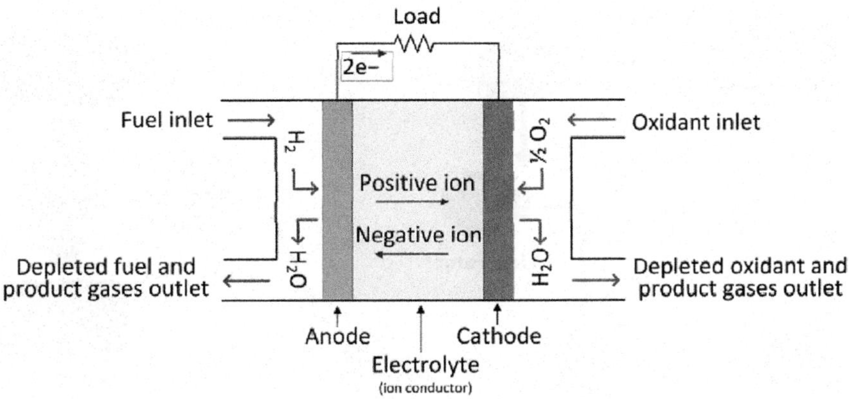

Figure 17.11 The working principle of a PEM fuel cell: Decomposition of H_2 at the anode and formation of H_2O at the cathode by the passage of protons through the membrane and DC in the external circuit.
From Bethoux, O. (2020). Hydrogen fuel cell road vehicles: State of the Art and perspectives. *Energies, 13* (21), 5843.

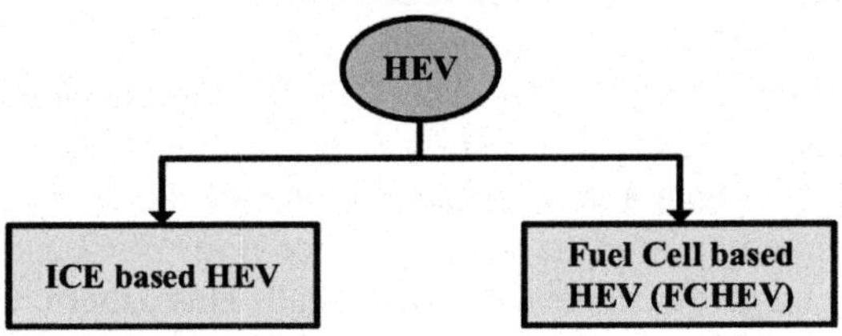

Figure 17.12 Classification of different HEV. *HEV*, Hybrid electric vehicle.
From Urooj, A., & Nasir, A. (2024). Review of hybrid energy storage systems for hybrid electric vehicles. *World Electric Vehicle Journal, 15* (8). https://doi.org/10.3390/wevj15080342.

17.11 Architectures and power electronics under uncertainty

In short, HEVs offer a high degree of flexibility in terms of propulsion power by combining multiple energy sources. As shown in Fig. 17.12, their prominent family has two branches: Internal combustion engine-based HEVs and fuel cell-based versions (FCHEV) (Hussan et al., 2022; Pan et al., 2025). At the same time, there has been an increasing trend toward the addition of a flying wheel (FESS), as these systems can deliver enormous amounts of energy in very short periods of time and are very efficient for power surges or energy recovery (Manjula et al., 2025; Shen et al., 2020).

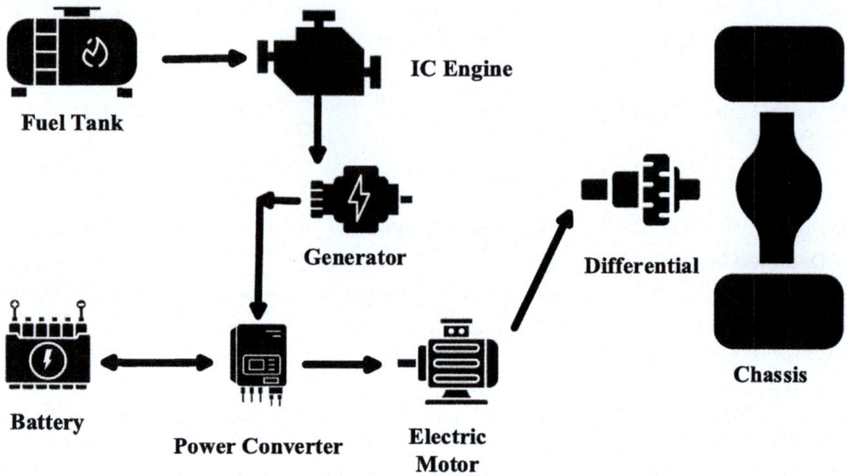

Figure 17.13 Series HEV. *HEV*, Hybrid electric vehicle.

17.12 Hybrid electric vehicle architectures: Series, parallel, series–parallel, complex

In the series architecture, as shown in Fig. 17.13, the electric motor is the only source of torque to the wheels, and the combustion engine is solely responsible for generating electricity through the generator. The internal combustion engine (ICE) + generator combination can directly power the traction motor (Fig. 17.13), so that in some cases there is no need to charge/discharge the battery at all. This pattern, which has been used for many years in diesel-electric locomotives, allows the combustion engine to operate close to its optimum consumption point by mechanically decoupling the ICE from the drive axle (Fig. 17.13) while at the same time managing the traction power with high flexibility (Khamesipour et al., 2022; Mohammadi & Saif, 2023). For this arrangement, sources have mentioned several specific operating modes, from direct traction from the generator to battery charging at constant ICE speed and all-electric motion, which are detailed in (Mohammadi & Saif, 2023).

1. Starting and acceleration: At the start of the movement, the combustion engine and electric motor provide torque simultaneously so that the vehicle accelerates smoothly and powerfully.
2. Light load: When the load on the engine is light, the excess ICE power is not wasted; it goes directly to charging the battery for later use, increasing the efficiency of the entire system.
3. Regenerative braking: When slowing down, the electric motor acts as a generator, recovering kinetic energy to charge the batteries, energy that would otherwise be lost.
4. Complete stop: When stationary and not moving, the ICE can keep the battery charged through the generator to be ready for the next move.

In short, the series arrangement creates a flexible and efficient platform for power sharing, where the combustion engine and electric motor intelligently switch between

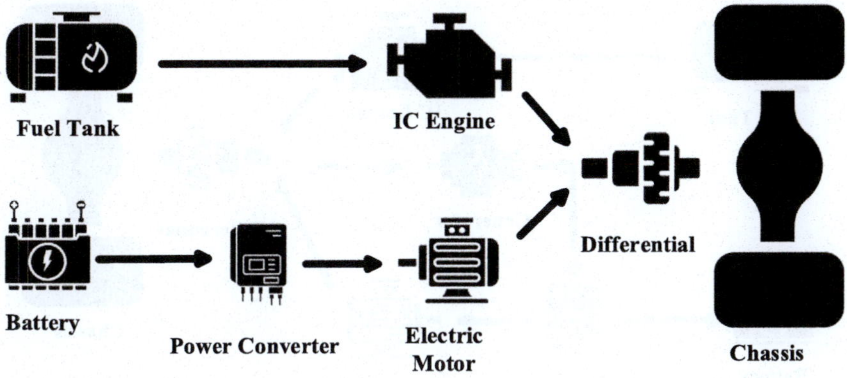

Figure 17.14 Parallel HEV. *HEV*, Hybrid electric vehicle.

propulsion and charging depending on the driving situation. The following are the most important advantages of series drivetrains (Cao et al., 2025; Luo et al., 2019):

Advantages:

- Because only one electric motor provides torque to the wheels, speed control becomes as precise and straightforward as operating the accelerator pedal.
- Mechanical decoupling of the ICE from the wheels allows the combustion engine to be kept within an optimal operating range.
- The nearly ideal torque-speed curve of the electric motor virtually eliminates the need for a multigear transmission.
- In terms of control, structure, and layout, the drivetrain is simple; the connection between the motor/generator and the battery and traction motor is primarily via electrical cables.

Disadvantages:

- Two electrical machines are required: A generator and a traction motor.
- There are two energy conversions (from mechanical in the ICE to electrical in the generator and back to mechanical in the traction motor), which increase losses.
- Because all of the drive torque must come from the electric motor, a larger/more powerful traction motor is usually required.

Due to its simple structure, ease of control, and convenient equipment placement, the series architecture has been widely used in electric propulsion systems for heavy-duty vehicles, including trucks, military vehicles, buses, and even locomotives (Kozachenko et al., 2014). One of the main reasons for this choice is the considerable space available in these vehicles, which allows for easy placement of relatively large components such as motor and generator sets. A well-known example of the industrial use of this configuration is hybrid city buses; manufacturers such as electric vehicles international, Ebus, ISE Research, and Thunder-Volt have been using this technology in production for many years (Shen et al., 2022).

In the parallel arrangement (Fig. 17.14), both the ICE and the electric motor deliver torque to the wheels directly through a mechanical coupling. This coupling can be a conventional gearbox, a belt pulley, a sprocket chain, or even a single axle; therefore

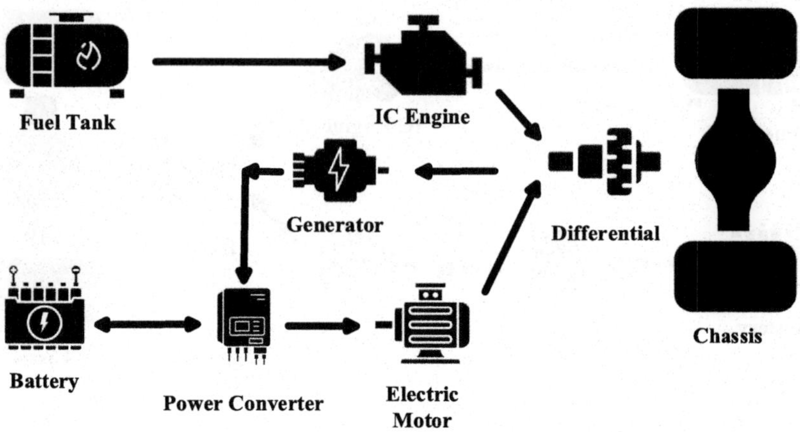

Figure 17.15 Series-parallel HEV. *HEV*, Hybrid electric vehicle.

the implementation is highly flexible. The result is a set of advantages reported in (Biswas et al., 2022; Zhang et al., 2023).

1. Because both the ICE and the electric motor directly deliver torque to the wheels, there is no energy going back and forth between mechanical and electrical; therefore there are fewer losses.
2. By eliminating the generator and shrinking the traction motor, the entire assembly becomes more compact, which is why parallel architecture is popular on small platforms (Biswas et al., 2022; Zhang et al., 2023).

Despite the advantages offered by the parallel configuration, this architecture is not without its weaknesses. Because the internal combustion engine is directly connected to the wheels, the ability to select the ideal operating point for it is reduced, and it is forced to operate in a narrower speed range, which will result in a loss of efficiency. On the other hand, the more complex control system and layout of this type of hybrid make the design process and maintenance operations more difficult. Nevertheless, this architecture works in practice and is very common in cars; from the Honda Civic and Honda Insight to the Ford Escape and other popular models, this arrangement has been used for many years to achieve stability and everyday use (Xu et al., 2022).

In the series-parallel architecture, the planetary gear set isolates the ICE speed from the wheel speed, allowing the engine speed to be adjusted independently of the vehicle speed (Fig. 17.15). In the "series flow" mode, power from the ICE is delivered to the planetary gear and then to the generator/motor to deliver it to the wheels with appropriate efficiency. When the speed of the motor/generator unit goes into negative values, the system enters the energy recovery mode. In this mode, the power generated by the internal combustion engine is directed into two paths: One part is used to charge or support the electrical path, and the other to provide motive power through the mechanical path. This control method helps the internal combustion engine to operate closer to its optimal operating point in variable driving conditions, thereby reducing fuel consumption.

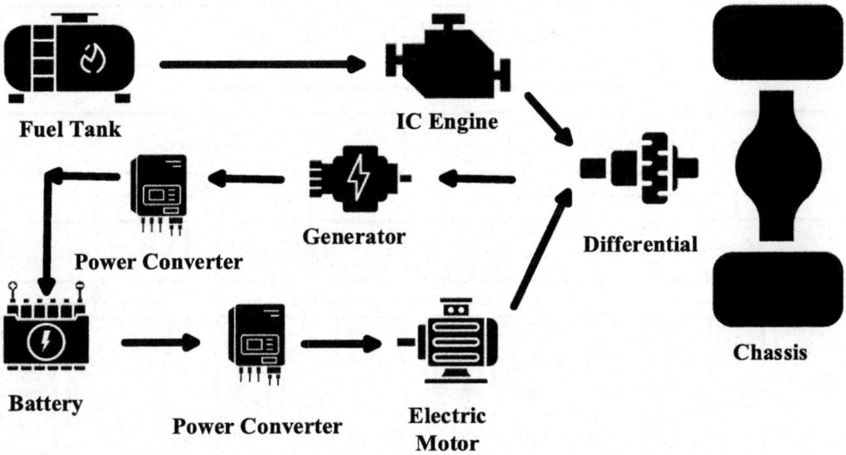

Figure 17.16 Complex HEV. *HEV*, Hybrid electric vehicle.

In the "complex hybrid," we are faced with a complex and nonsimple arrangement that does not fit easily into the previous three architectures. Fig. 17.16 shows a structure similar to the series-parallel one, with both a motor and a generator, but here is the main difference: The electric motor operates bidirectionally and can return power to both the wheels and the bus/generator; unlike series-parallel, where the generator power flow is assumed to be practically unidirectional. This bidirectional maneuverability provides the system with more control options and enables more diverse energy paths. In the complex hybrid, because power can flow in both directions, the designer has more freedom to define various operating modes, including the "triple drive" mode, which is not available in the series-parallel architecture. Of course, achieving this level of flexibility requires more expense and more complex system management. Despite these challenges, in recent years, a significant number of hybrid vehicles, especially those with two-axle configurations, have adopted this architecture and are taking advantage of its control advantages in various driving conditions.

The main difference is in the simplicity and efficiency of the regenerative braking: In the complex architecture, the return energy from the braking is transferred more directly and efficiently to the electrical path due to the bidirectional flow of power; but in series-parallel, there are usually more components involved, making the regenerative process less straightforward and possibly less efficient.

17.13 Hybrid energy storage system placement and DC-link interfacing under uncertainty

In practice, most HEVs use HESS to both provide a sharper transient response and extend FC/battery life: The FC or battery covers the base/average power load of the propulsion system, the SC compensates for instantaneous power surges during acceleration and attitude changes; the role of the DC–DC converter is to properly couple these sources on the DC bus and enable power distribution between them

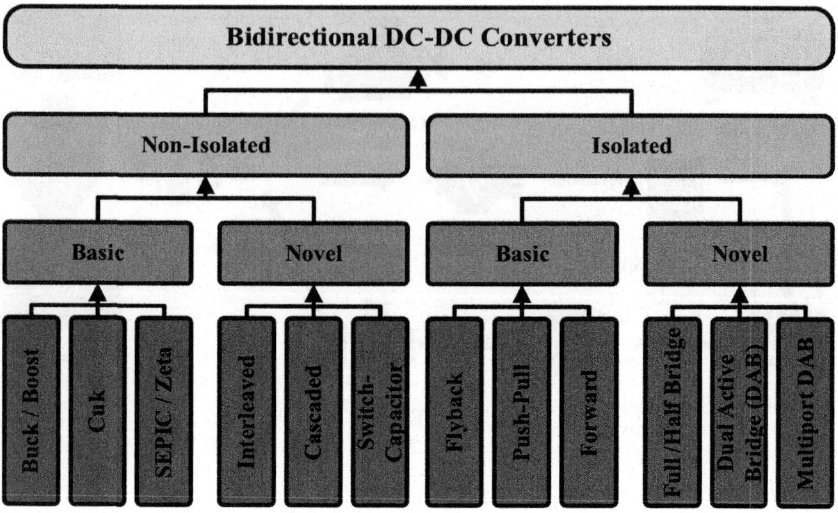

Figure 17.17 Novel DC–DC converter configuration.
From Urooj, A., & Nasir, A. (2024). Review of hybrid energy storage systems for hybrid electric vehicles. *World Electric Vehicle Journal, 15* (8). https://doi.org/10.3390/wevj15080342.

(Wang et al., 2022). The technical point here is that since the output voltage of the SC is not aligned with the vehicle DC bus voltage, a bidirectional DC–DC converter is unavoidable for equalization and recovery of regenerative braking energy. The efficiency of the bidirectional converter directly impacts the performance of the hybrid energy storage system and, consequently, the efficiency of the entire power transmission chain. Therefore if there is no capable and low-loss BDC in the system, HEVs cannot be seriously considered as an alternative to fossil fuel-based vehicles and play their expected role in reducing carbon emissions and moving toward green transportation (Wang et al., 2022). In the nonisolated branch, it is enough to replace the diodes with active switches (such as MOSFET/IGBT) to create a bidirectional BDC; this is the synchronous rectification that enables the power path in both directions. Then, by choosing the appropriate topology and tuning the control, it is possible to increase both the power density and the conversion efficiency. In general, this line of research on bidirectional BDC topologies is the driving force behind innovation in HEVs and is practically pushing their performance boundaries.

At a glance, BDCs have two branches: Nonisolated and isolated. In the nonisolated branch, direct DC-to-DC conversion is performed with an LC network and power switches. If we add a transformer between the two DC links, the DC→AC→DC path is formed, and high voltage gains can be achieved with the help of an adjustable turn ratio. Fig. 17.17 illustrates this classification and summarizes the differences in application of each group (Wang et al., 2022).

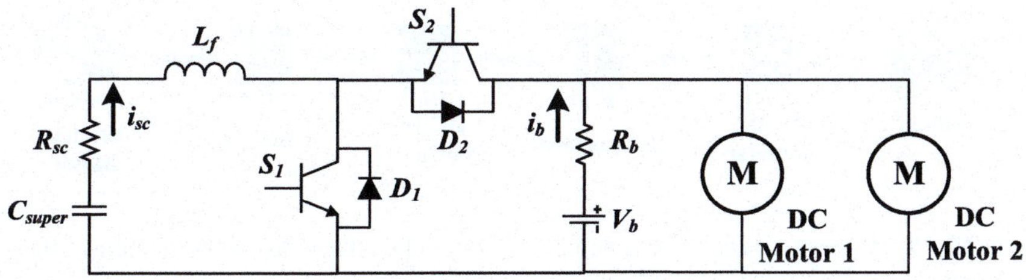

Figure 17.18 Supercapacitor bank with bidirectional DC–DC converter and two motor drivers.

Table 17.4 In wheel motor characteristics.

Characteristic	Value
τ_{max}	405 Nm
Wheel circumference	1.7687 m
$\Delta speed$	166.8 m/mim
Angular velocity	6.18 rad/s
Maximum power	2.5 KW

A bidirectional buck-boost converter controls the energy of an SC to charge it rapidly during regenerative braking without exceeding the maximum allowable current, and to discharge the stored energy during acceleration. Fig. 17.18 shows a schematic of an SC bank with a DC–DC converter and two motor loads (Mamun et al., 2022).

The size of the SC bank depends on the in-wheel motor ratings and the total torque required to accelerate the vehicle. The sizing of the SC bank is done by considering two situations:

1. Initial acceleration scenario: The power that must be provided at the beginning of the movement for the vehicle to overcome the inertia of rest and reach the desired speed.
2. Stopping scenario: The power that must be absorbed during the braking phase to bring the vehicle to a safe and complete stop.

Since the peak power demand occurs at the beginning of acceleration and is usually greater than the power demand during braking, we calculate the capacity of the SC bank based on the acceleration scenario. For numerical estimation, we use (17.7) and (17.8): Where μ is the coefficient of static friction of the surface-tire (for asphalt-tire: 0.72); R_{wheel} is the effective wheel radius, and η_d is the drivetrain efficiency to the wheels; and f_r is the static load fraction on the rear axle; the exact parameters of the in-wheel motors are also specified in Table 17.4 to complete the sizing calculations.

$$\tau_{max} = \frac{\left[\mu \times weight \times f_r \times R_{wheel} \times \eta_d\right]}{2} \tag{17.7}$$

$$Totals\ lack\ capacitance = \frac{C_{cell} \times No.parallel}{No.series} \tag{17.8}$$

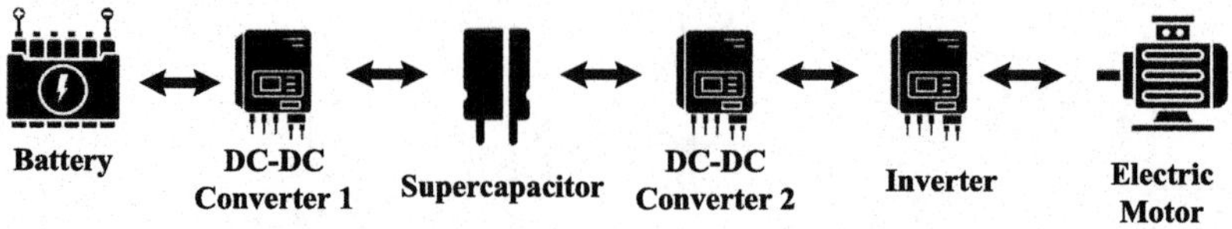

Figure 17.19 "Passive cascade" arrangement with two DC–DC converters to connect the battery and supercapacitor to the DC bus.

To meet or exceed 75 V to drive the electric motor with the following specifications of a SC:

$$Cell\,voltage = 2.7\ V$$

$$Rated\,capacitance = 650\ F$$

$$Cell\,resistance = 0.8\ m\Omega$$

$$Cell\,srequired = 75\ V/2.7\ V = 28\,inseries$$

The total capacity is calculated using a SC with the following specifications (Mamun et al., 2022).

$$Parallel = 1(singlestring)$$

$$Totals\,lack\,capacitance = \frac{650\ \times\ 1}{28} = 23.2\ F$$

In the two-converter arrangement, as shown in Fig. 17.19, the SC can practically transfer all of its stored energy to the bus. The reason is that the three voltage levels V_{battery}, V_{UC}, and $V_{\text{dc-link}}$ are not at the same level and each is coupled to the bus with its own converter; therefore UC does not have to be kept at a constant voltage and can be deeply discharged up to its operating range ($V_{battery} \neq V_{UC} \neq V_{dc-link}$) (Urooj & Nasir, 2024).

In this system, the boost converter plays a key role in reducing the stress on the battery by keeping the battery current fluctuations at moderate and controlled levels. The result is that the battery can be used more efficiently and effectively, with less wear/deterioration of components. The dual-converter SC battery configuration is well-suited for storage systems that require high energy density, long cycle life, and fast charge/discharge (Urooj & Nasir, 2024).

Fig. 17.20 shows the architecture of a multiinput converter, a solution that reduces the cost of the entire system by aggregating the inputs rather than using multiple separate converters and, at the same time, compensates for the weaknesses of previous topologies. In this arrangement, the three voltage levels V_{battery}, V_{UC}, and $V_{\text{dc-link}}$ are

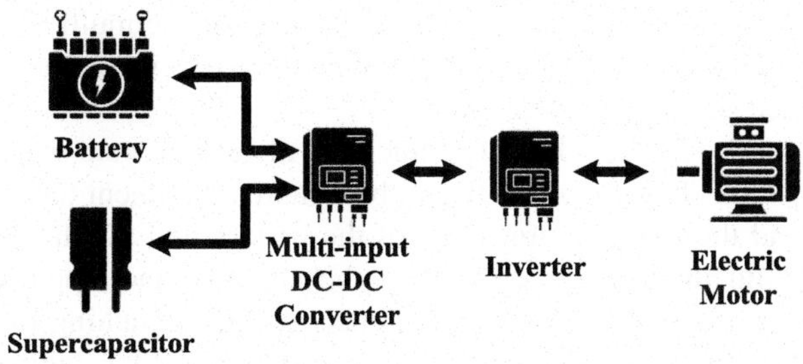

Figure 17.20 Multiinput/multiconverter DC–DC architecture for connecting battery and supercapacitor to the DC bus.

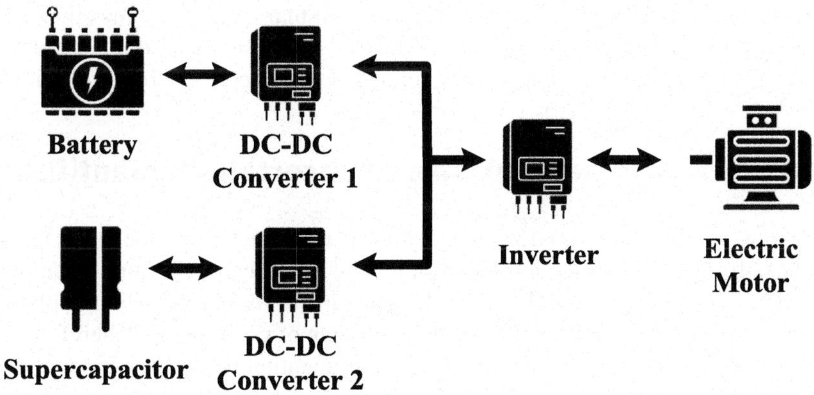

Figure 17.21 Multiple configurations of DC–DC converters.

not at the same level; this voltage freedom allows for better adaptation to the operating conditions and increases the overall efficiency of the system ($V_{battery} \neq V_{UC} \neq V_{dc-link}$) (Urooj & Nasir, 2024).

Implementing this arrangement simultaneously reduces the cost and mass of the system and improves the overall efficiency; therefore, it is an outstanding choice for designing a low-cost and high-efficiency HESS and is usually prioritized when the goal is to reduce costs (Sinha et al., 2022).

In a multiconverter arrangement, as shown in Fig. 17.21, each energy source is connected to the bus with an independent DC–DC converter. The clear advantage is that the three voltages of the battery, SC, and DC bus do not have to be at the same level; instead, two full-capacity converters must be used, increasing cost and size and increasing system complexity ($V_{battery} \neq V_{UC} \neq V_{dc-link}$) (Urooj & Nasir, 2024).

In such circumstances, the size and cost of the system increase, and its implementation becomes less operationally justifiable. Therefore, although this topology has some potential advantages, the existing limitations have made it less popular in real applications. Therefore, in the field of energy storage systems, there is a high need for alternative topologies that mitigate the disadvantages of the multiswap configuration discussed in (Wang et al., 2022; Zimmermann et al., 2016) while maintaining comparable performance (Urooj & Nasir, 2024).

In the real world, HESS and DC bus interface always operate under uncertainties: Driving patterns that change constantly, ambient and component temperatures that go up and down, and the health/charge state of the battery and SC that does not remain constant. These fluctuations directly affect the stability and ripple of the bus, the peaks of regeneration and acceleration currents, and the switching losses. Therefore, the choice of topology (nonisolated/isolated, multiinput, high gain) must be made simultaneously with the control design, where a rule-based layer for simplicity, a predictive layer (model predictive control [MPC]) to keep the current/voltage within safe limits, and a hierarchical layer to separate the fast dynamics of the converter from the slow energy management objectives sit side by side. The result of this "co-design of sizing + EMS + reliability constraints" is a system that both maintains efficiency and component life, remains resilient in variable driving scenarios, and ultimately delivers a clean and reliable DC bus with adequate control of power peaks (Urooj & Nasir, 2024).

17.14 Scenario generation and uncertainty quantification

The design and operation of HESS in electric/hybrid vehicles operate under numerous variables, from driver behavior and traffic and weather conditions to component parameter fluctuations and charging infrastructure availability. Relying solely on a single reference cycle does not reflect the true range of performance; therefore, generating representative scenarios and incorporating quantitative uncertainty analysis into decision indicators is at the core of evaluating and adjusting the EMS and sizing. The HESS literature has shown that the advantages of active topologies such as bidirectional converters and frequency splitting of power between the battery (low frequency) and the SC (high frequency) become apparent if the control settings are calibrated and stress-tested under a diverse set of operating conditions, not just under a few standard cycles (Nkembi et al., 2024; Urooj & Nasir, 2024). At the same time, fast charging infrastructure planning models show that ignoring exogenous uncertainties (traffic, demand shocks, weather conditions) leads to biases in the placement/capacity and, through queuing and routing, alters the vehicle operation patterns and load spectrum of the HESS (Pal et al., 2021).

To guide this process in practice, the scenario development and UQ framework are summarized in Table 17.5 . The table brings together the main scenario categories, proxy variables, HESS/EMS objective, sampling method/uncertainty analysis, and evaluation indicators to provide a concise overview of the design and evaluation process. References to key sources for each category are provided in the last column (Li et al., 2023; Mamun et al., 2022; Nkembi et al., 2024; Pal et al., 2021; Urooj & Nasir, 2024).

Following Table 17.5, scenario generation is defined in two complementary paths. The first path is scenario design based on real operation: Urban stop and go cycles, extra-urban cruising, sections with positive/negative gradients, and side loads are

Table 17.5 Summary of scenario building and UQ for HESS/EMS.

Scenario category	Representative variable/construct	Objective in HESS/EMS	UQ/sampling method	Main key performance indicators (KPIs)	References
Operation/driving	Urban stop-and-go cycles, highway cycles, road gradient, auxiliary loads.	Calibration of frequency-domain power split (supercapacitor for fast pulses, battery for slow energy).	Sampling over cycle descriptors (left-hand side (LHS)/Monte Carlo).	Battery current peak and RMS, SOC occupancy of battery/SC, converter losses.	Li et al. (2023), Urooj & Nasir (2024)
Data-augmented (synthetic)	Speed profiles generated by "random sampling over velocity transition probability surface."	Improving control generalization to unfamiliar cycles; pretraining then transfers.	Random sampling over speed transition matrix.	Inter-cycle performance stability, reduced control policy brittleness.	Li et al. (2023), Urooj & Nasir (2024)
Component stress	Short power steps, high-frequency wavelets, thermal sweeps.	Testing frequency-band limits and switching/thermal losses.	Parametric sweep (R_{int}, $\tau_{polarization}$, temperature coefficients).	Current peak reduction, thermal margin, efficiency, aging indicators.	Urooj & Nasir (2024)
Infrastructure-Coupled	Availability/unavailability of fast charging, demand shock, congestion.	Aligning vehicle design with real grid and queuing realities.	Stochastic traffic/demand scenarios, queuing models.	Waiting/detour time, entry/exit SOC, unmet energy.	Pal et al. (2021)
On-board multisource	PV/wind/fuel cell/alternator + battery/SC and role sequences.	Stable role/peak allocation, battery protection, source co-regulation.	Discrete source-role branches, SOC/power constraints.	Source duty cycle, battery stress, transient efficiency.	Mamun et al. (2022)

EMS, energy management system; *HESS*, hybrid energy storage system; *SC*, supercapacitor; *SOC*, state of charge; *UQ*, uncertainty-quantified.

chosen so that the power spectral content is separable, and the contribution of the volatile components is distinguished from the slow energy load. This separation is in line with the division of labor logic in active HESS: The SC absorbs fast transient loads, and the battery provides the slower energy demand (Urooj & Nasir, 2024). The second path is data-boosted scenario generation: Pretraining EMS policies on a set of artificial speed profiles constructed by random sampling on the speed transition probability surface increases the coverage of scenarios before transitioning to the destination route/fleet and improves the stability of performance in unfamiliar cycles (Li et al., 2023). The combination of these two paths creates a compact and reliable package of scenarios that is suitable for control tuning, topology comparison, and aging-sensitive evaluations.

In the UQ dimension, it is necessary to attribute distributional meanings to the performance. Indicators such as peak and root mean square (RMS) battery current, SOC occupancy distribution (battery and SC) with emphasis on residence time in stressed regions, converter losses, and power processing penalties in different paths, and aging indicators based on equivalent/electrothermal models should be modeled as stochastic and reported with confidence or plausibility intervals (Nkembi et al., 2024). Variance-based sensitivity analyses or screening are recommended to determine the contribution of uncertainty to model parameters (internal resistance, polarization times, temperature coefficients), operational/environmental descriptors (stop density, slope statistics, ambient temperature) and supervisor thresholds (SOC limits, current ceiling, converter constraints); such analysis prevents excessive specialization of the control policy to a specific cycle and reveals leverage points for robust calibration (Nkembi et al., 2024).

An upstream layer integrates infrastructure uncertainty into vehicle scenarios. Fast charging location/capacity formulations that consider stochastic traffic and demand patterns lead to different solutions than deterministic models; the difference in queuing and detour paths affects the fleet usage pattern and ultimately changes the HESS load spectrum (Pal et al., 2021). The inclusion of "with/without fast charging access" branches and demand shock scenarios according to the "infrastructure-coupled" row in Table 17.5 facilitates the alignment of the component/control level with network realities.

From an implementation perspective, a low-margin and repeatable framework is recommended: Collecting a base set of natural cycles in the target area and enriching it with artificial profiles obtained by random sampling on the velocity transition probability surface; pretraining EMS policies on this subset and then fine-tuning on the target data (Li et al., 2023; Urooj & Nasir, 2024). Designing component stress scenarios according to the corresponding row in Table 17.5 in active topologies significantly reduces battery peak currents and improves thermal behavior (Urooj & Nasir, 2024). Running sampling schemes (Latin Hyperbox/Monte Carlo) on uncertain component parameters and operational descriptors and reporting the results in a distributed manner allows attributing the variance of outputs to sources of uncertainty (Nkembi et al., 2024). Combining infrastructure/policy scenarios with vehicle scenarios makes sizing and economic decisions more resilient to external risks (Pal et al., 2021).

Reported evidence is consistent with this path: At the device level, tuning active HESS based on a portfolio of representative scenarios has led to the containment of battery peak currents, improved transients, and reduced effective aging rates (Urooj & Nasir, 2024); at the component modeling level, incorporating valid ranges for internal resistance, polarization dynamics, and temperature effects reduces false optimism in lifetime/performance estimates and identifies key risk drivers (Nkembi et al., 2024); at the network level, incorporating traffic/demand uncertainty into fast-charge location avoids capacity underestimation and keeps the user experience aligned with the technical constraints of the device (Pal et al., 2021). In addition, pretraining based on random sampling on the probability level of speed transition as a data-augmented scenario-building mechanism fills the data gap and enhances the generalizability of the EMS strategy to new cycles (Li et al., 2023). In multisource systems used in vehicles, the use of rule-based control frameworks has yielded positive results. These methods effectively specify the role of each energy source, such that a SC is responsible for absorbing momentary power, a more stable source, such as a fuel cell, provides continuous energy, and the battery SoC is maintained within a safe range, even in the face of unstable and real-world driving conditions (Mamun et al., 2022). As a result, the combination of scenario-building and uncertainty quantification is considered the data-model-decision connection loop in HESS, providing a transition from single-point assessments to distributed scales for robust decision-making (Nkembi et al., 2024; Urooj & Nasir, 2024).

17.15 Stochastic/risk-aware energy management and sizing

Energy management and sizing of HESS in electric/hybrid vehicles is defined by a reality that does not fully overlap with the "reference" laboratory conditions. Urban and suburban driving cycles, sudden changes in traffic and queueing patterns, climatic and temperature differences, variations in electrical side loads, as well as drift of battery and SC model parameters over time, cause the instantaneous load spectrum and performance indicators in the field of operation to be scattered and heterogeneous. Although deterministic models can provide valuable results for typical and average conditions, they are inherently unable to guarantee that the energy management strategy or the selected component size will continue to perform satisfactorily against rare but costly events that occur later in the distribution chain. The HESS literature has repeatedly shown that the advantages of active topologies (the presence of bidirectional DC–DC converters and frequency splitting of power between the battery for the low-frequency energy component and the SC for the high-frequency power component) are observed in a stable and repeatable manner when the design and calibration are performed under a variety of scenarios, not just under a few standard cycles (Nkembi et al., 2024; Urooj & Nasir, 2024). On the other hand, infrastructure studies on fast charging location and capacity have revealed that ignoring exogenous uncertainties (traffic, demand shocks, weather conditions) can lead to biased solutions; solutions that affect the fleet usage pattern through queuing and detours, and ultimately shift the spectrum of load imposed on the HESS (Pal et al., 2021). This device-network co-evolution requires that energy management and sizing be formulated from the outset in a stochastic/risk-aware framework so that uncertainties are present as first-order components of the problem alongside decision variables and physical constraints (Mamun et al., 2022; Nkembi et al., 2024; Pal et al., 2021; Urooj & Nasir, 2024).

In such a framework, decision variables are defined in two complementary layers: In the energy management layer (online), policy parameters such as SOC thresholds, power frequency division range, weighting of objective function components (loss reduction, peak current suppression, battery protection), and policy structure (rule-based, predictive, data-driven) play a role. In the sizing layer (offline), battery and SC nominal capacities and powers, converter capacity/efficiency, and related engineering constraints are determined. Uncertainties come in three complementary families: Operational/environmental (speed-acceleration profiles, stop density, path gradient, ambient temperature, and side load), technological/model (internal resistance dispersion, polarization time constants, adequate SC capacity, temperature coefficients, converter efficiency, and aging parameters), and infrastructure/network (fast charging availability and capacity, demand shocks, queueing patterns). Mapping these three categories to decision layers means that in energy management, the first and second categories are decisive, while in sizing, all three categories must be considered simultaneously to ensure that the size selection remains robust to external shocks (Nkembi et al., 2024; Pal et al., 2021; Urooj & Nasir, 2024).

To translate these intuitions into solvable decisions, risk measures are defined alongside classical objectives. At the constraint level, chance constraints keep the probability of violating safety limits such as SOC windows or converter temperature/power constraints below a specified threshold (e.g., SOC remains within the safe range with at least a 95% probability). At the objective level, tail measures such as VaR and CVaR focus on rare but costly events; CVaR of "battery peak current" or "ageing equivalent cost" at a given confidence level avoids designs that are good in the mean but fragile in the tail. This structure is complemented by sensitivity analysis and distributional reporting of indicators (percentiles and confidence intervals) to make cost-effectiveness risk trade-offs visible (Nkembi et al., 2024). Since real-world data does not always provide sufficient coverage of various driving conditions, the generation of augmentation data is of particular importance. In this approach, by using random sampling based on the probability distribution of speed changes, artificial speed profiles are created, and control policies are pretrained on this larger set. This process increases the generalization power of the model in the face of unfamiliar cycles and, by using the "separate training, validation, and testing" method, overfitting can be avoided (Li et al., 2023).

To see the relationship between decision layers, uncertainty sources, risk measures, and corresponding methods at a glance, Table 17.6 provides a concise summary. This table shows that in the online layer (EMS), SOC thresholds, frequency division bands, and objective function weights are adjusted under the uncertainty of driving cycles, temperature, and model parameters, along with chance constraints and CVaR targeting,

Table 17.6 Summary of risk-aware EMS and HESS sizing.

Decision layer	Key variables	Sources of uncertainty	Risk criteria/ constraints	Method/approach	Evaluation KPIs	References
EMS (online)	SOC thresholds; frequency-split bands; objective weights.	Driving cycles, temperature, auxiliary load; model parameters (R_int, τ_pol).	Chance constraints on SOC/temperature/ power; CVaR of current peaks.	Rule-based risk-aware control; scenario-based/Gaussian MPC; RL with scenario pretraining.	Battery current peak and RMS; SOC occupancy (battery/SC); converter losses; inter-cycle stability.	Nkembi et al. (2024), Li et al. (2021), Li et al. (2023), Urooj & Nasir (2024), Zheng et al. (2018)
Sizing (offline)	Battery and SC capacity/power; converter capacity/efficiency.	Operation scenarios; charging availability; aging parameters.	Chance constraints on performance/safety; CVaR of aging or unmet energy cost.	Two-stage scenario-based (SAA); multiobjective with distributional evaluation.	Cost/mass; CVaR of current peak; thermal margin; equivalent lifetime/SoH.	Nkembi et al. (2024), Li et al. (2021), Pal et al. (2021), Urooj & Nasir (2024), Zheng et al. (2018)

EMS, energy management system; *CvaR*, conditional value-at-risk; *HESS*, hybrid energy storage system; *SAA*, sample average approximation; *SC*, supercapacitor; *SoH*, state of health; *MPC*, model predictive control.

and methods such as risk-aware rule-based, scenario-based/Gaussian MPC, and pre-trained data-driven policies are applied; while in (offline) sizing, the capacities/powers and specifications of the converters are evaluated under operating and charging access scenarios with chance constraints and CVaR objectives. Two-stage scenario-oriented approach (SAA) and multiobjective optimization with distributed evaluation is used (Li et al., 2021; Nkembi et al., 2024; Pal et al., 2021; Urooj & Nasir, 2024; Zheng et al., 2018).

Based on this compact map, a detailed explanation of the design path can be provided. First, a package of scenarios is developed by combining the natural cycles of the target area and artificial profiles obtained by random sampling on the velocity transition probability surface to cover the range of speed-acceleration, stop-and-go, and path gradient well (Li et al., 2023). This package is the everyday basis for EMS calibration and evaluation of sizing options. In energy management, the rule-based policy is defined by the power frequency division logic (SC contribution in fast pulses and battery contribution in the slow component of energy) and SOC windows; the risk-aware version, by solving a scenario-based multiobjective problem, adjusts the thresholds and weights so that the battery peak current CVaR and the probability of SOC window violation are minimized, while also considering the power path losses and converter efficiency (Nkembi et al., 2024; Urooj & Nasir, 2024). If a predictive policy (MPC) is used, the multicriteria objective function is solved over short horizons with chance constraints on SOC/temperature/power. To be solvable, either the load/parameter uncertainty is sampled as a scenario and introduced into the SAA, or the chance constraints are transformed into deterministic equivalent constraints by Gaussian approximations and linearization; it is obvious that the accuracy of these approximations depends on the model validity and the quality of the online parameter estimation (Nkembi et al., 2024). In data-driven approaches, policies are first pretrained on a set of natural/artificial scenarios to encounter a range of velocity-acceleration patterns, then transferred to the destination path and fine-tuned. In this space, incorporating risk measures into the reward/value function (e.g., peak-flow CVaR) makes the policy sensitive to the tails of the distribution. It moves away from mean-centered solutions (Li et al., 2021, 2023).

Sizing in the risk-aware version is naturally expressed in a two-stage formulation (design–operation): In the first stage, the nominal battery and SC sizes and the converter capacities/efficiency are selected; in the second stage, the operation under a given EMS is evaluated. Uncertainty of the operation scenarios and model parameters is present in both stages, and performance indicators (peak/RMS battery current, SOC residence time in stressed regions, power electronics losses, and cyclic/calendar aging indicators) are reported in a distributed manner. Chance constraints ensure that the probability of violating safety limits remains low, and targeting the CVaR at the "equivalent cost of aging" or "unsupplied energy" (in the absence of fast charging) prevents the emergence of fragile designs. The problem is usually multiobjective, and a Pareto analysis is performed between the aging efficiency and the risk cost. In practice, the multiobjective frameworks introduced in recent reviews (although often definitive) are improved by replacing point evaluation with distributional evaluation and adding chance constraints to the risk-aware version (Li et al., 2021; Nkembi et al., 2024; Urooj & Nasir, 2024; Zheng et al., 2018).

At the heart of both decisions (EMS and sizing) are aging models and the definition of the equivalent cost of risk. In the rigorous version, current, SOC, and temperature under scenarios are transformed into stress profiles and then mapped to capacity loss/resistance gain with equivalent circuit/electrothermal models and cyclic/calendar regularities. This mapping allows the definition of a loss function that can be directly included in the objective function (as the aging cost CVaR) or in constraints (as a chance). In the lighter version when the whole model or necessary data is not available aging proxies such as peak/RMS current, SOC residence time in the stress regions, and cycle depth are used and personnel or their CVaR are targeted; Although these surrogates do not have the physical accuracy of full aging models, in practice they free the decision from reliance on single-point averages and provide a more conservative picture (Nkembi et al., 2024).

At the scale of multisource architectures on a vehicle, the role rule requires a risk-aware explanation. The SC should absorb short-term peaks, and the battery should take on the slow load, while a more uniform source (such as a fuel cell or alternator) provides the base power. Applying chance constraints on SOC/temperature ensures that safety margins are met in a high percentage of scenarios, and targeting CVaR for "peak current" or "ageing cost" avoids policies that occasionally perform very poorly. Proof of concept reports on rule-based control in multisource architectures have shown that this same logic can be implemented with risk-aware parameter-based tuning (temperature-dependent SOC thresholds, high-confidence current limiters, "peak-shutting" logic under thermal scenarios), leading to robust SOC and thermal response in the face of realistic disturbances (Mamun et al., 2022).

Despite the above clear framework, practical and research challenges remain. First, the computational cost of scenario-based solutions in high dimensions can be heavy; scenario dimensionality reduction strategies (clustering, adaptive sampling, allocation of sample budgets to the tails of the distribution) and the use of analytical approximations for chance constraints (in Gaussian-linearized forms) are part of the answer. Second, there is uncertainty in the aging models themselves: Coefficients and life functions often depend on specific laboratory conditions, and their transfer to the field of operation can be ambiguous; here, the use of distribution uncertainty sets and "robust-distribution" forms for chance constraints can provide controlled conservatism. Third, the lack of data is problematic for rare events; Relying on artificial scenarios derived from random sampling on the velocity transition probability surface (Li et al., 2023) and designing targeted tests for the tail of the distribution improves coverage.

The fourth point concerns the integration of software and hardware in the ECU, where computational power constraints and safety requirements play a decisive role. To manage this challenge, it is proposed to design the control blocks in a modular manner, to use precomputed maps in paths that require fast response, and to employ an online parameter estimation layer, such as continuous evaluation of the internal resistance of the battery to gradually update the model and compensate for drift (Nkembi et al., 2024). Finally, at the final selection level, the decision should be examined in the Pareto space of efficiency–aging–cost–risk; the approaches introduced in recent HESS reviews, although often deterministic (Li et al., 2021; Zheng et al., 2018), are easily improved by distributing the indicators and adding CVaR chance/objective constraints to the risk-aware version. This approach reduces the gap between conventional methods and field requirements (Nkembi et al., 2024; Urooj & Nasir, 2024).

The conclusion of this section emphasizes that the stochastic/risk-aware approach is not just "adding a risk term" to an existing problem; it is a redefinition of the entire decision model data chain. Starting with representative and data-enhanced scenario building, continuing with the formulation of objectives and constraints in a distributional and risk-based manner, and ending with out-of-sample assessment and sensitivity analysis, energy management and sizing move from relying on single-point numbers to decision-making distributional evidence. This transition is the missing link that elevates the HESS design from "good performance in the mean" to "stability in the tails of the distribution," making the chosen policy defensible not only on the laboratory track but also on busy streets and hot/cold days (Mamun et al., 2022; Nkembi et al., 2024; Urooj & Nasir, 2024). Finally, the systematic connection between the vehicle and infrastructure levels through the "with/without fast charging" and "normal/peak demand" scenario branches allows the final decision to be placed in the Pareto space of efficiency-aging-cost-risk. The output is a strategy that maintains safety and efficiency margins in both the average and worst-case scenarios (Nkembi et al., 2024; Pal et al., 2021; Urooj & Nasir, 2024).

17.16 On-board renewables and sustainability impacts

The discussion of "on-board renewables" makes sense when three questions are answered simultaneously: How much electrical power can realistically be obtained from on-board sources? How is this power integrated and managed alongside the HESS (battery/SC)? Moreover, finally, what are the sustainability implications of this integration from a lifecycle, critical materials, and safety perspective? Recent literature on multisource on-board systems that employ combinations of solar panels, slight wind, and fuel cells in conjunction with HESS provides plausible engineering answers, but also highlights physical limitations and design tradeoffs (Mamun et al., 2022; Nkembi et al., 2024; Urooj & Nasir, 2024).

From an architectural perspective, a systematic study has modeled a photovoltaic (PV) + slight wind + fuel cell combination alongside a battery and a SC on a vehicle. It has been shown that a rule-based logic for assigning roles to the fuel cell for steady power, PV/wind as complementary oscillatory inputs, the SC for short peaks, and the battery for the slow energy component can suppress transients and protect the battery from peak currents (Nkembi et al., 2024). This architecture is significant when the energy management is based on power frequency division, meaning that the high-oscillating loads are mounted on the SC and the slow loads on the battery to control power electronics losses and thermal stress (Urooj & Nasir, 2024). However, the implementation of a wind source on a vehicle faces aerodynamic and noise constraints and is mostly conceptual; in contrast, rooftop PV is much more established and has a detailed measurement/modeling literature (Mamun et al., 2022).

For rooftop PV, field measurements and detailed modeling have shown that the actual power is a combined function of the angle of incidence, shadow damage, cell temperature, and even the vehicle's own motion and vibration. Accurate estimates depend, in addition to the module and interface electronics (DC–DC converters with maximum power point tracking), on the road's radiation profile and the stop/go pattern. The practical result is that in sunny urban cycles, on-board PV can cover part of the side load (ventilation, lighting, cabin electronics). However, it has a limited contribution to the role of "propulsion pull source" and should not be considered a substitute for the battery or grid charging (Fathabadi, 2017). On the other hand, hybrid PV-battery designs for PHEVs have shown that proper integration of the converter and power management improves the efficiency of the PV→ storage path and smooths out PV fluctuations without imposing additional stress on the battery (Fathabadi, 2017). These findings are consistent with the logic of HESS:The more the share of transient loads is transferred to the SC and the shorter and more efficient the power paths are, the greater the sustainability and durability benefits will be (Urooj & Nasir, 2024).

Sustainability/life cycle implications are a second focus of discussion. Even if rooftop PV has a limited power contribution, integration of HESS with on-board renewables can mitigate grid energy requirements and charging peaks at the fleet scale; however, the net benefit must be assessed at a life cycle level. LCA studies have shown that recycling in the lithium-ion battery chain can significantly reduce the energy embodied and greenhouse gas emissions; in life cycle balance models, scenarios with higher recycling rates result in lower carbon intensity per kilometer, especially if the carbon intensity of grid electricity is also decreasing (Jiya et al., 2018). This message is directly relevant to HESS design: Energy management policies that lower the peak/RMS battery current and move the SOC residence time away from stress zones not only increase durability but also reduce the frequency of battery replacement/refreshment, which has a high impact on LCA (Urooj & Nasir, 2024). Furthermore, recent reviews of automotive storage systems, alongside the functional benefits of HESS, point to the challenges of critical materials, thermal safety, and waste management; these challenges, if not addressed in the design, will offset operational gains from a sustainability perspective (Nkembi et al., 2024).

From a fuel cell and hydrogen perspective, the "low-fluctuation base power" option is attractive and pairs well with HESS in multi-source architectures on the vehicle; the fuel cell can provide a more uniform power level, and the SC can absorb acceleration pulses (Mamun et al., 2022). However, the sustainability of this path depends on the hydrogen source and its logistics; if low-carbon hydrogen and a safe/efficient supply chain are available, the FC–HESS link will become more attractive from an emissions and sustainability perspective, otherwise it is simply a shift of emissions rather than a net reduction (Nkembi et al., 2024; Urooj & Nasir, 2024). This dependency on the "energy system context" once again highlights the necessity of a life cycle and full value chain view.

In engineering summary, the literature shows that the most significant sustainability benefits are achieved when the role of on-board resources is realistically defined: PV on the roof mainly for low-power loads and to help reduce peak charging, FC for uniform baseload power, and HESS for transient smoothing and battery protection. In this framework, sustainability-conscious energy management should incorporate life-cycle indicators into the decision criteria in addition to instantaneous efficiency: Reducing peak current CVaR and minimizing SOC residence time in stressed areas (as a proxy for aging costs), along with LCA metrics such as embodied energy and CO_2 equivalent emissions.

At the design level, adding PV on the body or FC module only makes sense if the increased mass/cost/complexity is offset by a significant reduction in battery or grid power stress; otherwise, the additional weight and losses may offset the benefits (Fathabadi, 2017; Nkembi et al., 2024). On the other hand, investing in recycling and recycling-friendly design (separable, low-critical materials) multiplies the sustainability gains. It shifts the big picture in favor of technology adoption by reducing the end-of-life environmental burden (Hao et al., 2017; Nkembi et al., 2024). Ultimately, the synergy of multisource architecture (Mamun et al., 2022), scenario-aware HESS and EMS logic (Urooj & Nasir, 2024), and LCA-driven policies (Hao et al., 2017; Nkembi et al., 2024) forms a pragmatic path that is compatible with both the physical constraints of the vehicle roof and the sustainability requirements of the value chain.

17.17 Conclusion

Modeling, conversion, and control were treated as a unified stochastic co-design problem for hybrid energy storage in electric platforms. Equivalent-circuit battery models identified from HPPC-style data reproduced lithium-ion polymer terminal voltage with maximum absolute errors of 15 to 21 mV across pulse-discharge tests at 80 to 640 mA, with runtime error not exceeding 0.12%. In broader profile validation, the same model achieved 15 mV and 0.395% for continuous discharge at 80 mA, 30 mV and 0.133% for pulse charge, and 20 mV and 0.338% for a four-step discharge sequence, indicating accurate capture of both steady-state and transient behavior. SC dynamics were represented with RC-family models selected for experiment-calibrated transient fidelity and tractable identification; higher-order variants maintained useful accuracy up to approximately 1 kHz when high-frequency effects and temperature dependence were included. Bidirectional DC–DC topology and control determined voltage gain, device stress, efficiency, and DC-link quality under uncertainty, while dual-converter or multiinput arrangements enabled deep SC utilization and reduced bus ripple. A scenario-driven, uncertainty-quantified workflow enabled risk-aware energy management and sizing using chance constraints and tail metrics, suppressing peak and RMS battery currents and protecting state-of-charge and temperature margins. Within this structure, frequency-split role assignment, with the SC handling fast transients, the battery supplying slow energy, and the fuel cell delivering baseload when present, improved thermal margins and mitigated effective aging. Overall, the results show that equivalent-circuit models provide the most favorable accuracy-to-complexity balance for the chapter's control and real-time objectives, while RC-based multi-branch SC models offer a practical compromise between fidelity and identification effort.

Acronym Corresponding term

BDC	Bidirectional DC–DC converter
BMS	Battery management system
CECM	Cauer equivalent circuit model
CVaR	Conditional value-at-risk
DAB	Dual active bridge
DoD	Depth of discharge
ECU	Electronic control unit
EDLC	Electric double-layer capacitor
EIS	Electrochemical impedance spectroscopy
ECM	Equivalent-circuit model
EMS	Energy management system
ESR	Equivalent series resistance
ESL	Equivalent series inductance
FC	Fuel cell
FCHEV	Fuel cell hybrid electric vehicle
FESS	Flywheel energy storage system
HEV	Hybrid electric vehicle
HESS	Hybrid energy storage systems
HF	High frequency
HPPC	Hybrid pulse power characterization
ICE	Internal combustion engine
MPC	Model predictive control
NREL	National renewable energy laboratory
OCV	Open-circuit voltage
ODE	Ordinary differential equation
PDE	Partial differential equations
PEMFC	Proton exchange membrane fuel cell
SAA	Sample average approximation
SAFT	Saft (battery manufacturer)
SC	Supercapacitor
SoC	State of charge
SoH	State of health
TLM	Transmission line model
UQ	Uncertainty-quantified
VaR	Value-at-risk

References

Argyrou, M. C., Christodoulides, P., Marouchos, C. C., & Kalogirou, S. A. (2018). Hybrid battery-supercapacitor mathematical modeling for PV application using Matlab/Simulink. In *2018 53rd International Universities Power Engineering Conference (UPEC)* (pp. 1–6). doi:10.1109/UPEC.2018.8541933.

Berrueta, A., San Martín, I., Hernández, A., Ursúa, A., & Sanchis, P. (2014). Electro-thermal modelling of a supercapacitor and experimental validation. *Journal of Power Sources, 259*, 154–165. doi:10.1016/j.jpowsour.2014.02.089.

Berrueta, A., Ursua, A., Martin, I. S., Eftekhari, A., & Sanchis, P. (2019). Supercapacitors: Electrical characteristics, modeling, applications, and future trends. *IEEE Access, 7*, 50869–50896. doi:10.1109/ACCESS.2019.2908558.

Bethoux, O. (2020). Hydrogen fuel cell road vehicles: State of the Art and perspectives. *Energies, 13*(21), 5843. doi:10.3390/en13215843.

Biswas, D., Ghosh, S., Sengupta, S., & Mukhopadhyay, S. (2022). Energy management of a parallel hybrid electric vehicle using model predictive static programming. *Energy, 250*, 123505. doi:10.1016/j.energy.2022.123505.

C. B. Corporation. (2019). Technical manual for lifeline batteries. https://lifelinebatteries.com/wp-content/uploads/2015/12/6-0101F-Lifeline-Technical-Manual-Final-5-06-19.pdf.

Cabrane, Z., Ouassaid, M., & Maaroufi, M. (2016). Analysis and evaluation of battery-supercapacitor hybrid energy storage system for photovoltaic installation. *International Journal of Hydrogen Energy, 41*(45), 20897–20907. doi:10.1016/j.ijhydene.2016.06.141.

CANBAT, AGM vs gel batteries. https://canbat.com/guide-agm-batteries-gel-batteries (Accessed 2025).

Cao, W., Goto, R., Yuno, T., Kawabe, T., & Mukai, M. (2025). Energy management method optimizing engine ignition on/off timing and operating points of powertrain components for series hybrid electric vehicle. *IEEJ Journal of Industry Applications, 14*(3), 366–375. doi:10.1541/ieejjia.23014068.

Chen, M., & Rincon-Mora, G. A. (2006). Accurate electrical battery model capable of predicting runtime and I–V performance. *IEEE Transactions on Energy Conversion, 21*(2), 504–511. doi:10.1109/TEC.2006.874229.

Cheng, Q., Zhang, R., Shi, Z., & Lin, J. (2024). Review of common hydrogen storage tanks and current manufacturing methods for aluminium alloy tank liners. *International Journal of Lightweight Materials and Manufacture, 7*(2), 269–284. doi:10.1016/j.ijlmm.2023.08.002.

Cultura, A., & Salameh, Z. (2015). Modeling, evaluation and simulation of a supercapacitor module for energy storage application. In *International Conference on Computer Information Systems and Industrial Applications* (pp. 876–882). Atlantis Press.

Devillers, N., Jemei, S., Péra, M. C., Bienaimé, D., & Gustin, F. (2014). Review of characterization methods for supercapacitor modelling. *Journal of Power Sources, 246*, 596–608. doi:10.1016/j.jpowsour.2013.07.116.

Du, L. (2009). Study on supercapacitor equivalent circuit model for power electronics applications. In *2009 2nd International Conference on Power Electronics and Intelligent Transportation System (PEITS): 2* (pp. 51–54). doi:10.1109/PEITS.2009.5406744.

Einhorn, M., Conte, V. F., Kral, C., Fleig, J., & Permann, R. (2010). Parameterization of an electrical battery model for dynamic system simulation in electric vehicles. In *IEEE Vehicle Power and Propulsion Conference, 1–3 Sept. 2010* (pp. 1–7). doi:10.1109/VPPC.2010.5729127.

Fathabadi, H. (2017). Novel standalone hybrid solar/wind/fuel cell power generation system for remote areas. *Solar Energy, 146*, 30–43. doi:10.1016/j.solener.2017.01.071.

Garche, J., Moseley, P. T., & Karden, E. (2015). Lead-acid batteries for hybrid electric vehicles and battery electric vehicles. In *Advances in battery technologies for electric vehicles* (pp. 75–101). Elsevier. doi:10.1016/B978-1-78242-377-5.00005-4.

Gómez, J. A., & Santos, D. M. F. (2023). The status of on-board hydrogen storage in fuel cell electric vehicles. *Designs, 7*(4), 97. doi:10.3390/designs7040097.

Hao, H., Qiao, Q., Liu, Z., & Zhao, F. (2017). Impact of recycling on energy consumption and greenhouse gas emissions from electric vehicle production: The China 2025 case. *Resources, Conservation and Recycling, 122*, 114–125. doi:10.1016/j.resconrec.2017.02.005.

He, H., Xiong, R., & Fan, J. (2011). Evaluation of lithium-ion battery equivalent circuit models for state of charge estimation by an experimental approach. *Energies, 4*(4), 582–598. doi:10.3390/en4040582.

Helseth, L. E. (2019). Modelling supercapacitors using a dynamic equivalent circuit with a distribution of relaxation times. *Journal of Energy Storage, 25*, 100912. doi:10.1016/j.est.2019.100912.

Hussan, U., Majeed, M. A., Asghar, F., Waleed, A., Khan, A., & Javed, M. R. (2022). Fuzzy logic-based voltage regulation of hybrid energy storage system in hybrid electric vehicles. *Electrical Engineering, 104*(2), 485–495. doi:10.1007/s00202-021-01315-4.

Hussein, A. A. (2021). An overview and practical considerations of common lithium-ion battery cell models. In *4th International Symposium on Advanced Electrical and Communication Technologies (ISAECT), 6–8 Dec. 2021* (pp. 1–5).

Jiya, I. N., Gurusinghe, N., & Gouws, R. (2018). Electrical circuit modelling of double layer capacitors for power electronics and energy storage applications: A review. *Electronics, 7*(11), 268. doi:10.3390/electronics7110268.

Johnson, V. H. (2002). Battery performance models in ADVISOR. *Journal of Power Sources, 110*(2), 321–329. doi:10.1016/s0378-7753(02)00194-5.

Kai, W., Baosen, R., Liwei, L., Yuhao, L., Hongwei, Z., & Zongqiang, S. (2017). A review of modeling research on supercapacitor 2017. In *Proceedings - 2017 Chinese Automation Congress, CAC 2017* (pp. 5998–6001). doi:10.1109/CAC.2017.8243857.

Khamesipour, M., Chitsaz, I., Salehi, M., & Alizadenia, S. (2022). Component sizing of a series hybrid electric vehicle through artificial neural network. *Energy Conversion and Management, 254*, 115300. doi:10.1016/j.enconman.2022.115300.

Kozachenko, V. F., Ostrirov, V. N., & Lashkevich, M. M. (2014). Electric transmission based on the switched reluctance motor with independent excitation. *Russian Electrical Engineering, 85*(2), 115–120. doi:10.3103/s1068371214020084.

Lee, S., Kim, J., Lee, J., & Cho, B. H. (2008). State-of-charge and capacity estimation of lithium-ion battery using a new open-circuit voltage versus state-of-charge. *Journal of Power Sources, 185*(2), 1367–1373. doi:10.1016/j.jpowsour.2008.08.103.

Li, F., Gao, Y., Wu, Y., Xia, Y., Wang, C., Hu, J., & Huang, Z. (2023). Incentive learning-based energy management for hybrid energy storage system in electric vehicles. *Energy Conversion and Management, 293*, 117480. doi:10.1016/j.enconman.2023.117480.

Li, M., Wang, L., Wang, Y., & Chen, Z. (2021). Sizing optimization and energy management strategy for hybrid energy storage system using multiobjective optimization and random forests. *IEEE Transactions on Power Electronics, 36*(10), 11421–11430. doi:10.1109/TPEL.2021.3070393.

Luo, C., Shen, Z., Evangelou, S., Xiong, G., & Wang, F. Y. (2019). The combination of two control strategies for series hybrid electric vehicles. *IEEE/CAA Journal of Automatica Sinica, 6*(2), 596–608. doi:10.1109/JAS.2019.1911420.

Mama, M., Solai, E., Capurso, T., Danlos, A., & Khelladi, S. (2025). Comprehensive review of multi-scale Lithium-ion batteries modeling: From electro-chemical dynamics up to heat transfer in battery thermal management system. *Energy Conversion and Management, 325*, 119223. doi:10.1016/j.enconman.2024.119223.

Mamun, K. A., Islam, F. R., Haque, R., Chand, A. A., Prasad, K. A., Goundar, K. K., Prakash, K., & Maharaj, S. (2022). Systematic modeling and analysis of on-board vehicle integrated novel hybrid renewable energy system with storage for electric vehicles. *Sustainability, 14*(5), 2538. doi:10.3390/su14052538.

Manjula, A., Kute, U. T., Reddy, C. V. K., & Mallala, B. (2025). Power quality improvement of microgrid for photovoltaic EV charging station with hybrid energy storage system using RPO-ADGAN approach. *Journal of Energy Storage, 108*. doi:10.1016/j.est.2024.114970.

Mohammadi, F., & Saif, M. (2023). A comprehensive overview of electric vehicle batteries market. *e-Prime - Advances in Electrical Engineering, Electronics and Energy, 3*, 100127. doi:10.1016/j.prime.2023.100127.

Naseri, F., Karimi, S., Farjah, E., & Schaltz, E. (2022). Supercapacitor management system: A comprehensive review of modeling, estimation, balancing, and protection techniques. *Renewable and Sustainable Energy Reviews, 155*, 111913. doi:10.1016/j.rser.2021.111913.

Navarro, G., Nájera, J., Torres, J., Blanco, M., Santos, M., & Lafoz, M. (2020). Development and experimental validation of a supercapacitor frequency domain model for industrial energy applications considering dynamic behaviour at high frequencies. *Energies, 13*(5), 1156. doi:10.3390/en13051156.

Nelms, R. M., Cahela, D. R., & Tatarchuk, B. J. (2001). Using a Debye polarization cell to predict double-layer capacitor performance. *IEEE Transactions on Industry Applications, 37*(1), 4–9. doi:10.1109/28.903119.

Nelms, R. M., Cahela, D. R., & Tatarchuk, B. J. (2003). Modeling double-layer capacitor behavior using ladder circuits. *IEEE Transactions on Aerospace and Electronic Systems, 39*(2), 430–438. doi:10.1109/TAES.2003.1207255.

New D.A. (2004). Double layer capacitors: Automotive applications and modeling. Massachusetts Institute of Technology.

Niu, R., & Yang, H. (2011). Modeling and identification of electric double-layer supercapacitors. In *2011 IEEE International Conference on Robotics and Automation* (pp. 1–4). doi:10.1109/ICRA.2011.5980586.

Nkembi, A. A., Simonazzi, M., Santoro, D., Cova, P., & Delmonte, N. (2024). Comprehensive review of energy storage systems characteristics and models for automotive applications. *Batteries, 10*(3). doi:10.3390/batteries10030088.

Pal, A., Bhattacharya, A., & Chakraborty, A. K. (2021). Placement of public fast-charging station and solar distributed generation with battery energy storage in distribution network considering uncertainties and traffic congestion. *Journal of Energy Storage, 41*, 102939. doi:10.1016/j.est.2021.102939.

Pan, M., Cao, S., Zhang, Z., Ye, N., Qin, H., Li, L., & Guan, W. (2025). Recent progress on energy management strategies for hybrid electric vehicles. *Journal of Energy Storage, 116*, 115936. doi:10.1016/j.est.2025.115936.

PV Education. Characteristics of lead acid batteries. https://www.pveducation.org/pvcdrom/lead-acid-batteries/characteristics-of-lead-acid-batteries (Accessed 2025).

Rafik, F., Gualous, H., Gallay, R., Crausaz, A., & Berthon, A. (2007). Frequency, thermal and voltage supercapacitor characterization and modeling. *Journal of Power Sources, 165*(2), 928–934. doi:10.1016/j.jpowsour.2006.12.021.

Şahin, M. E., Blaabjerg, F., & Sangwongwanich, A. (2021). Modelling of supercapacitors based on simplified equivalent circuit. *CPSS Transactions on Power Electronics and Applications, 6*(1), 31–39. doi:10.24295/CPSSTPEA.2021.00003.

Saidani, F., Hutter, F. X., Scurtu, R. G., Braunwarth, W., & Burghartz, J. N. (2017). Lithium-ion battery models: A comparative study and a model-based powerline communication. *Advances in Radio Science, 15*, 83–91. doi:10.5194/ars-15-83-2017.

Saldaña, G., Martín, J. I. S., Zamora, I., Asensio, F. J., & Oñederra, O. (2019). Analysis of the current electric battery models for electric vehicle simulation. *Energies, 12*(14). doi:10.3390/en12142750.

Seim, L. H. (2012). Modeling, control and experimental testing of a supercapacitor/battery hybrid system: passive and semi-active topologies. *Master's thesis, Department of Mathematical Sciences and Technology, Norwegian University of Life Sciences, Ås, Norway.*

Shen, L., Cheng, Q., Cheng, Y., Wei, L., & Wang, Y. (2020). Hierarchical control of DC microgrid for photovoltaic EV charging station based on flywheel and battery energy storage system. *Electric Power Systems Research, 179*, 106079. doi:10.1016/j.epsr.2019.106079.

Shen, Z., Luo, C., Dong, X., Lu, W., Lv, Y., Xiong, G., & Wang, F. Y. (2022). Two-level energy control strategy based on ADP and A-ECMS for series hybrid electric vehicles. *IEEE Transactions on Intelligent Transportation Systems, 23*(8), 13178–13189. doi:10.1109/TITS.2021.3121550.

Shi, M., Ren, Y., Cao, J., Kuang, Z., Zhuo, X., & Xie, H. (2024). Current situation and development prospects of discharge pretreatment during recycling of lithium-ion batteries: A review. *Batteries & Supercaps, 7*(2). doi:10.1002/batt.202300477.

Shin, H. K., & Ha, S. K. (2023). A review on the cost analysis of hydrogen gas storage tanks for fuel cell vehicles. *Energies, 16*(13). doi:10.3390/en16135233.

Sinha, V. K., Mahia, R. N., & Mahela, O. P. (2022). Power management in electric vehicles using a hybrid energy storage system. In *Active electrical distribution network: Issues, solution techniques, and applications* (pp. 201–215). Elsevier. doi:10.1016/B978-0-323-85169-5.00014-9.

Sun, K., & Shu, Q. (2011). China overview of the types of battery models. In *Proceedings of the 30th Chinese Control Conference, CCC 2011* (pp. 3644–3648).

Thanagasundram, S., Arunachala, R., Makinejad, K., Teutsch, T., & Jossen, A. (2012). A cell level model for battery simulation. In *European Electric Vehicle Congress* (pp. 1–13).

Tjandra, R., Thanagasundram, S., Tseng, K. J., & Jossen, A. (2014). Improved lithium-ion battery model with hysteresis effect. In *2014 IEEE Transportation Electrification Conference and Expo (ITEC)* (pp. 1–8).

Tran, M. K., Dacosta, A., Mevawalla, A., Panchal, S., & Fowler, M. (2021). Comparative study of equivalent circuit models performance in four common lithium-ion batteries: LFP, NMC, LMO. *NCA Batteries, 7*(3). doi:10.3390/batteries7030051.

Urooj, A., & Nasir, A. (2024). Review of hybrid energy storage systems for hybrid electric vehicles. *World Electric Vehicle Journal, 15*(8). doi:10.3390/wevj15080342.

Vural, B., Uzunoglu, M., Erdinc, O., & Onar, O. C. (2009). Turkey a dynamic ultra-capacitor model for vehicular applications. In *International Conference on Clean Electrical Power, ICCEP 2009* (pp. 595–598). doi:10.1109/ICCEP.2009.5211994.

Wang, J., Wang, B., Zhang, L., Wang, J., Shchurov, N. I., & Malozyomov, B. V. (2022). Review of bidirectional DC–DC converter topologies for hybrid energy storage system of new energy vehicles. *Green Energy and Intelligent Transportation, 1*(2), 100010. doi:10.1016/j.geits.2022.100010.

Xu, X., Zhao, J., Zhao, J., Shi, K., Dong, P., Wang, S., Liu, Y., Guo, W., & Liu, X. (2022). Comparative study on fuel saving potential of series-parallel hybrid transmission and series hybrid transmission. *Energy Conversion and Management, 252*, 114970. doi:10.1016/j.enconman.2021.114970.

Zhang, F., Xiao, L., Coskun, S., Pang, H., Xie, S., Liu, K., & Cui, Y. (2023). Comparative study of energy management in parallel hybrid electric vehicles considering battery ageing. *Energy, 264*, 123219. doi:10.1016/j.energy.2022.123219.

Zhang, L., Hu, X., Wang, Z., Sun, F., & Dorrell, D. G. (2018). A review of supercapacitor modeling, estimation, and applications: A control/management perspective. *Renewable and Sustainable Energy Reviews, 81*, 1868–1878. doi:10.1016/j.rser.2017.05.283.

Zhang, X., Zhang, W., & Lei, G. (2016). A review of Li-ion battery equivalent circuit models. *Transactions on Electrical and Electronic Materials, 17*(6), 311–316. doi:10.4313/TEEM.2016.17.6.311.

Zheng, C., Li, W., & Liang, Q. (2018). An energy management strategy of hybrid energy storage systems for electric vehicle applications. *IEEE Transactions on Sustainable Energy, 9*(4), 1880–1888. doi:10.1109/TSTE.2018.2818259.

Zhou, W., Zheng, Y., Pan, Z., & Lu, Q. (2021). Review on the battery model and SOC estimation method. *Processes, 9*(9), 1685. doi:10.3390/pr9091685.

Zimmermann, T., Keil, P., Hofmann, M., Horsche, M. F., Pichlmaier, S., & Jossen, A. (2016). Review of system topologies for hybrid electrical energy storage systems. *Journal of Energy Storage, 8*, 78–90. doi:10.1016/j.est.2016.09.006.

Zubieta, L., & Bonert, R. (2000). Characterization of double-layer capacitors for power electronics applications. *IEEE Transactions on Industry Applications, 36*(1), 199–205. doi:10.1109/28.821816.

A stochastic, Adaptive, and nature-inspired electric distribution grids architecture: Data-driven futuristic power grids through emergent intelligence-based operational mechanism

18

Neelofar Shaukat[a], *Md. Rabiul Islam*[b], *Syed Ayaz Ali Shah*[a],
Jamil Ahmad Khan[a] *and Frede Blaabjerg*[c]
[a] Electrical Engineering Department, Abbottabad Campus, COMSATS University Islamabad, Abbottabad, Khyber Pakhtunkhwa, Pakistan, [b] School of Engineering, University of Wollongong, Wollongong, New South Wales, Australia, [c] Department of Energy Technology, Aalborg University, Aalborg East, Denmark

18.1 Introduction

Modern power distribution systems confront unprecedented challenges due to integrating a complex mix of distributed energy resources (DERs) such as renewable energy resources (RERs), electric vehicles (EVs), and prosumers (Khalid, 2024). Conventional grids were developed for one-way power and communication flow; however, modern grids have evolved into a decentralized and dynamic system with bidirectional flows of power and communication. This evolution surfaces various challenges, including the need for real-time ambient-aware decision-making, flexibility under fluctuating conditions, and scalability to manage intermittent and unpredictable diverse resource bases (Kataray et al., 2023). The variability and dependency of RERs on weather conditions complicate the reliability and stability of power distribution grids (Wang et al., 2022). For example, intermittent wind or solar power generation outputs require conventional power plants to fill the gap, and the grid needs to adjust quickly. Thus, this approach introduces delays and strains grid resources because conventional grids are not designed to demonstrate flexibility to address rapid and random responses (Shen et al., 2021). Furthermore, EVs add another layer of complexity due to their unpredictable charging behavior. Therefore, preventing localized overload while maintaining stability requires real-time adaptation (Mohammed et al., 2023). Another core challenge due to the growth of decentralized generation is scalability, i.e., power grids must manage variable energy resources and accommodate more connections (Rodriguez-Calvo et al., 2018).

Stochastic Planning and Modeling for Energy Systems: Methods, Applications, and Developments.
DOI: https://doi.org/10.1016/B978-0-443-45298-7.00017-3

The traditional distribution grid structures rely on centralized and rigid hierarchical control schemes. Therefore conventional infrastructure requires huge computational resources to handle vast and diverse data (Ali Kadhem et al., 2017). The inclusion and integration of diverse energy resources call for adaptive and intelligent schemes for flexibility and real-time decision-making at every grid level (Machlev et al., 2022). Conventional systems struggle to adapt to switching, which is required for dynamic balance, because these systems were designed for power flow (steady state) from a few resources. Therefore conventional operational architectures struggle to accommodate real-time resource fluctuations and dynamic demand shifts (Couraud et al., 2023; Rani, 2024; Trivedi et al., 2022)

These transformations have shifted the power distribution grid from a predictable to a highly complex and dynamic environment characterized by bidirectional power and communication flows. Control strategies, either centralized, distributed, or decentralized, face inherent limitations in the context of being simultaneously intelligent, adaptive, and resilient (Jafari et al., 2024). The study of biological species self-organization in nature thus provides valuable inspiration in the abovementioned regard. Biological swarms such as fish schools, ant colonies, or bird flocks display emergent behavior, i.e., swarm-intelligence, achieving adaptive and robust behavior without a central authority to command (Paul et al., 2024; Priyadarshi & Kumar, 2025). The agents in the swarms follow simple rules, yet collectively they can organize themselves to absorb environmental shocks by adapting to external changes. These biological traits, i.e., adaptability, resilience, and scalability, are relevant to modern distribution grid systems because of variable RERs, demand shifts due to dynamic load characteristics, and unexpected weather contingencies. Thus, swarm intelligence can represent a methodological foundation for next-generation power grid governance. Moreover, the exponential energy demand results in highly complex and environmentally impactful power distribution grid infrastructures. Conventional centralized grids mostly require extensive transmission networks, land use, and backup generation (fossil-fuel-based), which disrupt native ecosystems, encroach upon ecological corridors, and increase greenhouse gas emissions (Kim, 2007; Luderer et al., 2019). The advent of smart grids continues to rely on large-scale grid infrastructures that disrupt natural habitats and introduce electromagnetic pollution that can disturb wildlife behaviors and migration patterns. Moreover, these frameworks are heavily dependent on fossil backup for grid stability, leading to terrestrial and aquatic pollution. Carbon emissions contribute to acid rain, damaging forest and aquatic ecosystems, and transmission networks interrupt wildlife migratory paths (Creutzig et al., 2019). Fig. 18.1 demonstrates the negative environmental consequences of centralized electric structures.

Therefore, these challenges raise another pressing research question: How can power grid systems achieve sustainability and resilience while minimizing environmental and ecological footprint and promoting harmony with natural ecosystems? Addressing this question requires a power grid architecture that embraces nature-aligned principles, decentralization, adaptiveness, and an ecologically harmonious structural framework. Therefore this chapter introduces the adaptive and intelligent power system (AIPS) framework for distribution grids, a new governance and operational framework that implements the swarm-inspired adaptive clustering for the coordination of the

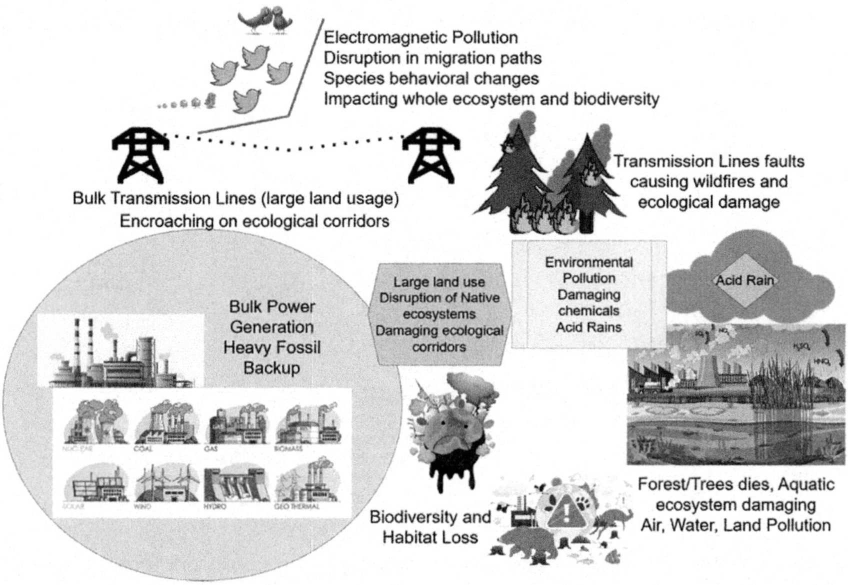

Figure 18.1 Impacts of conventional power structures. Negative ecological and environmental impacts of centralized power infrastructures.

system. Unlike traditional power system operational frameworks that predetermine the hierarchy of control zones, the proposed AIPS framework dynamically forms and dissolves clusters according to real-time operating conditions. The core of AIPS is intelligent adaptive hubs (IAHs), a regional supervisory entity managing a collection of cognitive energy units (CEUs). These CEUS represent the operational blocks of the system and oversee the energy adaptivity units (EAUs), which are the fundamental data blocks/physical entities. Under normal operation, IAHs operate autonomously to manage the local resources and loads. However, during disturbances, IAHs under the lightweight supervision of the coordinator of the central energy alliance (CEA) activate the adaptive clustering of IAHs, temporarily binding multiple IAHs into a coordinated cluster that redistributes resources until stability is restored. Fig. 18.2 illustrates the core concept of the proposed framework. The introduction of AIPS represents a fundamental departure from static operational frameworks because, in traditional approaches, zones or clusters of operation are fixed, leading to inefficiencies when the system state diverges from predetermined conditions. Conversely, the AIPS incorporates swarm-inspired clustering, in which clusters emerge dynamically in response to local system states, i.e., only during contingency periods and dissolve once the disturbance subsides. This framework allows the distribution grid system to reconfigure organically with minimal computational burden. Additionally, the AIPS is a modular system, i.e., IAHs operate locally with adaptive clustering invoked only in fault times, reducing unnecessary processing and communication.

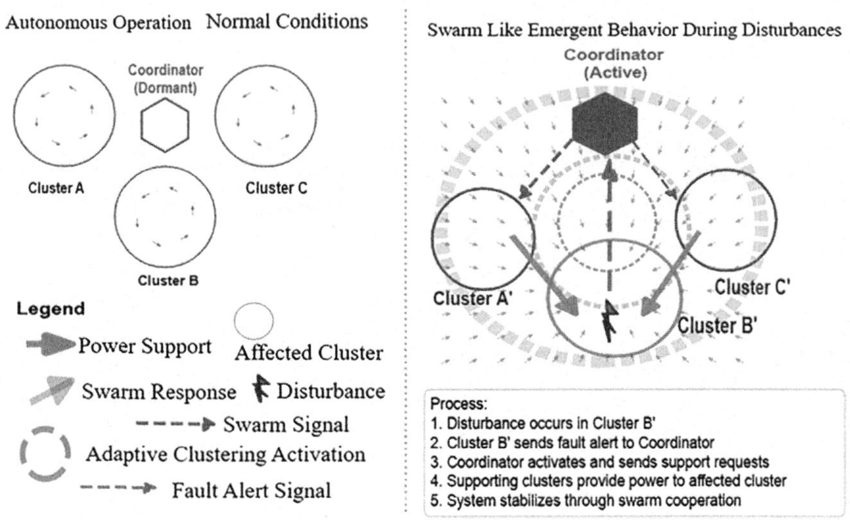

Figure 18.2 Graphical abstract. Core concept of the proposed framework for the power distribution system.

The distribution grids are increasingly dynamic, and uncertainty emerges as a defining characteristic of both load demand and generation. Variability in RERs, fluctuating prosumer behaviors, and other operational disturbances introduce stochastic elements that traditional deterministic control frameworks are not designed to address. Handling these uncertainties requires approaches that can address randomness and quantify the performance of the system under various and diverse possible states. Thus, in this chapter, the governance model is examined from both operational and stochastic perspectives, integrating probabilistic evaluation to assess the robustness under stochastic conditions. This uncertainty-based foundation strengthens the relevance of the proposed architecture for future electric distribution grids that need to operate reliably in stochastic and unpredictable environments.

In literature, various studies applied swarm intelligence to optimization tasks in electric systems such as unit commitment, scheduling, economic dispatching, etc., but these efforts treat swarm behavior as computational solvers rather than a guiding principle for system structure (Tang et al., 2021; Tiwari & Kumar, 2023). Algorithms based on ant colony or particle swarm are widely used in studies, but generally focus on algorithmic efficiency for particular problems in power systems. However, our research work in this chapter elevates swarm intelligence from an algorithmic or solver technique to a power system architectural principle, fundamentally reshaping the governance of power distribution systems. Therefore, this research work represents a step forward in bridging the gap between bio-inspired intelligence and power grid structural design.

Thus, in the light of the above-stated issues and bio-inspired principles, this work contributes to the State of the Art in the following three major ways.

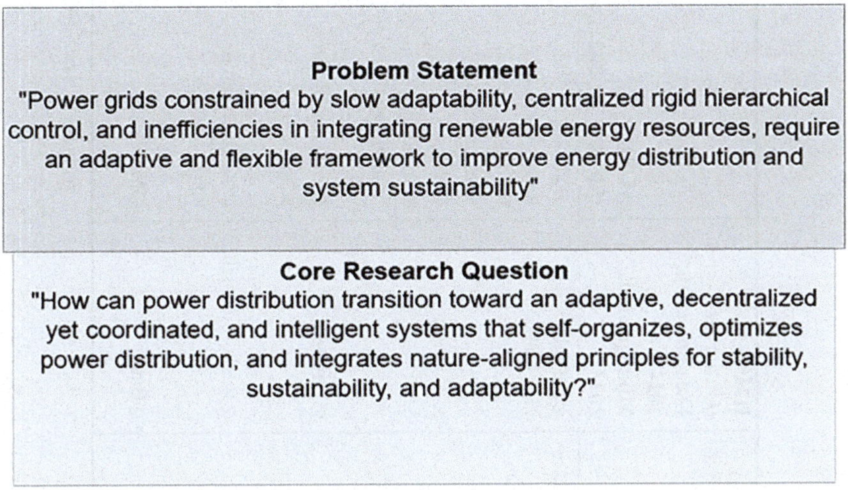

Figure 18.3 Guiding statements for the research chapter. The problem statement and core research question guiding this study.

Introduction of novel AIPS framework, which redefines distribution grid governance and intelligence through adaptive clustering, offering a nature-inspired framework.

Development of a fault-aware optimization model, extending the formulations with adaptive penalties and dynamic resource reconfiguration.

Analytical and statistical (through 1000 Monte Carlo simulations) insights into system modularity, demonstrating how swarm-inspired adaptive clustering enables effective operation under normal and dynamic conditions.

The problem statement and core research question that guide this work are demonstrated in Fig. 18.3.

18.1.2 Comparison with contemporary adaptive architectures

The accelerating complexity of modern distribution grid networks demands intelligent and adaptive architectures that are capable of managing diverse and distributed sources of energy generation and consumption and are flexible and robust (Alvarez-Alvarado et al., 2024; Li et al., 2024; Liang et al., 2024; Machlev et al., 2022; Nyangon, 2024). Several architectures, such as dynamic virtual power plants, autonomous energy grids by the National Renewable Energy Laboratory (NREL) (Kroposki et al., 2017), hierarchical microgrids (Mbungu et al., 2023), and transactive energy systems (Lin & Wang, 2022), etc., introduce advancements in distributed energy management, but they exhibit certain limitations in achieving real-time adaptivity, user-centric customization, and comprehensive decentralization. Generally, such architectures rely on centralized control schemes or are constrained by fixed/specific functional scope. Table 18.1 compares key characteristics of developed AIPS with some existing

Table 18.1 Comparison of AIPS with some existing partially adaptive distributed architectures.

Feature	AIPS	AEGs (Kroposki et al., 2017)	TES (Lin & Wang, 2022; Tooki & Popoola, 2024)	VPPs	HMGs (Gutiérrez-Escalona et al., 2024; Mbungu et al., 2023)	AI-DCS (Stock et al., 2021)
Local governance (autonomy)	√Strong local governance	∅Partial local governance relies on the center	∅Price driven local governance	∅Partial local governance	∅Partial local governance	∅Mostly partial local governance (varies by applications)
Adaptive clustering	√Swarm intelligence-based adaptive clustering.	× Not found	× Not found	× Not found	× Not found	∅Not so commonly implemented
Adaptive decision-making	√Highly adaptive	∅Limited	∅Limited	∅Partial (economy-driven)	∅Limited	∅Partially in some structures
Predictive analytics	√Implemented	∅Limited (predictive models)	∅Partial Price-based forecasts	∅Basic forecasting	∅Basic forecasting	∅Variable and in some cases
Anomaly detection	√Highly responsive	∅Limited	∅Limited	∅Limited	∅Basic	∅Varies

(*continued on next page*)

Data sovereignty	√Sovereign	× Not found	× Not found	× Rarely	× Not found	× Rarely
Adaptive policies	√Customer-centric adaptive	× Not found	× Only price-driven	∅Varies	× limited	∅Partial
Implicit analytics	√Implemented	∅Limited	∅Only market-based observability	× Not found	× Limited	× Limited (direct observability)
Scalability and flexibility	√Highly scalable and flexible (buildings to microgrids to larger grids).	∅Partial (larger segments/cells)	× Only the prosumer market-level	∅ Partial	× Limited	∅ Partially
Physic informed ML	√ Implemented	× Not found	× Not found	× Not found		

Symbols: √= Feature is present. ∅= Feature is partially present or limited implementation. × = Absence of feature or rarer/limited implementation.*AEGs*, autonomous energy grids (by NREL); *AI-DCS*, AI-enhanced distributed control systems; *AIPS*, adaptive and intelligent power system (proposed architecture: blue color); *HMGs*, hierarchical micro grids; *ML*, machine learning; *TES*, transactive energy systems; *VPPs*, virtual power plants.

distributed intelligent architectures, underscoring the unique contributions of AIPS.

This tabular comparison highlights that AIPS can address some critical gaps left by contemporary architectures, particularly in the areas of adaptive clustering, data sovereignty, and implicit analytics. The developed AIPS by employing swarm intelligence for clustering and de-clustering fosters resilience and improves load balancing. Moreover, AIPS revolves around customer centricity, i.e., supporting customizable policies for customers. This approach can provide operational flexibility and empower end-users to dynamically manage energy costs based on ambient insights.

Additionally, implicit analytics allows for nuanced system insights without overwhelming computational resources and preserving privacy (Lore & Reddy, 2019; Rowe et al., 2015). Thus, AIPS is the illustration of the sustainable, versatile, and adaptive structural framework developed to accelerate the green energy transition and support the next generation of data-driven futuristic grid networks. Table 18.2 presents the nature-positive dimension of AIPS and demonstrates how AIPS translates natural ecological processes into an energy system architecture that promotes environmental stability, aligning with natural world efficiencies.

Thus, AIPS is nature-inspired and incorporates natural principles of adaptability, balance, and resilience. Therefore, with its foundations based on ecological principles, AIPS is a forward-looking energy infrastructure that functions harmoniously with Mother Nature. In subsequent sections, we detail AIPS's nature-inspired design that can pave the way toward a future where power systems not only harmoniously coexist but also actively preserve the natural ecosystems. Ultimately, achieving a balance between environmental sustainability and human energy needs.

The rest of the chapter is structured as follows: Section 18.2 details the architecture of AIPS, explaining its key components and their functionalities. Section18.3 demonstrates the methodology of the case study as an application of AIPS within a microgrid. Results are discussed inSection 18.4, and future projections/directions are illustrated in Section 18.5. Section 18.6 concludes the article.

18.2 Adaptive and intelligent power system description

18.2.1 Components of the proposed adaptive and intelligent power system framework

The AIPS consists of flexible and modular IAHs, functioning as decentralized control units within the distribution grid. These IAHs manage local CEUs, i.e., clusters of EAUs within their local jurisdiction. This structural framework aims to develop localized units responsible for stabilizing and managing the specific geo-electric unit of the grid network. Each IAH has three core capabilities that include localized decision-making, learning and adaptation, and communication. The two main components of IAH are as follows.

a. EAUs = Individual physical energy units like generators or resources, homes, EVs, etc. EAUs represent entities with specific consumption or generation profiles.

Table 18.2 Translation and projection of natural phenomena in AIPS.

Principle inspired by nature	Application in AIPS	Positive impact on nature
Swarm intelligence (Duan et al., 2023; Xu et al., 2023) (Collective/community intelligence) of biological colonies.	IAHs mimic natural swarm behaviors of bees, ants, flocks of birds, etc., to self-organize and achieve efficient distribution, i.e., adaptive clustering emulating the natural way to optimize resources and movements.	Energy across IAHs and reduced dependency on centralized energy resources (usually fossil-based) accelerate environmentally friendly green energy transition.
Circular utilization (Sharma & Malaviya, 2023).	Real-time adjustments mirror natural ecosystems' resource allocation (predictive analytics leads to conservation of resources within IAHs).	Prevents energy wastage and decreases overgeneration. Reducing unnecessary energy consumption and carbon footprints.
Biodiversity	AIPS modules incorporate diverse resources and adaptive policies mimicking the biodiversity phenomenon of natural ecosystems.	Diversity supports grid flexibility and stability, reducing reliance on fossils. Also, diversity introduces resilience in the face of variable supply or weather conditions, i.e., grid robustness is introduced.
Community autonomy (ambient intelligence)	Like groups of organisms adapting to their niches in ecosystems, IAHs independently adapt, and decisions are made locally in response to environmental/ambient changes.	Improved response time results in reduced large-scale interventions, cutting environmental costs of distant resource management and transmission losses.
Collaborative learning	Local and collective learning without central authority. Similar to naturally distributed cooperative learning behavior. Secure learning across IAHs without data centralization and computational burden, like distributed intelligence in Neural systems.	Efficient data handling conserves resources while preserving privacy and autonomy, ultimately, less resource use and strain on natural resources.

(*continued on next page*)

Table 18.2 Translation and projection of natural phenomena in AIPS—cont'd

Principle inspired by nature	Application in AIPS	Positive impact on nature
Anomaly detection (Chalapathy & Chawla, 2019; Minias, 2023).	AIPS mimics the natural immune system response, handling and addressing the faults or anomalies at the local IAH level.	Early detection of faults reduces the risk of large-scale outages. Minimal disruptions and environmental impact by reducing the need for fossil-based resources to counter large-scale power disruptions.
Flexible ecosystems with fluid boundaries.	Adaptive clustering of AIPS works like ecosystems responding to resource flux, i.e., reconfiguration for optimized resource distribution and usage.	Responsiveness to supply and demand changes, thus reducing the negative environmental impacts of the shortage of overproduction.

AIPS, adaptive and intelligent power system; *IAH*, intelligent adaptive hub.

b. CEUs = Groups of EAUs that can act as single decision-making sub-units within IAHs, e.g., residential CEU represents aggregated 100 homes or wind farms with various wind turbines aggregated together.

Therefore, IAH is a cluster of various CEUs performing energy management tasks collectively and collaboratively. The atomic units of data are EAUs, and the functional grouping of EAUs with similar roles results in CEUs. Thus, IAH manages data aggregation, local optimization, and controls and shares data summaries to interact with neighboring IAHs. The division of the system into various IAHs is based on a mix of parameters, such as infrastructure type (i.e., residential, industrial, or commercial), geographical proximity, demand patterns/profiles, or resource availability. This framework enables localized control and optimization because IAHs are the modules that are responsible for intelligent local energy management.

Thus, this approach reduces the need for extensive and overloaded centralized control mechanisms. IAH represents geographically or electrically close energy resources and loads, while CEUs represent a logical aggregation of EAUs. The selection can be based on three criteria: Physical proximity of resources (geographically), sharing the same electric connectivity, and a sufficient resource mix (energy balance). Fig. 18.4 illustrates a concise overview of the IAH concept.

To avoid the overwhelming raw data exchange while ensuring a streamlined structure, the data is aggregated at the CEU level instead of the IAH. Individual EAUs report to the CEU level; for example, individual house meter reports energy usage to residential CEUs periodically/at set intervals. CEU then summarizes the critical insights/analytics from this data to report key insights to the IAH level. These insights include average load, cumulative generation profile, or abnormal alerts. IAH collects

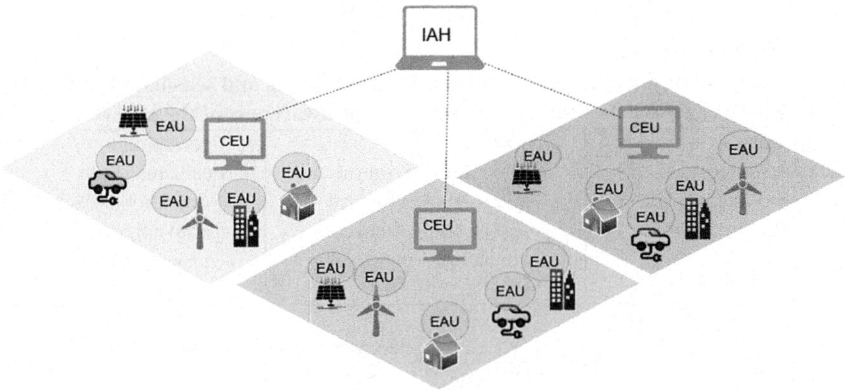

Figure 18.4 The IAH structure. The concise overview of the intelligent adaptive hubs (IAH).

summaries from its CEUs and performs its aggregation to represent broader system states that include cumulative demand, net generation, available storage, renewable resources portfolio, etc.

This analysis can be used to optimize the resource allocation locally and assist in the assessment of external energy support from other IAHs. There is a lightweight coordinator, the CEA, to enable adaptive clustering. Thus, each IAH shares data summaries and anomaly reports to the CEA only during significant events, minimizing communication and computational burden and data exchange. Therefore, aggregating data at different layers reduces data traffic but still captures the critical system insights. Fig. 18.5 illustrates the summarized view and structure of AIPS.

18.2.2 Description of the layered structure of adaptive and intelligent power system

AIPS operates through a structured, multilayered framework that enables adaptive decision-making and ambient coordination (real-time). Each layer contributes to optimizing and self-regulating the system. AIPS's core idea revolves around IAHs, which balance energy generation, consumption, and storage in localized hubs. The IAHs intelligently collaborate for system-wide optimized performance through swarm intelligence. The AIPS framework is structured into six interlinked and interdependent layers, which are demonstrated below in Table 18.3 and Fig. 18.6

18.2.3 Nature-inspired swarm approach

An intelligent and adaptive clustering mechanism is developed in AIPS to address network-wide events such as sudden demand spikes or drops in generation. The affected IAH can form a temporary cluster adaptive CEU (ACEU) with neighboring IAHs or within the IAH inter-CEU. These clusters work together to find solutions, such as redirecting power or sharing load to stabilize the area in their jurisdiction. After the

Table 18.3 Six interlinked layers of the proposed AIPS framework.

Physical/perception layer. **The foundational layer of AIPS. Physical real-world assets and sensors operate in this layer. This layer establishes the situational/ambient awareness of the grid conditions.**		
Components	Function	Interaction
Energy resources, storage systems, grid infrastructure, sensors, etc.	Collect data, monitor, and manage physical assets. Implements control action on assets.	Feeds real-time operational data to the above tier for processing.
Cognitive processing layer. The localized operational brain layer, i.e., adaptive/learning layer of AIPS.		
Components	Function	Interaction
Functional-focused (scope specific) data-driven controllers, predictive analytics, and anomaly detection modules.	Optimize power flows, detect potential anomalies, and proactively manage the system.	Executes decisions from the above tier and reports back to the upper-tier feedback/anomalies.
Strategic brain layer. The decision intelligence layer for system-wide optimization and operation.		
Components	Function	Interactions
Swarm intelligence, federated learning modules, energy agents (CEUs), data-driven decision making, predictive analytics, and anomaly detection.	Optimal global performance, implicit analytics, and IAH collaboration.	The lower tier feeds the data, i.e., localized analytics and insights. It is also guided by ecological constraints to refine decisions and strategies.
Energy alliance layer.		
Components	Function	Interactions
Central energy alliance (CEA)	Multi-IAH scenario, global federated model, and emergency intervention.	Event-triggered coordination and communication framework.
Communication layer. The backbone layer for secure real-time data sharing and decision distribution.		
Components	Function	Interactions
Inter and Intra IAH data sharing and coordination protocols and cybersecurity measures.	Enables inter-IAH data exchange and exchange with the strategic brain layer.	Maintains an efficient and secure communication network supporting all layers.

(*continued on next page*)

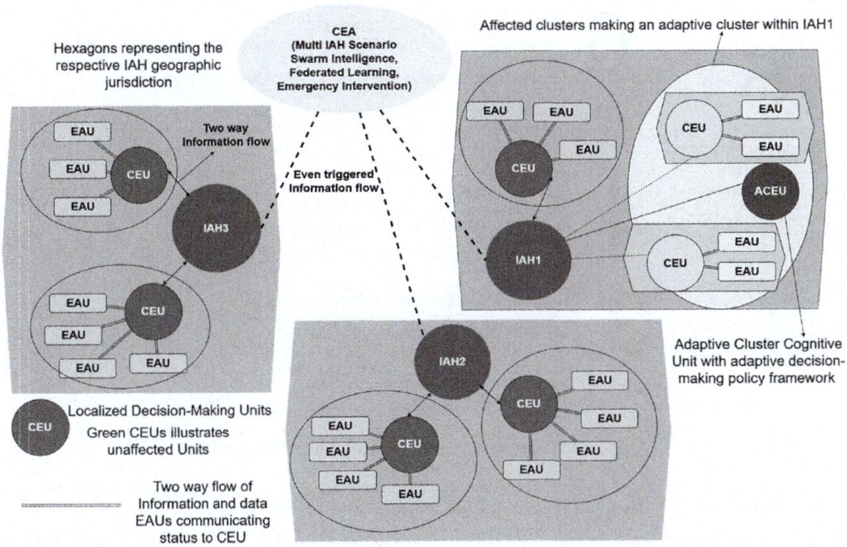

Figure 18.5 Proposed architecture. Summarized overview of the proposed AIPS framework. *AIPS*, Adaptive and intelligent power system.

Table 18.3 Six interlinked layers of the proposed AIPS framework—cont'd

Physical/perception layer. **The foundational layer of AIPS. Physical real-world assets and sensors operate in this layer. This layer establishes the situational/ambient awareness of the grid conditions.**		
Monitoring and an ecologically inspired layer. A user interface to provide operators with real-time analytics and sustainability-oriented insights and protocols.		
Components	Function	Interactions
Implicit environmental observability and carbon impact assessments.	Implicit and direct monitoring of ecosystem health to ensure energy operations with sustainability goals.	Coordinates and informs the decision layer about environmental constraints—validation of the ecological impact of grid operations.

AIPS, adaptive and intelligent power system; *CEUs*, cognitive energy units *IAH*, intelligent adaptive hub.

issue is resolved, de-clustering resumes the autonomous operation of IAH or CEUs. Therefore, this mechanism enables dynamic cooperation among IAHs or intra-IAHs during major grid events. The IAH or CEU that detects the issue broadcasts a swarm signal to nearby hubs via CEA, as shown. Hence, swarm behavior lets IAHs flexibly respond to grid-wide issues without continuous centralized command. The swarm intelligence phenomenon in AIPS falls under the category of adaptive collaboration, i.e., energy alliance through peer-support mechanisms. IAH periodically assesses its

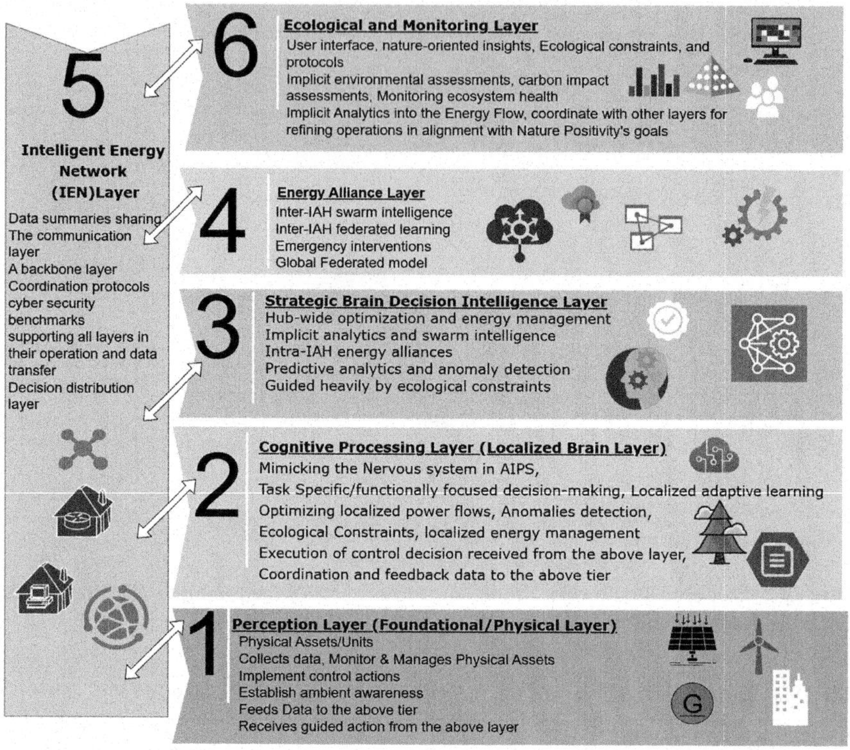

Figure 18.6 Layers of AIPS. Detailed description of the proposed AIPS framework layered structure. *AIPS*, Adaptive and intelligent power system.

local conditions and attempts to improve the resilience of the distribution grid by an inter-hub collaboration strategy if required.

Thus, the AIPS approach draws inspiration from natural ecosystems like bees, schools of fish, and ant colonies, where self-regulatory behaviors result in adaptive and efficient solutions. In such systems, principles of stigmergy and emergent behaviors are adapted for dynamic clustering. In AIPS, each IAH or CEU acting as an individual in a swarm forms clusters based on local energy conditions. Each IAH operates through decentralized intelligence but contributes to overall grid stability, enhancing energy flow and resource optimization akin to the natural ecosystem. Additionally, this approach improves fault tolerance and scalability, like ant colonies expanding the foraging networks. Fig. 18.7 demonstrates the ACEU formation within AIPS.

18.2.4 Holistic adaptive and intelligent power system framework demonstration

This section synthesizes the components and layers of AIPS into a comprehensive and unified perspective. Stitching together the previously discussed features in this

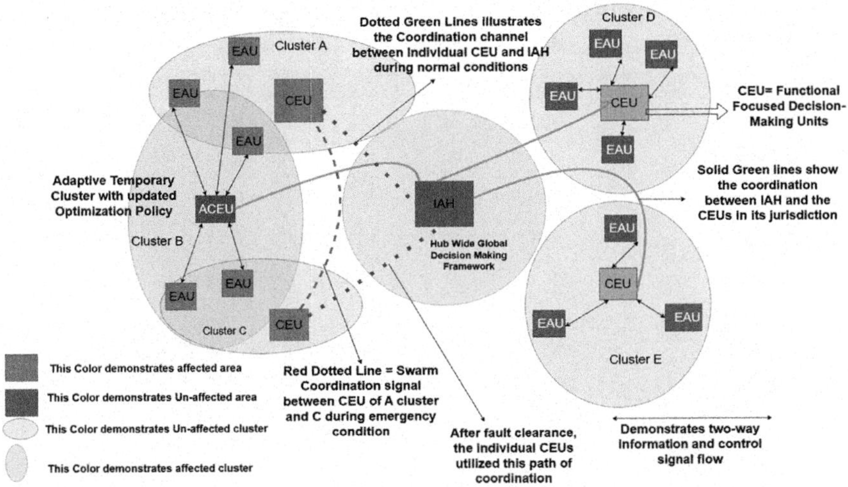

Figure 18.7 AECU formation. Demonstration of the formation of ACEU within AIPS. *ACEU*, Adaptive cognitive energy unit; *AIPS*, adaptive and intelligent power system.

section, holistic figures to visualize and illustrate how AIPS components interact to achieve adaptive and intelligent systems are demonstrated. The developed architecture operates as a multilayered adaptive ecosystem of diverse micro and macro energy assets. Each layer performs distant yet interconnected coordinated roles. The AIPS integrates layers and incorporates principles of decentralized decision-making and adaptive collaboration to manage the complexities of distribution grid networks. The modular architecture is presented in Figs. 18.8 and 18.9 illustrates the stitched AIPS, which illustrates the flow of data, energy decision-making, and collaboration across the system. Through localized governance, each IAH governs energy management within its locale, reducing latency and central control dependencies. The AIPS directly supports distributed RERs for global carbon reduction goals. Through localized control policies, AIPS introduces tailored governance mechanisms to ensure alignment with regional energy objectives. The adaptive policy concept enables AIPS to comply with various regulatory frameworks, ensure the ownership and privacy of data, and respect data sovereignty.

In AIPS, a modular layered structure enables seamless integration of new technologies and hubs, i.e., mechanisms of plug-and-play IAHs, making the architecture feasible for grid expansions. Therefore, AIPS introduces the concept of global-local governance harmony, where system-wide objectives are maintained via coordinated policies. AIPS, via its distributed setup, minimizes vulnerabilities from a single point of failure and centralized governance.

The AIPS, through an adaptive policy framework, allows IAHs to align with changing market environments and regulatory frameworks. Intra-IAH communication, i.e., CEUs can communicate with other CEUs and IAH's main decision-making through lightweight protocols such as Constrained Application Protocol (CoAP) and Message

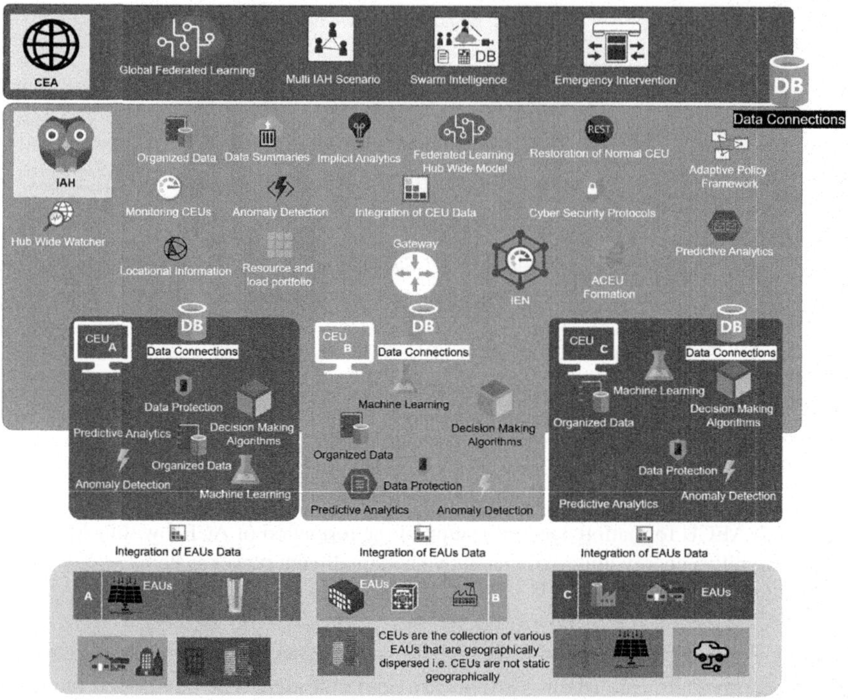

Figure 18.8 Modular AIPS. The proposed detailed modular overview of the AIPS framework with various capabilities for future extension. *AIPS*, Adaptive and intelligent power system.

Queuing Telemetry Transport (MQTT), ensuring real-time updates and minimal overhead. Inter-IAH communication can be done through high-bandwidth protocols such as ransmission Control Protocol/Internet Protocol (TCP/IP).

18.3 Case study methodology

18.3.1 System description

The case study considers AIPS consisting of one CEA and two IAHs. The CEA remains dormant under normal conditions and activates the adaptive clustering during disturbances. For this chapter, a 24-hour horizon is considered, with IAH1 comprising thermal, RERs, battery storage, and IAH2 consisting of distributed prosumers with rooftop photovoltaics (PVs) and battery storage. A fault is introduced in IAH1 between 08:00 AM and 12:00 PM, reducing the generation by 15%. Fig. 18.10 demonstrates the system architecture for validating AIPS.

The proposed AIPS framework, as shown in Fig. 18.8, operates within a microgrid environment consisting of two IAHs that manage different aspects of power distribution

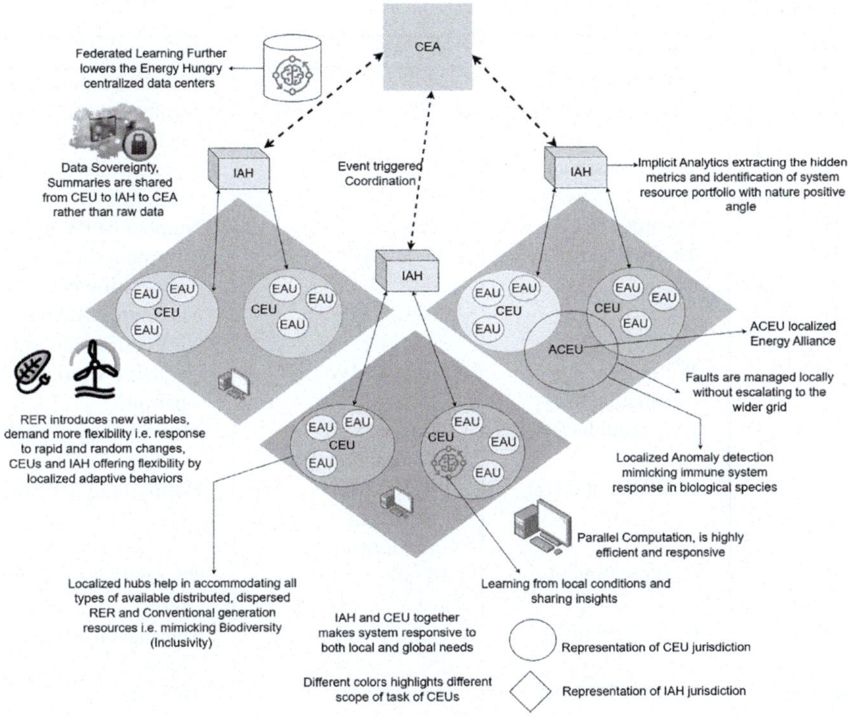

Figure 18.9 Full AIPS demonstration. Holistic AIPS framework description for power distribution systems. *AIPS*, Adaptive and intelligent power system.

systems. IAH1 handles thermal and utility-scale RER generation plants, i.e., IAH1 is an economic dispatching cluster. IAH2 manages distributed energy resources, which means residential-scale PVs and battery storage, i.e., IAH2 is a prosumer management cluster. CEA acts as a coordinator that orchestrates cooperation between both IAHs and implements adaptive clustering during disturbance periods. A disturbance scenario is configured as a persistent generation capacity constraint, and it affects the IAH1 only. Detailed system parameters are shown in Table 18.4.

To evaluate the system performance, metrics such as (1) frequency deviation, (2) power imbalance, and (3) violation deviation (number of hours frequency deviation >0.1 Hz) were used. Performance was evaluated through (1) focused assessment during the disturbance window, (2) time-domain analysis (hourly performance), (3) statistical comparison, and (4) visual comparison. Two comparative scenarios (1) noncooperative operation, where IAHs operate autonomously, and (2) adaptive clustering in which IAHs cooperate during disturbances, were compared. The approach is validated through (1) scenario testing and (2) comparative analysis.

Table 18.4 System parameters.

Category	Parameter	Value/range	Description
IAH1	Time horizon(T)	24 h	Simulation duration
IAH1	Thermal units(N_T)	1	Number of thermal generators
IAH1	Wind units(N_W)	1	Number of wind turbines
IAH1	PV units(N_{PV})	1	Number of PV units
IAH1	Thermal min(P_{min})	20–40 MW	Minimum thermal generation per unit
IAH1	Thermal max($P_{\max}$)	80–120 MW	Maximum thermal generation per unit
IAH1	Wind available(W_{av})	50–100 MW	Available wind power per unit
IAH1	PV available(PV_{av})	Time-varying Gaussian profile	Max 85 MW at peak (13 h)
IAH1	Load demand (D)	90–100 MW (uniform random)	Hourly load demand
IAH1	Cost coefficients (a)	10–20	Thermal cost constant term
IAH1	Cost coefficients (b)	1–2	Thermal cost linear term
IAH1	Cost coefficients (c)	0.1–0.2	Thermal cost quadratic term
IAH1	Valve point (e, f)	0.01–0.05	Thermal valve-point effect coefficients
IAH1	Frequency coefficient(kf)	0.1–0.5	Frequency response cost coefficient
IAH2	Number of houses	10	Residential prosumers
IAH2	Peak load per house	10 kW	Maximum load per house
IAH2	PV capacity per house	15 kW	Maximum PV generation per house
IAH2	Battery capacity	5 kWh	Storage capacity per house
IAH2	Grid import cost	0.20 $/kWh	Cost to import from the grid
IAH2	Grid export price	0.15 $/kWh	Revenue for exporting to the grid
IAH2	Frequency weight	100	Penalty weight for frequency deviations
IAH2	Load profile	kW per time window	Load values for 4 time windows

(*continued on next page*)

Table 18.4 System parameters—cont'd

Category	Parameter	Value/range	Description
IAH2	Time windows	Hours	Duration of each load segment
IAH2	PV generation profile	Gaussian curve (max at midday)	Peak generation ≈15 kW per house at noon
Fault scenario	Fault start hour	8	Fault begins at 8:00
Fault scenario	Fault end hour	12	Fault ends at 12:00
Fault scenario	Fault type	Economic dispatching cluster IAH1	Fault affects IAH1 only
Fault scenario	Severity	15% reduction	Generation reduced by 15% during the fault
CEA	Battery capacity	400 MWh (20 prosumers × 20 MWh)	Total storage capacity
CEA	Max charge/discharge Rate	100 MW (5 MW/prosumer × 20)	Maximum power flow rate
CEA	Initial SOC	80%	Initial state of charge
CEA	Min/Max SOC	20%/90%	Safe operating limits

CEA, central energy alliance; *IAH1*, intelligent adaptive hub 1; *IAH2*, intelligent adaptive hub 2; *PV*, photovoltaic; SOC, State of Charge.

18.3.2 Mathematical modeling

18.3.2.1 System notation

The system consists of N_t thermal units, N_w wind units, N_{pv} PVs units, and N_h prosumers houses. The scheduling horizon is divided into T time intervals indexed hourly by $t \in \{1, 2, \ldots, T\}$. Table 18.5 summarizes the key notation used in the formulation, and detailed notations are described after equations where necessary.

18.3.2.2 Intelligent adaptive hub 1 mathematical modeling

The IAH1 model implements the economic dispatching scenario, i.e., to minimize the total generation cost considering wind, PV, and thermal units, while satisfying unit constraints and load balancing.

Thermal units incur a quadratic fuel cost as described in

$$C_{i,t}^{fuel} = a_i + b_i P_{i,t}^{th} + c_i \left(P_{i,t}^{th} \right)^2$$

Table 18.5 Key notations used in mathematical formulation.

Symbol	Definition	Units
${P_{i,t}}^{th}$	Thermal generator *i* output at time *t*	MW
$P_i^{\min}, P_i^{\max}$	Minimum and maximum thermal generator capacity	MW
$P_{j,t}^{w}$	Wind generator *i* output at time *t*	MW
$W_{j,t}^{av}$	Available wind generator *i*output at time*t*	MW
$P_{k,t}^{pv}$	PV generator *i* output at time*t*	MW
$PV_{k,t}^{av}$	Available PV generator *i* output at time *t*	MW
P_t^{ch}, P_t^{dis}	Charging and discharging power of regional storage at time *t*	MW
SOC_t	State of charge of storage at time *t*	MWh
$SOC^{\min}$, $SOC^{\max}$	Minimum/maximum permissible SOC	MWh
D_t	Aggregate demand at time *t*	MW
Δf_t	Frequency deviation at time *t*	Hz
$P_t^{imp,n}, P_t^{exp,n}$	Grid import/export for prosumer *n* at time *t*	kW
$L_{n,t}^{h}$	Household load of prosumer *n* at time *t*	kW
η_c, η_d	Charging and discharging efficiency	....

PV, photovoltaic.

where a_i is fixed no load cost, b_i is the linear fuel coefficient, and c_i is the quadratic fuel coefficient. The valve point loading effect is captured through adding a sinusoidal ripple term, as demonstrated in

$$C_{i,t}^{valve} = e_i sin\left(f_i\left(P_i^{min} - P_{i,t}^{th}\right)\right)$$

where e_i and f_i are valve point coefficients.

Therefore, the total thermal cost can be defined as

$$C_{i,t}^{thermal} = C_{i,t}^{fuel} + C_{i,t}^{valve}$$

This new term reflects fuel consumption and irregularities in thermal generators.

The wind power curtailment penalty is used to penalize over-scheduling or not using available wind, as wind has zero fuel cost. Therefore, the wind cost is formulated as

$$C_{j,t}^{wind} = \phi k_j^{w} \max\left(0, W_{j,t}^{av} - P_{j,t}^{w}\right) + (1-\phi)k_j^{r}\max\left(0, P_{j,t}^{w} - W_{j,t}^{av}\right)$$

In the first term, penalizes curtailment, and the second term is for the over-dispatch penalty, with $\phi \in [0, 1]$ controlling the relative importance.

Similarly, the PV cost function is defined in

$$C_{k,t}^{PV} = \phi k_k^{PV} \max\left(0, PV_{k,t}^{av} - P_{k,t}^{PV}\right) + (1-\phi)k_k^{r}\max\left(0, P_{k,t}^{PV} - PV_{k,t}^{av}\right)$$

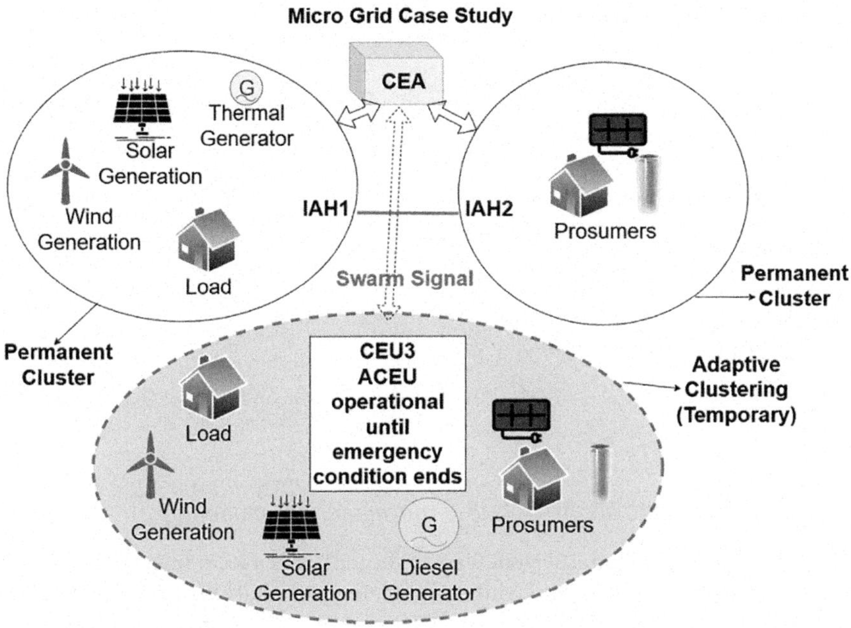

Figure 18.10 Case study. The proposed AIPS framework implementation within a microgrid. *AIPS*, Adaptive and intelligent power system.

With first term penalizes solar generation curtailment, and the second term penalizes the infeasible over-scheduling.

Storage cost is defined as

$$C_t^{storage} = \aleph\left(P_t^{ch} + P_t^{dis}\right)$$

In $\aleph$ is the cost coefficient for cycling and reflects the penalty for efficient use rather than wear and degradation due to excessive charging and discharging.

A quadratic penalty is used for any mismatch between generation, discharging, demand, and charging as

$$C_t^{imbalance} = \rho\left(\sum_i P_{i,t}^{th} + \sum_j P_{j,t}^{w} + \sum_k P_{k,t}^{PV} + P_t^{dis} - \left(D_t + P_t^{ch}\right)\right)^2$$

where ρ is a large positive constant to enforce balance. Physically, this means that the imbalance is reflected as frequency deviations.

This leads to the IAH1 system-level objective function as

$$J_{IAH1} = \sum_{t=1}^{T}\left(\sum_i C_{i,t}^{thermal} + \sum_j C_{j,t}^{wind} + \sum_k C_{k,t}^{PV} + C_t^{storage} + C_t^{imbalance}\right)$$

So, the IAH1 optimization problem is defined as

$$\min_{\{P^{th}_{i,t},P^{w}_{j,t},P^{PV}_{k,t},P^{ch}_{t},P^{dis}_{t},SOC_t\}} J_{IAH1} = \sum_{t=1}^{T}\left(\sum_{i\in\mathcal{G}} C^{thermal}_{i,t} + \sum_{j\in\mathcal{W}} C^{wind}_{j,t} + \sum_{k\in\mathcal{V}} C^{PV}_{k,t} + C^{storage}_{t} + C^{imbalance}_{t}\right)$$

Subject to the following constraints:

1. Power balance $\sum_{i\in\mathcal{G}} P^{thermal}_{i,t} + \sum_{j\in\mathcal{W}} P^{wind}_{j,t} + \sum_{k\in\mathcal{V}} P^{PV}_{k,t} + P^{dis}_{t} - (D_t + P^{ch}_{t}) = \in_t$, where $\in_t$ softened with a penalty.
2. Thermal limits $P^{\min}_{i} \le P^{th}_{i,t} \le P^{\max}_{i} \forall i, t$.
3. RER availability $0 \le P^{w}_{j,t} \le W^{av}_{j,t}, 0 \le P^{PV}_{k,t} \le PV^{av}_{k,t}$.
4. Storage dynamics $SOC_{t+1} = SOC_t + \eta_c P^{ch}_{t} - \left(\frac{1}{\eta_d}\right)P^{dis}_{t} and\ SOC^{\min} \le SOC_t \le SOC^{\max}, 0 \le P^{ch}_{t}, P^{dis}_{t}$.

18.3.2.3 Intelligent adaptive hub 2 mathematical modeling

IAH2 represents a residential system with rooftop PV and a local battery. The objective is to maintain frequency stability while minimizing electricity cost. The cost of buying power from the grid is $C^{import}_{n,t} = C^{imp}P^{imp,n}_{t}$ and export revenue is indicated with a negative sign as $C^{export}_{n,t} = -C^{export}P^{export,n}_{t}$

Deviation from local generation and demand is penalized to ensure that each prosumer contributes to frequency stability as

$$C^{freq}_{n,t} = \lambda\left(L^{h}_{n,t} - P^{PV,n}_{t} - P^{dis,n}_{t} - P^{imp,n}_{t} + P^{ch,n}_{t} + P^{exp,n}_{t}\right)^2$$

Thus, the cost function for IAH2 is defined as

$$J_{IAH2} = \sum_{t=1}^{T}\left(C^{import}_{n,t} + C^{export}_{n,t} + C^{freq}_{n,t}\right)$$

The IAH2 minimizes the household energy bills while supporting the local grid frequency stability.

The optimization problem for IAH2 is as follows.

$$\min_{\{P^{imp,n}_{t},P^{export,n}_{t},P^{ch,n}_{t},P^{dis,n}_{t},SOC^{n}_{t}\}} J_{IAH2} = \sum_{t=1}^{T}(C^{imp}P^{imp,n}_{t} - C^{export}P^{export,n}_{t}$$

$$+\lambda\left(L^{h}_{n,t} - P^{PV,n}_{t} - P^{dis,n}_{t} - P^{imp,n}_{t} + P^{ch,n}_{t} + P^{export,n}_{t}\right)^2)$$

Subject to the following constraints:

1. Local balance $L^{h}_{n,t} = P^{PV,n}_{t} + P^{dis,n}_{t} + P^{imp,n}_{t} - P^{ch,n}_{t} - P^{export,n}_{t}$.

2. Battery dynamics $SOC_{t+1}^{n} = SOC_t^n + \eta_c P_t^{ch,n} - \left(\frac{1}{\eta_d}\right) P_t^{dis,n} and SOC^{\min} \leq SOC_t^n \leq SOC^{\max}, 0 \leq P_t^{ch,n}, P_t^{dis,n}$.

18.3.2.4 Adaptive clustering mechanism mathematical modeling

During a fault window $[t_f^{start}, t_f^{end}]$, thermal generation is reduced $P{\prime}_{i,t}^{thermal} = (1-\gamma) P_{i,t}^{thermal} \forall t \in \left[t_f^{start}, t_f^{end}\right]$ with $\gamma \in [0, 1]$ as fault severity. This imbalance then forces cooperation from IAH2 to stabilize the frequency and support IAH1. Each IAH region updates its optimization strategy during the fault time period through swarm-based adaptive clustering with its neighbors as

$$P_{i,t}^{new} = P_{i,t}^{old} + \emptyset_1 . r_1 \left(P_{i,t}^{localbest} - P_{i,t}^{old}\right) + \emptyset_2 . r_2 \left(P_{global,t}^{best} - P_{i,t}^{old}\right)$$

In $P_{i,t}^{localbest}$ is the best local feasible solution, after local optimization, $P_{global,t}^{best}$ is the shared global solution across IAHs, $\emptyset_1$ and $\emptyset_2$ are cognitive/social learning weights, and r_1 and r_2 are stochastic exploration terms like swarm dynamics. This formulation mimics natural agents like birds or fish to adapt based on self-experience and collective emergent global intelligence.

The optimization problem during adaptive clustering means the CEA level is defined in

$$\min_{IAH1vars, IAH2vars} J_{AdaptiveclusteringorCEA} = J_{IAH1} + \sum_{n=1}^{N_h} J_{IAH2}^{n}$$

$$\min_{J_{IAH1} J_{IAH2}} J_{CEA} = \sum_{t=t_f^{start}}^{t_f^{end}} \left(J_{IAH1}(t) + \sum_{n=1}^{N_h} J_{IAH2}(t) + \alpha . Imbalance(t)^2\right)$$

$$Imbalance\,(t) = \sum_{i \in \mathcal{G}} P{\prime}_{i,t}^{th} + \sum_{j \in \mathcal{W}} P_{j,t}^{w} + \sum_{k \in \mathcal{V}} P_{k,t}^{PV} + \sum_{n} P_t^{dis,n}$$

$$-\left(D_t + \sum_{n} P_t^{ch,n} + P_t^{ch} - P_t^{dis} + \in_t\right)$$

Subject to coupling constraints:

1. Regional balance

$$\sum_{i \in \mathcal{G}} P{\prime}_{i,t}^{th} + \sum_{j \in \mathcal{W}} P_{j,t}^{w} + \sum_{k \in \mathcal{V}} P_{k,t}^{PV} + \sum_{n} P_t^{dis,n} = D_t + \sum_{n} P_t^{ch,n} + P_t^{ch} - P_t^{dis} + \in_t$$

2. Frequency deviation model $\Delta f_t = k_f(generation_t - demand_t)$

$$f_0 = 50Hz, \text{ and} \Delta f_{\max = } 0.1Hz,$$

$$|f_0 + \Delta f_t - f_0| \leq \Delta f_{\max}$$

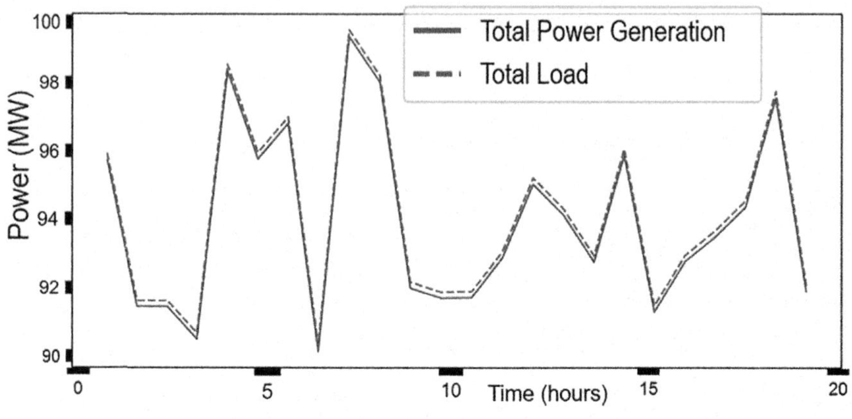

Figure 18.11 Power profile of IAH1. The total load profile and power generation of IAH1 under normal operating conditions. *IAH1*, Intelligent adaptive hub 1.

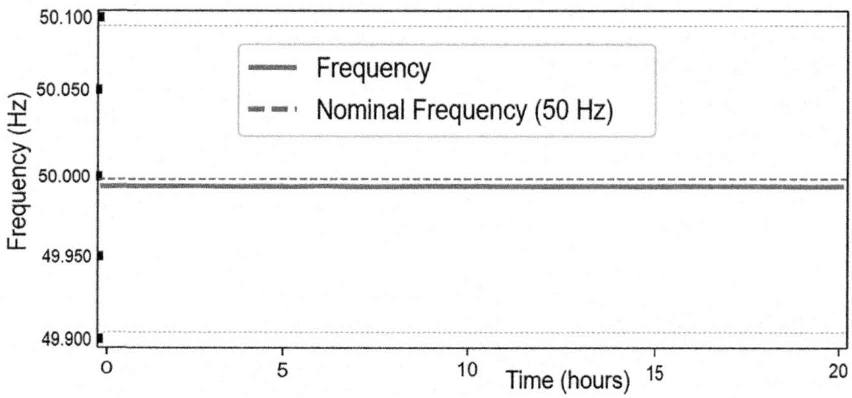

Figure 18.12 Frequency response of IAH1. The IAH1 system frequency response under normal operating conditions. *IAH1*, Intelligent adaptive hub 1.

Table 18.6 presents the adaptive clustering-based energy management algorithm.

18.4 Results and discussion

18.4.1 Normal operating conditions

Fig. 18.11 demonstrates the total load profile and power generation of IAH1 under normal operating conditions without any fault, and Fig. 18.12 illustrates the corresponding IAH1 system frequency response. The dispatch outcome in Fig. 18.10 shows that the scheduled generation closely follows the load profile over the complete time horizon (24 hours). The load profile fluctuates within the range of 90–100 MW, and the

Table 18.6 Swarm-inspired adaptive clustering algorithm.

Input:
IAH1 data, IAH2 data
System parameters: Generation limits, load profiles, cost coefficients
Fault scenario parameters and severity
battery parameters, capacity, and charge/discharge rates
Output:
Optimal dispatching
Regional and system frequency profile
Performance metrics
Stage I. Initialization:
1. Set time horizon T = 24 h
2. Generate RER profiles:
Wind availability
PV Availability
3. Thermal generation bounds
4. Initialize battery SOC = 0.8 Battery capacity
5. Configure fault scenario parameters:
Start hour, end hour, severity
6. Set swarm parameters
Stage II: IAH1 system dispatching
6. For each hour t [0, T]
a. Define variables: $\mathbf{G_{thermal(t)}}, \mathbf{G_{wind(t)}}, \mathbf{G_{pv(t)}}$
b. Constraints:
$\mathbf{P^{min}} \le \mathbf{G_{thermal}} \le \mathbf{P^{max}} 0 \le \mathbf{G_{wind}} \le \mathbf{W_{av}^{max}} 0 \le \mathbf{G_{PV}} \le \mathbf{PV_{av}^{max}}$ Power Balance
c. Cost Function $\mathbf{J_{IAH1}}$
d. Optimize to obtain dispatch
Stage III. IAH2 system household prosumer management
7. For each prosumer house $\mathbf{i} \in [1, \mathbf{N}]$
a. Variables: Grid import, grid export, battery charge, battery discharge
b. Constraints
power balance
SOC dynamics
c. Objective function $\mathbf{J_{IAH2}}$
d. Optimize prosumer-level operation
Stage IV: Adaptive Clustering
8. If cooperative operation is enabled
a. Construct a new optimization with joint variables for faulty windows
equation 12
b. Apply fault severity
c. System-wide new constraints (battery SOC + power balance)
d. Fault-aware cost function
J_{CEA}e. Solve optimization

(*continued on next page*)

Table 18.6 Swarm-inspired adaptive clustering algorithm—cont'd

Input: **IAH1 data, IAH2 data** **System parameters: Generation limits, load profiles, cost coefficients** **Fault scenario parameters and severity** **battery parameters, capacity, and charge/discharge rates**
Stage V: Frequency stability For each hour **t** ∈ [0, **T**] a. Compute total generation b. compute total load Calculate frequency deviation Stage VI: Performance analysis 10. Compute performance metrics a. Maximum, average frequency deviation b. Maximum, average power imbalance c. Frequency violations count d. Generates plots and comparison tables

IAH1, intelligent adaptive hub 1; *IAH2*, intelligent adaptive hub 2.

contribution from thermal, PV, and wind units is adjusted to track the load demand. The near overlap of power generation and load profile in Fig. 18.11 confirms that the economic dispatching optimization framework maintains an effective and optimal supply-demand balance with marginal deviations for a few hours. This behavior illustrates that IAH1 can autonomously sustain reliable operation when no disturbance occurs. The frequency of IAH1 in Fig. 18.12 remains tightly regulated around a nominal 50 Hz, with marginal deviation well within the operational limits. Therefore, this ability of IAH1 validates the efficacy of the developed economic dispatching strategy and its potential to integrate variable RERs while respecting reliability requirements.

Fig. 18.13 demonstrates the power balance of the IAH2 region that maintains a close match between aggregated demand and power generation. In the IAH2, neighborhood load oscillates between 0.02 MW during off-peak hours and around 0.09 MW during morning and evening peaks, obtained from a perhome demand pattern over four 6-hour windows scaled to MW. The prosumer management optimization enforces nodal balance each hour; as a result, total power supplied matches the aggregated load. The frequency of IAH2, as shown in Fig. 18.14, remains effectively anchored at 50 Hz during normal operation.

18.4.2 Fault condition of intelligent adaptive hub 1 without adaptive clustering

Fig. 18.15 shows the power balance degradation during the disturbance period. The disturbance is introduced in IAH1, which is the 15% reduction in available generation capacity from 8:00 AM to 12:00 PM. In Fig. 18.15 without an adaptive clustering mechanism, a visible supply-demand mismatch is observed, i.e., a deficit of several megawatts during the disturbance period, reaching 82.5 MW from 92.5 MW load

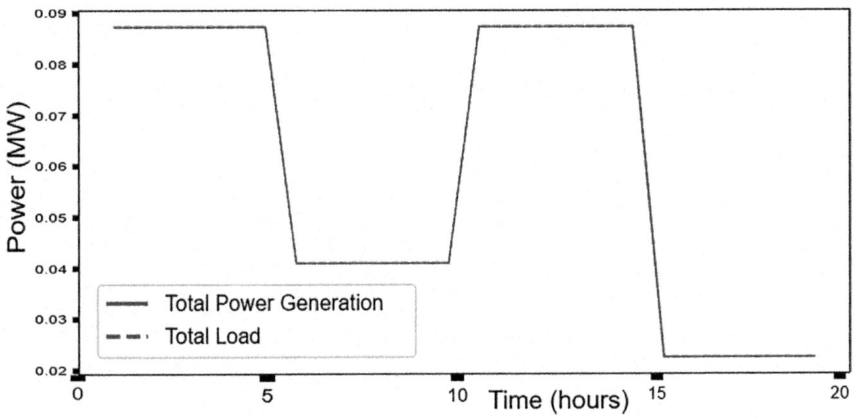

Figure 18.13 Power profile of IAH2. The power balance of IAH2 region under normal operating conditions. *IAH2*, Intelligent adaptive hub 2.

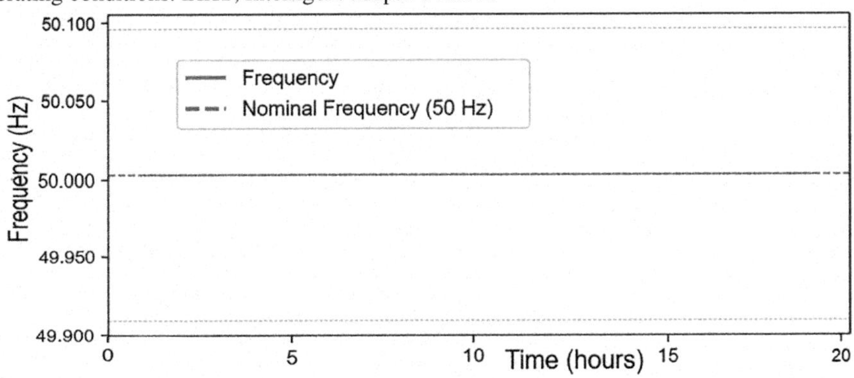

Figure 18.14 Frequency response of IAH2. The frequency response of IAH2 region under normal operating conditions. *IAH2*, Intelligent adaptive hub 2.

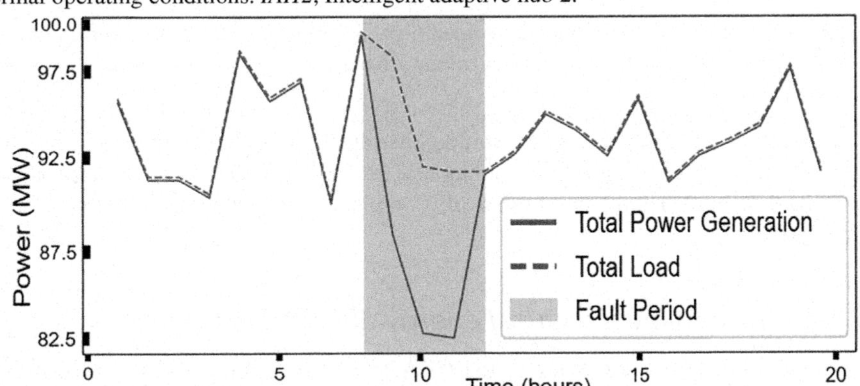

Figure 18.15 Power profile of IAH1 during disturbance. IAH1 region power profile under fault condition. *IAH1*, Intelligent adaptive hub 1.

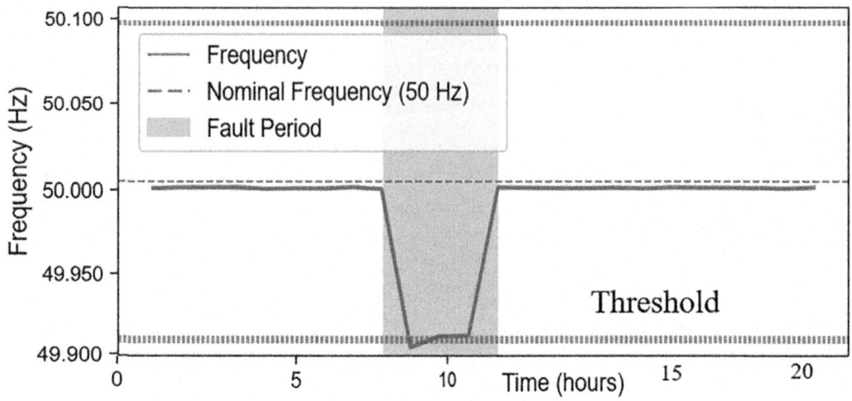

Figure 18.16 Frequency response of IAH1 during disturbance. IAH1 frequency response under fault condition. *IAH1*, Intelligent adaptive hub 1.

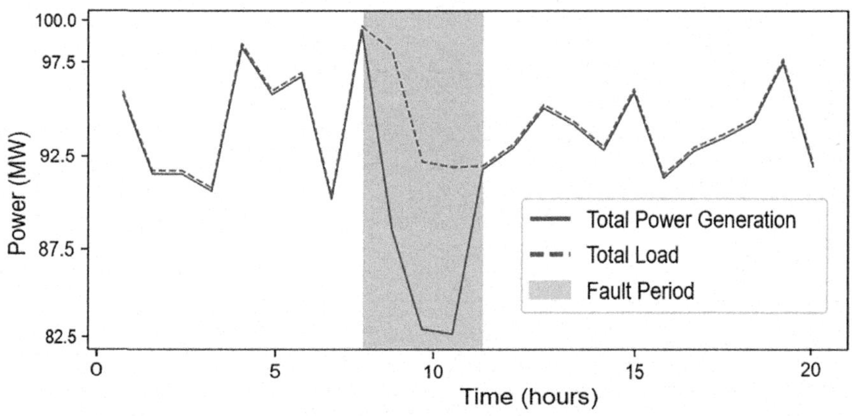

Figure 18.17 CEA power profile without adaptive clustering concept. CEA power profile without adaptive clustering. *CEA*, Central energy alliance.

demand around 10:00 AM. Fig. 18.16 demonstrates the frequency trajectory of IAH1 under the same scenario and shows noticeable deviations with excursions approaching tolerance threshold. Therefore, these results emphasize the vulnerability of noncoordinated operations.

18.4.3 System level (central energy alliance) response

Figs. 18.17 and 18.18 demonstrate the CEA level, i.e., aggregated system response under disturbance conditions without an adaptive clustering mechanism. The fault is localized (only introduced in IAH1), and the IAH2 continues to operate normally. The CEA-wide balance is directly impacted due to the reduced output of IAH1 during the

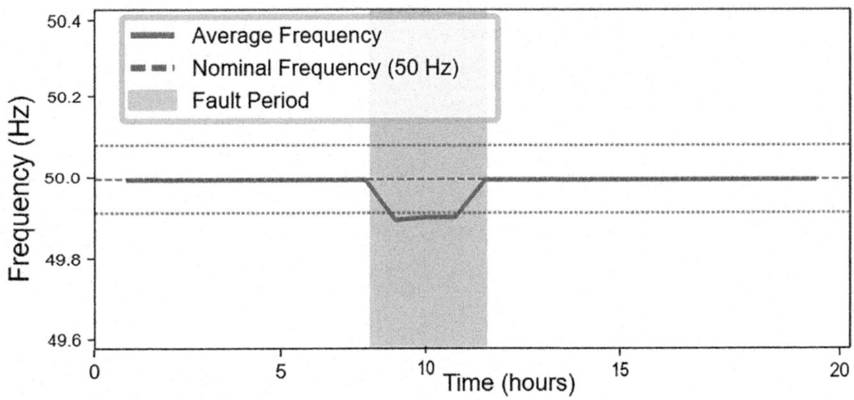

Figure 18.18 Frequency response of CEA without adaptive clustering concept. CEA frequency response without adaptive clustering. *CEA*, Central energy alliance.

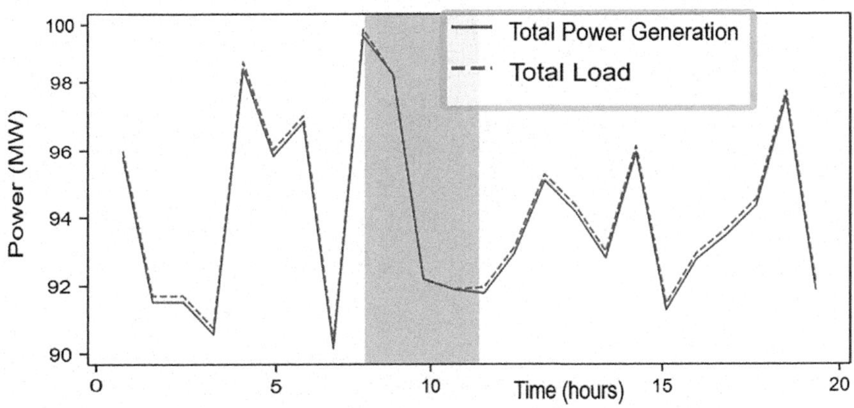

Figure 18.19 CEA power profile with adaptive clustering. CEA power profile with adaptive clustering. *CEA*, Central energy alliance.

fault window. Fig. 18.17 shows that the total CEA level power generation falls short of demand, producing a noticeable imbalance, and this effect is mirrored in Fig. 18.18, with frequency reaching the lower tolerance limit of 49.9 Hz. These results illustrate that the CEA region inherits the instability of the IAH1 region, which may lead to degraded reliability of the system-wide operation.

Figs. 18.19 and 18.20 depict the CEA level response under the same disturbance scenario with adaptive clustering. Unlike the nonadaptive clustering case, the optimizer reallocates available resources across IAH1 and IAH2. Fig. 18.19 demonstrates that total generation matches load demand during the fault period, with adaptive clustering, i.e., cooperative optimization compensating for the lost 15% generation capacity of IAH1. Correspondingly, Fig. 18.20 shows that frequency deviation remains confined

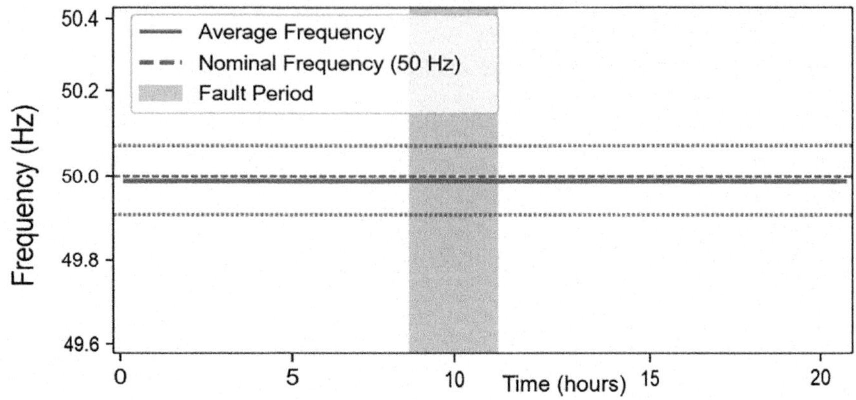

Figure 18.20 CEA frequency response with adaptive clustering concept. CEA frequency response with adaptive clustering. *CEA*, Central energy alliance.

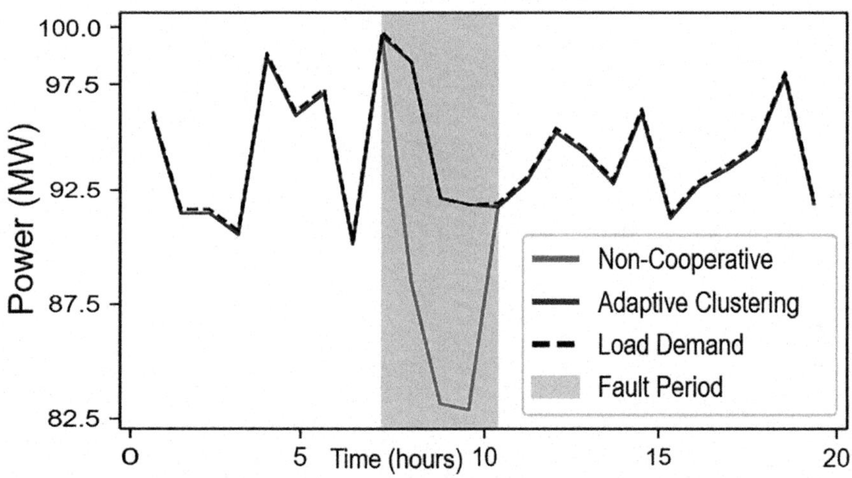

Figure 18.21 Comparison of CEA power profile with and without adaptive clustering. Power profile comparison at the CEA with and without adaptive clustering. *CEA*, Central energy alliance.

well within operational bounds. These results indicate the self-organizing and self-healing behavior of the adaptive-clustering strategy mimicking the natural swarm intelligence: The CEA region absorbs localized fault, preserving stable system operation through resource coordination.

18.4.4 Comparative analysis of adaptive clustering versus noncooperative strategies

Figs. 18.21 and 18.22 show the system-level power profile and frequency response for the adaptive-clustering and noncooperative (without adaptive-clustering)

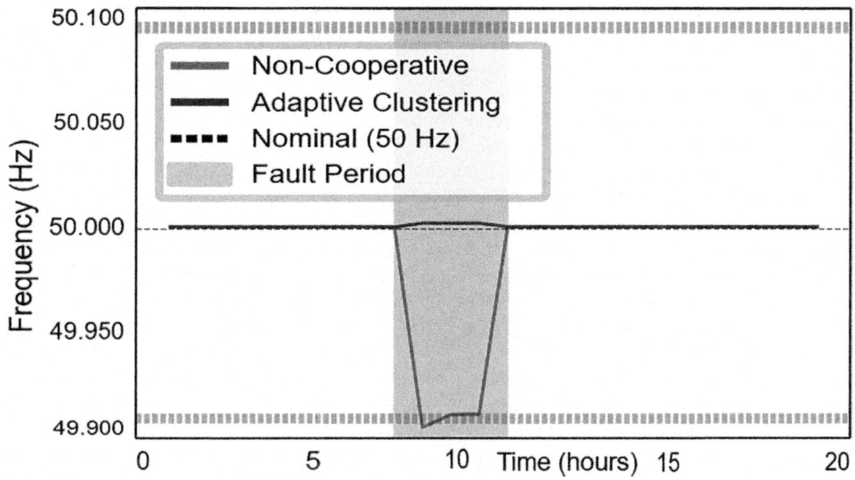

Figure 18.22 Comparison of frequency response of CEA with and without adaptive clustering. Frequency profile comparison at the CEA with and without adaptive clustering. *CEA*, Central energy alliance.

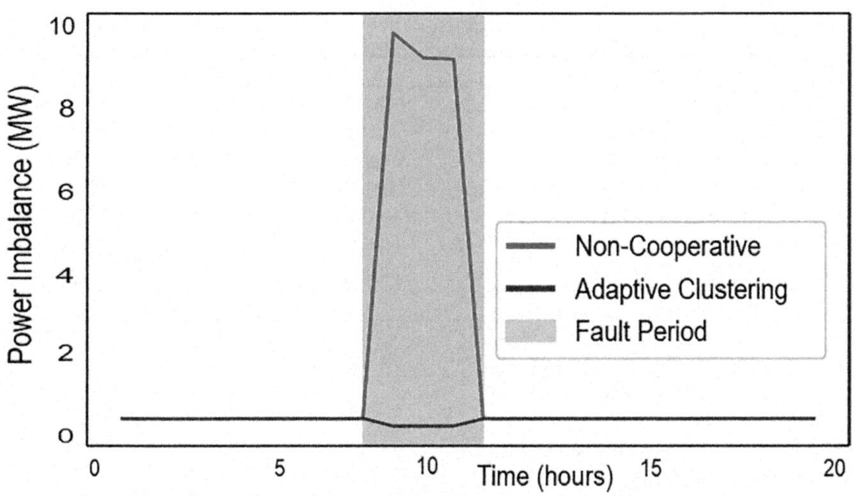

Figure 18.23 Absolute power deviation comparison at the CEA. Absolute power deviation comparison at the CEA with and without adaptive clustering. *CEA*, Central energy alliance.

strategies, respectively. Figs. 18.21 and 18.22 clearly demonstrate the comparison of both strategies between disturbance hours, illustrating a significant improvement in dynamic stability with the adaptive-clustering concept. While Figs. 18.23 and 18.24 quantify these improvements and effects through absolute deviations. In Fig. 18.23, during the fault window without adaptive clustering, a pronounced power imbalance is

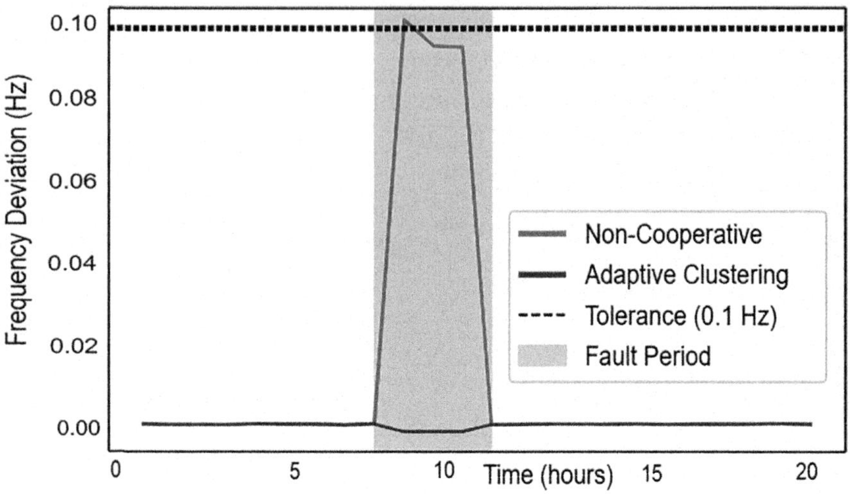

Figure 18.24 Absolute frequency deviation comparison at the CEA. Absolute frequency deviation comparison at the CEA level with and without adaptive clustering. *CEA*, Central energy alliance.

Table 18.7 Comparative performance analysis.

Performance analysis		
Metric	Noncooperative	Adaptive clustering
Max frequency deviation (Hz)	0.1001	0.0020
Avg frequency deviation (Hz)	0.0136	0.0017
Max power imbalance (MW)	10.0096	0.1995
Avg power imbalance (MW)	1.3627	0.1654
Frequency violation (>0.1 Hz)	1/24 h	0/24 h
Improvement analysis		
Metric	Value	
Max frequency deviation improvement with adaptive clustering	98.01%	
Max power imbalance improvement with adaptive clustering	98%	
Frequency stability improvement during a fault	99%	

observed, reaching approximately 10 MW, and frequency deviations nearly crossing the threshold. In the same figures, with the adaptive-clustering mechanism, both deviations contract sharply, showing the improved mean performance of the system response.

Table 18.7 shows the quantitative metrics in tabular form. In the nonadaptive clustering case, the system recorded a maximum frequency deviation of 0.1001 Hz with an average deviation of 0.0136 Hz, alongside a maximum power imbalance of 10.0096 MW and an average power imbalance of 1.3627 MW. In the case of adaptive clustering, these values were reduced to 0.0020 Hz, 0.0017 Hz, 0.1995 MW, and 0.1654 MW, respectively. The count of frequency violations, i.e., deviations greater than threshold 0.1 Hz, is one in the case of noncooperative operation and zero with adaptive clustering. Therefore, findings quantitatively establish the better performance of the adaptive-clustering framework.

The integrated evidence from results and Table 18.5 illustrates that adaptive clustering markedly enhances frequency regulation, power balance, and resilience of the system. With an adaptive clustering mechanism, i.e., swarm-inspired cooperation between IAH1 and IAH2, the CEA transforms a fragile, fault-escalating configuration into a stable and reliable architecture. Table 18.8 presents a robustness analysis of the system.

Table 18.8 demonstrates the robust results for both adaptive clustering and noncooperative strategies of the system for the complete time horizon. In terms of frequency stability, with the adaptive clustering mean deviation was reduced to 0.00165 Hz from 0.0136 Hz (the case of the noncooperative strategy). Thus, with adaptive clustering, a 95% reduction in RMSE is observed. Similarly, power balance decreased from 1.36 MW to 0.165 MW, demonstrating around 88% improvement. The most striking improvement during fault-focused assessment in Table 18.8 further illustrates the self-healing capability of the proposed clustering, with a stability index of 1.0, which is 0.913 in the noncooperative case.

To validate robustness under uncertainty, a Monte Carlo analysis (1000 simulations) was performed. The Monte Carlo analysis clearly shows that these enhancements are not case-specific because across 1000 randomized simulations, the results illustrate that adaptive clustering consistently outperformed noncooperative methodology. The mean gain of 57.7%, low variability (standard deviation of 4.3%), and a 100% probability of improvement showed the best performance of the system with the AIPS concept. These Monte Carlo analysis results confirmed the generalizability with a 95% confidence interval of (48.9, 66.3). The Monte Carlo Analysis provides quantitative evidence of the proposed adaptive clustering framework over the noncooperative benchmark under all simulated stochastic conditions.

18.5 Future projection/research direction of adaptive and intelligent power system

18.5.1 CEU to X

The concept of CEU to X allows AIPS to adapt to various types of energy environments or systems. X represents any functional grouping or application of EAUs. For example, industrial CEUs (aggregating EAUs in industrial zones, i.e., factories or warehouses with tight voltage and frequency regulation due to machinery sensitivity), commercial CEUs aggregating EAUs in commercial areas like office complexes, shopping malls with peak demands during business hours and load variability, agricultural CEUs

Table 18.8 Robustness analysis of the proposed AIPS system.

Category	Metric	Noncooperative	Adaptive clustering	Improvement/notes
Frequency stability	Mean deviation (Hz)	0.0136	0.00165	↓ ~88%
	Standard deviation (Hz)	0.0311	0.00063	Much tighter distribution
	RMSE (Hz)	0.0339	0.00177	↓ ~95%
	Violations >0.1 Hz	1/24 (4.2%)	0/24 (0%)	Eliminated
Power balance	Mean imbalance (MW)	1.36	0.165	↓ ~88%
	Generation efficiency (%)	98.55	99.82	+1.3%
	Peak generation (MW)	99.54	99.54	–
	Peak load (MW)	99.74	99.74	–
Fault-period analysis	Mean freq. deviation (Hz)	0.096	0.000	100% reduction
	Stability index	0.913	1.000	Perfect stability
Monte Carlo robustness (1000 runs)	Mean improvement (%)	–	57.71%	Consistent average gain
	Standard deviation (%)	–	4.32	Low variability
	95% Confidence Interval (%)	–	(48.9, 66.3)	Strong reliability
	Probability of improvement (%)	–	100%	Always better

AIPS, adaptive and intelligent power system; RMSE, Root Mean Square Error.

aggregating EAUs for farming operations with unique seasonal demand patterns, EVs CEUs aggregating EAUs like charging stations, or hybrid CEUs combining different or multiple functional EAUs, etc. Therefore, CEU to X generalizes the framework of CEUs to accommodate a wide range of systems. This approach will be able to accommodate new CEUs as the energy landscape evolves, such as the inclusion of hydrogen CEUs, datacenter CEUs, etc. Thus, it allows functional upgradation as well without reengineering the entire system. Moreover, tailored analytics and optimization for various CEUs enhance system governance and performance in the face of the growing availability of data.

18.5.2 Federated learning and data-sovereignty

Furthermore, collaborative and collective learning, i.e., federated learning (Zhan et al., 2025), may be enabled across AIPS while maintaining the system's privacy. Individual CEUs use their own functional and operational data to train the machine learning model. Each CEU then sends model updates periodically to an aggregator at IAH, and IAH sends updates to CEA, where the CEA updates the global model and shares the global model with each IAH. In this way, IAHs improve the local model based on global/collective insights. Thus, this collective learning strategy allows IAHs and CEUs to collaboratively learn without sharing their raw data. Each IAH and CEU uses its raw data, i.e., frequency and voltages, to train its local models. Afterward, central nodes aggregate the individual IAH model updates to refine the global model that then reflects the trends across the entire system. Thus, each IAH learns from the network-wide experiences without accessing private and raw data from other hubs. Fig. 18.25 presents the CEU to X, Data Governance, and federated learning concept within IAH.

18.5.3 Implicit analytics

Implicit Analytics allows IAHs to correlate and infer operational states using indirect indicators, enhancing system efficiency. IAHs detect issues indirectly using correlation rather than explicitly monitoring details or relying on granular details. The hubs can infer by correlating the measurements and send signals for help during the time window of significant deviations. This reduces data transmission and makes the system more private and resilient.

AIPS leverages this strategy to gain significant and critical insights into the system operations without exhaustive data collection or direct measurement. Analyzing the energy flow deviations, voltage and frequency fluctuations, and DER responses assists AIPS in inferring the health and performance of the system. This minimizes the burden of direct observability and complex instrumentation. Implicit analytics concentrate on actionable intelligence by correlating secondary indicators with the system underlying states. For example, changes in the frequency response of IAH might indicate either mechanical degradation of connected assets or disruption in power supply.

Similarly, changes in IAH energy storage patterns might signal shifting renewable resource availability or load profiles. Through implicit analytics, AIPS can detect anomalies in system behaviors without directly accessing energy components. Thus, AIPS maintains a comprehensive understanding of the system's behaviors. Likewise,

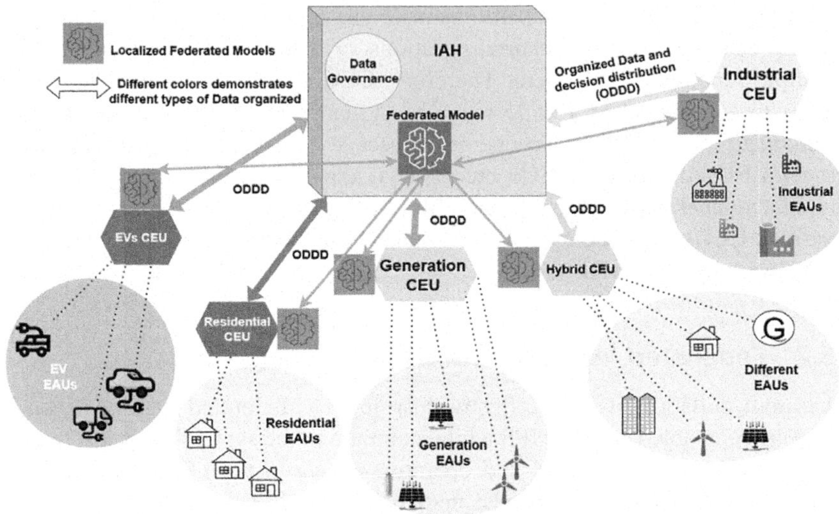

Figure 18.25 Future projection of AIPS. Federated learning, CEU to X, and data governance concept. *AIPS*, Adaptive and intelligent power system; *CEU*, cognitive energy unit.

implicit metrics like temperature variations in transmission lines can trigger predictive maintenance before failure occurs. Temporal correlation in supply-demand mismatches can help AIPS to analyze and enhance load forecasting and fine-tune energy dispatch policies. Through the mechanism of implicit analytics, AIPS is more cost-effective by reducing the need for intrusive sensors. Ambient data and indirect correlations may greatly assist in achieving observability in resource-constrained distributed grid environments such as rural electrification and community microgrids. This approach is beneficial in scenarios where installing direct measurement equipment is environmentally disruptive or impractical.

18.6 Conclusion

This chapter proposed AIPS, a swarm-inspired power distribution system governance framework, motivated by the need for cooperative, adaptive, and resilient operation of the system. Inspired by the emergent behavior of natural swarms, i.e., collective intelligence of simple agents working locally to achieve global stability, the AIPS enables distributed IAH regions to adaptively cluster, reallocate resources, and de-cluster without continuously relying on centralized authority. Such event-driven coordination and self-organizing/healing principles are imperative and feasible for modern smart distribution grids, which integrate decentralized RERs and distributed prosumers. Through a microgrid case study involving IAH1 (economic dispatch) and IAH2 (prosumer management), this work demonstrated that during the disturbance period, isolated operations may lead to instability with noticeable power and frequency excursions. Conversely, with swarm-inspired adaptive clustering, cross IAH support and coordination significantly reduce the power and frequency imbalances. Comparative and Monte Carlo analysis confirmed the improvements, thus establishing AIPS' self-organizing and self-healing capability while containing localized faults without escalating them to the wider grid. The stochastic evaluation conducted in this chapter reinforces that addressing uncertainty must be treated as an integral part of future power distribution grid design. The simulation results, by assessing the architecture under randomized operating conditions, confirm that the proposed adaptive mechanism remains effective when confronted with uncertain system conditions. This uncertainty-aware perspective highlights the architecture's robustness and further suggests that power grid architectures will require stochastic and probabilistic tools to ensure reliability and resilience. Therefore, this research study demonstrates the potential of bio-inspired cooperative mechanisms for ensuring adaptability and reliability in next-generation cyber-physical power distribution systems.

Declaration of Competing Interest

The authors declare no known competing interests

References

Ali Kadhem, A., Abdul Wahab, N. I., Aris, I., Jasni, J., & Abdalla, A. N. (2017). Computational techniques for assessing the reliability and sustainability of electrical power systems: A review. *Renewable and Sustainable Energy Reviews, 80*, 1175–1186. doi:10.1016/j.rser.2017.05.276.

Alvarez-Alvarado, M. S., Apolo-Tinoco, C., Ramirez-Prado, M. J., Alban-Chacón, F. E., Pico, N., Aviles-Cedeno, J., Recalde, A. A., Moncayo-Rea, F., Velasquez, W., & Rengifo, J. (2024). Cyber-physical power systems: A comprehensive review about technologies drivers, standards, and future perspectives. *Computers and Electrical Engineering, 116*, 109149. doi:10.1016/j.compeleceng.2024.109149.

Chalapathy, R., & Chawla, S. (2019). Deep learning for anomaly detection: A survey. *arXiv*. doi:10.48550/arxiv.1901.03407.

Couraud, B., Andoni, M., Robu, V., Norbu, S., Chen, S. i, & Flynn, D. (2023). Responsive FLEXibility: A smart local energy system. *Renewable and Sustainable Energy Reviews, 182*, 113343. doi:10.1016/j.rser.2023.113343.

Creutzig, F., Breyer, C., Hilaire, J., Minx, J., Peters, G. P., & Socolow, R. (2019). The mutual dependence of negative emission technologies and energy systems. *Energy and Environmental Science, 12*(6), 1805–1817. doi:10.1039/c8ee03682a.

Duan, H., Huo, M., & Fan, Y. (2023). From animal collective behaviors to swarm robotic cooperation. *National Science Review, 10*(5). doi:10.1093/nsr/nwad040.

Gutiérrez-Escalona, J., Roncero-Clemente, C., Husev, O., Matiushkin, O., & Blaabjerg, F. (2024). Artificial Intelligence in the hierarchical control of AC, DC, and hybrid AC/DC microgrids: A review. *IEEE Access, 12*, 157227–157246. doi:10.1109/ACCESS.2024.3486382.

Jafari, M., Gharehpetian, G. B., & Anvari-Moghaddam, A. (2024). On the role of virtual inertia units in modern power systems: A review of control strategies, applications and recent developments. *International Journal of Electrical Power and Energy Systems, 159*. doi:10.1016/j.ijepes.2024.110067.

Kataray, T., Nitesh, B., Yarram, B., Sinha, S., Cuce, E., Shaik, S., Vigneshwaran, P., & Roy, A. (2023). Integration of smart grid with renewable energy sources: Opportunities and challenges – A comprehensive review. *Sustainable Energy Technologies and Assessments, 58*, 103363. doi:10.1016/j.seta.2023.103363.

Khalid, M. (2024). Smart grids and renewable energy systems: Perspectives and grid integration challenges. *Energy Strategy Reviews, 51*, 101299. doi:10.1016/j.esr.2024.101299.

Kim, S. H. (2007). Evaluation of negative environmental impacts of electricity generation: Neoclassical and institutional approaches. *Energy Policy, 35*(1), 413–423. doi:10.1016/j.enpol.2005.12.002.

Kroposki, B. D., Dall-Anese, E., Bernstein, A., Zhang, Y., & Hodge, B. S. (2017). *Autonomous energy grids*. National Renewable Energy Laboratory (NREL), Golden.

Li, Y., Ding, Y., He, S., Hu, F., Duan, J., Wen, G., Geng, H., Wu, Z., Beng Gooi, H., Zhao, Y., Zhang, C., Mei, S., & Zeng, Z. (2024). Artificial intelligence-based methods for renewable power system operation. *Nature Reviews Electrical Engineering, 1*(3), 163–179. doi:10.1038/s44287-024-00018-9.

Liang, K., Wang, H., Pozo, D., & Terzija, V. (2024). Power system restoration with large renewable Penetration: State-of-the-Art and future trends. *International Journal of Electrical Power & Energy Systems, 155*, 109494. doi:10.1016/j.ijepes.2023.109494.

Lin, Y., & Wang, J. (2022). Realizing the transactive energy Future with local energy market: An overview. *Current Sustainable/Renewable Energy Reports, 9*(1), 1–14. doi:10.1007/s40518-021-00198-0.

Lore, K. G., & Reddy, K. K. (2019). Implicit context-aware learning and discovery for streaming data analytics. *arXiv*. doi:10.48550/arxiv.1910.08438.

Luderer, G., Pehl, M., Arvesen, A., Gibon, T., Bodirsky, B. L., de Boer, H. S., Fricko, O., Hejazi, M., Humpenöder, F., Iyer, G., Mima, S., Mouratiadou, I., Pietzcker, R. C., Popp, A., van den Berg, M., van Vuuren, D., & Hertwich, E. G. (2019). Environmental co-benefits and adverse side-effects of alternative power sector decarbonization strategies. *Nature Communications, 10*(1). doi:10.1038/s41467-019-13067-8.

Machlev, R., Heistrene, L., Perl, M., Levy, K. Y., Belikov, J., Mannor, S., & Levron, Y. (2022). Explainable artificial intelligence (XAI) techniques for energy and power systems: Review, challenges and opportunities. *Energy and AI, 9*, 100169. doi:10.1016/j.egyai.2022.100169.

Mbungu, N. T., Ismail, A. A., AlShabi, M., Bansal, R. C., Elnady, A., & Hamid, A. K. (2023). Control and estimation techniques applied to smart microgrids: A review. *Renewable and Sustainable Energy Reviews, 179*. doi:10.1016/j.rser.2023.113251.

Minias, P. (2023). The effects of urban life on animal immunity: Adaptations and constraints. *Science of The Total Environment, 895*, 165085. doi:10.1016/j.scitotenv.2023.165085.

Mohammed, A. S., Atnaw, S. M., Salau, A. O., & Eneh, J. N. (2023). Review of optimal sizing and power management strategies for fuel cell/battery/super capacitor hybrid electric vehicles. *Energy Reports, 9*, 2213–2228. doi:10.1016/j.egyr.2023.01.042.

Nyangon, J. (2024). Climate-proofing critical energy infrastructure: Smart grids, artificial intelligence, and machine learning for power system resilience against extreme weather events. *Journal of Infrastructure Systems, 30*(1). doi:10.1061/jitse4.iseng-2375.

Paul, S., De, S., & Bhattacharyya, S. (2024). *Emerging trends in computational swarm intelligence: A comprehensive overview recent trends in swarm intelligence enabled research for engineering applications* (pp. 1–40). Elsevier. doi:10.1016/B978-0-443-15533-8.00004-7.

Priyadarshi, R., & Kumar, R. R. (2025). Evolution of swarm intelligence: A systematic review of particle swarm and ant colony optimization approaches in modern research. *Archives of Computational Methods in Engineering, 32*(6), 3609–3650. doi:10.1007/s11831-025-10247-2.

Rani, P., Parkash, V., & Sharma, N. K. (2024). Technological aspects, utilization, and impact on power system for distributed generation: A comprehensive survey. *Renewable and Sustainable Energy Reviews, 192*, 114257.

Rodriguez-Calvo, A., Cossent, R., & Frías, P. (2018). Scalability and replicability analysis of large-scale smart grid implementations: Approaches and proposals in Europe. *Renewable and Sustainable Energy Reviews, 93*, 1–15. doi:10.1016/j.rser.2018.03.041.

Rowe, E., Asbell-Clarke, J., & Baker, R. S. (2015). *Serious games analytics to measure implicit science learning serious games analytics: Methodologies for performance measurement, assessment, and improvement* (pp. 343–360). Springer International Publishing. doi:10.1007/978-3-319-05834-4_15.

Sharma, R., & Malaviya, P. (2023). Ecosystem services and climate action from a circular bioeconomy perspective. *Renewable and Sustainable Energy Reviews, 175*, 113164. doi:10.1016/j.rser.2023.113164.

Shen, L., Tang, Y., & Tang, L. C. (2021). Understanding key factors affecting power systems resilience. *Reliability Engineering & System Safety, 212*, 107621. doi:10.1016/j.ress.2021.107621.

Stock, S., Babazadeh, D., & Becker, C. (2021). Applications of artificial intelligence in distribution power system operation. *IEEE Access, 9*, 150098–150119. doi:10.1109/access.2021.3125102.

Tang, J., Liu, G., & Pan, Q. (2021). A review on representative swarm intelligence algorithms for solving optimization problems: Applications and trends. *IEEE/CAA Journal of Automatica Sinica, 8*(10), 1627–1643. doi:10.1109/jas.2021.1004129.

Tiwari, S., & Kumar, A. (2023). Advances and bibliographic analysis of particle swarm optimization applications in electrical power system: Concepts and variants. *Evolutionary Intelligence, 16*(1), 23–47. doi:10.1007/s12065-021-00661-3.

Tooki, O. O., & Popoola, O. M. (2024). A comprehensive review on recent advances in transactive energy system: Concepts, models, metrics, technologies, challenges, policies and future. *Renewable Energy Focus, 50*, 100596. doi:10.1016/j.ref.2024.100596.

Trivedi, R., Patra, S., Sidqi, Y., Bowler, B., Zimmermann, F., Deconinck, G., Papaemmanouil, A., & Khadem, S. (2022). Community-based microgrids: Literature review and pathways to decarbonise the local electricity network. *Energies, 15*(3), 918. doi:10.3390/en15030918.

Wang, C., Ju, P., Wu, F., Pan, X., & Wang, Z. (2022). A systematic review on power system resilience from the perspective of generation, network, and load. *Renewable and Sustainable Energy Reviews, 167*, 112567. doi:10.1016/j.rser.2022.112567.

Xu, B., Bai, G., Liu, T., Fang, Y., Zhang, Y., & Tao, J. (2023). An improved swarm model with informed agents to prevent swarm-splitting. *Chaos, Solitons & Fractals, 169*, 113296. doi:10.1016/j.chaos.2023.113296.

Zhan, S., Huang, L., Luo, G., Zheng, S., Gao, Z., & Chao, H.-C. (2025). A Review on federated learning architectures for privacy-preserving AI: Lightweight and secure cloud–edge–end collaboration. *Electronics, 14*(13), 2512. doi:10.3390/electronics14132512.

Index

Page numbers followed by "*f*" and "*t*" indicate, figures and tables respectively.

Printed and bound by CPI Group (UK) Ltd, Croydon, CR0 4YY
14/09/2026
02216396-0002